# BROADBAND OPTICAL MODULATORS

Science, Technology, and Applications

# BROADBAND OPTICAL MODULATORS

## Science, Technology, and Applications

Edited by
Antao Chen
Edmond J. Murphy

CRC Press is an imprint of the
Taylor & Francis Group, an **informa** business

**Cover Image:** From Chapter 7, "InP-Based Electro-Optic Modulators" (Ian Betty).

CRC Press
Taylor & Francis Group
6000 Broken Sound Parkway NW, Suite 300
Boca Raton, FL 33487-2742

First issued in paperback 2020



Version Date: 20110818

ISBN-13: 978-0-367-57683-7 (pbk)
ISBN-13: 978-1-4398-2506-8 (hbk)



**Library of Congress Cataloging-in-Publication Data**

Broadband optical modulators : science, technology, and applications / edited by Antao Chen and Edmond J. Murphy.
p. cm.
Includes bibliographical references and index.
ISBN 978-1-4398-2506-8 (hardback)
1. Light modulators. 2. Optoelectronic devices. I. Chen, Antao. II. Murphy, Edmond J. III. Title.

TK8360.L5B76 2011
621.3815'2--dc23 2011026883

**Visit the Taylor & Francis Web site at**
**http://www.taylorandfrancis.com**

**and the CRC Press Web site at**
**http://www.crcpress.com**

*To the scientists and engineers whose hard work contributed to the advancement of modulator technology, and to the people who have the courage to stand up against totalitarianism and injustice.*

**—A.C.**

*To Kathy, Ed, Colleen and Matt, Dan, Tim, and Sean, with love.*

**—E.J.M.**

# Contents

# Foreword

Optical modulators, specifically waveguide optical modulators, provide the information encoding engines of most metro and all long-haul terrestrial and undersea fiber-optic transmission systems. The Internet that connects the people and businesses of the planet would not operate without them. The full benefit of optically amplified systems that allows information-bearing photons to travel even across oceans without the need for electrical regeneration can only be achieved because of the pure manner in which optical modulators impress high bit rate information on an optical carrier. As researchers seek to increase the capacity on a single fiber by increasing the spectral efficiency and lowering the cost per bit of these systems, the modulation methods needed require ever increasing complexity and performance. In response, modulator devices have evolved to highly complex modulator circuits. Who would have thought?

Optical modulators have been a subject of research since the 1960s. Bulk modulators using electro-optic or Franz-Keldysh effects in bulk crystals have been used to modulate the output of gas lasers since then. With the advent of semiconductor lasers—the natural light source for fiber-optic communication systems—the early common wisdom was that modulators would not be required because, unlike large gas lasers, the output power of a semiconductor laser can be directly turned on and off by modulating the driving current—thus providing a single device that was essentially a modulated light source. The primary concerns of directly modulated lasers in the early years was whether they could be modulated at the multigigabit rates needed for high capacity optical communication systems and whether they could maintain their spectral properties under modulation. Both issues were addressed by large research activities.

The prospects for fiber-optic systems strongly drove research on optical waveguide devices and integrated optic circuits. It was clear early on that waveguide devices would likely be single mode and therefore only be compatible with single mode fiber even thought initially fiber-optic systems used multimode fiber. Waveguide electro-optic and electroabsorptive modulators would have the benefit, compared to their bulky counterparts, of having much lower drive voltage because of the small lateral dimensions and the fact that both effects are field strength dependent. The challenge to the researcher was to achieve low optical loss in the waveguides, in waveguide bends, and in coupling to the single mode fiber.

Early work focused on a search for the right material system to form low-loss waveguides in an active material. The III–V materials, GaAs and InP, the materials used to make lasers and detectors for the shorter and longer wavelength regions, respectively, were natural choices given the desire to make modulators that could be integrated with laser sources. The demonstration of low-loss waveguides in lithium niobate via titanium in-diffusion provided a relatively simple, flexible, and reproducible fabrication method in a material with a strong electro-optic effect. Soon after its invention, a fairly large family of integrated optic devices, including modulators, switches, and tunable wavelength filters, were demonstrated.

Even in the early days, it was quite clear that the likely first application of waveguide devices if any—there were many skeptics—would be the modulator. A friendly and exciting competition began between the III–V and lithium niobate-based modulator research communities to be the first with a practical device that could provide significant value to high-capacity optical transmission systems. At the same time, the high-speed laser researchers were convinced that direct laser modulation would do just fine.

The first real indication that waveguide modulators might have a chance as an alternate to the much simpler and presumably more cost-effective direct modulation came when it was shown that waveguide modulators, particularly interferometer-based modulators, could be designed and driven such that information-bearing bits exhibited no wavelength "chirp" as the bit went from "zero" to "one." This is to be contrasted with the dynamic behavior of directly modulated lasers that chirp as the driving current is changed. Low chirp is essential in order to maximize transmission system performance. Ultimately, the capacity of optical transmission systems to transmit more information over a given distance, or to increase the distance for a given information capacity, is limited by signal intensity or distortion at the receiver. The result of wavelength chirp coupled with finite fiber dispersion is pulse spreading that can limit the ability to recover individual bits with high fidelity. Zero-chirp modulation offered by optical modulators thus offers substantial potential systems advantage relative to the unavoidable chirp of direct modulation.

For waveguide modulators, the path from possible to practical required much work and perseverance, together with a passionate belief that devices would be important commercially. First, the demonstration of low-propagation loss and efficient fiber to waveguide coupling had to be achieved. Low-drive voltage that could be delivered at multigigabit/sec rates by available electronics was required. For the lithium niobate modulators whose electro-optic effect is not as strong as the electroabsorptive effect in III–V materials, traveling wave electrodes are required to achieve sufficient bandwidth and low-drive voltage. Optimized electrode design was essential. Careful package design was critical for electrical performance and to maintain efficient fiber coupling under environmental extremes.

With practical devices in hand, the real test was to demonstrate the value of modulators in transmission systems. In the early 1990s, there was a global research competition on "hero" transmission experiments with the goal of sending the most bits per second over the longest distance. Initially, these systems were unamplified, with an unregenerated reach of 50 km to 100 km. Clearly to show value, modulators had to demonstrate system performance that was superior to directly modulated lasers. That milestone was first met using lithium niobate waveguide interferometer modulators at a rate of 1 Gb/s in 1985. The advantage of modulators increased as the research systems moved to 2 Gb/s.

Still, as long as commercial terrestrial systems were regenerated and the distance between regenerator sites was ~50 km, many believed that it would require very high bit rates before the low-chirp advantage of modulators would be material. The value proposition for modulators got its second big break about this time as fiber amplifiers became practical. With the advent of optical amplifiers, wavelength division multiplexed (WDM) systems became very attractive as a way of increasing the capacity on a fiber without having to move to extremely high bitrates per wavelength. While, at first glance, one may have concluded that such systems would delay the need for modulators because they allow high capacity without going to higher bitrates, that effect was countered by fact that, for long-haul systems, the value proposition for amplified WDM transmission increases as the total transmission distance without regeneration can be increased. For long-haul lightwave systems that distance could be as large as 600 km–1000 km before detection, electrical regeneration, and demultiplexing were required. Over such distances signal degradation due to chirp, even at several gigabits/sec, was significant. This was particularly true due to fiber dispersion in the 1550 nm wavelength region in which fiber amplifiers operate. The situation was even more obvious for transoceanic amplified systems that were early adopters of external modulators.

The case for modulators improved further as the feasibility and value proposition of reconfigurable wavelength routed networks was demonstrated in the late 1990s. In such networks, signals can be added, dropped, or routed at the wavelength level using optical switching technology. The result is that there is less need to return to the electrical domain to do any switching. Keeping signals in the optical domain as long as possible leads to lower-cost networks. In such networks, the return on investment that low-chirp modulation affords in enabling transcontinental transmission distances without the need to regenerate each individual wavelength on the fiber at multiple points along the way makes the use of these devices extremely compelling. Add to that the move from 2.5 Gb/s to 10 Gb/s and then to 40 Gb/s, and it is easy to see why virtually every long-haul system—both transoceanic and terrestrial—and the vast majority of metro systems today depend upon broadband optical modulators.

The story does not end there. Research on optical modulators continues to be very active, fruitful, and impactful. As the appetite for high-bandwidth services, including video, increases, the need for cost-effective bandwidth continues to grow. Commercial systems have just begun to offer 100 Gb/s per wavelength systems. In research, the race is on for 400 Gb/s and 1 Tb/s on a single wavelength. The role of the external modulator will continue to be to provide encoded signals that allow the minimum of propagation distortion at those high rates while traveling over the maximum possible distance. To do that, advanced modulations formats, currently used in wireless systems at much lower rates, are essential. These modulation formats require both intensity and phase modulation in increasingly complex combinations that require nested interferometers. Also, coherent techniques will be required. To achieve these complex modulation characteristics, highly integrated modulator circuits will be required. The value proposition of complex integration offered by waveguide technology becomes absolutely essential.

This book, *Broadband Optical Modulators: Science, Technology, and Applications*, edited by Antao Chen and Edmond Murphy, provides the full, exciting story of optical modulators. As the title indicates, it is a comprehensive review from the fundamental science to the material and processing technology to the optimized device design to the multitude of applications for which broadband optical modulators bring great value. While application to digital optical transmission systems and networks has served as the focal point for this foreword, applications extend to analog optical transmission for cable TV networks, optical pulse generation, optical analog to digital conversion, optical beam forming, and optical sensors, to name several. All are covered in this book. Especially valuable in my view is that the authors are internationally known researchers, developers, and systems people who are experts in their field, writing now, with the perspective that time offers, about their groundbreaking work. The book should prove valuable to a broad spectrum of readers, including graduate students, new entrants to the field, and systems researchers and engineers who want either an overview or an in-depth understanding of this exciting and highly enabling field of broadband optical modulators.

**Rodney C. Alferness**
*Alcatel Lucent*

# Preface

An optical modulator is the component at the beginning of a fiber-optic link used to encode electrical data on a lightwave before transmitting it down the optical fiber. It is the counterpart of an optical detector, which converts optical pulses back to electrical data at the receiving end of the fiber. In the early days of optical communications, optical modulation was achieved simply by the direct modulation of a semiconductor laser's drive current. As technology advanced and bit rates increased, researchers encountered fundamental problems with direct modulation of the laser—problems such as frequency chirp and wavelength excursion. Consequently, external modulation using optical modulators has become a standard approach. Today, all broadband fiber-optic links use optical modulators in their transmitters; some novel communication schemes even incorporate modulators in their receivers. Tens of millions of modulators have been deployed worldwide. As the trend toward ever-increasing bandwidths in fiber-optic communications continues, the optical modulator will remain a key component for transmitting information. Traditional lithium niobate electro-optic modulators have found wide application in commercial fiber-optic telecommunication systems and continue to push performance limits, while new modulator technologies are emerging and demonstrating great potential. For analog fiber-optic links, external modulators are the key to achieving the required linearity and dynamic range.

This book is designed to provide a comprehensive overview of modulator-related topics from the fundamental to the practical, from materials to systems, and from the historical to the present. Underlying physical effects such as electro-optic effects in nonlinear crystals and semiconductors and the electroabsorptive effect in semiconductor quantum well structures are described in detail. The reader will find descriptions of the optical and electro-optical properties for the commonly used single crystalline lithium niobate and III–V compound semiconductor materials. Also covered in this book are emerging new materials that show promise for new modulators with broad bandwidths and low drive voltages, such as silicon, electro-optic polymers, transparent ferroelectric oxides, and organic nonlinear optical crystals. For each type of modulator, principal factors important to modulator performance, typical modulator designs, fabrication techniques, and state-of-the-art performance are discussed in detail. Measurement methods for key performance parameters are presented. Electronic devices required for driving and controlling modulators are described. The use of modulators in current and emerging systems is also covered. Applications of high-speed optical modulators beyond those most relevant to the telecommunications industry, such as analog fiber-optic radio frequency links, fiber-optic gyroscopes, optoelectronic oscillators, and millimeter wave imaging, are also described in their respective chapters. In some cases, the subject matter from one chapter overlaps with that of other chapters. In these cases, we intentionally allowed overlapping treatments of the same matter under the belief that the reader will benefit from varied approaches to the material from two or more subject matter experts. To complement the text, each chapter contains an exhaustive list of references to aid in further study.

Although high-speed optical modulators have been key components widely used in fiber-optic telecommunication systems for decades, the technical literature on modulators is spread over multiple scientific and engineering journals and thus is not easy to access or comprehend. To our knowledge, no comprehensive treatment has been published. A few books on photonics and optoelectronics have chapters on optical modulators, but the details and depth of technical discussion are limited. A book entirely focused on high-speed optical modulators is needed by people who work with these devices. This need was the motivation for us to undertake this book project. We and the chapter authors try to provide a comprehensive account of high-speed optical modulators with a balanced and thorough treatment of the fundamentals, design, fabrication, practical issues, technology trends, and a broad range of applications. The chapters are written by internationally recognized leaders in their fields. Some of them are pioneers in their technical field. We are proud of our team of authors.

Two major groups of readers will find this book valuable. One group is comprised of engineers and other professionals directly working in the telecommunication industry, whose work requires a good understanding of optical modulation techniques. This group includes system and subsystem designers, product development engineers, and engineers exploring new modulator architectures who need to know the various modulation technologies available and the strengths and weaknesses of each technology. Technical managers and project managers in the telecommunications and defense industries will also find this book informative and helpful for gaining a high-level understanding of the advantages, trade-offs, and other relevant issues related to different optical modulator technologies. The other major group of readers for whom this book was written is scientific researchers, including graduate students, involved in fiber optics and fiber-optic communication research, especially in the fields of optical transmitters and optical modulators. Senior undergraduate students who are interested in fiber optics and telecommunication, and those students working on relevant class projects, will also find this book beneficial.

The editors extend a sincere thank you to the many contributing authors. It was truly our pleasure to work with this outstanding group of experts. Through your efforts, we are able to offer our readers a truly comprehensive treatise on the history and the state of the art of optical modulators. We hope our readers enjoy using the text as much as we have enjoyed compiling it.

Antao Chen would like to thank Dr. Philip Colosimo and Mr. Christopher Fuller for their help with editing the book. Their efforts made it possible for us to complete this project in a timely manner.

**Antao Chen**
*University of South Florida and*
*University of Washington*

**Edmond J. Murphy**
*JDSU*

# Editors

**Antao Chen** is a research professor at Physics Department of University of South Florida. He also has joint appointments as senior scientist at the Applied Physics Laboratory and associate professor at Department of Electrical Engineering, University of Washington. He received his BS and MEng degrees from Beijing Institute of Technology, Beijing, China, in 1983 and 1989, respectively, and his MS and PhD degrees from the University of Southern California, Los Angeles, in 1995 and 1998, respectively. He has worked in both the optical industry and academia since 1983.

Dr. Chen has worked on electro-optic waveguide modulators based on electro-optic polymer, lithium niobate, and silicon nanostructures. In 1997 he demonstrated the first electro-optic polymer modulator that operates at frequencies above 100 GHz. At Lucent Bell Laboratories, Dr. Chen developed lithium niobate optical cross-connect switches. These switches were the key to the success of the DARPA MONET project, which won the 1999 Bell Labs President Gold Award. He also led the development of a high-performance lithium niobate electro-optic polarization controller product, which won the 2001 Photonics Circle of Excellence Award as one of the top 15 new products of the photonics industry of the year. His current areas of research include free space, integrated, and fiber optics for information transmission and processing, chemical and environmental sensing using photonics and nanoelectronics, micro- and nanofabrication, and propagation and scattering of terahertz waves.

**Edmond J. Murphy** currently works for JDSU's Communication and Commercial Optical Products Division where he focuses on Advanced Technology Planning and Intellectual Property matters. He has been with JDSU since 1999, previously serving in various roles such as Vice President of Development, Electro-optic Products Division; General Manager, Electro-Optic Products Division; and Group Chief Technology Officer, Components and Modules Division.

Prior to joining JDSU, Dr. Murphy worked at AT&T/Lucent Bell Labs where he held a variety of technical staff and management positions. At Bell, he worked in a variety of technology areas, including high-speed modulation, optical switching, optical packaging, wafer processing, planar lightguide circuits, optical transmitters and receivers, and optical amplifiers.

Dr. Murphy received a BS in chemistry from Boston College and a PhD in chemical physics from MIT. He is a fellow of the IEEE and a fellow of the Optical Society of America, as well as a member of the Connecticut Academy of Science and Engineering. He has authored over 50 technical talks and papers and several book chapters; in addition, he has been awarded 13 patents and has edited a book on integrated optics.

# Contributors

**Edward I. Ackerman**
Photonic Systems, Inc.
Billerica, Massachusetts

**Rodney C. Alferness**
Alcatel-Lucent, Bell Laboratories
Holmdel, New Jersey

**Gary E. Betts**
Photonic Systems, Inc.
Escondido, California

**Ian Betty**
Ciena
Ottawa, Ontario, Canada

**John C. Cartledge**
Queen's University
Kingston, Ontario, Canada

**Pak S. Cho**
University of Maryland
College Park, Maryland

**Charles H. Cox III**
Photonic Systems, Inc.
Billerica, Massachusetts

**Larry R. Dalton**
University of Washington
Seattle, Washington

**John J. DeAndrea**
Finisar Corporation
New Hope/Solebury, Pennsylvania

**Danny Eliyahu**
OEwaves Inc.
Pasadena, California

**René-Jean Essiambre**
Alcatel-Lucent, Bell Laboratories
Holmdel, New Jersey

**Harold R. Fetterman**
University of California at Los Angeles
Los Angeles, California

**Matthew R. Fetterman**
University of California at Los Angeles
Los Angeles, California

**Frederic Y. Gardes**
University of Surrey
Guildford, United Kingdom

**Peter Günter**
ETH Zurich
and
Rainbow Photonics AG
Zurich, Switzerland

**John Heaton**
u2t Photonics UK Ltd
Durham, United Kingdom

**Fred Heismann**
JDSU Optical Networks Research Lab
Robbinsville, New Jersey

**Mojca Jazbinsek**
ETH Zurich
and
Rainbow Photonics AG
Zurich, Switzerland

**Seongku Kim**
University of California at Los Angeles
Los Angeles, California

**Karl Kissa**
JDSU
Bloomfield, Connecticut

**Brian H. Kolner**
University of California
Davis, California

**Steven K. Korotky**
Alcatel-Lucent, Bell Laboratories
Holmdel, New Jersey

**Jeffery E. Lewis**
Honeywell International Inc.
Phoenix, Arizona

**Ling Liao**
Intel Corporation
Santa Clara, California

**Ansheng Liu**
Intel Corporation
Santa Clara, California

**Yalin Lu**
U.S. Air Force Academy
Colorado Springs, Colorado

**Lute Maleki**
OEwaves Inc.
Pasadena, California

**Andrey B. Matsko**
OEwaves Inc.
Pasadena, California

**David Moilanen**
EOSPACE, Inc.
Redmond, Washington

**Ken Morito**
Fujitsu Laboratories Ltd.
Kanagawa, Japan

**Hiroshi Murata**
Osaka University
Osaka, Japan

**Hirotoshi Nagata**
JDSU
Tokyo, Japan

**Kazuto Noguchi**
NTT Photonics Laboratories
Kanagawa, Japan

**Mario Paniccia**
Intel Corporation
Santa Clara, California

**Vittorio M. N. Passaro**
Politecnico di Bari
Bari, Italy

**Yao Peng**
University of Delaware
Newark, Delaware

**Dennis W. Prather**
University of Delaware
Newark, Delaware

**Graham Reed**
University of Surrey
Guildford, United Kingdom

**Rebecca Schaevitz**
Intel Corporation
Santa Clara, California

**Christopher Schuetz**
University of Delaware
Newark, Delaware

**Haruhisa Soda**
FiBest Limited
Tokyo, Japan

**William H. Steier**
University of Southern California
Los Angeles, California

**Suwat Thaniyavarn**
EOSPACE, Inc.
Redmond, Washington

**Robert G. Walker**
u2t Photonics UK Ltd
Durham, United Kingdom

**Peter J. Winzer**
Alcatel-Lucent, Bell Laboratories
Holmdel, New Jersey

**Minyu Yao**
Tsinghua University
Beijing, China

**Hongming Zhang**
Tsinghua University
Beijing, China

# I

# Fundamentals

# I

# Fundamentals

# 1

# Role and Evolution of Modulators in Optical Fiber Communication

Steven K. Korotky
*Alcatel-Lucent,*
*Bell Laboratories*



*For*
*Pat, Kikki, Kris, Steve, Tina, Kate, Rusty, Abby, Jeff, Stevie, and Louis*
*With love.*

*"The secret of success is constancy to purpose."*—Benjamin Disraeli (1804–1881)

## 1.1 Introduction

Imagine a journey—a journey over great distances and over even greater time, a journey too challenging and long for one person and one lifetime, an epic journey of many, taken together, and spanning many generations. A journey sometimes with friends, but more often with strangers—some never to know. Such is the journey that has borne today's optical fiber communication networks and the optical technologies that enable them. This book is an update on the journey to realize and refine optical modulators. Yet we recognize that it necessarily represents the perspectives of but a few who have participated in and continue the pursuit, and so conveys only a portion of the saga. As a guide, the reader interested in the details of optical modulators for communication beyond those surveyed on these pages is referred to the cited literature, which includes a substantial body of earlier and complementary reviews [1–15].

The phenomenal success of fiber optics as a vehicle for communication is in large part attributable to its being a fertile technological platform—one that has supported high research and development (R&D) creativity and productivity and has manifested sustained, geometric, year-over-year increases in bandwidth capacity and, simultaneously, lower unit cost to transport information over large distances

[16]. Arguably, it was for this reason that one of optical communication's pioneers was recently recognized with a Nobel Prize for his achievements, which include the groundbreaking scientific analysis that concluded that purified silica glass might serve as a low-loss optical transmission medium for communication [17–21]. Among the many other key discoveries, inventions, and innovations that have contributed to sustaining the tremendous productivity of optical fiber communication for more than four decades are the optical detector, or photodetector, and the optical maser, or laser [22–25]. Less well known to the general public yet equally essential to the performance of contemporary long-distance optical communication are other physical devices such as optical modulators, optical amplifiers, optical multiplexer/demultiplexers, and optical switches [26,27]. Several of these basic building blocks have broad application and a long history, and have been pursued independently of the low-loss cylindrical dielectric optical waveguide transmission medium—otherwise known as the optical fiber. Some of these devices have then been adapted to the specific requirements of optical fiber communication.

The focus of this book is on the optical modulator, and in this introductory chapter, we concentrate on its role in optical fiber communication. The use of optical modulators in long-distance optical fiber communication became critical and widespread shortly following the demonstration of reliable, compact, low-loss, high-speed optical modulators that could exceed the dispersion-limited performance obtained using the direct switching of the electrical pump current of semiconductor injection lasers [28–30]. Yet optical modulators are fundamental to optical communication and have a long history with archetypes extending back to ancient and prehistoric times [31]. Optical modulator technology has also progressed significantly during the last two decades and, as described and illustrated in the chapters of this book, serves an expanding role in the continued viability of contemporary optical communications systems. In the remainder of this chapter, we sketch the milestones in the histories of optical modulators and optical fiber communication, highlight the interactions and impact of one upon the other, and summarize the key technological drivers of optical modulators, the performance demanded of them, and the milestone demonstrations and deployments achieved using them.

The high-level architecture and abstract components of a communication system employing electromagnetic waves are depicted in Figure 1.1. This representation encompasses the essence of communication systems ranging from those based on smoke signals to the most sophisticated lightwave systems yet envisioned [32]. The generic building blocks include a source of the electromagnetic wave, a coder to convert the user's native alphabet representing the abstract message into the symbols used for the physical system, a modulator to impress the symbols of the coded message onto the electromagnetic wave, a medium over which the modulated electromagnetic signal propagates, a demodulator to extract the symbols from the received signal, a detector(s) to collect the demodulated electromagnetic signal, and a decoder to convert the symbols of the message back into the user's native alphabet. In this section, we will describe the interwoven histories of optical fiber communication systems, or simply lightwave systems, and the optical technologies that comprise them, with particular emphasis on the role and evolution of optical modulators.

Some of the many milestones in the history and evolution of optical fiber communications and optical modulators are listed in Table 1.1. For reference and to help place the events important to optical

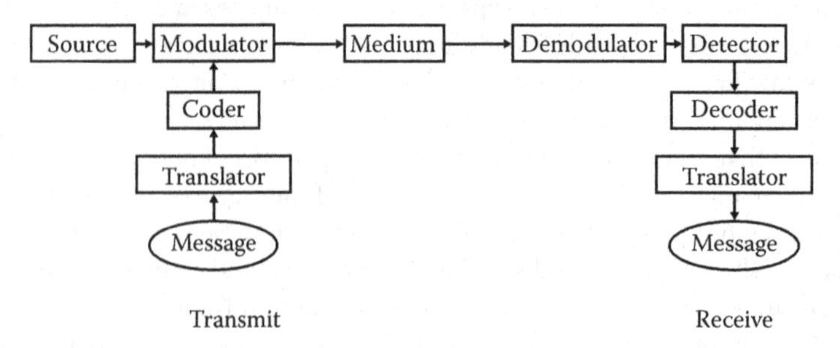


**FIGURE 1.1** The high-level architecture and basic components of a communication system employing electromagnetic waves.

**TABLE 1.1** Some Milestones in the Evolution of Optical Communications and Optical Modulators

| Year | Milestone | References |
|---|---|---|
| BC | Smoke signals, fire beacons, optical telegraph, shuttered torch, and heliograph | [31] |
| 1785 | Coulomb's law | |
| 1792 | Optical semaphore | [20] |
| 1809 | Electrical telegraph | [49] |
| 1813 | Photoelastic effect | [58] |
| 1831 | Faraday's law of induction | |
| 1845 | Magneto-optic effect | [56] |
| 1862 | Electromagnetic theory of light | [39,45] |
| 1866 | Wireless telegraphy | [47] |
| 1867 | Modern signal lamp | [238] |
| 1873 | Photoconduction | [22] |
| 1875 | Quadratic electro-optic effect | [51,57] |
| 1877 | Modern heliograph | [31] |
| 1880 | Photophone | [60] |
| 1886 | Radio waves | [38,46] |
| 1887 | Photoelectric effect | [38,53] |
| 1891 | Mach-Zehnder interferometer | [40,41] |
| 1893 | Linear electro-optic effect | [52,57] |
| 1895 | Radio communication | [48] |
| 1897 | Discovery of the electron | [239] |
| 1905 | Quantum nature of light | [42,43] |
| 1907 | Electroluminescence | [54,59] |
| 1910 | Electromagnetic theory of dielectric waveguides | [240] |
| 1927 | Light-emitting diode | [55,59] |
| 1933 | Microwave communication | [50] |
| 1942 | Contemporary signal lamp | [241,242] |
| 1948 | Communication theory | [243] |
| 1949 | Ferroelectric nature of lithium niobate | [62–64,244–245] |
| 1957 | Optical maser/laser | [20,23,25] |
| 1960 | Ruby laser | [68–69] |
| 1960 | Microwave modulation of light | [1,2,80–82,84,89] |
| 1962 | Practical light-emitting diode | [71] |
| 1962 | Laser diode | [70–73,77] |
| 1964 | High-speed avalanche photodiode | [74] |
| 1964 | Fiber amplifier | [20,78] |
| 1964 | Semiconductor modulator | [83,85] |
| 1965 | Growth of lithium niobate | [65,66,244,245] |
| 1965 | Low-loss fiber analysis | [17–20] |
| 1967 | Electro-optic and optoelectronic materials | [62,63] |
| 1969 | Integrated optics conception | [90] |
| 1970 | Continuous, room-temperature laser diode | [20,75–77] |
| 1970 | Low-loss fiber demonstration | [20,92] |
| 1971 | Distributed-feedback laser | [246] |
| 1972 | Semiconductor waveguide devices | [98] |
| 1972 | Traveling-wave modulator | [3] |
| 1973 | Fiber laser | [79] |
| 1974 | Ti:$LiNbO_3$ waveguides and modulator | [4,99,101] |

(*continued*)

**TABLE 1.1 (Continued)** Some Milestones in the Evolution of Optical Communications and Optical Modulators

| Year | Milestone | References |
|---|---|---|
| 1975 | Optical Fiber Transmission Conference | [96] |
| 1976 | Long wavelength laser diodes | [115] |
| 1977 | Integrated-optic traveling-wave modulator | [5,104,112,126] |
| 1977 | FT3 (45 Mb/s) multimode fiber transmission trial | [24,114] |
| 1978 | Low-loss, 1.3 μm/1.5-μm single-mode optical fiber | [20,116] |
| 1982 | Low-loss, 1.3-μm Ti:$LiNbO_3$ waveguide devices | [161,176] |
| 1983 | ASK, FSK, PSK, and DPSK transmission experiments | [247] |
| 1983 | FT3C (90 Mb/s) 0.8-μm multimode fiber system | [95] |
| 1985 | Electroabsorption modulator | [140,145,146,164] |
| 1985 | Low-loss, 1.5-μm Ti:$LiNbO_3$ waveguide devices | [161,180] |
| 1985 | External modulation outperforms direct OOK laser modulation | [28,29,151] |
| 1985 | Erbium-doped fiber amplifier | [182,183] |
| 1985 | FTG (417 Mb/s) 1.3-μm single-mode fiber system | [95] |
| 1985 | WDM transmission experiment over single-mode fiber | [186] |
| 1986 | 1.5-μm dispersion shifted fiber | [248] |
| 1988 | Transoceanic optical fiber cable system (TAT-8) | [172] |
| 1989 | FTG-1.7 WDM (1.3/1.5 μm) × 1.7-Gb/s system | [95] |
| 1990 | Arrayed waveguide grating multiplexer and router | [249–251] |
| 1990 | Dual-drive optical modulator permitting AM/PM modulation | [169,170] |
| 1991 | Ultra-broadband modulators | [14] |
| 1992 | Long-term lithium niobate modulator bias stability | [210,213] |
| 1993 | Fiber dispersion management and compensation | [26] |
| 1994 | Field trial of 10-Gb/s modulators | [8,192] |
| 1994 | Duobinary modulation | [205,252,253] |
| 1994 | Forward error correction deployment | [254–256] |
| 1994 | Undersea system using LN modulator and optical amplifiers | [33] |
| 1995 | DWDM × 2.5-Gb/s system using integrated laser/EA modulators | [95] |
| 1996 | 1-Tb/s fiber capacity | [20] |
| 1998 | PASS code modulation | [207] |
| 1999 | DWDM × 10-Gb/s systems using LN modulators | [10,11,95] |
| 1999 | M-ary ASK modulation | [229] |
| 2000 | Large port count single-mode optical cross-connects | [257,258] |
| 2001 | Data traffic exceeds voice traffic | [16] |
| 2001 | TAT-14 WDM × 10-Gb/s C-RZ using LN modulators | [215] |
| 2002 | DQPSK modulation | [259] |
| 2004 | DWDM × 40Gb/s DPSK system using LN modulators | [225] |
| 2006 | Optical OFDM | [260] |
| 2008 | 1-Tm fiber deployed worldwide | [227] |
| 2008 | 100-Gb/s system trial using LN modulators | [228] |

communication into the broader context of more general technological progress, also included in the list are some milestones for other modes of communication, namely, electrical wireline communication and radio and microwave communication. In considering the extent of the history, for the purpose of this chapter we have chosen to describe the advances by grouping them chronologically into the periods of BC–1750, 1750–1950, 1950–1970, and thereafter by decades. Following, we expand upon some of the major events in each period. For the benefit of the reader, we have selected references from among both original works and retrospective reviews.

## 1.2 Conception and Presages: Historical Archetypes (BC–1750)

The most ubiquitous optical modulator found in today's long-distance optical fiber communication systems is the lithium niobate (LN), waveguide, Y-branch Mach-Zehnder (MZ) interferometer, traveling-wave, electro-optic modulator, or LN-MZ modulator for short [8,10,11]. This type of optical modulator was the first external modulator of any kind to be deployed in commercial lightwave communication systems and was put into service on December 1, 1994, in the cable sections of the Americas-1/Columbus-2B undersea lightwave systems connecting St. Thomas in the U.S. Virgin Islands to Florida. This undersea cable system made use of the then-also-new optical amplifier technology and using external modulation operated at a bit rate of 2.5 Gb/s over a distance of 2000 km [33,34]. However, over the course of two or more millennia preceding the advent of optical fiber, various external optical modulators had been used for what is now referred to as free-space optical communication. These ancient and early modes of optical communication used mechanical means for modulating light and took the forms of smoke signals (ancient America, China, Egypt, Greece), fire beacons (ancient Greece and Rome), optical telegraph (150 BC Greece) and semaphore (eighteenth-century France), shuttered torch (ancient) and signal lamp (nineteenth to twentieth century), and reflected sunlight (405 BC Greece) and heliograph (nineteenth to twentieth century) [20,31,35,36].

Because the earliest forms of optical communication depended upon mechanical means for modulation, the physical data rates were correspondingly slow. Communication at higher data rates and longer distances than those that could be achieved by these primitive mechanical means would be made possible once electricity, magnetism, and light became better understood. In the next section, we review these modern advances, which in turn set the stage for the contemporary period of optical communication based on optoelectronics.

## 1.3 Modern Exploration and Discovery (1750–1950)

Today's wireline and wireless communication systems are based on the propagation of electromagnetic waves, and consequently the origins of contemporary optical communication and optical modulators may be traced directly to the extraordinary period of scientific exploration and discovery spanning the latter half of the eighteenth century through the first half of the twentieth century. As summarized in Table 1.1, this period includes the systematic, seminal experimental investigations on electricity and magnetism, the unifying electromagnetic theory of light, the invention and use of precision instrumentation for investigating the properties of light, the generation and detection of radio waves, the quantum theory of matter, and light, including the concept of the stimulated emission of radiation [37–46]. Leveraging this fundamental work, this period also gave rise to electrical and electromagnetic telegraph communications systems, and later to radio and microwave communications [47–50].

Of particular significance to the evolution of optical communication and optical modulators, during this period, pioneering works resulted in significant discoveries regarding the interaction and propagation of light in materials, including the detection and generation of light; the effects of mechanical strain and stress, and of electrical and magnetic fields, applied to the materials; and the theory of the propagation of light in dielectric media. Among these key discoveries were photoconduction, the photoelectric effect, electroluminescence, the light-emitting diode, the photoelastic effect, the magneto-optic effect, and the quadratic and linear electro-optic effects [22,38,40,41,50–59]. The latter phenomenon, also known as the Pockels effect, was discovered in 1893, and together with the MZ interferometer, which was conceived in 1891, has come to serve as the mechanism used within the high-speed modulators of present-day, high-capacity, long-distance fiber-optic transmission systems to impress information carried by electrical signals onto optical carrier waves [40,41,52]. The discovery of photoconduction formed the basis of the first optical detectors, which allowed analog information carried on optical waves (e.g., sun light) to be converted into electrical signals and facilitated the invention of the photophone, a wireless optical communication system for transporting voice, in 1880 [22,60,61]. At the transmitter, the photophone used a mechano-optical modulator in the form of a vibrating reflector to transform audio vibrations into variations in optical intensity.

## 1.4 Contemporary Science and Invention (1950–1970)

Although there had been seminal and substantial work on materials and elemental devices for generating, detecting, and transforming the intrinsic properties of light over the preceding 200 years, there was an explosive growth of research activity and results in these fields following World War II, and in particular, during the period of 1950–1970. An alphabet soup of naturally occurring and man-made electro-optic and optoelectronic materials were grown and studied throughout these two decades. Among these were both exotic and more common crystalline materials. Investigated for their electro-optic properties and potential use for optical modulators were materials such as ammonium dihydrogen phosphate (ADP; $NH_4H_2PO_4$), barium titanate ($BaTiO_3$), cadmium sulfide (CdS), gallium arsenide (GaAs), gallium phosphide (GaP), lithium niobate ($LiNbO_3$), lithium tantalite ($LiTaO_3$), potassium dihydrogen phosphate (KDP; $KH_2PO_4$), quartz ($SiO_2$), and zinc telluride (ZnTe) [62,63]. Already by 1949, the ferroelectric nature of lithium niobate ($LiNbO_3$), a man-made (i.e., not naturally occurring) crystalline dielectric, which is currently the material of choice for high-speed optical modulators, had been established, and by the mid-1960s, techniques for pulling this crystal from the melt had been developed and many of its properties had been extensively studied [64–67].

In parallel with research on new materials and novel devices for measuring and modifying the properties of light, during this period, groundbreaking progress was made on the effort to generate and detect light controllably at visible and near-infrared wavelengths. Among the former advances was the revolutionary invention and realization of the laser—the 50th anniversary of the solid-state ruby laser being celebrated this year [23,25,43,68,69]. Within only a few years of the demonstration of the optically pumped ruby laser, advances on semiconductor optoelectronic materials and processing resulted in the demonstration of the first practical electrically pumped (i.e., electrical current-driven) semiconductor, light-emitting laser diode, which emitted at visible and near-infrared wavelengths, and the first high-speed avalanche photodiode detectors [59,70–74]. By 1970, laser diodes were fabricated that could be operated continuously at room temperature, which further raised the prospect for a new generation of communication technology based on light [75–77]. The first fiber amplifiers and lasers were also demonstrated during this period [78,79].

The milestone invention and achievement of the laser at the beginning of the 1960s immediately spurred an intense period of pioneering work in related areas, including optical modulators, optical gain and laser media, and optical transmission media. In fact, by the time the demonstration of the ruby laser had been published, the first electro-optic modulators for modulating light at microwave frequencies had been demonstrated [80–82]. These early solid-state optical modulators were fabricated from bulk, large aperture crystalline samples. By the early 1970s, a solid foundation for optical modulation using both dielectric oxide and semiconductor crystals had been laid [1–3,83–89]. The innate advantages of the electro-optic effect to achieve ultra-high modulation frequencies and optical waveguides to permit low voltages were strong drivers of waveguide optical modulators from the onset. Already by 1970 some of the devices based on single crystal oxides incorporated traveling-wave electro-optic interaction and supported ultra-high-frequency modulation, while some of the devices based on reverse-biased *p–n* junctions in semiconductors exhibited aspects of optical waveguiding and boasted low-drive power [3,86,89].

To propagate light over short distances for the purpose of constructing miniaturized optical components suitable for commercial high-speed optical communication systems, such as lasers, modulators, splitters, and filters, the pioneers of this period conceived planar optical circuitry based on photolithographically defined channel waveguides [90,91]. In analogy to the integrated electronic circuits of the time, the miniature optical circuits envisioned for processing light were described as integrated optics. The technological foundations of optical waveguide devices necessary to realize integrated optical circuits, such as the choices of substrate and methods for producing the index change required for guiding light, which were begun in the 1960s, were systematically built during the 1970s.

Rounding out the activity and advances of this period was the important work on suitable media for optical transmission. One approach to long-distance optical transmission was based on periodically focused optical beams using gaseous lenses, another was based on hollow waveguides, and another was based on optical fiber waveguides [20]. Each had significant shortcomings or challenges, and possibilities were eliminated one by one. By 1970, persistence of vision and purpose led to the now celebrated proposal and demonstration of the low-loss, silica optical fiber waveguide [17–20,92,93].

## 1.5 Technological Foundation and Anticipation (1970–1980)

With the advances made on the enabling technologies of sources, detectors, and transmission media during the 1960s, the stage was set for the research, prototyping, development, and trials of optical fiber transmission systems during the 1970s [20,93–95]. Indicative of the concerted, worldwide effort to realize optical fiber transmission systems, the first topical meeting organized by the Optical Society of America on this specific subject was held in 1975—coincidentally the same year this author had just graduated from college [96].

To realize the vision of integrated optical components and circuits, first, techniques for producing low-loss optical waveguides in substrates supporting a means to affect the properties of light, for example, phase or polarization, were essential; and, second, concepts and demonstrations of efficient (e.g., low-voltage) waveguide versions of basic, discrete devices, such as phase, polarization, and amplitude modulators, were necessary. Among the materials considered as candidates for the substrate for optical waveguide modulators were those crystals that were investigated for bulk-optical modulators, are transparent in the visible and/or near-infrared spectrum, and exhibit a large linear electro-optic effect, as quantified by the figure of merit formed by the mathematical product of the index of refraction of the material cubed and the applicable electro-optic coefficient [1–3,85,97]. Additionally, it was recognized that the crystals should be relatively easy to grow and should be mechanically, thermally, optically, and electro-optically robust. The small number of candidates satisfying all the criteria expected of a practical technology included the crystalline dielectric oxides lithium niobate and lithium tantalate and crystalline compound semiconductors such as gallium arsenide. Methods for fabricating waveguides in these material systems were systematically pursued in the early to mid-1970s, and by the latter half of the decade, researchers had succeeded in demonstrating the first generation of high-speed (~1 GHz) lumped and traveling-wave optical modulators and switches, based on integrated optical waveguide directional couplers, interferometers, and attenuators in both dielectric and semiconductor crystals [4,5,98–113].

Ironically, in the late 1970s, after more than 2000 years of the use and refinement of optical modulators—and just as the first generation of waveguide optical modulators spurred by the realization of room-temperature semiconductor laser diodes were demonstrated—the success of the laser diode as the source of light for the first long-distance fiber-optic communication systems nearly resulted in the technological demise of optical modulators. With the ability to modulate its light emission directly by varying the electrical current passing through it, the compact laser diode—no bigger than a grain of sand—could simultaneously serve as both the source and modulator of light. In much the same way that a mechanical switch can be used to modulate the light emitted from a flashlight, the application of a varying electrical current to a laser diode results in modulation of the electron population inversion—referred to as gain modulation—and subsequently to the output light intensity through the processes of spontaneous and stimulated emission. Indeed, the first generation of optical fiber communication systems used direct current modulation of laser diodes to implement optical on/off keying, as exemplified by the trial link tested in 1977 between two Illinois Bell central offices in Chicago separated by 2.4 km. This fiber transmission (FT) system, designated FT3, operated at an optical wavelength of 0.8 μm and carried data at the T3 bit rate of 44.7 Mb/s over multimode optical fiber [24,114]. It was followed in 1983 and 1984 by the large-scale deployment along the U.S. Northeast corridor of commercial 90-Mb/s (FT3C) optical transmission systems—also based on direct laser modulation and multimode fiber—to connect Boston and Washington [95]. The FT3/FT3C systems had a reach of 7 km before signal regeneration was required.

The first optical transmission systems also presented a second dilemma for researchers pursuing the first generation of optical waveguide modulators: the first transmission systems used multimode fibers, as these were easier to fabricate, and it was not yet clear that single-mode fibers could be spliced or connectorized without incurring high optical loss. In contrast, the integrated-optical modulators provided high light extinction ratio and required low drive voltage only if the waveguides used to fabricate the integrated-optical circuits were single mode. Fortunately, the researchers working on integrated-optical modulators were not discouraged, but rather motivated, by the success of optical transmission systems using direct modulation of laser diodes and multimode fiber. Indeed, by the late 1970s, the limitations to the transmission capacity of multimode fiber, such as modal dispersion, were abundantly evident and work on low-loss single-mode optical fiber, single-mode connector technology, and single-transverse-mode laser diodes at 1.3-μm wavelength had progressed to the point where it was clear that the first multimode fiber systems for long-distance communication would also be the last for the conceivable future [20,95,115–117]. It was also already appreciated that while the direct modulation of lasers was architecturally simple, it was not without its own limitations. In particular, it was recognized early on that the interaction between the electron and photon populations within the semiconductor laser diode—each with their own characteristic lifetimes—gives rise to coupled oscillations, referred to as relaxation oscillations, resulting in a strongly peaked frequency response, which could limit the speed of direct modulation [94]. So high are the stakes to achieve the best optical transmission system performance at the lowest cost that direct modulation and external modulation have been considered competing options ever since the consideration of the first generation of fiber-optic systems [94]. A result of the potential of integrated-optical circuits, the success of these first fiber-optic systems, the possibility of a limit to practical direct modulation bandwidth, and the shift to single-mode optical fibers was that global R&D investments in single-mode integrated-optical modulators and switches would see a dramatic increase in the 1980s, as corporations, institutes, universities, and government agencies scrambled not to be left behind in the unfolding optical communication revolution.

## 1.6 Innovation—Realization and Application (1980–1990)

The progress and successes of semiconductor lasers and detectors, the steady advances on waveguide optical devices, and the realization of single-mode optical fiber and connector technology in the late 1970s sparked an explosion of research and development on integrated-optical devices in the 1980s—especially in the area of broadband optical amplitude and phase modulators, switches, and filters. The emphasis during this period was on the realization of robust device designs and fabrication processes driven by the requirements of practical applications [118]. In addition to substantially increased efforts on integrated-optical devices for analog and digital optical fiber communication, activity on waveguide optical components increased in anticipation of a diverse array of applications of optical modulators including signal processing, such as frequency shifting, spectrum analysis, and analog-to-digital conversion; optical pulse generation and optical sampling; sensing, such as electric field sensing, chemical sensing, and gyroscopes; and metrology [6,29,30,104,119–159]. These potential applications spanned optical wavelengths from the visible (~0.6 μm) through the near-infrared (~2 μm) and in nearly all cases simultaneously required low optical insertion loss, low-drive voltage, and high modulation speed.

To address the needs of the many applications, research on integrated-optical devices during the decade of the 1980s strove not only to demonstrate proof-of-principle functionality but also to establish practical fabrication processes and the relationships between the controllable design parameters and performance, which depends upon the wavelength of operation. For electroabsorption and electro-optic optical modulators in both crystalline dielectrics and semiconductors, this meant investigating and optimizing not only the optical waveguide, but also the driving electrodes and the optoelectronic interaction. For example, in the case of waveguide optical modulators based on lithium niobate (LN), these detailed studies led from basic first-generation designs to nonobvious, second-generation designs that made use of very thick gold-plated electrodes and a very thick insulating silicon dioxide buffer layer, which was placed between the surface of the crystalline substrate and the metallic electrodes, to produce a traveling-wave electric field within the crystal that was nearly velocity matched to the optical wave [5,6,104,130,135,147,148,160–162]. This period also saw advances in high-speed traveling-wave modulators and electro-absorption modulators based on semiconductor crystals [7,9,145,153,163–165]. In the remainder of this section, we specifically focus on the application of optical modulators to optical fiber communication. And, in particular, we consider the evolution of optical fibers and optical transmission systems during the 1980s and their effects on external modulator R&D, as researchers and developers sought to apply optical modulators to long-distance fiber-optic communication.

Fundamentally, the interest in optical fiber communication stems from its ability to support high-capacity data transport over long distances at the lowest cost, and it is this same overarching motivation that drives interest in single-mode optical fibers and waveguide external optical modulators. In the case of the wired optical transmission medium, the optical fiber, high performance translates into keeping the distortion (impairments) of the transported signal low. In other words, the fiber should have low optical loss, low dispersion, and low nonlinearities. The loss of the fiber is determined by the properties of the constituent materials and waveguide, including absorption by impurities, scattering by imperfections and fundamental inhomogeneities, and radiation in bends; the main advantage of optical fiber is that the propagation loss can be very low even for very high modulation frequency. For optical fibers based on silica, the losses are lowest in the 1.3 μm to 1.5-μm wavelength region. While the magnitude of nonlinearities is influenced by the material, impairments caused by nonlinearites can be ameliorated by keeping the optical power density low and by frustrating phase matching.

The dispersion of the fiber, which causes pulse spreading, is dependent on both the properties of the materials (chromatic dispersion) and the characteristics of the optical waveguide and the guided mode(s) it supports (waveguide dispersion). In general, the transverse modes of a multimode waveguide have different propagation constants, and so signal pulses carried by multiple fiber modes can quickly spread. This is why early optical fiber communication using directly modulated lasers quickly evolved from using multimode optical fiber to using single-mode optical fiber—even though variations in modal propagations could be reduced to some extent by tailoring the index distribution in the multimode fiber [117]. It had also been observed that the opposing effects of chromatic and waveguide dispersion of single-mode waveguides determined the wavelength of zero dispersion.

In addition to the characteristics of the fiber, the degree of the impairment caused by dispersion is dependent upon the properties of the signal waveform itself, and generally, the magnitude of the impairment for a linear system increases superlinearly as the signal bandwidth increases [117]. The optoelectronic transducer used to impart the modulation on the optical carrier may cause additional degradation if it further broadens the signal bandwidth through other processes. In the case of a directly modulated laser, not only can the relaxation oscillation distort the signal, but in a process referred to as laser chirp, the change in the real part of the index of refraction within the laser volume caused by the varying electrical charge density and changing gain acts to dynamically shift the laser wavelength (and corresponding carrier frequency), resulting in spectral broadening [166,167]. Like the other sources of signal distortion, impairments caused by dispersion can be managed in a variety of ways, and one way is to use a method of optical modulation that maintains high signal fidelity in the process of encoding (upconverting) the high-speed baseband signal onto the optical carrier wave. Electroabsorption modulators, for example, although they necessarily cause some chirp, avoid the larger wavelength variations that occur within a laser diode [140,145,146,150,153]. External optical modulators based on refractive mechanisms, such as the electro-optic effect, not only can provide high linearity and modulation bandwidth but, through unparalleled control of the amplitude and phase of the optical waveform, also can achieve very low chirp and, hence, narrow spectral width [94,137,168–171]. However, as suggested by Figure 1.2, although the advantages of external modulation were recognized by its early practitioners, it would not be until the 1990s that commercial optical fiber communication systems would require the demonstrated capabilities of electroabsorptive and electrorefractive external modulators.

The researchers and developers of external modulators at industrial laboratories during the 1980s carefully followed and were strongly influenced by the trends in the evolution of optical fiber systems using directly modulated lasers. These trends included the shift from systems using multimode fibers and 0.8-μm wavelength lasers to single-mode fibers and 1.3-μm wavelength lasers, and, of course, the

**FIGURE 1.2** A key benefit of external modulation—signal quality. As a result of the coupling of the electron and photon densities in semiconductor lasers, using direct current modulation, the repeated ignition and quenching of the laser action are accompanied by relaxation oscillations and optical frequency chirp. External modulation of the laser emission avoids these problems. Left Panel: Her: "There must be a better way to send smoke signals." Him: "I risk being burned or drowned with every puff." Right Panel: Her: "Remember the way we used to do this?" Him: "Yes, external modulation is sure to catch on." Cartoon by C. M. Korotky, 1994. (Reproduced from Korotky, S.K., *Proc. Conf. Optical Fiber Commun.*, 21–52, 1994. With permission of Optical Society of America.)

raison d'etre-increases in communication system bit-rate capacity [95,172–175]. Each generation of optical fiber system brought about a corresponding effort to design integrated-optical waveguides and optical waveguide modulators and switches optimized for those systems (e.g., see [5,176–179]). However, the higher modulation fidelity provided by external modulation did not translate into significantly increased performance in relation to direct laser modulation at 1.3 µm, as this was the zero-dispersion wavelength of the fiber, and the modulator added loss to the loss-limited systems.

As the shift to 1.3-µm optical transmission systems (based on direct laser modulation) was taking place, it was also established that the lowest loss window in silica fiber would be in the region near 1.55-µm wavelength, and so researchers shifted their attention to open that new ground. Again, integrated-optical waveguides and devices were optimized in anticipation of new opportunities (e.g., see [180]). The shift to a 1.55-µm wavelength would turn out to be another revolutionary change for long-distance communications—second perhaps only to the realization of low-loss optical fiber. First, if the systems were to operate at 1.55 µm, but the zero-dispersion wavelength remained at 1.3 µm, then the relative importance of dispersion management and external modulation would increase. Again, a unique opportunity for integrated-optical devices and optical fiber system refinement was not lost on researchers. Indeed, within two years of the deployment of the first 90-Mb/s laser diode-based optical transmission systems, strongly driven by the advent of low-loss single-mode optical fiber, integrated-optical modulators were demonstrated at 1.55 µm that could provide longer system reach for multigigabit per second data rates than could be achieved using directly modulated lasers [28–30,143,148,151,181]. Even with researchers' contemporaneous achievement to shift the zero-dispersion wavelength of single-mode optical fiber from 1.3 µm to the lower-loss 1.55-µm wavelength, the potential need for external optical modulators was increasing because the larger modulation bandwidth sampled the fiber dispersion away from the zero-dispersion wavelength [117,174]. From the perspective of researchers working on external optical modulators, the experimental demonstration that external modulators could increase system performance at multigigabit per second data rates launched an endurance race – a race to sustain technological advances, complete the development, establish the reliability, and lower the cost of high-performance, single-mode integrated-optical modulators that could outperform the moving target of state-of-the-art, directly modulated lasers in anticipation of the need for thirty times higher data rates a decade later. However, at the pace of advances in fiber and laser technology, a decade was far in the future, and so a place for external optical modulators was not yet assured.

As fate would have it, the attractiveness of external modulation suddenly and dramatically increased as a consequence of another advance—and it was an enormous one—namely, the demonstration of an optical fiber amplifier that could ameliorate optical fiber loss in both the 1.3-μm and 1.55-μm wavelength windows [182–184]. With the ability to provide signal amplification over a wide wavelength range, the emergence of a practical optical fiber gain medium immediately meant that the potential reach of high-bit-rate optical systems would no longer be limited by optical loss, but rather would be determined by other factors, such as dispersion and nonlinearities. And thus it was that—driven by the need of increased system capacity and lower cost for longer system reach—external modulators, fiber amplifiers, and wavelength-division multiplexed systems, which had been part of the collective consciousness of the R&D community since shortly after the realization of the laser, would come of age during the same period [68,78,80,95,139,182–187].

## 1.7 Optimization, Stabilization, Commercialization, and Deployment (1990–2000)

Owing to significant broad-based and sustained research investment, the decade of the 1980s was a productive period of proof-of-principle demonstrations for integrated-optical devices [188,189]. Among the wide variety of functions that were realized, high-speed optical modulation for communication emerged as a likely first, high-volume commercial application [161]. The need for high-bandwidth modulation with high spectral purity reached criticality following the advent of the erbium-doped fiber amplifier, which facilitated dense wavelength-division multiplexing as a means to enormously increase fiber capacity at relatively low cost. For example, with the broad optical spectrum potentially utilized by dense wavelength-division multiplexing, only a very few wavelength channels could be situated near the zero-dispersion wavelength of conventional or dispersion-shifted optical fiber. More significant, to curb performance-degrading distortion and cross talk in dense wavelength-division multiplexing transmission caused by the Kerr nonlinearity in optical fiber, a modest amount of fiber dispersion used in combination with dispersion compensation and a low-chirp transmitter proved both beneficial and practical [190]. Indeed, as illustrated in Figure 1.3, where the percentage of optical transmission experiments at gigabit-per-second data rates reported at the annual Conference on Optical Fiber Communication that used external modulators during the course of the 1990s is plotted, electroabsorptive and electrorefractive external modulators became major enabling technologies of contemporary high-capacity optical communication during this decade. In addition to optical amplifiers and optical modulators, other new optical technologies of this period included optical filters, multiplexer/demultiplexers, and switches [27,191].

The projected imminent convergence of the need for high-bandwidth modulation with high spectral purity for analog and digital fiber communication and the demonstrated capabilities of optical waveguide modulators—combined with a strong desire for both a technological and commercial success—very naturally resulted in focused R&D activities to bring waveguide modulators to the marketplace during the 1990s. However, displacing the incumbent technology, direct laser modulation, would prove difficult, as external modulation required both adding an additional device to the transmission system and lowering the total cost. Meeting these requirements would only be possible if the combination of a laser operated as a continuous-wave light source and a high-speed external optical modulator used as a data encoder could be manufactured with an overall higher yield for equal or improved performance and reliability. Hence, the R&D efforts on electroabsorptive and electrorefractive external modulators during the 1990s targeted continued performance optimization, reproducibility, hardening, stability,

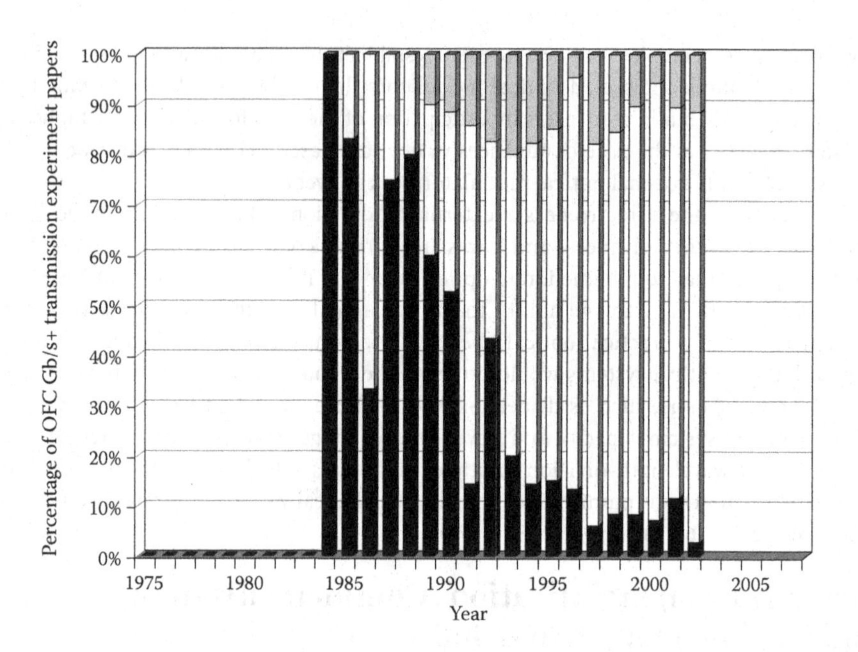


**FIGURE 1.3** Distribution of Gb/s+ transmission experiments at OFC. Graphed using histograms are the relative percentages of Gb/s+ transmission experiment papers presented at the Conference on Optical Fiber Communication using direct laser modulation and external modulation versus year for the period 1975–2002. Black fill—direct laser modulation. White fill—external modulation using lithium niobate modulators. Gray fill—external modulation using all other modulators.

reliability, high yield, medium volume manufacturing, and system field trials with the goal of system deployment [8,10,11,95,181,192,193].

In the case of optical amplitude modulators based on LN, which provided the highest level of chirp control, designs based upon both X-cut and Z-cut crystals, and initially upon both directional coupler switches and MZ structures, were pursued [5,102,103,130,135,147,148,161,194–197]. Each configuration had technical advantages and disadvantages, as well as historical—and sometimes emotional and mythical—preferences, associated with it, as pioneering device fabrication is often a mixture of art and science. For example, early on, many amplitude modulators based on X-cut crystals were designed using Y-branch (also known as 1 × 1) MZ modulators, whereas many amplitude modulators based on Z-cut crystals were originally based on 2 × 2 directional coupler switches. Both configurations utilized the Z-directed component of the applied electrical field to take advantage of the largest of LN's electro-optic coefficients ($r_{33}$) and were intended to be operated in a push–pull mode to minimize the required drive voltage.

The Y-branch MZ modulator in X-cut LN was preferred because of the perceived simplicity, as it did not require directional couplers or a buffer layer, the lower required optical phase shift, and the natural symmetries of the optical and electrical waveguides resulted in low sensitivity to environmental variations (e.g., temperature changes). However, the MZ X-cut configuration without a buffer layer has relatively low electrical impedance and a relatively high-velocity mismatch between the electrical and optical waves, which limit high-speed performance. The directional coupler in Z-cut LN was attractive because it could serve as a spatial switching element in addition to a modulator. However, for high-speed operation, the Z-cut implementations based on a single electrical transmission line tended toward asymmetric electrode designs, which heightened the device's sensitivity to environmental variations [5–198]. Also, the directional coupler with uniform optical coupling is intrinsically sensitive to variations of the electro-optically induced index change along the length of the device, and an observed consequence was that the attenuation of the electrical drive signal, as it propagated along the high-speed electrode, limited the achievable on/off extinction ratio [8,196,199].

The goal of multigigahertz optical modulators and switches requiring low-drive voltage and providing low chirp, high extinction ratio, and low environmental sensitivity led multiple research groups to focus independently on waveguide MZ interferometers in Z-cut LN with balanced drive electrodes to maintain high symmetry between the arms of the interferometer [169,200–202]. Given the importance of managing system penalties arising from fiber dispersion and nonlinearity, it was realized that an interferometric modulator with fully independent drive electrodes for each arm (dual drive) could not only provide zero chirp but could also provide precisely and fully adjustable amplitude (AM) and phase (PM) modulation and thereby practical control of the time–bandwidth product of the optical data pulses [169]. The commercial availability of the dual-drive AM/PM modulator not only would enable multigigabit/second undersea and terrestrial lightwave systems, but it would become a vehicle for stimulating, investigating, and implementing generations of advanced modulation schemes, beginning with prechirp, soliton, duo-binary, and phased amplitude-shift signaling code modulation [13,159,170,171,203–208].

As nature would have it, a thick silicon dioxide buffer layer would prove beneficial to simultaneously optimize LN integrated-optical modulators for the required high modulation bandwidth and low-voltage operation, as well as serve its original purpose to reduce optical propagation loss [162]. However, from the time of the fabrication of the earliest bulk-optical LN devices, drift of the optical phase operating point, referred to as bias drift, had been observed to varying degrees [2,209,210]. Obtaining low-bias drift in a multilayer dielectric system, such as the combination of a $SiO_2$ buffer layer and a titanium in-diffused LN waveguide device, creates additional challenges because of the possibility of time-varying voltage division across the highly resistive dielectric layers [210]. Consequently, before integrated-optical modulators based on crystalline $LiNbO_3$ and amorphous $SiO_2$ could be deployed in carrier-grade systems, it was necessary to develop repeatable crystal growth, titanium diffusion, and buffer layer deposition processes, and to establish the long-term bias stability of the devices. The latter was a critically important but often vexing task throughout the 1980s, and the outcome at that time was far from assured.

Early system level tests of the stability of LN amplitude modulators were encouraging though, and spurred by the better transmission performance achieved using external modulation, within a few years, researchers and developers had succeeded in demonstrating that the desired modulator bias point could be maintained for at least 15 years at elevated system temperature [181,211–213]. Shortly thereafter, the remaining hurdles, such as establishing thermal and mechanical reliability, conducting field trials, and reaching a price point lower than a solution based on direct modulation, were cleared for both LN and semiconductor modulators—thereby opening the path for system deployment. By 1995, LN electro-optic modulators and indium gallium arsenide phosphide electroabsorption modulators had been deployed in the land-based transmitters of optically amplified undersea and terrestrial long-distance fiber communication systems at a data rate of 2.5 Gb/s and optical reaches of 2000 km and 360 km, respectively [33,95]. Thereafter, virtually all long-distance optical fiber communication systems used external modulation. For example, the Transatlantic telecommunications 12/13 (TAT-12/13) undersea system deployed in 1996, which is a single wavelength system spanning more than 6000 km and operating at 5 Gb/s, used LN external modulators [214]. The TAT-12/13 upgrade in 1999 supported four 2.5-Gb/s wavelength-division multiplexed channels based on integrated laser–electroabsorption modulators. By 1999, terrestrially deployed wavelength-division multiplexed systems supporting many tens of channels and using lithium niobate (LN) external modulators operated at data rates as high as 10 Gb/s per channel [95]. The TAT-14 undersea system deployed in 2001, which supports 16 wavelength channels over a reach of more than 7600 km, also used LN external modulators operating at 10 Gb/s [215].

In parallel with the efforts to introduce external modulators into long-distance digital optical fiber communication systems at the required data rates during this period, device researchers and developers also pursued innovations to meet the raw bandwidth and corresponding drive voltage requirements of future systems. Through the exploration of advanced device fabrication techniques, such as shielding planes and etched substrates to further improve the characteristics of the high-speed traveling-wave electrodes, substantial progress on low-voltage, ultra-broadband modulators was also achieved [216–220].

## 1.8 Sophistication, Integration, and Variations (2000–2010)

Our historical review now brings us to the past decade. As the remainder of this book and other works cover the corresponding advances on optical modulators in detail, here, we briefly highlight some of the more notable trends [14,15]. First, the traffic carried over optical fibers has continued to grow at a very fast pace. Data traffic crossed voice traffic near the beginning of this millennium, and at present, the total traffic carried over global Internet backbones has been assessed to be growing at a compound annual growth rate of approximately 45%, corresponding to a doubling in slightly less than two years [16,221–224]. In absolute terms, today, the two largest carriers in North America each transport approximately 50 petabytes per day (PB/day) across their optical backbone networks [224]. To carry this traffic, as of 2006, ultra-long-haul optical transmission systems supporting 40-Gb/s wavelength channels and 1-Tb/s total capacity had been put into service; and as of 2008, more than 1 terameter (Tm) of optical fiber had been deployed worldwide [225–227]. By 2010, systems supporting 100 Gb/s per wavelength were tested in the field and deployed [228]. All of these high-capacity, long-reach optical transmission systems make use of LN external modulators.

Driven by the demand for more—and also flexibly reconfigurable—optical network capacity, another trend of the past decade has been an evolution resulting in increased capability and adaptability of high-speed optical modulators to encode the optical field of the source laser and thereby control the transmitted spectrum. While the initial motivation to tailor the spectral width of the transmitted signal using optical modulators capable of both amplitude and phase modulation had been to ameliorate dispersion and curb the corresponding system penalty, over time the emphasis has changed to achieve higher spectral density, and thereby reduce filtering penalties, such as incurred in reconfigurable optical add/drop multiplexers, and to increase the total capacity of a single fiber, optically amplified transmission system [13,159,169–171,203,205–208,229–234]. As a result, the design of integrated AM/PM optical modulators has become more sophisticated—evolving from the dual-drive MZ modulator to the nested dual-drive MZ modulator configuration, which bears a structural resemblance to the integrated single-sideband optical modulator and to the dual parallel optical modulator used for linearized analog modulation demonstrated decades earlier [13,126,169,235]. Discussions of advanced modulation formats and the corresponding modulator designs may be found in Chapter 2 of this book.

Other areas of major advances in optical modulator technology during the past decade have been in the further improvement of high-speed performance, drive voltage requirements, modulator integration and arrays, and new or improved material systems [14,158,236,237]. These topics are covered in Chapters 5–12.

## 1.9 Much Better, Much Faster, Much Cheaper—Limits and New Frontiers (2010–2020 and Beyond)

As we have discussed in this chapter, the essential utility of optical modulators stems from the need for additional affordable communication bandwidth, and hence future directions and advances in optical modulators will likely continue to be dictated by these considerations. Substantiating this is a macroeconomic perspective on the contemporary and future growth of overall network traffic. From that view, traffic growth occurs to the extent that the technological productivity of the value chain lowers the cost for transporting a bit of information while the total cost—such as measured as a fraction of all available, but limited, resources—remains roughly constant. In such a scenario, observed traffic growth reflects past increases in productivity, and a projection of future traffic growth presumes continued increases in productivity or a substantial shift in user behavior. An exception is when previously untapped resources become available to permit growth without necessarily increasing productivity. By extrapolating the historical trends in backbone network traffic, we therefore anticipate that there are drivers in the form that would tend to increase the volume of the traffic approximately 30-fold over the decade 2010-2020 [224]. While these expected increases are slightly less than actually achieved in the past, they are still staggering, especially when considering the implications on the reduction in cost and power consumption. They become even more daunting when we consider that practical and fundamental limits are being approached [231]. As in the past, today, we would be hard pressed to identify exactly what fantastic innovations will occur to achieve the expected improvements 10 years from now. We expect that further miniaturization and increased levels of integration will be critical to increase capacity and reduce cost. However, we also recognize that while smaller devices can provide higher modulation bandwidth, it is challenging to increase the bandwidth per unit drive power, for example, to lower the switching energy. Inspired by the challenges and accomplishments of those who have gone before us, perhaps informed by the insights summarized in this book, and driven with a constancy of purpose, we trust that we will find our way on this amazing journey.

# References


1. I. P. Kaminow and E. H. Turner, 1966, Electro-optic light modulators, *Proc. IEEE,* vol. 54, no. 10, pp. 1374–1390.
2. F.-S. Chen, 1970, Modulators for optical communication, *Proc. IEEE,* vol. 58, no. 10, pp. 1440–1457.
3. G. White and G. M. Chin, 1972, Traveling-wave electro-optic modulators, *Opt. Commun.*, vol. 5, no. 4, pp. 374–379.
4. I. P. Kaminow, 1975, Optical waveguide modulators, *IEEE Trans. Microwave Theory Tech.*, vol. 23, no. 1, pp. 57–70.
5. R. C. Alferness, 1982, Waveguide electro-optic modulators, *IEEE Trans. Microwave Theory Tech.*, vol. 30, no. 8, pp. 1121–1137.
6. S. K. Korotky and R. C. Alferness, 1988, Waveguide electro-optic devices for optical fiber communication, in *Optical Fiber Telecommunications II*, S. E. Miller and I. P. Kaminow, Eds., New York: Academic Press.
7. R. G. Walker, 1991, High-speed III–V semiconductor intensity modulators, *IEEE J. Quantum Electron.,* vol. 27, no. 3, pp. 654–667.
8. F. Heismann, S. K. Korotky, and J. J. Veselka, 1997, Lithium niobate integrated optics: Selected contemporary devices and system applications, in *Optical Fiber Telecommunications IIIB*, I. P. Kaminow and T. L. Koch, Eds., New York: Academic Press.
9. K. Wakita, 1998, *Semiconductor Optical Modulators*, Norwell, MA: Kluwer Academic Publishers.
10. E. L. Wooten, K. M. Kissa, A. Yi-Yan, E. J. Murphy, D. A. Lafaw, P. F. Hallemmeier, D. Maack, D. V. Attanasio, D. J. Fritz, G. J. McBrien, and D. E. Bossi, 2000, A review of lithium niobate modulators for fiber-optic communications systems, *IEEE J. Sel. Top. Quantum Electron.*, vol. 6, no. 1, pp. 69–82.
11. A. Mahapatra and E. J. Murphy, 2002, Electro-optic modulators, in *Optical Fiber Telecommunications IV-A*, New York: Academic Press, pp. 258–294.
12. G. T. Reed and C. E. J. Png, 2005, Silicon optical modulators, *Materials Today,* vol. 8, no. 1, pp. 40–50.
13. T. Kawanishi, T. Sakamoto, and M. Izutsu, 2007, High-speed control of lightwave amplitude, phase, and frequency by use of electrooptic effect, *IEEE J. Sel. Area Quantum Electron.*, vol. 13, no. 1, pp. 79–91.
14. K. Noguchi, 2007, Ultra-high-speed $LiNbO_3$ modulators, in *Ultrahigh-Speed Optical Transmission Technology*, Berlin, Germany: Springer, pp. 1–13.
15. L. Thylen, U. Westergren, P. Holmstrom, R. Schatz, and P. Janes, 2008, Recent developments in high-speed optical modulators, in *Optical Fiber Telecommunications V A: Components and Subsystems*, I. P. Kaminow, T. Li, and A. E. Willner, Eds., New York: Academic Press.

16. R. W. Tkach, 2010, Scaling optical communications for the next decade and beyond, *Bell Labs Tech. J.*, vol. 14, no. 4, pp. 3–9.
17. K. C. Kao and G. A. Hockham, 1966, Dielectric-fibre surface waveguides for optical frequencies, *Proc. IEEE*, vol. 113, no. 7, pp. 1151–1158.
18. K. C. Kao and G. A. Hockham, 1966, Dielectric-fibre surface waveguides for optical frequencies, in *Proc. URSI Symp. Electromagnetic Wave Theory*, Gent, Belgium: Radio Science Press, pp. 441–444.
19. K. C. Kao and G. A. Hockham, 1986, Dielectric-fibre surface waveguides for optical frequencies, *IEEE Proc. J. Optoelectronics*, vol. 133, no. 3, pp. 191–198.
20. J. Hecht, 1999, *City of Light: The Story of Fiber Optics*, New York: Oxford University Press.
21. The Royal Swedish Academy of Sciences, 2009, The masters of light. *Press Release* [Online]. Available: http://nobelprize.org/nobel_prizes/physics/laureates/2009/press.html (accessed December 1, 2010).
22. W. Smith, 1873, Effect of light on Selenium during the passage of an electric current, *Nature*, vol. 7, no. 173, p. 303.
23. A. L. Schawlow and C. H. Townes, 1958, Infrared and optical masers, *Phys. Rev.*, vol. 112, no. 6, pp. 1940–1949.
24. I. Jacobs, 1979, Lightwave communications begins regular service, *Bell Laboratories Record* vol. 57, no. 11, pp. 299–304.
25. M. Bertolotti, 2005, *The History of the Laser*, London, England: Institute Physics Pub.
26. I. P. Kaminow and T. L. Koch, Eds., 1997, *Optical Fiber Telecommunications IIIB*, New York: Academic Press.
27. I. P. Kaminow and T. Li, Eds., 2002, *Optical Fiber Telecommunications IV-A—Components*, New York: Academic Press.
28. A. H. Gnauck, B. L. Kasper, R. A. Linke, R. W. Dawson, T. L. Koch, T. J. Bridges, E. G. Burkhardt, R. T. Yen, D. P. Wilt, J. C. Campbell, K. C. Nelson, and L. G. Cohen, 1985, 4 Gb/s transmission over 103 km of optical fiber using a novel electronic multiplexer/demultiplexer, in *Proc. Conf. Opt. Fiber Commun.*, San Diego, CA, p. PD2. Washington, DC: Optical Society of America.
29. S. K. Korotky, G. Eisenstein, A. H. Gnauck, B. L. Kasper, J. J. Veselka, R. C. Alferness, L. L. Buhl, C. A. Burrus, T. C. D. Huo, L. W. Stulz, K. C. Nelson, L. G. Cohen, R. W. Dawson, and J. C. Campbell, 1985, 4 Gb/s transmission experiment over 117 km of optical fiber using a Ti:$LiNbO_3$ external modulator, in *Proc. Conf. Optical Fiber Commun.*, Washington, DC: Optical Society of America, p. PD1.
30. S. K. Korotky, G. Eisenstein, A. H. Gnauck, B. L. Kasper, J. J. Veselka, R. C. Alferness, L. L. Buhl, C. A. Burrus, T. C. D. Huo, L. W. Stulz, K. C. Nelson, L. G. Cohen, R. W. Dawson, and J. C. Campbell, 1985, 4 Gb/s transmission experiment over 117 km of optical fiber using a Ti:$LiNbO_3$ external modulator, *J. Lightwave Technol.*, vol. 3, no. 5, pp. 1027–1031.
31. R. W. Burns, 2004, *Communications: An International History of the Formative Years*, London, United Kingdom: Institution Electrical Engineers.
32. D. R. Smith, 1999, *Digital Transmission Systems*, 2nd Edition, Norwell: Kluwer Academic Publishers.
33. Business Wire, 1994, AT&T launches service on Columbus II cable system. [Online]. Available: http://findarticles.com/p/articles/mi_m0EIN/is_1994_Dec_1/ai_15928522/ (accessed December 1, 2010).
34. N. S. Bergano, 1997, Undersea amplified lightwave systems design, in *Optical Fiber Telecommunications IIIA*, I. P. Kaminow and T. L. Koch, Eds., New York: Academic Press, pp. 302–335.
35. Wikipedia (Retrieved: 2010), Semaphore line. [Online]. Available: http://en.wikipedia.org/wiki/Smoke_signal, http://en.wikipedia.org/wiki/Semaphore_line (accessed December 1, 2010).
36. Wikipedia (Retrieved: 2010), Heliograph. [Online]. Available: http://en.wikipedia.org/wiki/Heliograph (accessed December 1, 2010).
37. A. A. Michelson, 1881, The relative motion of the earth and the luminiferous ether, *Am. J. Sci.*, vol. 22, no. 128, pp. 120–129.
38. H. Hertz, 1887, Über einen Einfluss des ultravioletten Lichtes auf die elektrische Entladung, *Annalen der Physik*, vol. 267, no. 8, pp. 983–1000.

39. M. I. Niven, 2003, *The scientific papers of James Clerk Maxwell, Vol. II*, New York: Cambridge University Press, 1890, Reprinted, Mineola: Dover Publications.
40. L. Zehnder, 1891, Ein neuer Interferenzrefraktor, *Z. Instrumentenkunde*, vol. 11, pp. 275–285.
41. L. Mach, 1892, Über einen Interfernzrefraktor, *Z. Instrumentenkunde*, vol. 12, pp. 89–93.
42. A. Einstein, 1905, Über einen die Erzeugung und Verwandlung des Lichtes betreffenden heuristischen Gesichtsspunkt, *Annalen der Physik*, vol. 322, no. 6, pp. 132–148.
43. A. Einstein, 1917, Zur Quantentheorie der Strahlung, *Physik. Zeitschr.*, vol. 18, pp. 121–128.
44. P. A. M. Dirac, 1958; reprinted 1999, *The Principles of Quantum Mechanics*, 4th ed., New York: Oxford University Press.
45. P. M. Harman, Ed., 1995, *The Scientific Letters and Papers of James Clerk Maxwell, Vol. II: 1862–1873*, New York: Cambridge University Press.
46. J. H. Bryant, 1998, Heinrich Hertz's experiments and experimental apparatus: His discovery of radio waves and his delineation of their properties, in *Heinrich Hertz: Classical Physicist, Modern Philosopher*, D. Baird, R. I. G. Huges, and A. Nordmann, Eds., Hingham: Kluwer Academic Publishers.
47. J. J. Fahie, originally published: 1902, *A History of Wireless Telegraphy*, Charleston: BiblioBazaar.
48. G. Marconi, 1967, Wireless telegraphic communication, *Nobel Lectures, Physics 1901–1921*, Amsterdam, The Netherlands: Elsevier Publishing Co.
49. R. W. Burns, 1988, Soemmering, Schilling, Cooke and Wheatstone, and the electric telegraph, *Meeting History Electrical Engineering*, Twickenham, UK, pp. 70–79.
50. K. D. Stephan, 2001, Experts at play: Magnetron research at Westinghouse, 1930–1934, *Tech. and Culture*, vol. 42, no. 4, pp. 737–749.
51. J. Kerr, 1875, A new relation between electricity and light: Dielectrified media birefringent, *Philos. Mag.*, series 4, vol. 50, no. 332, pp. 337–348.
52. F. K. Pockels, 1893, Über den Einfluß deselektrostatischen Feldes auf das optische Verhalten piezoelektrischer Krystalle, *Abh. König. Ges. Wiss. Göttingen, Math. Phys. Klasse*, vol. 39, pp. 1–204.
53. P. Lenard, 1967, On cathode rays, *Nobel Lectures, Physics 1901–1921*, Amsterdam, The Netherlands: Elsevier Publishing Co.
54. H. J. Round, 1907, A note on carborundum, *Electr. World*, vol. 49, p. 308.
55. O. V. Losev, 1927, Luminous carborundum detector and detection with crystals, *Telegrafiya i Teliefoniya bez Provodov*, vol. 44, pp. 485–494.
56. O. Knudsen, 1976, The Faraday effect and physical theory, 1845–1873, *Archive for History of Exact Sciences*, vol. 15, no. 3, pp. 235–281.
57. W. A. Wooster, 1990, Brief history of physical crystallography, *Historical Atlas of Crystallography*, J. Lima-de Faria, Ed., Norwell: Kluwer Academic, pp. 61–76.
58. H. Aben, 2007, On the role of T. J. Seebeck in the discovery of the photoelastic effect in glass, *Proc. Estonian Acad. Sci. Eng.*, vol. 13, no. 4, pp. 283–294.
59. N. Zheludev, 2007, The life and times of the LED—a 100-year history, *Nature Photonics*, vol. 1, no. 4, pp. 189–192.
60. A. G. Bell, 1880, On the production and reproduction of sound by light: the photophone, *Proc. Am. Assoc. Adv. Sci., 29th Meeting*, Boston, pp. 115–136.
61. A. G. Bell, 1880, The Photophone, *Science*, vol. OS-1, no. 12, pp. 130–134.
62. I. P. Kaminow and E. H. Turner, 1971, Linear electrooptical materials, in *Handbook of Lasers*, R. J. Pressley, Ed., Cleveland: Chemical Rubber Co., pp. 447–459.
63. M. E. Lines and A. M. Glass, 1977, *Principles and applications of ferroelectrics and related materials*, Oxford, England: Clarendon Press.
64. B. T. Matthias and J. P. Remeika, 1949, Ferroelectricity in the ilmenite structure, *Phys. Rev.*, vol. 76, no. 12, pp. 1886–1887.
65. A. A. Ballman, 1965, Growth of piezoelectric and ferroelectric materials by the Czochralski technique, *J. Am. Ceramic Soc.*, vol. 48, no. 2, pp. 112–113.

66. K. Nassau, H. J. Levinstein, and G. M. Loiacono, 1966, Ferroelectric lithium niobate: 1. Growth, domain structure, dislocations and etching, *J. Phys. Chem. Solids*, vol. 27, nos. 6–7, pp. 983–988.
67. K. Nassau, H. J. Levinstein, and G. M. Loiacono, 1966, Ferroelectric lithium niobate: 2. Preparation of single domain crystals, *J. Phys. Chem. Solids*, vol. 27, nos. 6–7, pp. 989–996.
68. T. H. Maiman, 1960, Stimulated optical radiation in ruby, *Nature*, vol. 187, no. 4736, pp. 493–494.
69. Laserfest: Celebrating 50 years of laser innovation, 2010, Available: http://www.laserfest.org (accessed December 1, 2010).
70. R. N. Hall, G. E. Fenner, J. D. Kingsley, T. H. Soltys, and R. O. Carlson, 1962, Coherent light emission from GaAs junctions, *Phys. Rev. Lett.*, vol. 9, no. 9, pp. 366–368.
71. N. Holonyak, Jr. and S. F. Bevacqua, 1962, Coherent (visible) light emission from $Ga(As_{1-x}P_x)$ junctions, *Appl. Phys. Lett.*, vol. 1, no. 4, pp. 82–83.
72. M. I. Nathan, W. P. Dumke, G. Burns, F. H. Hill, Jr., and G. J. Lasher, 1962, Stimulated emission of radiation from GaAs p–n junctions, *Appl. Phys. Lett.*, vol. 1, no. 3, pp. 62–64.
73. T. M. Quist, R. H. Rediker, R. J. Keyes, W. E. Krag, B. Lax, A. L. McWhorter, and H. J. Zeiger, 1962, Semiconductor maser of GaAs, *Appl. Phys. Lett.*, vol. 1, no. 4, pp. 91–92.
74. K. M. Johnson, 1965, High-speed photodiode signal enhancement at avalanche breakdown voltage, *IEEE Trans. Electron Dev.*, vol. 12, no. 2, pp. 55–63.
75. Zh. I. Alferov, V. M. Andreev, D. Z. Garbuzov, Yu. V. Zhilyaev, E. P. Morozov, E. L. Portnoi, and V. G. Trofim, 1970, Investigation of the influence of the AlAs-GaAs heterostructure parameters on the laser threshold current and the realization of continuous emission at the room temperature, *Fiz. Tekh. Poluprovodn.*, vol. 4, pp. 1826–1829.
76. I. Hayashi, M. Panish, P. Foy, and S. Sumski, 1970, Junction lasers which operate continuously at room temperature, *Appl. Phys. Lett.*, vol. 17, no. 3, pp. 109–111.
77. J. W. Orton, 2004, *The Story of Semiconductors*, Oxford, UK: Oxford University Press, pp. 184–194.
78. C. J. Koestler and E. Snitzer, 1964, Amplification in a fiber laser, *Appl. Opt.*, vol. 3, no. 10, pp. 1182–1186.
79. J. Stone and C. A. Burrus, 1973, Neodymium-doped silica lasers in end-pumped geometry, *Appl. Phys. Lett.*, vol. 23, no. 7, pp. 388–389.
80. N. Bloembergen, P. S. Pershan, and L. R. Wilcox, 1960, Microwave modulation of light in paramagnetic crystals, *Phys. Rev.*, vol. 120, no. 6, pp. 2014–2023.
81. I. P. Kaminow, 1961, Microwave modulation of the electro-optic effect in $KH_2PO_4$, *Phys. Rev. Lett.*, vol. 6, no. 10, pp. 528–530.
82. S. E. Harris, B. J. McMurty, and A. E. Siegman, 1962, Modulation and direct demodulation of coherent and incoherent light at a microwave frequency, *Appl. Phys. Lett.*, vol. 1, no. 2, pp. 37–39.
83. D. F. Nelson and F. K. Reinhart, 1964, Light modulation by the electro-optic effect in reverse biased GaP p–n junctions, *Appl. Phys. Lett.*, vol. 5, no. 7, pp. 148–150.
84. R. T. Denton, F. S. Chen, K. Nassau, and A. A. Ballman, 1966, Optical modulators with low drive power requirements, *IEEE J. Quantum Electron.*, vol. 2, no. 4, p. 129.
85. F. K. Reinhart, 1968, Reversed-biased gallium phosphide diodes as high-frequency light modulators, *J. Appl. Phys.*, vol. 39, no. 7, pp. 3426–3434.
86. F. K. Reinhart, D. F. Nelson, and J. McKenna, 1968, Waveguide and electro-optic properties of reverse-biased GaP p–n junction, *IEEE J. Quantum Electron.*, vol. 4, no. 5, pp. 364–365.
87. K. K. Chow and W. B. Leonard, 1970, Efficient octave-bandwidth microwave light modulators, *IEEE J. Quantum Electron.*, vol. 6, no. 12, pp. 789–793.
88. H. V. Hance, R. C. Ohlmann, D. G. Peterson, R. B. Ward, and K. K. Chow, 1970, Ultra-wide bandwidth laser communications: Part II—An operating laboratory system, *Proc. IEEE*, vol. 58, no. 10, pp. 1714–1719.
89. I. P. Kaminow, T. J. Bridges, and M. A. Pollack, 1970, A 964-GHz traveling-wave electrooptic light modulator, *Appl. Phys. Lett.*, vol. 16, no. 11, pp. 416–418.
90. S. E. Miller, 1969, Integrated optics: An introduction, *Bell Syst. Tech. J.*, vol. 48, no. 7, pp. 2059–2069.

91. E. A. J. Marcatili, 1969, Dielectric rectangular waveguide and directional coupler for integrated optics, *Bell Syst. Tech. J.*, vol. 48, no. 7, pp. 2071–2102.
92. F. P. Kapron, D. B. Keck, and R. D. Maurer, 1970, Radiation losses in glass optical waveguides, *Appl. Phys. Lett.*, vol. 17, no. 10, pp. 423–425.
93. S. E. Miller, E. A. J. Marcatili, and T. Li, 1973, Research toward optical-fiber transmission systems, Part I: The transmission medium, *Proc. IEEE*, vol. 61, no. 12, pp. 1703–1726.
94. S. E. Miller, T. Li, and E. A. J. Marcatili, 1973, Research toward optical-fiber transmission systems, Part II: Devices and systems considerations, *Proc. IEEE*, vol. 61, no. 12, pp. 1726–1751.
95. R. C. Alferness, T. H. Wood, and H. Kogelnik, 2000, The evolution of optical systems: Optics everywhere, *Bell Labs Tech. J.*, vol. 5, no. 1, pp. 188–202.
96. Optical Society of America, 1975, *Tech. Dig. Top. Meet. Optical Fiber Transmission (Williamsburg, VA)*, 1975, Washington, DC: Optical Society of America.
97. I. P. Kaminow, 1974, *An Introduction to Electrooptic Devices*, New York: Academic Press.
98. F. K. Reinhart and B. I. Miller, 1972, Efficient GaAs-AlxGa1-xAs double-heterostructure light modulators, *Appl. Phys. Lett.*, vol. 20, no. 1, pp. 36–38.
99. R. V. Schmidt and I. P. Kaminow, 1974, Metal-diffused optical waveguides in $LiNbO_3$, *Appl. Phys. Lett.*, vol. 25, no. 8, pp. 458–460.
100. J. C. Campbell, F. A. Blum, D. W. Shaw, and K. L. Lawley, 1975, GaAs electro-optic directional coupler switch, *Appl. Phys. Lett.*, vol. 27, no. 4, pp. 202–204.
101. I. P. Kaminow, L. W. Stulz, and E. H. Turner, 1975, Efficient strip-waveguide modulator, *Appl. Phys. Lett.*, vol. 27, no. 10, pp. 555–557.
102. M. Papuchon, Y. Combemale, X. Mathieu, D. B. Ostrowsky, L. Reiber, A. M. Roy, B. Sejourne, and M. Werner, 1975, Electrically switched optical directional coupler: COBRA, *Appl. Phys. Lett.*, vol. 27, no. 5, pp. 289–291.
103. R. V. Schmidt and H. Kogelnik, 1976, Electro-optically switched coupler with stepped $\Delta\beta$ reversal using Ti-diffused $LiNbO_3$ waveguides, *Appl. Phys. Lett.*, vol. 28, no. 9, pp. 503–506.
104. M. Izutsu, Y. Yamane, and T. Sueta, 1977, Broad-band traveling-wave modulator using a $LiNbO_3$ optical waveguide, *IEEE J. Quantum Electron.*, vol. 13, no. 4, pp. 287–290.
105. P. K. Tien, 1977, Integrated optics and new wave phenomena in optical waveguides, *Rev. Mod. Phys.*, vol. 49, no. 2, pp. 361–420.
106. W. K. Burns, T. G. Giallorenzi, R. P. Moeller, and E. J. West, 1978, Interferometric waveguide modulator with polarization-independent operation, *Appl. Phys. Lett.*, vol. 33, no. 11, pp. 944–947.
107. V. Ramaswamy, M. D. Divino, and R. D. Standley, 1978, A balanced bridge modulator switch using Ti-diffused $LiNiO_3$ strip waveguides, *Appl. Phys. Lett.*, vol. 32, no. 10, pp. 644–646.
108. R. V. Schmidt and P. S. Cross, 1978, Efficient optical waveguide switch/amplitude modulator, *Opt. Lett.*, vol. 2, no. 2, pp. 45–47.
109. P. S. Cross and R. V. Schmidt, 1979, A 1 Gbit/s integrated optical modulator, *IEEE J. Quantum Electron.*, vol. 15, no. 12, pp. 1415–1418.
110. M. Minakata, 1979, Efficient $LiNbO_3$ balanced bridge modulator/switch with ion-etched slot, *Appl. Phys. Lett.*, vol. 35, no. 1, pp. 40–42.
111. A. Neyer and W. Sohler, 1979, High-speed cutoff modulator using a Ti-diffused $LiNbO_3$ channel waveguide, *Appl. Phys. Lett.*, vol. 35, no. 3, pp. 256–258.
112. K. Kubota, J. Noda, and O. Mikami, 1980, Traveling wave optical modulator using a directional coupler $LiNbO_3$ waveguide, *IEEE J. Quantum Electron.*, vol. 16, no. 7, pp. 754–760.
113. P. Thioulouse, A. Carenco, and R. Guglielmi, 1981, High-speed modulation of an electro-optic directional coupler, *IEEE J. Quantum Electron.*, vol. 17, no. 4, pp. 535–541.
114. I. Jacobs, 1980, Lightwave communications—yesterday, today, and tomorrow, *Bell Laboratories Record*, vol. 58, pp. 210.
115. J. J. Hsieh, 1978, GaInAsP/InP lasers and detectors for fiber optics communications at 1.1–1.3μm, *1978 Int. Electron Devices Meeting*, vol. 24, pp. 628–629.

116. N. Niizeki, 1981, Recent progress in glass fibers for optical communication, *Jpn. J. Appl. Phys.*, vol. 20, no. 8, pp. 1347–1360.
117. P. Henry, 1985, Lightwave primer, *IEEE J. Quantum Electron.*, vol. 21, no. 12, pp. 1862–1879.
118. L. D. Hutcheson, Ed., 1987, *Integrated Optical Circuits and Components: Design and Applications*, New York: Marcel Dekker.
119. P. LeFur and D. H. Auston, 1976, A kilovolt picosecond optoelectronic switch and Pockel's cell, *Appl. Phys. Lett.*, vol. 28, no. 1, pp. 21–23.
120. M. Papuchon and C. Puech, 1978, Integrated optics: A possible solution for the fiber gyroscope, *Proc. Soc. Photo-Opt. Instrum. Eng.*, vol. 157, pp. 218–219.
121. F. J. Leonberger, C. E. Woodward, and D. L. Spears, 1979, Design and development of a high-speed electro-optic A/D converter, *IEEE Trans. Circuits Syst.*, vol. 26, no. 12, pp. 1125–1131.
122. R. C. Alferness, N. P. Economou, and L. L. Buhl, 1980, Picosecond optical sampling technique for measuring the speed of fast electro-optic switch/modulators, *Appl. Phys. Lett.*, vol. 37, no. 7, pp. 597–599.
123. H. I. Bassen, C. H. Bulmer, and W. K. Burns, 1980, An RF field strength measurement system using an integrated optical linear modulator, *1980 MTT-S Int. Microwave Symp., Dig.*, vol. 80, pp. 317–318.
124. H. Haus, S. Kirch, K. Mathyssek, and F. J. Leonberger, 1980, Picosecond optical sampling, *IEEE J. Quantum Electron.*, vol. 16, no. 8, pp. 870–880.
125. E. A. J. Marcatili, 1980, Optical picosecond gate, *Appl. Opt.*, vol. 19, no. 9, pp. 1468–1476.
126. M. Izutsu, S. Shikama, and T. Sueta, 1981, Integrated optical SSB modulator/frequency shifter, *IEEE J. Quantum Electron.*, vol. 17, no. 11, pp. 2225–2227.
127. F. Heismann and Ulrich, 1982, Integrated-optical single-sideband modulator and phase shifter, *IEEE J. Quantum Electron.*, vol. 18, no. 4, pp. 767–771.
128. M. Kondo, Y. Ohta, M. Fujiwara, and M. Sakaguchi, 1982, Integrated optical switch matrix for single-mode fiber networks, *IEEE Tran. Microwave Theory Tech.*, vol. 30, no. 10, pp. 1747–1753.
129. J. A. Valdmanis, G. Mourou, and C. W. Gabel, 1982, Picosecond electro-optic sampling system, *Appl. Phys. Lett.*, vol. 41, no. 3, pp. 211–212.
130. C. M. Gee, G. D. Thurmond, and H. W. Yen, 1983, 17-GHz bandwidth electro-optic modulator, *Appl. Phys. Lett.*, vol. 43, no. 11, pp. 998–1000.
131. R. Kist and W. Sohler, 1983, Fiber-optic spectrum analyzer, *J. Lightwave Technol.*, vol. 1, no. 1, pp. 105–110.
132. B. H. Kolner, D. M. Bloom, and P. S. Cross, 1983, Electro-optic sampling with picosecond resolution, *Electron. Lett.*, vol. 19, no. 15, pp. 574–575.
133. R. C. Alferness, S. K. Korotky, L. L. Buhl, and M. D. Divino, 1984, High-speed, low-loss, low-drive-power traveling-wave optical modulator for $\lambda = 1.32$ μm, *Electron. Lett.*, vol. 20, no. 8, pp. 354–355.
134. R. C. Alferness, S. K. Korotky, and E. A. J. Marcatili, 1984, Velocity-matching techniques for integrated-optic traveling-wave switch/modulators, *IEEE J. Quantum Electron.*, vol. 20, no. 3, pp. 301–309.
135. R. A. Becker, 1984, Traveling-wave electro-optic modulator with maximum bandwidth-length product, *Appl. Phys. Lett.*, vol. 45, no. 11, pp. 1218–1170.
136. R. A. Becker, C. E. Woodward, F. J. Leonberger, R. C. Williamson, 1984, Wide-band electro-optic guided-wave analog-to-digital converters, *Proc. IEEE*, vol. 72, no. 7, pp. 802–819.
137. S. K. Korotky, G. Eisenstein, B. L. Kasper, R. C. Alferness, J. J. Veselka, and L. L. Buhl, 1984, Error-free external modulation of a single-frequency injection laser at 1.5 Gbit/s using a Ti:$LiNbO_3$ waveguide switch, *Electron. Lett.*, vol. 20, no. 21, pp. 878–879.
138. M. Shikada, E. Emura, S. Fujita, M. Kitamura, M. Arai, M. Kondo, and K. Minemura, 1984, 100-Mbit/sec ASK heterodyne detection experiment using 1.3-μm DFB laser diodes, *Tech. Dig. Conf. Optical Fiber Commun.*, New Orleans, LA, paper TUK6, pp. 62–64.

139. S. K. Korotky, G. Eisenstein, R. C. Alferness, J. J. Veselka, L. L. Buhl, and G. T. Harvey, 1985, Fully connectorized high-speed Ti:$LiNbO_3$ switch/modulator for time-division multiplexing and data encoding, *J. Lightwave Technol.*, vol. 3, no. 1, pp. 1–6.
140. T. H. Wood, C. A. Burrus, D. A. B. Miller, D. S. Chemla, T. C. Damen, A. C. Gossard, and W. Wiegmann, 1985, 131-ps optical modulation in semiconductor multiple quantum wells (MQW's), *IEEE J. Quantum Electron.*, vol. 21, no. 2, pp. 117–118.
141. R. C. Alferness, L. L. Buhl, M. D. Divino, S. K. Korotky, and L. W. Stulz, 1986, Low-loss, broad-band Ti:$LiNbO_3$ waveguide phase modulators for coherent systems, *Electron. Lett.*, vol. 22, no. 6, pp. 309–310.
142. P. J. Duthie, M. J. Wale, I. Bennion, and J. Hankey, 1986, Bidirectional fibre-optic link using reflective modulation, *Electron. Lett.*, vol. 22, no. 10, pp. 517–518.
143. A. H. Gnauck, S. K. Korotky, B. L. Kasper, J. C. Campbell, J. Talman, J. J. Veselka, and A. McCormick, 1986, Information-bandwidth-limited transmission at 8 Gb/s over 68.3 km of single-mode optical fiber, *Proc. Conf. Opt. Fiber Commun.*, Atlanta, GA, paper PDP9.
144. R. A. Linke, B. L. Kasper, N. A. Olsson, and R. C. Alferness, 1986, Coherent lightwave transmission over 150 km fibre lengths at 400 Mbits/s and 1 Gbit/s data rates using phase modulation, *Electron. Lett.*, vol. 22, no. 1, pp. 30–13.
145. Y. Noda, M. Suzuki, Y. Kushiro, and S. Akiba, 1986, High-speed electroabsorption modulator with strip-loaded GaInAsP planar waveguide, *J. Lightwave Technol.*, Washington, DC: Optical Society of America, vol. 4, no. 10, pp. 1445–1453.
146. T. H. Wood, E. C. Carr, B. L. Kasper, R. A. Linke, C. A. Burrus, and K. L. Walker, 1986, Bidirectional fibre-optical transmission using a multiple-quantum-well (MQW) modulator/detector, *Electron. Lett.*, vol. 22, no. 10, pp. 528–529.
147. S. K. Korotky, G. Eisenstein, R. S. Tucker, J. J. Veselka, and G. Raybon, 1987, Optical intensity modulation to 40 GHz using a waveguide electro-optic switch, *Appl. Phys. Lett.*, vol. 50, no. 23, pp. 1631–1633.
148. T. H. Okiyama, H. Nishimoto, T. Touge, M. Seino, and H. Nakajima, 1987, Optical transmission over 132 km at 4 Gb/s using a Ti:$LiNbO_3$ Mach-Zehnder modulator, *Proc. European Conf. Optical Commun.*, Helsinki, vol. 3, London: Institution of Electrical Engineers, postdeadline pp. 55–58.
149. W. E. Stephens and T. R. Josephs, 1987, System characteristics of directly modulated and externally modulated RF fiber-optic links, *J. Lightwave Technol.*, vol. 5, no. 3, pp. 380–387.
150. M. Suzuki, Y. Noda, H. Tanaka, S. Akiba, Y. Kushiro, and H. Isshiki, 1987, Monolithic integration of InGaAsP/InP distributed feedback laser and electroabsorption modulator by vapor phase epitaxy, *J. Lightwave Technol.*, vol. 5, no. 9, pp. 1277–1285.
151. T. Okiyama, H. Nishimoto, I. Yokota, and T. Touge, 1988, Evaluation of 4-Gbit/s optical fiber transmission distance with direct and external modulation, *J. Lightwave Technol.*, vol. 6, no. 11, pp. 1686–1692.
152. W. J. Minford, F. T. Stone, B. R. Youmans, and R. K. Bartman, 1989, Fiber optic gyroscope using an eight-component $LiNbO_3$ integrated optic circuit, *Fiber Optic and Laser Sensors VII, Proc. Soc. Photo-Opt. Instrum. Eng.*, vol. 1169, pp. 304–309.
153. H. Soda, T. Okiyama, M. Furutsu, K. Sato, M. Matsuda, I. Yokota, H. Nishimoto, and H. Ishikawa, 1989, 5 Gb/s transmission experiment using a monolithic electroabsorption modulator/DFB laser light source, *Proc. Conf. Optical Fiber Commun.*, Houston, TX, paper PD1.
154. R. B. Childs and V. A. O'Byrne, 1990, Multichannel AM video transmission using a high-power Nd:YAG laser and linearized external modulator, *IEEE J. Select. Areas Commun.*, vol. 8, no. 7, pp. 1369–1376.
155. C. H. Cox, III, G. E. Betts, and L. M. Johnson, 1990, An analytic and experimental comparison of direct and external modulation in analog fiber-optic links, *IEEE Trans. Microwave Theory Tech.*, vol. 38, no. 5, pp. 501–509.

156. P. G. Suchoski, T. K. Findakly, and F. J. Leonberger, 1990, Integrated optical devices for fiber gyroscope applications, in *Proc. Conf. Optical Fiber Commun.*, San Francisco, CA, paper FB3, Washington, DC: Optical Society of America, p. 203.
157. R. L. Jungerman and D. W. Dolfi, 1991, Frequency domain network analysis using integrated optics, *IEEE J. Quantum Electron.*, vol. 27, no. 3, pp. 580–587.
158. T. Tsang and V. Radeka, 1995, Electro-optical modulators in particle detectors, *Rev. Sci. Instrum.*, vol. 66, no. 7, pp. 3844–3854.
159. J. J. Veselka and S. K. Korotky, 1996, Pulse generation for soliton systems using lithium niobate modulators, *IEEE J. Sel. Top. Quantum Electron.*, vol. 2, no. 2, pp. 300–310.
160. N. J. Parsons, A. C. O'Donnell, and K. K. Wong, 1986, Design of efficient and wideband traveling-wave modulators, in *Proc. SPIE Integrated Optical Circuit Engineering III,* vol. 651, paper 24, Bellingham, Washington: International Society for Optics and Photonics, p. 148.
161. S. K. Korotky and R. C. Alferness, 1987, The Ti:$LiNbO_3$ integrated-optic technology: Fundamentals, design considerations, and capabilities, in *Integrated Optical Circuits and Components: Design and Applications*, L. D. Hutcheson, Ed., New York: Marcel Dekker.
162. S. K. Korotky, 1989, Optimization of traveling-wave integrated-optic modulators, in *Workshop Numerical Simulation Analysis Guided-Wave Optics Optoelectronics*, Houston, TX, paper SF2.
163. S. Y. Wang and S. H. Lin, 1988, High-speed III–V electro-optic waveguide modulators at $\lambda = 1.3$ μm, *J. Lightwave Technol.*, vol. 6, no. 6, pp. 758–771.
164. T. H. Wood, 1988, Multiple quantum well (MQW) waveguide modulators, *J. Lightwave Technol.*, vol. 6, no. 6, pp. 743–757.
165. K. Kawano, T. Kitoh, H. Jumonji, T. Nozawa, and M. Yanagibashi, 1989, New traveling-wave electrode Mach-Zehnder optical modulator with 20 GHz bandwidth and 4.7 V driving voltage at 1.52 μm wavelength, *Electron. Lett.*, vol. 25, no. 20, pp. 1382–1383.
166. T. L. Koch and J. E. Bowers, 1984, Nature of wavelength chirping in directly modulated semiconductor lasers, *Electron Lett.*, vol. 20, no. 25, pp. 1038–1039.
167. J. C. Cartledge and G. S. Burley, 1989, The effect of laser chirping on lightwave system performance, *J. Lightwave Technol.*, vol. 7, no. 3, pp. 568–573.
168. F. Koyama and K. Iga, 1988, Frequency chirping in external modulators, *J. Lightwave Technol.*, vol. 6, no. 1, pp. 87–93.
169. S. K. Korotky, J. J. Veselka, C. T. Kemmerer, W. J. Minford, D. T. Moser, J. E. Watson, C. A. Mattoe, and P. L. Stoddard, 1991, High-speed, low-power optical modulator with adjustable chirp parameter, in *Tech. Dig. Top. Meet. Integrated Photonics Research,* Monterey, paper TuG2.
170. A. H. Gnauck, S. K. Korotky, J. J. Veselka, J. Nagel, C. T. Kemmerer, W. J. Minford, and D. T. Moser, 1991, Dispersion penalty reduction using an optical modulator with adjustable chirp, in *Proc. Conf. Optical Fiber Commun.*, Washington, DC: Optical Society of America, paper PDP17.
171. A. H. Gnauck, S. K. Korotky, J. J. Veselka, J. Nagel, C. T. Kemmerer, W. J. Minford, and D. T. Moser, 1991, *IEEE Photonics Technol. Lett.*, vol. 3, no. 10, pp. 916–918.
172. P. K. Runge and P. R. Trischitta, 1986, The SL undersea lightwave system, in *Undersea Lightwave Communications*, P. K. Runge and P. R. Trischitta, Eds., New York: IEEE Press.
173. P. K. Runge and N. S. Bergano, 1988, Undersea cable transmission systems, in *Optical Fiber Telecommunications II*, New York: Academic Press.
174. P. S. Henry, 1988, Introduction to lightwave systems, in *Optical Fiber Telecommunications II*, S. E. Miller and I. P. Kaminow, Eds., New York: Academic Press.
175. Y. Niiro, Y. Ejiri, and H. Yamamoto, 1989, The first transpacific optical fiber submarine cable system, *Proc. IEEE Int. Conf. Communications,* Piscataway, NJ: Institute of Electrical and Electronic Engineers, vol. 3, pp. 1520–1524.
176. R. C. Alferness, V. Ramaswamy, S. K. Korotky, M. D. Divino, and L. L. Buhl, 1982, Efficient single-mode fiber to titanium diffused lithium niobate waveguide coupling for $\lambda = 1.32$ μm, *IEEE J. Quantum Electron.*, vol. 18, no. 10, pp. 1807–1813.

177. S. K. Korotky, W. J. Minford, L. L. Buhl, M. D. Divino, and R. C. Alferness, 1982, Mode size and method for estimating the propagation constant of single-mode Ti:LiNbO$_3$ strip waveguides, *IEEE J. Quantum Electron.*, vol. 18, no. 10, pp. 1796–1801.
178. W. J. Minford, S. K. Korotky, and R. C. Alferness, 1982, Low-loss Ti:LiNbO$_3$ waveguide bends at $\lambda$ = 1.3 µm, *IEEE J. Quantum Electron.*, vol. 18, no. 10, pp. 1802–1806.
179. F. Auracher, D. Schicketanz, and K. H. Zeitler, 1984, High-speed $\Delta\beta$-reversal directional coupler modulator with low insertion loss, *J. Opt. Commun.*, vol. 5, pp. 7–9.
180. J. J. Veselka and S. K. Korotky, 1986, Optimization of Ti:LiNbO$_3$ optical waveguides and directional coupler switches for 1.56 µm wavelength, *IEEE J. Quantum Electron.*, vol. 22, no. 6, pp. 933–938.
181. D. A. Fishman, 1993, Design and performance of externally modulated 1.5-µm transmitter in the presence of chromatic dispersion, *J. Lightwave Technol.*, vol. 11, no. 4, pp. 624–632.
182. S. B. Poole, D. N. Payne, R. J. Mears, M. E. Fermann, and R. I. Laming, 1986, Fabrication and characterization of low-loss optical fibers containing rare-earth ions, *J. Lightwave Technol.*, vol. 4, no. 7, pp. 870–876.
183. E. Desurvire, J. R. Simpson, and P. C. Becker, 1987, High-gain erbium-doped traveling-wave fiber amplifier, *Opt. Lett.*, vol. 12, no. 11, pp. 888–890.
184. R. J. Mears, L. Reekie, I. M. Jauncey, and D. N. Payne, 1987, Low-noise erbium-doped fibre amplifier operating at 1.54 µm, *Electron. Lett.*, vol. 23, no. 19, pp. 1026–1028.
185. W. J. Tomlinson, 1977, Wavelength multiplexing in multimode optical fibers, *Appl. Opt.*, vol. 16, no. 8, pp. 2180–2194.
186. N. A. Olsson, J. Hegarty, R. A. Logan, L. F. Johnson, K. L. Walker, L. G. Cohen, B. L. Kasper, and J. C. Campbell, 1985, 68.3 km transmission with 1.37 Tbit/s capacity using wavelength division multiplexing of ten single-frequency lasers at 1.5 µm, in *Proc. Conf. Optical Fiber Commun.*, Washington, DC: Optical Society of America, paper WB6, pp. 88-89.
187. N. A. Olsson, J. Hegarty, R. A. Logan, L. F. Johnson, K. L. Walker, L. G. Cohen, B. L. Kasper, and J. C. Campbell, 1985, 68.3 km transmission with 1.37 Tbit/s capacity using wavelength division multiplexing of ten single-frequency lasers at 1.5 µm, *Electron. Lett.*, vol. 21, no. 3, pp. 105–106.
188. J. T. Boyd, Ed., 1991 *Integrated Optics: Devices and Applications*, New York: IEEE Press.
189. E. J. Murphy, Ed., 1999, *Integrated Optical Circuits and Components: Design and Applications*, New York: Marcel Dekker.
190. A. R. Chraplyvy, A. H. Gnauck, R. W. Tkach, and R. M. Derosier, 1993, 8 × 10 Gb/s transmission through 280 km of dispersion-managed fiber, *IEEE Photon. Technol. Lett.*, vol. 5, no. 10, pp. 1233–1235.
191. N. A. Jackman, S. H. Patel, B. P. Mikkelsen, and S. K. Korotky, 1999, Optical cross-connects for optical networking, *Bell Labs Tech. J.*, vol. 4, no. 1, pp. 262–281.
192. C. D. Chen, J.-M. Delavaux, B. W. Hakki, O. Mizuhara, T. V. Nguyen, R. J. Nuyts, K. Ogawa, Y. K. Park, R. E. Tench, L. D. Tzeng, and P. D. Yeates, 1994, Field experiment of 10 Gb/s, 360 km transmission through embedded standard (non-DSF) fibre cables, *Electron. Lett.*, vol. 30, no. 14, pp. 1159–1160.
193. R. A. Jensen, R. E. Tench, D. G. Duff, C. R. Davidson, C. D. Chen, O. Mizuhara, T. V. Nguyen, L. D. Tseng, and P. D. Yeates, 1995, Field measurements of 10-Gb/s line-rate transmission on the Columbus-IIB submarine lightwave system, *IEEE Photon. Technol. Lett.*, vol. 7, no. 11, pp. 1366–1368.
194. F. J. Leonberger, 1980, High-speed operation of a LiNbO$_3$ electro-optic interferometric waveguide modulators, *Optics Lett.*, vol. 5, no. 7, pp. 312–314.
195. S. K. Korotky and R. C. Alferness, 1983, Time- and frequency-domain response of directional-coupler traveling-wave optical modulators, *J. Lightwave Technol.*, vol. 1, no. 1, pp. 244– 251.
196. S. K. Korotky, 1986, Three-space representation of phase-mismatch switching in coupled two-state optical systems, *IEEE J. Quantum Electron.*, vol. 22, no. 6, pp. 952–958.
197. J. P. Donnelly and A. Gopinath, 1987, A comparison of power requirements of traveling-wave LiNbO$_3$ optical couplers and interferometric modulators, *IEEE J. Quantum Electron.*, vol. 23, no. 1, pp. 30–41.

198. J. J. Veselka, S. K. Korotky, C. T. Kemmerer, W. J. Minford, D. T. Moser, and R. W. Smith, 1992, Sensitivity to RF drive power and the temperature stability of Mach-Zehnder modulators, in *Tech. Dig. Top. Meeting Integrated Photonics Research*, Washington, DC: Optical Society of America, paper TuG4.
199. H. Kogelnik, 1979, Theory of dielectric waveguides, in *Integrated Optics*, 2nd Edition, T. Tamir, Ed., New York: Springer-Verlag.
200. J. J. Veselka and S. K. Korotky, 1989, Velocity-matched Ti:$LiNbO_3$ switch for 16 Gb/s optical time-division multiplexing, in *Tech. Dig. Top. Meeting Photonic Switching*, Washington, DC: Optical Society of America, paper ThA2.
201. T. Namiki, H. Hamano, T. Yamane, M. Seino, and H. Nakajima, 1989, Perfectly chirpless and low drive voltage Ti:$LiNbO_3$ Mach-Zehnder modulator with two traveling-wave electrodes, in *Tech. Dig. Seventh Int. Conf. Integrated Optics and Optical Fiber Commun.*, Kobe, paper 19D4-2.
202. T. Namiki, N. Mekada, H. Hamano, T. Yamane, M. Seino, and H. Nakajima, 1990, Low-drive-voltage Ti:$LiNbO_3$ Mach-Zehnder modulator using a coupled line, in *Proc. Conf. Optical Fiber Commun.*, Washington, DC: Optical Society of America, paper TUH4.
203. J. C. Cartledge and R. G. McKay, 1992, Performance of 10 Gb/s lightwave systems using an adjustable chirp optical modulator and linear equalization, *IEEE Photon. Tech. Lett.*, vol. 4, no. 12, pp. 1394–1397.
204. S. K. Korotky, P. B. Hansen, L. Eskildsen, and J. J. Veselka, 1995, Efficient phase modulation scheme for suppressing stimulated Brillouin scattering, in *Proc. Conf. Integrated Optics and Opt. Commun.*, Hong Kong, paper WD2-1.
205. A. J. Price, L. Pierre, R. Uhel, and V. Havard, 1995, 210 km repeaterless 10 Gb/s transmission experiment through nondispersion-shifted fiber using partial response scheme, *IEEE Photon. Technol. Lett.*, vol. 9, no. 10, pp. 1219–1221.
206. L. F. Mollenauer, P. V. Mamyshev, and M. J. Neubelt, 1996, Demonstration of soliton WDM transmission at up to 8 × 10 Gbit/s, error-free over transoceanic distances, in *Proc. Conf. Optical Fiber Commun.*, Washington, DC: Optical Society of America, paper PD22.
207. J. B. Stark, J. E. Mazo, and R. Laroia, 1998, Phased amplitude-shift signaling (PASS) codes: Increasing the spectral efficiency of DWDM transmission, in *Proc. 24th European Conf. Optical Commun.*, London: Institution of Electrical Engineers, vol. 1, pp. 373–374.
208. S. K. Kim, O. Mizuhara, Y. K. Park, L. D. Tzeng, Y. S. Kim, and J. Jeong, 1999, Theoretical and experimental study of 10 Gb/s transmission performance using 1.55 μm $LiNbO_3$-based transmitters with adjustable extinction ratio and chirp, *J. Lightwave Technol.*, vol. 17, no. 8, pp. 1320–1325.
209. S. Yamada and M. Minakata, 1981, DC drift phenomena in $LiNbO_3$ optical waveguide devices, *Japanese J. Appl. Phys.*, vol. 20, no. 4, pp. 733–737.
210. S. K. Korotky and J. J. Veselka, 1996, RC network model of long term Ti:$LiNbO_3$ bias stability, *J. Lightwave Technol.*, vol. 14, no. 12, pp. 2687–2697.
211. C. R. Giles and S. K. Korotky, 1998, Stability of Ti:$LiNbO_3$ waveguide modulators in an optical transmission system, in *Dig. Top. Meeting Integrated Guided-Wave Optics*, Washington, DC: Optical Society of America, paper ME5.
212. D. A. Fishman, J. A. Nagel, and S. M. Bahsoun, 1991, Roaring Creek field trial: Transmission results, internal memorandum, unpublished; See [180].
213. M. Seino, T. Nakazawa, Y. Kubota, M. Doi, T. Yamane, and H. Hakogi, 1992, A low dc-drift Ti:$LiNbO_3$ modulator assured over 15 years, *Proc. Conf. Optical Fiber Commun.*, San Jose, CA, paper PD3.
214. P. Trischitta, M. Colas, M. Green, G. Wuzniak, and J. Arena, 1996, The TAT-12/13 cable network, *IEEE Commun. Mag.*, vol. 34, no. 2, pp. 24–28.
215. N. S. Bergano and H. Kidorf, 2001, Global undersea cable networks, *Optics Photonics News*, vol. 12, no. 3, pp. 32–35.
216. K. Noguchi, K. Kawano, T. Nozawa, and T. Suzuki, 1991, A Ti:$LiNbO_3$ optical intensity modulator with more than 20 GHz bandwidth and 5.2V driving voltage, *IEEE Photonics Technol. Lett.*, vol. 3, no. 4, pp. 333–335.

217. D. W. Dolfi and T. R. Ranganth, 1992, 50 GHz velocity-matched broad wavelength $LiNbO_3$ modulator with multimode active section, *Electron. Lett.*, vol. 28, no. 13, pp. 1197–1198.
218. G. K. Gopalakrishnan, W. K. Burns, R. W. McElhanon, C. H. Bulmer, and A. S. Greenblatt, 1994, Performance and modeling of broadband $LiNbO_3$ traveling wave optical intensity modulators, *J. Lightwave Technol.*, vol. 12, no. 10, pp. 1807–1819.
219. K. Noguchi, O. Mitomi, and H. Miyazawa, 1998, Millimeter-wave Ti:$LiNbO_3$ optical modulators, *J. Lightwave Technol.*, vol. 16, no. 4, pp. 615–619.
220. W. K. Burns, M. M. Howerton, R. P. Moeller, R. Krahenbuhl, R. W. McElhannon, and A. S. Greenblatt, 1999, Low drive voltage, broad-band $LiNbO_3$ modulators with and without etched ridges, *J. Lightwave Technol.*, vol. 17, no. 12, pp. 2551–2555.
221. P. K. Mutooni, 1997, Telecommunications @ crossroads: The transition from a voice-centric to a data-centric communication network, M.S. thesis, Dept. E. E. Comp. Sci., MIT, Cambridge, MA.
222. K. G. Coffman and A. Odlyzko, 2008, The size and growth rate of the Internet, *First Monday*, vol. 3, no. 10.
223. C. Labovitz, S. Iekel-Johnson, D. McPherson, J. Oberheide, F. Jahanian, and M. Karir, 2009, ATLAS Internet Observatory 2009 Annual Report, *Meet. N. Amer. Netw. Operators' Group, NANOG47*, Dearborn, MI.
224. D. C. Kilper, G. Atkinson, S. K. Korotky, S. Goyal, P. Vetter, D. Suvakovic, and O. Blume, 2011, Power trends in communication networks, *J. Sel. Top. Quantum Electron.*, vol. 17, pp. 275–284.
225. D. A. Fishman, W. A. Thompson, and L. Vallone, 2006, LambdaXtreme® transport system: R&D of a high capacity system for low cost, ultra long haul DWDM transport, *Bell Labs Tech. J.*, vol. 11, no. 2, pp. 27–53.
226. D. A. Fishman, D. L. Correa, E. H. Goode, T. L. Downs, A. Y. Ho, A. Hale, P. Hofmann, B. Basch, and S. Gringeri, 2006, The rollout of optical networking: LambdaXtreme® national network deployment, *Bell Labs Tech. J.*, vol. 11, no. 2, pp. 55–63.
227. R. Mack, 2008, Global landscape in broadband: Politics, economics, and applications, in *Optical Fiber Telecommunications V B: Systems and Networks*, I. P. Kaminow, T. Li, and A. E. Willner, Eds., New York: Academic Press, pp. 437–476.
228. P. J. Winzer, G. Raybon, H. Song, A. Adamiecki, S. Corteselli, A. H. Gnauck, D. A. Fishman, C. R. Doerr, S. Chandrasekhar, L. L. Buhl, T. J. Xia, G. Wellbrock, W. Lee, B. Basch, T. Kawanishi, K. Higuma, and Y. Painchaud, 2008, 100-Gb/s DQPSK transmission: From laboratory experiments to field trials, *J. Lightwave Technol.*, vol. 26, pp. 3388–3402.
229. S. Walklin and J. Conradi, 1999, Multilevel signaling for increasing the reach of 10 Gb/s lightwave systems, *J. Lightwave Technol.*, vol. 17, no. 11, pp. 2235–2248.
230. P. Winzer and R.-J. Essiambre, 2006, Advanced optical modulation formats, *Proc. IEEE*, vol. 94, no. 5, pp. 952–985.
231. R.-J. Essiambre, G. J. Foschini, G. Kramer, and P. J. Winzer, 2008, Capacity limits of information transport in fiber-optic networks, *Phys. Rev. Lett.*, vol. 101, article 163901.
232. D. J. Krause, J. C. Cartledge, and K. Roberts, 2008, Demonstration of 20-Gb/s DQPSK with a single dual-drive Mach-Zehnder modulator, *IEEE Photon. Technol. Lett.*, vol. 20, no. 16, pp. 1363–1365.
233. F. Buchali, R. Dischler, and X. Liu, 2009, Optical OFDM: A promising high-speed optical transport technology, *Bell Labs Tech. J.*, vol. 14, no. 1, pp. 125–146.
234. P. J. Winzer and R.-J. Essiambre, Evolution of digital optical modulation formats, Chapter 2, in this book.
235. S. K. Korotky and R. M. de Ridder, 1990, Dual parallel modulation schemes for low-distortion analog optical transmission, *IEEE J. Sel. Area. Commun.*, vol. 8, no. 7, pp. 1377–1381.
236. M. G. Young, U. Koren, B. I. Miller, M. A. Newkirk, M. Chien, M. Zirngibl, C. Dragone, B. Tell, H. M. Presby, and G. Raybon, 1993, A 16 × 1 wavelength division multiplexer with integrated distributed Bragg reflector lasers and electroabsorption modulators, *IEEE Photonics Technol. Lett.*, vol. 5, no. 8, pp. 908–910.

237. R. Nagarajan, C. H. Joyner, R. P. Schneider, Jr., J. S. Bostak, T. Butrie, A. G. Dentai, V. G. Dominic, P. W. Evans, M. Kato, M. Kauffman, D. J. H. Lambert, S. K. Mathis, A. Mathur, R. H. Miles, M. L. Mitchell, M. J. Missey, S. Murthy, A. C. Nilsson, F. H. Peters, S. C. Pennypacker, J. L. Pleumeekers, R. A. Salvatore, R. K. Schlenker, R. B. Taylor, H.-S. Tsai, M. F. Van Leeuwen, J. Webjorn, M. Ziari, D. Perkins, J. Singh, S. G. Grubb, M. S. Reffle, D. G. Mehuys, F. A. Kish, and D. F. Welch, 2005, Large-scale photonic integrated circuits, *IEEE J. Sel. Top. Quantum. Electron.*, vol. 11, no. 1, pp. 50–65.
238. C. H. Sterling, Ed., 2007, *Military Communications: From Ancient Times to the 21st Century*, Santa Barbara: ABC-CLIO, Inc.
239. J. J. Thomson, 1967, Carriers of negative electricity, in *Nobel Lectures, Physics 1901–1921*, Amsterdam, The Netherlands: Elsevier Publishing Co.
240. C. R. Doerr and H. Kogelnik, 2008, Dielectric waveguide theory, *J. Lightwave Technol.*, vol. 26, no. 9, pp. 1176–1187.
241. A. C. W. Aldis, 1944, Signaling apparatus, U.S. Patent 2 362 333.
242. A. C. W. Aldis, 1944, Signaling lamp, U.S. Patent 2 363 566.
243. C. E. Shannon, 1948, A mathematical theory of communication, *Bell Syst. Tech. J.*, vol. 27, nos. 3 and 4, pp. 379–423 and 623–656.
244. K. K. Wong, Ed., 2002, *Properties of Lithium Niobate*, London, England: INSPEC, The Institution of Electrical Engineers.
245. K. Nassau, 2004, Early history of lithium niobate: personal reminiscences, in *50 years Progress in Crystal Growth: A Reprint Collection*, R. S. Feigelson, Ed., Amsterdam, The Netherlands: Elsevier, pp. 155–160.
246. H. Kogelnik and C. V. Shank, 1971, Stimulated emission in a periodic structure, *Appl. Phys. Lett.*, vol. 18, no. 4, pp. 152–154.
247. T. Okoshi and K. Kikuchi, 1988, *Coherent Optical Fiber Communications*, Norwell, MA: Kluwer Academic Publishers.
248. W. A. Reed, L. G. Cohen, and H.-T. Shang, 1986, Tailoring optical characteristics of dispersion-shifted lightguides for applications near 1.55 μm, *AT&T Tech. J.*, vol. 65, pp. 105–122.
249. M. Smit, 1988, New focusing and dispersive planar component based on an optical phased array, *Electron. Lett.*, vol. 24, no. 7, pp. 385–386.
250. H. Takahashi, S. Suzuki, K. Kato, and I. Nishi, 1990, Arrayed-waveguide grating for wavelength division multi/demultiplexer with nanometer resolution, *Electron. Lett.*, vol. 26, no. 2, pp. 87–88.
251. C. Dragone, 1991, An N × N optical multiplexer using a planar arrangement of two star couplers, *IEEE Photon. Technol. Lett.*, vol. 3, no. 9, pp. 241–243.
252. G. May, A. Solheim, and J. Conradi, 1994, Extended 10 Gb/s fiber transmission distance at 1538 nm using a duobinary receiver, *IEEE Photon. Technol. Lett.*, vol. 6, no. 5, pp. 648–650.
253. J. Conradi, 1922 Bandwidth-efficient modulation formats for digital fiber transmission systems, in *Optical Fiber Telecommunications IV-B*, I. P. Kaminow and T. Li, Eds., New York: Academic Press, pp. 862–901.
254. J. L. Pamart, E. Lefranc, S. Morin, G. Balland, Y. C. Chen, T. M. Kissell, and J. L. Miller, 1994, Forward error correction in a 5 Gbit/s 6400 km EDFA based system, *Electron. Lett.*, vol. 30, no. 4, pp. 342–343.
255. S. Yamamoto, H. Takahira, and M. Tanaka, 1994, 5 Gbit/s optical transmission terminal equipment using forward error correcting code and optical amplifier, *Electron. Lett.*, vol. 30, no. 3, pp. 254–255.
256. T. Mizuochi, 2006, Recent progress in forward error correction and its interplay with transmission impairments, *IEEE J. Sel. Top. Quantum Electron.*, vol. 12, no. 4, pp. 544–554.
257. B. H. Lee and R. J. Capik, 2000, Demonstration of a very low-loss, 576 × 576 servo-controlled, beam-steering optical switch fabric, in *Proc. European Conf. Optical Commun.*, London: Institution of Electrical Engineers, vol. 4, pp. 95–96, paper 11.2.3.

258. R. Ryf, J. P. Hickey, A. Gnauck, D. Carr, F. Pardo, C. Bolle, R. Frahm, N. Basavanhally, C. Yoh, D. Ramsey, R. Boie, R. George, J. Krause, C. Lichtenwalner, R. Papazian, J. Gates, H. R. Shea, A. Gasparyan, V. Muratov, J. E. Griffith, J. A. Prybyla, S. Goyal, C. D. White, M. T. Lin, R. Ruel, C. Nijander, S. Arney, D. T. Neilson, D. J. Bishop, P. Kolodner, S. Pau, C. Nuzman, A. Weis, B. Kumar, D. Lieuwen, V. Aksyuk, D. S. Greywall, T. C. Lee, H. T. Soh, W. M. Mansfield, S. Jin, W. Y. Lai, H. A. Huggins, D. L. Barr, R. A. Cirelli, G. R. Bogart, K. Teffeau, R. Vella, H. Mavoori, A. Ramirez, N. A. Ciampa, F. P. Klemens, M. D. Morris, T. Boone, J. Q. Liu, J. M. Rosamilia, and C. R. Giles, 2001, 1296-Port MEMS transparent optical cross-connect with 2.07 Petabit/s switch capacity, *Proc. Conf. Optical Fiber Commun.*, Anaheim, CA, postdeadline paper PD28.
259. R. A. Griffin, R. I. Johnstone, R. G. Walker, J. Hall, S. D. Wadsworth, K. Berry, A. C. Carter, M. J. Wale, J. Hughes, P. A. Jerram, and N. J. Parsons, 2002, 10 Gb/s optical differential quadrature phase shift key (DQPSK) transmission using GaAs/AlGaAs integration, *Proc. Conf. Optical Fiber Commun.*, Washington, DC: Optical Society of America, paper FD6.
260. A. J. Lowery, D. Liang, and J. Armstrong, 2006, Orthogonal frequency division multiplexing for adaptive dispersion compensation in long haul WDM systems, *Proc. Conf. Optical Fiber Commun.*, Anaheim, CA, postdeadline paper PDP39. Washington, DC: Optical Society of America.
261. S. K. Korotky, 1994, External modulators for lightwave systems, *Proc. Conf. Optical Fiber Commun.*, Washington, DC: Optical Society of America, paper TuP, pp. 21–52.

# 2

# Evolution of Digital Optical Modulation Formats

Peter J. Winzer
*Alcatel-Lucent, Bell Laboratories*

René-Jean Essiambre
*Alcatel-Lucent, Bell Laboratories*



## 2.1 Role of Fiber Optics in Digital Optical Modulation

The transition from analog to digital technologies over the past ~50 years has enabled universal processing of all kinds of information, with very low loss of quality [1]. Breakthroughs in digital semiconductor technologies and their enormous ability to scale [2] have enabled cost-effective mass production of highly functional, reliable, and power-efficient microchips. Closely coupled to the generation, processing, and storage of digital information is the need for data transport, ranging from short on-chip [3] and board-level [4,5] interconnects all the way to long-haul transport networks spanning the globe [6,7] and to deep-space probes transmitting collected data back to Earth [8]. Each of these applications has its unique set of system characteristics that lead to different trade-offs regarding *capacity*, *sensitivity*, and *implementation* [9]. Despite the wide diversity of applications, a common theme unites the evolution of almost all digital communications applications: the demand for exponentially increasing system capacities. For example, data traffic in carrier networks is exponentially growing at about 60% per year [10], and emerging cloud computing applications are expected to generate an annual growth of close to 90% in machine-to-machine traffic across the network.*

* This high growth rate is based on Amdahl's rule [12,13], which linearly relates the exponentially growing computing power of microprocessors, close to 90% per year [11], with the processor interface bandwidth in a balanced computer architecture.

To meet these growing bandwidth demands, optical communications solutions are gradually replacing baseband electronics and radio frequency (RF) systems, owing to the large absolute bandwidth available at optical carrier frequencies.* This process started on a large scale in the late 1970s and 1980s with the most demanding high-bandwidth/long-distance applications of terrestrial [6] and submarine [7] transport. With massive fiber-to-the-home deployments now underway worldwide, optics is currently capturing the access space [15], and rack-to-rack interconnects are starting to become optical [3]. Despite the continuing improvement in electronic transmission techniques [16], optical solutions are expected to enter backplanes, paving the way to optical chip-to-chip and, eventually, on-chip communications once electronic transmission can no longer keep pace with the growing need for communication capacity [3–5]. At the same time, areas where optical communication techniques are already well established have to continue to support exponentially increasing capacity demands. Fiber-optic networks are therefore playing a pioneering role in pushing the boundaries of high-capacity digital data transport. Many concepts and technologies originally developed by the fiber-optics industry have been entering (and will most likely continue to enter) other areas of digital communications at a later point in time, adapted and augmented to meet the respective digital communication applications' needs. For example, while optical interconnects are still largely based on single-wavelength (or at the most on few-wavelength) transmission, dense wavelength-division multiplexing (WDM) with up to ~100 wavelengths per fiber, well established in fiber-optic networking today, is expected to enter the interconnect space once the amount of parallel fiber needed to support the growing interconnect bandwidth demands becomes a heavier economical burden than the cost of WDM transponders. Note in this context that the local area networking standards developed for 100G Ethernet interconnects are exclusively based on multiple parallel fibers or few-wavelength WDM [17]. Going to higher Ethernet rates (400 Gb/s and, eventually, 1 Tb/s) will likely require even more wavelengths for cost-effective interconnect solutions [18].

Due to the pioneering role of fiber optics in high-capacity transmission, we will review the evolution of optical modulation formats and modulator technologies from a fiber-optic point of view, anticipating that most of the concepts used in fiber-optic systems today will eventually become important for other digital optical communications applications as well. This chapter can only provide a brief review of the underlying concepts. More details on various aspects of advanced digital optical modulation techniques and technologies can be found in other chapters within this book and in the many references cited therein.

## 2.2 Fundamentals of Digital Optical Modulation

### 2.2.1 Primer on the Anatomy of Digital Modulation Formats

The basic structure of digital communication signals (optical or electrical alike) resembles the structure of many languages in several ways, as illustrated in Figure 2.1. A finite set or *alphabet* of *letters* is used to form words and sentences, whereby letters are written in series, one after the other. The countably discrete nature of letters contained in a (sometimes language-specific) alphabet motivates the term *digital*.† Note that letters are abstract concepts that need to be mapped into the analog reality of the world we live in. This is done by representing each letter by some kind of *analog waveform* that bears the key features of the letter. For example, the letter "A" in the 26-ary Roman alphabet can be represented by the "analog waveforms" A, *A*, $\mathcal{A}$, **A**, or ᴀ. As long as writer (transmitter) and reader (receiver) use the

* A 5-THz bandwidth at an optical carrier frequency of 193 THz (1.55-μm wavelength) corresponds to a mere 2.5% of relative system bandwidth. Low relative system bandwidths are generally advantageous for electromagnetic system design [14].

† The word *digital* is derived from the Latin word "digitus" (finger) and alludes to the basic way of counting members of discrete sets.

| **Language** | **Digital communications** |
|---|---|
| Alphabets of letters<br>{A,B,C,...,Z}, {α,β,γ,...,ω}, {0,1,2,...,9} | Symbol alphabets (constellations) 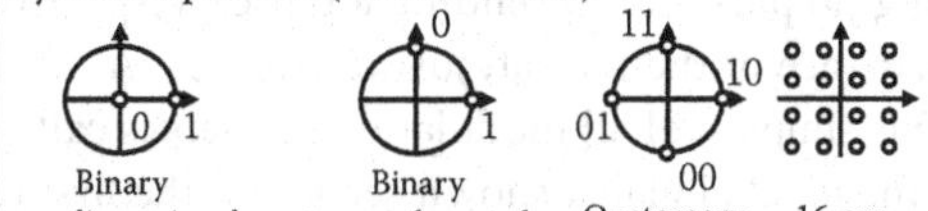<br>0 1 / 0 1 / 11 10 01 00<br>Binary one-dimensional, Binary two-orthogonal, Quaternary, 16-ary |
| Analog letter representations<br>'A' →A, A, **A**, A, *𝒜*, ... | Analog waveform representations (ex: "1101") 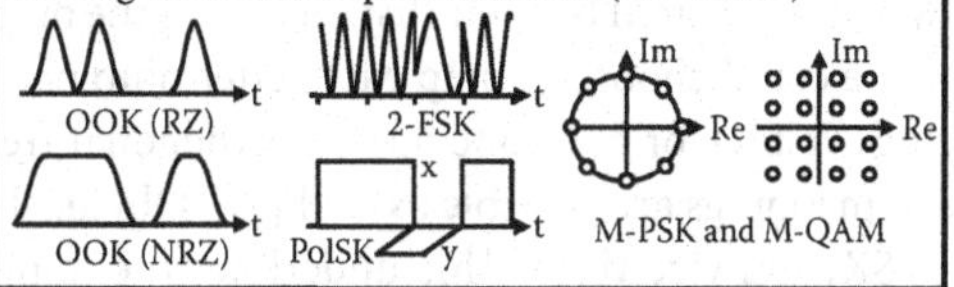<br>OOK (RZ), 2-FSK, OOK (NRZ), PolSK (x, y, t), M-PSK and M-QAM (Im, Re) |
| Letters arranged in series | One symbol transmitted per symbol period |
| Redundancy by words or sentences<br>"laguage" or "langyage" → "language" | Error correcting codes<br>Overhead in time (forward error correction, FEC)<br>or in symbol alphabet (coded modulation) |
| Synonym expressions<br>(ex: use "ponder" instead of "think" to avoid confusion with "sink") | Line coding<br>Overhead in time or in symbol alphabet<br>(ex: replace "11011" by {+1,+1,0,−1,−1}: DB) |

**FIGURE 2.1** The structure of digital communications in many ways resembles the structure of language.

same alphabet *and* are able to establish the correct mapping between the analog waveforms and the set of letters, communication may take place.

On top of using a well-defined alphabet of letters, each language also uses a considerable amount of *redundancy*. This is done by forming *words* and *sentences* from letters. By allowing a much smaller number of words and sentences than what would be mathematically possible by arbitrarily arranging letters within words and words within sentences, the receiver is put in the position to *correct* for spelling errors. For example, the words "laguage" or "langyage" are immediately identified as misspelled versions of the word "language." Being able to see and correct these typos shows the ability of our brain to act as an efficient real-time error correction device.

Language redundancy in the form of *synonyms* can also be used by the transmitter to avoid the use of certain words that are known to cause trouble in conveying a message. For example, many nonnative English speakers mispronounce a "th" as an "s," which can lead to uncorrectable confusion if both resulting words are legitimate English words, such as the words "think" and "sink." Error correction may be still possible on a sentence level by identifying the word's most likely meaning in a given context, but this kind of error correction is much more prone to errors than the one on the individual word level.* It may thus prove advantageous to substitute words containing "th" by suitable synonyms to *avoid* likely errors in the first place rather than having to correct them after they have occurred.

The above outline of basic language structure illustrates many important concepts of digital modulation, as summarized in Figure 2.1. In digital communications, the alphabet of letters becomes an *alphabet* or *constellation* of discrete communication *symbols*. Importantly, these symbols can be viewed as an abstraction that does not yet assign an analog physical meaning to the intended modulation. Before transmitting symbols over a physical channel, a set of analog waveforms has to be chosen to map a symbol constellation onto physical reality. For example, the simple binary symbol constellation shown

* This problem is well reflected in a popular language joke, in which a ship in distress radios the German coastguard with the message "Mayday, mayday, we are sinking."—The coastguard officer, pronouncing his "th" as an "s," answers with "What are you thinking about?"

first in Figure 2.1 consists of the two symbols {0,1}. The key feature of the two letters is that one of them carries no power and one of them does carry power. One may then choose to represent the two symbols by "sending no pulse" and "sending a pulse," respectively. The exact shape of the pulse adds detail to the performance of the format, just as A, *A*, $\mathcal{A}$, **A**, or A show key similarities and differ only in their details. This simplest of all modulation formats is called on/off keying (OOK). Another example is the binary orthogonal alphabet shown second in the first row of Figure 2.1. The key feature of this alphabet is that the two letters are orthogonal (in whatever physical dimension!), and that both letters have equal power, as measured by their distance from the origin in the symbol constellation. To transmit this alphabet, one could choose to map each letter onto one of two orthogonal polarizations (polarization-shift keying, PolSK), or onto one of two orthogonal frequencies* (frequency-shift keying, FSK); other orthogonal mappings are possible as well [19]. Obviously, a receiver built to detect PolSK will be unable to detect FSK, and vice versa. This underlines the importance of properly specifying *both* the symbol alphabet *and* the set of corresponding analog waveform representations to make digital communications possible. An important set of two orthogonal dimensions used to construct symbol alphabets in digital communications is the *quadrature space*, composed of real and imaginary components (also called "sine" and "cosine" components or in-phase (I) and quadrature (Q) components) of a bandpass signal, such as a digital communication signal at an optical carrier frequency. For example, while the quaternary (4-ary) or the 16-ary alphabets shown in Figure 2.1 could be equally well constructed using two orthogonal polarizations or two orthogonal frequencies, they are usually mapped onto real and imaginary parts of the complex optical field, where they are then called quadrature phase-shift keying (QPSK) and 16-ary quadrature amplitude modulation (16-QAM), respectively.

Once mapped to analog waveform representations, the symbols are sequentially transmitted at rate $R_S$, one symbol per symbol period $T_S = 1/R_S$. The symbol rate† is measured in baud, where 1 baud = 1 symbol/s. If all symbols in an $M$-ary alphabet (i.e., an alphabet of $M$ letters) occur with equal probability, and all symbols are used to carry information, each symbol can convey up to $\log_2 M$ bits of information, and bit rate $R_B$ and symbol rate $R_S$ are related by

$$R_B = R_S \log_2 M. \tag{2.1}$$

For example, QPSK has $M = 4$ symbols and can hence carry up to $\log_2(4) = 2$ bits per symbol, cf. Figure 2.1.

If information is transmitted on $p$ parallel channels (e.g., independently on $x$ and $y$ polarization of the optical field, with $p = 2$), the maximum achievable *aggregate* bit rate of the multiplex is given by the single-channel bit rate multiplied by the number of parallel channels,

$$R_B = pR_S \log_2 M. \tag{2.2}$$

As with language, redundancy in digital communications can be introduced either to *avoid* certain symbol combinations that are known to cause trouble on a specific communication channel (*line coding*) or to *correct* errors at the receiver (*forward error correction*, FEC). Both techniques introduce the required overhead in a variety of ways. The most common dimensions to include the overhead are the time domain (by transmitting at a higher symbol rate than what would be required by the client application) the size of the symbol constellation itself (by inserting additional symbols to carry redundancy

* As discussed in more detail in Section 2.4.2.3, signals modulated onto two different carrier frequencies are called *orthogonal* if the carrier frequencies differ by integer multiples of $1/T_S$, where $T_S$ is the symbol period. If the two carrier frequencies are separated by much more than $1/T_S$, the signals are approximately orthogonal even if the carrier frequencies do not differ by integer multiples of $1/T_S$ [19].

† Note that one sometimes encounters the term "baud rate" instead of "symbol rate," which is strictly speaking incorrect; "baud" is the *unit* of the symbol rate, just as "bits per second" is the unit of the bit rate.

as opposed to user information) [19]. The most prominent example of line coding in digital optical communication is *duobinary* (DB) modulation [20], in optical communications sometimes also referred to as *phase-shaped binary transmission* [21]. Redundancy for FEC is typically implemented in the time domain. An important quantity associated with codes is the *code rate** $\tilde{R}_c \le 1$, defined as the ratio of information bits to coded bits. In optical communications, the term *coding overhead* (OH), i.e., the percentage of bits that are added to the information bits for coding redundancy, is more popular. Coding overhead and code rate are related by

$$OH = (1 - \tilde{R}_c)/\tilde{R}_c. \tag{2.3}$$

To retain a certain information bit rate with a fixed modulation format, the redundancy carried by coding requires transmission at a higher line rate (e.g., 10.7 Gb/s are required to transport a 10-Gb/s information bit stream with 7% FEC-OH). Especially at higher OHs, this increase in the transmitted symbol rate results in a proportionally wider signal bandwidth, which can pose problems associated with the modulation and detection hardware. Furthermore, the large number of reconfigurable optical add/drop multiplexers (ROADMs) present in today's optically routed networks, repeatedly filter the signal and narrow its bandwidth. The penalties from such filtering processes act against the coding benefit from large OH, which results in system-specific optimum (temporal) coding OHs. A way to avoid having to increase the symbol rate while increasing the FEC-OH is to put redundancy into the symbol constellation by adding more symbols than required for pure information transport. This technique (often referred to as *coded modulation*) is well studied in electronic and wireless communications and is now starting to enter optical communications as well [22–25], as discussed in more detail in Section 2.4.4.

For more details on the basics of digital communications and coding, the interested reader is referred to classic textbooks such as [26–28]. Detailed tutorials on the basics of digital *optical* communications can be found in, e.g., [19,23,29].

## 2.2.2 Brief Overview of Optical Modulator Technologies

Owing to the high bit rates transported by optical communication signals, the design of digital optical transponders is concerned with hardware feasibility aspects as much as it is concerned with the application of classical communication techniques to the optical channel. Therefore, *communication technologies* are as important as *communication techniques* in optical communications. This strong interplay between digital communications engineering and high-speed optoelectronic and electronic device design continues to make digital optical communications a highly interdisciplinary and exciting field of research. In the context of this book, we will mainly focus on techniques and technologies for digital optical *modulation*, keeping in mind that similar, equally important considerations apply to *receiver, digital signal processing (DSP)*, and *coding* subsystems of modern optical transponders. Figure 2.2 gives a basic overview of the most important technology options used for modulation and pulse shaping in digital optical communications. Detailed accounts of advanced integrated optical modulator technologies can be found in the remainder of this book as well as in tutorial reviews [30–32].

### 2.2.2.1 Direct Laser Modulation

The simplest optical modulation concept is based on directly varying the laser drive current $I$ to change the intensity of the emitted optical wave, as shown in Figure 2.2a. Since light generation and light modulation are functionally combined into a single device, this technique yields the simplest and most compact transmitter element. Directly modulated lasers (DMLs) are primarily used for binary intensity modulation (OOK), where a drive current swing of typically a few tens of milliamperes (a drive voltage

* Note that the code rate is a dimensionless quantity as opposed to a bit rate or a symbol rate, which carries units of "per second."

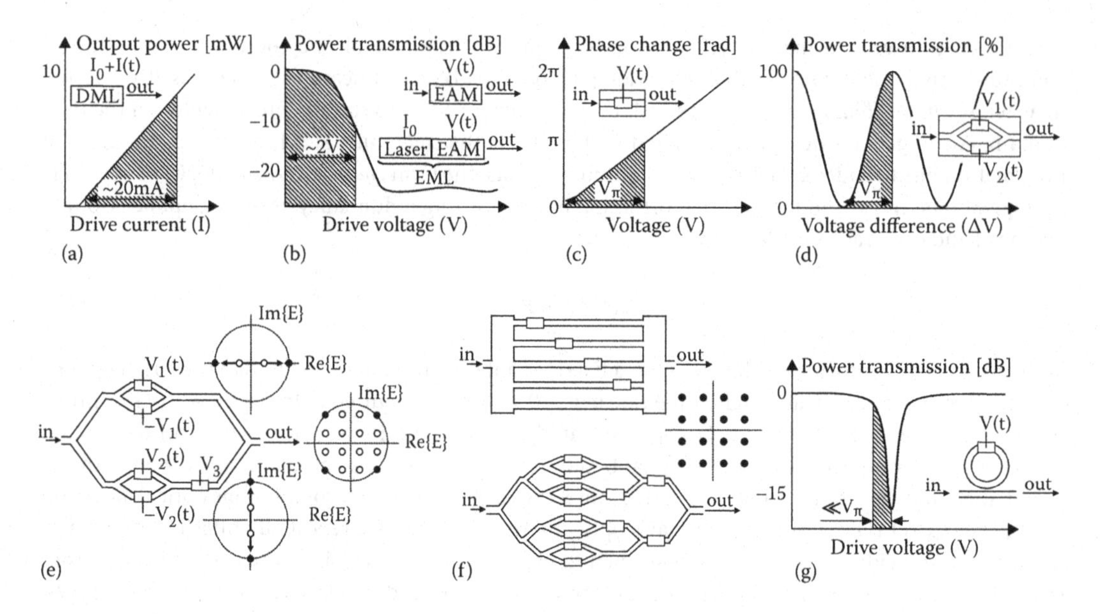


**FIGURE 2.2** Overview of optical modulator technologies.

swing of less than 1 V) is needed in conjunction with a bias current $I_0$ of about the same magnitude to achieve extinction ratios of around 8 dB. DMLs are available today for modulation up to 10 Gb/s, with research demonstrations up to 40 Gb/s [33,34]. The main drawback of DMLs for high bit rate intensity-modulated transmission over longer distances is their inherent and device-specific *chirp*, i.e., a residual data-dependent phase modulation accompanying the desired intensity modulation [35–37]. Laser chirp broadens the optical signal spectrum, which impedes dense WDM channel packing and can lead to increased signal distortions caused by the interaction with fiber chromatic dispersion and fiber nonlinearity [38]. Electronic equalization and FEC can be used to combat these impairments to some extent [39], but the associated hardware adds its own complexity, power consumption, and cost, which must be taken into account when designing DML-based transponders.

Through careful device engineering, the DML's phase modulation characteristics can be deliberately exploited. Examples are FSK [40], dispersion-supported transmission [41] or the application of a DML as a *chirp-managed laser* [42]. Using carefully preprogrammed analog electrical drive waveforms that are matched to a specific DML's frequency modulation characteristics, DML-based electronic predistortion systems have also been demonstrated [43].

### 2.2.2.2 Modulation by Absorption

The absorption of light in semiconductor structures can be controlled by applying an electrical field to the structure. As shown in Figure 2.2b, the resulting electroabsorption modulators (EAMs) feature relatively low drive voltages (typically 2 V) and are available for binary intensity modulation up to 40 Gb/s, with research demonstrations up to 80 Gb/s [44] and 100 Gb/s [45]. Similar to DMLs, EAMs produce some residual and component-specific *chirp*. Through careful device engineering, this chirp can be controlled and exploited to counteract chromatic dispersion, which increases the uncompensated reach of EAM-based transmitters. On the downside, EAMs have limited power-handling capabilities and exhibit wavelength-dependent performance. Nevertheless, EAMs are attractive because of their small size, integration capabilities with lasers to form *electroabsorption-modulated lasers* (EMLs), as well as their potential for integration into multibranch interference structures for advanced modulation formats [46], as discussed in more detail below.

#### 2.2.2.3 Optical Phase Modulation

Advanced optical modulation formats generally require modulation of the optical phase. Examples are multilevel phase-shift keying (PSK), or combined amplitude-shift keying (ASK) and PSK constellations ($M$-ASK/$N$-PSK), also known as "multiring PSK" constellations, cf. Section 2.4.4. Transmitters for the latter may be implemented by combining a pure optical phase modulator with a subsequent (chirp-free) intensity modulator. As shown in Figure 2.2c, optical phase modulators are based on materials that let the modulated optical phase be a (preferably linear) function of the electrical drive voltage or current. The drive voltage that generates a $\pi$-phase shift in an electro-optic phase modulator is usually referred to as $V_\pi$. Most widely used materials are $LiNbO_3$ [47–49], gallium arsenide [50], and indium phosphide (InP) [51–54]. $LiNbO_3$-based devices are bulkier than their semiconductor equivalents, but offer easier fiber-to-chip coupling and no residual intensity modulation accompanying the desired phase modulation. On the other hand, semiconductor devices, and in particular InP, lend themselves to much denser integration [53,55], which is particularly important for the multitude of phase modulator–based interference structures discussed below.

#### 2.2.2.4 Modulation by Interference

In contrast to modulating light through *absorption*, one can also exploit phase modulation followed by optical *interference*. An interference-based optical modulator first splits the incoming light into two or more branches. It then uses an electro-optic phase modulator to convert an electrical drive signal into an optical phase shift in one or more of these branches. In a third step, phase modulation is turned into intensity modulation or into combined amplitude/phase modulation using an appropriate interference structure that recombines the individual branches. Usually, these functions are integrated within the same device.

*Mach–Zehnder Modulator*

The simplest and most widely used interference structure used for modulation is the two-branch Mach–Zehnder interferometer, resulting in the Mach–Zehnder modulator (MZM) depicted in Figure 2.2d together with its sinusoidal optical power transmission characteristics. A periodic transmission characteristic is due to the $2\pi$ ambiguity of the modulated optical phase and is indicative of interference-based modulators. Modulating the optical phase difference between the two MZM branches by $\pi$ switches the modulator output from constructive to destructive interference. Hence, $V_\pi$ is also referred to as the "switching voltage" of a MZM. The MZM output signal is readily determined as (see, e.g., [19,29,56])

$$E_{out}(t) = E_{in}\frac{1}{2}\{e^{j\pi V_1(t)/V_\pi} + e^{j\pi V_2(t)/V_\pi}\} = e^{j\pi[V_1(t)+V_2(t)]/(2V_\pi)}\cos[\pi[V_1(t)-V_2(t)]/(2V_\pi)], \tag{2.4}$$

where we assume that the phase modulation depends linearly on the drive voltage. The drive voltages $V_1(t)$ and $V_2(t)$ may also incorporate a time-independent bias voltage $V_0$ to set the desired operating point on the MZM's transfer function [29]. Most notably, it can be seen from Equation 2.4 that the *amplitude* of the modulated optical signal is a sinusoidal function of the *difference* between the two drive signals $\Delta V = V_1(t) - V_2(t)$, while the *phase* of the modulated output signal depends linearly on the *sum* of the two drive voltages. Therefore, an MZM is able to produce pure amplitude modulation without residual chirp if its two branches are driven with opposite-sign ("differential") drive signals ($V_1(t) = - V_2(t)$), also referred to as "push–pull" operation. In some modulators, e.g., for $x$-cut $LiNbO_3$ MZMs, the electrodes are laid out to automatically achieve push–pull operation with a single electrical drive signal. The well-controlled nature of optical intensity and phase modulation by an MZM is widely exploited for advanced digital modulation [29]:

- *Chirp-free intensity modulation* is obtained by differentially driving the MZM with a data signal between its transmission minimum and its transmission maximum. Due to the sinusoidal power transfer function, drive signal overshoot and oscillatory ripple is effectively suppressed by the MZM.

- *Chirp-free binary phase modulation* is achieved by driving the MZM with a data signal symmetrically around its transmission minimum,* with a drive voltage swing of up to $2V_\pi$ if one wants to reach both adjacent transmission maxima [57,58].
- *Duobinary modulation* is obtained by driving the MZM differentially as in the case of chirp-free binary phase modulation, but using either a heavily filtered electrical drive signal or the frequency roll-off of the MZM itself as an electrical filter [59,60].
- *Pulse generation* ("pulse carving") is most widely performed by sinusoidally modulating the MZM, either at the full desired pulse rate between its transmission maximum and minimum, or at half the desired pulse rate between its two transmission minima or maxima. These three methods generate 50% and 33% duty cycle return-to-zero (RZ), and 67% duty cycle carrier-suppressed RZ modulation [29].
- *Arbitrary optical waveform generation* can be achieved using a MZM by independently driving the sum and the difference drive voltages, i.e., the common mode and the differential mode of $V_1(t)$ and $V_2(t)$. However, to cover the entire complex plane of the optical field, a drive voltage swing of $2V_\pi$ on each arm is needed [61], which can be difficult to achieve in practice. Furthermore, modulation of the two orthogonal quadratures (real and imaginary parts) of the optical field is not accomplished independently of each other, which can complicate the generation of the required high-speed electrical drive waveforms.

From an implementation point of view, MZMs rely on materials that yield efficient (low-$V_\pi$) and compact electro-optic phase modulators, as discussed in Section 2.2.2.3. $LiNbO_3$-based devices are the most widely used ones today, featuring wavelength-independent modulation characteristics, excellent extinction performance (typically 20 dB), and low insertion loss (typically 5 dB). The required high-speed switching voltages of up to about 6 V (and the required modulation voltages of up to $2V_\pi$ if the modulator is driven around a transmission null), however, require broadband driver amplifiers, which can be challenging to build at data rates in excess of 10 Gb/s. Today, $LiNbO_3$-based MZMs are widely available for modulation up to 40 Gb/s, and have been demonstrated at bit rates up to 100 Gb/s [62]. InP-based modulators, available for 10-Gb/s modulation, feature lower switching voltages (typically 2 V) but exhibit residual absorption in addition to phase modulation in the modulator arms, which can lead to chirp unless counteracted by proper electrode design [54]. InP-based MZMs have been demonstrated up to 80 Gb/s [63].

***Nested Mach–Zehnder Structures***

To independently modulate real and imaginary parts of the optical field with reduced drive voltages and still be able to cover the entire complex plane of the optical field, one typically uses the nested MZM structure, also referred to as "quadrature modulator," "I/Q modulator," or "Cartesian modulator" [49]. The structure was initially proposed for FSK and single-sideband (SSB) modulation [64,65] and was reintroduced for QPSK in 2002 [66,67]. Today, this modulator structure is widely used to generate advanced optical modulation formats. With reference to Figure 2.2e, describing its application as a modulator for QPSK (solid symbols) and 16-QAM (open symbols), the incoming light is first split into two arms. For QPSK, the sub-MZMs in each arm are driven by binary electric drive signals to generate two binary PSK signals using chirp-free (push–pull) modulation. For 16-QAM, the sub-MZMs are driven by four-level electric drive signals to generate 4-ASK. A third electrode in the outer MZM structure ($V_3$) introduces a static 90-degree phase shift between the two modulated optical signals, which puts them in quadrature to each other. Upon combination in the output coupler, the two optical fields add up to create QPSK [66] and 16-QAM [68], respectively.

Being a true I/Q modulator, the nested MZM can be used for a variety of other applications [49]: By driving one sub-MZM with the data sequence and the quadrature sub-MZM with the sequence's

* Note from Equation 2.4 that adjacent transmission maxima have opposite optical phase.

Hilbert transform, one generates SSB modulation [65]. Allowing the third electrode to be RF-modulated as well ($V_3 \rightarrow V_3(t)$), one can implement FSK and minimum-shift keying (MSK), a line-coded version of FSK with continuous optical phase [26,27,69,70]. Furthermore, a quadrature modulator can be used for complex optical pulse shaping [71] or to generate electronically predistorted optical waveforms [61] with independent control of real and imaginary parts of the (analog) optical field. Note that this modulator structure allows coverage of the entire complex plane even when using arbitrarily small drive voltage swings, albeit at the expense of an increased optical insertion loss. Quadrature modulators are commercially available in $LiNbO_3$ technology for modulation speeds of 28 Gb/s on each sub-MZM, and highly integrated InP-based devices have been reported [72]. Research prototypes have achieved up to 56-Gbaud modulation using QPSK [73–75] and 16-QAM [76].

***Multibranch Interference Structures***

While nested MZM structures with just two sub-MZMs offer full coverage of the complex plane of the optical field with independent control of real and imaginary optical field components, they require multilevel electronic drive signals to scale to symbol constellations beyond QPSK, as schematically represented for 16-QAM in Figure 2.2e. Since electrical drive amplifiers typically operate under saturation, it is generally not trivial to amplify a high-speed multilevel electrical transmit signal to the required modulator drive voltage swings without significant waveform distortion. One solution uses electrical digital-to-analog converters that can resolve more bits than what would be necessary to generate the desired modulation format and employs these extra bits for modulation distortion compensation [77]. Another solution uses multiparallel optical modulator structures with easily generated *binary* electrical drive waveforms. Figure 2.2f sketches examples for two such structures. In one, a 1 × 5 (not necessarily equal-power) input splitter and output coupler is used, with four modulating elements in between. These could either be optical intensity modulators, e.g., EAMs) or optical phase modulators, depending on the desired modulation functionality. A fifth unmodulated waveguide can be used when necessary to shift the constellation to the origin of the complex plane [46]. The other structure shown in Figure 2.2f, referred to as quad-parallel MZM [78], is a direct extension of the quadrature modulator of Figure 2.2e. This structure has been implemented monolithically in $LiNbO_3$ [78] as well as in hybrid designs with $LiNbO_3$ phase modulators and silica splitting and combining networks [79,80].

***Resonant Modulators***

Some modulator structures employ an *infinite* number of interferences, such as found in a Fabry–Perot cavity or in a ring resonator. Figure 2.2g depicts a ring resonator modulator as an example. Resonant modulators are based on the observation that the conditions for resonance of high-$Q$ cavities are typically very sensitive to perturbations [81]. In the case of a ring resonator, small changes in the coupling parameters between the ring and the straight waveguide, or small changes in the attenuation or in the propagation constant of the ring waveguide can significantly shift the ring off resonance. Thus, resonant modulators hold the promise of requiring much lower drive voltages and much shorter lengths of the active modulation region than nonresonant modulators to achieve a certain intensity modulation. This translates into potentially smaller modulator sizes and lower operating power compared to MZMs, which makes resonant devices particularly attractive candidates for complementary metal-oxide semiconductor (CMOS) integrated silicon photonics. On the downside, however, a high cavity $Q$ also implies a long cavity lifetime with long associated modulation time constants, which in turn limits achievable modulation speeds [81]. Furthermore, the transmission characteristics of resonant modulators can exhibit significant chirp, with likely similar implications on system performance as DML or EAM chirp. Finally, the high parameter sensitivity of resonant structures that forms the basis for their small switching voltages lets resonant modulators be highly sensitive to wavelength and temperature, which complicates their operation in many practically relevant environments. As a consequence, MZMs have so far been considered the preferred solution, even for CMOS integrated silicon photonics [82–84].

## 2.3 Pulse Shaping in Digital Optical Communications

As discussed in Section 2.2.1, every digital modulation format requires the specification of an *analog symbol waveform* in addition to a *digital symbol constellation*. Electrical and RF communication systems typically operate at low enough symbol rates to allow for sophisticated pulse shaping, which is used to achieve spectral compactness or other desired properties such as a high spectral side-lobe suppression ratio. Pulse shapes used to that end are, e.g., raised-cosine pulses or Gaussian-shaped phase transitions in Gaussian MSK. Optical modulation formats, on the other hand, usually operate at the highest symbol rates enabled by the current generation of electronics and optoelectronics, and therefore have to make use of pulse shapes that are available at these speeds. This is particularly true since most electro-optic modulators require drive voltage swings of several volts; the required high-bandwidth driver amplifiers typically exhibit significant nonlinearities and hence produce a substantial amount of nonlinear intersymbol interference (ISI). Figure 2.3 shows a measured binary electrical drive waveform and spectrum at 56 Gb/s. Such a drive waveform is referred to as *non return-to-zero** (NRZ). Also shown is a calculated waveform and spectrum, composed of pulses $p(t) = \sin(\pi t/T_S)/(\pi t/T_S)$, where $T_S$ is the symbol period. This sinc-pulse shape, sometimes also referred to as the Nyquist pulse shape, has a perfectly rectangular spectrum of full-width $1/T_S$ but at the same time is free of ISI, i.e., $p(t) = 0$ at $t = kT_S$. Using such sinc-pulses, the modulated waveform is obtained as†

$$s(t) = \sum_{k=-\infty}^{\infty} a_k p(t - kT_S), \tag{2.5}$$

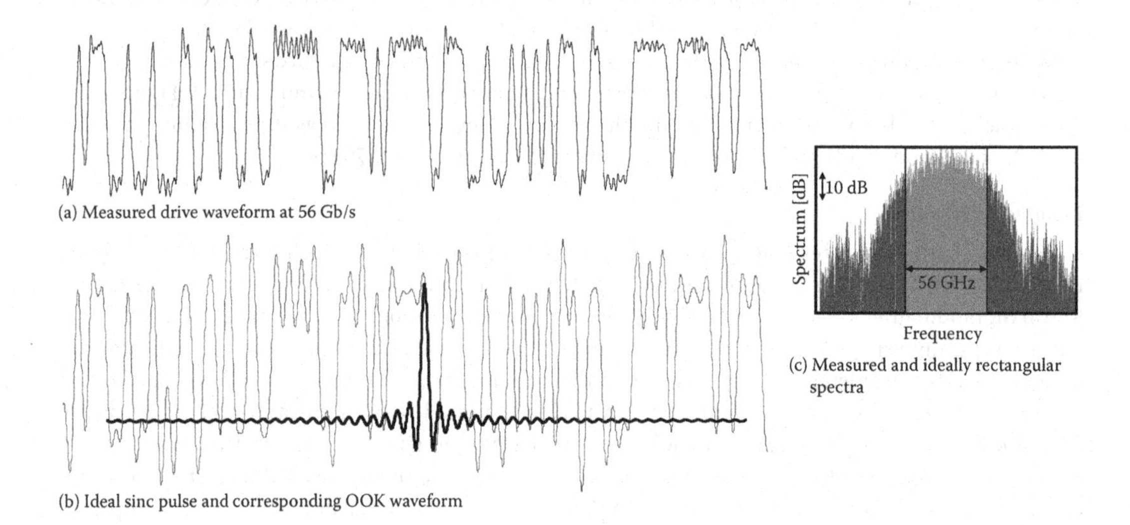


**FIGURE 2.3** (a) Measured binary drive waveform at 56 Gb/s; (b) ideal waveform for the same data sequence, based on sinc pulses; (c) measured and ideally rectangular spectra.

* The name is based on the observation that the drive signal level stays "high" for a pair of consecutive "1" bits, "not returning to zero" in between.

† In principle, it is possible to compress the spectrum even further by using more narrow-band pulses at the expense of ISI. To undo this controlled ISI at the receiver, more complicated digital signal processing is needed, though. The result of this process is "modulation with memory," DB being its most prominent example in optics; other "faster-than-Nyquist" transmission schemes have also been investigated in optical communications [85].

where $a_k$ is the data symbol from the modulation alphabet transmitted at time $kT_S$. In the example of Figure 2.3, $a_k \in \{0,1\}$. Ideally, the analog pulse shape $p(t)$ is identical for each symbol. However, due to nonlinear driver ISI, technically realizable waveforms lack this structure and hence cannot be fully transformed into a sequence of ideal Nyquist pulses using linear equalization.

To mitigate the nonlinear ISI of high-speed NRZ drive waveforms, some optical transmitters place a pulse carver as their pulse shaping element after the NRZ modulator, resulting in the RZ version of the underlying modulation format. The pulse carver (usually implemented as a sinusoidally driven MZM, cf. Section 2.2.2) optically samples the NRZ waveform at symbol center, where the nonlinear driver ISI is at its minimum, and hence provides a closer approximation to the ideal pulse sequence given by Equation 2.5. The narrower the RZ pulses, the more confined will be the temporal sampling of the NRZ waveform. Depending on the underlying receiver structure and the dominant noise processes, RZ signals can exhibit several dB better sensitivities than their NRZ equivalents [86], and will often exhibit higher tolerances to distortions from fiber nonlinearity [87]. On the downside, RZ signals have wider optical spectra than NRZ signals, which makes them less suitable for dense WDM applications. In some research experiments, the carved RZ signals are therefore converted back into an NRZ-like shape through appropriate optical filtering. Since optical filters are linear and can have well defined transfer characteristics, the resulting NRZ-like signal is a closer approximation to Equation 2.5 than the original NRZ signal generated by high-speed electronic circuits, and is hence very well suited for high-spectral-efficiency experiments [88,89]. Note that the distinction between NRZ and RZ is not restricted to OOK; almost all modulation formats can be generated either as NRZ or RZ versions [29].

Ultimately [23], one would like to implement perfect pulse shaping for spectrally compact yet ISI-free optical pulses in the spirit of Equation 2.5, which requires oversampled digital-to-analog converters in combination with compensation circuits to counteract the nonlinear driver ISI. This approach has been investigated in some multi-Gb/s research experiments [71,77].

In addition to shaping pulses for maximum spectral compactness, one can also electronically pre-distort the optical waveform to compensate for some (mostly linear) transmission impairments, such as chromatic dispersion or narrow-band optical filtering. Predistortion application-specific integrated circuits (ASICs) at 20 GSamples/s with 6-bit resolution digital-to-analog converters and finite impulse response (FIR) filters to compute the predistorted pulse shape have been reported [43]. Note, however, that in stark contrast to most electrical communication systems, the optical fiber channel is inherently a strongly nonlinear channel with significant memory, which limits the range of applicability of electronic predistortion techniques [90]. This topic is further addressed in Section 2.4.4.

## 2.4 Modulation and Multiplexing in Fiber-Optic Transport Networks

Equipped with a basic understanding of digital communication techniques (Section 2.2.1) as well as some of the key aspects associated high-speed optical modulation hardware (Section 2.2.2), we will now review the evolution of digital optical communication systems, as illustrated in Figure 2.4. The figure shows record research experiments reported in the postdeadline sessions of the annual Optical Fiber Communications Conference (OFC) and the European Conference on Optical Communication (ECOC). We will refer to this figure throughout this section.

### 2.4.1 Electronically Generated Single-Channel Bit Rates

Prior to the year 2000, digital optical communication systems almost exclusively used *binary intensity modulation* (OOK), where a logical "0" is represented by the absence of light, and a logical "1" by the presence of light, cf. symbol diagram in Figure 2.1. The success of OOK lies in the ease of its generation and detection at the forefront of technologically feasible modulation speeds. The performance of OOK

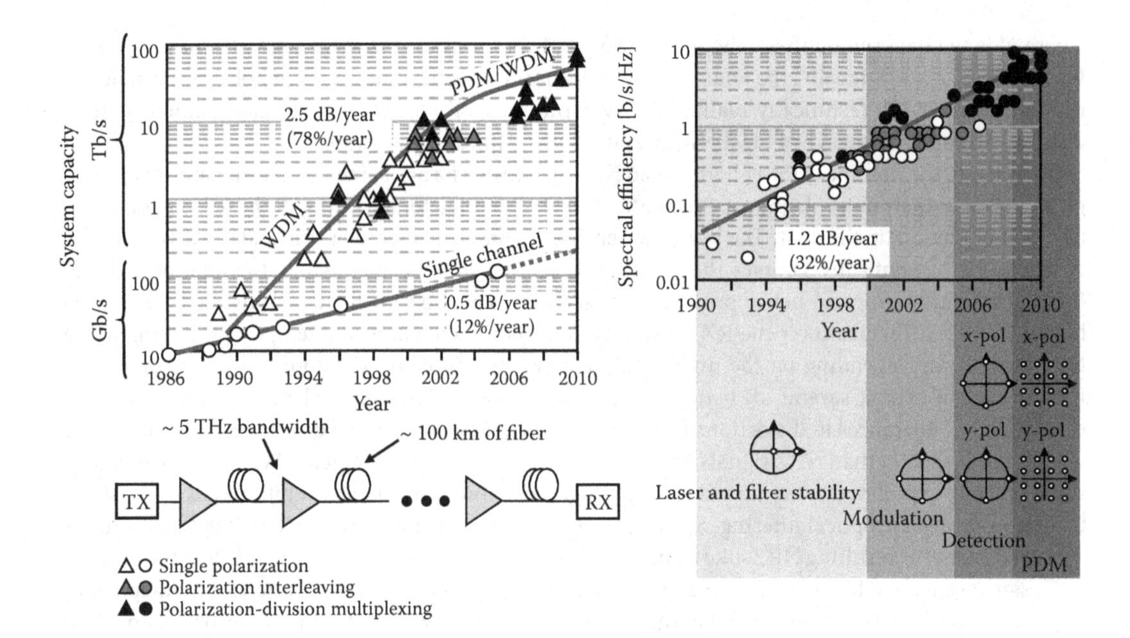


**FIGURE 2.4** Record single-channel bit rates, WDM capacities, and spectral efficiencies. (P. J. Winzer, Modulation and Multiplexing in Optical Communication Systems, *IEEE-LEOS Newsletter*, http://photonicssociety.org/newsletters/feb09/modulation.pdf. © 2009 IEEE. With permission.)

was sufficient to cover most geographically relevant communication distances up to 10 Gb/s, and the comparatively large spectral extent of OOK, especially when using RZ, was still significantly below the frequency stability range of lasers and optical filtering elements suitable for building optical networking elements at 100-GHz and 50-GHz WDM channel spacings. Hence, the evolution of optical communication systems using OOK was primarily an evolution of physical device design and engineering. Owing to its simplicity, OOK was also the first modulation format used in early 40-Gb/s products [91] and was the first modulation format taken to 100 Gb/s in research experiments, using passive optical equalization [62,92] and vestigial sideband (VSB) filtering [93] to make up for the limited bandwidth of optoelectronic components. For reasons of WDM spectral efficiency (SE) discussed in Section 2.4.2, subsequent research at 100 Gb/s and beyond used QPSK at up to 56 Gbaud [73–75,94] or other higher-order modulation formats, cf. Figure 2.5 [18].

Traditionally dominated by the Synchronous Optical Network (SONET), the Synchronous Digital Hierarchy (SDH), and the Optical Transport Network (OTN) standards, commercial optical

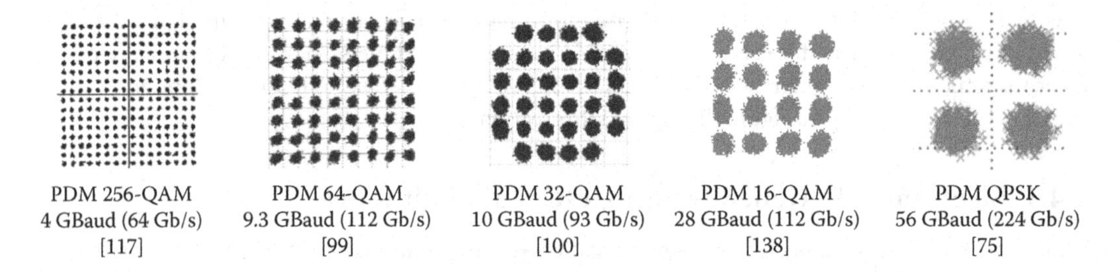


**FIGURE 2.5** State of the art in serial QAM at multi-Gbaud rates as of March 2010. (P. J. Winzer, Beyond 100G Ethernet, *IEEE Comm. Mag.*, 48:26–30. © 2010 IEEE. With permission.)

single-channel bit rates have increased in factors of four (2.5 Gb/s, 10 Gb/s, and 40 Gb/s). Ethernet, on the other hand, has shown historic bit rate increases in factors of 10 (100 Mb/s, 1 Gb/s, and 10 Gb/s). At the first common SONET/Ethernet bit rate of 10 Gb/s, Ethernet started to enter optical networking around the year 2000 [95]. The success of Ethernet and the dominance of data traffic in today's networks then triggered the desire to further scale optical networks with a focus on Ethernet. As a result, both the Institute of Electrical and Electronic Engineers (IEEE) and the International Telecommunications Union—Telecommunication Standardization Sector (ITU-T) are now finalizing the standardization of 100-Gb/s optical transport standards. Notably, the OTN standard has departed from the classical 4× scaling and has abandoned its next logical step of 160 Gb/s in favor of 112 Gb/s to natively support 100-Gb/s Ethernet in optical transport networks [17]. Ethernet, on the other hand, has standardized 40 Gb/s in addition to 100 Gb/s for economic and technology maturity reasons, departing from its so far strictly observed 10× scaling. This compromise suggests that the next optical channel bit rate to be standardized may well be 400 Gb/s, followed by 1 Tb/s, both in Ethernet and in OTN [18].

Generation and detection of electronic time-division multiplexed (ETDM) serial bit rates of ~200 Gb/s, allowing for polarization-division multiplexed (PDM) bit rates of ~400 Gb/s, have been reported in research in the first half of 2010, using 56-Gbaud 16-QAM [76]. Extrapolating the progress of serial bit rate research (Figure 2.4, circles), we may expect single-channel ETDM bit rates of around 500 Gb/s (allowing for ~1 Tb/s serial interfaces using polarization-division multiplexing using PDM) towards 2020 [18]. Note in this context that the exponential scaling of serial bit rates at about 12% (or 0.5 dB)* per year follows the speed increase of semiconductor devices ($f_T$), which are currently able to provide logic circuits suitable for electronically demultiplexing bit streams close to 200 Gb/s [96]. Note that the use of higher-order modulation formats ($M$-QAM) is unlikely to change this scaling, since richer constellations also require a higher effective number of bits (ENoB) in digital-to-analog and analog-to-digital conversion (DAC, ADC) at transmitter and receiver. Comparing the increased ADC requirements for $M$-QAM [97] with Walden's observations on the scaling of ADCs [98], which improve at about 1/3 ENoB per year at constant converter speed, we find little room for an extra increase in serial bit rates by going to higher-order modulation in addition to adopting higher symbol rates. The trade-off between symbol rate and the required resolution of ADCs is clearly visible in Figure 2.5, summarizing the current state-of-the-art in QAM research experiments. From a modulator perspective, optical parallelism could be used to reduce DAC resolution requirements, as discussed in Section 2.2.2.

## 2.4.2 Multiplexing Options

To exploit the full bandwidth supported by optical amplifiers and low-loss optical fiber, aggregate channel bit rates far beyond the capabilities of ETDM technologies are desirable, which asks for additional multiplexing techniques in the time, frequency, polarization, or spatial dimension [19], as illustrated in Figure 2.6.

### 2.4.2.1 Optical Time-Division Multiplexing

One option to go beyond the speed limitations of ETDM technologies is *optical time-division multiplexing* (OTDM) [101]. An OTDM transmitter, sketched in Figure 2.6b, generates several substreams ("OTDM tributaries") that each consists of short optical pulses at 1/Nth of the target symbol rate. The modulated substreams are then temporally interleaved in the optical domain to generate the full-rate serial data stream. Using this technique, single-channel bit rates of up to 1.28 Tb/s using OOK with temporal polarization interleaving of adjacent pulses [102] and up to 5.1 Tb/s using 16-QAM [103] have been generated. However, as evident from Figure 2.6, OTDM transmitters rely on *parallel* optical transmitter (and receiver) architectures, which are comparatively complex to implement. As a result, ETDM

* Exponential growth rates can be conveniently expressed in decibels, e.g., 12% equals $10 \log_{10}(1.12)=0.5$ dB [10].

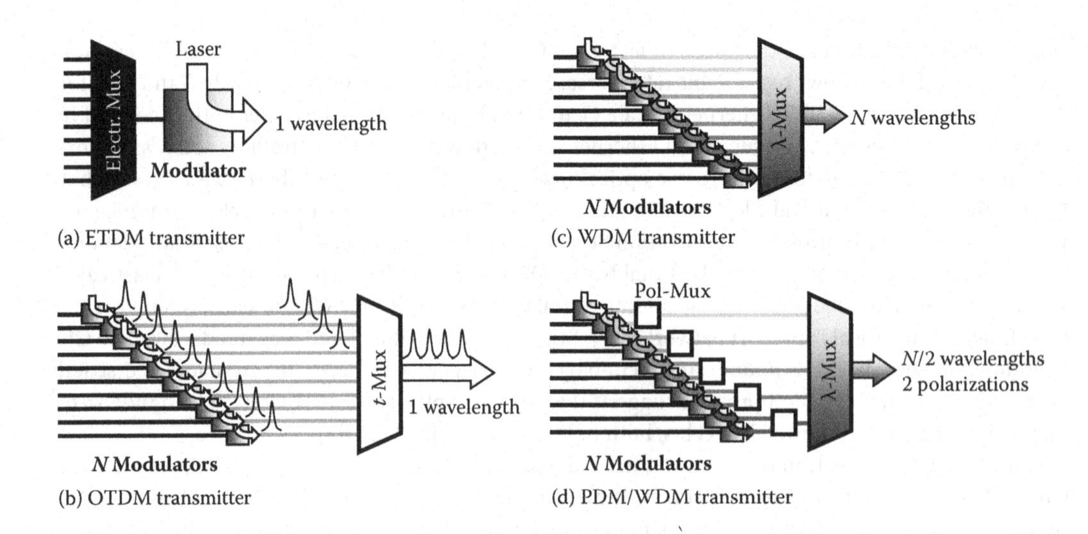


**FIGURE 2.6** Basic transmitter architectures using (a) ETDM, (b) OTDM, (c) WDM, and (d) WDM combined with PDM.

transponders have so far been economically superior to their OTDM counterparts in product development and commercialization. Furthermore, and more fundamentally, OTDM systems inherently rely on *short optical pulses* and sometimes even on temporal polarization interleaving to avoid crosstalk between adjacent OTDM tributaries. Short pulses result in broader modulation spectra compared to the corresponding ETDM signals, which stands against the desire to scale system capacity by dense WDM channel packing. Nevertheless, OTDM systems continue to be a valuable research tool to study the propagation properties of signals at bit rates beyond the available electronics and optoelectronics.

#### 2.4.2.2 WDM and PDM

Notwithstanding the great progress that has been made in single-channel serial transmission systems over the decades, it is clear that the modest single-channel capacity scaling of around 12% per year, cf. Figure 2.4a, would not have been sufficient to support the ~60% annual growth in capacity demand of modern data applications, including the build-out of the Internet. By the mid-1990s, the erbium-doped fiber amplifier (EDFA) became commercially viable [104]. This amplifier (as well as other amplification schemes, such as distributed Raman amplification [105]) could simultaneously amplify many WDM channels, enabling transparent optical transport of hundreds of individually modulated channels. The transmission of these channels over thousands of kilometers without the need for costly and power-hungry optoelectronic regeneration was enabled by the invention of optical dispersion management to combat fiber nonlinearity [87,106]. Both techniques allowed the capacity of fiber-optic communication systems to scale in the wavelength domain by two orders of magnitude compared to single-channel systems, as indicated by the triangle symbols in Figure 2.4a.

The key to WDM capacity scaling is the ability to pack as much information as possible into the amplification bandwidth of the underlying optical amplification technology, e.g., into the ~5-THz bandwidth of EDFAs. This implies increasing the system's SE, defined as the per-channel information bit rate divided by the WDM channel spacing,

$$\mathrm{SE} = \frac{R_B}{B}. \tag{2.6}$$

Up until ~2000, achieving a closer WDM channel spacing was primarily a matter of improving the stability of lasers and of building highly frequency selective yet stable and economically manufacturable optical filters; the increase in SE at about 32% per year, represented by the data points in Figure 2.4b, was therefore mostly due to improvements in *device physics* and *device technologies*. The additional ~35% of capacity increase of WDM systems is due to the simultaneous expansion of the optical amplification band, which comprised about 100 GHz in the early WDM experiments and has been expanded to between 10 Thz and 15 THz in more recent experiments using several parallel amplification bands, e.g., dual-band (C+L band) or triple-band (S+C+L band) systems [107]. Around 2000, the polarization dimension also started to be noticeably exploited to increase WDM capacity, first by polarization interleaving of adjacent WDM channels to suppress WDM crosstalk (cf. Figure 2.7b) and then by true PDM (cf. Figure 2.7c), i.e., by sending different symbol streams on each polarization at the same carrier wavelength. Capacity records using polarization interleaving and PDM are indicated by the different grayscales of the symbols in Figure 2.4. As evident from the figure, WDM capacity growth has started to considerably slow down over the past years, owing to fundamental capacity limits of the fiber-optic channel discussed in Section 2.4.4. It remains to be seen which technologies will allow to further scale the capacity of optical networks. One possibility is spatial multiplexing, either by using multiple parallel optical fibers or fiber ribbons, multiple modes of a single fiber [108], or multiple cores of a multicore fiber [109].

### 2.4.2.3 Orthogonal Frequency–Division Multiplexing

Multiplexing digital signals in the frequency domain requires establishing frequency-orthogonal channels, i.e., frequency channels that can ideally be separated without any crosstalk or interference. There are two ways of forming such channels in the frequency domain [19]. In a first approach, one can place different signals into spectrally nonoverlapping frequency bins. This is referred to as WDM and is illustrated in Figure 2.7a. Pulse shaping can be used to increase the spectral compactness of the WDM channels to a minimum ISI-free bandwidth of $R_S$. If energy spills from one channel into another in an

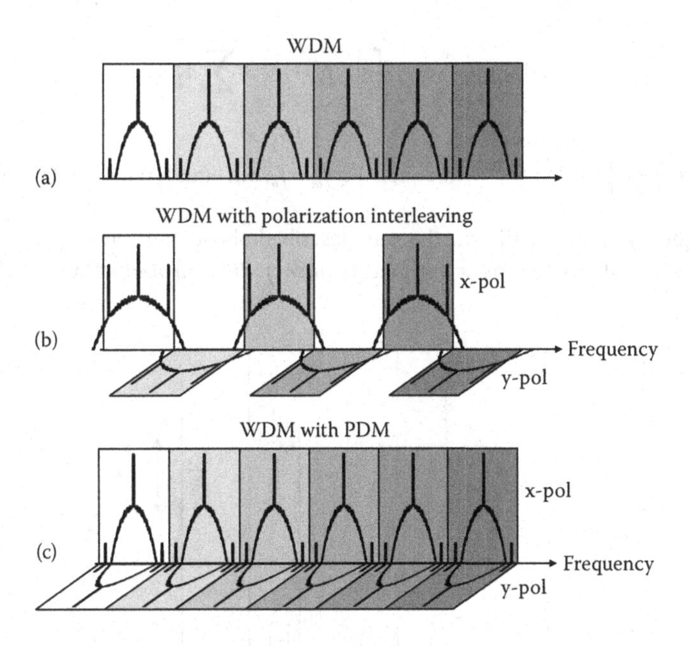


**FIGURE 2.7** Multiplexing on the optical fiber channel using WDM, polarization interleaving to reduce WDM crosstalk, and PDM to send two independent signal streams at the same carrier wavelength. (P. J. Winzer, Modulation and Multiplexing in Optical Communication Systems, *IEEE-LEOS Newsletter*, http://photonicssociety.org/newsletters/feb09/modulation.pdf. © 2009 IEEE. With permission.)

uncontrolled manner orthogonality is degraded and WDM crosstalk is observed; orthogonality can be reestablished using polarization interleaving, as introduced above.

In a second approach, one can establish frequency orthogonality in the following sense [26]: Two (RF or optical) complex carriers $e^{j2\pi f_1 t}$ and $e^{j2\pi f_2 t}$ are called "orthogonal" over a time interval of duration $T$ if they differ by an integer number of oscillations within that time interval. In other words, two carriers are orthogonal over $T$ if they are spaced in frequency by an integer multiple of $1/T$. Applied to digital communications, $T$ is identified as the symbol duration $T_S$. One can hence independently transmit digital information on two subcarriers if they are spaced in frequency by integer multiples of $1/T_S$ and if the symbol intervals are synchronized across the subcarriers. Even though the individual subcarrier signals overlap both in time and in frequency, no crosstalk is observed at the receiver. To see this, we consider the simple example shown in Figure 2.8, assuming three subcarriers modulated by Equation 2.5 using the pulse shapes $p_i(t) = e^{j2\pi f_i t}\text{rect}(t, T_S)$, with $\text{rect}(t,T_S)$ being a synchronized, rectangular pulse envelope of duration $T_S$. We write the multiplex as

$$\begin{aligned} s(t) &= \sum_{k=-\infty}^{\infty} a_{1,k}p_1(t-kT_S)+a_{2,k}p_2(t-kT_S)+a_{3,k}p_3(t-kT_S) \\ &= \sum_{k=-\infty}^{\infty} \left\{a_{1,k}e^{j2\pi f_1 t}+a_{2,k}e^{j2\pi f_2 t}+a_{3,k}e^{j2\pi f_3 t}\right\}\text{rect}(t-kT_S,T_S), \end{aligned} \tag{2.7}$$

where $a_{1,k}$, $a_{2,k}$, and $a_{3,k}$ are the digital modulation symbols in time slot $k$ that are to be transmitted. In our example, binary phase modulation is used ($a_k = \pm 1$), but any other complex quadrature modulation format is possible here as well. The three subcarrier frequencies $f_1$, $f_2$, and $f_3$ are spaced by $1/T_S$.

The receiver, in parallel, forms the "inner product" between the received multiplex $s(t)$ and all the subcarriers that it wants to detect. For example, detection of the subcarrier at $f_2$ requires processing of

$$\left\langle s(t)\middle|p_2(t)\right\rangle = \frac{1}{T_S}\int_0^{T_S} s(t)p_2^*(t)\text{d}t = \sum_{k=-\infty}^{\infty} a_{2,k}, \tag{2.8}$$

owing to the fact that $\frac{1}{T_S}\int_0^{T_S} e^{j2\pi f_m t}e^{-j2\pi f_n t}\,dt$ equals 1 if $f_m = f_n$, but equals 0 if $f_m$ and $f_n$ differ by an integer multiple of $1/T_S$.

Exploiting frequency orthogonality in the way described above, with an arbitrarily large number of subcarriers, is generally known as *orthogonal frequency division multiplexing* (OFDM) in digital

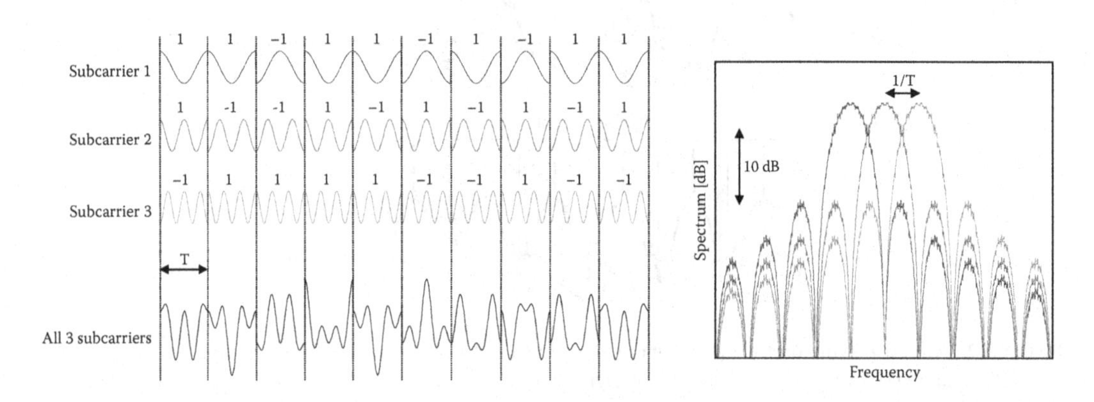


**FIGURE 2.8** An OFDM signal consists of a temporal as well as spectral superposition of orthogonal subcarriers.

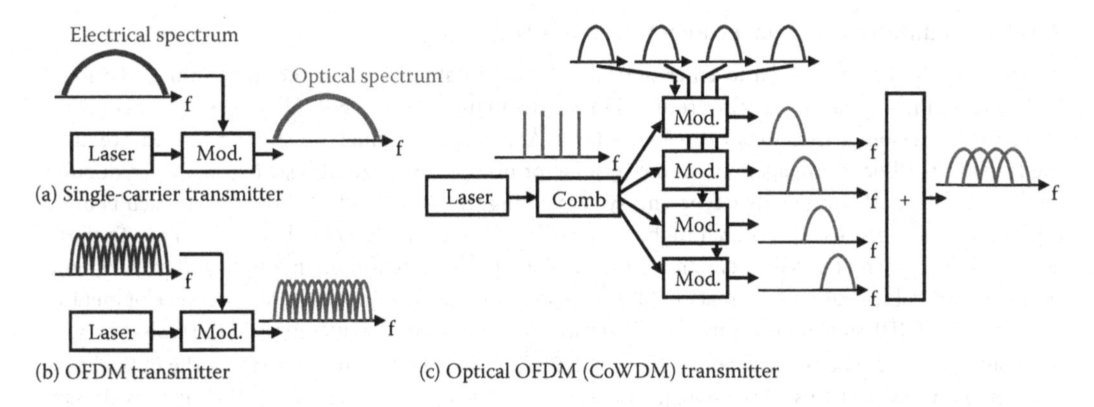


**FIGURE 2.9** Transmitter implementations of OFDM and oOFDM/CoWDM. (P. J. Winzer, Beyond 100G Ethernet, *IEEE Comm. Mag.*, 48:26–30. © 2010 IEEE. With permission.)

communications. Note that OFDM is not a modulation format but a subcarrier multiplexing technique; each subcarrier in OFDM is modulated using a certain digital modulation format.

In optical communications, we distinguish between two approaches of generating and detecting the orthogonal frequency multiplex, as illustrated for the transmitter in Figure 2.9. One typically speaks of "OFDM" if the subcarriers are generated *electronically*, as shown in Figure 2.9b. Since the operation described in Equation 2.7 represents a discrete Fourier transform, the most efficient OFDM systems are implemented based on the fast Fourier transform [110]. In contrast, Figure 2.9c shows a parallel optics implementation, where the orthogonal subcarriers are first generated optically, e.g., from a single laser source using a comb generator [111], and are then modulated individually using a set of parallel optical modulators. In this case, one typically speaks of "optical OFDM" (oOFDM) or "coherent WDM" (CoWDM). Note that an oOFDM/CoWDM signal can use either single-carrier modulation on each of its subcarriers [112] or electrically generated OFDM on each subcarrier [113,114]. Terabit superchannels have been obtained this way [111].

Irrespective of how it is generated, an orthogonal frequency multiplex using $N$ subcarriers transports an aggregate bit rate of up to $R_B = NR_S\log_2 M$ in a channel bandwidth of $B_{Ch} \approx NR_S = R_B/\log_2 M$ per WDM channel, just like a single-carrier format with proper pulse shaping. Hence, the native SE as well as the signal-to-noise ratio (SNR) requirements of single-carrier modulation and OFDM is the same when using the same modulation format. However, OFDM typically requires pilot carriers and a cyclic prefix, which constitutes a substantial overhead and reduces the transmitted information rate at fixed line rate [116]; this can make OFDM less spectrally efficient than single-carrier formats in practice.

Regarding the complexity of implementation, OFDM has been shown to require about the same DSP power than single-carrier modulation [116]. On the other hand, OFDM requires a high-speed (~$R_S$) and high-resolution DAC at the transmitter, which, together with a reasonably linear drive amplifier, can pose serious practical problems. For single-carrier modulation, a high-resolution DAC is only required for high-order QAM [117], or for pulse shaping [71,77] and predistortion [61]. Perhaps the only real advantage of OFDM in optical transport agreed upon today is its ability to dynamically adapt both the modulation format of each subcarrier and the number of subcarriers to the transmission channel, which could allow for software-defined, modem-like optical transponders in future bit-rate flexible optical networks.

### 2.4.3 Advanced Optical Modulation in WDM Networks

Having looked at the evolution of single-channel bit rates and at the multiplexing options available in digital optical communications, we will now study the evolution of optical modulation formats in the context of optically routed network over the last 10 years.

#### 2.4.3.1 Modulation for Longer Reach at 40 Gb/s

When 40-Gb/s systems started to enter optical networking at the turn of the millennium, the fourfold increase in per-channel bit rates resulted in a 6-dB-higher optical SNR (OSNR) requirement compared to the previous generation of 10-Gb/s systems. At fixed per-channel transmit powers, this reduces the reach in the linear propagation regime by a factor of four. In the nonlinear regime, the reduction depends significantly on the fiber type and system configuration and a factor of two in reach is more typical since one can usually launch higher per-channel powers at 40 Gb/s than at 10 Gb/s. To make up for the lost reach at 40 Gb/s, enhanced FEC [118,119], distributed Raman amplification [105], and advanced optical modulation formats [29,120,121] were introduced. In particular, the development of *differential PSK* (DPSK), more accurately, differential *binary* phase keying, improved the OSNR margin by about 3 dB, since the two symbols {+1, −1} of DPSK are spaced $\sqrt{2}$ further apart in the optical field than the symbols {0, +1} used by OOK for the same average optical power* [58]. Note in this context that PSK and DPSK are the same modulation format (since they have the same symbol alphabet and the same analog symbol waveforms), but they differ in the way symbols are encoded at the transmitter and are decoded at the receiver. PSK encodes bits directly onto the phase of the respective symbols, which requires a constant phase reference at the receiver to decode the symbols. The constant phase reference is provided by a local oscillator (LO) laser in a coherent receiver setup, as described below. DPSK, on the other hand, encodes information on the *difference* of the phase between adjacent symbols (thus the prefix "D" for "differential"), and each symbol is used as the phase reference for the immediately following symbol. Differential phase demodulation is performed by a *delay interferometer as* part of a DPSK receiver front end. While being significantly simpler than a coherent receiver, going to DPSK has undoubtedly increased the receiver complexity compared to OOK, as shown in Figure 2.10.

From a modulator point of view, DPSK can be generated either by a straight-line phase modulator that modulates the optical field by $\pi$ along the unit circle, or by a MZM that modulates the optical field between +1 and −1 along the real axis, as shown in Figure 2.2e. As detailed in [58], the MZM option is generally preferable for DPSK. When using a phase modulator, any overshoot or ISI in the drive signal directly translates into distortions of the information-bearing phase. Using a MZM, electrical drive imperfections are greatly suppressed due to the sinusoidal modulator transfer function, and the remaining overshoot and ripple only affects the signal amplitude and not the information-bearing optical phase. As a result, DPSK transmitters are almost exclusively based on MZMs today.

#### 2.4.3.2 Modulation for Higher SE at 40 and 100 Gb/s

While the reduced reach in going from 10 Gb/s to 40 Gb/s was partially addressed by DPSK, the spectral extent of 40-Gb/s binary signals was filling the bandwidth available to a WDM channel on a 50-GHz grid. Achieving an SE of 0.8 b/s/Hz (40 Gb/s / 50 GHz) required narrow-band modulation formats such as DB [123], VSB-OOK [124,125], or optically overfiltered DPSK that employed a delay interferometer with a slightly smaller delay than the symbol period [126], coining the name "partial DPSK" (P-DPSK). In a further step, reducing the symbol rate from 40 Gbaud for binary formats to 20 Gbaud using differential QPSK (DQPSK) [66] enabled enough spectral compression to comfortably accommodate a large number of ROADMs at almost the same transmission performance as DPSK [29,127]. On the downside, the hardware complexity at both transmitter and receiver is significantly higher for DQPSK compared to DPSK, as shown in Figure 2.10. A DQPSK receiver uses two delay interferometers followed by balanced detection, which yields twice the receiver complexity compared to DPSK. On the transmitter side, DQPSK is most conveniently generated with an I/Q modulator, as described along with Figure 2.2e, although transmitter implementations using a dual-drive MZM and nonbinary drive signals are also possible [61,128]. While the number and complexity of hardware elements increases for DQPSK, their

* In typical fiber-optic communication systems, the average optical power is usually the quantity of interest, owing to the average-power limitation of optical amplifiers [122]. In systems where the *peak* optical power is the quantity of interest, some statements made in this chapter would have to be revisited.

optoelectronic bandwidth requirements are reduced to half the bit rate, which offsets some of the added transponder complexity.

Following its success at 40 Gb/s, research on electronically multiplexed 100-Gb/s transmission systems also adopted DQPSK at symbol rates exceeding 50 Gbaud [74], demonstrating optically routed networking over 1200 km including 6 ROADMs at a SE of 1 b/s/Hz in the laboratory and in the field [73]. In research laboratories, DQPSK was used together with PDM to establish fiber transmission capacity records of up to 25.6 Tb/s at a SE of 3.2 b/s/Hz (1.6 b/s/Hz per polarization) [89]. These record capacities were the last ones performed with real-time, direct-detection receivers and marked the transition of high-capacity optical transport to *coherent detection*, which in research experiments has to resort to off-line processing of comparatively small chunks of the received signal (typically about 1 million symbols).

### 2.4.3.3 Coherent Detection and PDM QPSK

The transition to 100-Gb/s systems asked for further spectral compression to be able to stick with the 50-GHz WDM grid and thereby increase the aggregate system capacity by a factor of 2.5 from 40-Gb/s systems. This quest for increased SE motivated the shift to PDM systems. For example, a PDM QPSK signal at a line rate of 112 Gb/s (and hence at a symbol rate of 28 Gbaud) can traverse about 5 ROADMs on a 50-GHz WDM grid [129,130]. To properly demultiplex the two signal polarizations at the receiver, dynamic polarization tracking is needed in a fiber-optic context, where the signal polarizations can vary at kHz rates or faster [131]. Although optical polarization controllers are in principle feasible [132], polarization-diversity digital coherent receivers are a more practical solution. Digitally phase-locking coherent receivers (also referred to as "intradyne receivers") convert both polarizations of the optical field into the electrical domain by using a free-running LO laser followed by a high-speed ADC. Once in digital form, DSP algorithms essentially invert the polarization rotation matrix of the fiber channel [133]. From a hardware perspective, and as shown in Figure 2.10, PDM doubles the transmitter

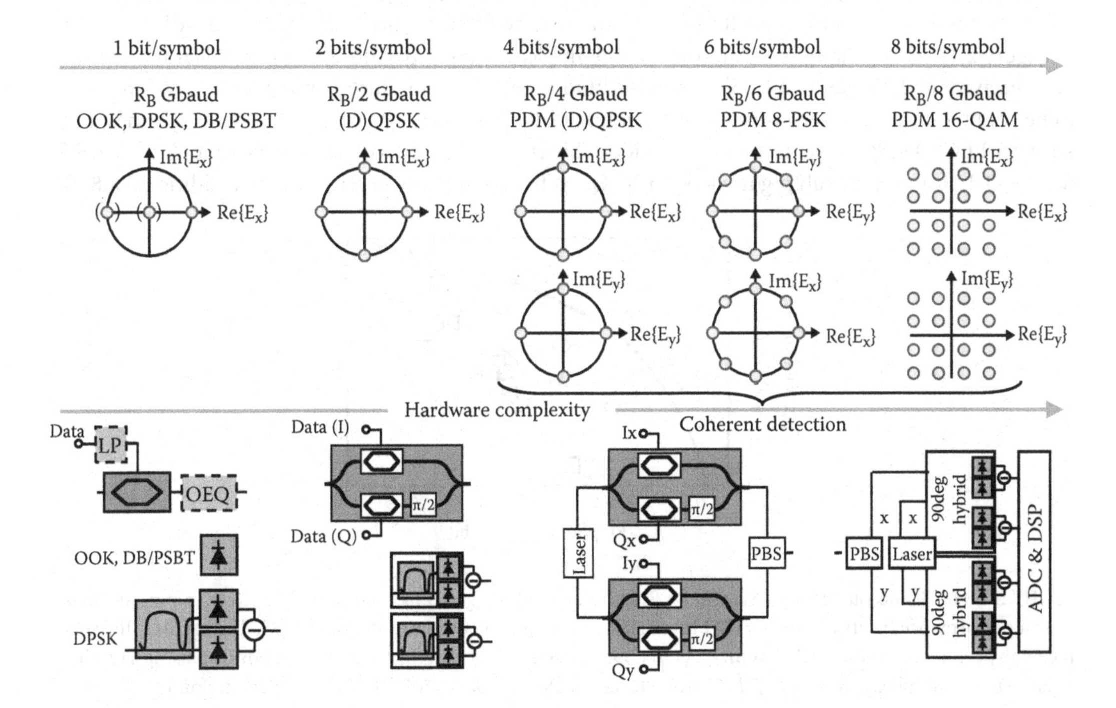


**FIGURE 2.10** Overview of hardware complexities associated with the generation and detection of various modulation formats. (G. Raybon and P. J. Winzer, 100Gb/s Challenges and Solutions, Tutorial at OFC/NFOEC 2008, Paper OTuG1. © 2008 IEEE. With permission.)

complexity and coherent detection significantly increases the receiver complexity, both through the required optical and optoelectronic components and through the DSP hardware that asks for dedicated ASIC developments [134]. Note, however, that the introduction of PDM and coherent detection opens the way to other complex quadrature modulation formats, with no further conceptual changes in the structure of the optoelectronic transmitter or receiver front ends. Different modulation formats are then enabled by the appropriate resolution DAC and ADC as well as by format- and application-specific DSP algorithms.

### 2.4.3.4 Beyond PDM QPSK: Higher-Order QAM

As discussed in Section 2.2, an $M$-ary symbol constellation with symbol-by-symbol detection* can theoretically support an SE of at most $\log_2 M$, or $2 \log_2 M$ if PDM is used, provided that perfect sinc pulse shaping is employed. With today's optical, electrical, and optoelectronic hardware, however, one can typically only achieve about 70% of the theoretical value for point-to-point systems and about 50% in an optically routed network with state-of-the-art ROADMs [135]. For example, QPSK with sinc pulse shaping theoretically supports an SE of 4 b/s/Hz but can only achieve 2 b/s/Hz in an optically routed network. If higher SEs are to be supported, more than four symbols per polarization are required. One possibility is the use of PSK with more than four levels. For example, 114-Gb/s 8-PSK has been demonstrated in a 662-km point-to-point experiment on a 25-GHz WDM grid at an SE of 4.2 b/s/Hz [136,137]. Going beyond $M$-ary PSK, one can use multiring PSK formats or QAM constellations, where the symbols are usually (but not necessarily) arranged on a rectilinear grid, representing a good trade-off between performance and implementation aspects. Using 16-QAM, one can perform 112-Gb/s optical networking with ROADMs on a 25-GHz WDM grid [68], or 228-Gb/s on a 50-GHz grid [138]. The most recent optical fiber capacity record at 69 Tb/s used 16-QAM at 171 Gb/s at a SE of 6.4 b/s/Hz over 240 km of fiber [139].

On the downside, introducing more symbols fundamentally reduces the robustness of a modulation format to noise, leading to an inherent SNR penalty. For example, going from QPSK (i.e., 4-QAM) to 16-QAM doubles the system capacity, but with the assumed FEC results in a penalty of 3.8 dB, as shown in Figure 2.11. This penalty is in addition to any implementation penalties from nonideal transmitter and receiver implementations [135] and is likely to manifest as a reduced system reach, which needs to be counteracted by lower-loss and lower-nonlinearity fiber, lower-noise amplification schemes, and stronger FEC. Owing to the fundamental trade-off between SE and SNR given by Shannon's theory (cf. Section 2.4.4), another doubling in SE from 16-QAM to 256-QAM would result in an additional 8.8 dB

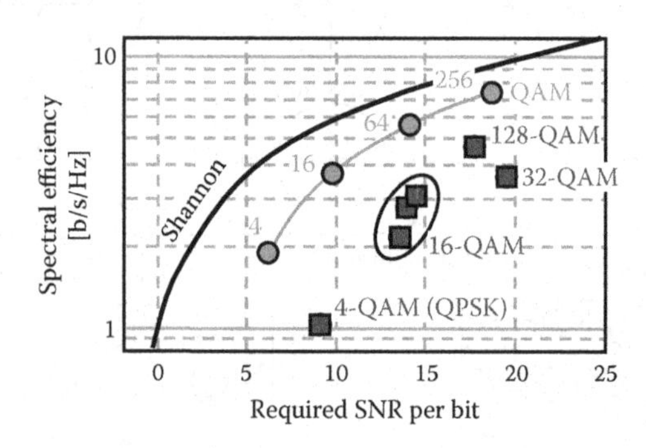


**FIGURE 2.11** Trade-off between SE and SNR as fundamentally given by Shannon. The circles represent theoretical performance limits of various QAM formats assuming a 7% OH hard-decision FEC. The squares illustrate recent experimental records. (P. J. Winzer, et al., Spectrally efficient long-haul optical networking using 112-Gb/s polarization multiplexed 16-QAM, *J. Lightwave Technol.*, 28:547–556. © 2010 IEEE. With permission.)

* Complex sequence detection at the receiver allows for stronger transmit filtering, inducing ISI [85] and leading to $>\log_2 M$ spectral efficiencies. In any case, the obtainable capacity remains limited by the modulation unconstrained Shannon limit.

of SNR penalty, again assuming state-of-the-art 7% OH hard-decision FEC. Furthermore, the robustness to laser phase noise and the requirements on the resolution of receiver ADCs becomes more severe for higher-order constellations [97].

### 2.4.4 Fundamental Limits to Capacity and Optical Modulation

The maximum capacity that can be achieved in the process of communicating information in a noisy environment (referred to as a noisy "channel") was determined by Shannon [1] in 1948. In his *theory of information*, Shannon derived the capacity $C$ of an additive white Gaussian noise (AWGN) channel [1,28,140] as

$$C = B\log_2(1 + \text{SNR}), \tag{2.9}$$

where $B$ is the channel bandwidth available for communication and the SNR is defined as the ratio of signal to noise within the channel.* At a fixed noise level, the capacity of the AWGN channel increases with signal power. Note that the capacity of a channel as given by Shannon assumes optimum coding [141], the use of Nyquist signaling (see Section 2.3 and [26,27,142]), and an optimum constellation, that follows a bidimensional Gaussian distribution.

Figure 2.12a shows the symbol constellation that yields optimum capacity on the AWGN channel, the bidimensional Gaussian constellation. This constellation is very challenging to implement as it contains a continuous (and infinite) set of constellation points, both for the in-phase and for the quadrature component of the optical field (or, alternatively, for radial and angular components). Practical implementations obviously require some level of quantization. A first way to quantize the bidimensional Gaussian distribution is to use ring constellations, such as the four-ring constellation shown in Figure 2.12b. From symmetry considerations, these ring constellations appear as a natural way to reduce the number of states of a bidimensional Gaussian distribution with minimum distortion. The number of levels of amplitude quantization determines the number of rings. Figure 2.13, shows ring constellations from one to eight rings.

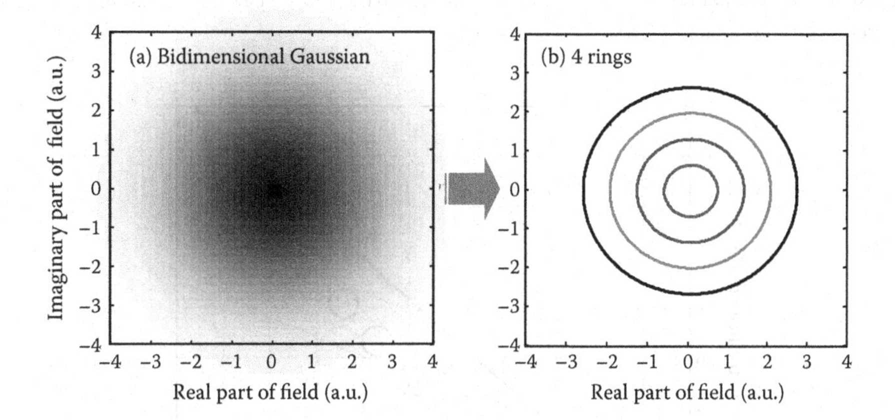


**FIGURE 2.12** A four-ring constellation (b) as an approximation of the bidimensional Gaussian constellation shown in (a). (R.-J. Essiambre, et al., Capacity limits of optical fiber networks, *J. Lightwave Technol.*, 28:662–701. © 2010 IEEE. With permission.)

* The SNR used in classical digital communications and the OSNR used in optics are universally related by OSNR = SNR $p\,R_S/(2B_{\text{ref}})$, where $p$ denotes the number of polarization states used to transmit information and $B_{\text{ref}}$ is a fixed reference bandwidth, commonly set to 12.5 GHz, independent of the bit rate [23].

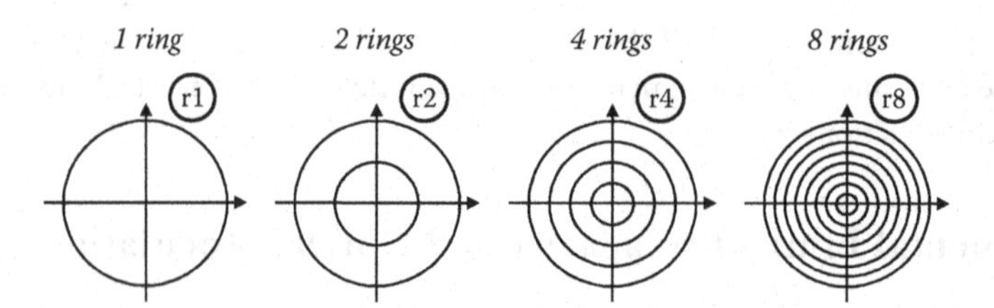


**FIGURE 2.13** Ring constellations with integer ring radii ratios. (R.-J. Essiambre, et al., Capacity limits of optical fiber networks, *J. Lightwave Technol.*, 28:662–701. © 2010 IEEE. With permission.)

Figure 2.14 shows the capacity per unit bandwidth for the ring constellations depicted in Figure 2.13, and larger constellations including up to 128 rings. We consider ring constellations with equally spaced radii in optical field amplitude and assume an equal frequency of occupation (equal probability for choosing a transmitted symbol) on each ring. Note that the phase state remains continuous, i.e., non-quantized, for our ring constellations. As can be seen from Figure 2.14, the process of quantizing the amplitude of the bidimensional Gaussian constellation into a finite set of rings leads to some loss of capacity compared to the Shannon limit. However, one can very closely approach the Shannon limit when using a sufficiently large number of rings. A sufficient number of rings can also reduce the SNR required to achieve a certain capacity, which is especially advantageous at high SNR values.

Calculating the capacity of a channel that incorporates optical fibers as a transmission medium represents a nontrivial extension of Shannon's results [23]. This is mainly due to the Kerr nonlinear effect found in fused silica optical fibers. The Kerr nonlinear effect produces a change of refractive index of optical fibers that increases with the optical signal power [37]. Figure 2.15 shows a capacity limit estimate of single-polarization modulation over standard single-mode fiber (SSMF) that takes into account the Kerr nonlinear effect. Also shown is a selection of the highest SE record experimental realizations from Figure 2.4, referenced to a single polarization. Different fiber types have different capacity limits, but variations in capacity are rather small across the wide range of commercially available fibers [143]. It is interesting to note that record research experiments have achieved capacities that are about a factor two to three from the capacity of SSMF.

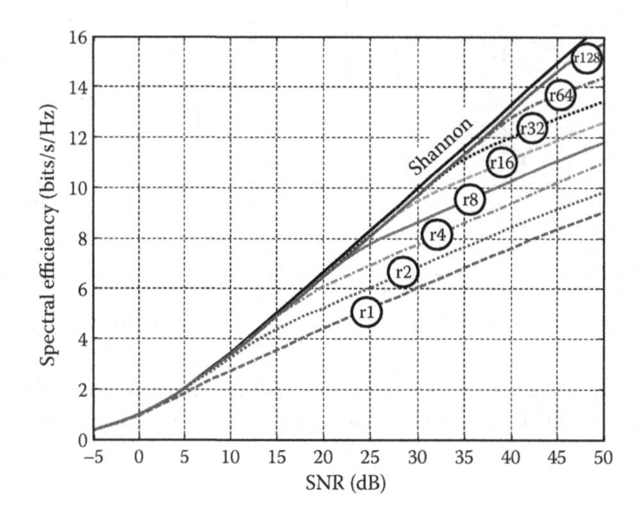


**FIGURE 2.14** Capacity per unit bandwidth (SE) for ring constellations with various numbers of rings. Nyquist signaling (making use of the sinc-pulse shape) is assumed here. (R.-J. Essiambre, et al., Capacity limits of optical fiber networks, *J. Lightwave Technol.*, 28:662–701. © 2010 IEEE. With permission.)

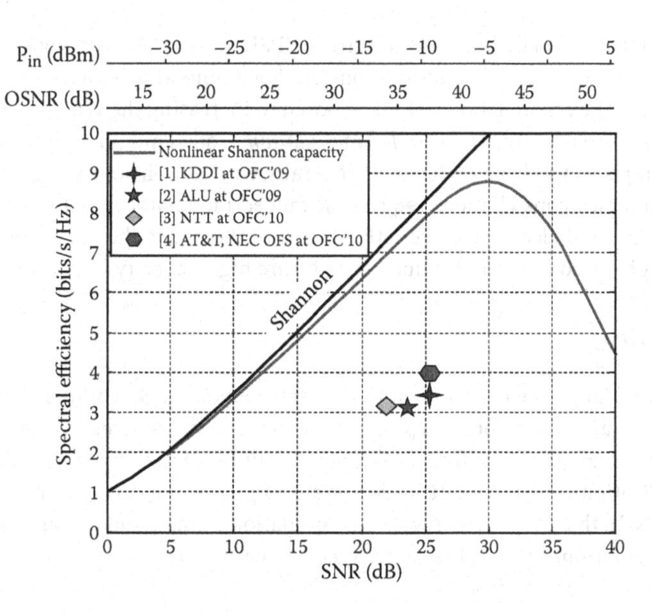


**FIGURE 2.15** SE for single-polarization modulation from a nonlinear Shannon capacity limit estimate and recent record experiments. Fiber and system parameters can be found in [23, Section XI]. (R.-J. Essiambre, et al., Capacity limits of optical fiber networks, *J. Lightwave Technol.*, 28:662–701. © 2010 IEEE. With permission.)

Ring constellations have been used to estimate the capacity of the nonlinear fiber-optic channel. However, one should also consider more practical advanced constellations using a finite set of symbols on a regular grid. The SE of such practical constellation for the single-polarization AWGN channel is shown in Figure 2.16. For the assumed Nyquist pulse shaping and symbol-by-symbol detection at the receiver, the SE of all formats saturates for high values of SNR at $\log_2(M)$, where $M$ is the number of points of the constellation.

One interesting feature of Figure 2.16 is the fact that at fixed SNR the SE always increases by using larger constellations (larger $M$). Consider a 10-dB SNR, for instance. The capacities increase from

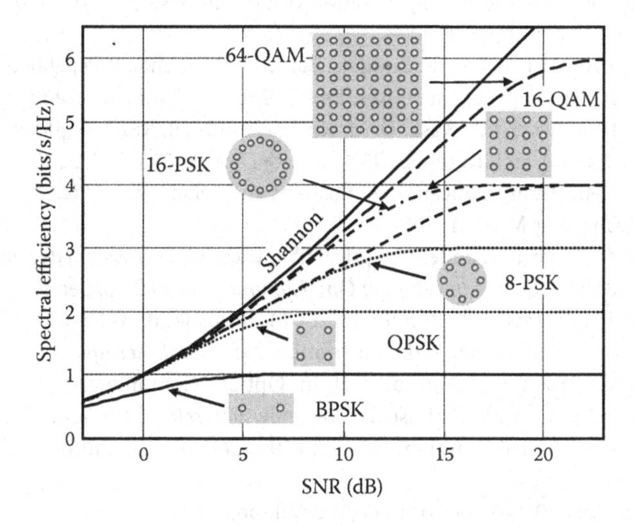


**FIGURE 2.16** SE for various commonly implemented single-polarization modulation formats. Nyquist signaling is assumed. (R.-J. Essiambre, et al., Capacity limits of Fiber-Optic Communication Systems, Tutorial at OFC/NFOEC 2009, Paper OthL1. © 2009 IEEE. With permission.)

1 bit/symbol to 3.3 bits/symbol for constellations from BPSK to 16-QAM. The Shannon limit is at $\log_2(11) \approx 3.46$. A capacity increase achieved by adding constellation points allows for a larger coding OH at fixed symbol rate. Growing the constellation size (as opposed to increasing the symbol rate) to accommodate a larger coding OH is often referred to as *coded modulation*. Coded modulation is a way to achieve high SE without incurring spectral broadening, and generally without sensitivity degradation (after decoding). Generating complex constellations requires advanced modulator structures with precise control of optical components and electrical drives. This is one of the major challenges ahead for designers of modulators and high-speed circuitry for upcoming future high-capacity transmission systems.

## 2.5 Conclusion

To meet the exponentially growing bandwidth demand of global data communications, optical communication solutions are continuously replacing electronic systems across a wide variety of communications applications ranging from free-space and fiber links all the way to on-chip communications. Within these applications, fiber-optic transport systems play a pioneering role in terms of aggregate transport capacities. In this chapter, we reviewed modulation formats and multiplexing techniques both from a conceptual and from a technological perspective, with a special emphasis on the past and future evolution of transmitter optoelectronics.

## Acknowledgment

We acknowledge many valuable discussions with S. Chandrasekhar, A. Chraplyvy, C. Doerr, J. Foschini, A. Gnauck, G. Kramer, M. Magarini, X. Liu, G. Raybon, J. Sinsky, R. Tkach, N. Weimann, and C. Xie.

## References

1. C. E. Shannon. A mathematical theory of communication. *Bell System Technical Journal*, 379–423 and 623–656, 1948.
2. G. E. Moore. Cramming more components into integrated circuits. *Electronics Magazine*, 38:114–117, 1965.
3. Y. Vlasov. Silicon photonics for next-generation computing systems. *Proc. European Conf. on Opt. Commun. (ECOC)*, paper Tu.1.A.1, 2008.
4. J. A. Kash, F. E. Doany, C. L. Schow, R. Budd, C. Baks, D. M. Kuchta, P. Pepeljugoski, P. et al. Terabus: Chip-to-chip board-level optical data buses. *Proc. IEEB/LEOS Annual Meeting*, paper WM1, 2008.
5. A. F. Benner, M. Ignatowski, J. A. Kash, D. M. Kuchta, and M. B. Ritter. Exploitation of optical interconnects in future server architectures. *IBM J. Res. & Dev.*, 49:755–775, 2005.
6. H. Kogelnik. On optical communication: Reflections and perspectives. *Proc. European Conf. on Opt. Commun. (ECOC)*, paper Mo1.1.1, 2004.
7. S. Abbott. Review of 20 years of undersea optical fiber transmission system development and deployment since TAT-8. *Proc. European Conf. on Opt. Commun. (ECOC)*, paper Mo.4.E.l, 2008.
8. S. A. Townes, B. L. Edwards, R. S. Bondurant, D. M. Boroson, B. A. Parvin, T. Roberts, A. Biswas, et al. The Mars laser communication demonstration. *Proc. Conf. Aerospace*, 1180–1195, 2004.
9. P. J. Winzer, Modulation and Multiplexing in Optical Communication Systems. *IEEE-LEOS Newsletter*, Feb. 2009, http://photonicssociety.org/newsletters/feb09/modulation.pdf.
10. R. W. Tkach. Scaling optical communications for the next decade and beyond. *Bell Labs Tech. J.*, 14:3–10, 2010.
11. http://top500.org/lists/2010/06/performance_development.
12. J. Gray and P. Shenoy. Rules of thumb in data engineering. *Microsoft Research Technical Report*, MS-TR-99-100, 2000.

13. J. L. Hennessy and D. A. Patterson. *Computer Architectures: A Quantitative Approach.* Morgan Kaufmann: San Francisco, 2003.
14. J. H. Sinsky and P. J. Winzer. 100-Gb/s optical communications. *IEEE Microw. Mag.*, 10:44–57, 2009.
15. R. E. Wagner. Fiber-based broadband access technology and deployment. In I. P. Kaminov, T. Li, and A. E. Winner, (eds.), *Optical Fiber Telecommunications V-B: Systems and Networks.* Academic Press: Burlington, MA, 2008.
16. A. Adamiecki, M. Duelk, and J. H. Sinsky. 25 Gbit/s electrical duobinary transmission over FR-4 backplanes. *Electron. Lett.*, 41:826–827, 2005.
17. J. D. Ambrosia. 100 Gigabit Ethernet and beyond. *IEEE Comm. Mag.*, 48:S6–S13, 2010.
18. P. J. Winzer. Beyond 100G Ethernet. *IEEE Comm. Mag.*, 48:26–30, 2010.
19. P. J. Winzer and R.-J. Essiambre. Advanced optical modulation formats. In I. P. Kaminov, T. Li, and A. E. Willner, (eds.), *Optical Fiber Telecommunications V-B: Systems and Networks.* Academic Press: Burlington, MA, 2008.
20. A. Lender. Correlative digital communication techniques. *IEEE Trans. Commun.*, 12:128–135, 1964.
21. D. Penninckx, M. Chbat, L. Pierre, J.-P. Thiery. The phase-shaped binary transmission (PSBT): A new technique to transmit far beyond the chromatic dispersion limit. *IEEE Photon. Technol. Lett.*, 9:259–261, 1997.
22. G. Kramer, A. Ashikhmin, A. J. van Wijngaarden, and X. Wei. Spectral efficiency of coded phaseshift keying for fiber-optic communication. *J. Lightwave Technol.*, 21:2438–2445, 2003.
23. R.-J. Essiambre, G. Kramer, P. J. Winzer, G. J. Foschini, and B. Goebel. Capacity limits of optical fiber networks. *J. Lightwave Technol.*, 28:662–701, 2010.
24. M. Magarini, R.-J. Essiambre, B. E. Basch, A. Ashikhmin, G. Kramer, and A. J. de Lind van Wijngaarden, et al. Concatenated coded modulation for optical communications systems. *IEEE Photon. Technol. Lett.*, 22:1244–1246, 2010.
25. H. Bulow, G. Thielecke, and F. Buchali. Optical trellis-coded modulation (oTCM). *Proc. Opt. Fiber Commun. Conf. (OFC)*, paper WM5, 2004.
26. J. G. Proakis and M. Salehi. *Digital Communications*, 5th edition. McGraw Hill: New York, 2007.
27. B. Sklar. *Digital Communications*, 2nd edition. Prentice Hall: Upper Saddle River, NJ, 2001.
28. T. M. Cover and J. A. Thomas. *Elements of Information Theory*, 2nd edition. John Wiley and Sons: New York, 2006.
29. P. J. Winzer and R. J. Essiambre. Advanced optical modulation formats. *Proc. IEEE*, 94:952–985, 2006.
30. C. R. Doerr. High-speed PIC transceivers for terabit transport networks. *Proc. European Conf. on Opt. Commun. (ECOC)*, paper 8.6.1, 2009.
31. C. R. Doerr. Photonic integrated circuits for high-speed communications. *Proc. Conf. on Lasers and Electro-optics*, paper CFE3, 2010.
32. C. R. Doerr. High performance photonic integrated circuits for coherent fiber communication. *Proc. Opt. Fiber Commun. Conf. (OFC)*, paper OWU5, 2010.
33. K. Sato, S. Kuwahara, Y. Miyamoto, and N. Shimizu. 40 Gb/s direct modulation of distributed feedback laser for very-short-reach optical links. *IEE Electron. Lett.*, 38:816–817, 2002.
34. B. Wedding, W. Pohlmann, H. Gross, and O. Thalau. 43 Gbit/s transmission over 210 km SMF with a directly modulated laser diode. *Proc. European Conf. on Opt. Commun. (ECOC)*, paper Mo4.3.7, 2003.
35. T. L. Koch. Laser sources for amplified and WDM lightwave systems. In I. P. Kaminow and T. L. Koch (eds.), *Optical Fiber Telecommunications III*, pp. 115–162. Academic Press: San Diego, 1997.
36. D. A. Ackermann, J. E. Johnson, L. J. P. Ketelsen, L. E. Eng, P. A. Kiely, and T. G. B. Mason. Telecommunication lasers. In I. Kaminow and T. Li (eds.), *Optical Fiber Telecommunications IV*, pp. 587–665. Academic Press: San Diego, 2002.
37. G. P. Agrawal. *Nonlinear Fiber Optics*, 4th edition. Elsevier Science & Technology: San Diego, 2006.

38. I. Tomkos, D. Chowdhury, J. Conradi, D. Culverhouse, K. Ennser, C. Giroux, B. Hallock, et al. Demonstration of negative dispersion fibers for DWDM metropolitan area networks. *IEEE J. Select. Topics Quantum Electron.*, 7:439–460, 2001.
39. P. J. Winzer, F. Fidler, M. J. Matthews, L. E. Nelson, H. J. Thiele, J. H. Sinsky, S. Chandrasekhar, et al. 10-Gb/s upgrade of bidirectional CWDM systems using electronic equalization and FEC. *J. Lightwave Technol.*, 23:203–210, 2005.
40. R. A. Linke and A. H. Gnauck. High-capacity coherent lightwave systems. *J. Lightwave Technol.*, 6: 1750–1769, 1988.
41. B. Wedding, B. Franz, and B. Junginger. 10-Gb/s optical transmission up to 253 km via standard single-mode fiber using the method of dispersion-supported transmission. *J. Lightwave Technol.*, 12:1720–1727, 1994.
42. D. Mahgerefteh, Y. Matsui, X. Zheng, and K. McCallion. Chirp managed laser and applications. *IEEE J. Select. Topics Quantum Electron.*, 16:1126–1139, 2010.
43. D. Walker, H. Sun, C. Laperle, A. Comeau, and M. O'Sullivan. 960-km transmission over G.652 fiber at 10 Gb/s with a laser/electroabsorption modulator and no optical dispersion compensation. *IEEE Photon. Technol. Lett.*, 17:2751–2753, 2005.
44. Y. Yu, R. Lewén, S. Irmscher, U. Westergren, L. Thylén, U. Eriksson, and T. W. Lee. 80Gb/s ETDM transmitter with a traveling-wave electroabsorption modulator. *Proc. European Conf. on Opt. Commun. (ECOC)*, paper OWE1, 2005.
45. C. Kazmierski, A. Konczykowska, F. Jorge, F. Blache, M. Riet, C. Jany, and A. Scavennec. 100 Gb/s operation of an AlGalnAs semi-insulating buried heterojunction EML. *Proc. Opt. Fiber Commun. Conf. (OFC)*, paper OThT7, 2009.
46. C. R. Doerr, P. J. Winzer, L. Zhang, L. Buhl, and N. J. Sauer. Monolithic InP 16-QAM modulator. *Proc. Opt. Fiber Commun. Conf. (OFC)*, paper PDP20, 2008.
47. F. Heismann, S. K. Korotky, and J. J. Veselka. Lithium niobate integrated optics: Selected contemporary devices and system applications. In I. P. Kaminow and T. L. Koch (eds.), *Optical Fiber Telecommunications III*, pp. 377–462. Academic Press: San Diego, 1997.
48. A. Mahapatra and E. J. Murphy. Electrooptic modulators. In I. Kaminow and T. Li (eds.), *Optical Fiber Telecommunications IV*, pp. 258–294. Academic Press: San Diego, 2002.
49. T. Kawanishi, T. Sakamoto, and M. Izutsu. High-speed control of lightwave amplitude and phase and frequency by use of electrooptic effect. *IEEE J. Select. Topics Quantum Electron.*, 13:79–91, 2007.
50. R. A. Griffin, R. G. Walker, and R. I. Johnstone. Integrated devices for advanced modulation formats. *IEEE/LEOS Workshop on Advanced Modulation Formats*, 39–40, 2004.
51. M. Erman, P. Jarry, R. Gamonal, P. Autier, J.-P. Chane, and P. Frijlink. Mach–Zehnder modulators and optical switches on III–V semiconductors. *J. Lightwave Technol.*, 6:837–846, 1988.
52. S. Akiyama, H. Itoh, T. Takeuchi, A. Kuramata, and T. Yamamoto. Mach–Zehnder modulator driven by 1.2 V single electrical signal. *IEE Electron. Lett.*, 41:40–41, 2005.
53. R. A. Griffin, B. Pugh, J. Fraser, I. B. Betty, K. Anderson, G. Busico, C. Edge, et al. Compact and high power and MQW InP Mach–Zehnder transmitters with full-band tunability for 10 Gb/s DWDM. *Proc. European Conf. on Opt. Commun. (ECOC)*, paper Th2.6.2, 2005.
54. R. A. Griffin, A. Tipper, and I. Betty. Performance of MQW InP Mach–Zehnder modulators for advanced modulation formats. *Proc. Opt. Fiber Commun. Conf. (OFC)*, paper OTuL5, 2005.
55. K. Tsuzuki, H. Kikuchi, E. Yamada, H. Yasaka, and T. Ishibashi. 1.3-Vpp push–pull drive InP Mach–Zehnder modulator module for 40 Gbit/s operation. *Proc. European Conf. on Opt. Commun. (ECOC)*, paper Th2.6.3, 2005.
56. J. Conradi. Bandwidth-efficient modulation formats for digital fiber transmission systems. In I. Kaminow and T. Li (eds.), *Optical Fiber Telecommunications IV B*, pp. 862–901. Academic Press: San Diego, 2002.

57. T. Chikama, S. Watanabe, T. Naito, H. Onaka, T. Kiyonaga, Y. Onoda, H. Miyata, et al. Modulation and demodulation techniques in optical heterodyne PSK transmission systems. *J. Lightwave Technol.*, 8:309–321, 1990.
58. A. H. Gnauck and P. J. Winzer. Optical phase-shift-keyed transmission. *J. Lightwave Technol.*, 23:115–130, 2005.
59. D. M. Gill, A. H. Gnauck, X. Liu, X. Wei, D. S. Levy, S. Chandrasekhar, and C. R. Doerr. 42.7-Gb/s cost-effective duobinary optical transmitter using a commercial 10 Gb/s Mach–Zehnder modulator with optical filtering. *IEEE Photon. Technol. Lett.*, 17:917–919, 2005.
60. P. J. Winzer, S. Chandrasekhar, C. R. Doerr, D. T. Neilson, S. Adamiecki, and R. A. Griffin. 42.7-Gb/s modulation with a compact InP Mach–Zehnder transmitter. *Proc. European Conf. on Opt. Commun. (ECOC)*, paper We1.6.1, 2006.
61. D. McGhan, M. O'Sullivan, C. Bontu, and K. Roberts. Electronic dispersion compensation. *Proc. Opt. Fiber Commun. Conf. (OFC)*, paper OWK1, 2006.
62. P. J. Winzer, G. Raybon, C. R. Doerr, M. Duelk, and C. Dorrer. 107-Gb/s optical signal generation using electronic time-division multiplexing. *J. Lightwave Technol.*, 24:3107–3113, 2006.
63. K. Tsuzuki, T. Ishibashi, T. Ito, N. Kikuchi, and F. Kano. 80 Gb/s InP Mach–Zehnder modulator module using liquid crystal polymer (LCP) transmission line. *Proc. European Conf. on Opt. Commun. (ECOC)*, paper 5.2.2, 2009.
64. B. Culshaw and M. G. F. Wilson. Integrated optic frequency shifter modulator. *IEE Electron. Lett.*, 17:135–136, 1981.
65. M. Itsuzu, T. Sueta, and S. Shikama. Integrated optical SSB modulator/frequency shifter. *IEEE J. Quantum Electron.*, QE-17:2225–2227, 1981.
66. R. A. Griffin and A. C. Carter. Optical differential quadrature phase shift key (oDQPSK) for high-capacity optical transmission. *Proc. Opt. Fiber Commun. Conf. (OFC)*, paper WX6, 2002.
67. R. A. Griffin, R. I. Johnstone, R. G. Walker, J. Hall, S. D. Wadsworth, K. Berry, A.C. Carter, et al. 10 Gb/s optical differential quadrature phase shift key (DQPSK) transmission using GaAs/AlGaAs integration. *Proc. Opt. Fiber Commun. Conf. (OFC)*, paper FD6, 2002.
68. P. J. Winzer, A. H. Gnauck, C. R. Doerr, M. Magarini, and L. L. Buhl. Spectrally efficient long-haul optical networking using 112-Gb/s polarization-multiplexed 16-QAM. *J. Lightwave Technol.*, 28:547–556, 2010.
69. T. Sakamoto, T. Kawanishi, and M. Izutsu. Continuous-phase frequency-shift keying with external modulation. *IEEE J. Select. Topics Quantum Electron.*, 12:589–595, 2006.
70. J. Mo, Y. J. Wen, Y. Dong, Y. Wang, and C. Lu. Optical minimum-shift keying format and its dispersion tolerance. *Proc. Opt. Fiber Commun. Conf. (OFC)*, paper JThB12, 2006.
71. X. Zhou, J. Yu, M. Huang, Y. Shao, T. Wang, L. Nelson, P. Magill, et al. 64-Tb/s (640×107-Gb/s) PDM-36QAM transmission over 320 km using both pre- and post-transmission digital equalization. *Proc. Opt. Fiber Commun. Conf. (OFC)*, paper PDPB9, 2010.
72. S. W. Corzine, P. Evans, M. Fisher, J. Gheorma, M. Kato, V. Dominic, P. Samra, et al. Large-scale InP transmitter PICs for PM-DQPSK fiber transmission systems. *IEEE Photon. Technol. Lett.*, 22:1015–1017, 2010.
73. P. J. Winzer, G. Raybon, H. Song, A. Adamiecki, S. Corteselli, A. H. Gnauck, D. A Fishman, et al. 100-Gb/s DQPSK transmission: From laboratory experiments to field trials. *J. Lightwave Technol.*, 2009.
74. M. Daikoku, I. Morita, H. Taga, H. Tanaka, T. Kawanishi, T. Sakamoto, T. Miyazaki, et al. 100Gbit/s DQPSK transmission experiment without OTDM for 100G Ethernet transport. *Proc. Opt. Fiber Commun. Conf. (OFC)*, p. PDP36, 2006.
75. A. H. Gnauck, P. J. Winzer, G. Raybon, M. Schnecker, and P. J. Pupalaikis. 10 × 224-Gb/s WDM transmission of 56-Gbaud PDM-QPSK signals over 1890 km of fiber. *IEEE Photon. Technol. Lett.*, 22:954–956, 2010.

76. P. J. Winzer, A. H. Gnauck, S. Chandrasekhar, S. Draving, J. Evangelista, and B. Zhu. Generation and 1,200-km transmission of 448-Gb/s ETDM 56-Gbaud PDM 16-QAM using a single I/Q modulator. *Proc. European Conf. on Optical Communications (ECOC'10)*, Torino (Italy), PD2.2, 2010.
77. Y. Kamio, M. Nakamura, and T. Miyazaki. ISI pre-equalization in a vector modulator for 5 Gsymbol/s 64-QAM. *Proc. European Conf. on Opt. Commun. (ECOC)*, paper Tu.1.D.3, 2008.
78. A. Chiba, T. Sakamoto, T. Kawanishi, K. Higuma, M. Sudo, and J. Ichikawa. 16-level quadrature amplitude modulation by monolithic quad-parallel Mach–Zehnder optical modulator. *Electron. Lett.*, 46:227–228, 2010.
79. H. Yamazaki, T. Yamada, T. Goh, Y. Sakamaki, and A. Kaneko. 64QAM modulator with a hybrid configuration of silica PLCs and $LiNbO_3$ phase modulators for 100-Gb/s applications. *Proc. European Conf. on Opt. Commun. (ECOC)*, paper 2.2.1, 2009.
80. T. Sakamoto, A. Chiba, and T. Kawanishi. 50-km SMF transmission of 50-Gb/s 16 QAM generated by quad-parallel MZM. *Proc. European Conf. on Opt. Commun. (ECOC)*, paper Tu.l.E.3, 2008.
81. A. Yariv. Critical coupling and its control in optical waveguide-ring resonator systems. *IEEE Photon. Technol. Lett.*, 14:483–485, 2002.
82. W. M. Green, M. J. Rooks, L. Sekaric, and Y. A. Vlasov. Ultra-compact, low RF power, 10 Gb/s silicon Mach–Zehnder modulator. *Opt. Express*, 15:17106–17113, 2007.
83. M. R. Watts, W. A. Zortman, D. C. Trotter, R. W. Young, and A. L. Lentine. Low-voltage, compact, depletion-mode, silicon Mach–Zehnder modulator. *IEEE J. Select-Topics Quantum Electron.*, 16:159–164, 2010.
84. A. Liu, L. Liao, D. Rubin, H. Nguyen, B. Ciftcioglu, Y. Chetrit, N. Izhaky, et al. Highspeed optical modulation based on carrier depletion in a silicon waveguide. *Opt. Express*, 15:660–668, 2007.
85. Y. Cai, D. G. Foursa, C. R. Davidson, J.-X. Cai, O. Sinkin, M. Nissov, and A. Pilipetskii. Experimental demonstration of coherent MAP detection for nonlinearity mitigation in long-haul transmissions. *Proc. Opt. Fiber Commun. Conf. (OFC)*, paper OTuEl, 2010.
86. P. J. Winzer and A. Kalmar. Sensitivity enhancement of optical receivers by impulsive coding. *J. Lightwave Technol.*, 17:171–177, 1999.
87. R.-J. Essiambre, G. Raybon, and B. Mikkelsen. Pseudo-linear transmission of highspeed TDM signals: 40 and 160 Gb/s. In I. Kaminow and T. Li (eds.), *Optical Fiber Telecommunications IV*, pp. 232–304. Academic Press: San Diego, 2002.
88. A. Sano, H. Masuda, Y. Kisaka, S. Aisawa, E. Yoshida, Y. Miyamoto, M. Koga, et al. 14-Tb/s (140×111-Gb/s PDM/WDM) CSRZ-DQPSK transmission over 160km using 7-THz bandwidth extended L-band EDFA. *Proc. European Conf. on Opt. Commun. (ECOC)*, paper Th4.1.1, 2006.
89. A. H. Gnauck, G. Charlet, P. Tran, P. Winzer, C. Doerr, J. Centanni, E. Burrows, et al. 25.6-Tb/s C+L-band transmission of polarization-multiplexed RZ-DQPSK signals. *Proc. Opt. Fiber Commun. Conf. (OFC)*, paper PDP19, 2007.
90. R.-J. Essiambre, P. J. Winzer, X. Q. Wang, W. Lee, C. A. White, and E. C. Burrows. Electronic predistortion and fiber nonlinearity. *IEEE Photon. Technol. Lett.*, 18:1804–1806, 2006.
91. M. Birk, D. Fishman, and P. Magill. Field trial of end-to-end OC-768 transmission using 9 WDM channels over 1000 km of installed fiber. *Proc. Opt. Fiber Commun. Conf. (OFC)*, paper TuS4, 2003.
92. J. H. Sinsky, A. Adamiecki, L. Buhl, G. Raybon, P. Winzer, O. Wohlgemuth, M. Duelk, et al. A 107-Gbit/s optoelectronic receiver utilizing hybrid integration of a photodetector and electronic demultiplexer. *J. Lightwave Technol.*, 26:114–120, 2008.
93. K. Schuh, E. Lach, B. Junginger, G. Veith, J. Renaudier, G. Charlet, and P. Tran. 8 Tbit/s (80 × 107 Gbit/s) DWDM ASK-NRZ VSB transmission over 510 km NZDSF with 1 b/s/Hz spectral efficiency. *Proc. European Conf. on Opt. Commun. (ECOC)*, paper PD1.8, 2007.
94. C. R. S. Fludger, T. Duthel, T. Wuth, and C. Schulien. Uncompensated transmission of 86Gbit/s polarisation multiplexed RZ-QPSK over 100 km of NDSF employing coherent equalisation. *Proc. European Conf. on Opt. Commun. (ECOC)*, paper Th4.3.3, 2006.

95. C. F. Lam and W. I. Way. Optical ethernet: Protocols and management and 1–100G technologies. In I. P. Kaminov, T. Li, and A. E. Willner (eds.), *Optical Fiber Telecommunications V-B: Systems and Networks*. Academic Press: Burlington, MA, 2008.
96. C. Monier, M. D'Amore, D. Scott, A. Cavus, E. Kaneshiro, S. Lin, P. C. Chang, et al. 172 GHz divide-by-two circuit using a 0.25-μm InP HBT technology. *Proc. IPRM*, 24–27, 2009.
97. T. Pfau, S. Hoffmann, and R. Noé. Hardware-efficient coherent digital receiver concept with feedforward carrier recovery for M-QAM constellations. *J. Lightwave Technol.*, 27:989–999, 2009.
98. R. H. Walden. Analog-to-digital converter survey and analysis. *J. Select. Areas in Commun.*, 17:539–550, 1999.
99. Y. Yu, X. Zhou, Y. Huang, S. Gupta, M. Huang, T. Wang, and P. Magill. 112.8-Gb/s PM-RZ-64QAM optical signal generation and transmission on a 12.5 GHz WDM grid. *Proc. Opt. Fiber Commun. Conf. (OFC)*, paper OThMl, 2010.
100. Y. Mori, C. Zhang, U. M. Igarashi, K. Katoh, and K. Kikuchi. 200-km transmission of 100-Gbit/s 32-QAM dual-polarization signals using a digital coherent receiver. *Proc. European Conf. on Opt. Commun. (ECOC)*, paper 8.4.6, 2009.
101. H.-G. Weber and R. Ludwig. Ultra-high speed OTDM transmission technology. In *Optical Fiber Telecommunications V.B*. Elsevier: Burlington, MA, 2008.
102. M. Nakazawa, T. Yamamoto, and K. R. Tamura. 1.28 Tbit/s70 km OTDM transmission using third and fourth-order simultaneous dispersion compensation with a phase modulator. *IEE Electron. Lett.*, 36:2027–2029, 2000.
103. C. Schmidt-Langhorst, R. Ludwig, D.-D. Gross, L. Molle, M. Seimetz, R. Freund, and C. Schubert. Generation and coherent time-division demultiplexing of up to 5.1 Tb/s single-channel 8-PSK and 16-QAM signals. *Proc. Opt. Fiber Commun. Conf. (OFC)*, paper PDPC6, 2009.
104. E. Desurvire. *Erbium-doped fiber amplifiers*. John Wiley and Sons, Inc.: New York, 1994.
105. J. Bromage. Raman amplification for fiber communications systems. *J. Lightwave Technol.*, 22:79–93, 2004.
106. F. Forghieri, R. W. Tkach, and A. R. Chraplyvy. Fiber nonlinearities and their impact on transmission systems. In I. P. Kaminov and T. L. Koch (eds.), *Optical Fiber Telecommunications III A*, pp. 196–264. Academic Press: San Diego, 1997.
107. A. Sano and Y. Miyamoto. Technologies for ultrahigh bit-rate WDM transmission. *Proc. IEEE/LEOS Annual Meeting*, paper TuP1, 2007.
108. H. R. Stuart. Dispersive multiplexing in multimode optical fiber. *Science*, 289:281–283, 2000.
109. Y. Kokubun and M. Koshiba. Novel multi-core fibers for mode-division multiplexing: proposal and design principle. *IEICE Electron. Exp.*, 6:522–528, 2009.
110. S. L. Jansen. Optical OFDM for long-haul transport networks. *Proc. IEEE/LEOS Annual Meeting*, paper MH1, 2008.
111. S. Chandrasekhar, X. Liu, B. Zhu, and D. W. Peckham. Transmission of a 1.2-Tb/s 24 carrier noguard-interval coherent OFDM superchannel over 7200-km of ultra-large-area fiber. *Proc. European Conf. on Opt. Commun. (ECOC)*, paper PD2.6, 2009.
112. A. Sano, E. Yamada, H. Masuda, E. Yamazaki, T. Kobayashi, E. Yoshida, Y. Miyamoto, et al. No-guard-interval coherent optical OFDM for 100-Gb/s long-haul WDM transmission. *J. Lightwave Technol.*, 27:3705–3713, 2009.
113. Y. Ma, Q. Yang, Y. Tang, S. Chen, and W. Shieh. 1-Tb/s single-channel coherent optical OFDM transmission with orthogonal-band multiplexing and subwavelength bandwidth access. *J. Lightwave Technol.*, 28:308–315, 2010.
114 X. Liu, S. Chandrasekhar, B. Zhu, P. J. Winzer, A. H. Gnauck, D. W. Peckham, 448-Gb/s reduced-guard-interval CO-OFDM transmission over 2000 km of ultra-large-area fiber and five 80-GHz-grid ROADMs. *J. Lightwave Technol.*, 29:483–490, 2011.
115. S. L. Jansen, I. Morita, K. Forozesh, S. Randel, D. Van Den Borne, and H. Tanaka. Optical OFDM, a hype or is it for real? *Proc. European Conf. on Opt. Commun. (ECOC)*, paper Mo.3.E.3, 2008.

116. B. Spinnler. Equalizer design and complexity for digital coherent receivers. *IEEE J. Select. Topics Quantum Electron.*, 16:1180–1192, 2010.
117. M. Nakazawa, S. Okamoto, T. Omiya, K. Kasai, and M. Yoshida. 256-QAM (64 Gb/s) coherent optical transmission over 160 km with an optical bandwidth of 5.4 Ghz. *IEEE Photon. Technol. Lett.*, 22:185–187, 2010.
118. F. Chang, K. Onohara, and T. Mizuochi. Forward error correction for 100 G transport networks, IEEE Communications Magazine, 48:S48-S55, 2010.
119. T. Mizuochi. Recent progress in forward error correction and its interplay with transmission impairments. *IEEE J. Select. Topics Quantum Electron.*, 12:544–554, 2006.
120. A. H. Gnauck. Advanced amplitude- and phase-coded formats for 40-Gb/s fiber transmission. *Proc. IEEE/LEOS Annual Meeting*, paper WR1, 2004.
121. S. Bigo. Multiterabit DWDM terrestrial transmission with bandwidth-limiting optical filtering. *IEEE J. Select. Topics Quantum Electron.*, 10:329–340, 2004.
122. D. O. Caplan. Laser communication transmitter and receiver design. In *Free-Space Laser Communications: Principles and Advances*, pp. 109–246. Springer: New York, 2007.
123. G. Charlet, J.-C. Antona, S. Lanne, P. Tran, W. Idler, M. Gorlier, S. Borne, et al. 6.4 Tb/s (159 × 42.7 Gb/s) capacity over 21 × 100 km using bandwidth-limited phase-shaped binary transmission. *Proc. European Conf. on Opt. Commun. (ECOC)*, paper PD4.1, 2002.
124. G. Raybon, S. Chandrasekhar, A. H. Gnauck, B. Zhu, and L. L. Buhl. Experimental investigation of long-haul transport at 42.7 Gb/s through concatenated optical add/drop nodes. *Proc. Opt. Fiber Commun. Conf. (OFC)*, paper ThE4, 2004.
125. A. Agarwal, S. Chandrasekhar, and R.-J. Essiambre. 42.7 Gb/s CSRZ-VSB for spectrally efficient meshed networks. *Proc. European Conf. on Opt. Commun. (ECOC)*, paper We3.4.4, 2004.
126. B. Mikkelsen, C. Rasmussen, P. Mamyshev, and F. Liu. Partial DPSK with excellent filter tolerance and OSNR sensitivity. *IEE Electron. Lett.*, 42:1363–1364, 2006.
127. A. H. Gnauck, P. J. Winzer, S. Chandrasekhar, and C. Dorrer. Spectrally efficient (0.8 b/s/Hz) 1-Tb/s (25×42.7 Gb/s) RZ-DQPSK transmission over 28 100-km SSMF spans with 7 optical add/drops. *Proc. Opt. Fiber Commun. Conf. (OFC)*, paper Th4.4.1, 2004.
128. D. J. Krause, J. C. Cartledge, and K. Roberts. Demonstration of 20-Gb/s DQPSK with a single dual-drive Mach–Zehnder modulator. *IEEE Photon. Technol. Lett.*, 20:1363–1365, 2008.
129. C. R. S. Fludger, T. Duthel, D. van den Borne, C. Schulien, E.-D. Schmidt, T. Wuth, J. Geyer, et al. Coherent equalization and POLMUX-RZ-DQPSK for robust 100-GE transmission. *J. Lightwave Technol.*, 26:64–72, 2008.
130. G. Charlet, J. Renaudier, H. Mardoyan, P. Tran, O. Bertran Pardo, F. Verluise, M. Achouche, et al. Transmission of 16.4 Tbit/s capacity over 2,550 km using PDM QPSK modulation format and coherent receiver. *Proc. Opt. Fiber Commun. Conf. (OFC)*, paper PDP3, 2008.
131. P. M. Krummrich, E.-D. Schmidt, W. Weiershausen, and A. Mattheus. Field trial results on statistics of fast polarization changes in long haul WDM transmission systems. *Proc. Opt. Fiber Commun. Conf. (OFC)*, paper OThT6, 2005.
132. B. Koch, V. Mirvoda, H. Grießer, H. Wernz, D. Sandel, and R. Noé. Endless optical polarization control at 56 krad/s, over 50 Gigaradian, and demultiplex of 112-Gb/s PDM-RZ-DQPSK signals at 3.5 krad/s. *IEEE J. Select. Topics Quantum Electron.*, 16:1158–1163, 2010.
133. S. Savory. Digital coherent optical receivers: Algorithms and subsystems. *IEEE J. Select. Topics Quantum Electron.*, 16:1164–1179, 2010.
134. S. Han, S., K.-T. Wu, and K. Roberts. Real-time measurements of a 40 Gb/s coherent system. *Opt. Express*, 16:873–879, 2008.
135. P. J. Winzer and A. H. Gnauck. High-speed coherent detection at high spectral efficiencies. *IEEE Photonics Society Summer Topicals*, paper WA2.1, 2010.

136. X. Zhou, J. Yu, D. Qian, T. Wang, G. Zhang, and P. Magil. 8 × 114 Gb/s, 25-GHz-spaced, PolMux-RZ-8PSK transmission over 640 km of SSMF employing digital coherent detection and EDFA-only amplification. *Proc. Opt. Fiber Commun. Conf. (OFC)*, paper PDP1, 2008.
137. J. Yu, X. Zhou, M. F. Huang, Y. Shao, D. Qian, T. Wang, M. Cvijetic, et al. 17 Tb/s (161 × 114 Gb/s) PolMux-RZ 8PSK transmission over 662 km of ultra-low loss fiber using C-band EDFA amplification and digital coherent detection. *Proc. Eur. Conf. Opt. Commun. (ECOC)*, paper Th.3.E.2, 2008.
138. A. H. Gnauck, P. J. Winzer, S. Chandrasekhar, X. Liu, B. Zhu, and D. W. Peckham. Spectrally efficient long-haul WDM transmission using 224-Gb/s polarization-multiplexed 16-QAM. *J. Lightwave Technol.*, 29:373-377, 2011.
139. A. Sano, H. Masuda, T. Kobayashi, M. Fujiwara, K. Horikoshi, E. Yoshida, Y. Miyamoto, et al. 69.1-Tb/s (432 × 171-Gb/s) C- and extended L-band transmission over 240 km using PDM-16-QAM modulation and digital coherent detection. *Proc. Opt. Fiber Commun. Conf. (OFC)*, paper PDPB7, 2010.
140. R. G. Gallager. *Information Theory and Reliable Coding*. John Wiley and Sons: New York, 1968. (Chapters 2 and 4.)
141. S. Lin and D. J. Costello. *Error Control Coding*. Prentice Hall, 2004.
142. H. Nyquist. Certain topics of telegraph transmission theory. *Trans. Am. Inst. Electr. Eng.*, 47:617–644, 1928.
143. R.-J. Essiambre. Impact of Fiber Parameters on Nonlinear Fiber Capacity. *Proc. Opt. Fiber Commun. Conf. (OFC)*, paper OTuJ1, 2011.

# 3

# Fiber-Optic Analog Radio Frequency Links

Charles H. Cox III
*Photonic Systems, Inc.*
Edward I. Ackerman
*Photonic Systems, Inc.*



## 3.1 Introduction

Since the development of low-loss optical fiber in the late 1960s [1], designers of all sorts of systems for transmitting, detecting, and processing electronic signals have sought to exploit its uniquely advantageous characteristics. (These properties of optical fiber are discussed in Chapters 1 and 2.) The interconnection of optical sources, modulators, amplifiers, photodetectors, and other components using optical fiber so as to convey an electronic signal from one point to another is what is known as a fiber-optic link. The subject of this chapter is the use of fiber-optic links to transport radio frequency (RF) signals.

Say "fiber optics" to people not intimately familiar with the technology and they will likely think of the internet or of the pervasive advertisements for fiber-to-the-home communication and entertainment services. It is true as of this writing, and may forever be, that most optical fiber in use carries digital rather than analog information, and therefore, the design of fiber-optic links for the transport of digital signals has been discussed in great detail in many publications (including in some other chapters of this book). Fewer publications discuss the design of links for analog RF signal transport.

To convey a high-frequency (or broadband) RF signal via fiber, one option is to first down-convert the high RF (or a narrow-band portion of the broadband RF signal) to a lower RF at which commercial analog-to-digital signal converters can operate (e.g., 750 MHz [2]), and then convey the signal in digital format using a digital fiber-optic link. There do exist situations, however, in which it is impractical or impossible to perform this down-conversion and analog-to-digital conversion prior to the fiber-optic link.

For example, fiber-optic links are useful tools for the practice of antenna remoting, which is the communication between an antenna and a remotely located signal transmitter and/or receiver. Figure 3.1 shows the advantages of using an analog rather than a digital fiber-optic link to transport a signal received by an RF antenna to a remote receiver and signal processor: that is, using the analog RF fiber-optic link situates fewer and less complex and DC power-hungry components at the antenna's location,

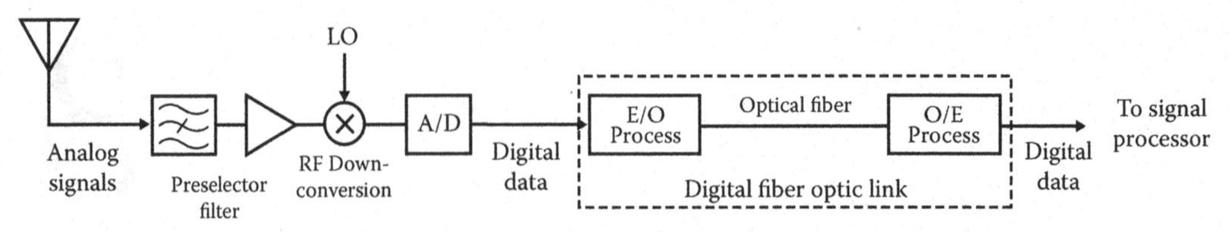


Disadvantages:

- Downconverter and A/D size and power consumption are major concerns; one set of this hardware is required at each element
- Requires phase synchronization of Local Oscillators (LOs) at all the sensor elements

(a)

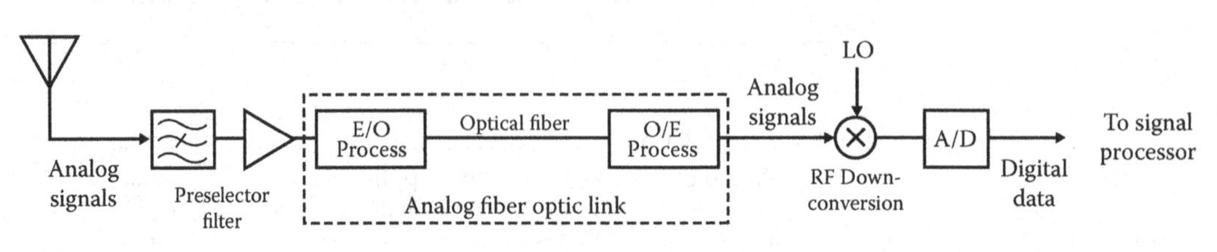


Advantage:

- Minimizes size, complexity, and power consumption of componentry at antenna element

(b)

**FIGURE 3.1** Comparison of the hardware needed to perform RF antenna remoting using (a) a digital fiber-optic link vs. (b) an analog fiber-optic link.

which is often remote from the remainder of the system. The complexity of the antenna-site hardware increases if the system consists of more than one antenna element, because all of the local oscillators needed for the pre–analog-to-digital RF down-conversion must be synchronized to one another.

Figure 3.1 gives no details about the devices that perform the electrical-to-optical (E/O) and optical-to-electrical signal conversion processes in an analog or in a digital fiber-optic link. The necessary configuration of devices depends on the choice of *modulation techniques* and *detection techniques.*

The light generated by an optical source can be modulated either directly or externally. In a *direct modulation* RF link, a single device—typically a semiconductor laser or light emitting diode (usually the former, unless inferior performance can be tolerated)—is responsible for generating and modulating the light. In an *external modulation* RF link, two separate devices perform the functions of generating light and allowing it to be modulated by the RF signal. In both direct and external modulation links, a small-signal RF modulation around a bias point set by a DC current or voltage is converted into a corresponding small-signal modulation of the emitted optical carrier's intensity, frequency, or phase (or, most likely, all of these) around the average values of these quantities that occur at this bias point. Unlike that of a direct modulation link, the RF performance of an external modulation link generally improves when the optical power being modulated by the small signal is increased [3].

Although external modulation is slightly more complex—and hence, costly—than direct modulation, it offers vastly greater design freedom, which in turn, enables better performance than is possible with direct modulation. For example, because the optical source does not have to be a semiconductor laser designed to enable high-frequency direct modulation, the designer of an external modulation link can choose instead a semiconductor laser designed for high-power continuous wave (CW) output, or can even choose a high-power solid-state or doped-fiber laser as the optical source. The advantages of external modulation when compared to direct modulation are explained later in this chapter, which focuses on the design, modeling, and reported performance of *external modulation* links.

Depending on what type of modulator is used at the input end of an external modulation link, the RF signal can be made to modulate the amplitude, phase, or intensity of the laser's CW output. At the output end of the link, two different methods can be used to recover the RF signal from the optical carrier it modulates: *direct detection* or *coherent detection*. Figure 3.2 [4] shows block diagrams of links using these two detection techniques.

In either detection process of Figure 3.2, photons incident upon a photodetector excite valence band electrons into a higher-energy conduction band, and therefore the photodetector responds to a modulation of incident photon intensity with a proportional modulation of its output current, provided that the detection process and the device in which it occurs are both sufficiently fast. This *optical-to-electrical* conversion in the photodetector is represented mathematically as

$$I_D = r_d \left| \sum_n E_n \right|^2, \tag{3.1}$$

where the proportionality constant $r_d$ is the detector's responsivity. Representing each optical field $E_n$ as $E_{0,n} e^{j(2\pi \nu_n t + \theta_n)}$, where $E_{0,n}$, $\nu_n$, and $\theta_n$ are, respectively, the amplitude, frequency, and phase of the optical field $E_n$, it is evident from Figure 3.2a that in a *direct detection* link, which involves a single optical carrier, the only property of the optical field to which the photodetector responds is its intensity (which is the square of its amplitude $E_0$), that is,

$$I_D = r_d |E_0|^2. \tag{3.2}$$

For a fiber-optic link's output current to be affected by modulation of optical phase or frequency requires that more than one optical carrier be incident upon the photodetector as shown in either of

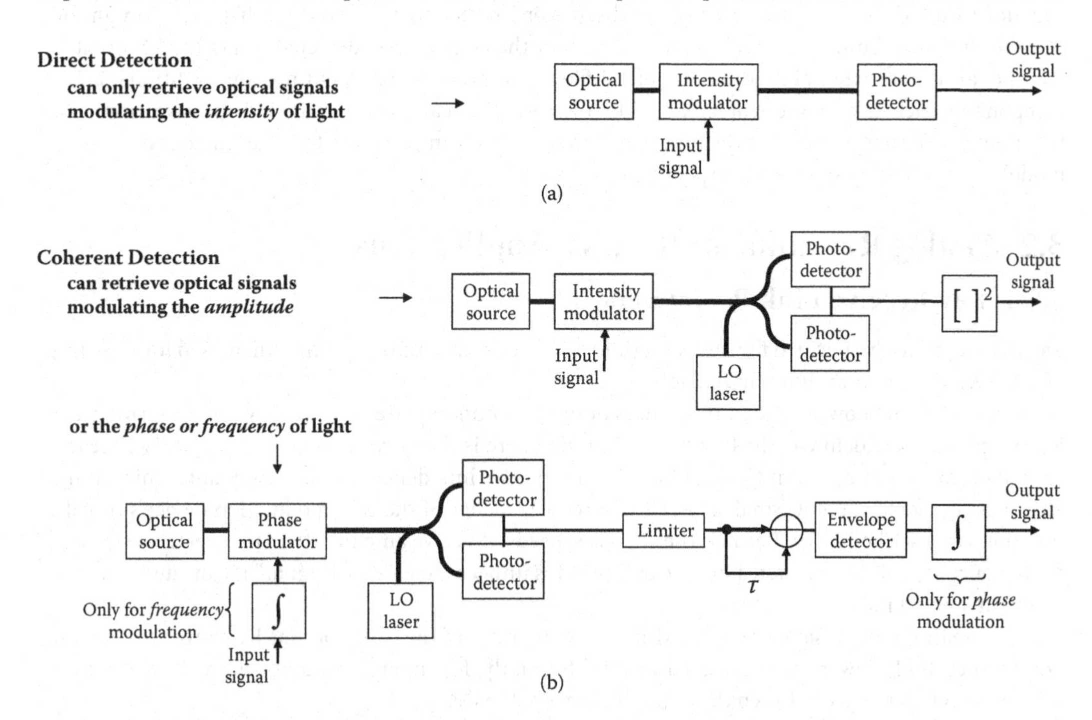


**FIGURE 3.2** Detection techniques. Direct detection (a) can recover an RF signal that modulates the intensity of an optical carrier, whereas amplitude, phase or frequency modulation of an optical carrier can only be recovered by a coherent detection technique (b). (Reproduced from R. Kalman et al., *J. Lightwave Technol.*, 12, 7, 1263–1277, © 1994 IEEE. With permission of Optical Society of America.)

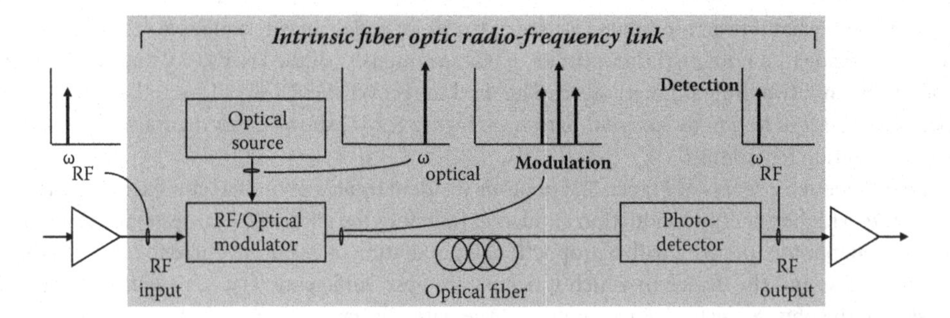


**FIGURE 3.3** Definition of an "intrinsic" RF fiber-optic link. (Reproduced with permission from C. Cox et al., *IEEE Trans. Microwave Theory Tech.*, 45, 1375–1383, © 1997 IEEE.)

the two coherent detection architectures of Figure 3.2b. Because of the increased cost of implementing either of these coherent detection link architectures or variations thereof, to the authors' knowledge, no fielded system exists at this time in which an RF signal modulating an optical carrier is retrieved using a coherent detection technique. The remainder of this chapter therefore describes only links that employ *direct detection*, and therefore also only links that employ *intensity modulation*—the one and only modulation technique that can be recovered via direct detection.

Figure 3.3 [5] depicts the direct detection link architecture of Figure 3.2a in slightly greater detail, and shows the spectral patterns of signals at various points throughout the link, for the assumed case where the link's input RF signal is fed to an external modulator which intensity modulates the optical carrier supplied by a CW laser. For the purposes of discussion, in the remainder of this chapter, a straightforward or "intrinsic" link is defined to include each of the components depicted in Figure 3.3—that is, the optical source, external modulator, optical fiber, and photodetector, but not any additional active components such as amplifiers or mixers. The intrinsic link can, however, include passive circuits at the input and/or output ports of the link that transform their impedances to better match those of the modulator and/or photodetector, respectively.

## 3.2 Analog RF Links for Receive Applications

### 3.2.1 Key Receive Link Parameters

The most commonly specified figures of merit for an analog fiber-optic RF link, when used for remoting of a receive antenna, are summarized here.

Figure 3.4 shows how several of the figures of merit commonly used to quantify the performance of RF components are defined. The first to be examined here is the *intrinsic small-signal gain*, $g_i$, because several of the other important figures of merit for a receive link depend in some way upon this parameter. A precise definition of "small signal" requires knowledge of the E-O modulation device's transfer function, as described in [3]. Because the signals sensed by receive antennas tend to be extremely weak, for the purposes of this chapter, a precise definition is unnecessary; "small signal" means just what you would imagine it means.

For a small input RF signal, $g_i$ is the dimensionless ratio of the link's output RF power to its input power when both powers are expressed in mW. Similarly, for input and output powers expressed in dBm, $G_i$ is the difference between these and is expressed in dB:

$$g_i\,(\text{dimensionless}) = \frac{s_{out}\,(\text{in mW})}{s_{in,a}\,(\text{in mW})} \tag{3.3a}$$

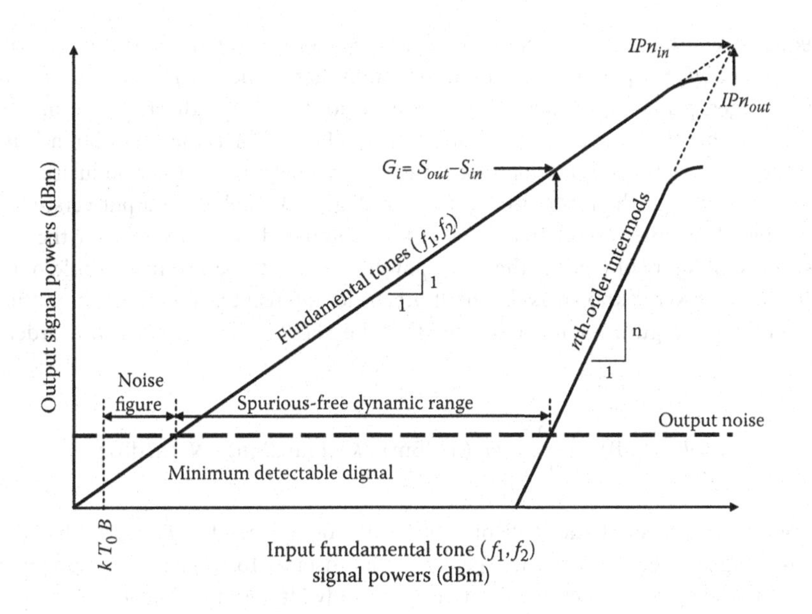


**FIGURE 3.4** Principal figures of merit for two-port analog devices, including radio-frequency fiber-optic links.

$$G_i(\text{in dB}) = 10\log(g_i) = S_{out}(\text{in dBm}) - S_{in,a}(\text{in dBm}). \tag{3.3b}$$

In Equation 3.3, the input signal power must be defined as that which is *available* (hence the "*a*" in the subscript) at the link input—that is, what that signal source would supply to the modulator if they were perfectly impedance-matched to one another [3].

Another important figure of merit for a receive link is its *noise figure* (*NF*), defined as the extent to which the signal-to-noise ratio degrades between the input and output ports of the link:

$$\begin{aligned} NF(\text{in dB}) &= 10\log\left[\frac{(s_{in}/n_{in})}{(s_{out}/n_{out})}\right]_{n_{in}=kT_0B} = 10\log\left[\frac{n_{out}}{kT_0Bg_i}\right] \\ &= N_{out}(\text{in dBm}) - 10\log(kT_0B)(\text{in dBm}) - G_i(\text{in dB}). \end{aligned} \tag{3.4}$$

In the definition for *NF*, the input noise is specified as only the noise arising from thermal sources. This input noise is the product of $k$ (Boltzmann's constant), $T_0$ (room temperature, standardized by the Institute of Electrical and Electronic Engineers as 290 K [6]) and $B$ (the instantaneous bandwidth of the electronic receiver, sometimes called the "resolution bandwidth" or "noise bandwidth"). Because the output noise term $n_{out}$ is also proportional to $B$, the noise figure is independent of $B$.

Just as the existence of noise sets a lower limit on the link's useful range of input signal powers, there is also an upper limit imposed by the nonlinear characteristics of the link components, which give rise to harmonic and/or intermodulation distortion products. A receive link's *spurious-free dynamic range (SFDR)* is defined as the ratio of the largest input power (expressed in mW) that it can convey, free of distortion products, to the smallest such input power in mW (or as the difference in dB between these two powers if they are both expressed in dBm). The first of these two signal powers, the upper limit to the *SFDR*, is the largest signal for which the strongest distortion products are weaker than the output noise; the second of the two powers, the lower limit to the *SFDR*, is the minimum detectable signal (see Figure 3.4).

For two equal-strength input signal tones at RF frequencies $f_1$ and $f_2$, the resulting intermodulation tones at $pf_1 \pm qf_2$ and $pf_2 \pm qf_1$ are stronger than the output harmonics at $(p + q)f_1$ and $(p + q)f_2$ for any given positive integers $p$ and $q$ for which $p \geq q$. Therefore, for $n = p + q$, $n$th-order intermodulation distortion products, rather than $n$th harmonics, are shown in Figure 3.4 as the upper bound on the *SFDR*. As the two-tone input signal power is increased, the output power at the intermodulation frequencies increases as input power to the $n$th power so that, on Figure 3.4's plot of output versus input power in dBm, the slope of the intermodulation product is $n$, compared to a slope of 1 for the fundamental signal tones. A useful figure of merit is the input power at which these two lines would hypothetically intersect. This input power is known as the input $n$th-order intercept power $IPn_{in}$. From the geometric arrangement of lines in Figure 3.4, it can be seen that the *SFDR* is limited by the $n$th-order distortion products to:

$$SFDR_n(\text{in dB}) = \frac{n-1}{n}\left[IPn_{in}(\text{in dBm}) - kT_0B(\text{in dBm}) - NF(\text{in dB})\right]. \tag{3.5}$$

Unlike $G_i$ or *NF*, at any instant, the *SFDR* of a receive link does depend on $B$, the width of the passband of frequencies being allowed to reach the receiver at that instant. To compare the performance of one link to another irrespective of receiver instantaneous bandwidth, and to enable the most straightforward scaling of *SFDR* for different values of $B$, *SFDR* is often quoted for a hypothetical 1-Hz instantaneous bandwidth. To indicate how it scales for more useful (larger) values of $B$, the 1-Hz *SFDR* is stated in units of dB × $\text{Hz}^{(n-1)/n}$.

### 3.2.2 Analytical Model of Analog Link Performance

We begin the description of how to model the RF performance of an external modulation link with an examination of the *intrinsic small-signal gain*, $g_i$, because several of the other important figures of merit for a receive link depend upon $g_i$. This important parameter is easily calculated from the square of the product of small-signal slope efficiencies of the external modulator and the detector. The modulator's slope efficiency, $s_m$, is the change in its optical output power for a given change in input voltage. Specifically,

$$s_m \equiv \left|\frac{dp_{M,O}(v_M)}{dv_M}\right|_{v_m=0}, \tag{3.6}$$

where $p_{M,O}$ is the modulator's output optical power, which is a function of $v_M$, the total voltage on the modulator, which is in turn the sum of a DC bias voltage $V_M$ and the modulation signal voltage $v_m$. If the modulator's transfer function (i.e., the mathematical relationship between its output optical power and its input voltage) is known, Equation 3.6 can be used to derive a useful expression for $s_m$. A Mach-Zehnder interferometric (MZI) modulator, for example, has the transfer function

$$p_{M,O} = \frac{T_{FF}P_I}{2}\left[1 + \cos\left(\frac{\pi v_M}{V_\pi}\right)\right], \tag{3.7}$$

where $T_{FF}$ is the modulator's fiber-to-fiber optical insertion loss, $P_I$ is the CW optical power supplied to its input fiber and $V_\pi$ is the voltage change necessary to shift the optical phase in one of the arms of the MZI by $\pi$ radians relative to the other. Given that $v_M = V_M + v_m$, substituting Equation 3.7 into Equation 3.6 yields

$$s_m = \frac{\pi T_{FF} P_I R_M}{2V_\pi} \sin\left(\frac{\pi V_M}{V_\pi}\right), \tag{3.8}$$

where $R_M$ is the modulator's resistance. The detector's slope efficiency is simply

$$s_d = r_d |H_d(f)|, \tag{3.9}$$

where $r_d$ is its responsivity and $H_d(f)$ represents the frequency response of the detector circuit (including any resistive or reactive matching circuitry). Therefore,

$$g_i = \left[\frac{\pi r_d T_{ff} P_I R_S}{2V_\pi(f)} |H_d(f)| \sin\varphi\right]^2, \tag{3.10}$$

where

$$V_\pi(f) = \frac{\alpha L}{1 - e^{-\alpha L}} V_\pi(DC), \tag{3.11}$$

and where $L$ is the length of the modulator electrodes and $\alpha$ is their frequency-dependent loss coefficient [7].

Equation 3.10, in which we have defined the ratio $\pi\, V_M / V_\pi$ as the MZI modulator's "bias angle" $\phi$, shows that an MZI modulator-based external modulation link gain is maximum at the so-called "quadrature" bias angle (90°). Also, notice that the slope efficiency depends on two independent parameters, $P_I$ and $V_\pi$, whose maximum ratio is set by practical, but not fundamental, constraints. This is a prime example of the expanded design space external modulation affords in comparison to direct modulation, since with direct modulation, the maximum slope efficiency is limited by energy conservation. The impact of higher slope efficiency is that one can increase the gain by increasing $P_I$, such that it is possible to achieve $g_i > 1$ so that $G_i$ exceeds 0 dB [3]. Figure 3.5 shows the measured gains and corresponding measured noise figures for nine different links reported in eight different publications [8–15]. The figure includes only those links whose reported gain exceeded 0 dB, and for which a corresponding noise figure was also reported.

Figure 3.6 shows an equivalent circuit for an external modulation link, including noise sources representing the photo-detected laser relative intensity noise (*RIN*), the shot noise, and several sources of thermal noise that contribute to the total output thermal noise $\overline{n_{out,th}}$. The noise figure of this link is

$$NF = \left.\frac{\overline{n_{out,th}} + \overline{n_{out,rin}} + \overline{n_{out,shot}}}{kT_0\, g_i}\right|_{\overline{n_{in}} = kT_0}, \tag{3.12}$$

where $\overline{n_{out,RIN}}$ and $\overline{n_{out,shot}}$, the output noise spectral densities due to the laser's *RIN* and the shot noise of the detection process, respectively, are defined in [16] and elsewhere. Recall from Equation 3.4 that noise figure is defined for input noise equal to the thermal noise generated at $T_0 = 290$ K. This input thermal noise is amplified (or attenuated) by the link's gain (or loss) $g_i$, and if $g_i$ is sufficiently large that the amplified input thermal noise term makes the dominant contribution to the total noise at the link output, then there is virtually no degradation in signal-to-noise ratio from the link's input to its output,

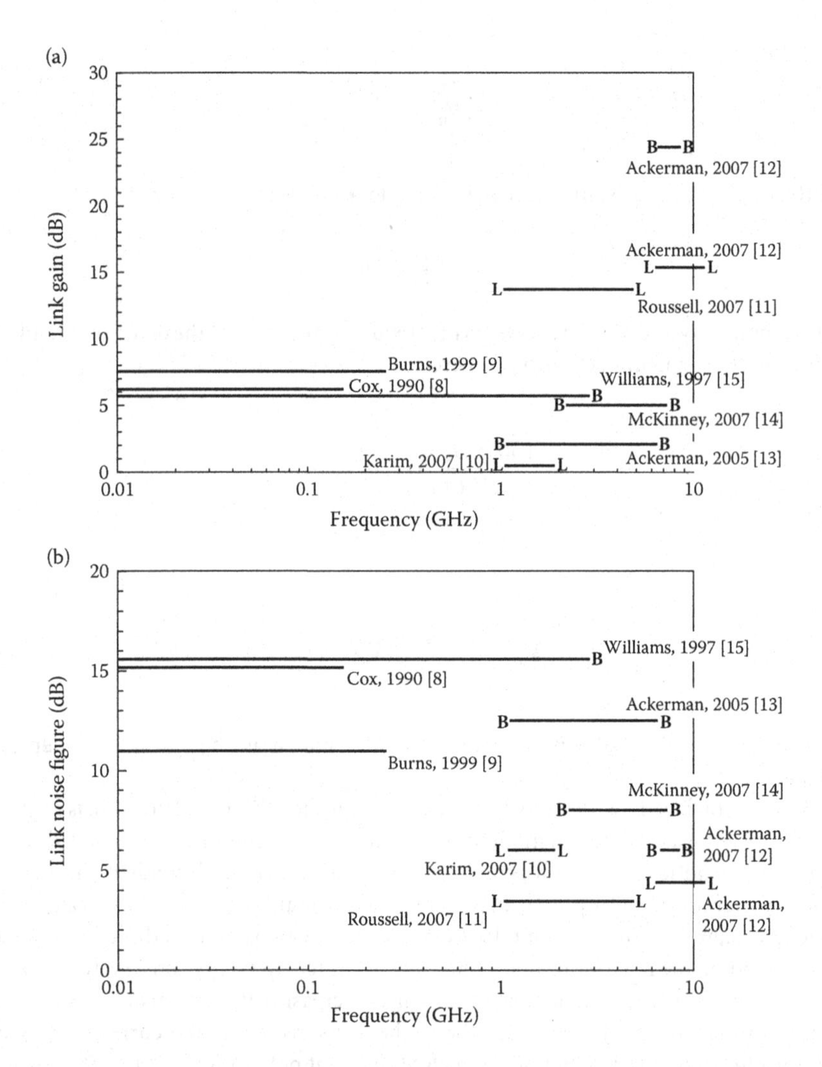


**FIGURE 3.5** (a) Gains and (b) noise figures of intrinsic RF fiber-optic links. Characters at line endpoints indicate links using the low-biasing (L) and balanced differential modulation and detection (B) techniques, which are discussed in Sections 3.2.3 and 3.2.4, respectively.

and *NF* will be minimized. This thermal-noise-limited situation yields the lower limit to *NF*, which we can express as $NF_{\min}$:

$$NF_{min} = \left.\frac{\overline{n_{out,th}}}{kT_0\, g_i}\right|_{\overline{n_{in}}=kT_0} = 10\log\left[1 + x + \frac{1}{g_i}\right], \tag{3.13}$$

where the three terms in the latter expression are the individual contributions of the input thermal noise source, of the modulation device and its optional matching circuit, and of the detector and its optional matching circuit. If we assume the modulator electrodes are *lossless* and that the RF and optical velocities have been successfully matched to one another in the modulator, then

$$x = x_{lossless} = \frac{\sin^2(\beta_{RF}L)}{(\beta_{RF}L)^2}, \tag{3.14}$$

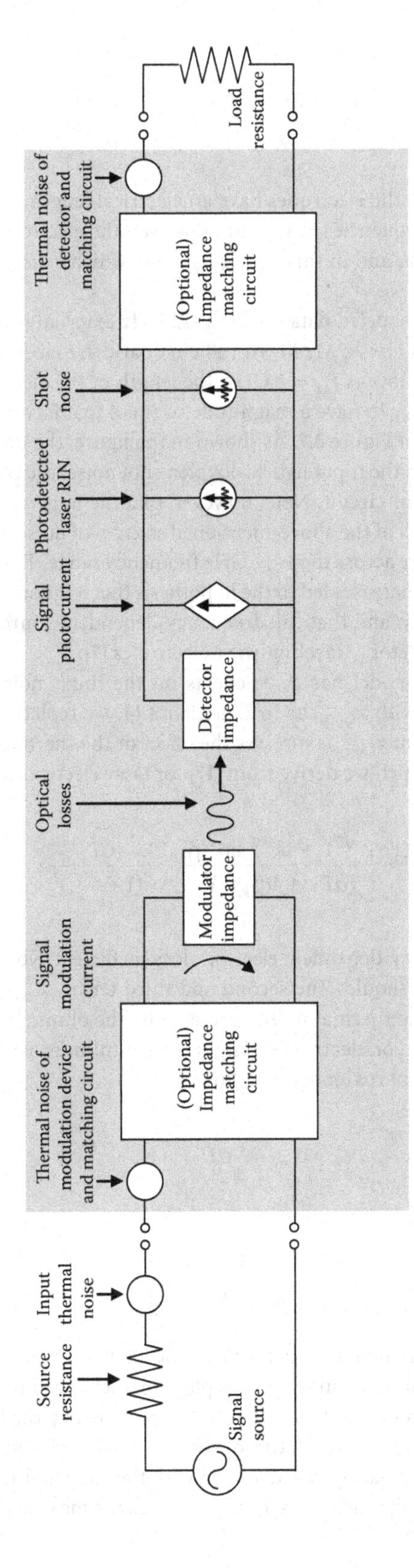


**FIGURE 3.6** Equivalent circuit of an intrinsic RF fiber-optic link using external intensity modulation and direct detection. (Reproduced from C. Cox, *Analog Optical Links: Theory and Practice*, Cambridge University Press, Cambridge, UK, 2004. With permission.)

where

$$\beta_{RF} = \frac{2\pi}{\lambda_{RF}} = \frac{2\pi f_{RF} n_{opt}}{c}. \tag{3.15}$$

At low RF frequencies, for which the electrodes have an electrical length <<360°, $x_{lossless} \to 1$, $NF \to$ 3 dB for $g_i >> 0$ [16]. At most RF frequencies of interest, however, the electrode length is likely at least a large fraction of an RF wavelength, and in this case $x_{lossless} \to 0$, which causes $NF$ to approach a value less than 3 dB.

Figure 3.7a shows measured vs. modeled data for the 1–12 GHz external modulation link described in [11], which featured an extremely low-$V_\pi$ MZM with 14 cm dual-drive electrodes fed by a directional coupler. Even for RF frequencies as low as $f_{RF}$ = 1 GHz, the length of the electrodes in this modulator was greater than 360°, causing $x_{1,lossless}$ to have a magnitude so small that its contribution would appear below the lowest line on the y-axis of Figure 3.7. As shown in the figure, the dominant sources of noise at the link output were due to loss in the input hybrid coupler, shot noise, the photo-detected laser *RIN*, and the thermal noise of the detector circuit. Note, however, that the uppermost curve in Figure 3.7a, which shows the modeled effect of all of the above-mentioned sources of noise, falls between 1.1 and 2.3 dB short of the measured noise figure across the 1–12 GHz frequency range. The degradation of the lossless model's accuracy as frequency increases led to the hypothesis that it ignores a source of noise whose magnitude increases with frequency, and that this frequency-dependent source of noise is that which arises from ohmic loss in the modulator's traveling-wave electrodes [7].

Including electrode loss in the model has *three* effects on the link's noise figure, so that rather than replacing $x$ in Equation 3.13 with $x_{lossless}$ as in Equation 3.14, we replace it instead with the sum $x_{1,lossy} + x_{2,lossy} + x_{3,lossy}$. The first term, $x_{1,lossy}$, expresses the effect of the thermal noise generated by the electrode termination resistance, which we derive from [17] for lossy electrodes:

$$x_{1,lossy} = \frac{(1-e^{-\alpha L})^2 + 4e^{-\alpha L}\sin^2(\beta_{RF} L)}{(\alpha L)^2 + 4(\beta_{RF} L)^2} \frac{(\alpha L)^2}{(1-e^{-\alpha L})^2}, \tag{3.16}$$

where $\alpha$ is the electrodes' RF frequency-dependent electrode loss coefficient. Note that, for $\alpha = 0$, Equation 3.16 reduces to Equation 3.14, as it should. The second and third terms, $x_{2,lossy}$ and $x_{3,lossy}$, express the effect of co- and counterpropagating thermal noise generated by the ohmic loss in the traveling-wave electrodes themselves, respectively. For electrodes designed for minimum RF loss, it has been shown that the extent of this effect can be approximated as follows [7]:

$$x_{2,lossy} \approx \frac{2}{3}\alpha L \tag{3.17}$$

and

$$x_{3,lossy} \approx 0. \tag{3.18}$$

The effect of modifying the noise figure model to account for loss in the modulator electrodes is shown in Figure 3.7b. The unlabeled solid curves in this plot are unchanged relative to Figure 3.7a, but the three additional effects of electrode loss factor into the total shown by the bold dashed curve. Two of the three effects are so small as to never reach the lowest y-axis value of –190 dBm/Hz, but the third effect—copropagating electrode thermal noise—does increase the expected noise figure to the point where the modeled data of Figure 3.7b much more nearly matches the measured data.

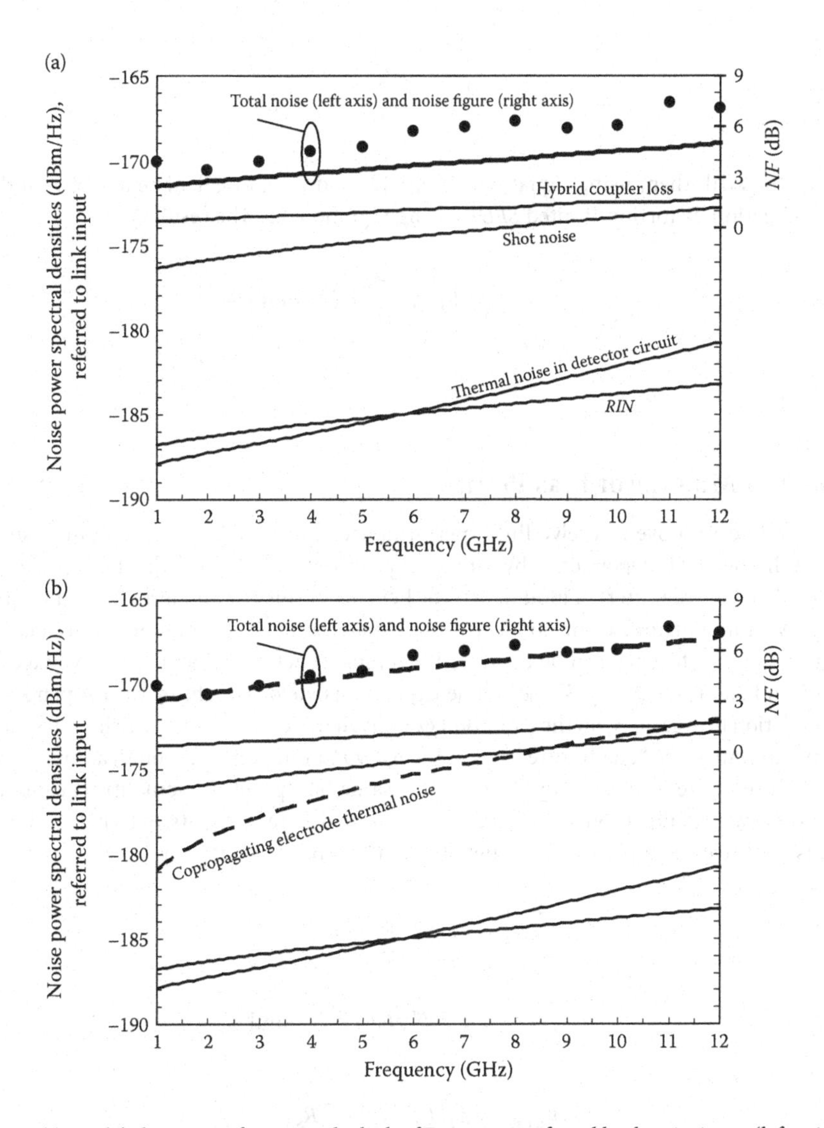


**FIGURE 3.7** (a) Modeled sources of noise in the link of Figure 3.6, referred back to its input (left axis) assuming *lossless* modulator electrodes, and the measured link noise figure, *NF* (right axis). Noise due to the electrode termination resistance, giving rise to the $x_{lossless}$ term, falls below −190 dBm/Hz. (b) Same model, assuming *lossy* modulator electrodes. The modified effect of termination resistance (accounted for in the $x_{1,lossy}$ term) is again negligible, as is counterpropagating thermal noise arising from the electrode loss ($x_{3,lossy}$). The effect of copropagating thermal noise from the electrode loss ($x_{2,lossy}$) is shown by the additional dashed line. With electrode loss accounted for, the modeled link *NF* more nearly matches the measured data. (Reproduced from E. Ackerman et al., *J. Lightwave Technol.*, 26, 15, 2441–2448, © 2008 IEEE. With permission of Optical Society of America.)

Once the noise figure of an external modulation receive link is predicted using the equations above, only one more parameter must be determined before the receive link's *SFDR* can also be predicted; as described above, this one remaining parameter is the input *n*th-order intercept power $IPn_{in}$. For links operated over less than one octave of RF bandwidth, and for links in which the modulator and photodetector are biased so as to suppress all second-order distortion products, the dominant intermodulation distortion products are usually third-order. Moreover, when an MZI modulator, the most prevalent type, is used, the sinusoidal nature of its transfer function allows for prediction of $IP3_{in}$ using a very succinct expression:

$$IP3_{in} = \frac{4V_\pi^2}{\pi^2 R_S}. \tag{3.19}$$

It is straightforward, then, using Equations 3.10, 3.12, and 3.19, to calculate a receive link's third-order intermodulation distortion-limited $SFDR_3$ using Equation 3.5. The result is

$$SFDR_3 = \frac{n-1}{n} \cdot 10 \log \left[ \frac{\left( r_d T_{ff} P_I |H_d(f)| \sin\varphi \right)^2 R_S}{\overline{n_{out,th}} + \overline{n_{out,rin}} + \overline{n_{out,shot}}} \right]. \tag{3.20}$$

### 3.2.3 Effect of Modulator Bias Point

One effective way to improve a receive link's performance is to reduce its noise figure using a technique that was discovered independently by three groups in 1993 [18–20] and that has generally become known as "low-biasing" the external modulator. The benefits of this technique are easiest to quantify in the case of an MZI modulator in a linear electro-optic material like lithium niobate, because its transfer function and slope efficiency can be expressed as simple functions of its DC bias voltage $V_M$ using Equations 3.7 and 3.8, respectively. These simple expressions yield straightforward dependence of the external modulation link gain, $g_i$, on the bias point $\varphi$ as given in Equation 3.10, which shows for example that gain is maximum at the "quadrature" bias point $\varphi = 90°$. Equation 3.12 would, at first glance, seem to imply that noise figure is always minimum at this same bias point. In actuality, because the three terms in the numerator of Equation 3.12 each depend on $\varphi$ to different extents, external modulation link noise figure is a relatively complicated function of modulator bias. Specifically,

$$\overline{n_{out,th}} = kT_0 g_i (1+x) + kT_0, \tag{3.21}$$

$$\overline{n_{out,RIN}} = \langle I_D \rangle^2 RIN |H_d(f)|^2 R_S, \text{ and} \tag{3.22}$$

$$\overline{n_{out,shot}} = 2q \langle I_D \rangle |H_D(f)|^2 R_S, \tag{3.23}$$

where

$$\langle I_D \rangle = \frac{r_d T_{FF} P_I}{2}(1+\cos\varphi). \tag{3.24}$$

Figure 3.8 shows, for a specific set of assumed values for the components in Equations 3.10–3.24, the effect of the low-biasing technique on the $NF$ of an MZI modulator-based external modulation link. From the curves showing the intrinsic link gain ($G_i$) and the total output noise power density $\overline{n_{out,total}} = \overline{n_{out,th}} + \overline{n_{out,RIN}} + \overline{n_{out,shot}}$, it is evident that, over a large range of the modulator bias point $\varphi$, the noise decreases more quickly than the signal as the modulator bias is increased from 90° towards the light-extinguishing bias of 180°. At some optimum low-biasing point between 90° and 180° that depends on component parameters such as the laser $RIN$, $NF$ is minimized. If the bias point is moved further towards 180°, the signal gain begins to decrease more quickly than the noise, and therefore $NF$ begins to increase relative to its value at the optimum low-biasing point.

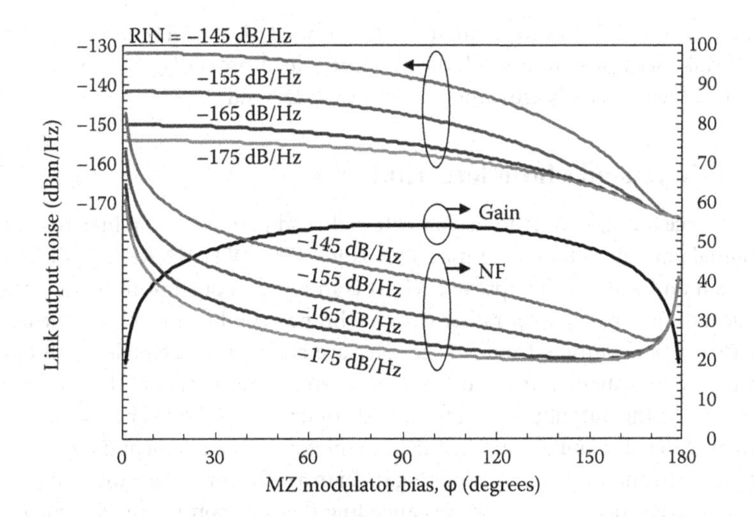


**FIGURE 3.8** Illustration of the "low-biasing" technique for reducing the noise figure of an external modulation link that uses a MZI modulator (assumptions: $V_\pi = 3$ V, $\langle I_D \rangle = 10$ mA at $\varphi = 90°$). The link's output noise (left axis) decreases more quickly than its gain $G_i$ (right axis), causing the optimum *NF* (right axis) to occur at a bias point between 90° and 180°. (Reproduced with permission after C. Cox et al., *IEEE Trans. Microwave Theory Tech.*, 54, 2, 906–920, © 2006 IEEE.)

An external modulator produces no second-order distortion products when biased where $G_i$ is maximum (e.g., at $\varphi = 90°$ for an MZI modulator). Therefore, the low-biasing technique for reducing *NF* has an adverse effect on the link's second-order distortion limited dynamic range $SFDR_2$, such that it tends not to be employed except in links with bandwidth of less than one octave, in which all second-order distortion products fall out of band [19].

The lowest noise figure ever demonstrated for an RF fiber-optic link at X-band frequencies (6–12 GHz) was described in [12] and is repeated in Figure 3.9. The link consisted of a high-power (2.5 W), low-*RIN* (approximately −175 dB/Hz) master oscillator power amplifier, a dual-drive MZI modulator with the lowest RF $V_\pi$ available—less than 1.4 V at 12 GHz when measured from the delta port of a broadband directional coupler connected to its two RF inputs [21], and a p-i-n photodetector with $r_d$ of nearly 1.0 A/W at $\lambda = 1.55$ μm. The modulator was biased at about 75° away from quadrature, resulting in an average photocurrent $\langle I_D \rangle$ of only 6.2 mA. For the 6–12 GHz band, the square points in Figure 3.9

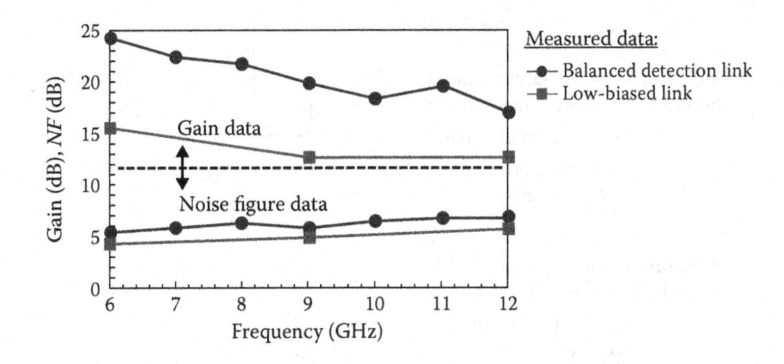


**FIGURE 3.9** Measured gain and noise figure of links using a single photodetector to detect one output of a low-biased MZI modulator (curves with square markers), and balanced photodetectors to differentially detect two anti-phase outputs of a quadrature-biased MZI modulator (curves with circle markers). (Reproduced with permission from E. Ackerman et al., *IEEE MTT-S Int. Microwave Symp. Dig.*, 51–54, © 2007 IEEE.)

show the measured $G_i$ of > 12.7 dB and record low *NF*—only 4.4 dB at 6 GHz and <5.7 dB at 12 GHz—of this "low-biased" link. Because of its low *NF*, the measured third-order distortion-limited *SFDR* for this link was impressive: 120 dB×Hz$^{2/3}$ across the entire 6–12 GHz band [12].

### 3.2.4 Effect of Balanced Photodetection

Another way to reduce a link's *NF* is to use an external modulation link architecture that employs a balanced differential photodetector configuration to cancel the CW laser's RIN. Figure 3.10 shows such a link configuration and why it has improved *NF*. This link uses a quadrature-biased MZI modulator in which an optical directional coupler rather than a y-branch combiner produces the necessary interferometry. The two optical outputs from this coupler are conveyed via equal-length optical fibers to two detectors whose photocurrents are made to subtract from one another at the link output port—for example, by connecting the output port to the cathode of one photodiode and the anode of the other. Because the two modulated outputs from the MZI modulator are complementary to one another—that is, 180° out of phase with one another—subtracting them in a differential detector configuration doubles the output signal, (thereby increasing $g_i$) while cancelling the common-mode component, which is the laser's *RIN* [22].

As was reported in [12], cancellation of the laser *RIN* using the balanced photodetection link architecture shown in Figure 3.10 also yielded impressively low *NF*. The round data points in Figure 3.9 show the measured gain and *NF* for a link using a balanced photodetector to demodulate both antiphase outputs of the same MZI modulator used in the "low-biased" link that had the measured performance shown by the square data points.

Note from Figure 3.9 that the balanced photodetection link exhibited much higher gain than the low-biased link because its modulator must be biased at quadrature ($\varphi = 90°$, where $g_i$ is maximum) for the *RIN* cancellation to succeed. Its measured *NF*, ranging from 5.5 dB at 6 GHz to 6.9 dB at 12 GHz, is not quite as low as that enabled by the low-biased configuration because, whereas balanced detection suppresses only the portion of the total link output noise that is photodetected laser *RIN*, low-biasing reduces both the *RIN* and shot noise components of the output noise, as shown in Equations 3.22 and 3.23. Other points of comparison between these two noise-reduction architectures are also

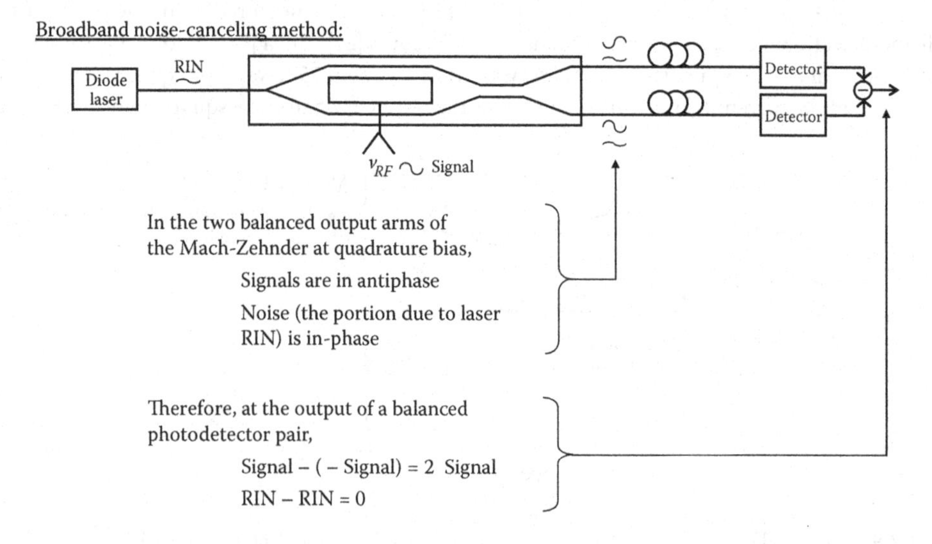


**FIGURE 3.10** External modulation link configuration using a dual-output MZI modulator and balanced photodetector to cancel the laser's *RIN*. (Reproduced with permission from E. Ackerman et al., *IEEE MTT-S Int. Microwave Symp. Dig.*, 51–54, © 2007 IEEE.)

worth noting. In addition to enabling slightly lower *NF*, an important advantage of the low-biasing approach is its relative simplicity—that is, the fact that it requires no more components than a conventional quadrature-biased external modulation link. The balanced photodetection link, by contrast, requires a dual-output modulator and specially configured pair of detectors, plus two lengths of optical fiber in which precise balancing of the amplitude and phase balance must be maintained. The increased complexity of the balanced photodetection architecture is sometimes worth the trouble, though, because it does have important advantages compared to the low-biasing architecture. Its higher gain (see Figure 3.9) is one advantage, but its principal benefit is the fact that quadrature biasing minimizes second-order distortion. Therefore, unlike a low-biased link, a balanced photodetection link will have high *SFDR* that is limited by third-order rather than second-order distortion even when operated over multioctave RF bandwidths.

### 3.2.5 Limitations of EA Modulator-Based Links

As was mentioned in Section 3.2.2, many of the performance advantages that modulator-based analog links offer over direct modulation links can be traced to the fact that it is possible to achieve higher RF-to-optical conversion efficiency with external modulation links [3]. The basis for this higher efficiency arises from the fact that the RF power required to drive the modulator is *independent* of the amount of optical power that is flowing through the modulator. It should be possible to realize this independence from voltage-driven modulators that are based on the linear electrooptic effect. In fact, this theoretically predicted independence has been confirmed experimentally for up to 2.5 W of optical power in an MZI modulator fabricated in lithium niobate [12].

Optical modulators based on the electroabsorption (EA) effect have demonstrated this independence, at least at low optical powers. However, recently, it has been shown by Betts et al. [23] that at higher optical powers, the independence breaks down, resulting in limits on the RF performance that can be achieved from EA-modulation-based links. The presentation below follows that of Betts et al. [23].

The intrinsic EA-modulator-based link is shown in Figure 3.11a and a simple circuit model of this link is shown in Figure 3.11b. It is important to note that there are two key limitations of this model. One is that this circuit model contains no frequency-dependent elements and so is only valid for low-frequency modulation where these frequency-dependent elements can be neglected. The other limitation is that this is a small signal model, where "small signal" is defined as the condition wherein the amplitude of the RF modulation voltage, $v_m$, is much less than $V_{\pi e}$, the equivalent $V_\pi$ of the EA modulator.

This simple model considers only two sources of loss: voltage-independent input and output coupling losses ($T_I$ and $T_O$), and voltage-dependent absorption loss; distributed scattering losses and voltage-independent absorption have been neglected. This model also assumes that there is no impact of electrorefractive index change. (Physically, this assumes that there is no reflection and that the

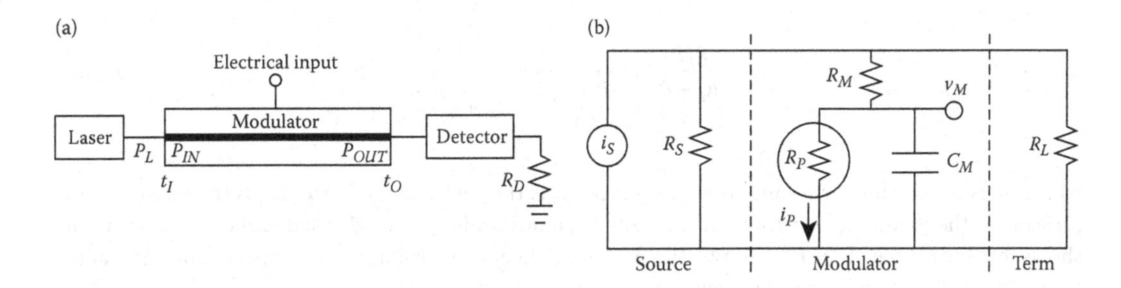


**FIGURE 3.11** (a) Block diagram of intrinsic, EA-modulator-based link. (b) Simplified schematic diagram of link shown in (a) that is limited to low-frequency, small-signal analysis. (Reproduced with permission from G. Betts, *IEEE Photon. Technol. Lett.*, 18, 19, 2065–2067, © 2006 IEEE.)

waveguide-substrate index difference is much larger than the electrorefractive index change.) The CW optical power in the modulator input waveguide is $P_{IN} = P_L T_I$, where $P_L$ is the input laser power. The total optical power, that is, the sum of the bias and modulated optical powers, in the modulator output waveguide is $p_{OUT}$. With these assumptions, the electrooptic transfer function can be approximated by the transmission at the bias point $T_B$:

$$p_{OUT} \cong P_{IN}\left(T_B - \frac{\pi}{2V_{\pi e}} v_m\right) \tag{3.25}$$

and the slope at the bias point is characterized by the equivalent $V_{\pi E}$:

$$V_{\pi E} = (\pi/2)(dT/dV)^{-1}, \tag{3.26}$$

where the derivative of the electrooptic transfer function $T$ is evaluated at the bias $T_B$. The photocurrent $i_P$ in the EA modulator is given by $(P_{IN} - p_{OUT})\eta_m$, where $\eta_m$ characterizes the dependency of modulator drive on optical power, where $\eta_m = 0$ represents independence. Thus, at the bias point, the dependency of photocurrent on optical power and modulation voltage can be written as

$$i_P = P_{IN}\eta_m(1 - T_B) + P_{IN}\eta_m \frac{\pi}{2V_{\pi E}} v_m. \tag{3.27}$$

We are interested in the small signal component of the total photocurrent $i_P$, which is represented by the second term in Equation 3.27. To assist in the intuitive understanding of the basis for the EA modulator limitations, we note that the coefficient of $v_m$ in Equation 3.27 has the units of conductance, or equivalently 1/resistance. Hence, we can write the small signal component of the photocurrent as

$$i_p = \frac{v_m}{R_p}, \tag{3.28}$$

where

$$R_P = \frac{2V_{\pi E}}{P_{IL} T_I \eta_m \pi}. \tag{3.29}$$

The equivalent circuit can be solved to give the modulator voltage $v_m$ in terms of the source current $i_s$ as

$$v_m = i_s \frac{R_L R_S}{R_L + R_S} \frac{1}{1 + \frac{P_L T_I \eta_m \pi}{2V_{\pi E}}\left(R_M + \frac{R_L R_S}{R_L + R_S}\right)}, \tag{3.30}$$

where $R_L$ is the modulator termination resistance, $R_S$ is the source impedance, $R_M$ is the resistance in series with the modulator junction, and the other quantities have been defined earlier. This equation shows that the source of the gain limit—that is, the modulator AC voltage—is inversely proportional to the optical power at high optical power.

The link gain is by Equation 3.31, where we have assumed no losses in the link except in the modulator. The input RF power is defined as the power delivered by the source to a matched load, which is the

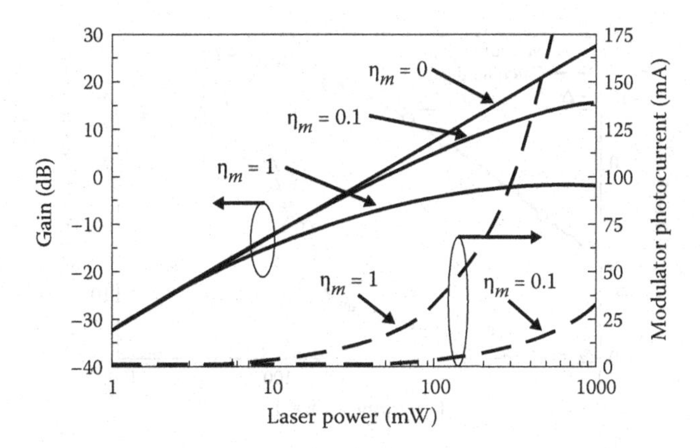


**FIGURE 3.12** Link gain as a function of laser power, for various values of the modulator responsivity $\eta_m$. The DC component of the modulator photocurrent is also plotted. The parameter values are $V_{\pi e} = 1$ V, $R_S = R_L = R_D = 50$ Ω, $R_M = 5$ Ω, $r_d = 0.8$ A/W, $T_I = -2$ dB, $T_O = -2$ dB, and $T_B = 0.5$. (Reproduced with permission from G. Betts, *IEEE Photon. Technol. Lett.*, 18, 19, 2065–2067, © 2006 IEEE.)

available power $\langle i_s^2 \rangle R_S/4$. The link output is the power delivered to the detector load resistance $R_D$. The link gain is given by Equation 3.31, where we have assumed no losses in the link except in the modulator.

$$g = \left[\left(\frac{P_L T_I T_O r_d \pi}{2V_{\pi E}}\right)^2 R_D R_L \left[\frac{4R_L R_S}{(R_L + R_S)^2}\right] \times \left[\frac{1}{1 + \frac{P_L T_I \eta_m \pi}{2V_{\pi E}}\left(R_M + \frac{R_L R_S}{R_L + R_S}\right)}\right]^2 \right. . \tag{3.31}$$

We see from Equation 3.31 that the gain expression can be written as the product of two terms. The first is the gain of an external modulation link with impedance matched input and assuming independence between the RF drive and modulator optical powers. The second term represents the gain compression that is introduced by the interdependence between the RF drive and modulator optical powers at high power in an EA-based modulator.

The effect of this gain limit is shown in Figure 3.12. The case of $\eta_m = 0$ is the standard external modulation result with no photocurrent effect. The case with $\eta_m = 1$ A/W approximates performance expected from a high-power electroabsorption modulator. For a high-performance modulator, the limiting value for the link gain in Equation 3.31 is near 0 dB. The limit can be increased if the modulator responsivity is reduced, but even at a low responsivity, such as 0.1 A/W, the photocurrent effect has an impact.

Betts et al. [23] have verified the gain limit by measuring the gain of a link using an electroabsorption modulator at high optical power levels. The modulator structure is similar to that described in [24]. The $V_{\pi e}$ was 0.85 V and the input and output losses were approximately $T_I = T_O = -2$ dB. The bias point was $T_B = 0.5$, which occurred at 1.5 V reverse bias. The AC input voltage was 0.063 V peak-to-peak. The modulator's apparent DC responsivity varied from 0.7 A/W to 1.5 A/W, indicating some mechanism creating additional photocurrent beyond simple absorption. An RF responsivity $\eta_m = 0.8$ A/W was used to fit the calculation to the measured data. The measurement frequency was 50 MHz, well below the RC bandwidth. The results are shown in Figure 3.13. The gain follows the theoretical prediction very closely. The gain deviates from the prediction of this model only at the highest powers used (>250 mW) due to heating.

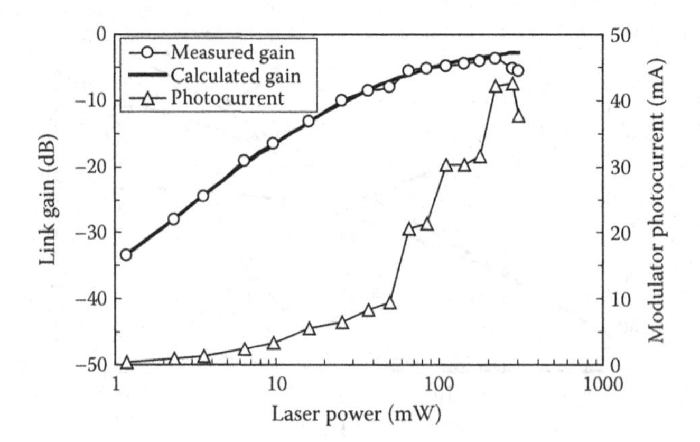


**FIGURE 3.13** Experimental measurement of a link using an electroabsorption modulator at 1550 nm, compared with the theoretical gain calculation. (Reproduced with permission from G. Betts, *IEEE Photon. Technol. Lett.*, 18, 19, 2065–2067, © 2006 IEEE.)

# 3.3 Transmit Links

## 3.3.1 Key Transmit Link Parameters

In addition to their receive (Rx) antenna remoting applications, discussed in Section 3.2 of this chapter, analog fiber-optic links can also be used in transmit (Tx) applications. This section describes and evaluates methods for delivering high-power RF signals to an antenna site for transmission by an antenna element or subarray.

The figures of merit that define a high-performance Tx link differ considerably from those that define a high-performance Rx link. In an Rx antenna remoting link, noise is a major concern because it limits the ability of the link to sense low-power signals; in a Tx antenna remoting link, however, noise is of little or no concern. Only a fraction of the transmitted signal and noise energy will impinge on the intended, distant, target—the antenna connected to a receiver front-end in the case of a communications application, or some sought-for entity (for example, inbound airliner, weather front, enemy vehicle) in the case of a radar application. Therefore, in the receiver that senses the transmitted signal, the noise that was transmitted with this signal will be negligible in comparison with the noise added by the receiver front-end. (This is especially true in the case of a radar, which senses the Tx signal only after it reflects off the distant object.)

In short, the primary goal in the design of a Tx link is to achieve some desired output RF power ($S_{out}$). It is additionally desirable to consume as little DC power as possible in the process of delivering the Tx signal to the antenna site. Subsections 3.3.2 and 3.3.3 compare some established methods.

## 3.3.2 Transmit Links with Post-Detector Power Amplification

We saw from Equation 3.10 and from Figures 3.5 and 3.9 that fiber-optic links can have positive (> 0 dB) gain without any amplifiers. This fact, coupled with the light's extremely low attenuation per unit length in optical fiber, means that it can be quite advantageous to perform Tx antenna remoting using an analog fiber-optic link. The clearest way to show this is with an illustrative example.

Figure 3.14 shows the loss as a function of frequency for a typical coaxial cable with an outer diameter of 3.15 mm [25]. Four plots of loss vs. frequency are shown for four different cable lengths in feet. Note, for example, that at about 5 GHz, the loss in just 20 feet of this coaxial cable is 10 dB, and is proportionately greater for longer cable lengths (~20 dB of loss at 50 feet, 40 dB at 100 feet, and 80 dB at 200 feet). The plot in Figure 3.14 also shows the analog signal gain or loss in up to 1,000 feet of optical fiber using

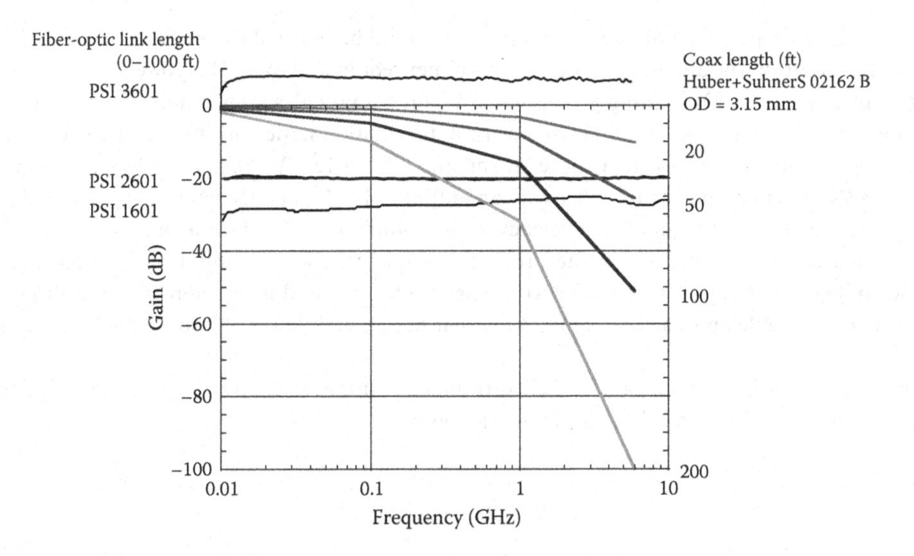


**FIGURE 3.14** Comparison of length- and frequency-dependence for a typical coax cable [25] and for three commercial fiber-optic links [26].

three different fiber-optic link products sold by Photonic Systems, Inc. [26]. Only one curve is shown per product because if one were to (for example) halve or double the fiber length, the gain or loss would be changed only negligibly because the attenuation of the 1,550-nm light in standard telecommunications-grade optical fiber is only ~0.3 dB/km.

In Figure 3.15, two Tx antenna remoting links are shown for comparison's sake. Figure 3.15a shows using a 20-foot length of the coax cable whose loss vs. frequency is shown in Figure 3.14 to deliver a 5-GHz Tx signal, hence incurring 10 dB of loss. Figure 3.15b shows using the PSI 3601 fiber-optic link whose gain vs. frequency is shown in Figure 3.14 to deliver the 5-GHz Tx signal via optical fiber of up to at least 1,000 feet in length. Rather than loss, this link has positive gain (~10 dB) without any amplification. Figure 3.15b shows how, using a power amplifier chain developed by Photonic Systems, Inc., the

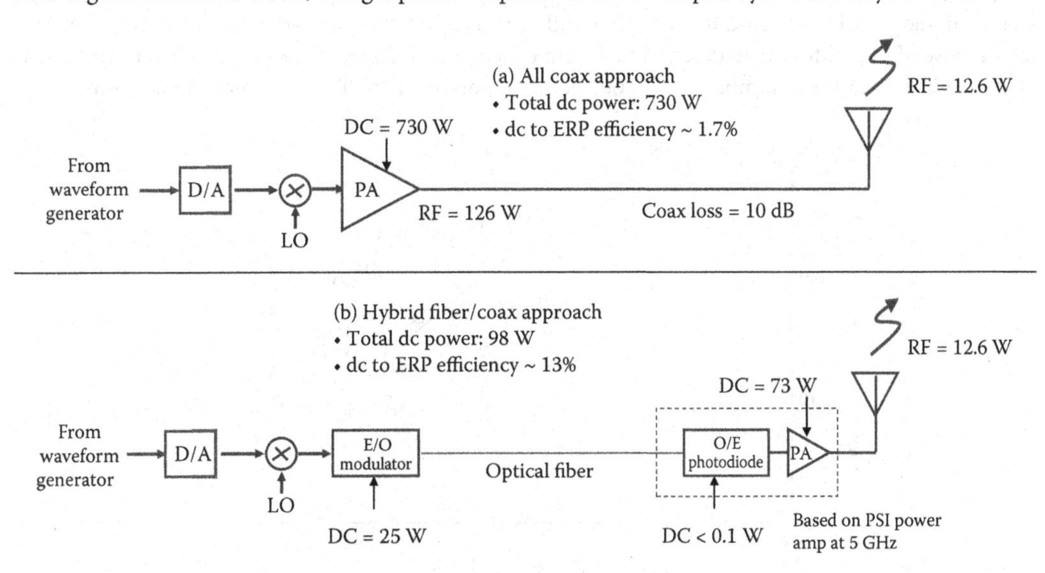


**FIGURE 3.15** Comparison of (a) all coax and (b) hybrid fiber/coax approaches to delivering a signal to a transmitting antenna.

5-GHz Tx signal output of a PSI 3601 fiber-optic link could be boosted to an effective radiated power (ERP) of 12.6 W from the antenna. This power amplifier requires 73 W of DC power. Additionally, the Tx signal link requires a DC biasing power of ~25 W, for a total DC supply power of ~98 W. In Figure 3.15a, we hypothesize a power amplifier at a central distribution node with the same DC efficiency as the PSI amplifier in Figure 3.15. To achieve an antenna ERP of 12.6 W for the hardware configuration in Figure 3.15a requires a power amplifier with an output of 126 W, and therefore a DC power supply of 730 W (again, assuming the same DC efficiency for this amplifier as for the one in Figure 3.15b).

To summarize the example from the previous paragraph, the postamplified fiber-optic Tx link approach of Figure 3.15b yields a DC-to-ERP efficiency of ~13%, and is therefore preferable to the pre-amplified coaxial cable approach of Figure 3.15a that has a much lower DC-to-ERP efficiency of only 1.7%.

To get a sense for how much gain is needed from a Tx link postamplifier, we can calculate the RF signal's output power from the photodetector as follows:

$$s_{out}(f)=\frac{1}{2}\left|i_d(f)\right|^2\left|H_d(f)\right|^2 R_S, \tag{3.32}$$

where $H_d$ and $R_S$ were defined in Section 3.2.2 and where $i_d(f)$ is the component of the detector photocurrent at the RF signal frequency of interest. If we assume for the moment that we could obtain a modulator for the Tx link that was perfectly linear, then the magnitude of $i_d(f)$ could be made as large as the average value of photocurrent—that is, 100% depth of modulation. In this case, the maximum RF output power of the intrinsic link would be

$$s_{out,\max}(f)=\frac{1}{2}I_D^2\left|H_d(f)\right|^2 R_S. \tag{3.33}$$

From Equation 3.33, it is clear that we desire a photodetector that can withstand high photocurrents while still being able to respond to light modulated at the RF frequency in question. For surface-illuminated photodetectors, for example, this comes down to choosing the diameter of the photosensitive area wisely. A large diameter will withstand higher $I_D$ but will also generally have a lower $|H_d(f)|$ at high frequencies because of its greater capacitance. Figure 3.16, which is a collection of saturation current and 3-dB RF bandwidth data for a number of photodetectors reported in the RF photonics literature over the

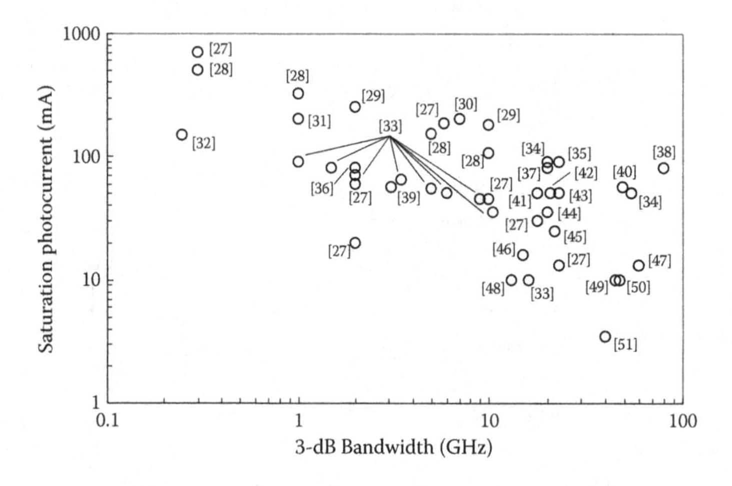


**FIGURE 3.16** Saturation current and 3-dB bandwidth reported in the literature [27–51] for many photodetectors.

last 15–20 years [27–51], shows the general trend of saturation current that decreases with increasing bandwidth.

If we stay with the example of a 5-GHz transmit signal, Figure 3.16 indicates that the best saturation current one can expect for a suitably fast photodetector is ~200 mA. From Equation 3.33, then, $s_{out,\max}$ works out to be about 1 W and therefore only about 11 dB of amplification would be needed to reach the ERP level shown in Figure 3.15. Using a Tx link like that shown in Figure 3.15b, however, would in practice result in much less than 1 W of signal power at the detector output because there is no E-O modulator sufficiently linear to allow $s_{out}$ to approach the maximum value given by Equation 3.33. The desire to deliver Tx signals to a remote antenna site with a greater depth of modulation, and therefore higher output signal power, led to research on the Tx signal delivery method discussed in Subsection 3.3.3.

### 3.3.3 Direct Photonic RF Power Generation

Instead of generating an RF Tx signal and then sending it to an antenna via a fiber-optic link, an alternative method is to use optical fibers to illuminate a photodetector at the antenna site with two optical carriers and cause a signal to be generated there at the RF "beat" frequency that is the difference between the two optical carrier frequencies. Figure 3.17 shows this arrangement. If, as shown in Figure 3.17 [52], an external modulator is used to modulate one or both of the two optical carriers, the RF output of the photodetector is correspondingly modulated.

An advantage of this "heterodyne" method is that the signal generated at the RF difference frequency can be made to have a maximum depth of modulation enabling much higher output powers without amplification than what is possible using a straightforward external modulation link as described in the previous subsection. Using Equation 3.1 for two optical carriers at frequencies $f_{opt1}$ and $f_{opt2}$,

$$I_D = r_d|E_{opt1}\cos(2\pi f_{opt1}t) + E_{opt2}\cos(2\pi f_{opt2}t)|^2. \tag{3.34}$$

If the Tx signal frequency of interest is $f_{RF} = f_{opt1} - f_{opt2}$, then

$$|i_d(f_{RF})| = r_d\sqrt{P_{opt1}P_{opt2}}, \tag{3.35}$$

where $P_{opt} = |E_{opt}|^2$. Substituting Equation 3.35 into Equation 3.32 yields

$$s_{out}(f_{RF}) = \frac{1}{2}r_d^2 P_{opt1}P_{opt2}|H_d(f_{RF})|^2 R_S. \tag{3.36}$$

If the detector in the architecture of Figure 3.17 is able to withstand a photocurrent ($r_d \cdot P_{opt}$) of 200 mA, Equation 3.36 dictates that 1 W of RF signal output power can be generated at any frequency $f_{RF}$ within the bandwidth of the detector—that is, for frequencies where $|H_d(f_{RF})|$ is approximately unity. Indeed, a conference paper by Itakura et al. showed exactly that: when their photodetector with a 3-dB bandwidth of 7 GHz was illuminated by two lasers to produce 200 mA of photocurrent, they measured an output power of 29 dBm (~800 mW) at 5 GHz, where their detector had about 1 dB of roll-off [30].

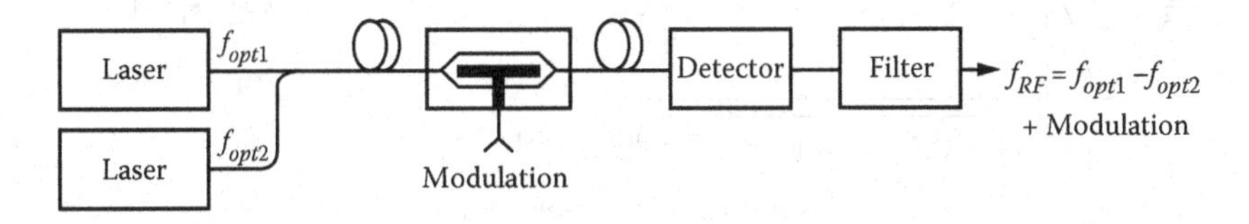


**FIGURE 3.17** Method of heterodyning two lasers at optical frequencies $f_{opt1}$ and $f_{opt2}$ to generate a Tx signal at frequency $f_{RF} = f_{opt1} - f_{opt2}$. (Reproduced from C. Cox et al., *J. Communications*, XLVIII, 22–25, 1997. With permission.)

For further details and discussion of this technique for high-power RF signal delivery via optical fiber, the reader is referred to the following articles by Gliese et al. [53], Helkey et al. [54], Wake et al. [55,56], Kitayama and Kuri [57,58], Huggard et al. [59], and Stohr et al. [60].

## 3.4 Combined Transmit and Receive Links

### 3.4.1 Key Transmit/Receive Link Parameters

Up to this point in the chapter, we have treated the receive and transmit functions separately. And while there are many applications that involve receive-only, and other applications—perhaps fewer—that involve transmit-only, by far the majority of antenna remoting applications involve systems that must both transmit and receive (Tx/Rx). If these Tx/Rx systems use separate antennas for transmit and receive, then remoting the signals to and from these antennas, respectively, can be accomplished using the appropriate transmit and receive photonic links that have already been discussed. However, it is often highly desirable—and usually required—to transmit and receive via the same antenna element. Hence, it is important to investigate and develop photonic links that can support transmitting and receiving via the same radiating element.

The key figures of merit for links that handle both transmit and receive are bandwidth and transmit-to-receive (T/R) isolation. Figure 3.18 illustrates these two figures of merit. While bandwidth is just the common 3-dB bandwidth, T/R isolation requires a slight extension to accommodate the unique properties of active solutions such as the photonic solution to be discussed below.

However, before we have the isolation discussion, it is important to distinguish two categories of isolation. When one refers to "T/R isolation," one is generally referring to a system in which the interface hardware is connected to the antenna. In this case, the T/R isolation is a function of both the interface and the return loss of the antenna. However, components such as the 3-port transmit–receive antenna interface shown in Figure 3.18 are often characterized individually, with 50 Ω terminations on their various ports. When an interface is characterized in this manner, the isolation will be that of just the 3-port component itself. In this case, we refer to the T/R isolation as the "port 1–3 isolation."

Traditionally, port 1–3 isolation is simply the amount of transmit signal that leaks from the transmit port, port 1, to the receive port, port 3. If we treat the generic antenna interface as a 3-port device with the port assignments shown in Figure 3.18, then port 1–3 isolation would simply be $S_{31}$.

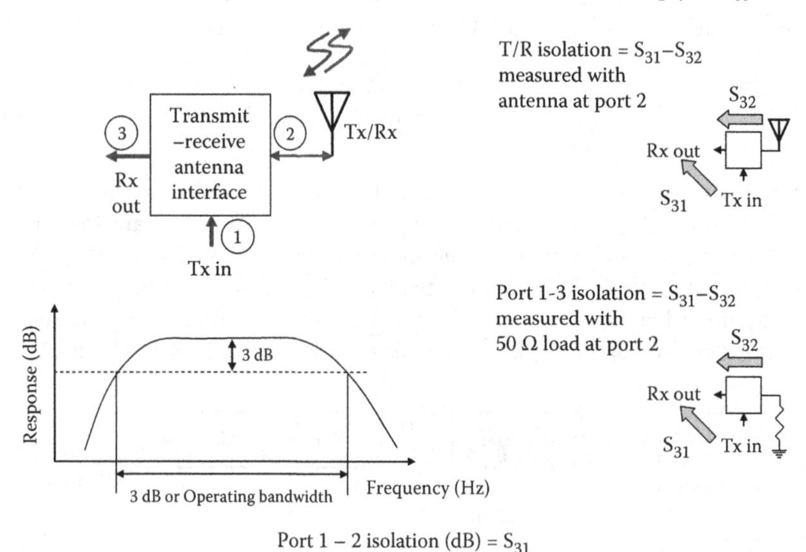


**FIGURE 3.18** Figures of merit for transmit/receive links.

While port 1–3 isolation defined in this manner has been sufficient for the passive electronic interfaces to which it has been applied in the past, it does not accommodate the newer active means, such as the photonic one to be discussed below. The primary limitation of the traditional definition of port 1–3 isolation is that it does not take into account the efficiency of the antenna-port-to-receive port path. With passive approaches, this path is typically very efficient, with losses less than 1 dB being typical. With active antenna interfaces, the loss may be replaced with power gain, which would go completely unrecognized by the present definition.

Further, the present definition allows for the possibility of realizing high port 1–3 isolation while not achieving the goal of actually conveying the receive signal from the antenna port to the receive port. To illustrate this case, consider the extreme situation where there is infinite loss in the antenna port to receive port path. Hence, there is no receive signal at port 3, independent of the level of the transmit signal. Therefore the transmit signal is totally isolated from the receive signal—which is desired—but there is no receive signal—which is not desired.

A simple way to avoid these issues is to modify the definition of port 1–3 isolation to be the ratio of the port 1–3 isolation to the port 2–3 transmission. When this definition is expressed in terms of S-parameters it becomes

$$\text{Port 1–3 Isolation (dB)} = S_{31} - S_{32}. \tag{3.37}$$

This definition of port 1–3 isolation avoids the previous issues because loss or gain in the receive path is explicitly accounted for. Hence, this is the definition of port 1–3 isolation that will be used in the remainder of this chapter.

There are two main approaches to implementing the transmit/receive antenna interface that was shown in Figure 3.18. Perhaps the most common approach is to use a transmit/receive (T/R) switch to alternately connect *either* the transmit path to the antenna *or* the antenna to the receive path; see Figure 3.19a. When the T/R switch is implemented with a PIN diode switch, the bandwidth can be broad, the switching speed can be less than a microsecond and the port 1–3 isolation can be on the order of 40 dB. Except for the fact that a switch cannot simultaneously connect the transmit and receive paths to the same antenna element, a PIN diode T/R switch comes reasonably close to the ideal transmit–receive antenna interface. If simultaneous connection of the transmit and receive paths to a single antenna is a dominant factor, then historically, a ferrite circulator is the only option, as shown in Figure 3.19b. However, the performance of ferrite circulators imposes some significant constraints on the system. For example, the maximum bandwidth is about an octave and the maximum port 1–3 isolation is about 20 dB, however isolations as high as 35 dB to 40 dB can be achieved over narrow fractional bandwidths, such as ~10%. These performance limitations stem from the fact that a ferrite circulator is basically a narrow bandwidth device, to which various impedance matching circuits are applied to broaden its bandwidth resulting in a tradeoff of reduced port 1–3 isolation.

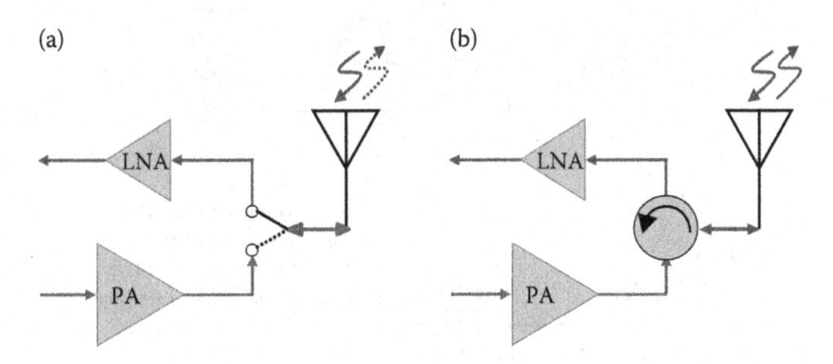


**FIGURE 3.19** Approaches for implementing a transmit/receive interface using (a) a transmit/receive (T/R) switch, and (b) a ferrite circulator.

### 3.4.2 Transmit-Isolating Photonic Receive (TIPRX) Link

Recently, it has been found to be possible to apply photonic techniques to the implementation of an antenna interface that permits simultaneous connection of the transmit and receive paths to a single antenna element. As will become clear from the discussion that follows, the photonic T/R antenna interface replaces more than just the ferrite circulator; it also replaces the low noise amplifier in the receive path. Hence, this interface has isolation like a circulator, but it has gain and low noise figure, like a low noise amplifier. For this reason, we elect to give this interface a new name that is an acronym for all of the functions it performs: TIPRX, for transmit-isolating photonic receiver.

The hexagon in Figure 3.20 encloses the photonic hardware comprising the TIPRX. The Tx signal enters the TIPRX at RF port 1. At RF port 2, where the antenna is connected, the Tx signal exits the TIPRX and the Rx signal enters. The Rx signal is routed to RF port 3.

The key component in the TIPRX is an electrooptic external modulator with traveling-wave electrodes. We begin the heuristic description of how a TIPRX works by describing the receive function, which is basically just a conventional external modulation receive link, but with one important distinction: the termination at the output end of the modulator's traveling wave electrode is replaced with the input port for the transmit signal. It will be helpful to recall that for maximum modulation sensitivity, as is desired for the receive signal, the receive signal wave on the modulator electrodes copropagates with, and ideally is velocity matched to, the CW optical wave that is propagating in the modulator's optical waveguides. Therefore, the Rx signal modulates the light efficiently. A high-speed photodetector retrieves the Rx signal from the modulated optical output of the modulator, and directs it out of the TIPRX at RF port 3.

The remaining Rx signal at the output end of the modulator's traveling wave electrode is dumped into the output impedance of the transmit signal source, which typically would be the output impedance of a power amplifier. The transmit signal that is injected at this port travels along the modulator's electrodes and is then connected to the antenna. So the transmit path is all-RF; that is, there is no conversion to and from optics as there was with the receive signal. However, by injecting the Tx signal at what would normally be the output end of the modulator electrodes, the Tx signal counterpropagates with respect to the CW optical traveling wave. Hence, in terms of velocity match between RF and optical waves, the Tx signal achieves a very poor one. But this means that the Tx signal is relatively inefficient at modulating the CW optical carrier, which is just what we want.

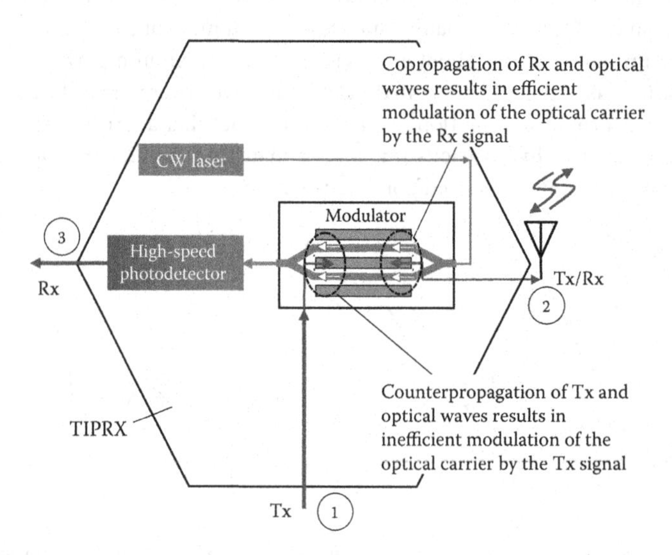


**FIGURE 3.20** Transmit isolating photonic receive link (TIPRX).

Formally, we may establish the response of a TIPRX to the simultaneous application of transmit and receive signals by applying the principle of superposition. Recall that in linear systems, the principle of superposition states that the response of a system to stimulation by multiple stimuli is simply the sum of the responses to each of the stimuli applied individually, with the remaining stimuli set to zero. Thus in the case of a TIPRX, we can establish its linear response to the simultaneous application of transmit and receive signals by summing its response to the receive and transmit signals applied individually with the other source set to zero.

Figure 3.21a is a TIPRX schematic augmented by the equivalent circuits for the receive and transmit signals. Figure 3.21b shows the TIPRX with the receive source active and the transmit source set to zero, which for an ideal voltage source is a short circuit. Figure 3.21c shows the complementary equivalent circuit when the transmit source is active and the receive source is set to zero.

To quantify the performance of a TIPRX over bandwidth and isolation, we need to develop an analytical model [61]. Models for the performance of the Rx signal channel through the TIPRX (port 2 to port 3) are well established both theoretically and experimentally [3,7,16], and the Tx signal channel (port 1 to port 2) can be modeled like any passive electrical transmission line with propagation constant $\gamma = \alpha + j\beta$. Therefore, in this chapter, we focus exclusively on the modeling of the port 1–3 isolation. At any RF frequency, the port 1–3 isolation afforded by the TIPRX can be modeled using expressions that have been derived previously for the efficiency with which light is modulated in an electro-optic modulator by co- vs. counterpropagating RF signals on its traveling-wave electrodes [17].

The most accurate expression of isolation as a function of frequency is quite complex because it accounts for many imperfections in the modulator, including attenuation of the RF signal on the electrodes, imperfect matching of the RF signal's velocity on the electrodes to that of the light in the modulator's optical waveguides, and one or more points at which a portion of the RF signal is reflected

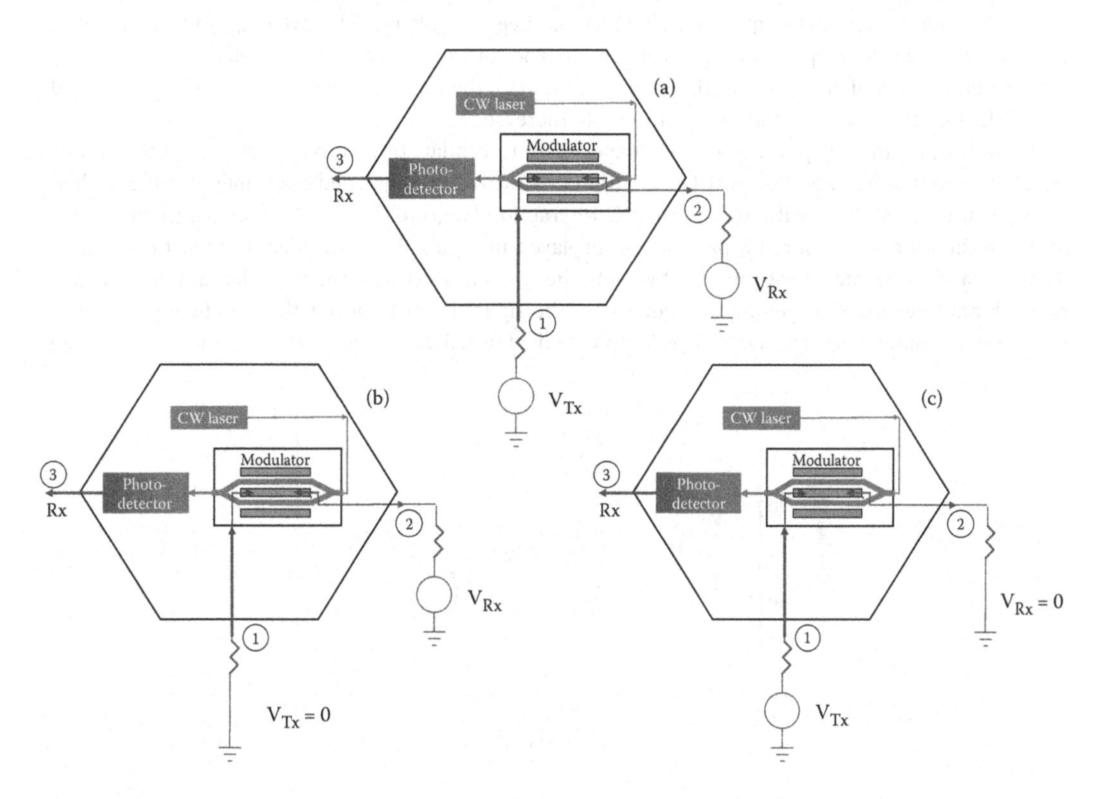


**FIGURE 3.21** (a) TIPRX schematic with equivalent circuits for Tx and Rx signal sources. (b) TIPRX schematic with Rx source active, Tx source set to zero. (c) TIPRX schematic with Tx source active, Rx source set to zero.

because of an impedance mismatch. Significant insight can be gained from a relatively simple model that neglects all of the above-mentioned imperfections in the modulator as second-order effects. Under this simplifying assumption, the resulting expression for port 1–3 isolation becomes

$$\text{Port 1-3 Isolation} \equiv \left|\frac{S_{31}}{S_{32}}\right|^2 = \frac{\sin^2(\beta_{RF}L)}{(\beta_{RF}L)^2}, \tag{3.38}$$

where $L$ is the electrical-optical interaction length in the modulator, and where we have assumed perfect matching of the RF traveling-wave velocity $v_{RF}$ to the optical wave's guided velocity $v_{opt}$ in the device, such that

$$\begin{aligned}\beta_{RF} &\equiv \frac{2\pi}{\lambda_{RF}} = \frac{2\pi f_{RF}}{v_{RF}} \\ &= \frac{2\pi f_{RF}}{v_{opt}} = \frac{2\pi f_{RF} n_{opt}}{c}.\end{aligned} \tag{3.39}$$

Note from Equations 3.38 and 3.39 that, in the limit as the RF frequency approaches DC and thus the distinction between co- and counterpropagating waves disappears, the port 1–3 isolation is defined such that it approaches unity (0 dB). As frequency increases, the isolation quickly improves, ostensibly reaching 0 ($-\infty$ dB) at the exact frequency for which its corresponding wavelength in the modulator's traveling-wave electrode structure is $2 \times L$. As dictated by Equations 3.38 and 3.39, the port 1–3 isolation will hit similar optimal frequencies wherever an integer number of RF wavelengths fits exactly into $2 \times L$. In between these optimal frequencies, frequencies of worst-case port 1–3 isolation occur whenever an odd number of half wavelengths fits into $2 \times L$, with the worst-case isolation becoming more and more tolerable as this integer number continues to increase.

We performed an early proof-of-concept experiment to validate this simplified version of the model. Figure 3.22 shows the measured port 1–3 isolation of a TIPRX that included a commercial off-the-shelf traveling-wave modulator with an electrooptic interaction length of half an RF wavelength at $f$ = 920 MHz. In the ratio of measured S-parameters displayed in Figure 3.22, the effect of reflections due to RF impedance mismatches was negated by using the "gating" feature in the time-domain mode of the network analyzer to examine only the portion of the signal input at port 1 that reached port 3 after traversing the modulator electrodes once. With the effect of reflections negated, the resulting measured

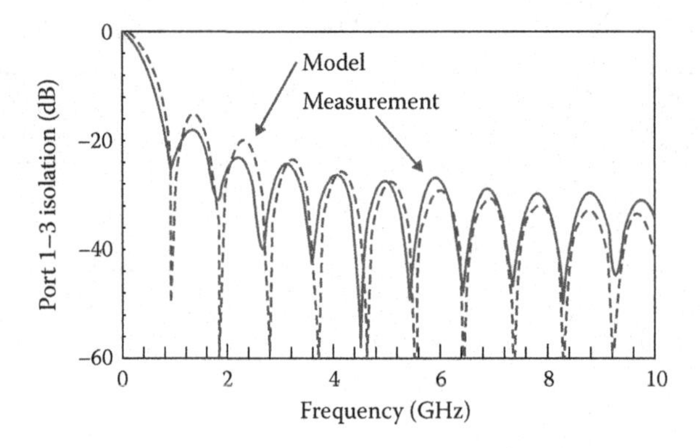


**FIGURE 3.22** Measured port 1–3 isolation of a TIPRX based on network analyzer measurements of $S_{31}$ and $S_{32}$ in an early experiment to validate the simple model expressed in (1) and (2).

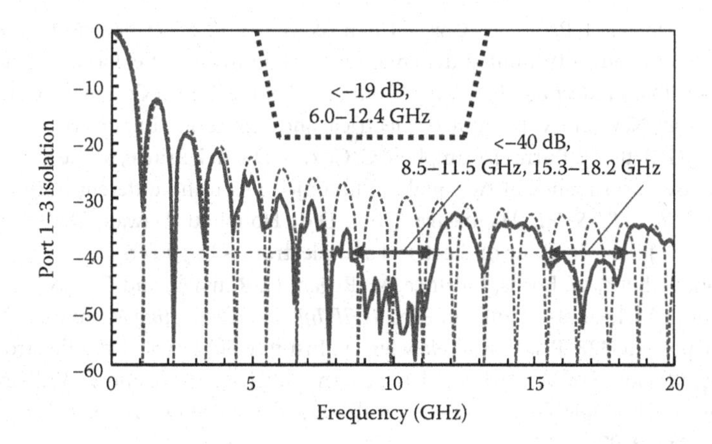


**FIGURE 3.23** Example of the measured port 1–3 isolation of a TIPRX (solid curve) compared to its modeled performance neglecting the effect of reflections (dashed curve on the bottom) and to the quoted performance of a ferrite circulator (dashed curve on the top).

behavior closely resembles the "model" curve in Figure 3.22, which was generated by substituting into Equation 3.38 the value of $\beta_{RF}L$ obtained by entering $c/2n_{opt}$ = 920 MHz × $L$ into Equation 3.39.

More recently, we have made additional measurements of the TIPRX port 1–3 isolation vs. frequency, a representative sample of which is shown in Figure 3.23. These data, which show ~30 dB of port 1–3 isolation over nearly a decade of bandwidth clearly indicate that a TIPRX has both wider bandwidth and higher isolation than a conventional ferrite circulator.

# References


1. F. Kapron, D. Keck, and R. Maurer, "Radiation losses in glass optical waveguides," *Appl. Phys. Lett.*, vol. 17, no. 10, pp. 423–425, Nov. 1970.
2. AD9480: 8 Bit, 250 MSPS, 3.3 V A/D Converter, Analog Devices, Inc., Norwood, Massachusetts, USA. http://www.analog.com/en/analog-to-digital-converters/ad-converters/ad9480/products/product.html, (accessed on July 1, 2011).
3. C. Cox, *Analog Optical Links: Theory and Practice,* Cambridge, UK: Cambridge University Press, 2004.
4. R. Kalman, J. Fan, and L. Kazovsky, "Dynamic range of coherent analog fiber-optic links," *J. Lightwave Technol.*, vol. 12, no. 7, pp. 1263–1277, Jul. 1994.
5. C. Cox, E. Ackerman, R. Helkey, and G. Betts, "Techniques and performance of intensity-modulation, direct-detection analog optical links," *IEEE Trans. Microwave Theory Tech.*, vol. 45, pp. 1375–1383, Aug. 1997.
6. H. Haus, et al., "IRE standards on methods of measuring noise in linear twoports, 1959," *Proc. IRE*, vol. 48, pp. 60–68, Jan. 1959.
7. E. Ackerman, W. Burns, G. Betts, J. Chen, J. Prince, M. Regan, H. Roussell, and C. Cox, "RF-over-fiber links with very low noise figure," *J. Lightwave Technol.*, vol. 26, no. 15, pp. 2441–2448, Aug. 2008.
8. C. Cox, D. Tsang, L. Johnson, and G. Betts, "Low-loss analog fiber-optic links," in *IEEE MTT-S Int. Microwave Symp. Dig.*, Dallas, Texas, USA, pp. 157–160, 1990. Piscataway, New Jersey: Institute of Electrical and Electronics Engineers.
9. W. Burns, M. Howerton, and R. Moeller, "Broad-band unamplified optical link with RF gain using a $LiNbO_3$ modulator," *IEEE Photon. Technol. Lett.*, vol. 11, no. 12, pp. 1656–1658, Dec. 1999.
10. A. Karim and J. Devenport, "Low noise figure microwave photonic link," in *IEEE MTT-S Int. Microwave Symp. Dig.*, Honolulu, Hawaii, USA, pp. 1519–1522, 2007. Piscataway, New Jersey: Institute of Electrical and Electronics Engineers.

11. H. Roussell, M. Regan, J. Prince, C. Cox, J. Chen, W. Burns, E. Ackerman, and J. Campbell, "Gain, noise figure, and bandwidth-limited dynamic range of a low-biased external modulation link," in *Proc. IEEE Int. Topical Meeting Microwave Photonics*, Victoria British Columbia, Canada, pp. 84–87, 2007. Piscataway, New Jersey: Institute of Electrical and Electronics Engineers.
12. E. Ackerman, G. Betts, W. Burns, J. Campbell, C. Cox, N. Duan, J. Prince, M. Regan, and H. Roussell, "Signal-to-noise performance of two analog photonic links using different noise reduction techniques," in *IEEE MTT-S Int. Microwave Symp. Dig.*, Honolulu, Hawaii, USA, 2007, pp. 51–54. Piscataway, New Jersey: Institute of Electrical and Electronics Engineers.
13. E. Ackerman, G. Betts, W. Burns, J. Prince, M. Regan, H. Roussell, and C. Cox, "Low noise figure, wide bandwidth analog optical link," in *Proc. IEEE Int. Topical Meeting Microwave Photonics*, Seoul, Korea, 2005, pp. 325–328. Piscataway, New Jersey: Institute of Electrical and Electronics Engineers.
14. J. McKinney, M. Godinez, V. Urick, S. Thaniyavarn, W. Charczenko, and K. Williams, "Sub-10-dB noise figure in a multiple-GHz analog optical link," *IEEE Photon. Technol. Lett.*, vol. 19, no. 7, pp. 465–457, Apr. 2007.
15. K. Williams, L. Nichols, and R. Esman, "Externally-modulated 3 GHz fibre-optic link utilizing high current and balanced detection," *Electron. Lett.*, vol. 33, no. 15, pp. 1327–1328, Jul. 1997.
16. C. Cox, E. Ackerman, G. Betts, and J. Prince, "Limits on the performance of RF-over-fiber links and their impact on device design," *IEEE Trans. Microwave Theory Tech.*, vol. 54, no. 2, pp. 906–920, Feb. 2006.
17. G. Gopalakrishnan, W. Burns, R. McElhanon, C. Bulmer, and A. Greenblatt, "Performance and modeling of broadband LiNbO3 traveling wave optical intensity modulators," *J. Lightwave Technol.*, vol. 12, no. 10, pp. 1807–1819, Oct. 1994.
18. G. Betts and F. O'Donnell, "Improvements in passive, low-noise-figure optical links," in *Proc. Photonic Systems Antenna Applications Conf.*, Monterey, California, USA, 1993. US Defense Advanced Research Projects Agency (DARPA).
19. M. Farwell, W. Chang, and D. Huber, "Increased linear dynamic range by low biasing the Mach-Zehnder modulator," *IEEE Photon. Technol. Lett.*, vol. 5, no. 7, pp. 779–782, Jul. 1993.
20. E. Ackerman, S. Wanuga, D. Kasemset, A. Daryoush, and N. Samant, "Maximum dynamic range operation of a microwave external modulation fiber-optic link," *IEEE Trans. Microwave Theory Tech.*, vol. 41, no. 8, pp. 1299–1306, Aug. 1993.
21. PSI 3600 MOD D1 Low-Vpi Modulator, Billerica, Massachusetts, USA: Photonic Systems, Inc., http://www.photonicsinc.com/pdfs/PSI-3600-MOD-D1.pdf (accessed on August 10, 2010).
22. E. Ackerman, S. Wanuga, J. MacDonald, and J. Prince, "Balanced receiver external modulation fiber-optic link architecture with reduced noise figure," in *IEEE MTT-S Int. Microwave Symp. Dig.* 1993, Atlanta, GA, pp. 723–726. Piscataway, New Jersey: Institute of Electrical and Electronic Engineers.
23. G. Betts, X. Xie, I. Shubin, W. Chang, and P. Yu, "Gain limit in analog links using electroabsorption modulators," *IEEE Photon. Technol. Lett.*, vol. 18, no. 19, pp. 2065–2067, Oct. 2006.
24. J. Chen, Y. Wu, W. Chen, I. Shubin, A. Clawson, W. Chang, and P. Yu, "High-power intrastep quantum well electroabsorption modulator using single-sided large optical cavity waveguide," *IEEE Photon. Technol. Lett.*, vol. 16, no. 2, pp. 440–442, Feb. 2004.
25. RF Cables, Huber+Suhner, Switzerland, http://www.hubersuhner.com/products, (accessed on July 1, 2011).
26. High Performance Fiber-optic Links, Billerica, Massachusetts, USA: Photonic Systems, Inc., July 1, 2011, http://www.photonicsinc.com/fiber_optics.html.
27. N. Li and J. Campbell, private communication, 2005.
28. D. Tulchinsky, K. Williams, X. Li, N. Li, and J. Campbell, "High power photodetectors for microwave photonic links," in *IEEE Conf. Avionics Fiber-Optics Photonics*, Minneapolis, MN, 2005, pp. 73–74. Piscataway, New Jersey: Institute of Electrical and Electronic Engineers.
29. N. Duan, X. Wang, N. Li, H. Liu, and J. Campbell, "Thermal analysis of high-power InGaAs-InP photodiodes," *IEEE J. Quantum Electron.*, vol. 42, no. 12, pp. 1255–1258, Dec. 2006.

30. S. Itakura, K. Sakai, T. Nagatsuka, T. Akiyama, Y. Hirano, E. Ishimura, M. Nakaji, and T. Aoyagi, "High-current backside-illuminated InGaAs/InP p-i-n photodiode," in *Proc. IEEE Int. Topical Meeting Microwave Photonics,* Valencia, Spain, paper #We2.6, pp. 4, 2009. Piscataway, New Jersey: Institute of Electrical and Electronic Engineers.
31. H. Ito, S. Kodama, Y. Muramoto, T. Furuta, T. Nagatsuma, and T. Ishibashi, "High-speed and high-output InP-InGaAs uni-traveling-carrier photodiodes," *IEEE J. Select. Topics Quantum Electron.*, vol. 10, no. 4, pp. 709–726, Jul./Aug. 2004.
32. G. Davis, R. Weiss, R. LaRue, K. Williams, and R. Esman, "A 920-1650-nm high-current photodetector," *IEEE Photon. Technol. Lett.*, vol. 8, no. 10, pp. 1373–1375, Oct. 1996.
33. X. Li, N. Li, X. Zheng, S. Demiguel, J. Campbell, D. Tulchinsky, and K. Williams, "High-saturation-current InP-InGaAs photodiode with partially depleted absorber," *IEEE Photon. Technol. Lett.*, vol. 15, no. 9, pp. 1276–1278, Sep. 2003.
34. N. Li, X. Li, S. Demiguel, X. Zheng, J. Campbell, D. Tulchinsky, K. Williams, T. Isshiki, G. Kinsey, and R. Sudharsansan, "High-saturation-current charge-compensated InGaAs-InP uni-traveling-carrier photodiode," *IEEE Photon. Technol. Lett.*, vol. 16, no. 3, pp. 864–866, Mar. 2004.
35. H. Pan, A. Beling, H. Chen, J. Campbell, and P. Yoder, "The influence of nonlinear capacitance and responsivity on the linearity of a modified uni-traveling carrier photodiode," in *Proc. IEEE Int. Topical Meeting Microwave Photonics, 2008,* Gold Coast, Australia, pp. 82–85, 2008, Piscataway, New Jersey: Institute of Electrical and Electronic Engineers.
36. Y. Fu, H. Pan, Z. Li, and J. Campbell, "High linearity photodiode array with monolithically integrated Wilkinson power combiner," in *Proc. IEEE Int. Topical Meeting Microwave Photonics, 2010,* Montreal, Quebec, Canada, pp. 111–113, 2010. Piscataway, New Jersey: Institute of Electrical and Electronic Engineers.
37. M. Chitioui, A. Enard, D. Carpentier, F. Lelarge, B. Rousseau, M. Achouche, A. Marceaux, A. Renoult, C. Feuillet, M. Queguiner, and T. Merlet, "High power UTC photodiodes design and application for analog fiber optic links," in *Proc. IEEE Int. Topical Meeting Microwave Photonics,* Valencia, Spain, paper #We2.4, pp. 4, 2009. Piscataway, New Jersey: Institute of Electrical and Electronic Engineers.
38. J. Klamkin, A. Ramaswamy, L. Johansson, N. Nunoya, J. Bowers, S. DenBaars, and L. Coldren, "Uni-traveling-carrier balanced photodiode with tunable MMI coupler for optimization of source laser RIN suppression," in *Proc. IEEE Int. Topical Meeting Microwave Photonics,* Valencia, Spain, paper #We2.2, pp. 4, 2009. Piscataway, New Jersey: Institute of Electrical and Electronic Engineers.
39. H. Pan, A. Beling, H. Chen, J. Campbell, and P. Yoder, "A high-linearity modified uni-traveling carrier photodiode with offset effects of nonlinear capacitance," *J. Lightwave Technol.*, vol. 27, no. 20, pp. 4435–4439, Oct. 2009.
40. L. Lin, M. Wu, T. Itoh, T. Vang, and R. Muller, D. Sivco, and A. Cho, "High-speed photodetectors with high saturation for high performance microwave photonic systems," in *Proc. IEEE Int. Topical Meeting Microwave Photonics,* Kyoto, Japan, 1996, pp. 313–316. Piscataway, New Jersey: Institute of Electrical and Electronic Engineers.
41. N. Li, H. Chen, N. Duan, M. Liu, S. Demiguel, R. Sidhu, A. Holmes, and J. Campbell, "High power photodiode wafer bonded to Si using Au with improved responsivity and output power," *IEEE Photon. Technol. Lett.*, vol. 18, no. 23, pp. 2526–2528, Dec. 2006.
42. A. Beling, H. Pan, H. Chen, and J. Campbell, "Measurement and modeling of a high-linearity modified uni-traveling carrier photodiode," *IEEE Photon. Technol. Lett.*, vol. 20, no. 14, pp. 1219–1221, Jul. 2008.
43. A. Beling, H. Pan, H. Chen, and J. Campbell, "High-power modified uni-traveling carrier photodiode with > 50 dBm third order intercept point," in *2008 IEEE MTT-S Int. Microwave Symp. Dig.*, Atlanta, GA, pp. 499–502, 2008. Piscataway, New Jersey: Institute of Electrical and Electronic Engineers.
44. J. Raring, E. Skogen, J. Barton, C. Wang, S. DenBaars, and L. Coldren, "Quantum well intermixing and MOCVD regrowth for monolithic integration of 40 Gbit/s UTC type photodiodes with QW based components," *Electron. Lett.*, vol. 41, no. 18, pp. 1033–1034, Sep. 2005.

45. T. Ishibashi, N. Shimizu, S. Kodama, H. Ito, T. Nagatsuma, and T. Furuta, "Uni-traveling-carrier photodiodes," *Tech. Dig. Ultrafast Electronics Optoelectronics*, pp. 83–86, Jun. 1997.
46. J. Paslaski, P. Chen, J. Chen, C. Gee, and N. Bar-Chaim, "High-power microwave photodiode for improving performance of RF fiber optic links," *Proc. SPIE*, vol. 2844, pp. 110–119, Aug. 1996.
47. http://www.u2t.de/system/files/sites/default/files/DS_BPDV2xxxR_2v2.pdf. AG, Berlin, Germany: u2t Photonics. (accessed on July 1, 2011).
48. H. Yang, C. Daunt, F. Gity, K. Lee, W. Han, B. Corbett, and F. Peters, "Zero-bias high-speed edge-coupled unitraveling-carrier InGaAs photodiode," *IEEE Photon. Technol. Lett.*, vol. 22, no. 23, pp. 1747–1749, Dec. 2010.
49. S. Demiguel, L. Giraudet, L. Jouland, J. Decobert, F. Blache, V. Coupe, F. Jorge, P. Pagnod-Rossiaux, E. Boucherez, M. Achouche, and F. Devaux, "Evanescently coupled photodiodes integrating a double-stage taper for 40-Gb/s applications—compared performance with side-illuminated photodiodes," *J. Lightwave Technol.*, vol. 20, no. 12, pp. 2004–2013, Dec. 2002.
50. H. Ito, T. Furuta, T. Ito, Y. Muramoto, K. Tsuzuki, K. Yoshino, and T. Ishibashi, "W-band uni-travelling-carrier photodiode module for high-power photonic millimeter-wave generation," *Electron. Lett.*, vol. 38, no. 22, pp. 1376–1377, Oct. 2002.
51. Y. Wu, J. Shi, J. Wu, F. Huang, Y. Chan, Y. Huang, and R. Xuan, "High-performance evanescently edge coupled photodiodes with partially p-doped photoabsorption layer at 1.55-μm wavelength," *IEEE Photon. Technol. Lett.*, vol. 17, no. 4, pp. 678–680, Apr. 2005.
52. C. Cox, E. Ackerman, R. Helkey, and G. Betts, "Applications of analog fiber-optic links," *J. Communications*, vol. XLVIII, pp. 22–25, Aug. 1997.
53. U. Gliese, T. Nielsen, M. Bruun, E. Christensen, K. Stubkjaer, S. Lindgren, and B. Broberg, "A wide-band heterodyne optical phase-locked loop for generation of 3–18 GHz microwave carriers," *IEEE Photon. Technol. Lett.*, vol. 4, no. 8, pp. 936–938, Aug. 1992.
54. R. Helkey, D. Derickson, A. Mar, J. Wasserbauer, and J. Bowers, "Millimeter-wave signal generation using semiconductor diode lasers," *Microwave Opt. Technol. Lett.*, vol. 6, no. 1, pp. 1–5, Jan. 1993.
55. D. Wake, C. Lima, and P. Davies, "Optical generation of millimeter-wave signals for fibre-radio systems using a dual-mode DFB semiconductor laser," *IEEE Trans. Microwave Theory Tech.*, vol. 43, no. 9, pp. 2270–2296, Sep. 1995.
56. D. Wake, C. Lima, and P. Davies, "Transmission of 60-GHz signals over 100 km of optical fiber using a dual-mode semiconductor laser source," *IEEE Photon. Technol. Lett.*, vol. 8, no. 4, pp. 578–580, Apr. 1996.
57. K. Kitayama and T. Kuri, "Dual lightwave technique for optical generation and transport of wireless signals," in *Proc. IEEE Int. Topical Meeting Microwave Photonics, 1997,* Duisburg, Germany, pp. 43–46, 1997. Piscataway, New Jersey: Institute of Electrical and Electronic Engineers.
58. T. Kuri and K. Kitayama, "Optical heterodyne detection of millimeter-wave-band radio-on-fiber signals with a remote dual-mode local light source," *IEEE Trans. Microwave Theory Tech.*, vol. 49, no. 10, pp. 2025–2029, Oct. 2001.
59. P. Huggard, B. Ellison, P. Shen, N. Gomes, P. Davies, W. Shillue, A. Vaccari, and J. Payne, "Efficient generation of guided millimeter-wave power by photomixing," *IEEE Photon. Technol. Lett.*, vol. 14, no. 2, pp. 197–199, Feb. 2002.
60. A. Stohr, A. Malcoci, A. Sauerwald, I. Mayorga, R. Gusten, and D. Jager, "Ultra-wideband traveling-wave photodetectors for photonic local oscillators," *J. Lightwave Technol.*, vol. 21, no. 12, pp. 3062–3070, Dec. 2003.
61. E. Ackerman and C. Cox, "Optimization of photonic transmit/receive module performance," in *Proc. IEEE Int. Topical Meeting Microwave Photonics,* paper #Th2.1, pp. 4, 2009, Valencia, Spain, 2009. Piscataway, New Jersey: Institute of Electrical and Electronic Engineers.

# 4

# Overview of Optical Modulators and the Properties That Affect Transmission System Performance

Gary E. Betts
*Photonic Systems, Inc.*

John C. Cartledge
*Queen's University*



A large variety of optical modulators has been invented, researched, and developed over the last 40 years. This research is still very active, and at present, there are many papers published every year describing new designs and new advances in existing modulators. On the other hand, only two modulator designs are presently used in large numbers in the fiber-optic communication market.

This chapter will give a brief description of the classes of modulators in existence and attempt to give a perspective on why one might be chosen over another for particular applications. It will also discuss performance measures that are important for most digital and analog communication applications. Later chapters will cover specific types of modulators in much more detail.

## 4.1 Practical Overview

This section lists some of the present applications that drive the development of modulators, and discusses some of the primary reasons for particular choices that have been made.

One restriction should be noted at the outset: this chapter is concerned with electro-optic modulators using single-mode optical waveguides as their input and output. This is the type of modulator compatible with single-mode fiber and high-speed operation.

### 4.1.1 Applications of Modulators

The primary application of optical modulators is digital fiber-optic communication. This consumes the largest volume of modulators and is perhaps the largest driver of modulator development. After the advent of erbium doped fiber amplifiers in the 1990s, fiber-optic communication has been almost exclusively at 1550 nm. This application was described in detail in Chapter 2. Digital fiber communication affects development in all other applications because it dominates the design of components. For example, the preferred wavelength for most applications has become 1550 nm due to the availability of cheap, high-quality components at that wavelength. Modulators have been made over the wavelength range 450 nm to 3.4 μm [1] and beyond; 1550 nm is just the dominant wavelength at present.

A second modulator application is analog fiber-optic communication. This is much smaller than digital fiber optics, but it has still driven a significant amount of modulator research. The one large commercial application in this category was analog cable TV transmission. This became practical about 1992 and a large number of systems were installed over the next decade or so. At present, however, much of the fiber-optic cable TV plant has been converted to digital (even for the signals that eventually appear as analog to the end user), so this is no longer a key driver of modulator development. The other driver of analog system development has been defense applications. None of these has been large in volume, but due to the large variety of end uses and the availability of substantial research funds for this purpose, defense has driven a significant amount of analog modulator development. Analog system considerations were described in Chapter 3. Some analog system applications that are primarily defense-related are described in Chapters 17 through 23.

There are several other niche applications that also use modulators. Perhaps the largest of these in volume at present is modulator chips to support fiber-optic gyroscopes (Chapter 23).

### 4.1.2 Why Modulators at All: Direct versus External Modulation

Semiconductor diode lasers are the laser sources most commonly used with optical modulators. These can be modulated directly [2], which is cheaper and simpler than adding the additional optical modulator component. Indeed, this is how many optical communication systems operate.

External modulation offers better performance than direct modulation in return for increased cost and complexity. External modulation separates optical power generation from optical modulation; the external modulator can be optimized for modulation while the laser source can be separately optimized for optical carrier generation.

For long-distance digital fiber-optic communication, a key parameter is chirp, which is a measure of the ratio of frequency modulation to amplitude modulation. Directly modulated lasers generate significant chirp and so, are limited to systems with lower speed-dispersion products [3]. The tradeoff point between when direct modulation is usable and when external modulation is required depends on many factors including both technology (e.g., how low a chirp is possible from a directly modulated laser) and cost (e.g., whether dispersion compensation is cheaper than an external modulator). To give a very rough benchmark, at the present time a data rate of 10 Gb/s over a distance of 80 km or greater of standard fiber (17 ps/(km·nm) dispersion) generally requires external modulation (this tradeoff point changes over time: see [2] versus [3]).

For analog communication, several performance measures may be involved. Chirp was a major issue for cable TV transmission. For analog defense systems, which tend to have short fiber lengths, the linearity and the sensitivity of the modulator are the prime factors [4]. The external modulator is a three-terminal device so RF signal gain may be achieved by increasing the optical power passing through the modulator; much smaller signals can be detected than for direct modulation. Second-order distortion at all frequencies and third-order distortion at high frequency (see Subsection 4.3.10) are generally better for external than direct modulation.

A factor important to both digital and analog modulation is bandwidth. Modulators can work at speeds of 100 GHz and more [5], while directly modulated lasers are limited by carrier dynamics to speeds of ~30 GHz [6]. The laser speed limit improves as more research is done and also depends on what bandwidth criterion is used so it is not a precisely defined number. Up to the present, this has not been the deciding factor for most systems, since most digital communication at present is 12.5 Gb/s or less and most analog communication is below 18 GHz. Speculating as to the future, it is likely that bandwidth needs will increase faster than direct modulation speed will improve, so external modulation will continue to dominate in the fastest communication links.

### 4.1.3 Two Modulators in Widespread Use

This book will describe dozens of modulator types that have been built or are being researched now. Several are commercially available. At the present time, however, only two are actually in widespread use: the lithium niobate ($LiNbO_3$) Mach-Zehnder (MZ) modulator, and the III-V semiconductor electroabsorption modulator (EAM). Of these, only the $LiNbO_3$ MZ modulator is widely used as a stand-alone external modulator. There are a number of factors, both technical and economic, that have contributed to this result.

The $LiNbO_3$ MZ modulator provides very good control of the chirp, better than the EAM, which is a major factor in the choice of the external modulator in the first place. It has low optical loss, a drive voltage within the capabilities of economical amplifiers, an electrical bandwidth in excess of that required for digital communication, a wide optical bandwidth, and it can be pigtailed with single-mode optical fibers. These things could be said of MZ modulators based on other inorganic crystals or on polymers as well. But in addition to these performance measures, the $LiNbO_3$ MZ modulator is a reliable device, able to operate for years over the temperature range required for practical system use. $LiNbO_3$ is a standard material, available from multiple suppliers as single-crystal, single-domain material. $LiNbO_3$ also has a good, low-loss waveguide fabrication technology.

The III-V semiconductor EAM has most of the same key advantages of the $LiNbO_3$ MZ modulator. However, it does not control chirp as well as the $LiNbO_3$ MZ modulator [7], nor does it have as low a loss or as wide an optical bandwidth (modern EAMs are reducing these penalties [8]). But it does have a key advantage over the $LiNbO_3$ MZ modulator: it can be monolithically integrated with a semiconductor laser. It is this integrability that has led to its widespread use, even though the $LiNbO_3$ MZ modulator provides superior modulation performance. The monolithically integrated electroabsorption-modulated laser provides better chirp control than a simple directly modulated laser, but at a lower cost than a laser-$LiNbO_3$ MZ modulator combination.

One point of this subsection is to briefly illustrate what is required for practical utility. It is not a single performance measure or even a heroic performance in several areas. It is a combination of high performance, high reliability, and economic factors. For new modulators to gain wide market acceptance, they must meet or exceed the performance of existing modulators in *all* dimensions important to the targeted application.

This is just a snapshot of how things are in 2010. Future development is needed to improve performance in existing applications and to open new applications. There are several areas of current research that are likely to provide new capabilities to the modulator market. Reduced drive voltage, as much as 10× below today's high-speed $LiNbO_3$ MZ modulators, is helpful for digital communication and critical for many analog communication applications; this is most likely to come from new materials. If the reliability of polymer modulators were to be improved, polymer MZ modulators could compete with $LiNbO_3$ MZ modulators in most analog and digital applications. The III-V semiconductor MZ modulator, like the EAM, can be monolithically integrated with a laser and can offer better chirp performance. Silicon modulators may be able to reduce costs 100× below those of today's modulators, which would expand optical communication into new applications such as intracomputer [9] and even intrachip optical communication.

## 4.2 Modulator Landscape: Materials, Design, Electro-Optic Effects, and Relative Performance

There are three basic elements that define an optical modulator: the material, the modulator design (waveguide layout), and the electro-optic effect used. Table 4.1 summarizes these elements for the range of modulators covered in this book. Not every effect or every design can be made in every material, but most materials support several possibilities. Parts II and III of this book are primarily organized by material; however, there are also common performance features across many materials that use the same modulator design or physical effect. This section will discuss each of these basic modulator elements.

This section will also give a few quantitative results comparing various materials and modulators. Much of the connection between performance and the basic elements depends on the electrical design, so that will also be discussed briefly.

### 4.2.1 Electro-Optic Effects

There are several electro-optic effects that allow an electrical signal to affect the propagation of light. The effects most relevant to electro-optic modulators that allow a "low" frequency (low compared to optical frequencies) electric field to affect light are listed in Table 4.1. (There are other effects, such as acousto-optic where a sound wave affects light, that can be used to make modulators, but these are outside the scope of this book.) The linear and quadratic electro-optic (EO) effects may be found in both insulating and semiconductor crystals. The electroabsorption (EA) and carrier density effects are only found in semiconductors.

The first three effects, linear and quadratic EO effects and the EA effect, are extremely fast, with subpicosecond response time. These effects arise from the distortion of the potential seen by electrons bound in the crystal. The speed of modulators based on these effects is always (at least up to the time of this writing) limited by technological restrictions, not by the speed of the basic effects themselves. Different materials do have different ultimate speeds, but these differences are at modulation frequencies well above 100 GHz and have not yet factored in to practical modulator design.

The fourth type of effect depends on carrier density. This does have a speed limit in the range where it affects practical modulators.

#### 4.2.1.1 Linear Electro-Optic Effect

The linear EO effect, or Pockels effect, is found in crystals that do not have inversion symmetry, and in EO polymers. This includes $LiNbO_3$, barium titanate ($BaTiO_3$), III-V semiconductors, and many other crystals but not silicon. In this effect, an electric field applied along a particular crystal axis will affect the refractive index along one or more of the three principal crystal axes. It is thus sensitive to the polarization of light and to the direction along which the field is applied. In some cases, the applied

**TABLE 4.1** Defining Characteristics of Modulators

| Material | Modulator Design | Electro-Optic Effect |
|---|---|---|
| Lithium niobate (Chapter 6) | Mach-Zehnder interferometer (MZ) | Linear electro-optic (Pockels) |
| Other inorganic crystals (Chapter 12) | Resonant (e.g., ring) | Quadratic electro-optic (Kerr) |
| Polymers (Chapter 9) | Directional coupler | Electroabsorption |
| Organic crystals (Chapter 11) | Compound (e.g., cascaded MZs) | Carrier density (plasma effect, band filling, bandgap shrinkage) |
| III-V semiconductors (Chapters 5, 7, and 8) | Electroabsorption (EAM) | |
| Silicon (Chapter 10) | | |

electric field can rotate the direction of the crystal axes and lead to polarization rotation. The complete description of this effect involves the product of an electro-optic tensor with the electric field vector to get a vector of refractive index changes. A good introductory description of the full calculation can be found in [10].

For the purposes of this overview chapter, the linear EO effect will be described by

$$\Delta n = -\frac{1}{2} n^3 r E. \tag{4.1}$$

Here, $n$ is the refractive index, $\Delta n$ refers to the electro-optically induced index change, $r$ refers to the relevant EO coefficient, and $E$ is the electric field component along the appropriate crystal direction. Although this is a simplification of the full tensorial effect, in many modulators, Equation 4.1 is a highly accurate description as long as the correct effective values of $r$ and $E$ are used.

Equation 4.1 shows the key characteristic of the linear EO effect: the refractive index, and through it the optical phase, depends linearly on the applied electric field.

#### 4.2.1.2 Quadratic Electro-Optic Effect

The quadratic EO effect, or Kerr effect, results in an index change that depends on the square of the applied electric field. Like the linear EO effect, the quadratic EO effect modulates the phase of the light.

The quadratic EO effect is universal in all transparent, nonmetallic crystals. An applied electric field creates a direction (breaks inversion symmetry), which in turn allows the EO effect. This effect is generally very small, so when the linear EO effect is present, the linear EO effect usually dominates. Some crystals, such as certain formulations of lead lanthanum zirconate-titanate (PLZT), can have a quadratic EO effect high enough that when biased with reasonable field, the effect can rival the strength of the linear EO effect.

A quadratic EO effect also accompanies electroabsorption. This will be described below.

#### 4.2.1.3 Electroabsorption

In the EA effect, an applied electric field changes the energy difference between the conduction and valence bands in a semiconductor. This changes the absorption of light for photon energies (wavelengths) near the band gap energy [11]. This occurs in bulk semiconductors including both silicon and the III-Vs, but it is strongest in quantum wells in III-V semiconductors.

The EA effect modulates the intensity of light directly. The relation of absorption to electric field is nonlinear and has no simple, general analytic representation.

Whenever there is an absorption change, there is also a refractive index change through the Kramers–Kroenig relations [12]. The refractive index change from this source has a quadratic dependence on the applied electric field. This results in optical phase modulation along with the optical intensity modulation.

The EA effect is strongly dependent on the wavelength of light, as well as on the temperature.

#### 4.2.1.4 Carrier Density Effects

There are several types of effect that relate the density of carriers in a semiconductor to the optical refractive index and absorption [13]: band filling, bandgap shrinkage, and free-carrier absorption (plasma effect). In all of these effects, a change in carrier density affects absorption, which in turn causes an index change through the Kramers–Kroenig relations.

The key characteristic of these effects is that they achieve modulation by varying the density of carriers. This can involve significant speed limitations due to carrier lifetime, which is typically in the 1 ns–10 ns range for pure silicon and III-V semiconductors. This speed limitation can be overcome by shortening the lifetime via material damage or by operating in depletion mode where carriers are moved instead of recombined, but these techniques lead to an increase in switching voltage [14].

Carrier effects are primarily of interest in silicon, where the faster electro-optic effects are absent or very weak.

## 4.2.2 Modulator Designs

The linear and quadratic EO effects produce phase modulation of the light. This can be used directly, but it requires a coherent optical receiver with an optical local oscillator, which is much more complicated and expensive than detection of intensity-modulated light. Therefore, several modulator designs have been developed to convert phase modulation into intensity modulation. For the linear EO effect, the design of the modulator is the primary factor determining the modulator's transfer function (dependence of transmission on voltage). For other EO effects, the phase versus voltage relationship and the modulator design together determine the transfer function.

### 4.2.2.1 Electroabsorption Modulator

The EAM works by modulating the intensity of light directly, so it is the one exception to the introductory comments above that will be discussed.

The EAM has a very simple design: it is a single waveguide with electrodes that provide the modulating electric field across the waveguide. A schematic version is shown in Figure 4.1a; Figure 4.1b shows the transfer function. It is monotonic, with absorption increasing with voltage. The EAM will be discussed in detail in Chapter 5.

Practical EA modulators are built in III-V material. These are very compact devices, with typical lengths of just a few hundred microns. They can achieve >10 dB absorption change with <2 V drive voltage, while achieving a bandwidth of >40 GHz [15]. These features are highly desirable in most applications.

Because of the index change that accompanies the absorption change in the EA effect, the EA modulator causes chirp. Another drawback of the EA modulator is that it absorbs light. One impact of this is that its optical power handling ability is limited due to heating (this limitation can be partly overcome by some designs [16]). Another impact is the generation of modulated photocurrent, which limits the link gain and noise figure in analog applications [17].

### 4.2.2.2 Mach-Zehnder Interferometric Modulator

Perhaps the most successful modulator design that converts phase to intensity modulation is the MZ modulator, shown in Figure 4.2 [18,19]. In this design, the input light is split into two arms of an interferometer. Through the EO effect, the refractive index of one arm is changed relative to the other, which changes the relative phase of the light in the two arms. At the output, the light from the two arms recombines. In the 1 × 1 design, the light is transmitted when the two arms are in phase and radiated away when they are out of phase [20]. In the 1 × 2 design, there is an optical 3-dB coupler at the output so the light is switched to one waveguide or another [21]. The transfer function is periodic, as shown in Figure 4.2b.

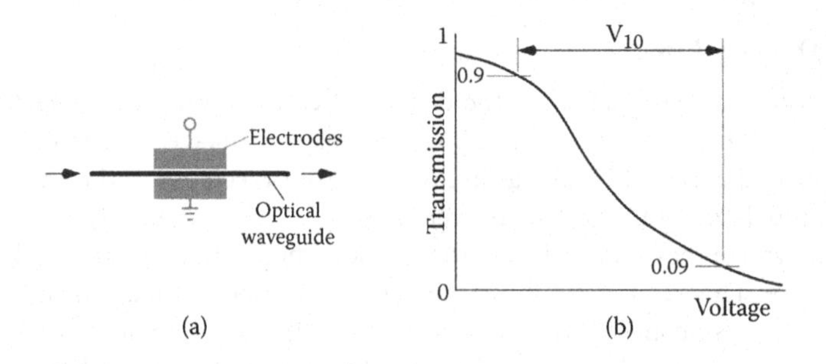


**FIGURE 4.1** Electroabsorption modulator. (a) Schematic view. (b) Transfer function.

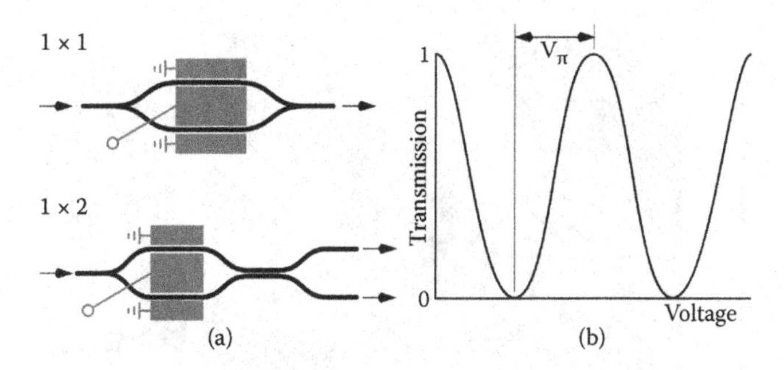


**FIGURE 4.2** Mach-Zehnder interferometric modulator. (a) Schematic view of 1 × 1 and 1 × 2 versions. (b) Transfer function of the 1 × 1 (or of a single output of the 1 × 2).

If the phase modulators in the interferometer arms are based on the linear EO effect and the optical power splits are ideal, the power transfer function is the simple cosine function

$$T = \frac{1}{2}\left[1 + \cos(aV)\right], \tag{4.2}$$

where $T$ is the optical transmission, $a$ is a constant and $V$ is the voltage applied to one arm of the modulator. This is the case, to a very good approximation, in $LiNbO_3$ and EO polymers. If the material has a quadratic EO effect, or if there is mixed EA and EO effects as in the III-V semiconductors, the transfer function is more complex and the simple cosine function can be severely distorted [22].

The output signal from a symmetric MZ modulator can be written as

$$\begin{aligned} E(t) &= 0.5E_0\left[\exp\left(j\Delta\beta(V_1(t))L\right) + \exp\left(j\Delta\beta(V_2(t))\right)\right] \\ &= E_0 \cos\left(\frac{\Delta\beta(V_2(t)) - \Delta\beta(V_1(t))}{2}L\right)\exp\left(j\left(\Delta\beta(V_2(t)) + \Delta\beta(V_1(t))\right)L\right), \end{aligned} \tag{4.3}$$

where $\Delta\beta(V)$ is the change in the phase constant due to the applied voltage, $V_1$ and $V_2$ are the voltages applied to arms 1 and 2, respectively, and $L$ is the interaction length of modulator arm. For a $LiNbO_3$ MZ modulator, the change in phase is a linear function of the applied voltage. In this case, Equation 4.3 can be written in the low-pass equivalent form as

$$\begin{aligned} E(t) &= A(t)\exp[j\phi(t)] \\ &= \frac{E_0}{\sqrt{2}}\left[1 + \cos\left(\frac{\pi}{V_\pi}\left(V_2(t) - V_1(t)\right)\right)\right]^{1/2} \exp\left[j\frac{\pi}{2V_\pi}\left(V_2(t) + V_1(t)\right)\right]. \end{aligned} \tag{4.4}$$

$V_\pi$ is the voltage required to switch the modulator between the off-state and on-state. It corresponds to the voltage required to obtain a phase change of $\pi$ between the two optical signals at the input to the 2 × 1 combiner. Typical values are in the range of 3 V–5 V for a $LiNbO_3$ MZ modulator. Over a range of wavelengths such as the C-band (1528.77 nm–1568.36 nm), $V_\pi$ does not depend appreciably on the wavelength of the input CW optical signal.

By varying the voltages applied to each arm of an MZ modulator, a contour plot for the normalized output power can be obtained. This is shown in Figure 4.3 for an MZ modulator with the bias voltages

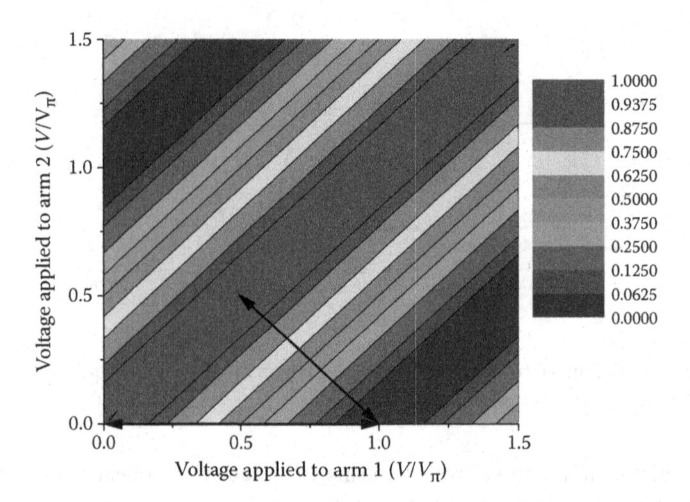


**FIGURE 4.3** Contour plot for the normalized optical power for a symmetric dual-drive Mach-Zehnder modulator as a function of the voltages applied to the arms of the modulator.

normalized by $V_\pi$. The arrowed straight lines illustrate two practical operating conditions for single-drive and dual-drive (push-pull) modulation. Depending on the design and use of an MZ modulator, the output signal may exhibit phase modulation (chirp) due to the applied voltage(s), as seen from Equation 4.4.

When the MZ modulator is built in a material with a well-developed, low-loss optical waveguide technology such as $LiNbO_3$, it has several useful features. Its switching voltage $V_\pi$ can be reduced by increasing its length. The chirp can be controlled by the distribution of the voltage-induced index change between the arms: for equal and opposite index changes ("push-pull," $V_2(t) = -V_1(t)$), the chirp is zero; for index change predominantly in one arm, the chirp can have a controlled positive or negative value. The optical extinction can be high, >25 dB. The optical loss can be low, since no absorption is involved in the modulation. The optical bandwidth is very wide, >40 nm.

The primary disadvantage of the MZ modulator is the long length often required for low switching voltage. For example, an EAM with 2 V switching voltage might be 300 μm long, while a $LiNbO_3$ MZ modulator with the same bandwidth and switching voltage is typically 50 mm long.

#### 4.2.2.3 Compound Modulators

Compound modulators refer to the series, parallel, or more complex connection of multiple modulators. The most useful of these compound modulators has been the connection of multiple MZ modulators, but combinations of the MZ modulator with other modulators have been made as well. The advantage of these compound modulators is that they perform some function better than a single modulator or they enable a function unavailable from a single modulator. The disadvantage is that they are more difficult to build and control.

One design is the dual-parallel (or quadrature) MZ modulator, which is comprised of two conventional zero-chirp MZ modulators nested within an MZ interferometer [23]. This can be used to produce single-sideband suppressed-carrier modulation [24]. The same arrangement can also be used to generate quadrature phase shift keying and differential quadrature phase shift keying [25] (also see Chapters 2 and 6). One modulator generates the in-phase component of the signal and the other modulator, followed by an optical phase shift of $\pi/2$, generates the quadrature component. Higher level modulation formats such as quadrature amplitude modulation can be achieved by cascading an MZ modulator and a phase modulator [26].

One major reason for a host of compound modulator designs has been linearization of modulation for analog communication [27]. Perhaps the most successful of these is the series connection of two MZ modulators, which has been used commercially for cable TV transmission. Linearized modulators can offer a substantial improvement in linear dynamic range over that of a single modulator, but they do require complex and precise control circuits.

Compound modulators account for a very small fraction of the total number of modulators produced so far, but they continue to be developed and used for a variety of special purposes.

#### 4.2.2.4 Resonant Modulators

Several modulator designs are derived from optical filters. The basic idea is that a small voltage-induced index change shifts the frequency of the passband of the filter. If the input light is at a frequency near the edge of the filter's passband, the filter's frequency shift will result in a change in optical transmission.

One way of implementing this idea is to use a resonant optical cavity. This has been done by coupling an input/output waveguide to a waveguide ring [28], as diagrammed in Figure 4.4a. The electrodes change the refractive index of the ring waveguide, which changes the optical path length, which changes the resonant frequency. The transfer function of the ring modulator is shown in Figure 4.4b. Many other optically resonant designs have been implemented also, including Fabry-Perot resonators, resonant disks, and even more complex structures.

A second way of implementing the filter idea is to use a grating. The grating is an optical filter with a center frequency that depends on the grating period. The applied electric field changes the effective grating period by changing the refractive index of the material, which changes the transmission at a frequency near the edge of the grating's passband.

The big advantage of resonant modulators is that the rate of change of transmission with optical phase change can be increased by increasing the Q of the resonance. This can be seen by comparing the transfer functions of Figures 4.2b and 4.4b. Both are periodic, with period $2\pi$ in optical phase. But as the ring modulator's Q is increased, the peaks in Figure 4.4b become narrower, resulting in a steeper slope.

The drawback of increasing the Q is that the optical bandwidth is decreased, which limits the electrical modulation bandwidth. The exact relationship is complex, depending on the modulator type and other parameters such as the free spectral range, but the overall effect limits the resonant advantage achievable at large electrical signal bandwidths. Resonant modulators are much shorter than MZ modulators with the same switching voltage, but resonant modulators do not necessarily achieve a lower switching voltage than that of a long MZ modulator in the same material [29].

It is also possible to make the modulator electrodes into an electrically resonant structure, which increases the applied voltage for a given electrical input power [30,31]. Only a moderate advantage over nonresonant electrodes can be achieved this way due to the low Q at microwave frequency of electrical resonators with small dimensions.

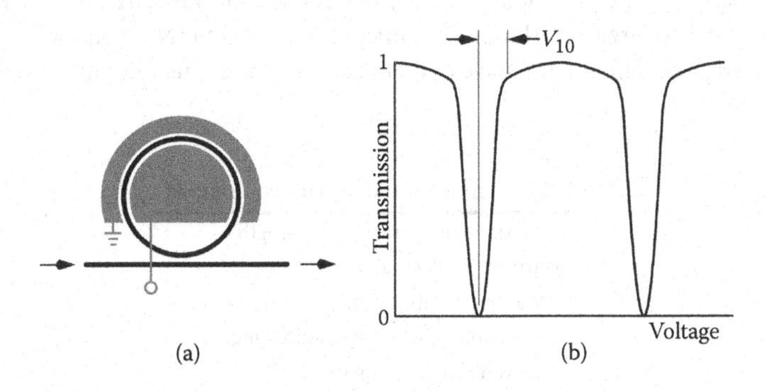


**FIGURE 4.4** Resonant ring modulator. (a) Schematic view. (b) Transfer function.

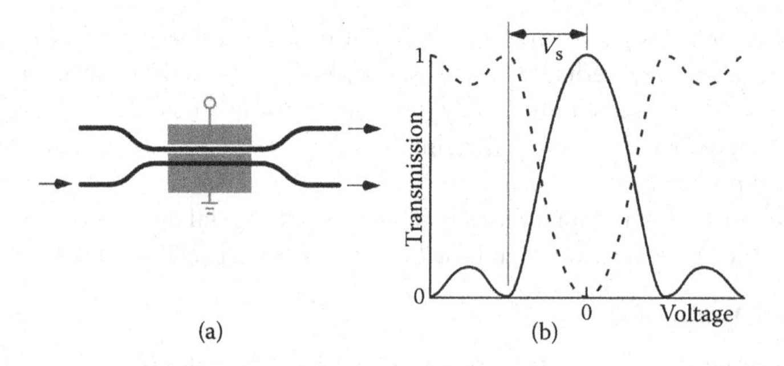


**FIGURE 4.5** Directional coupler modulator. (a) Schematic view (b) Transfer function solid line is the cross (upper) channel, dotted line is the bar (lower) channel.

#### 4.2.2.5 Directional Coupler and Other Designs

In addition to the modulator types discussed above, there have been many other waveguide electro-optic modulator designs developed [27,32]. This multiplicity of designs came about mostly in the early days of modulators, as people tried to find the best way to modulate light, but some of these designs reappear for special purposes. One of these devices, the directional coupler, deserves some discussion here.

The basic directional coupler modulator is shown in Figure 4.5. Two identical waveguides are spaced closely enough that their modes overlap. Optical power couples from one to the other and the amount of coupling can be controlled by the applied voltage. The transfer function is not periodic. A split electrode [33] is commonly used to compensate for fabrication problems.

The directional coupler does not make as good a high-speed modulator as the MZ modulator because it is difficult to scale up the length to reduce the switching voltage, and it also has a switching voltage $\sqrt{3}$ larger than the MZ modulator at the same length.

The directional coupler does, however, make a good, compact, high-speed switch. It has been used in demonstrations of large (8 × 8) integrated optic switch arrays [34]. It is also used as an actively controlled input and/or output coupler on the MZ modulator when precise control is needed.

### 4.2.3 Materials

In many ways, the modulator material determines the characteristics of the modulator. It determines which of the electro-optic effects are possible, what waveguide fabrication processes can be used, and the reliability and other characteristics of the modulator.

There are several characteristics a material must have in order to be useful for waveguide electro-optic modulators. These are listed in Table 4.2. Only the first two of these are basic material properties; the rest are technological. The reasons that III-V semiconductors and $LiNbO_3$ are widely used and other materials are regarded as developmental have to do mostly with these technological factors.

**TABLE 4.2** Modulator Material Requirements

| |
|---|
| Transparent at operating wavelength |
| Electro-optic effect (any from Table 1) |
| Waveguide fabrication method |
| Stable over operating temperature range |
| Available in adequate size |
| High quality (e.g., single-crystal) |

This section will give a brief introduction to the major classes of material and some specific materials that are used for making modulators.

#### 4.2.3.1 Inorganic Crystals

From a modulator perspective, there are two types of inorganic crystal: lithium niobate and everything else.

##### *4.2.3.1.1 Lithium Niobate*

$LiNbO_3$ is the material used for the mainstream commercial MZ modulator as well as many developmental modulator designs. It meets all the criteria in Table 4.2: it is transparent from ~400 to ~5000 nm, it has a strong linear EO effect, it has two good waveguide fabrication methods, its Curie temperature and melting point are both above 1100°C [35], and it is available as single-crystal, single-domain wafers (up to 125 mm in diameter at present). $LiNbO_3$ is also used for surface acoustic wave devices and for nonlinear optical applications, which has helped material quality and availability. It is not the material with the highest EO coefficient or any other particular standout property, but its suitability in all key requirements is what has made it a material of choice.

There are two waveguide fabrication technologies in lithium niobate, titanium indiffusion and annealed proton exchange. Both have low optical loss, as low as 0.1 dB/cm at 1550 nm. Ti-indiffused waveguides guide both polarizations (although they are birefringent), while annealed proton exchange waveguides only guide a single polarization. Ti-indiffused waveguides are more stable against long-term drift under DC bias. The single-polarization property made annealed proton exchange waveguides the preferred choice for cable TV transmission where elimination of the second polarization was required to meet the linearity specifications. The better stability of the Ti waveguides made them the preferred choice for digital fiber-optic communication [36].

Lithium niobate is not without its problems. It is a ferroelectric so it has a built-in electric dipole moment that is temperature dependent, which causes modulator bias point shifts proportional to the rate of change of temperature [37]. It is mildly photorefractive and photoconductive, so high optical power can also cause bias shifts; this is not a significant problem at 1550 nm [38], but can become severe at wavelengths <1000 nm [39]. It is piezoelectric so modulator electrodes launch sound waves. This causes changes in the electrode impedance with frequency due to the acoustic radiation impedance and due to acoustic resonances in the substrate [40]. It is acousto-optic so these sound waves modulate the light as well. The acoustic effects are primarily a problem below 1 GHz and in analog applications where even small ripples in frequency response can be a problem. All of these problems have been addressed with technological fixes, but one must be aware they exist because not every modulator is built with everything fixed.

##### *4.2.3.1.2 Other Inorganic Crystals*

There are a variety of other inorganic crystals with strong EO effects. Some such as lithium tantalate ($LiTaO_3$) and potassium titanyl phosphate (KTP) are very roughly similar to $LiNbO_3$ in their general properties. These crystals are used sometimes due to specific advantages, such as higher optical power handling. These are not far behind $LiNbO_3$ in availability and several of the other technological criteria.

There are several crystals that have been investigated due to their very high EO coefficients, such as strontium barium niobate [41], lead lanthanum zirconate-titanate (PLZT), and barium titanate. Barium titanate is a good example of this type of material. It has $n^3r \approx 18000$ pm/V, much higher than $LiNbO_3$'s 306 pm/V, which is very advantageous in making modulators with low switching voltage [42]. But it has several drawbacks: its dielectric constant drops with frequency and is ~1000 at 20 GHz, which makes broadband modulator design difficult; its Curie temperature is only 120°C, which causes severe temperature dependencies over normal device operating ranges; and it is difficult at present to get large single-crystal, single-domain wafers. So while barium titanate is very attractive for its high EO effect, its other physical properties and its present poor crystal quality make it impractical to use.

### 4.2.3.2 Semiconductors

Semiconductors offer a much richer variety of physical effects than insulating materials. In addition to the linear and quadratic EO effects, they have EA and carrier-density effects that modulate light. They offer the possibility of integrating active electronic components such as transistors with electro-optic devices. Direct-gap semiconductors also offer the possibility of integration with lasers. There are two classes of semiconductor that are of primary relevance to modulators: silicon and III-V semiconductors.

#### *4.2.3.2.1 III-V Semiconductors*

The III-V semiconductors are perhaps the most important electro-optic materials in existence. Lasers, modulators, detectors, and transistors are all made in these materials. However, integration of more than one of these functions on the same substrate involves fabrication difficulties and performance compromises. It has been proven economically viable to integrate a laser and modulator; the cost and size advantages more than offset the reduced performance compared to discrete components. Other integrations have been proven feasible in the laboratory but have not yet become viable in the market. The large III-V electronic and optical industries have made III-V material technology mature and widely available.

III-V semiconductors are compounds of group III and group V elements in the periodic table. Binaries such as gallium arsenide and indium phosphide are available as wafers, and ternaries such as indium gallium arsenide and quaternaries like indium gallium arsenide phosphide are grown as epitaxial layers on the wafer surfaces. Waveguides are generally made by using the layer structure for vertical confinement of the mode, and etching a ridge to provide horizontal confinement. For wavelengths far from the band edge, where losses are dominated by scattering instead of absorption, losses can be <1 dB/cm.

The EA modulator is the most commonly built modulator in III-V semiconductors. Several other modulator designs have also been built that are based on EO effects, most notably the MZ modulator. The MZ modulator may operate far (>150 nm) from the band edge and use primarily the material's linear EO effect [43], or it may operate closer to the band edge and use a combination of this and the quadratic EO effect induced by the EA effect [44]. The induced quadratic EO effect is stronger than the material linear EO effect, but it is accompanied by absorption, which can lead to photocurrent, high losses, and a nonperiodic transfer function [22,44].

#### *4.2.3.2.2 Silicon*

Silicon is the basis of a vast electronics industry, but it lacks the strong high-speed electro-optic effects present in other materials. It would not be a good choice as a modulator material except for the possibility of integration with silicon circuitry and fabrication in silicon foundries. If components such as lasers, modulators, and detectors could be successfully made in silicon using fabrication processes compatible with the existing large-scale processes used for electronic integrated circuit fabrication, these electro-optic devices and circuits based on them could be integrated with electronic circuits and potentially be made for a small fraction of their present cost [9].

The idea of using silicon for integrated-optical circuits has existed for many years [13], but it became much more interesting within the last decade when lithography advanced to the point where $Si/SiO_2$ waveguides with dimensions <1 μm could be made with losses <1 dB/cm [45]. These waveguides have a high index difference ($n = 3.5$ for silicon and 1.45 for $SiO_2$) so they can be just a few hundred nm wide and they can be bent very sharply, with curve radii of <10 μm. This enables complex, compact integrated optical circuits such as multipole filters. One drawback of these extremely small waveguides is that their power handling capability is limited: two-photon absorption can cause problems with just several mW at 1550 nm [46].

There is no linear EO effect in silicon due to its crystal symmetry. The quadratic EO effect and the EA effect are both very weak. So silicon modulators have had to use carrier density effects to achieve modulation. These effects and their problems were discussed in Subsection 4.2.1.4. However, the very small dimensions of Si waveguides can offset these drawbacks by producing very high electric fields at low voltages, by enabling resonant modulators (e.g., a ring cavity) with practical bandwidths [47], and by enabling new modulator structures [48].

#### 4.2.3.3 Polymers

Electro-optic polymers are insulating materials that use the linear EO effect for optical modulation. These are usually prepared in liquid form and then spun on to a substrate such as silicon or glass and then cured to form thin films. Optical waveguides are usually made by fabricating layers of different material for vertical confinement and then using a ridge or channel for horizontal confinement. Waveguides are moderately low loss, 1 dB/cm to 3 dB/cm at 1550 nm. Waveguide losses are both from absorption by the polymer and from scattering.

Polymers offer several advantages over other materials. They are spin-on films so they can be applied to any substrate material to form an active optical layer. They provide a variety of materials for cladding and waveguide layers so waveguides with moderate waveguide-substrate index difference (similar in magnitude to III-V waveguides, less than silicon but more than typical in inorganic crystals) can be made, enabling tighter bends and smaller mode sizes than in inorganic crystals such as $LiNbO_3$. Polymers are a whole class of materials, allowing chemical engineers to optimize properties like the electro-optic coefficient; values of $n^3r$ as high as 1600 pm/V (five times higher than $LiNbO_3$) have been reported [49].

There are some drawbacks to polymers as well. A primary issue is thermal stability [49,50]. There is a tradeoff between high EO coefficient and high thermal stability, so the most impressive EO effects have been in materials that had limited lifetimes for operation above room temperature. Polymer technology is constantly improving, though, so the EO coefficient that can be achieved at a given level of stability is improving. Polymers must be poled after layer fabrication, and the EO coefficient achievable in devices falls short of the EO coefficient achievable in bulk material [50]. Optical power handling has been an issue in the past, but by sealing the modulator in an oxygen-free environment, it can handle power levels >100 mW at 1550 nm [51].

The potential advantages of polymers have kept research on them active for over 20 years. Their drawbacks have prevented them from producing practical commercial modulators in the past, but present polymer technology is on the verge of changing this.

### 4.2.4 Modulator Performance

So far, this overview of modulators has been almost entirely qualitative. This section will give a few quantitative comparisons of materials and devices.

Before moving on to the performance comparisons, a discussion of modulator electrical design is appropriate. This will show how a number of different material and waveguide properties combine with engineering design to determine switching voltage and bandwidth, two of the most important modulator performance criteria.

#### 4.2.4.1 Electrical Design

The MZ modulator using the linear EO effect is a good example to show how the switching voltage depends on material and design parameters. In this case, the switching voltage $V_\pi$ is

$$V_\pi = \frac{\lambda_o}{n^3 r}\frac{1}{L}\frac{\iint dx\,dy\,I(x,y)}{\iint dx\,dy\left(\frac{E(x,y)}{V}\right)I(x,y)}, \tag{4.5}$$

where $\lambda_o$ is the vacuum wavelength of the modulated light, $L$ is the length of the modulator electrodes, $I(x,y)$ is the intensity distribution of the optical waveguide mode, and $E(x,y)/V$ is the electric field per volt applied to the electrodes (e.g., if the electrodes were ideal parallel plates, $E(x,y)/V$ would equal $1/d$ where $d$ is the distance between the plates) [27]. The material parameter $n^3r$ and the modulator length are equally important. A material with a modest $n^3r$ value can still yield modulators with low switching voltage if the length can be made long enough (this is what is responsible for the success of the $LiNbO_3$ MZ modulator). The integral factor is reduced by a small electrode gap (more field per volt) and by placement of the waveguide mode in a high-field region, both of which are key design factors.

One figure of merit is the *VL product*, which is the product of switching voltage and electrode length. Equation 4.5 shows this to be a measure of $n^3r$, the integral factor, and the wavelength.

Other types of modulators have more complex descriptions of the switching voltage. For all, though, small cross-sectional dimensions and good confinement of the optical mode to the high-field regions are important. The EA modulator has a significant difference from Equation 4.5: its switching voltage cannot be arbitrarily reduced by increasing the length [52].

The modulator bandwidth is determined by the electrical properties of the modulator electrodes. There are two general types of electrode design: lumped element and traveling wave.

In the lumped-element design, the electrodes are driven as a capacitor. This is the design most often used with the EA modulator. The frequency-dependent switching voltage $V_s(\omega)$ is

$$V_s(\omega) = V_{s,dc}(1 + j\omega R_p C), \tag{4.6}$$

where $V_{s,dc}$ is the switching voltage at DC, $C$ is the capacitance, and $R_p$ is the parallel resistance (including the input source impedance). The 3-dB bandwidth is $1/(2\pi R_p C)$. There is a tradeoff between bandwidth and switching voltage: shrinking the electrode gap or increasing the electrode length increases the capacitance.

In the traveling-wave electrode structure, the electrodes form a transmission line so the electrical signal and the optical signal travel together along the device. This is the design used on almost all high-speed $LiNbO_3$, polymer, and III-V MZ modulators.

The electrical velocity must be matched to the optical velocity to achieve long lengths at high bandwidth; the modulator will work with mismatched velocities but the bandwidth will be reduced. The optical velocity is set by the index of refraction. The electrical velocity can be varied by design (see Chapter 6). When the modulator material has a high dielectric constant, a high electrical velocity can still be achieved by keeping a large fraction of the electric field in the air or other low-dielectric-constant materials instead of the modulator material. There is a tradeoff with switching voltage in that these velocity-matching designs often result in weakening the electric field in the optical waveguide. In general, materials with high dielectric constant make this velocity-matching more difficult, but the relationship is not a simple one.

When the electrical and optical velocities are matched and there is a lossless impedance match to the source, the frequency-dependent switching voltage, as measured at the source impedance, is

$$V_s(\omega) = V_{s,dc}\sqrt{\frac{Z_i}{Z_o}}\,\frac{\left[\dfrac{\alpha_p(\omega)L}{2}\right]}{1-\exp\left[-\dfrac{\alpha_p(\omega)L}{2}\right]}, \tag{4.7}$$

where $Z_i$ is the input system impedance, $Z_o$ is the modulator transmission line impedance, $\alpha_p(\omega)$ is the frequency-dependent electrical power loss per unit length on the electrode, and $V_{s,dc}$ is the switching voltage at DC as measured directly on the electrodes (this is given by Equation 4.5 for a linear EO material). Equation 4.7 shows that the bandwidth is determined by the electrode loss. This results in a tradeoff with switching voltage, since large cross-sectional dimensions give low electrode loss but also give higher switching voltage. Note also that the switching voltage is increased if the modulator has low impedance relative to the system, a problem that can occur with high-dielectric-constant modulator materials.

#### 4.2.4.2 Relative Performance

Two of the most important modulator performance measures are switching voltage and bandwidth. These must be considered together because there is a tradeoff between them, as was discussed in the previous subsection. The performance versus frequency of several modulators is shown in Figure 4.6. Most of these are the best published performance available for each type of modulator at the time of this writing, for example, the widest bandwidth or the lowest switching voltage in a particular frequency range.

Figure 4.6 includes a variety of modulators with different designs and transfer functions, so some conversions were used. For modulators that were not MZ modulators, an effective $V_\pi$ was calculated from the transfer function using Equation 4.8. For MZ modulators with $V_\pi$ measured at wavelengths shorter than 1550 nm, the effective $V_\pi$ was scaled linearly with wavelength (see Equation 4.5). For MZ modulators with separate electrodes on the two arms, the $V_\pi$ of one arm was divided by $\sqrt{2}$. For modulators with impedance other than 50 ohms, the $V_\pi$ from 50 ohms with an ideal impedance match was calculated. These latter two adjustments make the effective $V_\pi$ represent the total radio frequency power required to drive the modulator.

MZ modulators built in $LiNbO_3$, polymer, and III-V materials all have approximately comparable performance. The one with significantly lower voltage at high frequency is a III-V MZ modulator [44] operated very close to the bandgap so the lower voltage comes at the expense of nonlinearity and optical absorption. All the MZ modulators have a tradeoff between $V_\pi$ and bandwidth; this is most obvious for the $LiNbO_3$ MZ modulators simply because more devices with different optimizations have been reported. The EAM [15] has only a weak tradeoff between bandwidth and voltage (which is why only one is plotted). It is comparable to a $LiNbO_3$ modulator with ~40 GHz bandwidth.

Silicon modulators show a substantial difference from the other modulator materials because they use the slow carrier-density effects. The injection-mode MZ modulator [14] has extremely low voltage but this advantage disappears by ~1 GHz. A depletion-mode MZ modulator [14] has much higher bandwidth, but with the penalty of a very high $V_\pi$. A depletion-mode resonant ring [47] trades off some bandwidth for improved effective $V_\pi$.

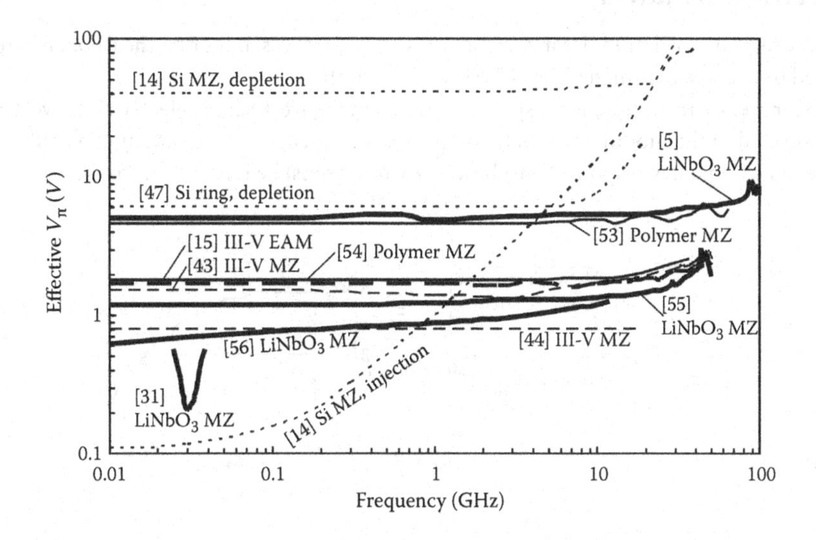


**FIGURE 4.6** Comparison of performance of several state-of-the-art modulators.

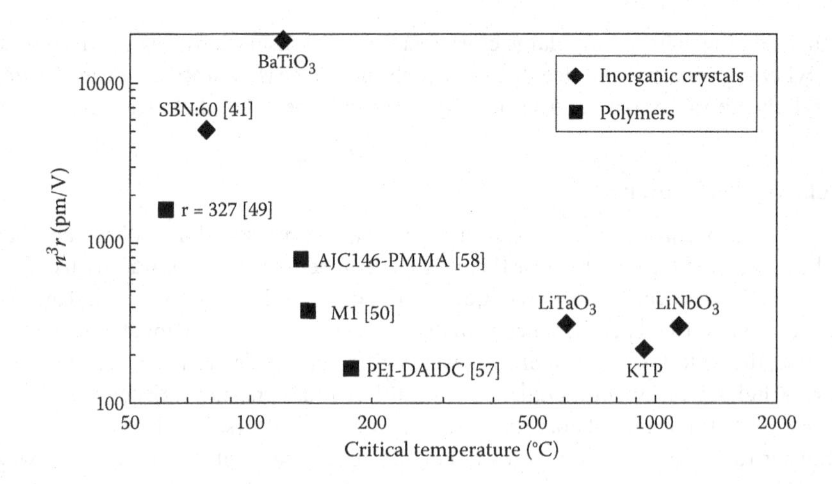


**FIGURE 4.7** Electro-optic strength versus critical temperature (Curie temperature for inorganic crystals, glass transition temperature for polymers) for several electro-optic materials.

For polymer and inorganic crystal materials, there is a rough tradeoff between thermal stability and $n^3r$. Figure 4.7 shows the material critical temperature (glass transition for polymers and Curie temperature for inorganics) versus $n^3r$. The relation of reliability to these temperatures is complex [50], but these give some indication of the tradeoff between temperature stability and electro-optic coefficient.

The performance shown in Figures 4.6 and 4.7 represents the present state of the art, and improvements can be expected. The relationships are not fundamental laws, but they do represent empirical realities that are likely to hold in the future.

## 4.3 Performance Criteria and Measurements

This section will define several common modulator measurements and performance criteria. Some are common to all applications and some are specific to only analog or digital applications.

### 4.3.1 Electrical Bandwidth

The frequency response of a modulator can contain several features, but often the frequency response is characterized by a single parameter, the electrical bandwidth.

The setup for measuring frequency response is shown in Figure 4.8a. An electrical network analyzer's port 1 is connected to the modulator input, and port 2 is connected to the output of a calibrated detector. The detector's frequency response must be known and it must be removed from the total response to

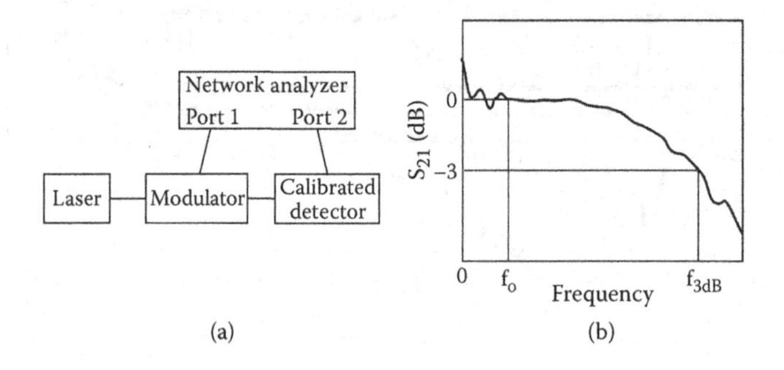


**FIGURE 4.8** Measurement of modulator electrical bandwidth. (a) Setup. (b) Frequency response plot.

get the modulator's frequency response. Some models of network analyzer include calibrated detectors to make this measurement easy. Note that the measurement is of the electrical power response, which is proportional to the square of the optical modulation depth or alternatively, to the inverse square of the switching voltage.

The definition of the electrical bandwidth is fairly obvious: the frequency where the response has dropped 3 dB from its low-frequency value (or, for bandpass modulators sometimes used in analog applications, where the response has dropped 3 dB from its peak value).

A second response specification is the ripple, which refers to the peak-to-peak variation in frequency response from the mean, over a bandwidth much smaller than the 3-dB bandwidth. This specification is important primarily in analog applications.

In practice, the frequency response is often more complicated than a simple rolloff. Figure 4.8b shows an example of a measurement of a $LiNbO_3$ MZ modulator frequency response. There is a rapid rolloff of a couple of dB and some significant ripple at very low frequency. Often, the bandwidth is measured relative to a "low-frequency" point that is above these low frequency effects, typically, 1 GHz.

Low-frequency anomalies occur in $LiNbO_3$, other inorganic crystals, polymers, and perhaps other materials for one or more of the following reasons: (1) Resistive and capacitive voltage division differing across the layers in the modulator structure. This usually affects frequencies from sub-Hz to ~1 MHz [59]. (2) Launching of acoustic waves, in $LiNbO_3$ and some other inorganics. This is a problem up to ~1 GHz and usually causes shallow (<1 dB) dips at frequencies where the electrode structure is a good transducer. (3) Acoustic resonances in the substrate, in $LiNbO_3$ and some other inorganics [40]. These are very sharp, often <100 kHz wide, can be >1 dB magnitude, and are most severe at frequencies where the acoustic launch is most efficient. (4) Basic crystal properties changing from stress-free ("unclamped") to strain-free ("clamped") as the frequency increases. For $LiNbO_3$, this occurs at <100 MHz [35] and causes a clear signature and 1-dB response drop in modulators on z-cut substrates [60]. If these problems are suspected to exist, measurements with adequate resolution and frequency span must be made in order to find them.

Finally, the group delay (or the phase response) can be measured with the setup in Figure 4.8a. For some modulators, the magnitude response alone may be misleading and the group delay measurement is necessary to verify that it is constant over the desired bandwidth [61]. Equivalent circuit models and detailed device simulators have been used to investigate the frequency response of optical modulators [62–65].

### 4.3.2 Switching Voltage

The switching voltage is the voltage required to switch the modulator from on to off. Although this seems like a clear definition, there are some variations.

When a modulator has a periodic transfer function, there is an unambiguous definition possible, which is the half-period of the transfer function. An example of this is the MZ modulator, whose transfer function is shown in Figure 4.2b. The switching voltage is called $V_\pi$, since a relative phase change of $\pi$ between interferometer arms changes the output from maximum to minimum.

When a modulator's transfer function is not periodic, the switching voltage can be defined as the voltage required to switch from an arbitrarily defined high transmission to some specific attenuation relative to the first state. This is shown in Figure 4.1b for the EAM, where the voltage "$V_{10}$" denotes the voltage required to switch from "on" to 10-dB extinction. There are obviously many choices of specific definition in this case.

In analog applications, the maximum slope of the transfer function is the key parameter. This can be expressed as an "equivalent $V_\pi$" ($V_{\pi e}$) [52]

$$V_{\pi e} = \frac{\pi}{2}\left|\frac{dT}{dV}\right|_{\max}^{-1}, \tag{4.8}$$

where $dT/dV$ is the slope of the transfer function at the point of maximum slope. This allows easy comparison between any modulator type and the widely used MZ modulator.

The switching voltage is most easily measured by applying a voltage ramp to the modulator electrodes, detecting the output with a photodetector connected to an oscilloscope, then measuring the voltage difference between the relevant points of the transfer function. This is generally done at low frequency, and for reasons discussed in Subsection 4.3.1, it may not represent the switching voltage at high frequency. The switching voltage must be measured at a frequency within the range of a network analyzer sweep to calibrate the relative response given by the network analyzer to an actual switching voltage.

In digital systems, a characterization related to the switching voltage is the "drive voltage." This is the peak-to-peak voltage of a pseudorandom bit sequence required to achieve adequate electrical extinction in an eye pattern (see Subsection 4.3.8). This drive voltage may be greater or less than the modulator switching voltage.

The switching voltage is important in all applications because it determines the amount of signal amplification required to drive the modulator (which often affects the cost). For unamplified analog applications, switching voltage is critically important because it is a key determinant of the optical link noise figure [4].

### 4.3.3 Optical Loss

The optical loss is the output optical power divided by the input optical power, when the modulator is biased for maximum transmission (sometimes this is referred to as excess optical loss). Since most modulators are used in a single-mode fiber environment, this measurement is usually done by measuring the optical power in the input and output fibers.

Optical loss is important for all applications, but it is important mainly as a cost factor or power efficiency factor, not as a performance factor. System performance is generally determined by modulator output power, so a modulator with higher loss requires a more powerful laser to achieve the same performance as a lower-loss modulator.

### 4.3.4 Optical Bandwidth

Optical bandwidth refers to the range of wavelengths over which a modulator can maintain its performance.

Measurement of optical bandwidth for a modulator is much more complicated than for a filter: it is not simply the optical transmission versus wavelength. The switching voltage, optical extinction, optical loss, chirp, and linearity (or the subset of these that is important to the application) must each be measured over wavelength to see where they degrade to an unacceptable level. In the special case where multiple wavelengths will be used simultaneously, the measurement must be made with at least the wavelength extremes simultaneously present (or other steps must be taken to avoid reoptimizing the modulator at different measurement wavelengths).

Optical bandwidth is only important in systems that use multiple wavelengths. Only in very few cases does this mean actually running multiple wavelengths through the modulator simultaneously. Normally, it is a requirement because it is desirable for a single standard modulator to be able to be used on any of the wavelength channels in a system.

### 4.3.5 Chirp

Chirp refers to the ratio of frequency modulation to intensity modulation that is produced by a laser or modulator. Chirp is a key parameter for any application, analog or digital, involving long-distance fiber transmission. Chirp is characterized by the parameter $\alpha$, where $\alpha$ is defined by [66,67]

$$\alpha = \frac{d\phi}{dt}\frac{2I(t)}{\frac{dI(t)}{dt}} = \frac{4\pi\Delta f(t)I(t)}{\frac{dI(t)}{dt}}. \tag{4.9}$$

Here $I(t)$ is the optical intensity, $dI(t)/dt$ is the change in intensity with time, $d\phi/dt$ is the change in optical phase with time, and $\Delta f(t)$ is the change in carrier frequency. As can be seen from Equation 4.4, chirp only occurs during the rising and falling edges of pulse when the time derivatives of the applied voltage are nonzero. The term *transient chirp* is used to distinguish it from *adiabatic chirp*, both of which occur for a directly modulated laser. In the case of a directly modulated laser, $\alpha$ can also be defined in terms of certain material parameters [68].

Figure 4.9a through c illustrate schematically the optical field for a chirp-free optical pulse, a positively chirped pulse for which the carrier frequency increases during the rising edge and decreases during the falling edge, and a negatively chirped pulse for which the carrier frequency decreases during the rising edge and increases during the falling edge. The optical pulse is not drawn to scale. For an optical carrier with a frequency of 193.1 THz (wavelength of 1548.42 nm), the period of the carrier

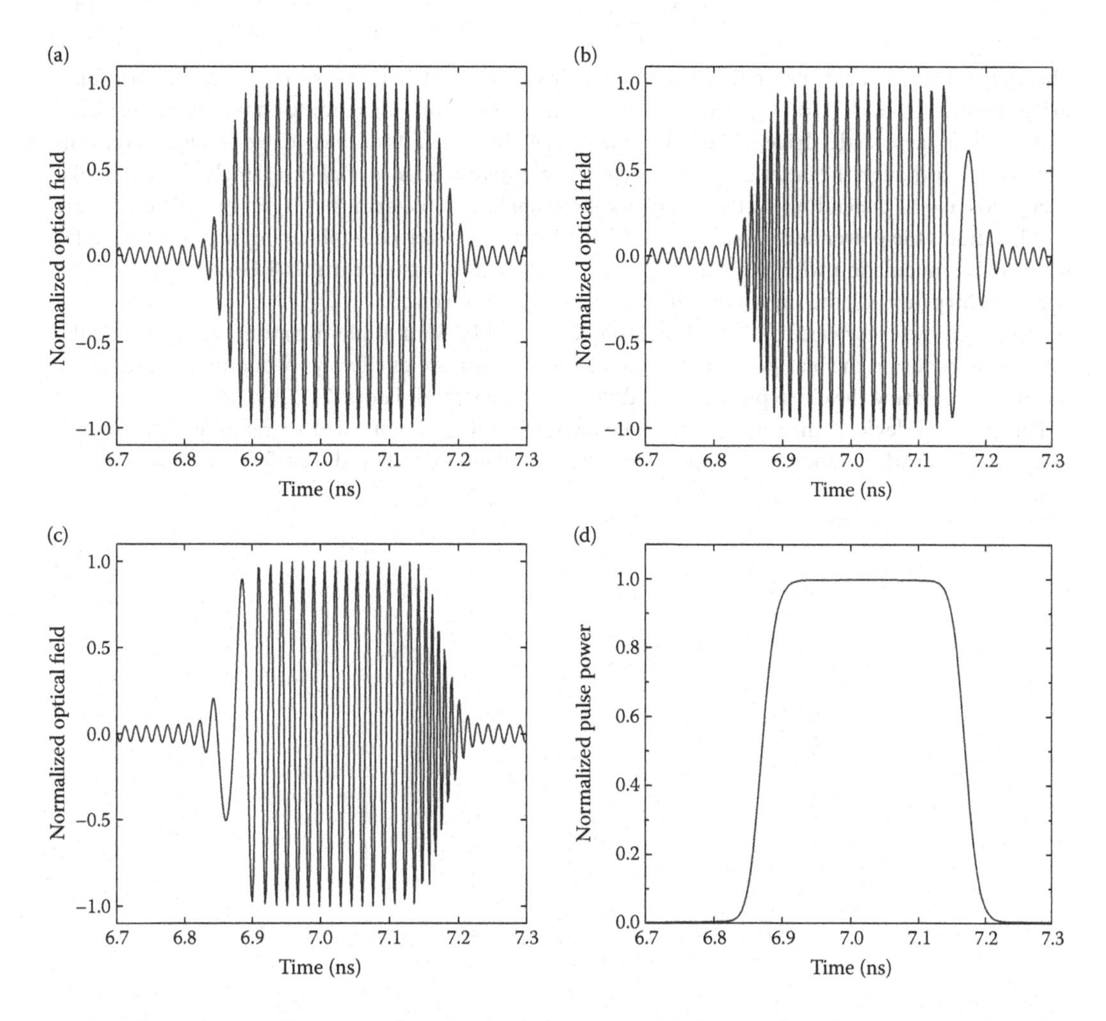


**FIGURE 4.9** Schematic illustration of an optical pulse. (a) Optical field for zero chirp. (b) Optical field for positive chirp. (c) Optical field for negative chirp. (d) Corresponding optical power.

is 5.18 fs. There would be 57,930 carrier periods in the pulse duration of 300 ps. Figure 4.9d shows the corresponding optical power.

The α-parameter depends on the bias point of the modulator and is often specified at the half-power point. For an ideal symmetric MZ modulator, the α-parameter at the half-power point takes a particularly simple form:

$$\alpha = \left[\frac{\Delta V_2 + \Delta V_1}{\Delta V_2 - \Delta V_1}\right] \tag{4.10}$$

for the operating region illustrated in Figure 4.3. The simplicity of this result is due to the fact that the change in the phase of the signal in an arm of the modulator is a linear function of the applied voltage. An advantage of the dual-drive MZ modulator is that the chirp can be adjusted simply by varying the drive voltage signals [69,70].

For a dual drive MZ modulator, the chirp from Equation 4.4 is

$$\Delta\nu(t) = \frac{1}{4V_\pi}\left[\frac{dV_2(t)}{dt} + \frac{dV_1(t)}{dt}\right]. \tag{4.11}$$

The chirp is zero for ideal push-pull modulation with $V_1(t) = -V_2(t)$. In this case $\alpha = 0$. A symmetric dual-drive modulator with $V_1(t) \neq -V_2(t)$ exhibits positive or negative chirp depending on the applied voltages. Out-of-phase (push-pull) drive voltage waveforms yield $|\alpha| < 1$ and in-phase drive voltage waveforms yield $|\alpha| > 1$, although the latter case is not of practical interest. As a special case, single-drive modulation yields $\alpha = \pm 1$ depending on the DC bias voltage applied to the arm that is not modulated. Figure 4.10 illustrates the power and chirp for a 10 Gb/s nonreturn-to-zero on–off keying signal obtained for single-drive modulation with positive chirp ($\alpha = 1$). The peak-to-peak chirp is 8 GHz. For the case of negative chirp ($\alpha = -1$), the sign of the chirp in Figure 4.10 is reversed.

The α-parameter for electro-absorption modulators and III-V multiple-quantum-well MZ modulators also depends on the bias point of the modulator. The concept of effective α-parameters provides a means of comparing the chirp properties of different types of modulators [71,72].

For practical devices, the change in frequency during modulation, and therefore α, is a complicated time-dependent function. It depends on the modulation depth and waveform as well as on the

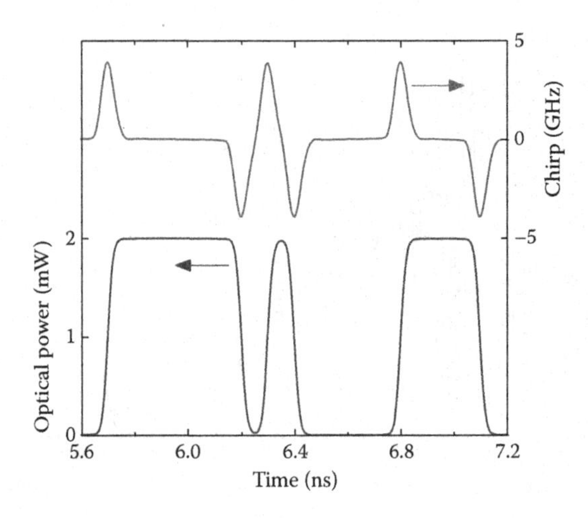


**FIGURE 4.10** Time dependence of the power and chirp for positive chirp ($\alpha = 1$). Transient chirp occurs during the rising and falling edges of optical power waveform.

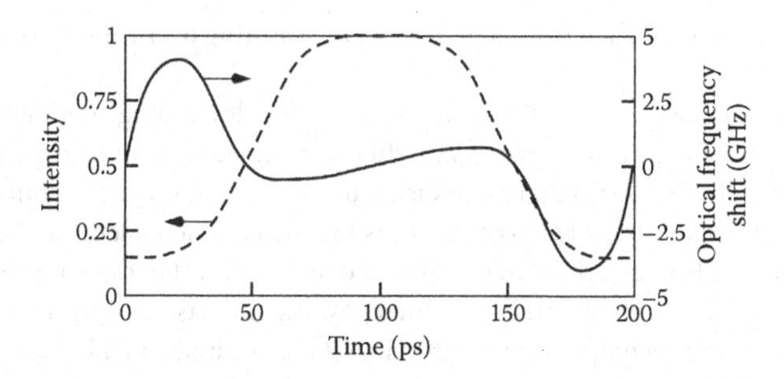


**FIGURE 4.11** Intensity vs. time (dashed line) and optical carrier frequency shift vs. time (solid line) for a $LiNbO_3$ MZ modulator with poor extinction (design chirp −0.7, extinction 8 dB), during a 0-1-0 data sequence at 10 Gb/s.

modulator itself. Many times $\alpha$ is quoted as a single number, which is an extreme oversimplification and is misleading except in carefully defined circumstances. Figure 4.11 shows an example of an MZ modulator designed for $\alpha = -0.7$ but with the poor optical extinction of 8 dB. It can be seen that $\alpha$ is nearly zero over large portions of the pulse and that there are significant tails with $\alpha > 0$.

A measurement of carrier frequency change under modulation ("time-resolved chirp") is the ideal, but it is difficult to accomplish due to high modulation frequencies and to the basic limit on resolving both frequency and time simultaneously. There have been many methods reported for this; two will be briefly described here.

The frequency discriminator method [67] is shown in Figure 4.12a. The output of the modulator is fed through a fiber MZ interferometer, one of whose arms contains a delay so the interferometer is frequency sensitive. By taking measurements at quadrature points with positive and negative slope and at a maximum-transmission point, the frequency and intensity modulation components can be resolved. The time resolution is accomplished by the high-speed oscilloscope.

The scanning filter method is shown in Figure 4.12b. The output of the modulator is fed through a narrowband filter (narrow enough to provide the desired frequency resolution; the filter is not used as a frequency discriminator). The filter is stepped across the optical frequency range and for each filter frequency, the oscilloscope records the time dependence. The measured time dependences of each frequency component can then be combined to give the full frequency versus time information [73]. (In the original version of this method [74], a sampling gate was used instead of an oscilloscope so the

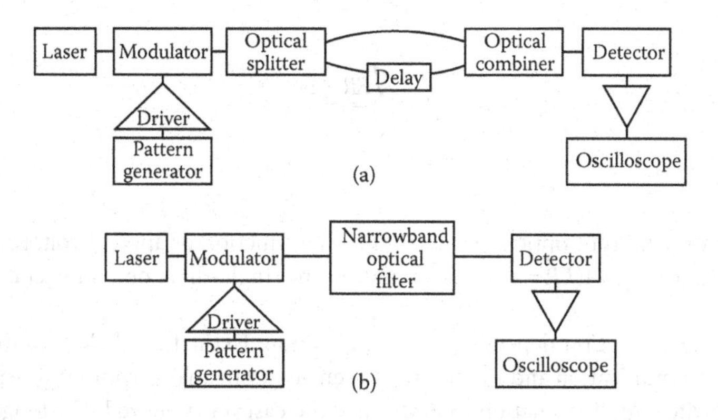


**FIGURE 4.12** Measurement of time-resolved chirp. (a) Frequency discriminator method. (b) Scanning filter method.

optical spectrum was scanned at each time step, instead of scanning over time at each frequency step as described above.)

These methods give detailed information that is useful for developing modulators. However, it is common for digital optical system engineers to skip this complex modulator test and instead do a simple "power penalty" test [75]. In this test, a digital link is set up with the modulator connected directly to the detector and the bit error rate versus received power is measured. Then a length of fiber with a known total dispersion is inserted between the modulator and detector and the test is repeated. Assuming the optical power is low enough to avoid optical nonlinearity, the increase in optical power required for a given bit error rate (the power penalty) is due to the chirp (and the unavoidable modulation bandwidth) interacting with the fiber dispersion. This provides a simple, direct way to evaluate the magnitude of the system degradation caused by chirp.

### 4.3.6 Impedance Match/Electrical Reflections

The modulator's electrical input port should not reflect any electrical power when driven from the system impedance. This can be accomplished by either making the modulator's internal impedance the same as that of the system, or by matching the modulator's internal impedance to that of the system.

Electrical reflections are generally specified in all applications, analog and digital. This is a purely electrical specification.

The electrical reflection is measured by connecting the modulator's input port to an electrical network analyzer and measuring $S_{11}$. Modulator specifications are usually in terms of $S_{11}$. The $S_{11}$ data is also useful for modulator development since it can be used to determine the modulator's impedance and the magnitude, type, and sometimes location of electrical parasitics.

### 4.3.7 Optical Extinction

The optical extinction is the ratio of the optical output power when the modulator is "on" to the output power when the modulator is "off." This is measured at DC or at low frequency.

For an ideal symmetric MZ modulator, the extinction ratio is infinite. In practice, asymmetry due to nonideal splitting and combining ratios of the 1 × 2 splitter and 2 × 1 combiner, respectively, and a difference in loss between the two arms of the interferometer lead to a finite extinction ratio. In this case, the output signal from the MZ modulator can be written as [76]

$$E(t) = 0.5E_0[\exp(j\Delta\beta(V_1(t))L) + \gamma \exp(j\Delta\beta(V_2(t)))], \tag{4.12}$$

where

$$\gamma = \frac{\sqrt{ER}-1}{\sqrt{ER}+1} \tag{4.13}$$

and $0 < \gamma \leq 1$.

The static L-V curve (output optical power (light) as a function of applied voltage) is illustrated in Figure 4.13. The result for $\gamma = 1 (ER = \infty)$ corresponds to the single-drive operating condition illustrated in Figure 4.3.

The importance of extinction depends on the application. Extinctions below 10 dB cause a power penalty in digital communication and a reduction in sensitivity in analog communication. If the modulator is used as a switch, high extinction (>40 dB in some cases) is required. For long-distance digital communication, extinction >20 dB can be required in an MZ modulator to avoid uncontrolled chirp (unbalanced arm intensity causes both poor extinction and chirp) [77].

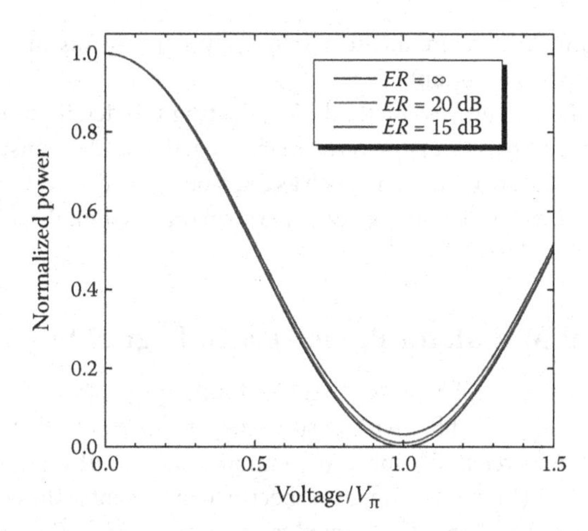


**FIGURE 4.13** Dependence of the normalized optical power for a symmetric single-drive Mach-Zehnder modulator as a function of the voltage applied to the arm of the modulator.

### 4.3.8 Electrical Extinction and Other Eye Pattern Measurements

Several measurements can be made on the eye pattern produced by the modulator. These measurements are all specific to digital applications, since it is a random binary bit stream that produces the eye. Eyes with multiple levels are generated by more complex modulation formats, but the binary format illustrates the key features.

Figure 4.14a shows the setup for generating the eye pattern. Nonideal effects are generated by each component in the link from pattern generator to oscilloscope, so care must be taken to separate how much eye degradation is due to the modulator and how much is due to other components.

The electrical extinction is the ratio of the maximum detected voltage to the minimum detected voltage, at the eye center. This is diagrammed in Figure 4.14b. The electrical extinction is analogous to the optical extinction, but it includes more factors than just the optical extinction. Electrical reflections and rolloff in the frequency response both contribute to reducing the electrical extinction below the optical extinction. If the eye is measured after transmission through dispersive fiber, chirp will also reduce extinction.

The jitter is a measure of timing errors. This usually is dominated by nonmodulator sources, but the modulator can contribute to jitter if it has electrical reflections (the modulator jitter is deterministic jitter, not random jitter).

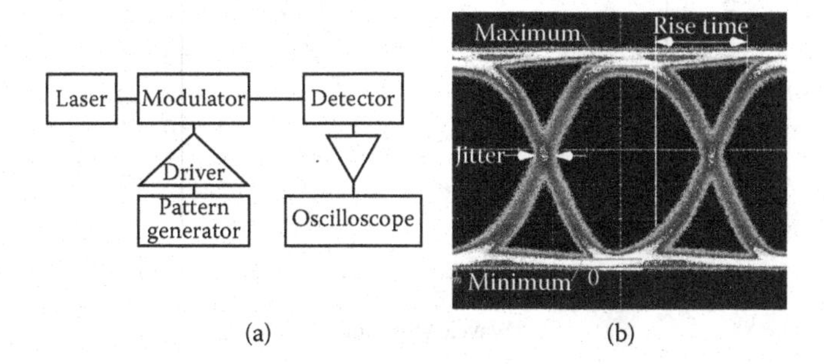


**FIGURE 4.14** Measurement of modulator eye pattern. (a) Setup. (b) Eye pattern with a few key measurements indicated (extinction is ratio of maximum to minimum).

The rise time is determined by the modulator's frequency response (assuming the modulator is the limiting factor in the measurement setup).

An eye pattern is usually not the best method of evaluating the details of modulator performance because it mixes several aspects of modulator performance together and because in practice it is difficult to separate out effects due to other components such as the modulator driver. However, the eye pattern is widely used for simple, quick and visual judgment of performance in digital applications so it is relevant to modulator evaluation.

### 4.3.9 Implications of Modulator Properties on Digital System Performance

Modulator chirp has important implications in terms of pulse propagation on a dispersive fiber. The group delay of a typical single-mode fiber with positive dispersion, as illustrated in Figure 4.15, causes the higher frequency (lower wavelength) spectral components of a pulse to propagate with a shorter delay than the lower frequency (higher wavelength) spectral components. The corresponding dispersion coefficient is also shown in Figure 4.15 and is related to the group delay $\tau_g(\lambda)$ per unit length by

$$D = \frac{d\tau_g(\lambda)}{d\lambda}. \tag{4.14}$$

The dispersion coefficient is usually expressed in units of ps/(km-nm). Positive chirp causes the rising edge of a pulse to propagate faster than it would for a zero-chirp pulse, and the falling edge to propagate slower. This enhances the pulse broadening. Conversely, a negatively chirped pulse is compressed as it propagates along a fiber with positive dispersion.

The combined implications of the modulator chirp and fiber dispersion are shown in Figure 4.16 by considering transmission over 40 km of standard single-mode fiber (dispersion coefficient of 17 ps/(km·nm)). The average power of the transmitted signal ($P_{tx}$) is 0 dBm in order to avoid fiber nonlinear effects [78]. For the case of zero chirp (Figure 4.16b), the pulse broadening is due to the modulated signal bandwidth. Pulse broadening and pulse compression occur for positive (Figure 4.16c) and negative chirp (Figure 4.16d), respectively. A key benefit of negative chirp is that the zero bit in the bit sequence . . . 101 . . .

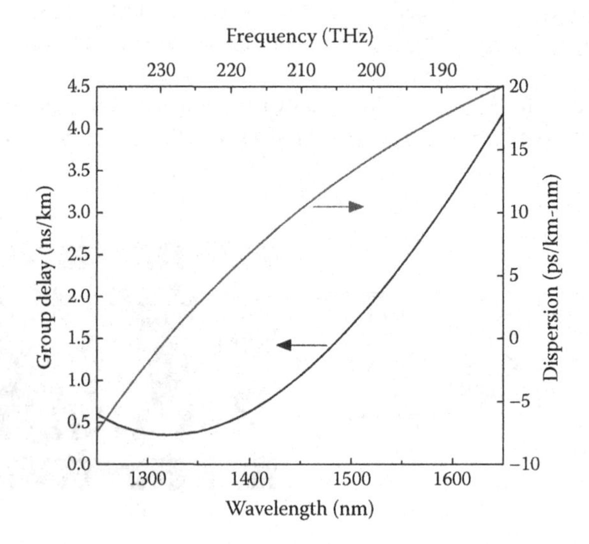


**FIGURE 4.15** Dependence of the group delay per unit length and dispersion of a standard single-mode fiber on wavelength and frequency.

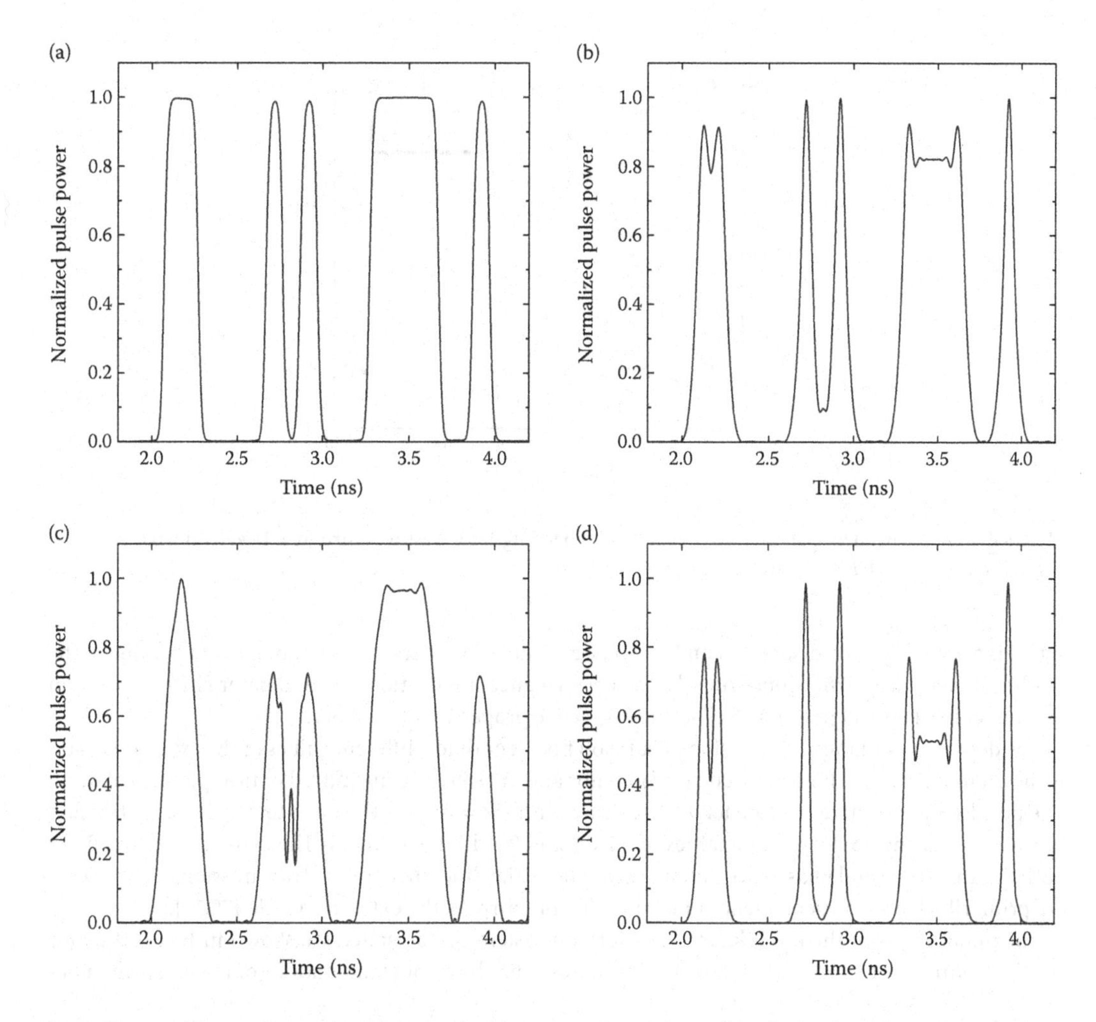


**FIGURE 4.16** The 10-Gb/s NRZ-OOK signals. (a) Output of modulator. (b) After 40 km of SMF for zero chirp ($\alpha = 0$). (c) After 40 km of SMF for positive chirp ($\alpha = 1$). (d) After 40 km of SMF for negative chirp ($\alpha = -1$).

remains distinct, whereas for zero chirp, and more so for positive chirp, intersymbol interference occurs due to the broadened pulses on either side.

The implications of the chirp are quantified in Figure 4.17 by showing the dependence of the penalty in system performance on fiber length for 10 Gb/s nonreturn-to-zero on–off keying modulation and different values of the α-parameter. The extinction ratio is infinite and the average transmitted power is 0 dBm. The penalty is relative to the back-to-back performance, which does not depend on the modulator chirp. Clearly, the chirp has a pronounced effect on the system performance. The pulse compression for $\alpha = -1$ initially yields a negative power penalty. These results were obtained for a symmetric dual-drive MZ modulator by varying the amplitudes of the two drive voltage waveforms according to Equation 4.10. Alternatively, the α-parameter can be varied by altering the design of the modulator [79–81]. This allows, for example, the realization of a zero-chirp, single-drive modulator [82]. A zero-chirp MZ modulator can be used to generate a binary phase modulated signal by biasing the modulator at the extinction point and using a drive voltage with amplitude $2V_\pi$ [83].

Figure 4.18 illustrates the dependence of the power penalty on the α-parameter for 80 km of standard single-mode fiber and different values of the extinction ratio and transmitted power. For each case, the penalty is relative to the back-to-back performance. For a transmitted power of 10 dBm, the optical

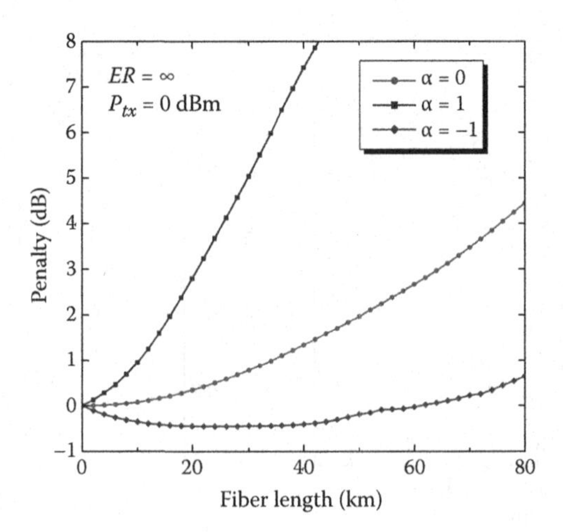


**FIGURE 4.17** Dependence of the power penalty on fiber length for positive chirp ($\alpha = 1$), zero chirp ($\alpha = 0$) and negative chirp ($\alpha = -1$), $ER = \infty$ and $P_{tx} = 0$ dBm.

signal is affected by fiber dispersion and self-phase modulation that occurs in an optical fiber due to the nonlinear Kerr effect [78]. Consequently, the performance implications of modulator chirp depend on the transmitted optical power as well as the dispersive properties of the fiber.

The dependence of the transmission penalty on fiber length for different values of the extinction ratio is shown in Figure 4.19 for a zero-chirp modulator and an average transmitted optical power of 0 dBm. The penalties are relative to the back-to-back case with $ER = \infty$. For the back-to-back case, extinction ratios of 20 dB and 15 dB yield penalties of 0.2 dB and 0.7 dB, respectively. The extinction ratio affects both the amplitude and phase of the modulated signal. The differences in the transmission performance are primarily due to the dependence of the modulator chirp on the extinction ratio [70,72].

The implications of the modulator bandwidth on system performance are shown in Figure 4.20 for 10 Gb/s nonreturn-to-zero on–off keying modulation and different combinations of the chirp and fiber

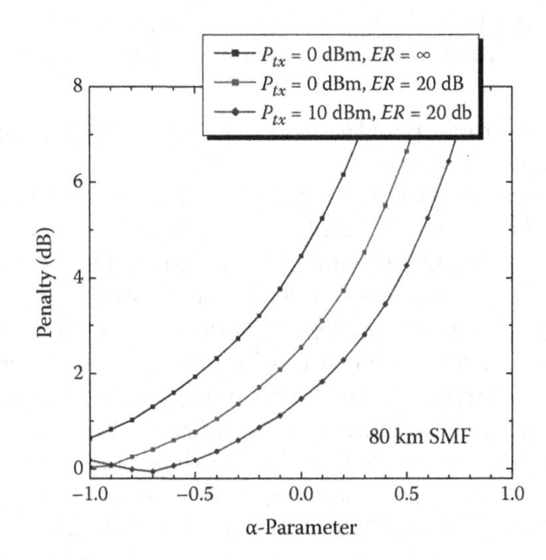


**FIGURE 4.18** Dependence of the power penalty on the α-parameter for different values of the transmitted power and extinction ratio. The fiber length is 80 km.

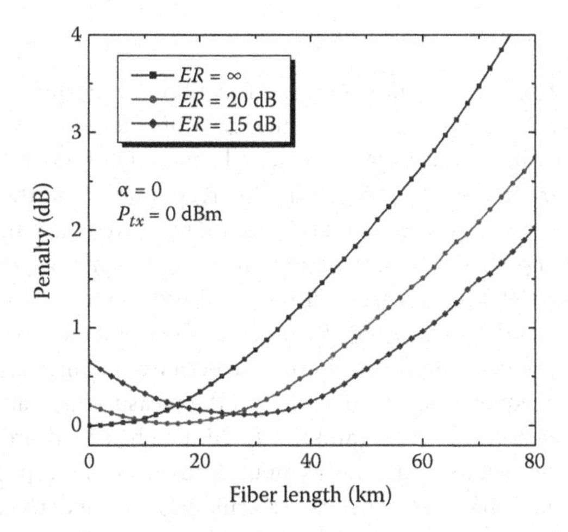


**FIGURE 4.19** Dependence of the power penalty on fiber length for different values of the extinction ratios. $\alpha = 0$ and $P_{tx} = 0$ dBm.

length. The extinction ratio is infinite and the average transmitted power is 0 dBm. Here, the penalty for each case is relative to the performance obtained for a modulator bandwidth of 20 GHz. The strictest requirement is for the back-to-back case, which is actually independent of the modulator chirp.

An important observation from these illustrative examples is that the impact that a particular modulator property has on system performance is dependent on other modulator properties and the specific system configuration. Also, what would appear to be a relatively small difference in the L–V curves (Figure 4.13) for extinction ratios of ∞ and 20 dB, has a significant effect on the transmission performance (Figure 4.19). In addition, modulator properties can affect system performance in other ways. For example, modulator chirp has an impact on the performance of nonideal dispersion compensating fiber Bragg gratings [84].

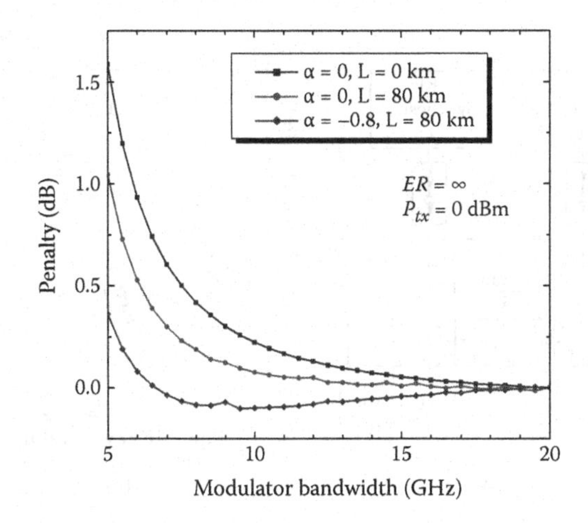


**FIGURE 4.20** Dependence of the power penalty on the modulator bandwidth for different α-parameters and fiber lengths, $ER = \infty$ and $P_{tx} = 0$ dBm.

### 4.3.10 Linearity

Measurement of the linearity of a modulator's transfer function is important for analog applications.

Quantifying the nonlinearity of a modulator is somewhat difficult because the normal electrical measures apply to a component with an electrical output, and a modulator has an optical output. The most straightforward measures are the input second-order intercept point and the input third-order intercept point. The 1-dB compression point ($P_{1dB}$) is also a useful measure. These measures are diagrammed in Figure 4.21b. The intercept points are extrapolations of measurements at small signal levels. The input third-order intercept point (IIP3) is a measure of the third-order intermodulation power, that is, the power at frequencies $2f_1 - f_2$ and $2f_2 - f_1$, where $f_1$ and $f_2$ are the fundamental input frequencies. Third-order intermodulation is used instead of third harmonic because the intermodulation frequencies are close to two closely spaced frequencies, which simplifies the measurement at high frequency. Second-order intermodulation frequencies are $f_1 - f_2$ and $f_1 + f_2$. Although second-order intercept point is not shown in Figure 4.21b, it is defined the same way as input third-order intercept point (IIP2) except that it uses the second-order intermodulation frequencies. Harmonics can be derived from the intermodulation results—the same nonlinearity that produces third-order intermodulation produces third harmonics—so they do not need to be separately measured to characterize the modulator.

These measures refer to the electrical power at the modulator input. This is directly useful to radio frequency system engineers. For modulator design, though, it is sometimes useful to convert these to a modulation depth, or to normalize these by switching voltage. Conversions such as these are specific to particular modulators. An example for the MZ modulator is given in [85].

The setup for making linearity measurements is shown in Figure 4.21a. Two frequencies are applied to the modulator, and the output electrical power spectrum from the detector is measured on an radio frequency spectrum analyzer. Extreme care must be taken that nonlinear distortion from the detector, from the synthesizers (due to imperfect isolation), from the spectrum analyzer, or from any amplifiers or other sources does not exceed the distortion from the modulator. Several measurements must be taken to verify that the slope of the third-order distortion line is actually three (two for second-order). The measured input third-order intercept point and IIP2 assume the intermodulation powers have slopes of three and two; if they do not, these measures are meaningless. (Linearized modulators may have different slopes than these. Intercept points can still be defined, but they require a modification of the usual cascade equations [86].)

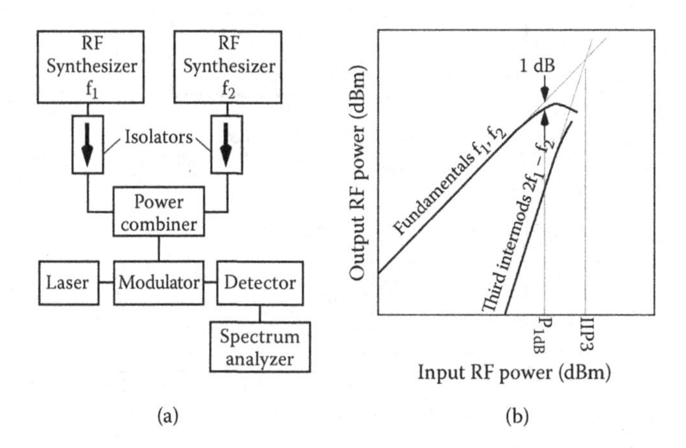


**FIGURE 4.21** Modulator linearity measurement. (a) Setup. (b) Determination of input 1-dB compression point and input third-order intercept point.

### 4.3.11 Optical Power Handling

Optical power handling is generally pass/fail: if the modulator can handle the power needed for the intended application, it is unimportant how much more it can handle; if it cannot, it is useless. Digital applications generally require 1 mW to 10 mW output power. Analog applications can require much more since performance depends on optical power; modulators have been used with up to 2.5 W input [56].

In many cases, the limit of optical power handling ability is catastrophic destruction. Lifetime measurements are not necessary with this mechanism, but one has to verify that the damage threshold is reliably above the operating power. In other cases, there is a power-dependent drift such as the photorefractive effect in lithium niobate at short wavelengths [38,39]. If a mechanism like this is present, a lifetime test is necessary to verify that the modulator will continue to operate properly at the power necessary.

### 4.3.12 Stability over Temperature and Reliability

Reliability is a key performance area, since a modulator is not much use in a practical application if it is unreliable. Reliability involves many aspects, so only a few brief comments can be made here; a comprehensive specification is [87].

Survival over temperature is perhaps the generic aspect of reliability most relevant to basic modulator operation. Other measures such as vibration, electrostatic discharge survival, fiber pull strength, humidity resistance, etc., are mostly controlled by packaging and fiber pigtailing methods. Survivability over temperature involves packaging and pigtailing as well, but it also can be limited by the modulator itself.

The temperature range defined by [87] is −40°C to 85°C, and survival involves 500 cycles −40°C to 70°C and a 5000 h soak at 85°C. Some applications can be as benign as −5°C to 50°C, but the preference is for a single design that satisfies all applications. Military applications vary, with a few specifications being as extreme as −80°C or +125°C. If a modulator relies on an electro-optic material or a waveguide fabrication process that cannot survive these types of temperature stresses, it is unlikely to win wide acceptance.

## References

1. R. H. Rediker, T. A. Lind, and B. E. Burke, "Optical wavefront measurement and/or modification using integrated optics," *J. Lightw. Technol.*, vol. 6, no. 6, pp. 916–932, Jun. 1988.
2. B. Huiszoon, R. J. W. Jonker, P. K. van Bennekom, G.-D. Khoe, and H. de Waardt, "Cost-effective up to 40 Gb/s transmission performance of 1310 nm directly modulated lasers for short- to medium-range distances," *J. Lightw. Technol.*, vol. 23, no. 3, pp. 1116–1125, Mar. 2005.
3. T. Okiyama, H. Nishimoto, I. Yokota, and T. Touge, "Evaluation of 4 Gbit/s optical fiber transmission distance with direct and external modulation," *J. Lightw. Technol.*, vol. 6, no. 11, pp. 1686–1692, Nov. 1988.
4. C. H. Cox III, *Analog Optical Links: Theory and Practice.*, Cambridge, UK: Cambridge University Press, 2004.
5. K. Noguchi, O. Mitomi, and H. Miyazawa, "Millimeter-wave Ti:$LiNbO_3$ optical modulators," *J. Lightw. Technol.*, vol. 16, no. 4, pp. 615–619, Apr. 1998.
6. K. Y. Lau, "Dynamics of quantum well lasers," in *Quantum Well Lasers*, P. S. Zory Jr., Ed., San Diego: Academic Press, 1993, ch. 5, pp. 217–276.
7. S. D. McDougall, B. C. Qiu, G. Ternent, D. A. Yanson, V. Loyo-Maldonado, O. P. Kowalski, and J. H. Marsh, "Monolithic integration of InGaAs/InAlGaAs-based semiconductor optical amplifiers and 10 Gb/s broadband electroabsorption modulators using quantum well intermixing technology," in *Int. Conf. Indium Phosphide Related Materials 2004 Conf. Proc.*, pp. 403–406. Piscataway, NJ: IEEE (Institute of Electrical and Electronic Engineers).

8. W. Kobayashi, M. Arai, and T. Yamanaka, "Design and fabrication of 10-/40-Gb/s, uncooled electroabsorption modulator integrated DFB laser with butt-joint structure," *J. Lightw. Technol.*, vol. 28, no. 1, pp. 164–171, Jan. 2010.
9. S. Hardy, "Luxtera launches single chip transceiver for motherboard deployment," *Lightwave* [Online], Nov. 2009. Available: http://www.lightwaveonline.com/top-stories/Luxtera-launches-single-chip-transceiver-for-motherboard-deployment-69773572.html (accessed June 27, 2011).
10. A. Yariv, *Introduction to Optical Electronics*, 2nd ed., New York: Holt, Rinehart and Winston, 1976.
11. D. S. Chemla, T. C. Damen, D. A. B. Miller, A. C. Gossard, and W. Wiegmann, "Electroabsorption by Stark effect on room-temperature excitons in GaAs/GaAlAs multiple quantum well structures," *Appl. Phys. Lett.*, vol. 42, no. 10, pp. 864–866, May 1983.
12. J. S. Weiner, D. A. B. Miller, and D. S. Chemla, "Quadratic electro-optic effect due to the quantum-confined Stark effect in quantum wells," *Appl. Phys. Lett.*, vol. 50, no. 13, pp. 842–844, Mar. 1987.
13. R. A. Soref and B. R. Bennett, "Electrooptical effects in silicon," *IEEE J. Quantum Electron.*, vol. 23, no. 1, pp. 123–129, Jan. 1987.
14. S. J. Spector, C. M. Sorace, M. W. Geis, M. E. Grein, J. U. Yoon, T. M. Lyszczarz, E. P. Ippen, and F. X. Kartner, "Operation and optimization of silicon-diode-based optical modulators," *IEEE J. Sel. Top. Quantum Electron.*, vol. 16, no. 1, pp. 165–172, Jan. 2010.
15. H. Fukano, T. Yamanaka, M. Tamura, and Y. Kondo, "Very-low-driving-voltage electroabsorption modulators operating at 40 Gb/s," *J. Lightw. Technol.*, vol. 24, no. 5, pp. 2219–2224, May 2006.
16. P. K. L. Yu, I. Shubin, X. B. Xie, Y. Zhuang, A. J. X. Chen, and W. S. C. Chang, "Transparent ROF link using EA modulators," in *Int. Topical Meeting Microwave Photonics 2005 Digest*, pp. 21–24. Piscataway, NJ: IEEE.
17. G. E. Betts, X. Xie, I. Shubin, W. S. C. Chang, and P. K. L. Yu, "Gain limit in analog links using electroabsorption modulators," *IEEE Photon. Technol. Lett.*, vol. 18, no. 19, pp. 2065–2067, Oct. 2006.
18. R. Madabhushi, "Lithium niobate optical modulators," in *WDM Technologies: Active Optical Components*, A. K. Dutta, N. K. Dutta, and M. Fujiwara, Eds., San Diego: Academic Press, 2002, pp. 207–248.
19. A. Mahapatra and E. J. Murphy, "Electrooptic modulators," in *Optical Fiber Telecommunications IV A: Components*, I. P. Kaminow and T. Li, Eds., San Diego: Academic Press, 2002, pp. 258–294.
20. W. E. Martin, "A new waveguide switch/modulator for integrated optics," *Appl. Phys. Lett.*, vol. 26, no. 10, pp. 562–563, May 1975.
21. K. M. Kissa, P. G. Suchoski, and D. K. Lewis, "Accelerated aging of annealed proton-exchanged waveguides," *J. Lightw. Technol.*, vol. 13, no. 7, pp. 1521–1529, Jul. 1995.
22. M. Ushigome, M. Fukuoka, T. Arakawa, and K. Tada, "Low voltage Mach-Zehnder modulator with InGaAs/InAlAs five layer asymmetric coupled quantum wells," in *IEEE LEOS Annual Meeting 2009 Conf. Proc.*, pp. 434–435. Piscataway, NJ: IEEE.
23. T. Kawanishi, T. Sakamoto, and M. Izutsu, "High-speed control of lightwave amplitude, phase, and frequency by use of electrooptic effect," *IEEE J. Sel. Top. Quantum Electron.*, vol. 13, no. 1, pp. 79–91, Jan. 2007.
24. M. Izutsu, S. Shikama, and T. Sueta, "Integrated optical SSB modulator/frequency shifter," *IEEE J. Quantum Electron.*, vol. 17, no. 11, pp. 2225–2227, Nov. 1981.
25. R. A. Griffin and A. C. Carter, "Optical differential quadrature phase-shift key (oDQPSK) for high capacity optical transmission," in *Optical Fiber Communication Conf. 2002 Tech. Dig.*, pp. 367–368. Washington, DC: OSA (Optical Society of America).
26. X. Zhou and J. Yu, "Multi-level, multi-dimensional coding for high-speed and high-spectral-efficiency optical transmission," *J. Lightw. Technol.*, vol. 27, no. 16, pp. 3641–1653, Aug. 2009.
27. G. E. Betts, "$LiNbO_3$ modulators and their use in high performance analog links," in *RF Photonic Technology in Optical Fiber Links*, W. S. C. Chang, Ed., Cambridge, UK: Cambridge University Press, 2002, ch. 4, pp. 81–132.
28. H. Tazawa, Y. H. Kuo, I. Dunayevskiy, J. Luo, A. K.-Y. Jen, H. R. Fetterman, and W. H. Steier, "Ring resonator-based electrooptic polymer traveling-wave modulator," *J. Lightw. Technol.*, vol. 24, no. 9, pp. 3514–3519, Sep. 2006.

29. I. L. Gheorma and R. M. Osgood, "Fundamental limitations of optical resonator based high-speed EO modulators," *IEEE Photon. Technol. Lett.*, vol. 14, no. 6, pp. 795–797, Jun. 2002.
30. R. Krähenbühl and M. M. Howerton, "Investigations on short-path-length high-speed optical modulators in $LiNbO_3$ with resonant-type electrodes," *J. Lightw. Technol.*, vol. 19, no. 9, pp. 1287–1297, Sep. 2001.
31. G. E. Betts, L. M. Johnson, and C. H. Cox III, "High-sensitivity lumped-element bandpass modulators in $LiNbO_3$," *J. Lightw. Technol.*, vol. 7, no. 12, pp. 2078–2083, Dec. 1989.
32. R. C. Alferness, "Waveguide electrooptic modulators," *IEEE Trans. Microwave Theory Tech.*, vol. 30, no. 8, pp. 1121–1137, Aug. 1982.
33. H. Kogelnik and R. V. Schmidt, "Switched directional couplers with alternating $\Delta\beta$," *IEEE J. Quantum Electron.*, vol. 12, no. 7, pp. 396–401, Jul. 1976.
34. J. E. Watson, M. A. Milbrodt, K. Bahadori, M. F. Dautartas, C. T. Kemmerer, D. T. Moser, A. W. Schelling, T. O. Murphy, J. J. Veselka, and D. A. Herr, "A low-voltage 8×8 $Ti:LiNbO_3$ switch with a dilated-Benes architecture," *J. Lightw. Technol.*, vol. 8, no. 5, pp. 794–801, May 1990.
35. K. K. Wong, Ed., *Properties of Lithium Niobate*, London, UK: INSPEC, 2002.
36. E. L. Wooten, K. M. Kissa, A. Yi-Yan, E. J. Murphy, D. A. Lafaw, P. F. Hallemeier, D. Maack, D. V. Attanasio, D. J. Fritz, G. J. McBrien, and D. E. Bossi, "A review of lithium niobate modulators for fiber-optic communication systems," *IEEE J. Sel. Top. Quantum Electron.*, vol. 6, no. 1, pp. 69–82, Jan. 2000.
37. J. Nayyer and H. Nagata, "Suppression of thermal drifts of high speed $Ti:LiNbO_3$ optical modulators," *IEEE Photon. Technol. Lett.*, vol. 6, no. 8, pp. 952–955, Aug. 1994.
38. G. E. Betts, F. J. O'Donnell, and K. G. Ray, "Effect of annealing on photorefractive damage in titanium-diffused $LiNbO_3$ modulators," *IEEE Photon. Technol. Lett.*, vol. 6, no. 2, pp. 211–213, Feb. 1994.
39. T. Fujiwara, S. Sato, and H. Mori, "Wavelength dependence of photorefractive effect in Ti-indiffused $LiNbO_3$ waveguides," *Appl. Phys. Lett.*, vol. 54, no. 11, pp. 975–977, Mar. 1989.
40. G. E. Betts, K. G. Ray, and L. M. Johnson, "Suppression of acoustic effect in lithium niobate integrated-optical modulators," in *Integrated Photonics Research 1990 Tech. Dig.*, pp. 37–38. Washington, DC: OSA.
41. J. M. Marx, Z. Tang, O. Eknoyan, H. F. Taylor, and R. R. Neurgaonkar, "Low-loss strain induced optical waveguides in strontium barium niobate ($Sr_{0.6}Ba_{0.4}Nb_2O_6$) at 1.3 μm wavelength," *Appl. Phys. Lett.*, vol. 66, no. 3, pp. 274–276, Jan. 1995.
42. J. Tang, S. Yang, and A. Bhatranand, "Electro-optic barium titanate waveguide modulators with transparent conducting oxide electrodes," in *CLEO Pacific Rim 2009*, Shanghai, pp. 1–2.
43. T. Yamanaka, K. Tsuzuki, N. Kikuchi, E. Yamada, Y. Shibata, H. Fukano, H. Nakajima, Y. Akage, and H. Yasaka, "High-performance InP-based optical modulators," in *Optical Fiber Communication Conf. 2006 Tech. Dig.*, Anaheim, CA, paper OWC1. Washington, DC: OSA.
44. M. Jarrahi, T. H. Lee, and D. A. B. Miller, "Wideband, low driving voltage traveling-wave Mach-Zehnder modulator for RF photonics," *IEEE Photon. Technol. Lett.*, vol. 20, no. 7, pp. 517–519, Apr. 2008.
45. L. C. Kimerling, D. Ahn, A. B. Apsel, M. Beals, D. Carothers, Y.-K. Chen, T. Conway, et al. "Electronic-photonic integrated circuits on the CMOS platform," *Proc. SPIE*, vol. 6125, paper 2, Mar. 2006.
46. B. Jalali, V. Raghunathan, D. Dimitropoulos, and O. Boyraz, "Raman-based silicon photonics," *IEEE J. Sel. Top. Quantum Electron.*, vol. 12, no. 3, pp. 412–421, May 2006.
47. P. Dong, S. Liao, D. Feng, H. Liang, D. Zheng, R. Shafiiha, C.-C. Kung, W. Qian, G. Li, X. Zheng, A. V. Krishnamoorthy, and M. Asghari, "Low $V_{pp}$, ultralow-energy, compact, high-speed silicon electro-optic modulator," *Opt. Exp.*, vol. 17, no. 25, pp. 22484–22490, Dec. 2009.
48. T. Baehr-Jones, B. Penkov, J. Huang, P. Sullivan, J. Davies, J. Takayesu, J. Luo, et al. "Nonlinear polymer-clad silicon slot waveguide modulator with a half-wave voltage of 0.25V," *Appl. Phys. Lett.*, vol. 92, no. 16, article 163303, Apr. 2008.
49. T.-D. Kim, J.-W. Kang, J. Luo, S.-H. Jang, J.-W. Ka, N. Tucker, J. B. Benedict, et al. "Ultralarge and thermally stable electro-optic activities from supramolecular self-assembled molecular glasses," *J. Am. Chem. Soc.*, vol. 129, no. 3, pp. 488–489, Jan. 2007.

50. D. Jin, H. Chen, A. Barklund, J. Mallari, G. Yu, E. Miller, and R. Dinu, "EO polymer modulators reliability study," *Proc. SPIE*, vol. 7599, paper 0H, Jan. 2010.
51. B. Li, R. Dinu, D. Jin, D. Huang, B. Chen, A. Barklund, E. Miller, et al. "Recent advances in commercial electro-optic polymer modulator," in *Optical Fiber Communications Optoelectronics Conf. Asia 2007*, Shanghai, pp. 115–117. Piscataway, NJ: IEEE.
52. W. S. C. Chang, "Multiple quantum well electroabsorption modulators for RF photonic links," in *RF Photonic Technology in Optical Fiber Links*, W. S. C. Chang, Ed., Cambridge, UK: Cambridge University Press, 2002, ch. 6, pp. 165–202.
53. J. Mallari, C. Wei, and D. Jin, "100 Gbps EO polymer modulator product and its characterization using real-time digitizer, " *Optical Fiber Communication Conf. 2010 Tech. Dig.*, paper OThU2. Washington, DC: OSA.
54. H. Chen, B. Chen, D. Huang, et al., "Broadband electro-optic ploymer modulators with high electro-optic activity and low poling-induced optical loss," *Appl. Phys Lett.*, vol. 93, p. 43507, July 2008.
55. M. Sugiyama, M. Doi, S. Taniguchi, et al., Driver-less 40 Gb/s LiNb03 modulator with sub-IV drive voltage," *Optical Fiber Communication Conf. 2002 Postdeadline papers*, FB6. Washington, DC: OSA.
56. E. I. Ackerman, G. E. Betts, W. K. Burns, et al., "Signal-to-noise performance of two analog photonic links using different noise reduction techniques," *IEEE/MTT-S Int. Microwave Symp. 2007 Tech. Dig.*, pp. 51–54. Piscataway, NJ: IEEE.
57. S. Park, J. J. Ju, J. Y. Do, et al., "Thermal stability enhancement of electrooptic polymer modulator," *IEEE Photon. Technol. Lett.*, vol. 16, pp. 93–95, Jan. 2004.
58. Y. Enani, C. T. Derose, D. Mathine, et al., "Hybrid polymer/sol-gel waveguide modulators with exceptionally large electro-optic coefficients," *Nature Photonics*, vol. 1 pp. 180–185, Mar. 2007.
59. S. K. Korotky and J. J. Veselka, "An RC network analysis of long term Ti:$LiNbO_3$ bias stability," *J. Lightw. Technol.*, vol. 14, no. 12, pp. 2687–2697, Dec. 1996.
60. W. K. Burns, M. M. Howerton, R. P. Moeller, R. Krahenbuhl, R. W. McElhanon, and A. S. Greenblatt, "Low drive voltage, broad-band $LiNbO_3$ modulators with and without etched ridges," *J. Lightw. Technol.*, vol. 17, no. 12, pp. 2551–2555, Dec. 1999.
61. M. Nazarathy, D. W. Dolfi, and R. L. Jungerman, "Velocity-mismatch compensation in traveling-wave modulators using pseudorandom switched-electrode patterns," *J. Opt. Soc. Am. A*, vol. 4, no. 6, pp. 1071–1079, June 1987.
62. G. K. Gopalakrishnan, W. K. Burns, R. W. McElhanon, C. H. Bulmer, and A. S. Greenblatt, "Performance and modeling of broadband LiNbO3 traveling wave optical intensity modulators," *J. Lightw. Technol.*, vol. 12, no. 10, pp. 1807–1819, Oct. 1994.
63. R. Krähenbühl and W. K. Burns, "Modeling of broad-band traveling-wave optical-intensity modulators," *IEEE Trans. Microwave Theory Tech.*, vol. 48, no. 5, pp. 860–864, May 2000.
64. T. Yamanaka, H. Fukano, and T. Saitoh, "Lightwave-microwave unified analysis of electroabsorption modulators integrated with RF coplanar waveguides," *IEEE Photon. Technol. Lett.*, vol. 17, no. 12, pp. 2562–2564, Dec. 2005.
65. C. G. Lim, "A passive broadband impedance equalizer for improving the input return loss of electro-absorption modulators," *J. Lightw. Technol.*, vol. 27, no. 8, pp. 1051–1058, Apr. 2009.
66. F. Koyama and K. Iga, "Frequency chirping in external modulators," *J. Lightw. Technol.*, vol. 6, no. 1, pp. 87–93, Jan. 1988.
67. C. Laverdiere, A. Fekecs, and M. Tetu, "A new method for measuring time-resolved frequency chirp of high bit rate sources," *IEEE Photon. Technol. Lett.*, vol. 15, no. 3, pp. 446–448, Mar. 2003.
68. D. Sands, *Diode Lasers*, Bristol, UK: Institute of Physics Publishing, 2005, pp. 278–279.
69. S. K. Korotky, J. J. Veselka, C. T. Kemmerer, W. J. Minford, D. T. Moser, J. E. Watson, C. A. Mattoe, and P. L. Stoddard, "High-speed low power optical modulator with adjustable chirp parameter," in *Top. Meet. Integrated Photon. Research 1991 Tech. Dig.*, Monterey, CA, paper TuG2. Washington, DC: OSA.

70. A. H. Gnauck, S. K. Korotky, J. J. Veselka, J. Nagel, C. T. Kemmerer, W. J. Minford, and D. T. Moser, "Dispersion penalty reduction using an optical modulator with adjustable chirp," *IEEE Photon. Technol. Lett.*, vol. 3, no. 10, pp. 916–918, Oct. 1991.
71. N. Suzuki and Y. Hirayama, "Comparison of effective α-parameters for multiquantum-well electroabsorption modulators," *IEEE Photon. Technol. Lett.*, vol. 7, no. 9, pp. 1007–1009, Sep. 1995.
72. J. C. Cartledge, "Comparison of effective α-parameters for semiconductor Mach-Zehnder optical modulators," *J. Lightw. Technol.*, vol. 16, no. 3, pp. 372–379, Mar. 1998.
73. Making time-resolved chirp measurements using the optical spectrum analyzer and digital communications analyzer. Agilent Application Note 1550-7 [Online], Mar. 2002. Available: http://cp.literature.agilent.com/litweb/pdf/5988-5614EN.pdf (accessed June 27, 2011).
74. R. A. Linke, "Modulation induced transient chirping in single frequency lasers," *IEEE J. Quantum Electron.*, vol. 21, no. 6, pp. 593–597, Jun. 1985.
75. S. Makino, K. Shinoda, T. Shiota, T. Kitatani, T. Fukamachi, M. Aoki, N. Sasada, K. Naoe, K. Uchida, and H. Inoue, "Wide temperature (15°C to 95°C), 80-km SMF transmission of a 1.55-μm, 10 Gbit/s InGaAlAs electroabsorption modulator integrated DFB laser," in *Optical Fiber Communication Conf. 2007 Tech. Dig.*, Anaheim, CA, paper OMS1. Washington, DC: OSA.
76. H. Kim and A. H. Gnauck, "Chirp characteristics of dual-drive Mach-Zehnder modulator with a finite DC extinction ratio," *IEEE Photon. Technol. Lett.*, vol. 14, no. 3, pp. 298–300, Mar. 2002.
77. J. C. Cartledge, "Performance of 10 Gb/s lightwave systems based on lithium niobate Mach-Zehnder modulators with asymmetric Y-branch waveguides," *IEEE Photon. Technol. Lett.*, vol. 7, no. 9, pp. 1090–1092, Sep. 1995.
78. G. P. Agrawal. *Nonlinear Fiber Optics*. San Diego: Academic Press, 2001.
79. M. M. Howerton, R. P. Moeller, A. S. Greenblatt, and R. Krahenbuhl, "Fully packaged, broad-band $LiNbO_3$ modulator with low drive voltage," *IEEE Photon. Technol. Lett.*, vol. 12, no. 7, pp. 792–794, Jul. 2000.
80. G. Ghislotti, S. Balsamo, and P. Bravetti, "Single-drive $LiNbO_3$ Mach-Zehnder modulator with widely DC tunable chirp," *IEEE Photon. Technol. Lett.*, vol. 15, no. 11, pp. 1534–1536, Nov. 2003.
81. J. Kondo, K. Aoki, A. Kondo, T. Ejiri, Y. Iwata, A. Hamajima, T. Mori, et al. "High-speed and low-driving-voltage thin-sheet X-cut $LiNbO_3$ modulator with laminated low-dielectric-constant adhesive," *IEEE Photon. Technol. Lett.*, vol. 17, no. 10, pp. 2077–2079, Oct. 2005.
82. J. Ichikawa, S. Oikawa, F. Yamamoto, T. Sakane, S. Kurimura, and K. Kitamura, "Zero chirp broadband Z-cut $LiNbO_3$ optical modulator using polarization reversal and branch electrode," in *Optical Fiber Communication Conf. 2004 Tech. Dig.*, paper MF56. Washington, DC: OSA.
83. P. J. Winzer and R.-J. Essiambre, "Advanced modulation formats for high-capacity optical transport networks," *J. Lightw. Technol.*, vol. 24, no. 12, pp. 4711–4728, Dec. 2006.
84. J. C. Cartledge and H. Chen, "Influence of modulator chirp in assessing the performance implications of the group delay ripple of dispersion compensating fiber Bragg gratings," *J. Lightw. Technol.*, vol. 21, no. 7, pp. 1621–1628, Jul. 2003.
85. B. H. Kolner and D. W. Dolfi, "Intermodulation distortion and compression in an integrated electro-optic modulator," *Appl. Opt.*, vol. 26, no. 17, pp. 3676–3680, Sep. 1987.
86. J. H. Schaffner and W. B. Bridges, "Intermodulation distortion in high dynamic range microwave fiber-optic links with linearized modulators," *J. Lightw. Technol.*, vol. 11, no. 1, pp. 3–6, Jan. 1993.
87. *Generic Reliability Assurance Requirements for Optoelectronic Devices Used in Telecommunications Equipment*, Telcordia document number GR-468, Sept. 2004. Piscataway, NJ: Telcordia.

# II

# Modulator Technology

# II

# Modulator Technology

# 5

# Electroabsorption Modulators




Haruhisa Soda
*FiBest Limited*

Ken Morito
*Fujitsu Laboratories*


## 5.1 Quantum-Confined Stark Effect in Multiple Quantum Well Structures

As the first generation of electroabsorption modulators, very simple bulk-type electroabsorption modulators were used. This modulator was based on the Franz–Keldysh effect. It comes from an electric field–assisted tunneling of electrons into the band-gap of the semiconductor material. When the electric field is applied to the semiconductor, the effective absorption edge moves to lower energies. Figure 5.1a shows the theoretical absorption characteristics of indium-gallium-arsenide-phosphide (InGaAsP) having 1.50-μm band-gap wavelength calculated by Aspnes' theory [1]. It was found that the light with lower energies than the band-gap of the semiconductor can be modulated by the applied electrical signal. Actually, high speed and low voltage operation were achieved successfully in semiinsulating buried heterostructure (BH) structure InGaAsP electroabsorption modulator [2]. Such bulk-type InGaAsP/InP electroabsorption modulators were put to practical use in long-distance 2.5-Gb/s transmission systems due to the lower chirp characteristics [3] compared to a directly modulated distributed feedback (DFB) lasers. However, it was very difficult to obtain a negative chirp parameter, which is indispensable

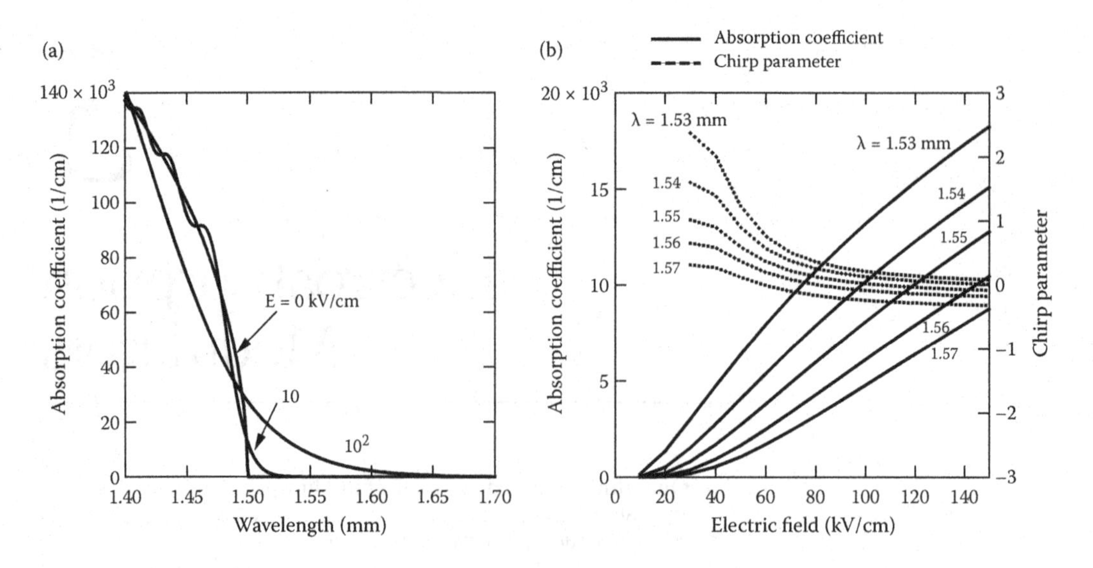


**FIGURE 5.1** (a) Theoretical absorption characteristics of InGaAsP having 1.50 μm band-gap wavelength calculated by Aspnes' theory. (b) Calculated chirp parameters of bulk-type InGaAsP/InP electroabsorption modulators having 1.50 μm band-gap wavelength.

for long-distance 10-Gb/s transmission applications. Figure 5.1b shows the calculated chirp parameters of such bulk-type InGaAsP/InP electroabsorption modulators. The chirp parameters were fixed to be almost positive values for low insertion loss regions. A multiple quantum well electroabsorption modulator resolved this engineering problem and realized current long-haul 10 Gb/s transmission system.

A multiple quantum well (MQW) electroabsorption modulator consists of MQWs sandwiched by barriers as shown in Figure 5.2a. The thicknesses of well and barrier are around several nanometers. In the quantum well structure, electrons and holes are localized in the well because of potential barriers of the conduction and valence bands resulting in discrete subband levels. When no electric field is applied, the wave function associated with each subband level is symmetrical about the center of the well as shown in Figure 5.2b. The exciton, due to localized electrons and holes in the MQW structure, strongly increases the binding energy and oscillator strength. This enhanced binding energy leads to stable exciton effects even at room temperature. The exciton transitions are easily observed at electric fields as high as over 100 kV/cm for 10 nm-wide quantum wells. The photon energy $h\nu$ absorbed in quantum well structure is basically given by Equation 5.1, taking the exciton absorption into account:

$$h\nu = E_g + E_{c1} + E_{hh1} - E_{ex}, \tag{5.1}$$

where $E_g$ is the band-gap energy of the well material, and $E_{c1}$ and $E_{hh1}$ are the electron and heavy hole ground state subband energies as shown in Figure 5.2b. $E_{ex}$ is the exciton biding energy. When the electric field applied to the quantum well structure is perpendicular to the layers, the potentials of the well are deformed as shown in Figure 5.2c. The wave functions of the electron and hole subbands are pushed to opposite sides of the well, and the ground state subband energies of the conduction band and the valence band shift downward and upward. Then $E_{c1}$ and $E_{hh1}$ are decreased with increasing electric fields. Simultaneously, the binding energy of the exciton decreased. But the subband shift is dominant for well localized practical structures. Thus, the effective band-gap energy reduces with increasing electric field. This effect is known as quantum-confined Stark effect. Figure 5.3 shows the calculated electric field dependence of the intersubband absorption energy wavelength ($E_g + E_{c1} + E_{hh1}$) for an unstrained quantum structure with various well thicknesses.

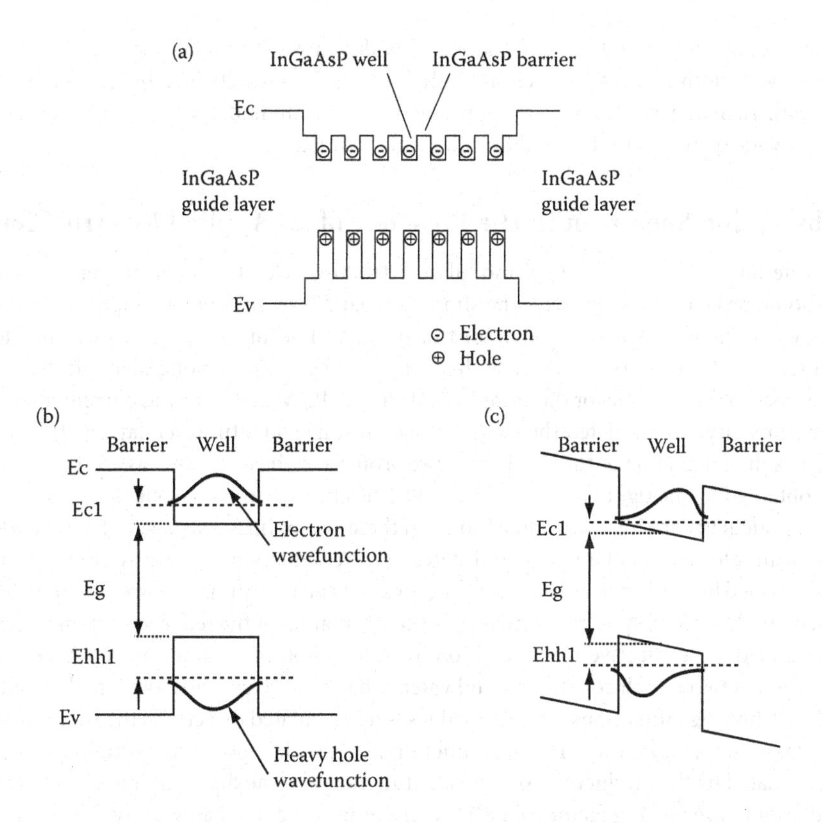


**FIGURE 5.2** (a) An MQW electroabsorption modulator structure. (b) Wave functions in conduction band and valence band for a flat band condition. (c) Deformed wave functions and changed subband levels due to quantum-confined Stark effect with applied electric field.

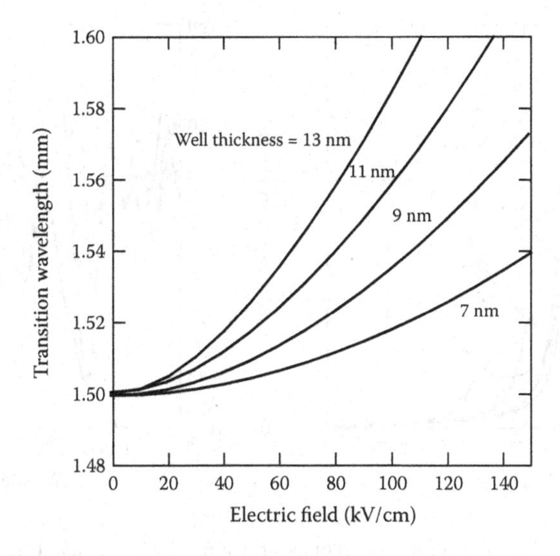


**FIGURE 5.3** Calculated electric field dependence of the intersubband absorption energy wavelength ($E_g + E_{c1} + E_{hh1}$) for an unstrained quantum structure with various well thicknesses.

From these results, it is found that the wider well width shows the higher modulation efficiency. But the increased well width reduces the exciton binding energy and results in deterioration of the absorption change [4]. Then, the well thickness is optimized for over-all modulation characteristics. Typically, around 9-nm-wide quantum well is used for practical modulators.

### 5.1.1 Absorption Spectrum in the Presence of an Applied Electric Field

In order to design the practical MQW modulator structure, electrooptical properties of the MQW structure should be clarified accurately. The simple subband level shift due to quantum-confined Stark effect as shown in Figure 5.3 is not enough for this purpose. Lots of studies were presented to simulate absorption spectra of the MQW structure with an electric field [5–7]. Among them, the exciton Green's function approach [8] is suitable for the complicated strain MQW structure and current numerical techniques. For simplicity, we simulated the structure with a single quantum well layer with barrier layers. The first step is the calculation of the band structure profile and the effective masses. The band discontinuities are obtained by the tight binding method [9]. The effective masses of electrons, heavy holes and light holes are calculated taking the composition and the strain of InGaAsP layers. The second step is the acquisition of all information of quasi-bound states in the conduction band, heavy hole band, and light hole band deformed by an electrical field. Each energy level and eigenfunction was obtained by using the finite element method technique. The third step is the calculation of the reduced exciton Green function matrix derived from the effective-mass equation for the exciton in a quantum well. Overlap integrals of eigenfunction between each conduction and valence bands were used to calculate this matrix. In the last step, the exciton eigenfunctions and eigenvalues were calculated directly as the solution of the eigen value problem of the reduced exciton Green function matrix. The optical susceptibility at each photon energy is calculated by the product of exciton eigenfunction and the dipole moment, the exciton eigenvalues, and exciton broadening factor over all the discretized exciton states. Only the exciton broadening factor is the given factor to fit the experimental results. Figure 5.4a shows the calculated absorption spectrum of an unstrained InGaAsP quantum well structure as the electric field increases from 0 kV/cm to 150 kV/cm. The operating wavelength is detuned from the band edge by approximately 40 nm to 50 nm for practical devices. So the thickness and λg of the well were set to be 9 nm and 1.585 μm to obtain

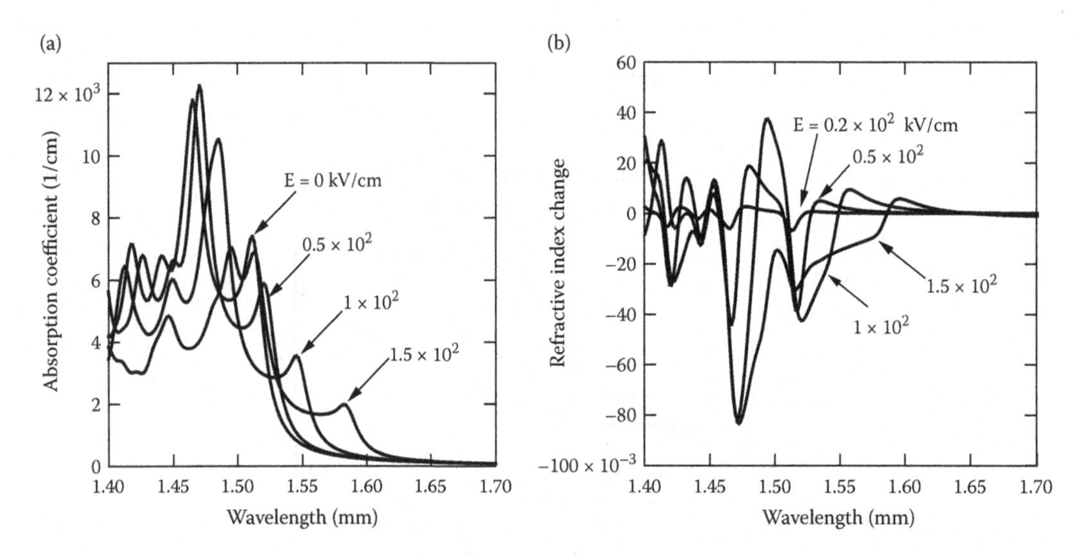


**FIGURE 5.4** (a) Calculated absorption spectrum of an unstrained InGaAsP quantum well structure as increasing an electric field from 0 kV/cm to 150 kV/cm. (b) Simulated refractive index change spectra of an unstrained InGaAsP quantum well structure as increasing an electric field from 0 kV/cm to 150 kV/cm.

the intersubband absorption wavelength of 1.50 μm with no field, when the composition of the barrier layer was set to be 1.15 μm. The electric field effects on the absorption spectrum can be clearly observed from this figure. As the electric field increases the excitonic peak shifts to longer wavelengths, there is also a substantial decrease in the absorption peak caused by the reduction in the overlap integral of the eigenfunctions between the conduction and the valence band and the binding energy of the exciton.

### 5.1.2 Refractive Index Change Due to Absorption Spectrum Change

The chirp parameter governs the transmission characteristic of the electroabsorption modulators [10], because the distortion of the received eye pattern is characterized by the chirp parameter after dispersive fiber transmission. So this is the most important parameter for this device. The chirp parameter is calculated as the ratio between the variations of the real and imaginary parts of the refractive index induced by the electric field.

$$\text{Chirp parameter} = \frac{\Delta n'}{\Delta n''}, \tag{5.2}$$

where $\Delta n'$ is the real part of the complex refractive index change and $\Delta n''$ is the imaginary part of the complex refractive index change. The imaginary part of the complex refractive index change $\Delta n''$ is calculated using the relation:

$$\Delta n'' = \frac{\lambda}{4\pi}\Delta\alpha, \tag{5.3}$$

where $\Delta\alpha$ is the attenuation coefficient change due to the electric field. The electrorefractive index spectra are inherently related to the electroabsorption spectra of the electroabsorption modulator through the Kramers–Kronig relations which can be derived from the causality between electrical flux and fields in a linear and stable medium. Then the real part of the refractive index change is calculated by the absorption change as Equation 5.4:

$$\Delta n' = \frac{\lambda_i^2}{2\pi^2} PV \int_0^\infty \frac{\Delta\alpha}{\lambda_i^2 - \lambda^2} d\lambda. \tag{5.4}$$

Here, $\lambda_i$ is an input light wavelength to the MQW modulator. The symbol *PV* stands for the Cauchy principal value. Figure 5.4b shows the simulated refractive index change spectra under the same condition of Figure 5.4a.

## 5.2 Basic Structure of Electroabsorption Modulator

Strong confinement of the electrical field induced by an input electrical signal and the input light in the MQW structure attains the highest performance electroabsorption modulator.

### 5.2.1 Layered Structure

For semiconductor devices, it is very easy to obtain p- and n-type material by using various doping materials. By applying a reverse bias to a p-i-n structure, it is possible to concentrate a high electric field in the nondoped MQW layer. The MQW structure is sandwiched by p- and n-type guide and clad layers. In general, the MQW structure is nondoped to maintain the good excitonic absorption and both of the guide layers are lightly doped to suppress the excess light absorption. It is also important to adjust the excitonic absorption peak wavelength for the required operating wavelength. Advanced metal-organic vapor phase epitaxy (MOVPE) epitaxy technology enables us to obtain enough accuracy for the absorption wavelength because the layer thickness, composition, and the interface between layers are well controlled. The detuning (= the operating wavelength–the excitonic absorption peak wavelength) is set to optimize the extinction ratio, the chirp parameter, and the insertion loss. The value of 40 nm to 50 nm is often used.

### 5.2.2 Waveguide Structure

The modulator waveguide is categorized as a passive waveguide. The modulator waveguide is designed to be a straight confined waveguide structure with a lossy core. Even an oversized waveguide structure with higher lateral modes works well when the input light is not an off-axis beam because the modulator length is less than 300 μm. But from the coupling efficiency point of view, a BH or a ridge waveguide structure are widely used. The BH structure is superior due to its circular radiation pattern of the output beam which is well coupled to the single mode fiber.

## 5.3 Electroabsorption Modulator Design

For long-haul optical fiber transmission application, the high speed, high power, and low chirp characteristics are absolutely necessary for the electroabsorption modulator. This section shows the design consideration for these characteristics.

### 5.3.1 Design for High-Speed Operation

Figure 5.5 shows the equivalent circuit model for the electroabsorption modulator and the drive circuit.

The modulator is expressed by the series circuit of the series resistance $R_s$ and the capacitance $C$. The reverse biased electroabsorption modulator exhibits a large resistance for no input light. Then the load resistance $R_L$ is connected parallel to the modulator for radio frequency matching. When the modulator is driven by the radio frequency signal with source resistance of $R_i$, the cut-off frequency $f_c$ is calculated as Equation 5.5:

$$f_C = \frac{1}{2\pi\left(R_S + \left(\frac{1}{R_i} + \frac{1}{R_L}\right)^{-1}\right)C}. \tag{5.5}$$

To increase the cut-off frequency, the reduction of $C$ is very important. The capacitance consists of the reverse biased p-n junction capacitance, the parasitic capacitance, and the bonding pad capacitance. Thus, the parasitic capacitance and bonding pad capacitance should be as small as possible. In practical

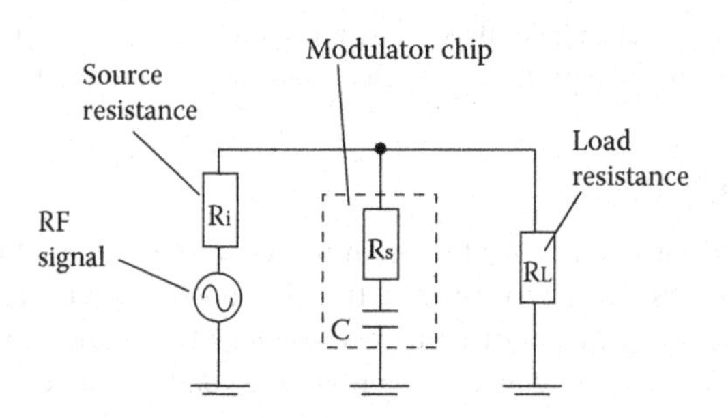


**FIGURE 5.5** An equivalent circuit model for the electroabsorption modulator and the drive circuit.

electroabsorption modulators, a semiinsulating InP BH structure is used to reduce the parasitic capacitance and the bonding pad capacitance. The junction capacitance of the modulator is determined by the total thickness of nondoped MQW layers and the length of the modulator. Thicker nondoped layers and shorter modulator length reduce the capacitance, but at the expense of a reduction of the applied electric field and the extinction ratio. So, in practical designs, the nondoped layer thickness and the modulator length are designed to optimize three characteristics of the cut-off frequency, the extinction ratio, and the operating voltage. The capacitance is typically 0.5 pF and the series resistance is several ohms, so over 12 GHz cut-off frequency is easily obtained for 50 ohm matching condition. For high output power operation, we observed the cut-off frequency reduction due to the carrier pile-up mechanism. This effect will be explained in the following section.

## 5.3.2 Design for Suppressing Carrier Pile-Up

Unstrained InGaAsP/InGaAsP MQW modulators suffer from a saturation of the absorption at moderate light intensity, which induces the deterioration of dynamic performance. In particular, the cut-off frequency is reduced and output eye is distorted. Such absorption saturation is attributed to the field screening in the quantum wells by accumulation of holes trapped by a large valence band discontinuity between the well and the barrier [11]. Then the careful design is required to control the valence band discontinuity.

As shown in Figure 5.6, two kinds of modulators are compared to investigate the carrier pile-up characteristics [12].

The type A structure has unstrained MQW structure and stepped separate confinement heterostructure. On the other hand, type B has strained MQW structure where the well has 0.5% compressive strain and the barrier has −0.3% tensile strain. Type B has graded separate confinement heterostructure. The number and the thickness of wells are 7 nm and 9 nm, and the barrier width is 5.1 nm for both type structures. For 1.55 μm wavelength operation, the PL wavelength of the MQW modulator is adjusted to be 1.49 μm. The valence band discontinuity of types A and B is estimated to be 150 meV and 74 meV by the tight-binding model. For the conduction band, discontinuity is kept to be 95 meV for both structures. Type B structure is expected to reduce the hole escape time from wells. In order to investigate experimentally the characteristics for the high photocurrent condition, a DFB laser is integrated. A photocurrent of 40 mA is obtained at the reverse bias of −2 V for both type of modulators. Figure 5.7 shows the cut-off frequency and reverse bias characteristics.

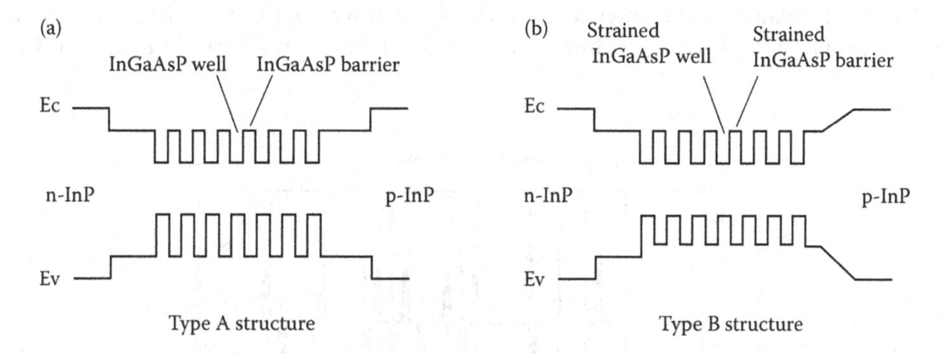


**FIGURE 5.6** Two kinds of modulator structures. Type A structure has unstrained MQW structure and step separate confinement hetero (SCH) structure. Type B has strained MQW structure, where the well has 0.5% compressive strain and the barrier has −0.3% tensile strain. Type B has graded SCH structure. The number and the thickness of wells are 7 nm and 9 nm, and the barrier width is 5.1 nm for both type structures. For 1.55 μm wavelength operation, the PL wavelength of the MQW modulator is adjusted to be 1.49 μm. The valence band discontinuity of Types A and B is estimated to be 150 meV and 74 meV by the tight-binding model. For the conduction band, discontinuity is kept to be 95 meV for both structures.

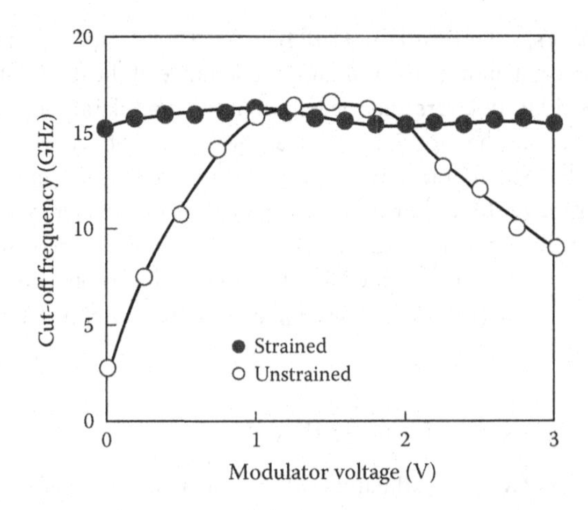


**FIGURE 5.7** Cut-off frequency and reverse bias characteristics for Types A and B structure obtained experimentally.

It is clearly shown there was a drop of the cut-off frequency at the high photocurrent condition for only type A unstrained structure. Type B strained structure showed no reduction of the cut-off frequency. The cut-off frequency was almost 15 GHz for any voltage. From these results, it is shown that the strained MQW structure is suited for practical devices.

### 5.3.3 Design for Negative Chirp

Koyama and Iga showed the wavelength chirp of the external modulator is determined by the chirp parameter and modulated waveform. The wavelength chirp $\Delta\lambda$ of the electroabsorption modulator is expressed in Equation 5.6:

$$\Delta\lambda = -\frac{\lambda_i^2}{2\pi c}\alpha_H \frac{1}{P}\frac{dP}{dt}, \tag{5.6}$$

where $P$ is the modulated output power, $\alpha_H$ is the chirp parameter and $c$ is the light velocity. Figure 5.8 shows the example of the wavelength chirp characteristics of the electroabsorption modulator with $\alpha_H$ of 0.4.

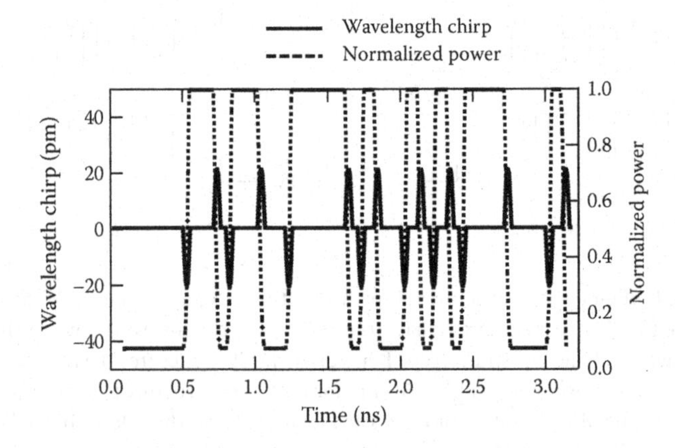


**FIGURE 5.8** An example of the wavelength chirp characteristics of the electroabsorption modulator with $\alpha_H$ of 0.4.

At the rising edge, the output wavelength is shifted to the shorter wavelength. On the other hand, the wavelength is shifted to the longer wavelength at the falling edge. Even such small wavelength changes seriously deformed the eye pattern after long haul transmission. The calculated waveforms after dispersive fiber transmission of the electroabsorption modulator with chirp parameters of 0.4 and −0.8 under 10 Gb/s modulation are shown in Figure 5.9.

For simplicity, the chirp parameter is fixed and the extinction ratio is set to be 10 dB. It is obvious that the negative chirp parameter is very important for the clear eye opening for 1600 ps/nm dispersion. For the positive chirp parameters, 0 level is shifted to 1 level due to the large pulse broadening. The chirp parameter $\alpha_H$ is expressed by Equation 5.7:

$$\alpha_H = \frac{2\lambda_i}{\pi\Delta\alpha}\lim_{\varepsilon\to 0}\left(\int_0^{\lambda_i-\varepsilon}\frac{\Delta\alpha}{\lambda_i^2-\lambda^2}d\lambda+\int_{\lambda_i+\varepsilon}^{\infty}\frac{\Delta\alpha}{\lambda_i^2-\lambda^2}d\lambda\right). \tag{5.7}$$

Thus, it is proportional to a shaded area minus two hatched areas as shown in Figure 5.10.

Then the negative chirp parameter is obtained in a large decrease of the shaded area and a large increase of the hatched areas. From viewpoint of this relation, the absorption from the low-order transition should be dominant. The design of strains in the quantum well and the barriers enables such condition and provides the negative chirp parameters without degradations of the insertion loss and the

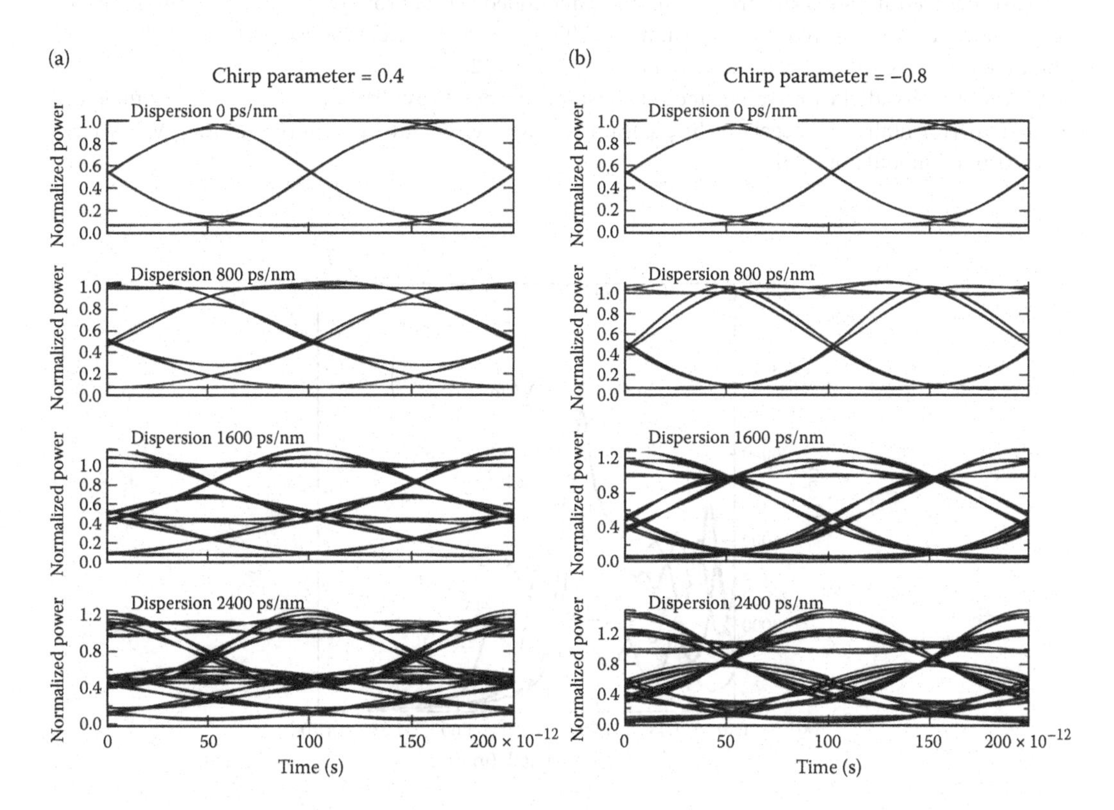


**FIGURE 5.9** Calculated waveforms after dispersive fiber transmission of the electroabsorption modulator with various chirp parameters under 10 Gb/s modulation.

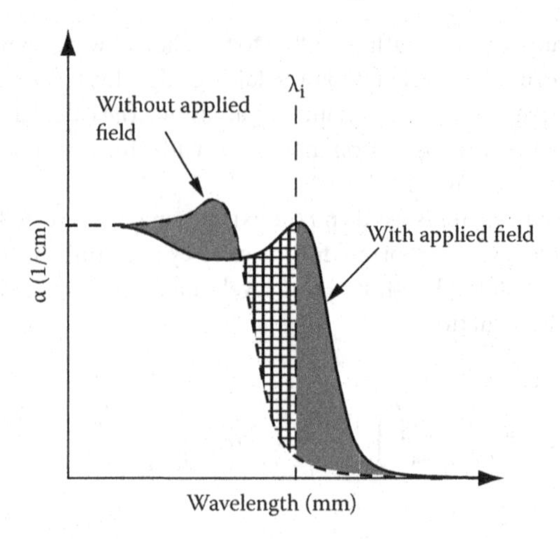


**FIGURE 5.10** Illustrated estimation of the chirp parameter from absorption spectra.

modulation efficiency. Figure 5.11 shows the calculated absorption spectrum for 0.8% strained well and −0.7% strained barrier quantum well structure.

The quantum well thickness is set to be 9 nm. In this case, the light hole transition and heavy hole higher-order transition disappeared because of unbounded states. From the figure, it is found that the shorter wavelength absorption reduces remarkably as the electric field increases. As a result, we can obtain the negative chirp parameters as shown in Figure 5.12a.

On the other hand, the unstrained modulator shows the positive chirp parameters as shown in Figure 5.12b. Experimentally, the negative chirp parameters were also confirmed in the strained MQW electroabsorption modulators [13].

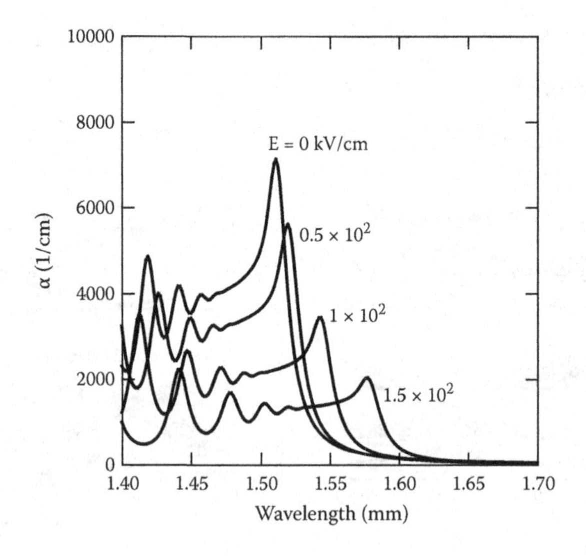


**FIGURE 5.11** Calculated absorption spectrum for 0.8% strained well and −0.7% strained barrier quantum well structure, having 9-nm-thick quantum well.

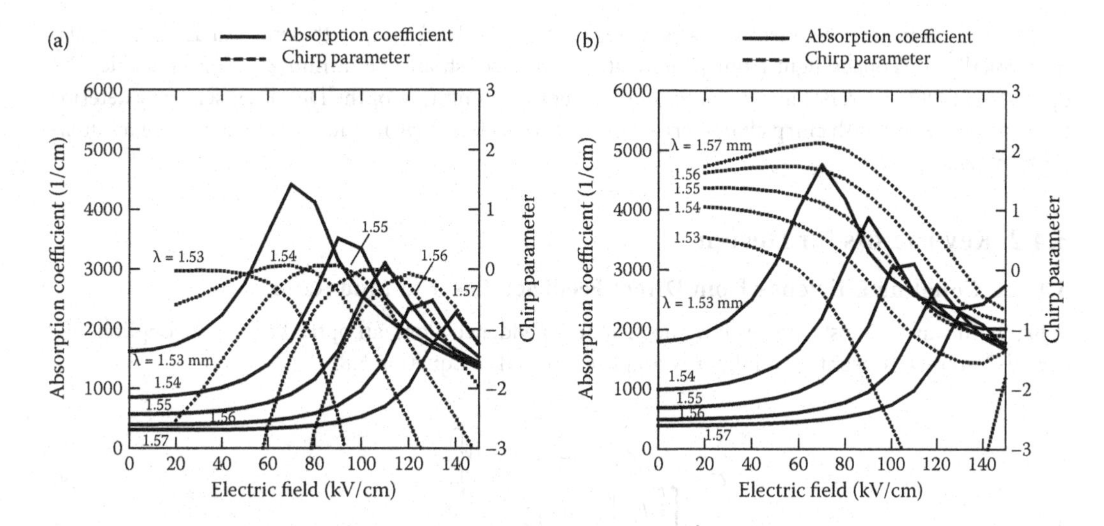


**FIGURE 5.12** (a) Calculated chirp parameters for 0.8% strained well and −0.7% strained barrier quantum well structure, having 9-nm-thick quantum well. (b) Chirp parameters for unstrained quantum well structure.

## 5.4 Electroabsorption Modulator Integrated with Laser

The discrete modulator is not suited for practical applications because it needs an expensive fiber coupling for both input and output side. Monolithic integration of electroabsorption modulator and DFB laser provided us the compact and high performance electroabsorption modulated laser (EML) suited for the practical long-haul 10 Gb/s fiber transmission systems.

### 5.4.1 Basic Structure

The EML is the first practical monolithic opto-semiconductor device. The functions of the continuous wave (CW) light source and the high speed modulator are integrated into one semiconductor chip. Figure 5.13 shows the schematic structure of EML.

Important issues between the DFB laser and the modulator for this integration are the coupling for the light and the isolation for the electrical signal. The emitted light from DFB laser should be guided to the modulator through the optical waveguide structure with a small loss. Modulation signals applied

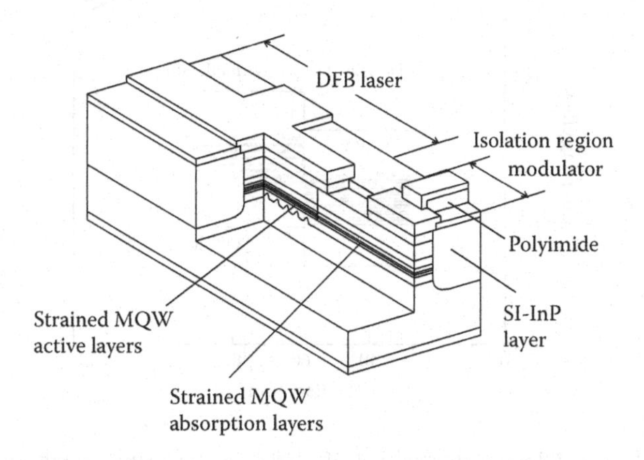


**FIGURE 5.13** Schematic view of EML.

to the modulator must be effectively isolated from the CW DFB laser with semiinsulating layer BH. Additionally, a feedback light from a modulator front facet should be minimized to keep stable CW operation of DFB laser. Because a change of the operation condition of the DFB laser seriously deteriorates the low wavelength chirp characteristics, an antireflection coating must be applied to the modulator front facet.

## 5.4.2 Key Factors for Design

### 5.4.2.1 Coupling Efficiency from Direct Feedback Laser to Modulator

When both waveguide structures have no spatial dependence along an optical axis, the coupling efficiency $C$ from DFB laser to modulator is simply expressed as Equation 5.8:

$$C = \frac{\left| \iint E_{dfb} E_{mod} {}^{*} \, dx\, dy \right|^2}{\iint \left| E_{dfb} \right|^2 dx\, dy \iint \left| E_{mod} \right|^2 dx\, dy}, \tag{5.8}$$

where $E_{dfb}$ is the fundamental mode field of DFB laser, $E_{mod}$ is the transmitted fundamental mode field of the modulator. The asterisk means the complex conjugated field. From this expression, it is clear that the optical field distribution of modulator should match that of DFB laser for high coupling efficiency. The MQW structure of the modulator and the DFB laser is not identical in thickness, composition, or well numbers. Nonetheless, guide layers and control of the waveguide offset can provide high coupling efficiency. Figure 5.14 shows the typical butt-joint structure of DFB laser and modulator and the calculated coupling efficiency dependence of the waveguide offset.

In the calculation, we assumed that the DFB MQW structure has seven compressively strained (1%) InGaAsP wells and lattice matched InGaAsP barriers ($\lambda g = 1.3$ μm) with thicknesses of 5 nm and 9 nm and the modulator MQW has seven compressively strained (0.8%) InGaAsP wells and tensile strained (−0.7%) InGaAsP barriers ($\lambda g = 1.15$ μm) with thicknesses of 9 nm and 5 nm. Both MQW are sandwiched by 0.1 μm thick InGaAsP separate confinement heterostructure layers ($\lambda g = 1.15$ μm). From this

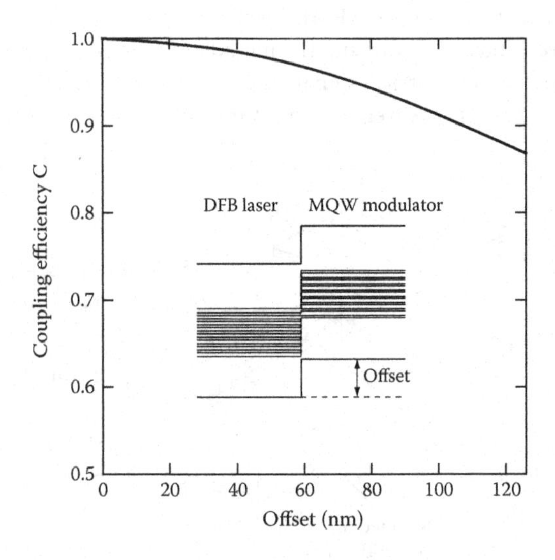


**FIGURE 5.14** Typical butt-joint structure model of DFB laser and modulator, and the calculated coupling efficiency dependence of the waveguide offset.

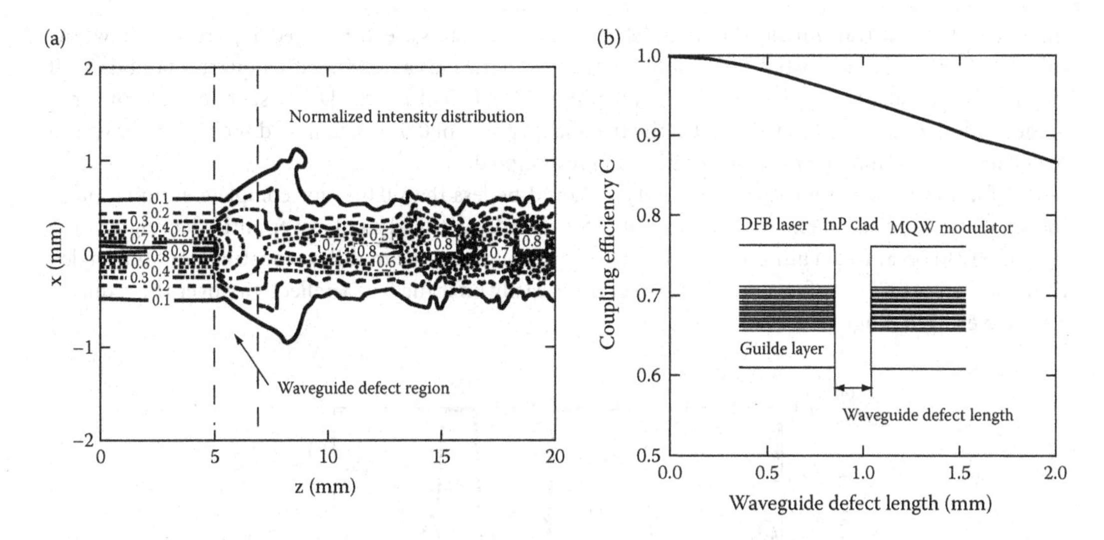


**FIGURE 5.15** (a) Calculated field propagation when a 2 μm long waveguide defect is introduced at the interface between the DFB laser and the modulator. (b) Waveguide defect length dependence of the coupling efficiency.

figure, there is no serious reduction in the coupling efficiency when the offset is less than around 50 nm. In current fabrication process, this offset is achievable. For further analysis of the coupling efficiency, a finite difference beam propagation mode analysis [14] is very useful. By using this analysis, we can consider the thickness and the refractive index distribution along optical axis. Figure 5.15a shows the calculated field propagation when a 2 μm long waveguide defect is introduced at the interface between the DFB laser and the modulator.

As DFB laser waveguide terminates at $z = 5$ μm, the optical field is radiated at $z = 5$ μm. But after the waveguide of the modulator appears from $z = 7$ μm, the optical field is guided again. Such a short waveguide defect remarkably reduces the coupling efficiency. Figure 5.15b shows the waveguide defect length dependence of the coupling efficiency. From this figure, it was found that the defect length should be less than 0.5 μm for good optical coupling. During the regrowth process, voids should be eliminated.

### 5.4.2.2 Electrical Isolation

The DFB laser is a forward bias device, while the modulator is a reverse bias device. If the electrical isolation is not enough, an unexpected current flows from the constant current source of the DFB laser to the modulator at reverse bias. This effect changes the operating condition of DFB laser and induces the dynamic wavelength change. In general, frequency modulation efficiency of DFB laser is around 0.3 GHz/mA–1 GHz/mA. Then for less than 0.1 GHz frequency change, the change of DFB laser current should be less than 0.1 mA. From these results, the isolation resistance should be larger than 20 kΩ. By using semiinsulating BH structure with about 100 μm long isolation region, the isolation resistance of 50 kΩ is realized.

### 5.4.2.3 Front Facet Reflectivity

So called adiabatic wavelength chirp occurs due to the residual front facet reflectivity in an EML. This wavelength chirp is expressed by the dynamic wavelength change for "on" and "off" level of modulation. Because the DFB laser region is modulated by the equivalent front facet reflectivity modulation, this adiabatic wavelength chip induces serious eye distortion after fiber transmission. For the wavelength chirp analysis, some dynamic large signal DFB laser and absorption modulator combined models are invented [15,16]. In these models, the EML is divided into multisection simple functions and analyzed by using rate equations of photons and carriers, and the connection equation of each section at each time step. Here, the transmission laser model base analysis results are introduced. Figure 5.16 shows the calculated chirp characteristics for various front reflectivities for an EML with a quarter lambda shift grating. In this calculation, the normalized coupling coefficient of 1.5, the DFB laser chirp parameter of 3, the modulator chirp parameter of 0.4, the front facet phase of 0.5*π radian, and the DFB laser length of 300 μm and the modulator length of 200 μm are assumed.

It is found that the front facet reflectivity $r_f$ should be less than 0.01% for complete elimination of the adiabatic wavelength chirp. It is impossible for the conventional single layer antireflecting coating technology to obtain such ultra-low front facet reflectivity. Thus, the multilayer coatings and the angled front facet structure are often used. Multilayer films easily offer the 0.01% reflectivity over the required wide wavelength range.

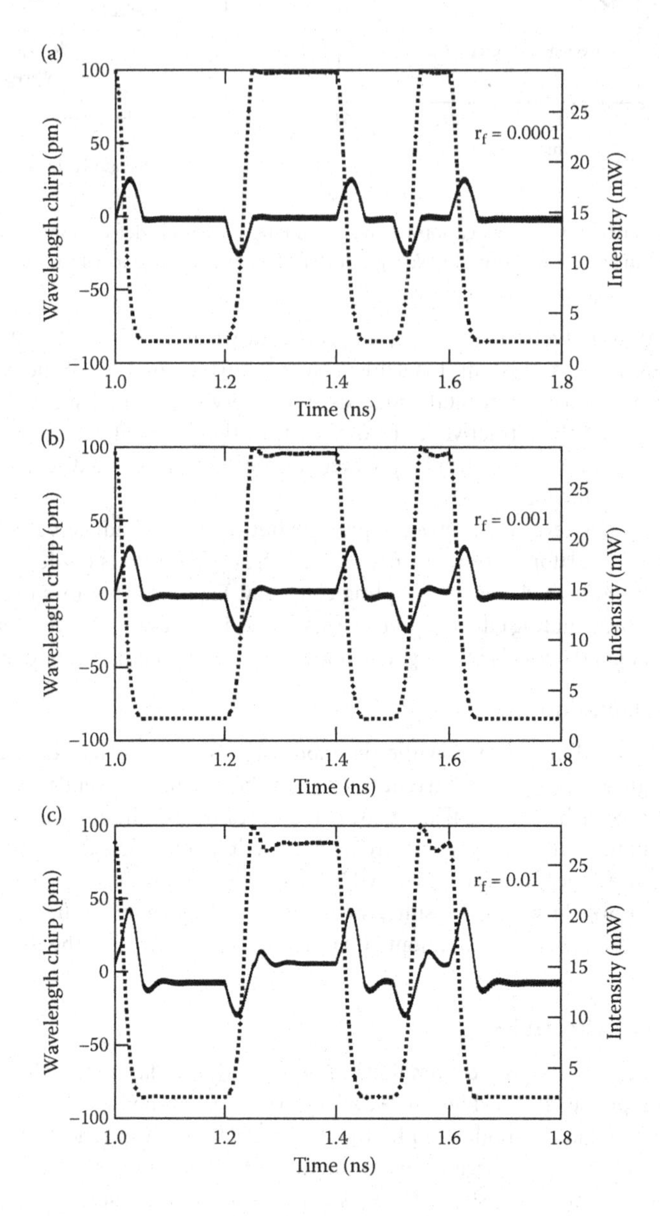


**FIGURE 5.16** Calculated chirp characteristics for various front reflectivity of EML having quarter lambda shift grating.

## 5.4.3 Integration Technology

Unlike electrical integrated circuits (ICs), there is a difficulty to make different layered structures and connect both structures optically. The special epitaxy technologies that have been developed to address this issue are discussed here.

### 5.4.3.1 Butt–Joint Coupling Technology

Butt–joint coupling between two waveguides is expected to be one of most suitable technologies because of complete mode matching and great flexibility in the design of both waveguide structures compared with other technologies [17,18]. There is a long development history from liquid phase epitaxy to MOVPE technology era. Basically, butt–joint coupling technology is very simple. After the base wafer growth, the dielectric etch mask is patterned on the base wafer and unnecessary base layers in open area are etched off. After that, the connected waveguide layers are grown on the etched area, while the deposition of epitaxial material elsewhere is inhibited by the dielectric mask. In early development, the obtained coupling efficiency was very low compared with the theoretical values because poor coupling resulted from the anomalous butt–joint coupling shape due to the undesirable regrowth shape and immature wet chemical etching technology. Slanted facets after wet chemical etching induce the anomalous butt–joint coupling shape during regrowth process. In addition, layer uniformity was not enough for practical production. To solve this problem, dry etching technology was introduced. The etched facet became almost vertical and the shape and depth were well controlled. By using such etched facets and an optimized growth rate enables uniform and low-loss butt–joint coupling structure.

### 5.4.3.2 Selective Area Growth (SAG) Technology

Butt–joint coupling technology requires two epitaxy steps at least. Epitaxy process is known as an expensive and time-consuming process. SAG technology was developed to grow both waveguide structures simultaneously, thus reducing the epitaxy steps [19,20]. The shorter bandgap required in the DFB laser region can be obtained by growing this part of the device between a pair of dielectric SAG masks. Just as in the butt coupling technology, the deposition of epitaxial material is inhibited by such dielectric SAG mask. Some of the material that would have been deposited on the masked area is deposited between the pair of dielectric masks. Thus, the growth rate increases in these regions. For quantum well structure, the equivalent bandgap (= quantum level) is directly related to the well thickness. Thus, the thickness changes caused by this SAG technology directly changes the band-gap in the quantum well structure. Fortunately, we can use quantum well structure for both DFB laser and modulator. So such SAG technology is applicable for EML. Compared with the butt–joint coupling process, SAG is superior in terms of ease of continuous bandgap control, almost 100% coupling efficiency, and high quality of the growth. However, it is inferior in the design flexibility because of its limitations in simultaneous growth of the epitaxy layers.

## 5.4.4 Characteristics

The typical EML device consists of a 300 μm long DFB laser, a 50 μm isolation region, and a 200 μm long modulator, as shown in Figure 5.13. We fabricated such EMLs by using a five step all-MOVPE growth technique. First, the distributed feedback grating was etched into an n-InP substrate by $C_2H_6$-RIE (reactive ion etching) to a depth consistent with the optimum κL value of 1.5. Next, the strained MQW laser was grown by MOVPE. The MQW structure has ten compressively strained (0.8%) InGaAsP wells and lattice matched InGaAsP barriers (λg = 1.3 μm) with thicknesses of 5.1 nm and 10 nm. Such MQW structure was sandwiched by InGaAsP separate confinement heterostructure layer (λg = 1.15 μm). The obtained photoluminescence wavelength of this MQW structure was 1.58 μm. After defining the DFB laser with a C2H6/H2/O2-RIE, the MQW modulator was grown by MOVPE. Here, the seven strained (0.5%) 9-nm- thick InGaAsP wells and tensile strained (−0.3%) 5.1 nm thick barriers were used. Then,

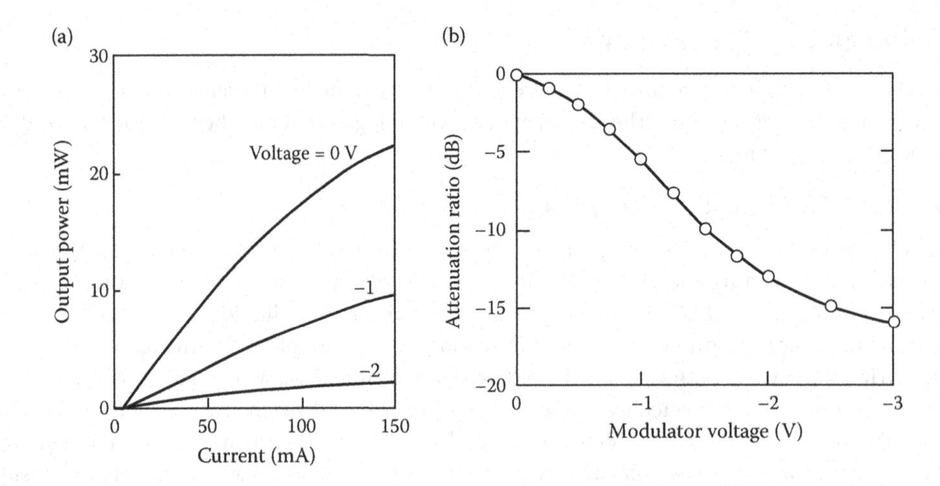


**FIGURE 5.17** (a) Light-output current characteristics for various modulation bias voltage. (b) Attenuation ratio dependence on modulator bias voltage.

the p-InP clad and p+–InGaAsP contact layer were grown. After a 1.5 μm wide waveguide mesa was defined by etching, an iron (Fe) doped InP semiinsulating layer was grown by MOVPE. To reduce the electrical cross talk between the DFB laser and the modulator, an isolation region was formed by removing the contact layer. Electrodes were formed on both sides of the wafer and a thick layer of polyimide was applied to minimize the modulator bonding pad capacitance. Finally, high reflection and antireflection coating were applied to the back and front facets.

The light-current curve in Figure 5.17a shows the 5.5 mA threshold current and 24 mW peak power with 140 mA laser driver current.

Since the output from a solitary 300 μm DFB laser is 33 mW driven with 140 mA, the insertion loss of the modulator is around 1.4 dB when the modulator is unbiased. The lasing wavelength is 1.556 μm. The resistance of the isolation region is typically 50 kΩ. For light coupled into an optical fiber, the attenuation characteristics were measured. Figure 5.17b shows the attenuation ratio dependence on modulator bias voltage when 100 mA is injected into the laser. An attenuation ratio of −10 dB for 2.0 V peak-to-peak operation was obtained when −2.0 V was applied to the modulator. Figure 5.18a shows

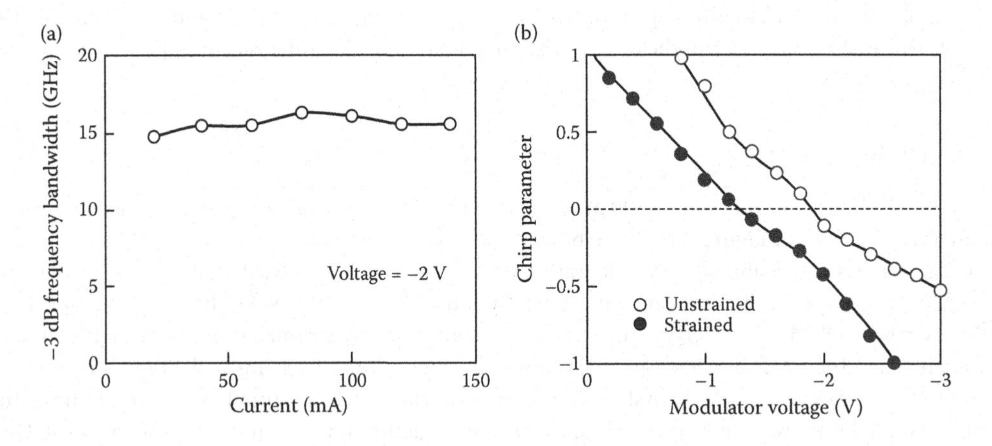


**FIGURE 5.18** (a) The −3-dB frequency bandwidth dependence on the laser injection current. (b) Experimental chirp parameters of fabricated strained MQW EML.

the –3 dB frequency bandwidth dependence on the laser injection current when –2.0 V is applied to the modulator.

From the output power of a 300 μm solitary DFB laser, light power coupled into the modulator was estimated to change from 3 mW to 30 mW for a change in the laser injection current of between 20 mA and 140 mA. The frequency bandwidth of 15 GHz was not sensitive to changes in the laser injection current. This means that fabricated InGaAsP MQW modulator does not suffer from frequency degradation caused by hole accumulation and that much higher speed modulation can be achieved by reducing its parasitic capacitance. The obtained symmetric, single lobe far field pattern indicates the absence of interference from scattered light, and is clear evidence of good optical coupling at the butt-joint. Vertical reactive ion etching and MOVPE with precise thickness control suppressed abnormal overgrowth and produced the high-quality butt-joint.

Figure 5.18b shows the experimental chirp parameters of fabricated strained MQW EML.

The operating temperature was 25°C. The chirp parameters were measured via a fiber response peak method. Chirp parameter characteristics of these strained layers differ from conventional unstrained ones. The chirp parameter reduced remarkably and crossed zero as the reverse bias voltage increased. These characteristics are similar to the theoretical prediction.

## 5.5 Packaging of Electroabsorption Modulator Integrated with Laser

Compared with other modulators, the EML is a very short device. Typically, the device length is about 0.5 mm. Thus, a new packaging technology was established to take advantage of such compactness for pluggable transceivers.

### 5.5.1 Compact Packaging Concept

EMLs were first assembled in a butterfly pig-tail module mainly for high end standard 300 pin multi-source agreement transceivers. This legacy transceiver module is as large as 114.3 × 88.9 × 13.5 mm, and consumes a lot of power. So a new type 10 Gigabit small form factor pluggable module is developed and installed in practical system. This 10 Gbit small form factor pluggable module converts serial electrical signals to external serial optical. The small form factor pluggable module is hot pluggable and supports SONET [OC192/STM-64], 10G Fiber Channel, G.709, and 10 Gigabit Ethernet with the same module. 10 Gbit/s miniature device-transmitter optical subassembly (TOSA) is one of the transmitter modules for these pluggable modules. In a TOSA, the EML is mounted on a miniature thermoelectric cooler in the small package (19 × 6 × 6 mm). The output light is coupled by a one or two lens system. The package is hermetically sealed for high reliable operations. Generally, 10 Gb/s input signals were fed through the flexible cable at the rear of the package. For radio frequency (RF) matching, 50 ohm termination was done in the package. Careful design of the radio frequency circuit of the package and some equalizer circuits are needed to obtain good eye openings. After 50 km transmission (800 ps/nm dispersion), the path penalty is less than 1 dB. For usage of high performance thermoelectric cooler and EML, the power consumption is around 1.1 W even at 75°C for 0 dBm average output power under 10 Gb/s modulation.

### 5.5.2 Transmitter Optical Sub-Assembly Integrated with Driver IC

To further reduce the power consumption of the entire transceiver and attain good extinction ratio and eye waveform, the connection between EML and driver IC should be as short as possible. Figure 5.19a shows a photograph of a driver integrated EML TOSA. The size is 25.7 × 6 × 6 mm.

In this TOSA, the driver IC was also assembled in the package and the output of driver IC is connected to the EML on the thermoelectric cooler through a microwave transmission line. As shown in Figure 5.19b, an alternating current (ac)-coupled circuit is used to reduce the power consumption of

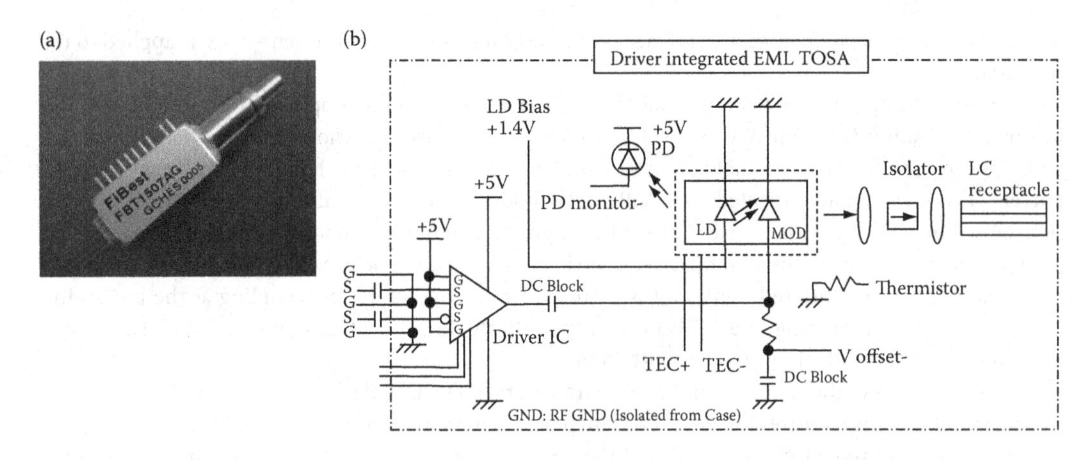


**FIGURE 5.19** (a) Photograph of a driver integrated EML TOSA. (b) Driver integrated EML TOSA internal circuit.

modulator direct current (DC) bias of EML. The driver IC works as a waveform shaper to make clear optical eye opening even with a small voltage swing. Figure 5.20a shows an example of nonreturn-to-zero eye patterns under 10 Gb/s modulation.

It was found that a small jitter of 2.1 ps and a large mask margin of 25% were obtained with an extinction ratio of 9.8 dB. The total power consumption is only 1.38 W at the case temperature of 75°C. After a 100-km transmission (1600 ps/nm dispersion), a clear eye opening was obtained as shown in Figure 5.20b. The bit-error rate characteristics are also shown in Figure 5.21. The path penalty after 100 km was only 1.7 dB under 10 Gb/s modulation.

For practical applications, the reliability of EML TOSAs is very important. Appropriate burn-in and process control enabled us to achieve around 20 failures per billion hours (also referred to as failure in time, FITs) for 35°C 20 years operation for EML chip. Application of yttrium aluminum garnet welding and assembly technique with high temperature soldering also realized very stable optical coupling of less than 100 FITs (40°C 20 years operation) for EML TOSAs. Many thousands of EML TOSAs have operated reliably in 10-Gb/s long-distance transmission system with a simple conventional automatic power control circuit.

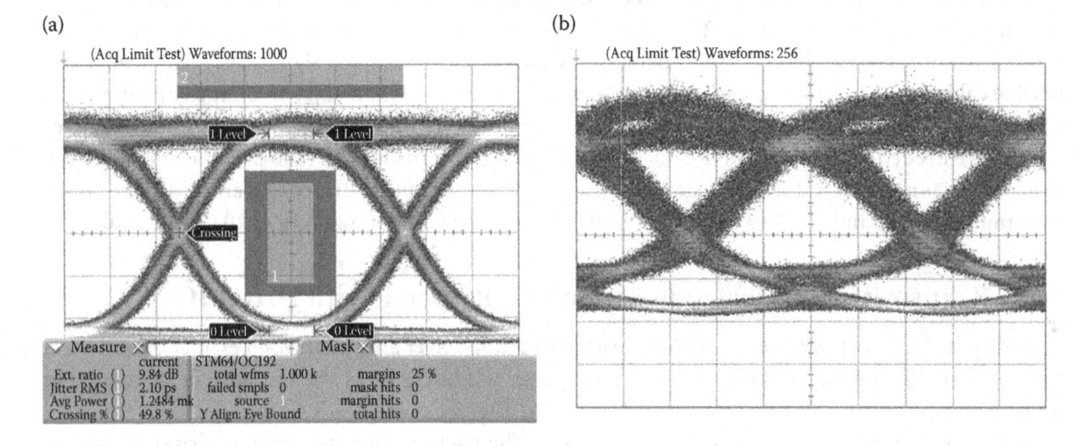


**FIGURE 5.20** (a) An example of eye patterns under 10-Gb/s NRZ modulation of a driver integrated EML TOSA. (b) Eye waveform after 100-km transmission (1600 ps/nm dispersion).

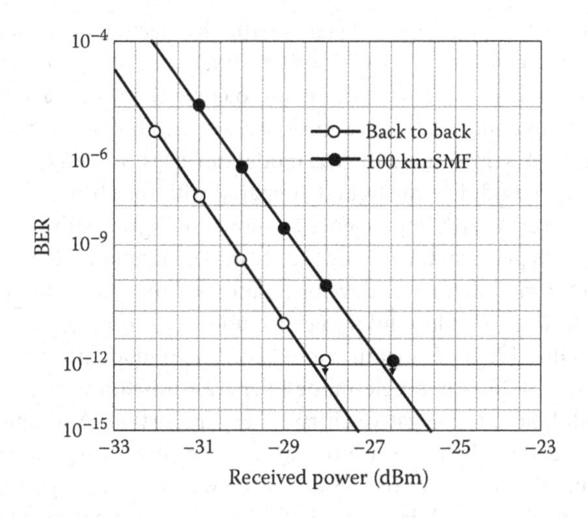


**FIGURE 5.21** Bit-error rate characteristics.

# 5.6 Comparison and Future Technology Trend

## 5.6.1 Comparison of State-of-the-Art EML

Table 5.1 shows reported state-of-the-art EML devices for 10 Gb/s applications. All EMLs adopted the MQW structure for both the DFB and modulator regions. Butt-joint coupling technology was preferred to optimize the structure [21,22].

## 5.6.2 Competing Technologies

From a compact modulation technology perspective, a semiconductor Mach-Zehnder (MZ) modulator technology and a chirp managed laser technology are competitive technology. Both are not widely used for the present practical transmission system due to an immature technology but they have some advantages for EML.

**TABLE 5.1** State-of-the-Art EML Devices for 10-Gb/s Applications

| Company | Opnext | SEDI | NEL | Mitsubishi | Cyoptics |
|---|---|---|---|---|---|
| Crystal system | InGaAsP/InP | InGaAsP/InP | InGaAsP/InP | InGaAsP/InP | InGaAsP/InP |
| DFB structure | MQW | MQW | MQW | MQW | MQW |
| Mod structure | MQW GRIN-SCH | MQW | MQW | Shallow MQW | MQW |
| Integration technology | Butt-joint | Butt-joint | Butt-joint | Butt-joint | Butt-joint |
| Waveguide structure | BH | BH | BH | Ridge | BH |
| Operating current (mA) | 100 | | | 85 | 110 |
| Operating temperature (°C) | 50 | 40 | 45 | 45 | 40 |
| Output power (dBm) | 1.6 | 0 | 2 | 3 | 2 |
| Extinction ratio (dB) | 10.9 | 9 | 10 | 10.1 | 10 |
| Bit rate (Gb/s) | 10.7 | 10 | 10 | 10.7 | 10 |
| Path penalty (dB) | 1.9 at 80 km | 2 at 100 km | 2 at 100 km | 1.6 at 96 km | 2 at 100 km |
| Reference | 19 | *TOSA ELD54010K-N | *TOSA NLK3C8FBUB | 20 | *TOSA 1625/26-Series |

As shown in Section 5.3.3, the chirp parameter is strongly affected by the detuning. For a required operating wavelength, the band-gap of the MQW modulator must be properly adjusted to obtain the negative chirp parameter. Thus, it is very difficult to cover the wide wavelength range of C-band (1530–1565 nm) by using one modulator structure. A wavelength insensitive modulator is required for a tunable laser source. A semiconductor MZ modulator uses the refractive index change due to an applied electrical field to modulate the light. Compared with the absorption change, the refractive index change has a weak wavelength dependence as shown in Figure 5.4b. To date, there was a lot of research to achieve the high performance semiconductor MZ modulator. Recently, the traveling wave electrode n-i-n MZ MQW modulator was developed and assembled in a compact module with tunable DFB laser array [23]. This module exhibits a continuous-wave output power of 6.5 dBm and a full C-band 10-Gb/s optical duo-binary transmission with 2.4 Vpp modulation. The path penalty of less than 1 dB was achieved in a 200 km single-mode fiber transmission over entire C-band range. For nonreturn-to-zero modulation, the negative chirp is very important. A lumped electrode InP-based MZ Modulator with Ru-doped SI-BH-structure showed a negative chirp parameter and a transmission distance of 103 km with a penalty of 2–3 dB in the wavelength range 1528–1600 nm [24]. The drive voltage was 3.8 Vpp. These modulators have the advantage of small size but the disadvantage of high insertion loss.

A chirp managed laser is based on the conversion technology from a frequency modulation signal to an amplitude modulation signal [25]. The frequency modulated signal from the DFB laser is coupled to the waveform shaper which is a detuned Fabry–Perot etalon. Due to edge filtering of Fabry–Perot etalon, the output from the waveform shaper is converted to the amplitude modulated signal with negative chirp characteristics. Optimization of the quantity of the frequency modulation, and the finesse and the detuning of Fabry–Perot etalon enables good transmission characteristics for over 100-km transmission span. These long-distance transmission characteristics are very attractive. However, the precise detuning is indispensable from the operation principal. So compared with EML, the complicated control circuit is necessary to adjust the wavelength detuning. Chirp managed lasers can exhibit extinction ratio penalties.

### 5.6.3 Advanced Electroabsorption Modulated Laser Technology

Most of today's applications for EMLs are in 2.5 Gb/s and 10 Gb/s WDM transmission systems. Here, we discuss some advanced EML technologies for future applications. A reduction on the power consumption of the transceiver module is strongly required for the densely packaging system. The TEC is one of the most power consuming devices for EML TOSAs. So elimination of the TEC realizes a noticeable power reduction. In uncooled EMLs, the DFB laser must operate at high temperatures of 70°C or 85°C, and the structure must be optimized for the changes in wavelength due to the temperature change from −40°C to 85°C. Recently, 1.55 μm uncooled electroabsorption/DFB-LD coaxial-type InGaAlAs TOSA was demonstrated successively for wide temperature range of −5°C to 75°C [26]. The InGaAlAs multiquantum well system has good temperature characteristics compared with the InGaAsP/InP system because of the larger conduction band offset. For the entire temperature range, the path penalty of about 2 dB was attained after a 10 Gb/s 40 km transmission. For a ridge-waveguide structure InGaAlAs uncooled EA modulator integrated DFB laser, an extended operating temperature range of 125°C (−25°C to 100°C) was demonstrated [27]. An 80-km single mode fiber transmission with a power penalty of below 2 dB was achieved by adjusting only the bias voltage of the EA modulator while maintaining the constant modulation voltage swing at all operating temperatures.

For higher bit rates of 40 Gb/s or 100 Gb/s operation, advanced EMLs are reported. The cut-off frequency of EML is determined by the capacitance of the modulator region. Thus, shortening of the modulator length and reduction of parasitic capacitance were applied to 1.55-μm InGaAlAs electroabsorption modulator integrated DFB laser. A 150-micron long modulator using ridge waveguide and a low dielectric constant benzocyclobutene layer achieved 40 Gb/s transmission [28]. A 3-dB frequency bandwidth of over 39 GHz was achieved at a laser current of 100 mA and a modulator bias voltage of –2.5 V. After 2-km transmission, the path penalty was less than 2 dB from –15°C to 80°C.

An even shorter InGaAlAs EML showed 100 Gb/s eye opening at room temperature [29]. This device has a 5-μm high active stripe buried in a semiinsulating iron doped regrown InP layer and a 50 micron long modulator. The EML showed a large extinction ratio of -12 dB at -4 V and a steep modulation slope owing to the efficient modulation properties of InGaAlAs material even though it has a 50 micron long modulator.

## References

1. D. E. Aspnes, "Electric field effects on the dielectric constant of solids," *Phys. Rev.*, vol. 153, no. 3, pp. 972–982, Jan. 1976.
2. H. Soda, K. Nakai, H. Ishikawa, and H. Imai, "High-speed GaInAsP/InP buried heterostructure optical intensity modulator with semi-insulating InP burying layers," *Electronics Lett.*, vol. 23, no. 23, pp. 1232–1234, Nov. 1987.
3. H. Soda, "10 Gb/s monolithic electro-absorption modulator/DFB laser light source," in *Third Optoelectronics Conf.*, 12B1-1, 1990. Chiba: IEICE.
4. K. Wakita, *Semiconductor Optical Modulators,* Norwell: Kluwer Academic Publishers, 1998.
5. P. J. Stevens, M. Whitehead, G. Parry, and K. Woodbridge, "Computer modeling of the electric field dependent absorption spectrum of multiple quantum well material," *IEEE J. Quantum Electron.*, vol. 24, no. 10, pp. 2007–2016, Oct. 1988.
6. K. Nakamura, A. Shimizu, K. Fujii, M. Koshiba, and K. Hayata, "Numerical analysis of the absorption and the refractive index change in arbitrary semiconductor quantum-well structures," *IEEE J. Quantum Electron.*, vol. 28, no. 7, pp. 1670–1677, Jul. 1992.
7. P. Debernardi and P. Fasano, "Quantum confined Stark effect in semiconductor quantum wells, including valence band mixing and Coulomb effects," *IEEE J. Quantum Electron.*, vol. 29, no. 11, pp. 2741–2755, Nov. 1993.
8. S. L. Chuang, S. Schmitt-Rink, D. A. B. Miller, and D. S. Chemla, "Exciton Green's-function approach to optical absorption in a quantum well with an applied electric field." *Phys. Rev. B*, vol. 43, no. 2, pp. 1500–1509, Jan. 1991.
9. M. Sugawara, N. Okazaki, T. Fujii, and S. Yamazaki, "Conduction-band and valence-band structures in strained InxGa1-xAs/InP quantum wells on (001) InP substrates," *Phys. Rev. B*, vol. 48, no. 11, pp. 8102–8118, Sep. 1993.
10. F. Koyama and K. Iga, "Frequency chirping in external modulators," *IEEE J. Lightwave Technol.*, vol. 6, no. 1, pp. 87–93, Jan. 1988.
11. R. Sahara, K. Morito, and H. Soda, "Engineering of barrier band structure for electro-absorption MQW modulators," *Electron. Lett.*, vol. 30, no. 9, pp. 698–699, Apr. 1994.
12. K. Morito, K. Sato, Y. Kotaki, and H. Soda, "High power modulator integrated DFB laser incorporating a strain-compensated MQW and graded SCH modulator for 10 Gb/s transmission," *Electron. Lett.*, vol. 31, no. 12, pp. 975–976, Jun. 1995.
13. M. Matsuda, K. Morito, K. Yamaji, T. Fujii, and Y. Kotaki, "A novel method for designing chirp characteristics in electro-absorption MQW optical modulators," *IEEE J. Photonics Technol. Lett.*, vol. 10, no. 3, pp. 364–366, Mar. 1988.
14. Y. Chung and N. Dagli, "An assessment of finite difference beam propagation method," *IEEE J. Quantum Electron.*, vol. 26, no. 8, pp. 1335–1339, Aug. 1990.
15. Y. Kotaki and H. Soda, "Analysis of static and dynamic wavelength shifts in modulator integrated DFB lasers," in *19th European Conference on Optical Communication (ECOC 93)*, paper WeP8.6, pp. 381–384, 1993. Montreux: SEV & EPFL/ETHZ.
16. Y. Kim, H. Lee, J. Lee, J. Han, T. W. Oh, and J. Jeong, "Chirp characteristics of 10-Gb/s electro-absorption modulator integrated DFB lasers," *IEEE J. Quantum Electron.*, vol. 36, no. 8, pp. 900–908, Aug. 2000.

17. H. Soda, M. Furutsu, K. Sato, N. Okazaki, S. Yamazaki, I. Yokota, T. Okiyama, H. Nishimoto, and H. Ishikawa, "High-power semi-insulating BH structure monolithic electro-absorption modulator/DFB laser light source operating at 10 Gb/s," in *Integrated Optics Optical Communications*, paper 20PDB-5, 1989. Kobe: IEICE.
18. M. Suzuki, H. Tanaka, M. Usami, H. Taga, and Y. Matsushima, "SI-InP buried planar-type λ/4-shifted DFB laser/EA modulator integrated light source and 2.4 Gb/s-118 km penalty-free conventional fiber transmission," in *Integrated Optics Optical Communications*, paper 20PDB-3, 1989. Kobe: IEICE.
19. T. Kato, T. Sasaki, N. Kida, K. Komatsu, and I. Mito, "Novel MQW DFB laser diode/modulator integrated light source using bandgap energy control epitaxial growth technique," in *17th European Conf. Optical Communication/8th Int. Cod. Optical Communication*, paper WeB7-1, 1991. Paris: SEE.
20. M. Aoki, M. Suzuki, H. Sano, T. Kawano, T. Ido, T. Taniwatari, K. Uomi, and A. Takai, "InGaAs/InGaAsP MQW electroabsorption modulator integrated with a DFB laser fabricated by bandgap energy control selective area MOCVD," *IEEE J. Quantum Electron.*, vol. 29, no. 6, pp. 2088–2096, Jun. 1993.
21. S. Sumi, K. Naoe, Y. Sakuma, R. Washino, K. Okamoto, K. Motoda, H. Sato, K. Shinoda, T. Kitatani, A. Taike, and K. Uomi, "Semi-cooled operation (TLD=50°C) of 10.7-Gbit/s 1.55-μm electro-absorption modulator integrated DFB laser for 80 km transmission," in *Semiconductor Laser Conference*, pp. 85–86, 2006. Hawaii: IEEE.
22. Y. Miyaazaki, T. Yamatoya, K. Matsumoto, K. Kuramoto, K. Shibata, T. Aoyagi, and T. Ishikawa, "High-power ultralow-chirp 10-Gb/s *electroabsorption* modulator integrated laser with ultrashort photocarrier lifetime," *IEEE J. Quantum Electron.*, vol. 42, no. 4, pp. 357–362, Apr. 2006.
23. K. Tsuzuki, Y. Shibata, N. Kikuchi, M. Ishikawa, T. Yasui, H. Ishii, and H. Yasaka, "Full C-band tunable DFB laser array co-packaged with InP Mach-Zehnder modulator for DWDM optical communication systems," *IEEE J. Selected Topics Quantum Electron.*, vol. 15, no. 3, pp. 521–527, May 2009.
24. M. L. Nielsen, K. Tsuruoka, T. Kato, T. Morimoto, S. Sudo, T. Okamoto, K. Mizutani, K. Sato, and K. Kudo, "Demonstration of 10-Gb/s C+L-band InP-based Mach-Zehnder modulator," *IEEE J. Photonics Technol. Lett.*, vol. 20, no. 14, pp. 1270–1272, Jul. 2008.
25. Y. Matsui, D. Mahgerefteh, X. Zheng, C. Liao, Z. F. Fan, K. McCallion, and P. Tayebati, "Chirp-managed directly modulated laser (CML)," *IEEE J. Photonics Technol. Lett.*, vol. 18, no. 2, pp. 385–387, Jan. 2006.
26. H. Yamamoto, M. Hirai, O. Kagaya, K. Nogawa, K. Naoe, N. Sasada, and M. Okayasu, "Compact and low power consumption 1.55-μm electro-absorption modulator integrated DFB-LD TOSA for 10-Gbit/s 40-km Transmission," in *OFC 2009*, San Diego, CA, paper OThT5, 2009. San Diego: IEEE.
27. W. Kobayashi, M. Arai, T. Yamanaka, N. Fujiwara, T. Fujisawa, M. Ishikawa, K. Tsuzuki, Y. Shibata, Y. Kondo, and F. Kano, "Extended operating temperature range of 125°C (-25°C to 100°C) of 10-Gbit/s, 1.55-μm electro-absorption modulator integrated DFB laser for 80-km SMF Transmission," in *OFC 2009*, San Diego, CA, paper OThT4, 2009. San Diego: IEEE.
28. W. Kobayashi, T. Yamanaka, M. Arai, N. Fujiwara, T. Fujisawa, K. Tsuzuki, T. Ito, Y. Kondo, and F. Kano, "40-Gbit/s, uncooled (-15°C to 80°C) operation of a 1.55-μm, InGaAlAs, electro-absorption modulated laser for very short reach applications," in *IEEE Int. Conf. InP Related Materials 2009*, Newport Beach, CA, paper ThB1.2, 2009. Newport Beach: IEEE.
29. C. Kazmierski, A. Konczykowska, F. Jorge, F. Blache, M. Riet, C. Jany, and A. Scavennec, "100 Gb/s operation of an AlGaInAs semi-insulating buried heterojunction EML," in *OFC 2009*, San Diego, CA, paper OThT7, 2009. San Diego: IEEE.

# 6

# Lithium Niobate Modulators




Kazuto Noguchi
*NTT Photonics Laboratories*


## 6.1 History of Lithium Niobate Modulator Development

After the Czochralski process for single-crystal growth had been established, research on lithium niobate ($LiNbO_3$) optical modulators started in the 1960s. $LiNbO_3$'s large electro-optic (EO) coefficient led to the development of bulk-type optical modulators, in which a $LiNbO_3$ crystal is sandwiched by two metal plates. The optical beam propagating in the crystal is modulated by applying a voltage of several hundred volts to the electrode plates. In the 1970s, fabrication technologies for optical waveguides in $LiNbO_3$ substrate were developed by exploiting the out-diffusion of lithium oxide ($Li_2O$), where the optical wave is confined and propagates in the high-index guiding layer and so the drive voltage (half-wave voltage) was reduced to 24 V [1]. A ridge waveguide with a cross section of several square micrometers was fabricated by ion-beam etching as shown in Figure 6.1 [2]. This advance made it possible to restrict the interaction region of the external electrical signal and optical waves near the optical waveguide and to keep the interaction length at several centimeters. As a result, the efficiency of the EO interaction increased and drive voltages were reduced to about 4 V.

From the late 1970s to the late 1980s, research focused on using $LiNbO_3$ for external modulators in high-speed coherent optical transmission systems [3–10] and for matrix switches for optical cross connects [11–16]. For those purposes, the insertion loss of the modulator had to be minimized to achieve long span systems or large-scale integration of switching elements. Much effort was then focused on optimizing the fabrication method for Ti-diffused optical waveguides to reduce propagation loss [17–19] and coupling loss with single-mode fiber [20–23]. During this decade, fiber-to-fiber insertion losses of

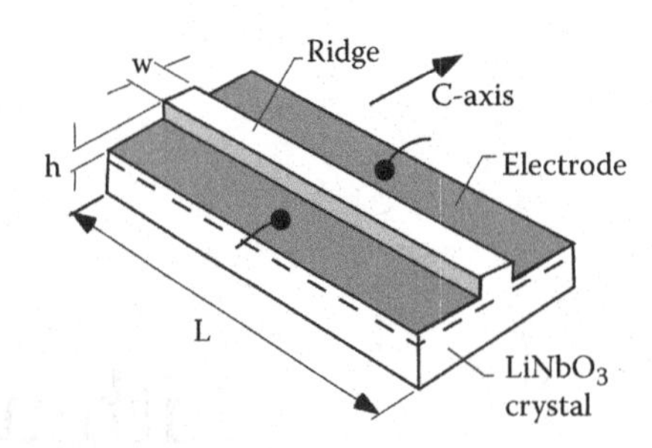


**FIGURE 6.1** Schematic diagram of a $LiNbO_3$ ridge waveguide modulator. (Reprinted with permission from I. P. Kaminow et al., *Appl. Phys. Lett.*, 24, 12, 622–624, Copyright, 1974, American Institute of Physics.)

$LiNbO_3$ devices were greatly reduced and the procedure for fabricating devices as optical components was established.

From the late 1980s to the early 1990s, the development of the erbium-doped fiber amplifier significantly advanced large-capacity and long-haul optical transmission systems [24,25]. $LiNbO_3$ optical modulators became one of the key devices for those systems operating at 5 Gbit/s–20 Gbit/s. To operate at such high speeds, $LiNbO_3$ modulators were developed into broadband microwave and millimeter-wave components. Various types of modulators with velocity-matched traveling-wave electrodes were developed [26–32] and modulation speeds reached more than 100 GHz [33]. Simultaneously, studies on the reliability of $LiNbO_3$ modulators and packaging technologies progressed. Thermal drift [34–36] and DC (direct current) drift [37–41,63,64] were fully suppressed. $LiNbO_3$ modulators were accepted into commercial trunk network systems between metropolitan areas and in submarine systems across the Pacific and Atlantic Oceans.

After 2000, the spread of broadband internet access and high-quality video applications caused continual explosive traffic growth and an ever-increasing demand for more bandwidth in trunk networks [54]. $LiNbO_3$ are now used in high-speed systems operating at 40 Gbit/s or 100 Gbit/s. As discussed in Chapter 2, new modulation technologies, such as differential quadrature phase shift keying, have been developed [42,43].

# 6.2 Properties of Single-Crystal Lithium Niobate and Design Considerations for Optical Modulators

## 6.2.1 Physical and Optical Properties of Lithium Niobate

$LiNbO_3$ is ferroelectric with large self-polarization along the z-axis and belongs to the *3m* crystallographic point group. Its physical, optical, and EO constants are summarized in Table 6.1 [44,45]. The phase-change temperature (Curie temperature) $T_c$ is as high as 1483 K, and the material is stable at room temperature. In addition to being an important EO material, $LiNbO_3$ has relatively large piezoelectric coefficients and is also used for surface acoustic wave components, such as intermediate frequency filters for televisions and radio frequency filters for cellular phones. $LiNbO_3$ is the third most produced single crystal material worldwide, after silicon and gallium arsenide. $LiNbO_3$ crystal growth technologies have matured and high-quality wafers with diameters up to six inches are commercially available at a relatively low price.

$LiNbO_3$ crystal is transparent in the wavelength range from 0.33 μm to 4.5 μm, that is, from visible light to infrared. The refractive index is almost constant and ranges from 2.1 to 2.2. Since the wavelength dependence of its optical properties is very small in the wavelength region used for optical communications systems, $LiNbO_3$ modulators are well-suited for use in wavelength division multiplexing transmission systems.

**TABLE 6.1** Characteristics of Lithium Niobate Crystal

| Characteristic | Numerical Value |
|---|---|
| Point group | 3m |
| Space group | R3c |
| Lattice constants (A) | $a = 5.15$, $c = 13.863$ |
| Melting point (C) | 1255 ± 5 |
| Curie point (C) | 1140 ± 5 |
| Density (g/cm$^3$) | 4.64 |
| Pyroelectric coefficient (C/C/m$^2$) | $-8.3 \times 10^{-5}$ (at 25°C) |
| Transparent range (nm) | 400–5200 |
| Absorption coefficient (%/cm) | 0.1 (at $\lambda = 1064$ nm) |
| Refractive indices | $n_o = 2.2967$, $n_e = 2.2082$ (at $\lambda = 632.8$ nm) |
| | $n_o = 2.200$, $n_e = 2.146$ (at $\lambda = 1300$ nm) |
| Electro-optic coefficients (pm/V) | $r_{33} = 30.8$, $r_{13} = 8.6$, $r_{22} = 8.4$, $r_{51} = 2.8$ |
| Relative dielectric constant | $\varepsilon_{33} = 29$, $\varepsilon_{11} = 44$ |

*Sources:* T. Yamada, *Numerical Data and Functional Relationships in Science and Technology*, 16, 149–163, Springer-Verlag, Berlin, 1980. With permission; A. M. Prokhorov and Y. S. Kuz'minov, *Physics and Chemistry of Crystalline Lithium Niobate*, 235, Adam Hilger, Bristol and New York, 1990. With permission.

The phase velocity of a plane wave in a transparent dielectric crystal $v$ is given as [46]

$$v = \frac{c}{n},$$

where $c$ is speed of light in vacuum and $n$ is the refractive index.

The phase velocity of an optical wave propagating in a crystal with optical anisotropy depends on the direction of its polarization. The index ellipsoid of the crystal is written as

$$\frac{x^2}{n_x^2} + \frac{y^2}{n_y^2} + \frac{z^2}{n_z^2} = 1,$$

where $n_x$, $n_y$, and $n_z$ are refractive indices along the $x$, $y$, and $z$ directions, respectively. $LiNbO_3$ is a uniaxial crystal, where $n_x = n_y = n_o$ (ordinary index) and $n_z = n_e$ (extraordinary index).

### 6.2.2 Electro-Optic Properties and Modulator Structure

When an electric field is applied to a material, the refractive index of the material changes with the strength of the field. This phenomenon is called EO effect. In a $LiNbO_3$ optical modulator, a linear EO effect (the Pockels effect) is utilized—the index change is proportional to the applied electrical field $E_j$ as

$$\Delta \frac{1}{n_i^2} = r_{ij} E_j,$$

where $i = [1, 6]$ and $j = [1, 3]$. The EO constant $r_{ij}$ is a third-rank tensor and it can be expressed as a matrix with six rows and three columns:

$$r_{ij} = \begin{bmatrix} 0 & -r_{22} & r_{13} \\ 0 & r_{22} & r_{13} \\ 0 & 0 & r_{33} \\ 0 & r_{51} & 0 \\ r_{51} & 0 & 0 \\ -r_{22} & 0 & 0 \end{bmatrix}.$$

Optical modulators are most commonly fabricated on X-cut and Z-cut single crystal $LiNbO_3$ wafers. Typical configurations of $LiNbO_3$ optical modulators are shown in Figures 6.2 and 6.3 for X-cut and Z-cut wafers, respectively. In both cases, the most effective modulation is obtained when the directions of the main components of the electric fields of both the modulating electrical signal and the optical wave coincide with the z-axis, the main axis of the crystal.

As shown in Figure 6.2, the coplanar electrodes are fabricated on an X-cut $LiNbO_3$ wafer and the optical wave propagates along y-axis. The optical waveguide is placed at the center of the gap between the two electrodes. In the optical waveguide, the direction of the field of the applied electrical wave is parallel to the substrate surface along the z-axis. The refractive index change $\Delta n_z$ is

$$\Delta n_z = \Delta n_e = -\frac{1}{2} n_e^3 r_{33} E_z,$$

where $E_z$ is the z-component of the electric field of the modulating electrical signal. In this case, to obtain a large optical phase change, the polarization of the optical wave in the waveguide is set to the transverse electric mode, where the main field component is along the z-axis. The value of $n^3r$ is one of the key performance criteria of the linear EO effect of a material.

In the case of a Z-cut $LiNbO_3$ substrate (Figure 6.3), light propagates along the x-axis. The optical waveguide is formed beneath one of the electrodes and the main component of the electric field of the modulating signal in the waveguide $E_z$ is perpendicular to the substrate surface. In this case, the optical wave in the transverse magnetic mode, where the main field component is perpendicular to the substrate surface along the z-axis, is modulated most effectively.

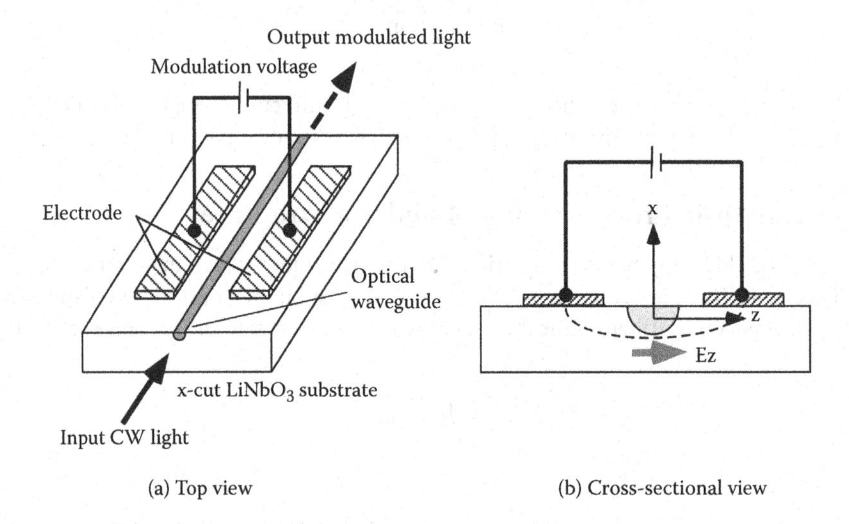


**FIGURE 6.2** Schematic diagram of an X-cut $LiNbO_3$ phase modulator. (a) Top view (b) Cross-sectional view.

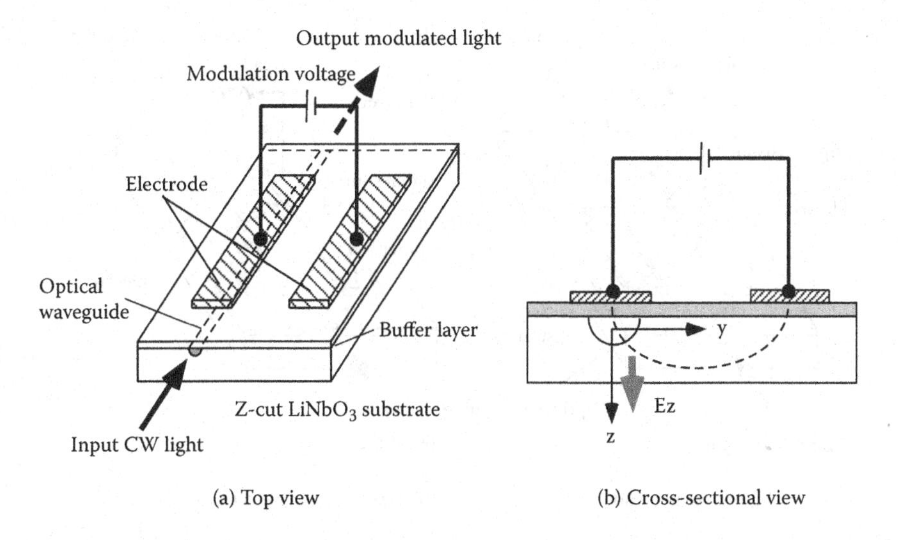


**FIGURE 6.3** Schematic diagram of a Z-cut $LiNbO_3$ phase modulator. (a) Top view (b) Cross-sectional view.

$LiNbO_3$ crystal has three major advantages as a material for EO modulators: First, its EO coefficient $r_{33}$ is one of the largest among inorganic compounds. This leads to a low-drive voltage, and for this reason, the $LiNbO_3$ crystal has attracted much attention. Second, the frequency response of the EO effect extends to the millimeter wave frequency region [47,48], enabling modulators to achieve broad bandwidths. Third, the relative dielectric constants of $LiNbO_3$ are relatively small. These characteristics are advantageous for obtaining velocity matching and broadband operation, as discussed in the following sections.

## 6.3 Design Considerations for High-Speed Modulation

### 6.3.1 Electrode Design as a Microwave Component

A broadband $LiNbO_3$ optical modulator works not only as an optical waveguide component but also as a microwave (millimeter-wave) component with electrodes. The EO operation results from the interaction between the optical wave and the microwave, and the interaction depends on the properties of the $LiNbO_3$ crystal in both the optical and microwave frequency regions.

#### 6.3.1.1 Lumped-Element Electrode

The typical configuration of a modulator with lumped-element electrodes is shown in Figure 6.4. The electrical field is launched between the two electrodes and the modulator behaves as a capacitor for the microwave signal source. The optical wave propagates in the microwave field between the capacitor plates. To obtain impedance matching between the internal characteristic impedance of the signal source and the electrodes, a load resistor is connected across the two electrodes. In this structure, the modulation bandwidth depends on the time constant of the charging and discharging of the capacitor, which is determined by the capacitance $C$ and the load resistance $R$. As the modulation frequency increases, the characteristic impedance of the electrodes decreases and thus the modulation voltage decreases. In this case, the modulation bandwidth $\Delta f$ is expressed [3] as

$$\Delta f = \frac{1}{\pi RC} = \frac{1}{n_m^2 C_0 RL},$$

where $n_m$ is the effective refractive index of the electrode, $C_0$ is the electrode capacitance per unit length, and $L$ is the electrode length.

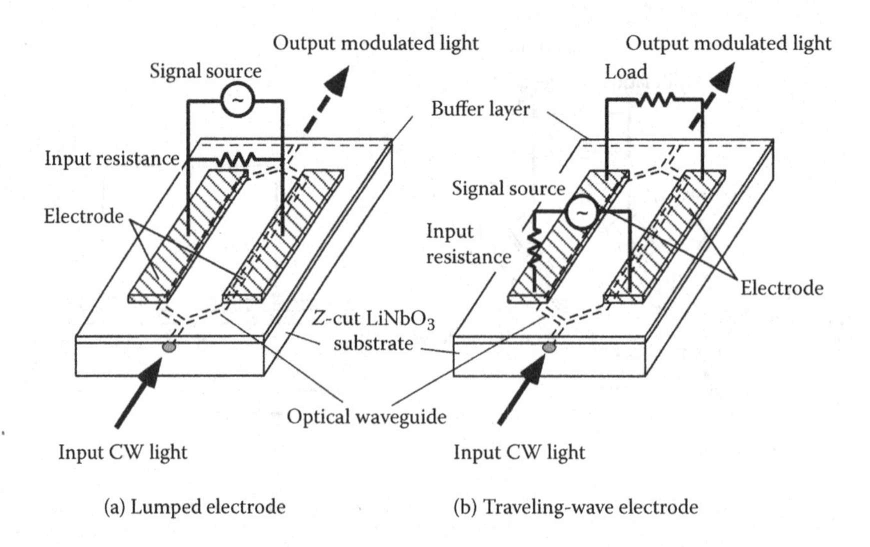


**FIGURE 6.4** Mach-Zehnder optical intensity modulators. (a) Lumped electrode (b) Traveling-wave electrode.

### 6.3.1.2 Traveling-Wave Electrode

An optical modulator with a traveling-wave electrode is shown in Figure 6.5, where the electrode works as a transmission line for the modulating electrical wave. The electrode and optical waveguide are parallel to each other and the modulating electrical wave and optical wave propagate along the same direction. The two waves interact in a distributed manner and the optical wave is modulated by the electrical wave through the EO effect. When the velocities of the two waves are equal (velocity matched), the wavefront of the electrical wave and that of the optical wave, launched simultaneously at the input, propagate to the output at the same velocity without a time lag. The change in the electrical wave is exactly transferred to the optical wave and theoretically, there exists no limitation on modulation speed or modulation bandwidth if the frequency-dependent electrode loss can be ignored.

However, the dielectric constants for the optical wave and electrical wave are different, as are their effective refractive indices. Therefore, the optical wave and the electrical wave usually propagate at different velocities in the modulator. As the optical wave and electrical wave proceed, the separation between the wavefronts of the two waves grows gradually. This leads to the degradation of modulation efficiency at higher frequencies and to a limitation of modulation bandwidth. This is called "velocity mismatch."

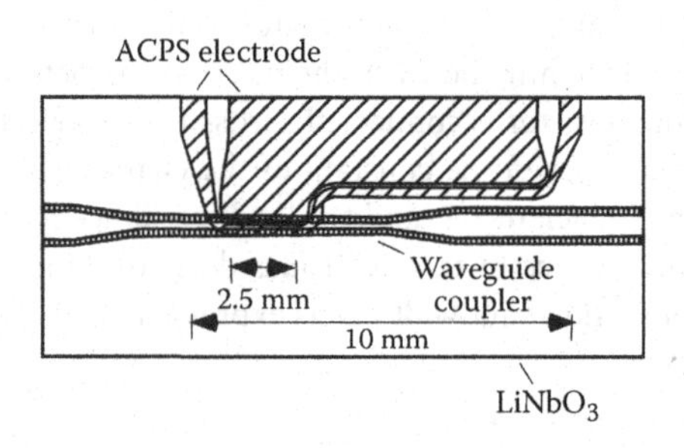


**FIGURE 6.5** Schematic diagram of a Ti:$LiNbO_3$ waveguide EO switch with a very short electrode. (Reprinted with permission from S. K. Korotky et al., *Appl. Phys. Lett.*, 50, 23, 1631–1632, Copyright 1987, American Institute of Physics.)

The modulation bandwidth $\Delta f$ with velocity mismatch [3] is

$$\Delta f = \frac{1.9c}{\pi |n_m - n_0| L},$$

where $L$ is electrode length, $c$ is the speed of light in vacuum, and $n_m$ and $n_o$ are the effective refractive indices of the electrical wave and optical wave, respectively.

Izutsu et al. compared a lumped modulator to a traveling-wave modulator and found that the latter is three times more efficient for $LiNbO_3$ [3].

### 6.3.2 Velocity Matching Technologies

The modulation bandwidth of a traveling-wave modulator depends on the microwave propagation characteristics of the electrode. The bandwidth-limiting factors are velocity mismatch between the microwave and optical wave, impedance mismatch between the driving circuit and electrode, and microwave propagation loss of the electrode. Considering all of these parameters, the optical response, $R(f)$, at frequency $f$ of a traveling-wave modulator is expressed [49] as

$$R(f) = \left| \frac{(\Psi_+ + \rho_2 \Psi_-)(1+\rho_1)}{\exp(2j\phi_+) + \rho_1 \rho_2 \exp(-2j\phi_-)} \right|,$$

$$\Psi_\pm = \exp(\pm j\phi_\pm)\frac{\sin\phi_\pm}{\phi_\pm},$$

$$\phi_\pm = \frac{\gamma_{m\pm} L}{2} - j\frac{\alpha_m L}{2},$$

$$\gamma_{m\pm} = \frac{2\pi(n_m \mp n_o) f}{c},$$

$$\alpha_m = \alpha_0 f^{1/2},$$

$$\rho_1 = \frac{Z_c - Z_s}{Z_c + Z_s},$$

$$\rho_2 = \frac{Z_l - Z_c}{Z_l + Z_c},$$

where $\alpha_0$ is conductor loss due to the skin effect and $Z_c$, $Z_s$, and $Z_l$ are the characteristic impedance of the electrode, the output impedance of the signal source, and the load impedance, respectively.

Korotky et al. fabricated a modulator with a very short (2.5-mm long) asymmetric coplanar strip electrode (Figure 6.5) to obtain small $\phi_+$. This configuration enabled them to reduce both the velocity mismatch and the microwave propagation loss. Although they obtained a wide modulation bandwidth of 40 GHz, the drive voltage was as high as 26 V because of the short electrode [50].

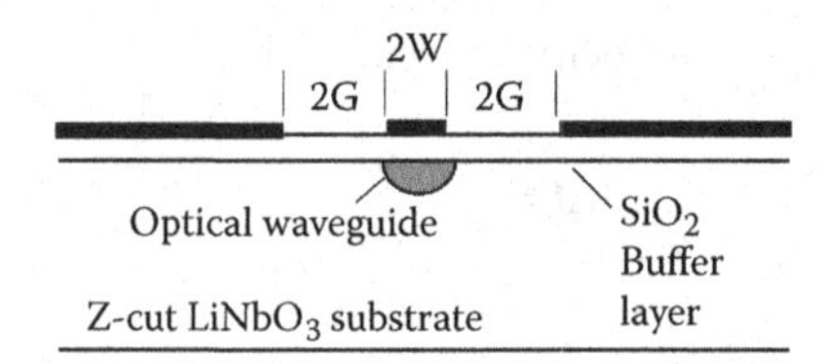


**FIGURE 6.6** Cross-sectional view of a Ti:$LiNbO_3$ phase modulator with a CPW electrode. (Reprinted with permission from K. Kawano et al., *IEEE Photon. Technol. Lett.*, 1, 2, 33–34, © 1989 IEEE.)

Kawano et al. were the first to adopt a relatively thick buffer layer of silicon oxide as a low-index material and a CPW electrode with a narrow center conductor and wide gap to achieve good velocity matching, as shown in Figure 6.6. The center conductor width was 8 μm, and the gap was 15 μm. Since the widths of the electrode and the optical waveguide were the same, they obtained a large interaction between the microwave and the optical wave and achieved a drive voltage of only 5 V with a 2.7-cm-long electrode [51]. These device parameter values became the standard ones for later modulator development.

Electrode parameters $n_m$, $Z_c$, and $\alpha_0$ can be calculated from a numerical analysis of the electrode cross section by using the finite element method [52,53]. Design parameters of the structure are buffer layer thickness $T_b$, electrode thickness $T_e$, center conductor width $W$, and the gap between the center conductor and ground electrode $G$.

The dependence of $n_m$, $Z_c$, and $\alpha_0$ on the above design parameters is shown in Figure 6.7 [53]. The characteristic impedance depends on the width-to-gap ratio of the electrode and on the dielectric constants of the surrounding materials. Electrode thickness affects the microwave index and microwave loss.

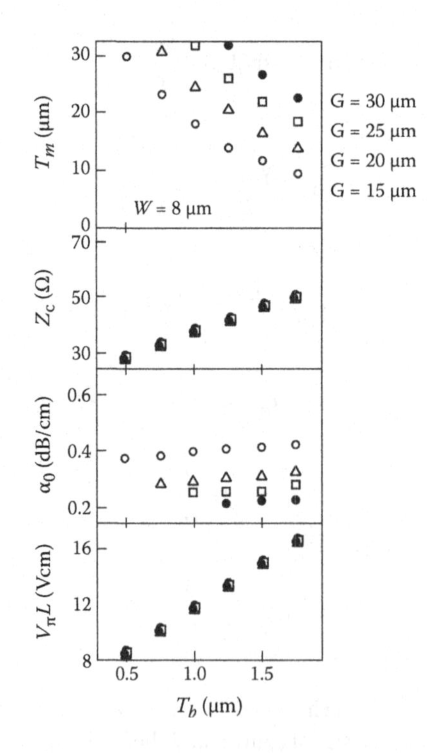


**FIGURE 6.7** Calculated characteristics of $LiNbO_3$ modulators for the electrode thickness, Tm, allowing a velocity-match of $n_m = 2.15$, and the consequent microwave characteristics of $Z_c$, $\alpha_0$ and $V_\pi L$ as a function of the buffer layer thickness $T_b$, $W$ is a constant 8 μm and $G$ is a parameter. (Reprinted from O. Mitomi et al., *IEE Proc.-Optoelectron.*, 145, 6, 360–364, 1998. With permission.)

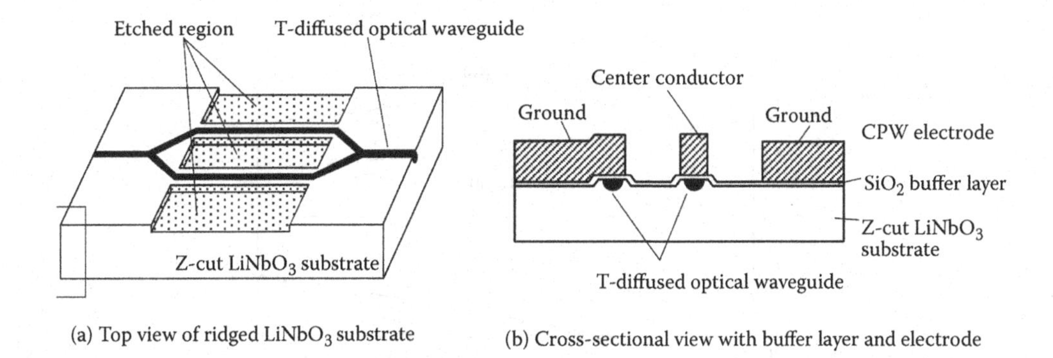


**FIGURE 6.8** Schematic diagram of a Ti:LiNbO$_3$ modulator with a ridge structure. (a) Top view of ridged LiNbO$_3$ substrate. (b) Cross-sectional view. (Reprinted with permission from K. Noguchi et al., *Electron. Lett.*, 34, 7, 661–662, © 1998 IEEE.)

Various electrodes and buffer layer structures have been proposed to obtain velocity matching, impedance matching, and low microwave loss. These include: shielding type electrode [26], thick gold asymmetric coplanar strip electrode with a narrow tip [27], separated buffer layer [28], thick CPW electrode with wide gap [29], buried buffer layer [30], mushroom-type electrode [31], and ridge-type modulator [32,33].

### 6.3.3 Ridge-Type Modulator

Replacing high-index LiNbO$_3$ with low-index silicon oxide SiO$_{20}$ and air in the interaction region of an electrode is effective for reducing the effective microwave refractive index and obtaining velocity matching. Figure 6.8 shows an optical intensity modulator with a ridge structure [33]. A Mach-Zehnder Ti-diffused optical waveguide is formed on a LiNbO$_3$ substrate and both sides of the waveguide in the interaction region are etched to form ridges where optical waveguides are. In this structure, the optical field is confined in the low-loss Ti-diffused waveguide, while the microwave field of the CPW electrode goes through the LiNbO$_3$ ridges, the SiO$_2$ buffer layer, and air. Therefore, conventional techniques for fabricating a Ti-diffused waveguide, SiO$_2$ buffer layer, and gold-plated electrode are applicable and thus there is little increase in waveguide propagation loss.

The ridge height dependence of electrode thickness $T_m$, characteristic impedance $Z_c$, and conductor loss $\alpha_0$, under the velocity-matching condition, are shown in Figure 6.9. The center conductor width is 8 μm and the gap $G$ is a design variable. As the ridge height increases, $T_m$ decreases, while $Z_c$ and $\alpha_0$ increase. A wider gap leads to a larger $T_m$, which results in a lower $\alpha_0$. Impedance matching at $Z_c$ of 45 Ω to 50 Ω can be achieved when $H_r = 3$ to 4 μm and $T_m = 25$ μm. With these parameters, optical modulators with a ridge structure have been fabricated. A fully packaged module has been used in a 40 Gbit/s transmission experiment [54] and a very wide modulation bandwidth of over 100 GHz has been confirmed. Figure 6.10 shows the frequency dependence of the EO response of the fabricated modulator [48].

## 6.4 Modulator Fabrication Technologies

### 6.4.1 Optical Waveguide Formation in Lithium Niobate

#### 6.4.1.1 Waveguide Formation Technologies

The formation of a low-loss optical waveguide in a LiNbO$_3$ substrate is crucial for obtaining modulators with both low loss and low drive voltage, because the drive voltage of a modulator is inversely proportional to electrode length. To date, three methods have been used to form optical waveguides LiNbO$_3$ substrate: out-diffusion of Li$_2$O [1], in-diffusion of metal [55], and proton exchange [56].

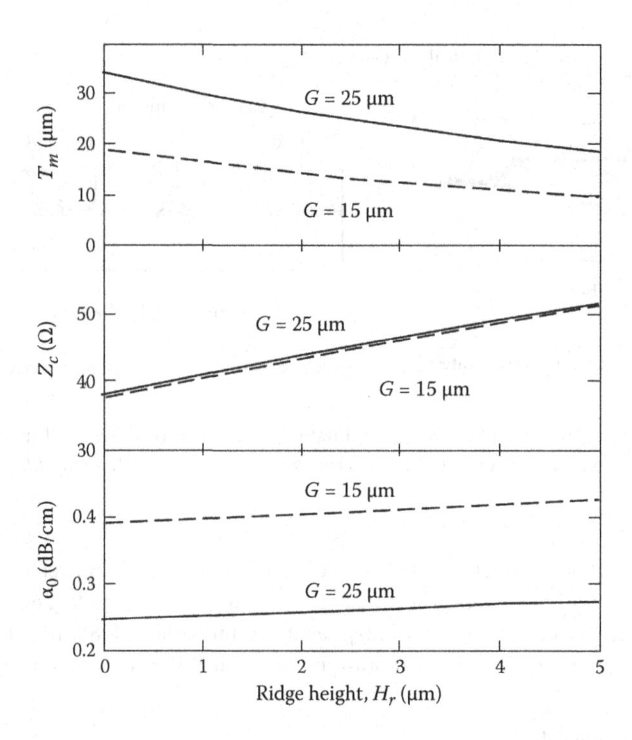


**FIGURE 6.9** Ridge height dependence of electrode characteristics. (Reprinted with permission from K. Noguchi et al., *J. Lightwave Technol.*, 16, 4, 615–619, © 1998 IEEE.)

*Out-diffusion of $Li_2O$:* Waveguide-type $LiNbO_3$ modulators were initially fabricated using out-diffusion of $Li_2O$ from the $LiNbO_3$ substrate. The $LiNbO_3$ crystal was heated in a furnace to 1000°C. $Li_2O$ out-diffused from the surface of the crystal, forming a high index layer. The channel waveguide was formed by ion etching of the out-diffused high-index layer [46]. The index change was limited to the extraordinary index ($n_e$) and the value was as small as $10^{-3}$.

*In-diffusion of transition metals:* Thermal in-diffusion of transition metals, such as Ti, V, Ni, and Cu, into $LiNbO_3$ crystal at about 1000°C causes a refractive index increase for both $n_o$ and $n_e$ [55]. The index change is controlled by the thickness of the metal layer and channel waveguide patterns are easily

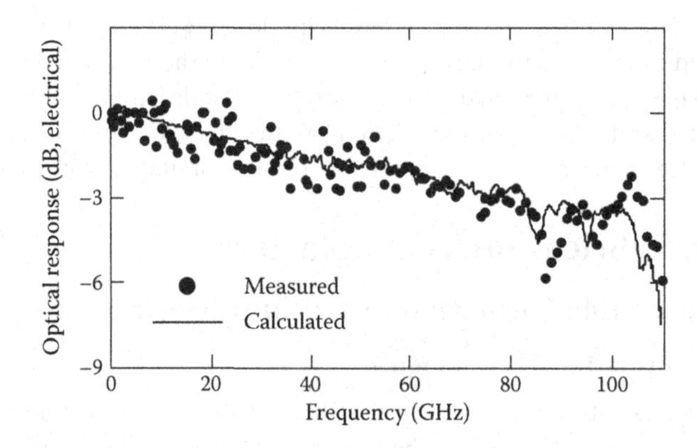


**FIGURE 6.10** Electro-optic response of a Ti:$LiNbO_3$ modulator with a ridge structure. (Reprinted with permission from K. Noguchi et al., *J. Lightwave Technol.*, 16, 4, 615–619, © 1998 IEEE.)

obtained by forming a metal pattern on the substrate by means of standard photolithography technology used in the manufacture of electronic integrated circuits. Ti in-diffused waveguides have become mainstream because they show excellent propagation characteristics and no degradation of the EO effect [7,17]. Although the $Li_2O$ out-diffusion method is not commonly used today, the knowledge gained from the past use of the process has been used for optimizing the atmosphere for Ti diffusion [57–60].

*Proton exchange waveguide:* When $LiNbO_3$ crystal is immersed in benzoic acid at 250°C, ion exchange between H+ in the benzoic acid and Li+ in the crystal occurs and a high-index layer is formed in the crystal [56]. In this case, a patterned Cr or Ta thin film is used as a mask to obtain a channel waveguide. The index change is limited to $n_e$ and is as large as 0.1. To obtain low-loss and stable waveguides, proton-exchange waveguides have to be annealed at 400°C for several hours. An annealed proton-exchange waveguide shows high resistance to the photorefractive effect and is therefore often used for fabricating second-harmonic generation devices that convert infrared light to visible light [61,62].

### 6.4.1.2 Waveguide Formation by Ti In-Diffusion

#### *6.4.1.2.1 Formation of Ti Pattern on the Substrate*

Since the melting point of Ti is as high as 1725°C, sputtering or electron beam evaporation is used to deposit a thin film of Ti on the $LiNbO_3$. The Ti film thickness should be 70 nm–100 nm to obtain a single-mode waveguide in the 1.3 μm to 1.5-μm wavelength range. The Ti pattern is formed by transferring a photoresist pattern to Ti thin film. For the transfer, wet and dry etching or the lift-off method are applicable. Though wet etching of Ti in hydrofluoric acid is easy and less damaging to the $LiNbO_3$ surface, it is difficult to precisely control Ti widths to less than 1 μm. Dry etching of Ti in $CF_4$ plasma may damage the $LiNbO_3$ surface which leads to increased propagation loss and dc drift.

The lift-off method is a relatively low-temperature process and allows patterning of fine features of less than 1 μm with little damage to the crystal. The procedure to form a Ti pattern with the lift-off method is shown in Figure 6.11 a–g. After the $LiNbO_3$ substrate has been coated with a photoresist and prebaked, the substrate is exposed using a photomask and immersed in a developer. Waveguide patterns are formed in the photoresist on the substrate. Next, a Ti thin film is deposited on the patterned photoresist. The photoresist is removed with a solvent, leaving a 5 μm to 10-μm-wide Ti pattern on the $LiNbO_3$ substrate.

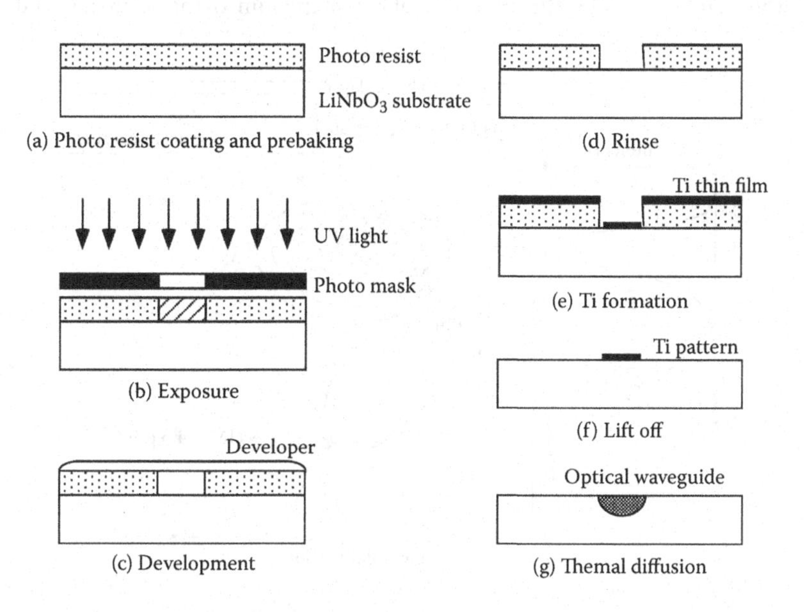


**FIGURE 6.11** Fabrication process for Ti-diffused waveguides.

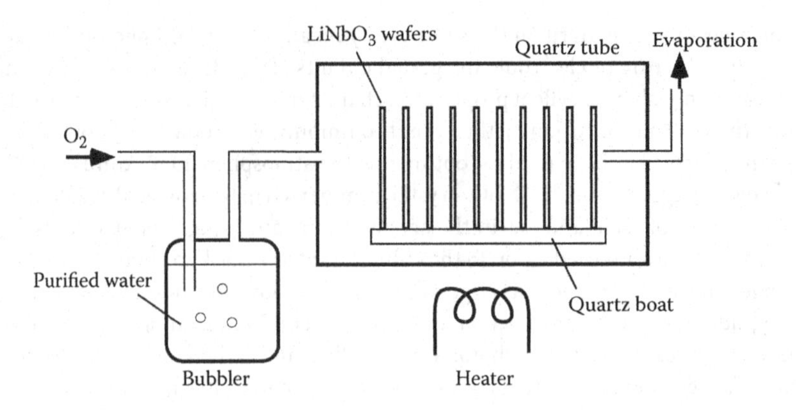


**FIGURE 6.12** Thermal diffusion apparatus.

#### *6.4.1.2.2 Thermal In-Diffusion*

Ti strip pattern on a $LiNbO_3$ substrate is in-diffused into the crystal surface in a furnace at about 1000°C to form a high index region. During the heating process for in-diffusion, $Li_2O$ out-diffuses from the crystal surface because the vapor pressure of Li is much higher than that of Nb. This leads to the formation of a high-index layer and a degradation of the waveguide confinement factor for extraordinary light.

To suppress out-diffusion of $Li_2O$, "wet" ambient technology, where water vapor is introduced into the furnace, has been developed [19]. A schematic diagram of the apparatus is shown in Figure 6.12. $LiNbO_3$ substrates with a Ti strip pattern are loaded into a quartz tube and the ambient air is replaced with flowing oxygen. The oxygen flows through the bubbler to introduce humidity. The amount of water vapor in the oxygen is controlled by the temperature of the bubbler. The quartz tube is heated to about 1000°C and the in-diffusion process takes 5–10 hours.

#### *6.4.1.2.3 Characteristics of Ti-Diffused Waveguides*

A typical output intensity pattern of an optical beam at 1.55-μm wavelength from a Ti-diffused $LiNbO_3$ waveguide is shown in Figure 6.13. The full widths at half maximum of the beam in lateral and vertical

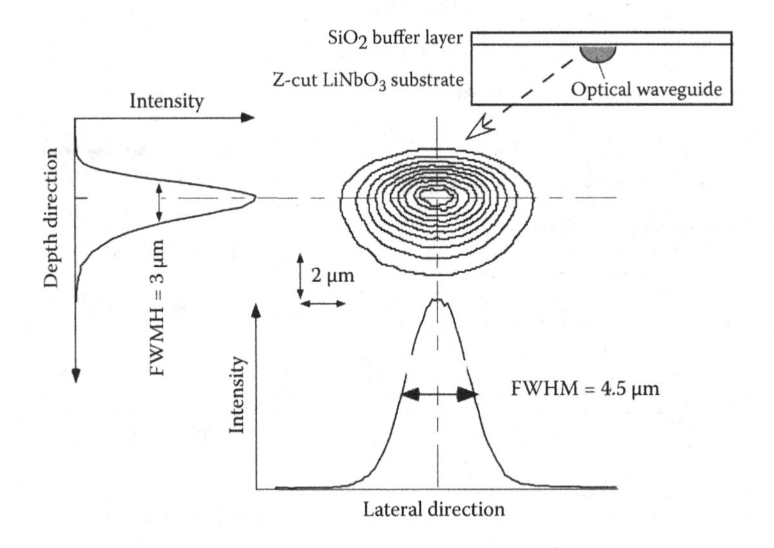


**FIGURE 6.13** Spatial profile of output optical beam from a Ti-diffused waveguide on z-cut $LiNbO_3$.

directions are 6 μm and 3.5 μm, respectively. The width and thickness of the Ti pattern are 6 μm and 90 nm, respectively, and the diffusion time is ten hours in a wet $O_2$ atmosphere. The propagation loss of the waveguide is 0.2 dB/cm, and the coupling loss with a single-mode fiber is less than 0.5 dB/facet.

When visible light is used as the optical wave, such as light from a Helium–Neon laser at 632.8 nm, the photorefractive effect occurs in the Ti-diffused waveguide [63]. The index change due to the optical signal in the waveguide causes a phase change of the optical signal and an increase of propagation loss. This phenomenon is called "optical damage." Note that optical damage is observed only in the visible-light region. In the near-infrared wavelength region above 1.3 μm, which is used for optical communications systems, optical damage in $LiNbO_3$ optical modulators has not been observed [64].

### 6.4.2 Buffer Layer Fabrication

The buffer layer is located between the optical waveguide and the electrode in a Z-cut $LiNbO_3$ modulator. It plays three major roles: First, it is used to avoid increases in the propagation loss of the waveguide due to the optical absorption by metal electrodes. Second, as described in the previous section, it affects the propagation characteristics of a traveling-wave electrode considerably and it is used to achieve velocity and impedance matching [51]. Third, the composition of the buffer layer and its fabrication method greatly affect the suppression of dc and thermal drifts in $LiNbO_3$ modulators [36,38,65].

To avoid optical absorption by the electrode metal, the refractive index of the buffer layer is lower than that of the $LiNbO_3$ substrate. To obtain velocity and impedance matching, the dielectric constant should be small. To ensure the reliability of the modulator, the material should be stable at operating temperature and have strong adhesion to the $LiNbO_3$ substrate.

$SiO_2$ thin films with thickness of 0.5 μm to 1 μm have been commonly used as the buffer layer. For the fabrication of silicon oxide thin films, plasma-enhanced chemical vapor deposition was initially used. However, high-energy plasma and heating the wafer to about 400°C in a vacuum may damage the surface of the $LiNbO_3$ substrate, causing an increase of propagation loss in the optical waveguide and dc drift. Therefore, electron-beam evaporation or radio frequency sputtering become preferred methods. An indium titanium oxide buffer layer or silicon coating is effective for obtaining high reliability [66].

### 6.4.3 Electrode Fabrication

As described in the previous section, the bandwidth of a $LiNbO_3$ modulator is determined by the microwave propagation characteristics of the traveling-wave electrode. The electrode characteristics depend on the shape of the electrode: the width of the center conductor, the gap between the center conductor and ground, the thickness and the length. They also depend on the dielectric constants of the surrounding materials: the $LiNbO_3$ substrate, $SiO_2$ buffer layer, and air.

Large interaction between the optical wave and the microwave is obtained by making the width of the center conductor the same as that of the optical waveguide because this increases the overlap integral of the applied electric field and the optical mode in the waveguide. This in turn reduces the drive voltage. The width of the center conductor of a CPW electrode is typically about 8 μm.

The thickness of the traveling-wave electrode is typically 10 μm to 20 μm for velocity matching and low conductor loss. Since drive voltage is inversely proportional to electrode length, a longer electrode is desirable but the length of the electrode is limited by velocity matching and wafer size. As a trade-off, electrode length of 2 cm to 5 cm is normally chosen. As a result, 8 μm wide, 20 μm thick, 5 cm long electrodes are common for $LiNbO_3$ wafers of three to four inch sizes.

Gold (Au) is chosen for the electrode material because its resistivity is the third smallest among all the metals and thick gold films can be easily formed by electroplating. The electrode fabrication procedure is shown in Figure 6.14. After Ti-diffused waveguides and the $SiO_2$ buffer layer have been formed, a thin film of Ti and a thin film of Au are deposited over the entire surface of the $LiNbO_3$ substrate in a single vacuum deposition run. The Au film is the seed layer for plating, and the Ti film acts as an adhesive

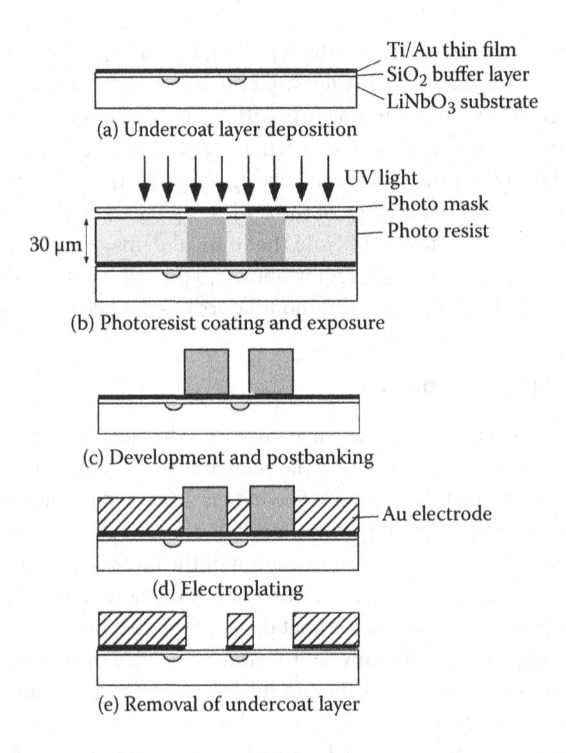


**FIGURE 6.14** Electrode fabrication process.

between the $SiO_2$ and Au. Next, the wafer is coated with a 30-μm-thick layer of photoresist and is prebaked. After photolithographic patterning and developing, the patterned resist is postbaked in an oven to form a plating template. The substrate is then electroplated for several tens of minutes to grow a thick gold electrode in the openings of the photoresist template. After the photoresist is removed, the remaining Au and Ti seed layer between the plated electrodes are removed by ion milling to separate the center conductor from the ground electrodes.

As shown in Figure 6.15, the plating apparatus consists of a bath filled with plating solution, a circulation system with a pump, a filter, and a heater, an electrical current source, an anode electrode made of a

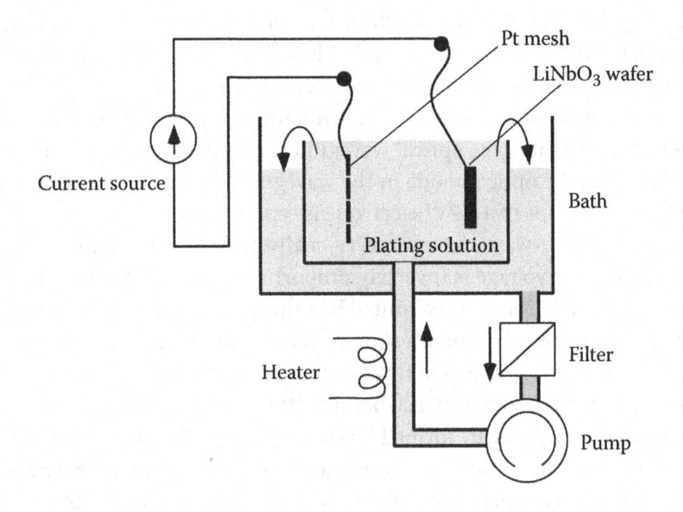


**FIGURE 6.15** Electroplating apparatus.

platinum mesh, and a wafer holder. A $LiNbO_3$ wafer clad with Ti/Au thin film seed layer and a patterned photoresist template is mounted on the holder and is then immersed along with the platinum electrode. When a dc current flows between the wafer and the electrode, gold ions in the solution deposit on the wafer with the patterned photoresist guiding the deposition. A 20-μm-thick gold electrode is typically obtained after two hours of plating at 60°C with a current density of 2 mA/cm$^2$.

## 6.5 Packaging Considerations

Optical modulators are fabricated on $LiNbO_3$ wafers, which are cut into chips of individual modulators several centimeters in length and several millimeters in width. To supply electrical signals from the drive circuit to the electrode, the electrode is connected to microwave connectors, such as Anritsu V-connectors. The connector connects the modulator to the drive circuit through a coaxial line. The width of the center conductor of a V-connector is 220 μm, while that of the interaction region of the traveling-wave electrode is 8 μm. Tapered regions are introduced between the interaction region and the large contact areas for the connectors, as shown in Figure 6.16.

Both an asymmetric coplanar strip line and a CPW have been used as a traveling-wave electrode. However, the former has not been used recently because it has two disadvantages, especially at frequencies higher than 20 GHz. First, standard microwave feeding connectors are symmetric. The configuration matches better with CPW electrodes which have a symmetric ground-signal-ground structure. Second, the substrate resonant frequency of asymmetric coplanar strip electrodes is only half of that of CPW electrodes [27].

When an electrical signal propagates through the joint points between the connectors and the electrodes, part of the energy easily radiates into substrate resonant modes. The resonant coupling frequency of the substrate mode $f_c$ is calculated [67] as

$$f_c = \frac{11.9}{d} (\text{GHz}),$$

where $d$ is the substrate thickness in millimeters. For example, to obtain high-speed optical modulators operating at 40 Gbit/s, we must eliminate undesired substrate resonances by using $LiNbO_3$ substrates thinner than 0.25 mm.

## 6.6 Future Development

### 6.6.1 Development in $LiNbO_3$ Modulator Technology

*New fabrication technology:* Although ridged $LiNbO_3$ modulators show excellent performance [32,48,68], the fabrication throughput is not high due to the very slow etching rate of $LiNbO_3$ crystal. Kondo et al. have developed a micromachining technology based on krypton fluoride excimer laser ablation [69].

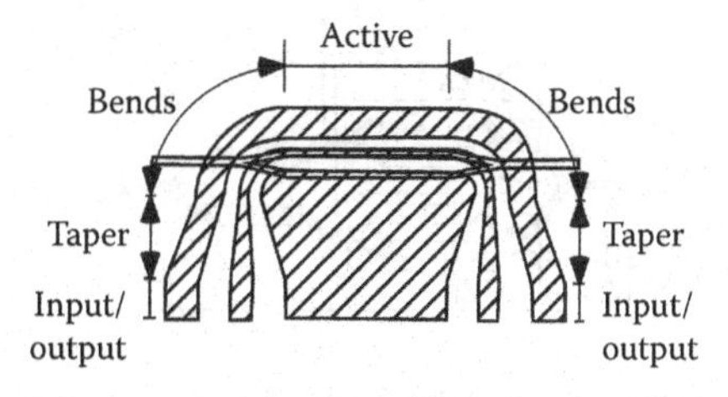


**FIGURE 6.16** Layout of modulator electrode. (Reprinted with permission from G. K. Gopalakrishnan et al., *IEEE J. Lightwave Technol.*, 12, 10, 1807–1818, © 1994 IEEE.)

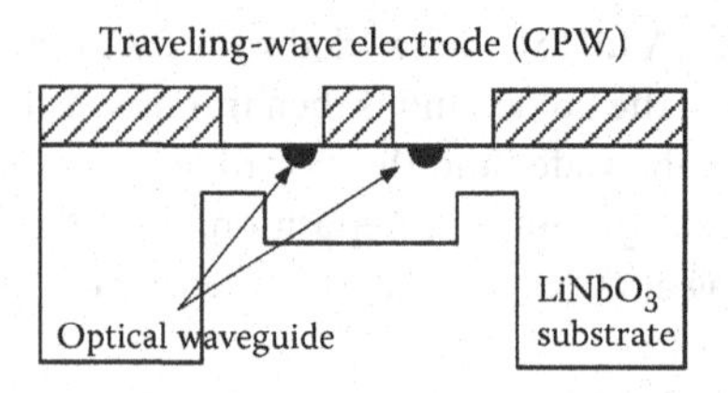


**FIGURE 6.17** Cross-sectional view of x-cut $LiNbO_3$ with a two-step back-slot structure. (Reprinted with permission from J. Kondo et al., *IEEE J. Lightwave Technol.*, 20, 12, 2110–2114, © 2002 IEEE.)

They have built thin-film $LiNbO_3$ modulators with a two-step back-slot structure, as shown in Figure 6.17. They use air as a low-index material at the back side of the substrate and an X-cut $LiNbO_3$ substrate to avoid thermal drift and have successfully developed 40-Gbit/s modulators.

*Modulators for new modulation formats:* Intensity modulation and on-off keying are commonly used in commercial optical communications systems. The key to obtain higher spectral efficiency or receiver sensitivity of the system is to control the lightwave amplitude, phase, and frequency by using $LiNbO_3$ modulators [43,70]. Figure 6.18 shows a frequency shift keying/single sideband modulator consisting of six optical phase modulators with traveling-wave electrodes, developed by Kawanishi et al. [42]. A pair of sub-Mach-Zehnder structures was incorporated in a main Mach-Zehnder structure. They demonstrated 40-Gbit/s frequency shift keying modulation and a 40-GHz optical frequency shift. The modulator was used for differential quadrature phase shift keying by supplying appropriate signals to electrodes A1, A2, B1, B2, C1, and C2. Differential quadrature phase shift keying systems are promising because they can double the symbol rate per time slot and halve the operation frequency of the drive circuit and modulator.

*Hybrid integration:* The refractive index contrast of Ti-diffused waveguides is typically below 0.5 %, where the optical mode field diameter in the waveguide matches that in the single-mode fiber, and the propagation loss can be as low as 0.2 dB/cm. To suppress excess bending loss in waveguides of low index contrast, Y-branches of Mach-Zehnder modulators need to be more than 5 mm long. In the differential quadrature phase shift keying modulator in Figure 6.18, four Y-branches are connected in series on one chip. The total length of the Y branches is 20 mm. Because the total length of the device is limited by the available size of the $LiNbO_3$ wafer, long Y branches limit the interaction length of modulators.

Yamada et al. introduced planar lightwave circuit (PLC) technologies to enhance the function of $LiNbO_3$ devices [71]. Figure 6.19 shows the hybrid-integration of a planar lightwave circuit $LiNbO_3$ modulator. Input/output and circuit silica waveguides with a monitor port were formed on planar lightwave circuit chips on silicon substrates. On the $LiNbO_3$ chip, two Mach-Zehnder waveguides and four traveling-wave electrodes were fabricated to form two parallel intensity modulators. In this device, low-loss high index-contrast waveguides made with planar lightwave circuit technology and high-speed modulators made with $LiNbO_3$ modulator technology were effectively integrated [72].

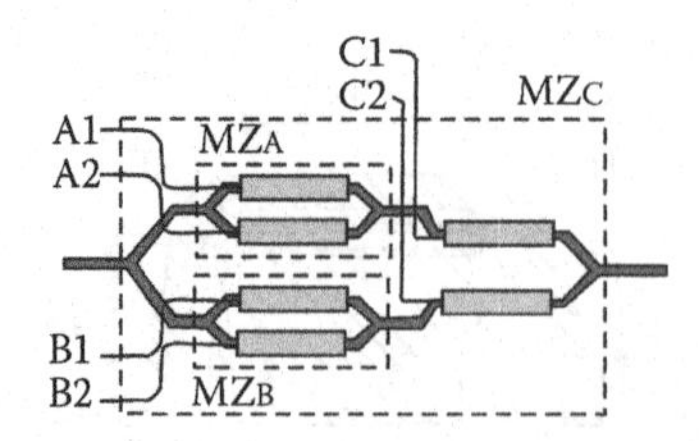


**FIGURE 6.18** Z-Cut dual-electrode FSK/SSB modulator, consisting of six optical phase modulators with traveling-wave electrodes. (Reprinted with permission from T. Kawanishi et al., *IEEE J. Selected Topics Quantum Electron.*, 13, 1, 79–91, © 2007 IEEE.)

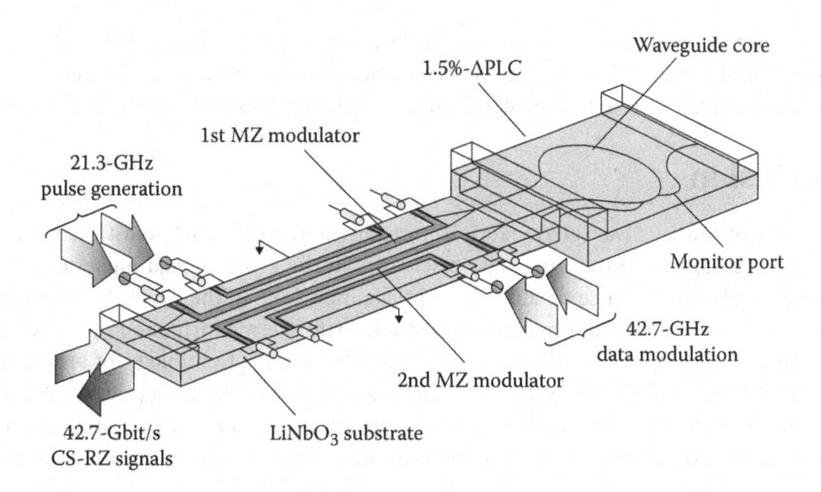


**FIGURE 6.19** A hybrid modulator consisting of two parallel Mach-Zehnder modulators and a PLC chip. (Reprinted from T. Yamada, *10th Intl. Symp. Microwave Optical Technology. ISMOT 2005*, 674–676. With permission.)

## 6.6.2 Other Materials

The EO properties of commonly used modulator materials are summarized in Table 6.2 [45,73,74]. For the fabrication of low drive-voltage devices, a large EO coefficient is preferable. Simultaneously, a small dielectric constant is desirable for high-speed operation. From the industrial viewpoint, the stability of the material and availability of high-quality crystals are important.

Semiconductor materials, such as gallium arsenide and indium phosphate, are also used in EO modulators operating at above 40 Gbit/s [75–77]. To cope with the relatively large coupling loss with single-mode fiber and the large propagation loss of the waveguide, indium phosphate electroabsorption modulators are integrated with laser diodes [78] or semiconductor optical amplifiers [79,80]. Electro-absorption and EO modulators based on semiconductor materials are discussed in detail in Chapters 5, 7, and 8, respectively.

Lead lanthanum-modified zirconate titanate ferroelectric ceramics have very large EO coefficients and the drive voltages of optical modulators can be reduced [81]. However, there are technical challenges to using these ceramics for high-speed modulators due to their very high dielectric constants. Lead lanthanum-modified zirconate titanate is therefore mostly used for optical switches [82]. Chapter 12 of this book focuses on lead lanthanum-modified zirconate titanate material and devices.

Electro-optic polymers are also promising for high-speed modulators [74,83,84]. Since their dielectric constants are small, EO polymers have great advantages for obtaining velocity matching. Modulation

**TABLE 6.2** Electro-Optic Coefficient and Dielectric Constants of Electro-Optic Materials

| Materials | Electro-Optic Coefficient (pm/V) | Refractive Index | $n^3r$ (pm/V) | Relative Dielectric Constant |
|---|---|---|---|---|
| $LiNbO_3$ | $r_{33} = 30.8$ | $n_e = 2.146$ (at $\lambda = 1300$ nm) | 304 | $\varepsilon_{33} = 29$, $\varepsilon_{11} = 44$ |
| GaAs | $r_{41} = 1.41$ | $n_o = 3.42$ (at $\lambda = 1250$ nm) | 56 | $\varepsilon = 13.2$ |
| InP | $r_{41} = 1.3$ | $n_o = 3.20$ (at $\lambda = 1350$ nm) | 43 | $\varepsilon = 12.6$ |
| PLZT | $r = 102$ | $n = 2.45$ | 1500 | $\varepsilon = 2382$ |
| EO Polymer | $r = 30$ | $n = 1.6$ | 123 | $\varepsilon = 3.5$ |

*Sources:* A. M. Prokhorov and Y. S. Kuz'minov, *Physics and Chemistry of Crystalline Lithium Niobate*, 235, Adam Hilger, Bristol and New York, 1990. With permission; T. Yamada et al., *10th Intl. Symp. Microwave Optical Technology. ISMOT 2005*, 674–676. With permission; G. H. Jin et al., *IEEE J. Lightwave Technol.*, 18, 6, 807–812, 2000. With permission.

bandwidths beyond 100 GHz can be achieved relatively easily [85]. Reducing dc drift and enhancing the thermal stability of these materials may be necessary before the modulators can be used in commercial systems [86–88]. Chapter 9 has a detailed description of EO polymers and EO polymer modulators.

## 6.7 Conclusion

The development of low-loss waveguide fabrication technology on $LiNbO_3$ substrate by Ti-diffusion in the 1970s and research into traveling-wave electrodes in the 1980s turned high-speed low-voltage modulators into high-performance reliable fiber-optic components in the 1990s. Practical issues such as instabilities of $LiNbO_3$ modulator operation due to dc drift, thermal drift, and optical damage were overcome by steady research in the 1990s. Today $LiNbO_3$ is the leading technology for broadband optical modulators and a large number of $LiNbO_3$ modulators are in service in fiber-optic telecommunication systems worldwide. $LiNbO_3$ technology is constantly progressing and new modulators are continually emerging to meet the ever increasing demand in performance and for new modulation formats.

## References

1. I. P. Kaminow, J. R. Carruthers, E. H. Turner, and L. W. Stulz, "Thin-film $LiNbO_3$ electro-optic light modulator," *Appl. Phys. Lett.*, vol. 22, no. 10, pp. 540–542, May 1973.
2. I. P. Kaminow, V. Ramaswamy, R. V. Schmidt, and E. H. Turner, "Lithium niobate ridge waveguide modulator," *Appl. Phys. Lett.*, vol. 24, no. 12, pp. 622–624, Jun. 1974.
3. M. Izutsu, Y. Yamane, and T. Sueta, "Broad-band traveling-wave modulator using a $LiNbO_3$ optical waveguide," *IEEE J. Quantum Electron.*, vol. 13, no. 4, pp. 287–290, Apr. 1977.
4. T. R. Ranganath and S. Wang, "Suppression of $Li_2O$ out-diffusion from Ti-diffused $LiNbO_3$ optical waveguides," *Appl. Phys. Lett.*, vol. 30, no. 8, pp. 376–379, Apr. 1977.
5. H. Sasaki, "Efficient intensity modulation in a Ti-diffused $LiNbO_3$ branched optical waveguide device," *Electron. Lett.*, vol. 13, no. 23, pp. 693–694, Nov. 1977.
6. O. Mikami, J. Noda, and M. Fukuma, "Directional coupler type light modulator using $LiNbO_3$ waveguides," *Trans. IEICE Japan*, vol. E-61, no. 3, pp. 144–147, Mar. 1978.
7. M. Minakata, S. Saito, M. Shibata, and S. Miyazawa, "Precise determination of refractive-index changes in Ti-diffused $LiNbO_3$ optical waveguides," *J. Appl. Phys.*, vol. 49, no. 9, pp. 4677–4682, Sep. 1978.
8. R. C. Alferness, R. V. Schmidt, and E. H. Turner, "Characteristics of Ti-diffused $LiNbO_3$ optical directional couplers," *Appl. Opt.*, vol. 18, no. 23, pp. 4012–4018, Dec. 1979.
9. F. J. Leonberger, "High-speed operation of $LiNbO_3$ electrooptic interferometric waveguide modulators," *Opt. Lett.*, vol. 5, no. 7, pp. 312–314, Jul. 1980.
10. R. E.Tench, J. M. P. Delavaux, L. D. Tzeng, R. W. Smith, L. L. Buhl, and R. C. Alferness, "Performance evaluation of waveguide phase modulators for coherent systems at 1.3 and 1.5 microns," *J. Lightwave Technol.*, vol. 5, no. 4, pp. 492–501, Apr. 1987.
11. M. Kondo, Y. Tanisawa, Y. Ohota, T. Aoyama, and R. Ishikawa, "Low-drive-voltage and low-loss polarization-independent $LiNbO_3$ optical waveguide switches," *Electron. Lett.*, vol. 23, no. 21, pp. 1167–1169, Oct. 1987.
12. D. J. Blumenthal, P. R. Prucnal, L. Thylen, and P. Granestrand, "Performance of and 8 × 8 $LiNbO_3$ switch matrix as a gigahertz self-routing switching node," *Electron. Lett.*, vol. 23, no. 25, pp. 1359–1360, Dec. 1987.
13. L. Thylen, "Integrated optics in $LiNbO_3$: Recent developments in devices for telecommunications," *J. Lightwave Technol.*, vol. 6, no. 6, pp. 847–861, Jun. 1988.
14. I. Sawaki, T. Shimoe, H. Nakamoto, T. Iwama, T. Yamane, and H. Nakajima, "Rectangularly configured 4 × 4 Ti: $LiNbO_3$ matrix switch with low drive voltage," *IEEE J. Selected Comm.*, vol. 6, no. 7, pp. 1267–1272, Aug. 1988.

15. H. Nishimoto, M. Iwasaki, S. Suzuki, and M. Konodo, "Polarization independent $LiNbO_3$ 8 × 8 matrix switch," *IEEE Photonics Technol. Lett.*, vol. 2, no. 9, pp. 634–636, Sep. 1990.
16. H. Miyazawa, T. Nozawa, and M. Yanagibashi, "A high-speed 1 × 4 Ti:$LiNbO_3$ optical switch," in *Technical Digest of the Third Optoelectron. Conf.*, 1990, vol. PD-2, pp. 8–9.
17. M. Fukuma, J. Noda, and H. Iwasaki, "Optical properties in titanium-diffused $LiNbO_3$ strip waveguides," *J. Appl. Phys.*, vol. 49, no. 7, pp. 3693–3698, Jul. 1978.
18. S. Fouchet, A. Carenco, C. Daguet, R. Guglielmi, and L. Rivere, "Wavelength dispersion of Ti induced refractive index change in $LiNbO_3$ as a function of diffusion parameters," *J. Lightwave Technol.*, vol. 5, no. 5, pp. 700–708, May 1987.
19. T. Nozawa, K. Noguchi, H. Miyazawa, and K. Kawano, "Water vapor effects on optical characteristics in Ti:$LiNbO_3$ channel waveguides," *Appl. Opt.*, vol. 30, no. 9, pp. 1085–1089, Mar. 1991.
20. M. Fukuma and J. Noda, "Optical properties of titanium-diffused $LiNbO_3$ strip waveguides and their coupling-to-a-fiber characteristics," *Appl. Opt.*, vol. 19, no. 4, pp. 591–597, Feb. 1980.
21. R. C. Alferness, L. L. Buhl, and M. D. Divino, "Low-loss fiber-coupled waveguide directional coupler modulator," *Electron. Lett.*, vol. 18, no. 12, pp. 490–491, Jun. 1982.
22. V. Ramaswamy, R. C. Alferness, and M. Divino, "High efficiency single-mode fibre to Ti:$LiNbO_3$ waveguide coupling," *Electron. Lett.*, vol. 18, no. 1, pp. 30–31, Jan. 1982.
23. K. Komatsu, S. Yamazaki, M. Konodo, and Y. Ohta, "Low-loss broad-band $LiNbO_3$ guided-wave phase modulators using titanium/magnesium double diffusion method," *IEEE J. Lightwave Technol.*, vol. 5, no. 9, pp. 1239–1245, Sept. 1987.
24. T. Kataoka, Y. Miyamoto, K. Hagimoto, and K. Noguchi, "20 Gbit/s long span transmission experiments using a 270 photon/bit optical preamplifier receiver," *Electron. Lett.*, vol. 30, no. 9, pp. 715–716, Apr. 1994.
25. K. Hagimoto, M. Yoneyama, A. Sano, H. Hirano, T. Kataoka, T. Otsuji, K. Sato, and K. Noguchi, "Limitations and challenges of single-carrier full 40-Gbit/s repeater system based on optical equalization and new circuit design," in *Conf. Optical Fiber Communication, OFC 97*, Dallas, TX, 1997, vol. ThC1, pp. 242–243.
26. K. Kawano, T. Kitoh, H. Jumonji, T. Nozawa, and M. Yanagibashi, "New traveling wave elecrode Mach-Zehnder optical modulator with 20 GHz bandwidth and 4.7 V driving voltage at 1.52 μm wavelength," *Electron. Lett.*, vol. 25, no. 20, pp. 1382–1383, Sep. 1989.
27. M. Seino, N. Mekada, T. Yamane, Y. Kubota, M. Doi, and T. Nakazawa, "20-GHz 3 dB-bandwidth Ti:$LiNbO_3$ Mach-Zehnder modulator," in *16th European Conf. Optical Communication, ECOC'90*, Amsterdam, The Netherlands, 1990, vol. PD3, pp. 999–1002.
28. D. Dofli and T. R. Ranganath, "50 GHz volocity-matched, broad wavelength $LiNbO_3$ modulator with multimode active section," *Electron. Lett.*, vol. 28, no. 13, pp. 1197–1198, Jun. 1992.
29. G. K. Gopalakrishnan, C. H. Bulmer, W. K. Burns, R. W. McElahanon, and A. S. Greenblatt, "40 GHz, low half-wave voltage Ti:$LiNbO_3$ intensity modulator," *Electron. Lett.*, vol. 28, no. 9, pp. 826–827, Apr. 1992.
30. H. Miyamoto, H. Ohta, K. Tabuse, and Y. Miyagawa, "Evaluation of $LiNbO_3$ intensity modulator using electrodes buried in buffer layer," *Electron. Lett.*, vol. 28, no. 11, pp. 976–977, May 1992.
31. R. Madabushi, Y. Uematsu, K. Fukuchi, and A. Noda, "Wide-band, low driving voltage Ti:$LiNbO_3$ optical modulators for 40 Gb/s applications," in *24th European Conf. Optical Communications, ECOC'98*, Madrid, Spain, 1998, vol. 1, pp. 547–548.
32. M. M. Howerton, R. P. Moeller, A. S. Greenblatt, and R. Krahenbuhl, "Fully packaged, broad-band $LiNbO_3$ modulator with low drive voltage," *IEEE Photon. Technol. Lett.*, vol. 12, no. 7, pp. 792–794, Jul. 2000.
33. K. Noguchi, H. Miyazawa, and O. Mitomi, "Frequency-dependent propagation characteristics of a coplanar waveguide electrode on a 100-GHz-Ti:$LiNbO_3$ optical modulator," *Electron. Lett.*, vol. 34, no. 7, pp. 661–662, Apr. 1998.

34. C. H. Bulmer, W. K. Burns, and S. C. Hiser, "Pyroelectric effects in $LiNbO_3$ channel-waveguide devices, *Appl. Phys. Lett.*, vol. 48, no. 16, pp. 1036–1038, Apr. 1986.
35. P. Skeath, C. H. Bulmer, S. H. Hiser, and W. K. Burns, "Novel electrostatic mechanism in the thermal instability of z-cut $LiNbO_3$ interferometers," *Appl. Phys. Lett.*, vol. 49, no. 19, pp. 1221–1223, Nov. 1986.
36. I. Sawaki, H. Nakajima, M. Seino, and K. Asama, "Thermally stabilized z-cut Ti:$LiNbO_3$ waveguide switch," in *Technical Digest of Conference on Lasers and Electro-Optics, CLEO '86*, Washington, DC, 1986, paper MF2, pp. 46–47. Washington, DC: Optical Society of America.
37. S. Yamada and M. Minakata, "DC drift phenomena in $LiNbO_3$ optical waveguide devices," *Jpn. J. Appl. Phys.*, vol. 20, no. 4, pp. 733–737, Apr. 1981.
38. M. Seino, T. Nakazawa, Y. Kubota, M. Doi, T. Yamane, and H. Hakogi, "A low DC-drift Ti:$LiNbO_3$ modulator assured over 15 years," in *Digest of Conf. Optical Fiber Communication, OFC'92*, San Jose, CA, 1992, paper PD-3, pp. 325–328. Washington, DC: Optical Society of America.
39. S. K. Korotky and J. Veselka, "An RC network analysis of long term Ti:$LiNbO_3$ bias stability," *J. Lightwave Technol.*, vol. 14, no. 12, pp. 2687–2697, Dec. 1996.
40. D. R. Maack, "Reliability of lithium niobate Mach Zehnder modulators for digital optical fiber communication systems" in *Reliability of Optical fibers and Optical Fiber Systems*, D. K. Paul and B. Javidi, Eds., Washington: SPIE Optical Engineering Press, 1999, pp. 197–230.
41. H. Nagata, "DC drift failure rate estimation on 10 Gb/s X-cut lithium niobate modulators," *IEEE Photon. Technol. Lett.*, vol. 12, no. 11, pp. 1477–1479, Nov. 2000.
42. T. Kawanishi, T. Sakamoto, and M. Izutsu, "High-speed control of lightwave amplitude, phase, and frequency by use of electrooptic effect," *IEEE J. Selected Topics Quantum Electron.*, vol. 13, no. 1, pp. 79–91, Jan. 2007.
43. A. Sano, T. Kobayashi, K. Ishihara, H. Masuda, S. Yamamoto, K. Mori, E. Yamazaki, E. Yoshida, Y. Miyamoto, T. Yamada, and H. Yamazaki, "240-Gb/s polarization-multiplexed 64-QAM modulation and blind detection using PLC-LN hybrid integrated modulator and digital coherent receiver," in *35th European Conference on Optical Communication, 2009. ECOC '09*, Vienna, 2009, Supplement, pp. 1–2.
44. T. Yamada, "$LiNbO_3$ family," in *Numerical Data and Functional Relationships in Science and Technology* 16, K.-H. Hellwege, Ed., Berlin: Springer-Verlag, 1980, pp. 149–163.
45. A. M. Prokhorov and Y. S. Kuz'minov, *Physics and Chemistry of Crystalline Lithium Niobate*, Bristol and New York: Adam Hilger, 1990.
46. I. P. Kaminow, *An Introduction to Electrooptic Devices*, New York: Academic Press, 1974.
47. W. M. Robertson, G. Arajavalingam, and G. Kopcsay, "Broadband microwave dielectric properties of lithium niobate," *Electron. Lett.*, vol. 27, no. 2, pp. 175–176, Jan. 1991.
48. K. Noguchi, H. Miyazawa, and O. Mitomi. "Millimeter-wave Ti:$LiNbO_3$ optical modulators," *J. Lightwave Technol.*, vol. 16, no. 4, pp. 615–619, Apr. 1998.
49. K. Kubota, J. Noda, and O. Mikami, "Traveling-wave optical modulator using directional coupler $LiNbO_3$ waveguide," *IEEE J. Quantum Electron.*, vol. 16, no. 7, pp. 754–760, Jul. 1981.
50. S. K. Korotky, G. Eisenstein, R. S. Tucker, J. J. Veselka, and G. Raybon, "Optical intensity modulation to 40 GHz using a waveguide electro-optic switch," *Appl. Phys. Lett.*, vol. 50, no. 23, pp. 1631–1632, Jun. 1987.
51. K. Kawano, T. Kitoh, O. Mitomi, T. Nozawa, and H. Jumonji, "A wide-band and low-driving-power phase modulator employing a Ti:$LiNbO_3$ optical waveguide at 1.5 μm wavelength," *IEEE Photon. Technol. Lett.*, vol. 1, no. 2, pp. 33–34, Feb. 1989.
52. G. K. Gopalakrishnan, W. K. Burns, R. W. McElhanon, C. H. Bulmer, and A. S. Greenblatt, "Performance and modeling of broadband $LiNbO_3$ traveling wave optical intensity modulators," *IEEE J. Lightwave Technol.*, vol. 12, no. 10, pp. 1807–1818, Oct. 1994.
53. O. Mitomi, K. Noguchi, and H. Miyazawa, "Broadband and low driving-voltage $LiNbO_3$ optical modulators," *IEE Proc.-Optoelectron.*, vol. 145, no. 6, pp. 360–364, Dec. 1998.

54. Y. Miyamoto, "40-Gbit/s transport system: its WDM upgrade," *Conference on Optical Fiber Communication, Technical Digest Series*, vol. 3, pp. 323–325, 2000.
55. R. V. Schmidt and I. P. Kaminow, "Metal diffused optical waveguides in $LiNbO_3$, *Appl. Phys. Lett.*, vol. 25, no. 8, pp. 458–460, Oct. 1974.
56. J. L. Jackel, C. E. Rice, and J. J. Veselka, "Proton exchange for high-index waveguides in $LiNbO_3$," *Appl. Phys. Lett.*, vol. 41, no. 7, pp. 607–608, Dec. 1982.
57. T. R. Ranganath and S. Wang, "Ti-diffused $LiNbO_3$ branched-waveguide modulators: Performance and design," *IEEE J. Quantum Electron.*, vol. 13, no. 4, pp. 290–295, Apr. 1977.
58. W. K. Burns, C. H. Bulmer, and E. J. West, "Application of $Li_2O$ compensation techniques to Ti-diffused $LiNbO_3$ planar and channel waveguides," *Appl. Phys. Lett.*, vol. 33, no. 1, pp. 70–72, Jul. 1978.
59. B.-U. Chen and A. C. Pastor, "Elimination of $Li_2O$ out-diffusion waveguide in titanium-diffused $LiNbO_3$," *Appl. Phys. Lett.*, vol. 30, no. 11, pp. 570–571, Jun. 1981.
60. J. L. Jackel, V. Ramaswamy, and S. Lyman, "Elimination of out-diffused surface guiding in titanium-diffused $LiNbO_3$," *Appl. Phys. Lett.*, vol. 38, no. 7, pp. 509–510, Apr. 1981.
61. O. Tadanaga, M. Asobe, H. Miyazawa, Y. Nishida, and H. Suzuki, "Variable optical frequency shifter using multiple quasi-phase-matched $LiNbO_3$ wavelength converters," in *Optical Fiber Communication Conference. OFC 2003*, Atlanta, GA, paper FP3. Washington, DC: Optical Society of America.
62. A. Enokihara, A. Suzuki, J. Adachi, T. Iwamoto, H. Murata, and Y. Okamura, "Fabrication and evaluation of $LiNbO_3$ periodic waveguide with etched grooves," *Electron. Lett.*, vol. 43, no. 11, pp. 629–610, May 2007.
63. R. V. Schmidt, P. S. Cross, and A. M. Glass, "Optically induced crosstalk in $LiNbO_3$ waveguide switches," *J. Appl. Phys.*, vol. 51, no. 1, pp. 90–93, Jan. 1979.
64. G. T. Harvey, G. Astfalk, A. Feldblum, and B. Kassahun, "The photorefractive effect in titanium indiffused lithium niobate optical directional couplers at 1.3 μm," *IEEE J. Quantum Electron.*, vol. 22, no. 6, pp. 939–946, Jun. 1986.
65. H. Nagata and J. Nayyer, "Stability and reliability of lithium niobate optical modulators," in *Integrated Photonics Research. IPR'95*, Dana Point, CA, 1995, paper ISaB1-1. Washington, DC: Optical Society of America.
66. M. Seino, T. Nakazawa, M. Doi, and S. Taniguchi, "The long-term reliability estimation of Ti:$LiNbO_3$ for DC drift," in *10th Int. Conf. Integrated Optics and Optical Fibre Communication. IOOC'95*, Hong Kong, 1995, pp. PD1-8.
67. Y. Shi, "Micromachined wide-band lithium-niobate electrooptic modulators," *IEEE Trans. Microwave Theory Tech.*, vol. 54, no. 2, pp. 810–815, Feb. 2006.
68. M. Sugiyama, M. Doi, S. Taniguchi, T. Nakazawa, and H. Onaka, "Driver-less 40 Gb/s $LiNbO_3$ modulator with sub-1 V drive voltage," in *Optical Fiber Communication Conf., 2002. OFC2002*, pp. FB6-1-4. Washington, DC: Optical Society of America.
69. J. Kondo, A. Kondo, K. Aoki, M. Imaeda, T. Mori, Y. Mizuno, S. Takatsuji, Y. Kozuka, O. Mitomi, and M. Minakata, "40-Gb/s X-cut $LiNbO_3$ optical modulator with two-step back-slot structure," *IEEE J. Lightwave Technol.*, vol. 20, no. 12, pp. 2110–2114, Dec. 2002.
70. S. Norimatsu, K. Iwashita, and K. Noguchi, "An 8 Gb/s QPSK optical homodyne detection experiment using external-cavity laser diodes," *IEEE Photon. Technol. Lett.*, vol. 4, no. 7, pp. 765–767, Jul. 1992.
71. T. Yamada, S. Mino, M. Ishii, T. Shibata, S. Suzuki, T. Ohara, S. Kuwahara, and Y. Miyamoto, "Highly functional hybrid modules using low loss direct attachment technique with planar lightwave circuit and $LiNbO_3$ devices," in *10th Intl. Symp. Microwave Optical Technology. ISMOT 2005*, pp. 674–676, Piscataway, NJ:IEEE.
72. K. Suzuki, T. Yamada, O. Moriwaki, H. Takahashi, and M. Okuno, "Polarization-insensitive operation of lithium niobate Mach-Zehnder interferometer with silica PLC-based polarization diversity circuit," *IEEE Photon. Technol. Lett.*, vol. 20, no. 10, pp. 773–775, May 2008.
73. K. Tada and N. Suzuki, "Linear electrooptic properties of InP," *Jpn. J. Appl. Phys.*, vol. 19, no. 11, pp. 2295–2296, Nov. 1980.

74. R. Lytel, "EO polymer materials and devices: From research to reality," in *Nonlinear Optics: Materials, Fundamentals, and Applications, 1994. NLO '94 IEEE*, Waikoloa, HI, pp. MA1:3–5. Piscataway, NJ: IEEE.
75. K. Tsuzuki, T. Ishibashi, T. Ito, S. Oku, Y. Shibata, T. Ito, R. Iga, Y. Kondo, and Y. Tohmori, "A 40-Gb/s InGaAlAs-InAlAs MQW n-i-n Mach-Zehnder modulator with a drive voltage of 2.3 V," *IEEE Photon. Technol. Lett.*, vol. 17, no. 1, pp. 46–48, Jan. 2005.
76. R. A. Griffin, A. Tipper, and I. Betty, "Performance of MQW InP Mach-Zehnder modulators for advance modulation formats," in *Optical Fiber Communication Conf., OFC 2005*, Anaheim CA, paper OTuL5, Washington, DC: Optical Society of America.
77. C. R. Doerr, G. Raybon, L. Zhang, L. L. Buhl, A. L. Adamiecki, J. H. Sinsky, N. J. Sauer, et al., "Low-chirp 85-Gb/s duobinary modulator in InP using electroabsorption modulators," *IEEE Photon. Technol. Lett.*, vol. 21, no. 17, pp. 1199–1201, Sep. 2009.
78. Y. Kawamura, K. Wakita, Y. Itaya, Y. Yoshikuni, and H. Asahi, "Monolithic integration of InGaAsP/InP DFB lasers and InGaAs/InAlAs MQW optical modulator," *Electron. Lett.*, vol. 22, no. 5, pp. 242–243, Feb. 1986.
79. D. F. Welch, F. A. Kish, S. Melle, R. Nagarajan, M. Kato, C. H. Joyner, J. L. Plemeekers, et al., "Large-scale InP photonic integrated circuits: Enabling efficient scaling of optical transport networks," *IEEE J. Selected Topics Quantum Electron.*, vol. 13, no. 1, pp. 22–31, Jan. 2007.
80. T. Yasui, Y. Shibata, K. Tsuzuki, N. Kikuchi, M. Ishikawa, Y. Kawaguchi, M. Arai, and H. Yasaka, "10-Gb/s 100-km SMF transmission using InP Mach-Zehnder modulator monolithically integrated with semiconductor optical amplifier," *IEEE Photon. Technol. Lett.*, vol. 20, no. 13, pp. 1178–1180, Jul. 2008.
81. G. H. Jin, Y. K. Zou, V. Fuflyigin, S. W. Liu, Y. L. Lu, J. Zhao, and M. Cronin-Golomb, "PLZT film waveguide Mach-Zehnder electrooptic modulator," *IEEE J. Lightwave Technol.*, vol. 18, no. 6, pp. 807–812, Jun. 2000.
82. A. L. Glebov, M. G. Lee, L. Huang, S. Aoki, K. Yokouchi, M. Ishii, and M. Kato, "Electrooptic plannar deflector switches with thin-film PLZT active elements," *IEEE J. Selected Topics Quantum Electron.*, vol. 11, no. 2, pp. 422–430, Mar. 2005.
83. M.-C. Oh, H. Zhang, C. Zhang, H. Erlig, Y. Chang, B. Tsap, D. Chang, A. Szep, W. H. Steier, H. R. Fetterman, and L. R. Dalton, "Recent advances in electrooptic polymer modulators incorporating highly nonlinear chromophore," *IEEE J. Selected Topics Quantum Electron.*, vol. 7, no. 5, pp. 826–835, Sep. 2001.
84. R. A. Norwood, "Electro-optic polymer modulators for telecommunications applications, in *Optical Fiber Communication Conf. Expo. Nat. Fiber Optic Engineers Conf., OFC/NFOEC 2008*, paper OMJ3, Washington, DC: Optical Society of America.
85. D. Chen, H. R. Fetterman, A. Chen, W. H. Steier, and L. R. Dalton, "Demonstration of 110 GHz electro-optic polymer modulators," *Appl. Phys. Lett.*, vol. 70, no. 25, pp. 3335–3337, Jun. 1997.
86. S. Park, J. J. Ju, J. Y. Do, S. K. Park, and M.-H. Lee, "Thermal stability enhancement of electrooptic polymer modulator," *IEEE Photon. Technol. Lett.*, vol. 16, no. 1, pp. 93–95, Jan. 2004.
87. Y.-H. Kuo, J. Luo, W. H. Steier, and A. K.-Y. Jen, "Enhance thermal stability of electrooptic polymer modulators using the Diels-Alder crosslinkable polymer," *IEEE Photon. Technol. Lett.*, vol. 18, no. 1, pp. 175–177, Jan. 2006.
88. R. Dinu, D. Jin, G. Yu, B. Chen, D. Huang, H. Chen, A. Barklund, E. Miller, C. Wei, and J. Vemagiri, "Environmental stress testing of electro-optic polymer modulators," *IEEE J. Lightwave Technol.*, vol. 27, no. 11, pp. 1527–1532, Jun. 2009.

# 7

# Indium Phosphide-Based Electro-Optic Modulators

Ian Betty
*Ciena*



## 7.1 History and Evolution of InP-Based Electro-Optic Modulators in Optical Communications

### 7.1.1 1990–2000: Innovation—Realization and Application

In the early 1990s, Rolland et al. [1–4] at Bell Northern Research developed a negative chirp Mach–Zehnder modulator (MZM) for 10 Gb/s optical communication systems that utilized the quantum confined Stark effect (QCSE) [5,6] in the InGaAsP/InP material system. They published early work on the means to generate the optimal negative chirp for maximum dispersion-limited reach over standard optical fiber with these InP-based MZMs [7], demonstrating >100 km of over-fiber reach in 1993 [8]. In parallel, Sano et al. [9,10] at Hitachi published early work for an MZM design in the InGaAs/InAlAs material system.

In 1996, Adams et al. [11] demonstrated monolithic integration of their MZM with a distributed feedback laser source enabling 100 km of reach at 10 Gb/s with <1 dB dispersion penalty. At 2.5 Gb/s an over-fiber reach of 1102 km was demonstrated for both the discrete modulator and the monolithic laser integrated device [12].

In 1996 to 1997, Delansay and Penninckx et al. [13,14] at Alcatel were the first to demonstrate optical communication system performance as a function of wavelength for an InP-based MZM design.

Commercial optical system deployments of InP-based MZM began in 1995 and were concentrated in 2.5 Gb/s applications, where the modulator was copackaged with the fixed wavelength distributed feedback laser laser source [15]. For 10 Gb/s applications, the maturity of the modulator designs and also of the required semiconductor processing technology, did not provide enough optical system performance uniformity to successfully compete with the linear electro-optic (LEO) $LiNbO_3$ MZM alternatives.

### 7.1.2 2000–2010: Optimization and Commercialization

In the early 2000s, copackaged distributed feedback laser/MZM modules became very successful in 10 Gb/s extended reach Metro applications [16], enabled by improved uniformity in InP semiconductor processing. The use of 3" InP substrates and i-line 5× stepper-based lithography allowed among other things, robust multimode interference (MMI) designs for optical splitter and combiner functions. Bookham (Oclaro) announced the shipment of its 100,000th InP-based MZM in 2007.

The research interest in InP-based MZM designs continued to increase, in part due to the growing commercial success, and in part due to the rise in standards-based 10 Gb/s transponder and transceiver footprints [17,18]. InP-based MZM are being increasingly seen as the technology of choice to enable long-haul, high performance, optical transmission in the smallest of these standards-based footprints.

In the 1990s, research had concentrated on p-i-n capacitive lumped element high speed designs for the electro-optic (EO) phase shifters; however, multiple InP-based MZMs with bandwidth in excess of 40 GHz were demonstrated in quick succession starting in 2002. They employed either independent conventional coplanar radio-frequency (RF) microstrip electrodes on each interferometric arm of the modulator [19], or they placed the p-i-n epitaxy in a series push–pull configuration to capacitively load a coplanar stripline electrode [20,22]. The InP-based MZM high-speed phase shifter design is addressed in Section 7.4.

Other research directions have focused on addressing the perceived shortcomings, or playing to the strengths of the InP-based MZM technology. The optical insertion losses have been reduced by adding optical mode spot-size converters [22,23], or integrating an optical amplifier [24]. Both monolithic [25–27] and hybrid [28] integration of the MZM with full C-Band (erbium doped fiber amplifier band) tunable laser sources have been shown. In-line and complementary evanescently coupled InGaAs power taps have been integrated [29]. Push–pull MZM designs for zero chirp modulation over the full C-Band have been demonstrated [30], and compatibility of InP-based MZM with maximum likelihood sequence estimation (MLSE) receivers has been proven [31]. The applicability of InP-based MZM designs for various advanced modulation schemes has also been shown, including duobinary [32,33], differential-phase-shift keying (DPSK) [34–36], and Differential-Quadrature-Phase-Shift Keying (DQPSK) [37,38]. Large scale monolithic integration of multiple InP-based MZMs has also been published (>40 devices) with intent for commercial system deployment [39,40].

This chapter is focused on InP-based MZM devices but it should be noted that during the late 1990s and early 2000s, there was work done to commercialize a linear electro-optic (LEO) III-V MZM in the AlGaAs/GaAs material system [41,42]. The commercial success was limited given the rapidly changing market in GaAs foundries, and the drop in the demand for optical components at that time. The current state of the art for GaAs-based MZM devices is detailed in Chapter 8.

### 7.1.3 2010–2015: Sophistication and Miniaturization

In this time frame, it is expected that dense-wavelength-division-multiplexed (DWDM) tunable transponders and pluggable XFP transceivers will become the dominant delivery vehicle for 10 Gb/s line side optical communications. Very competitive multisupplier agreement (MSA) transponders that use InP-based MZMs already exist from several vendors. Nontunable electroabsorption modulated laser-based

DWDM XFP transceivers will start to be challenged by full C-Band tunable XFPs, that are enabled, by transmitters, consisting of a hybrid or monolithic integrated tunable laser and InP-based MZM. Full C-Band tunable XFP transceivers have the potential to displace their nontunable DWDM variants, as occurred earlier in the decade with discrete optical components, since the same wavelength sparing, inventory carrying, and market pricing challenges exist.

The silicon industry semiconductor roadmap has enabled analog-to-digital converters at speeds that allow the suite of digital signal processing techniques developed for radio to be applied in optical communications [43]. This is revolutionizing the performance available from spectrally efficient coherent modulation formats at 40 Gb/s and beyond [44] and should also be transformative for 10 Gb/s applications.

A transponder using a tunable InP-based MZM transmitter with a MLSE receiver has already been demonstrated for 10 Gb/s applications [31]. This technology will likely be integrated into an existing pluggable transceiver footprint, or a new transceiver standard with a linear output receiver will emerge, to allow high performance line-side 10 Gb/s optics to move from larger MSA transponder designs to transceivers having much smaller and pluggable form factors. InP-based MZMs are the technology of choice for the smallest form factor tunable transmitters, therefore their commercial future appears bright, with the biggest downside risk being a move away from DWDM line-side optical communications at 10 Gb/s.

The base modulator of choice for the coherent optical systems being developed for 40 Gb/s applications and beyond is a Cartesian MZM design providing access to the full complex plane for encoding data. The use of polarization diversity in a coherent optical receiver is favored, so the use of polarization multiplexing in the transmitter offers a low-cost method for increasing spectral efficiency, and thus total system capacity. The most versatile coherent modulator is, therefore, a dual polarization Cartesian MZM design providing access to the full complex plane for encoding data on two output optical fields along orthogonal polarization axes. InP-based dual polarization Cartesian MZM designs will need to compete directly with $LiNbO_3$ designs, much like was done in the mid-1990s, for these high performance system applications.

## 7.2 Basic Modulator Structures and Models

The wavelength range of interest for InP-based MZMs is within either the C or L erbium doped fiber amplifier band. InP-based MZM are typically designed for optimal performance with only transverse electric (TE) optical mode transmission.

### 7.2.1 Material Systems and the Quantum Confined Stark Effect

InP is a semiconductor material with a bandgap energy of 1.344 eV and a lattice constant of 5.8697 Å at 300 K. Substrates of InP are commercially available, with various dopants, and in diameters ranging from two to four inches. The refractive index ($n$) of undoped InP can be calculated using the modified Sellmeier coefficients determined by Martin [45]. In doped semiconductor materials the principle carrier-induced effects on refractive index have been calculated by Bennett et al. [46] and also experimentally determined for InP [47–49]. The doping-induced free carrier plasma in a semiconductor reduces the real refractive index, and in III-V compounds the influence is particularly strong for electrons since their effective mass is generally small. In InP, the p-doping level induces a factor of 7–8 greater contribution to optical loss and a negligible real refractive index change relative to a comparable n-doping level—a significant consideration for modulator waveguide design with this material.

The phase shift functions in InP-based MZM are typically implemented in deeply etched ridge waveguides, through p-i-n doped epitaxial layers, grown on (100) InP substrates using thin film epitaxy. The intrinsic waveguide core region is most often a multiquantum well (MQW) stack made from thin films of typically quaternary semiconductor $In_{1-x}Ga_xAs_yP_{1-y}$. The $In_{1-x}Ga_xAs_yP_{1-y}$ bandgap and lattice constant are a function of the group III and group V composition, denoted by $x$ and $y$, respectively. Strained, or even strain compensated, MQW thin films are rarely used in InP-based MZM design to date, due to the large number of layers in the MQW required for optimal phase efficiency. The choice of $x$ and $y$ is restricted in alloys that are lattice-matched to InP and the bandgap energies ($E_g$) of these lattice-matched alloys range from 0.750 eV to 1.344 eV (300 K.) In InP-based optical waveguide design the bandgap energy ($E_g$) is often expressed by the corresponding peak photoluminescence of the material in microns Q, where $E_g$(eV) = 1.2398/Q. In an MQW structure, this photoluminescence peak will be an exciton resonance. It is the wavelength range well above $Q$ that is of most interest in EO modulators, therefore the refractive index of the $In_{1-x}Ga_xAs_yP_{1-y}$ thin films can be calculated using the modified Sellmeier formula coefficients determined by Fiedler and Schlachetzki [50].

To enable efficient numerical calculations of waveguide properties, the MQW structure must be replaced by a single homogeneous layer, with an equivalent refractive index $n_{eq}$, and equivalent thickness $t_{eq}$. A linear perturbation of the TE solutions for the wave equation [51,52] leads to a root-mean square approximation, with $n_{eq}$ given by a weighted average of the dielectric constants of the individual MQW layers and $t_{eq}$ equal to the as-grown thickness of the MQW. It is critical to incorporate the well quantum size effect when determining $n_{eq}$. For an MQW with uncoupled quantum wells, it is valid to assume the barrier layer index is given by its bulk value, and the well layer index is given by the value of a bulk layer with the same $Q$ composition as the peak exciton absorption in the MQW.

The MQW waveguide core design proposed by C. Rolland and W. Bardyszewski is used for illustrative purposes throughout this chapter because it has the most fully characterized phase and absorption characteristics of the MQW core designs in the literature [3,4,30,53–55]. This reference MQW contains $N$ layers of 95 Å thick $Q$ = 1.47 μm unstrained InGaAsP wells and $N$ + 1 layers of 80 Å thick $Q$ = 1.1 μm unstrained InGaAsP barriers.

Figure 7.1a presents the absorption spectra α, as a function of the applied bias voltage $V$, for the reference MQW design with $N$ = 20. The room temperature exciton photoluminescence peak $Q$ is approximately 140 nm detuned from the operating wavelength. This defines the MQW wavelength detuning $\lambda_D$. The change in α with bias $V$ in Figure 7.1a is the QCSE and it is discussed in depth in Chapter 5. The phase shift given by the Kramers–Krönig transformation of this QCSE absorption change is the basis for the phase shift function in InP-based MZMs,

$$\Delta n(\hbar\omega, F_z) = \frac{\hbar c}{\pi} \lim_{\delta \to \infty} \left[ \int_0^{\hbar\omega-\delta} \frac{\Delta\alpha(E, F_z)}{E^2 - (\hbar\omega)^2} dE + \int_{\hbar\omega+\delta}^{\infty} \frac{\Delta\alpha(E, F_z)}{E^2 - (\hbar\omega)^2} dE \right], \tag{7.1}$$

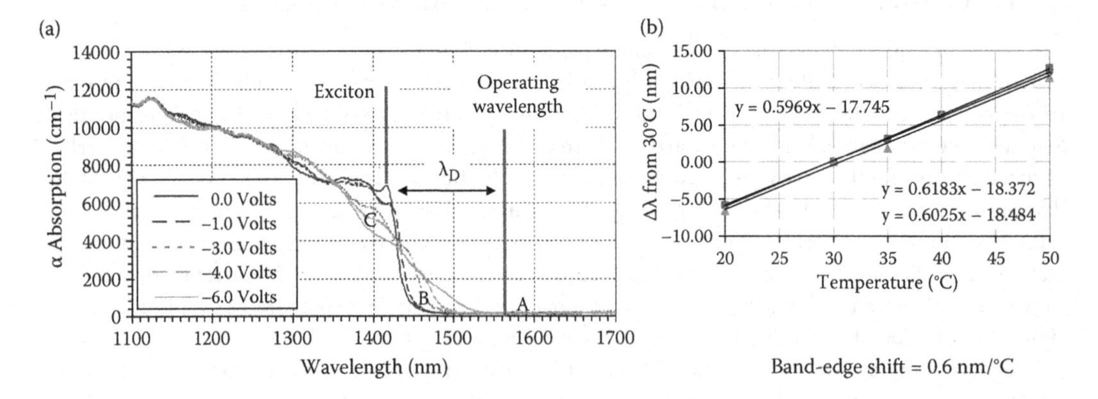


**FIGURE 7.1** (a) Measured absorption spectra α for the reference MQW core with N = 20 wells. (b) Exciton absorption band-edge shift versus temperature for the reference MQW core with N = 20 wells.

where $\Delta n$ and $\Delta\alpha$ are the index change and absorption change, respectively, due to the presence of the electric field $F_z$ relative to that at zero electric field [56]. From Eq. 7.1, it is clear that the $\Delta\alpha$ in regions $A$ and $C$ of Figure 7.1 contribute to negative $\Delta n$, whereas $\Delta\alpha$ in region $B$ contributes to positive $\Delta n$. Due to the weighting of $\Delta\alpha(E, F_z)$ by $[E^2-(\hbar\omega)^2]$ the net $\Delta n$ for any given reverse bias voltage is positive for typical MZM $\lambda_D$ offsets. The ideal QCSE for MZM operation would cause a large red shift of the absorption band edge, maintain the exciton confinement and its associated absorption peak, and cause minimal broadening [56]. Deep wells for both electrons and holes are desired, which with all else being equal, would tend to favor InGaAlAs/InAlAs materials since they offer stronger electron confinements than InGaAsP/InGaAsP MQW materials in use today. MQW bandgap engineering for InP-based MZMs has not been revisited in many years [53,57,58] and is a potential area for future research.

The absorption band edge is shown in Figure 7.1b, to undergo translation without distortion to longer wavelengths at the rate of 0.6 nm/°C. MQW materials containing aluminum show similar absorption band edge translations [59]. A change in wavelength detuning $\lambda_D$ can, therefore, be obtained by either directly changing the operating wavelength, or by varying the operating temperature at a fixed wavelength. This correspondence between temperature change and wavelength change is an important consideration for uncooled applications of InP-based MZMs.

InP-based semiconductors are isotropic Zinc Blende ($\bar{4}3m$) crystals that present a LEO effect due to the absence of inversion symmetry [60]. The thin film epitaxy growth of the materials proceeds perpendicular to the (100) InP surface along the z-axis. For TE modes in optical waveguides that run along the [011] axis, the LEO index modulation is given by $n_o + \frac{1}{2} n_o^3 r_{41} F_z$, and it *adds* to the QCSE induced phase shift [56]. Transverse magnetic (TM) modes polarized along the z-axis are not modulated by the LEO effect. Here $n_o$ is the ordinary refractive index, $r_{41}$ is an EO tensor component having a value around 1.7 pm/V [61] in the reference MQW design at 1550 nm, and $F_z$ is the applied electric field along the $z$ axis perpendicular to the MQW layers. For waveguides that run along the $[01\bar{1}]$ axis, the LEO index modulation is given by $n_o - \frac{1}{2} n_o^3 r_{41} F_z$, and it opposes the QCSE induced phase shift. To maximize the phase efficiency of the phase shift electrode, the MZM optical waveguide should run parallel to the major flat of the (100) InP wafer, along the [011] axis.

Without bandgap engineering, the bias induced absorption and the nonlinearity of the phase shift in the operating wavelength range of interest, can be minimized through a combination of the following: decrease the applied electric field for a given voltage bias by increasing the thickness of the waveguide cross-section intrinsic region; increase the waveguide mode confinement to the phase change region by increasing the number of layers in the MQW core; increase the $\lambda_D$ detuning by reducing the $Q$ of the quantum wells; and increase the EO interaction length. Obviously, this optimization must be achieved without compromising the required EO electrical bandwidth and return loss from the phase shift electrode as discussed in Section 7.4. Two such optimizations with the reference MQW design in optical waveguides along the [011] axis are discussed in Section 7.2.2.1 and shown in Figure 7.3.

## 7.2.2 General Mach–Zehnder Device Model and Operating Conditions

### 7.2.2.1 Mach–Zehnder Modulator Device Model

The example InP-based MZM shown in Figure 7.2 is a 0th order interferometer with a direct current (DC) transfer function for the output electric field $E_0(V_L, V_R)$ that can be mathematically expressed as a function of the left and right arm biases by

$$E_o(V_L, V_R) = \sqrt{S_{li} S_{lo}} \cdot e^{-[\alpha(V_L, \lambda_D)/2 \cdot L_L]} \cdot e^{i[-\Delta\phi(V_L, \lambda_D) \cdot L_L - \delta_L]}$$
$$+ \sqrt{(1 - S_{li})(1 - S_{lo})} \cdot e^{-[\alpha(V_R, \lambda_D)/2 \cdot L_R]} \cdot e^{i[-\Delta\phi(V_R, \lambda_D \cdot L_R - \delta_R)]}, \tag{7.2}$$

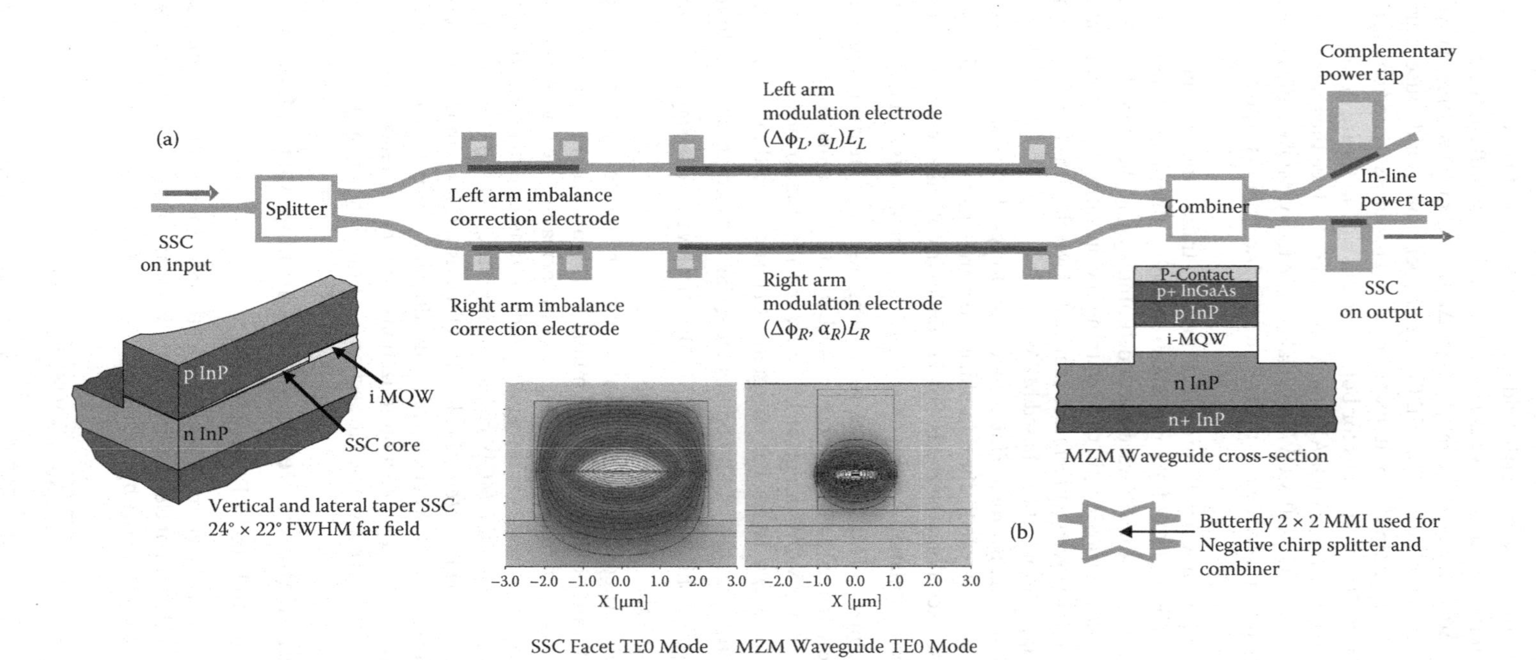


**FIGURE 7.2** (a) Example of zero chirp MZM design. (b) Example of negative chirp MZM design.

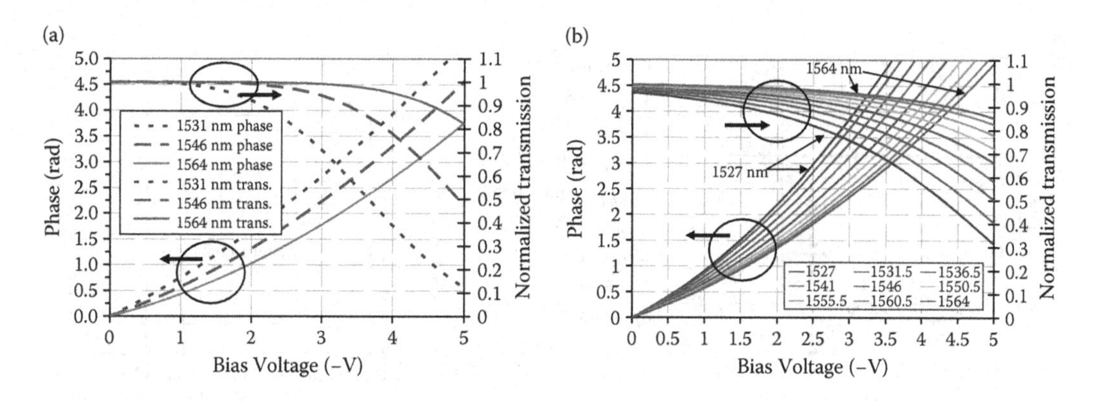


**FIGURE 7.3** Measured optical phase and absorption characteristics versus voltage and wavelength for two phase shifter design variants using the reference MQW core: (a) 600 μm long, 2.0 μm wide, one 20-well core with $Q$ = 1415 nm, (b) 1500 μm long, 1.5 μm wide, 31-well core with $Q$ = 1390 nm.

assuming a unity input field amplitude. The MZM output power transfer function can be determined from $P_o = [E_o \cdot E_o^*]$. Here, $S_{li}$ is the input MMI power split ratio into the left modulator arm (0–1); $S_{lo}$ is the left modulator arm's MMI power split ratio into the output guide (0–1); $L_L$ and $L_R$ are the left and right arm phase shift electrode optical interaction lengths; $V_L$ and $V_R$ are the reverse bias voltages applied to the electrodes; and $\delta_L$ and $\delta_R$ are the phase offsets (0–2π) in the left and right arm paths introduced by the MMI splitter and combiner functions or the separate imbalance control electrodes. $\Delta\phi(V, \lambda_D)$ is the phase shift per micron for an applied negative voltage bias to the phase shift electrodes. An example of this relationship in radians corresponding to a 1550 nm operating wavelength in Figure 7.3a is given by $\Delta\phi(V,\lambda_D) = \frac{1}{600}\left[p_1|V| + p_2|V|^2\right]$ where $p_1 = 0.5055$ and $p_2 = 0.0817$. $\alpha(V, \lambda_D)$ is the optical power absorption coefficient in inverse microns for an applied negative voltage bias. An example of this relationship, also corresponding to a 1550 pm operating wavelength in Figure 7.3a, is given by $\alpha(V,\lambda_D) = \frac{1}{600}\left[a_1|V|\right]^{a_2}$ with $a_1 = 0.2079$ and $a_2 = 4.5285$.

In Figure 7.3 the measured optical phase and absorption characteristics versus voltage and wavelength are shown for two phase shifter design variants using the reference MQW core. In Figure 7.3a the phase shift electrode is 600 μm long on a 2.0 μm wide deeply etched waveguide and it uses a 20-well undoped MQW core with $Q$ = 1415 nm. This phase shifter design achieves a 58%–60% overlap between the $TE_0$ optical mode in C-Band and the bias induced electrical field, it has a capacitance per unit length of ≈0.65 pF/mm, and a 3 dB EO electrical bandwidth of 14 GHz [54]. In Figure 7.3b, the phase shift electrode is 1.5 mm long on deeply etched waveguide and it uses a 31-well undoped MQW core with $Q$ = 1390 nm. This phase shifter design achieves an 80% overlap between the $TE_0$ optical mode in C-Band and the bias induced electrical field, it has a capacitance per unit length of ≈0.4 pF/mm, and a 3 dB EO electrical bandwidth of 14 GHz [30]. The optical group index for the waveguide in Figure 7.3b is ≈3.6 in C-Band. To determine the phase versus voltage characteristic, a nonlinear fit routine based on Eq. 7.2 is used to fit both the left and right arm light-versus-voltage sweeps, using calculated absorption data obtained from the measured photocurrent in the individual MZM arms [54]. Note the voltage required on one arm to switch the modulator from the off-to-on state ($V_\pi$) may be different from the voltage bias from 0V required for a material π phase shift in the modulation arm ($V_\pi$). This is due to the nonlinearity of $\Delta\phi(V, \lambda_D)$ and the distortion introduced by $\alpha(V, \lambda_D)$ to the light transmission versus reverse bias voltage curve.

### 7.2.2.2 Chirped MZM Designs

A $P_o$ output intensity versus operating voltage surface is shown in Figure 7.4a for a negatively chirped InP-based MZM design using independently driven phase shifter electrodes as provided in Figure 7.3a,

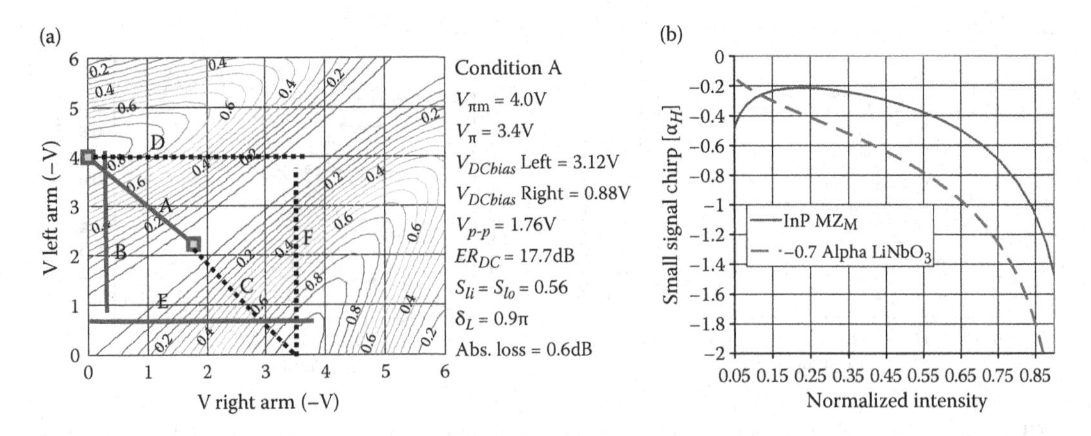


**FIGURE 7.4** (a) A modeled $P_o$ output intensity versus voltage operating surface for the negative chirp InP-based MZM design. (b) Measured variation of the small signal chirp parameter $\alpha_H$ through a bit transition with drive condition "A" and also for a reference $-0.7\alpha_H$ $LiNbO_3$ MZM.

where $V_{\pi m} = -4.0$ V. This MZM design can be realized using butterfly MMI splitters and combiners with balanced modulator arms lengths, (nominally $\delta_L = \pi$ and $\delta_R = 0$) as given by Figure 7.2b. Since the MZM arms are independently addressable, there is a large set of possible bias conditions, with examples labeled A–F in Figure 7.4a; however, the differential condition "A" is preferred for dispersion-limited transmission on standard single mode optical fiber [2]. This drive configuration exploits the super-linear phase-voltage characteristic of the device. At quadrature, the left arm is biased more deeply than the right; hence its dynamic phase shift is higher, contributing to negative chirp at the output. The negative chirp is increased further by unbalancing the optical split ratios to favor the more deeply biased left arm with $S_{li} = S_{lo} = 0.56$. This unbalancing limits the DC extinction ratio to the 17.7 dB indicated.

This tailored optical imbalance MZM design, originally proposed by Cartledge and Rolland et al. [7], provides superior dispersion-limited transmission performance [62] compared with the RF drive imbalance design more traditionally used in z-cut $LiNbO_3$. The optical unbalancing produces a chirped nonreturn-to-zero data stream without broadening the power spectrum relative to a zero chirp device [63] and the more uniform small signal chirp parameter ($\alpha_H$) profile, as shown in Figure 7.4b, provides lower group velocity variation between different points on a bit transition than a $LiNbO_3$ MZM.

The optical imbalance ($\delta_L$, $\delta_R$), the DC bias depth, and the relative detuning wavelength $\lambda_D$, are all minor contributors to the over-fiber system performance [28,54] dominated by the selected values for $S_{li}$ and $S_{lo}$. Since $S_{li}$ and $S_{lo}$ define the easily measured DC extinction ratio, and since this ratio can be independent of the optical input fiber alignment with optimized waveguide geometries, it is possible to screen system performance at the chip level, enabling high transmitter module yields. For this negatively chirped InP-based MZM design, a $\lambda_D$ range between 120 nm and 155 nm, with a nominal 18 dB target for the DC extinction ratio, offers excellent tolerance to the typical specification allowing 2 dB maximum penalty for 1600 ps/nm of fiber dispersion [54].

This design approach using differentially driven, short, lumped element, phase shift electrodes is favored for negatively chirped InP-based MZMs due to the resulting small chip size, and <3 $V_{pp}$ RF drive requirements. The use of traveling wave designs for the phase shift RF electrodes is of limited interest, given the lack of applications for >10 Gb/s negative chirp MZM devices, and the focus of negative chirp MZM designs on space sensitive applications, such as in monolithic integration with tunable laser sources as discussed in Section 7.6.

#### 7.2.2.3 Zero Chirp MZM Designs

A near zero chirp InP-based MZM must be operated at quadrature output with equal DC bias on the left and right arm RF phase shift electrode; have a continuous wave optical input signal that is split and combined equally ($S_{li} = S_{lo} = 0.5$ nominally); and it must be driven differentially. This operating mode allows each arm to counteract the frequency chirp introduced by the other under modulation. Any deviations from zero net frequency chirp during a transition are caused by mismatched electrical signals to the two RF phase shift electrodes, or by the nonlinearities in the $\Delta\phi(V, \lambda_D)$ relationship. The waveguide design for this modulator can be realized using a 1 × 2 MMI splitter and a 2 × 2 MMI combiner with balanced modulator arms lengths (nominal $\delta_L = \pi/2$ and $\delta_R = 0$), and ideally requires separate imbalance control electrodes to optimize ($\delta_L$,$\delta_R$), as shown in Figure 7.2a. The use of a p-i-n waveguide cross section allows the use of short, current driven, free-carrier plasma-based, and imbalance control electrodes [30,37]. At the expense of additional chip length, QCSE imbalance control electrodes can also be used [38]. With this MZM design, neither arm RF phase shifter need receive an absolute voltage under modulation greater than $V_\pi/2$, thereby greatly improving the effective linearity of the phase shifters.

Figure 7.5 provides the system performance of a 10 Gb/s zero chirp InP-based MZM, with *independently driven* left and right arm RF phase shift electrodes, in a tunable transmitter module [30]. Here the phase shifter nonlinearities and absorption versus bias are reduced, relative to the chirped design in

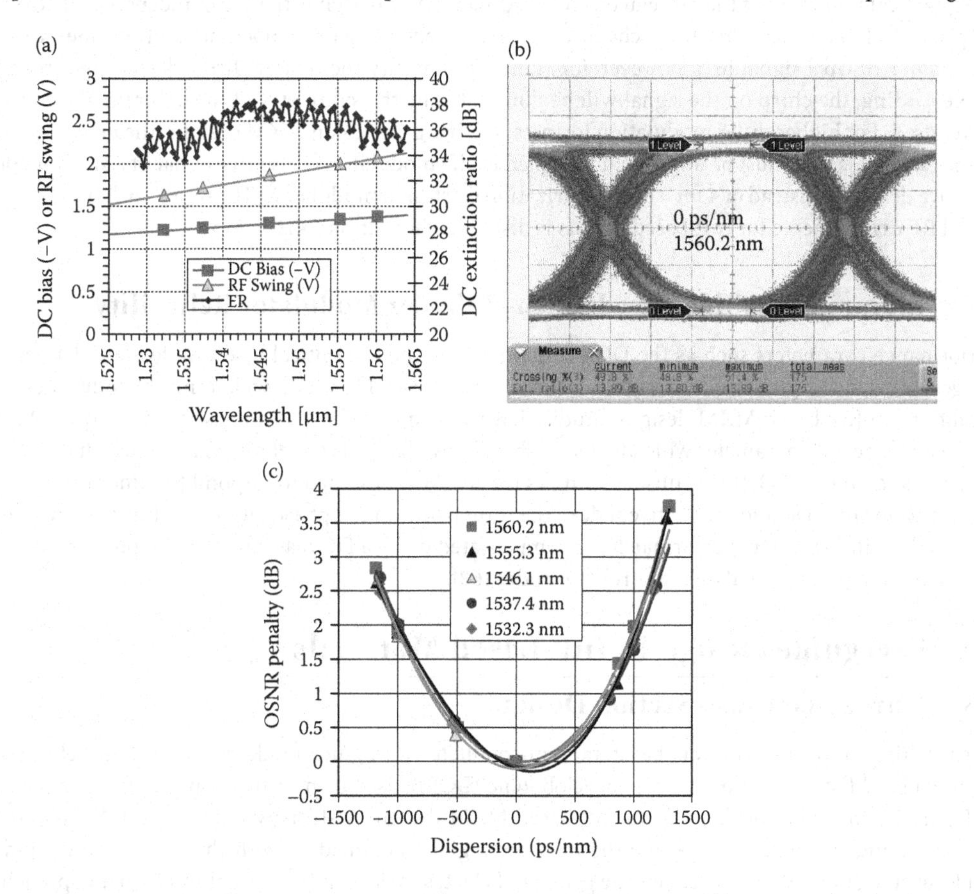


**FIGURE 7.5** (a) Measured zero chirp transmitter module operating conditions over wavelength. (b) Transmitter module back-to-back eye diagram at 1560.2 nm. (c) Transmitter module OSNR penalty versus dispersion for 5 channels across C-band. (Reprinted from I. Betty et al., "Zero Chirp 10 Gb/s MQW InP Mach-Zehnder Transmitter," Optical Fiber Communications Conference, 25-29 Mar., 2007, OWH6. With permission of Optical Society of America.)

Section 7.2.2.2, using the phase shifter design given in Figure 7.3b. In Figure 7.5a, the transmitter module operating bias conditions over C-B and are shown. The electrical drive is applied differentially to the modulator by a 10.7 Gb/s $2^{31}$–1 pseudo-random-bit sequence signal from an Advantest D3186 Pulse Pattern Generator. In Figure 7.5b, a representative back-to-back eye for this module at 1560.2 nm is shown; all channels had dynamic extinction ratio values above 13.8 dB. The optical signal-to-noise ratio penalty at a bit error rate of $6 \times 10^{-5}$ exhibits symmetric behavior for positive and negative dispersion, shown in Figure 7.5c, with <2 dB penalty for +1000 ps/nm, confirming that there is minimal residual chirp for the transmitter over C-Band.

A fixed RF driving voltage over wavelength can be achieved by adapting the phase shift electrode DC bias depth to maintain $\lambda_D$, the relative difference between the operating wavelength and the QCSE absorption edge (higher DC bias at higher wavelengths). This has been demonstrated experimentally with both InGaAsP and InGaAlAs MQW core designs [64,65]. If a constant DC bias depth is used then the modulation efficiency is a function of wavelength, with the highest efficiency at the shortest wavelength.

The zero chirp modulator design presented here has been shown to provide equivalent performance to linear electro-optic MZM in instances where both are combined with an 8-State MLSE receiver [31]. Duobinary, Binary Phase Shift Keying (BPSK), and differential-quadrature-phase-shift keying modulation formats have also been demonstrated with variants of this MZM design [34,37,66]. The most significant practical limitation of the presented zero chirp modulator design is the use of independently driven RF phase shift electrodes. This drive scheme is optimal for encoding precompensation of the fiber dispersion onto a 10 Gb/s signal [67]; however, for standard intensity modulated direct detect nonreturn to zero encoding, the chirp on the signal will be dominated by the symmetry of the differential electrical driver used. For higher order modulation formats requiring Cartesian MZM designs, the number of RF electrical signals that must be aligned and coupled to the modulator, is also twice that of a single-ended RF drive design (8 instead of 4 for a dual polarization Cartesian modulator.) The series push–pull single-ended RF drive scheme for InP-based MZMs is discussed in Section 7.4.2.

### 7.2.3 Indium Phosphide-Based Mach–Zehnder Modulator Reliability

Performance parameters such as the DC extinction ratio and the optical insertion loss are defined by the geometry of the processed waveguide structure and are stable over a wide range of endurance test conditions in InP-based MZM designs. Studies have shown that the MZM phase shift electrode $V_{\pi m}$ is also stable. The only parameter which is susceptible to any change is the phase offset in the left and right arm paths ($\delta_L$ and/or $\delta_R$) This is unsurprising as the MZM is designed to respond to minor differential perturbations in waveguide refractive index. The minor shift in the phase offset over life must be either budgeted for in the system performance, or compensated using a DC bias on either the phase shift electrodes, or on separate imbalance control electrodes [68].

## 7.3 Waveguide Design in InP-Based Materials

### 7.3.1 Waveguide Cross-Section Design

Deep etching to strongly confine the lateral semiconductor waveguide modes with a low dielectric constant material (i.e., silicon oxide, benzocyclobutene (BCB)), as shown in the waveguide cross section of Figure 7.2, has multiple benefits in InP-based MZM design. A strongly-guided p-i-n diode optical waveguide under reverse bias, where the intrinsic region is coincident with the higher optical index guide layer, enables overlap between the generated electric field and the optical mode that approaches 80%. The confinement of the electric field lines by the low dielectric constant cladding also enables lower waveguide capacitance per unit length and reduced dielectric RF losses than in weakly-guided structures, where p- and n-doped regions surround the core over the full area of the device. The number of modes supported and the properties of the fundamental modes in deeply-etched waveguides are not significantly dependent on the etch depth, thus enabling the use of the conventional dry etching techniques. Strong lateral confinement also provides more ideal self-imaging in splitter and combiner MMI functions. Any light launched or scattered into the higher-order modes of a deeply-etched waveguide will leak vertically into the substrate rather than horizontally; thus, inter-waveguide crosstalk and background scatter into the output optics are reduced. In addition, strong-guiding greatly reduces the radiation loss from the waveguide in low radius of curvature bends, and allows the spot-size of the lateral mode field to be changed over short distances using waveguide tapers.

In a 2.0 μm or 1.5 μm wide waveguide, it is possible to obtain <2 dB/cm or <3.5 dB/cm propagation loss, respectively, using conventional 5× stepper-based i-line photolithography [23,69] Unfortunately, in optical modulator applications, it is not usually possible to obtain true single lateral mode operation using a deep-etched waveguide with a ridge width near 2.0 μm. Reducing the core composition wavelength and/or the core thickness will increase the vertical evanescent field overlap with the substrate and, therefore, increase the loss of the higher-order lateral modes for a given ridge width; however, in InP-based MZM applications, a thick core is required to maximize the mode confinement to the MQW region, and an effectively high composition index is required to make use of the QCSE. The problems introduced by higher-order lateral modes and the means of suppressing them are a large focus of InP-based MZM waveguide design [23].

Many different technologies exist for etching III-V semiconductors. Those that have been demonstrated for deeply etched waveguide structures include $CH_4$:Ar [70,71] or $CO_2$:$CH_4$:$H_2$ [72] reactive ion etching, HBr [73] or $Cl_2$ [74,75] inductive coupled plasma etching, and $Br_2N_2$ [76] self-passivating reactive ion etching. The ridge width uniformity on a 3" wafer is typically better than ±0.1 μm using the HBr inductive coupled plasma etch technology with i-line 5× stepper-based photolithography [74]. This level of uniformity is critical to yield the MMI splitters and combiner waveguide functions discussed in Section 7.3.2.

The InP doping concentrations used in the n- and p-sides of the intrinsic MQW core in the Figure 7.2 waveguide cross section must be selected to favor either low optical loss or low RF loss. The profile for the as-grown p-doping concentration must also be offset to account for spatial diffusion of zinc when using metal organic chemical vapor deposition epitaxial growth [77], and the inevitable hydrogen passivation of zinc during the fabrication process [78]. For the integration of the InP-based MZM with a laser source, a common p-overclad layer stack is often used, requiring a trade-off between laser resistance and modulator optical loss. The use of a n-p-i-n or n-i-n doping profile has been demonstrated as a means to reduce the optical and RF losses inherent with Zn-doping, as discussed in Section 7.4.1; however, these doping profiles also prevent the use of on-chip carrier-based functions, such as those discussed in Section 7.5, including optical power detectors, free-carrier plasma-based DC imbalance control electrodes, and/or low-loss MMI variable optical attenuators.

There are several design factors that also influence the thickness of the overclad InP layers in the Figure 7.2 waveguide cross section: The minimization of the overlap between the fundamental optical $TE_0$ waveguide mode and the highly lossy InGaAs p-contact layer; the minimization of the stress-induced refractive index change in the intrinsic region caused by the presence of the metallization on the ridge [79]; and the design of the optical output mode spot-size converter, as discussed in Section 7.3.4.

## 7.3.2 MMI Splitters and Combiners

MMI waveguide components are used for the splitter and combiner functions in InP-based MZM designs. MMI devices are based on the principle of self-imaging [80,81]. In multimode waveguides, self-imaging is the mechanism by which an input field profile is reproduced, in single and multiple images, at periodic intervals along the $z$ propagation direction of the waveguide. MMI self-imaging can be analytically determined from the propagation of the waveguide normal modes in only the $(x,z)$ dimensions [81,82].

An arbitrary MMI input field $\Psi(x, 0)$ can be decomposed into the set of $m$ guided modes, $\Psi(x,0)=\sum_{v=0}^{m-1} c_v\Psi_v(x)$ with $c_v$ the mode expansion coefficients, and $\Psi_v(x)$ the orthonormal mode field distributions. Here, the set of radiation modes are neglected as a valid decomposition in strongly guided InP-based MZM waveguides. The mode expansion coefficients $c_v$ express rigourously the decomposition of the input field onto the guided modes of the multimode section, but do not take into account the reflected electric field. As the refractive index difference between the single-mode waveguide and the multimode section is low, this last phenomenon can generally be neglected. Three different operating regimes for MMI devices can be defined based on the subset of modes excited in the multimode waveguide:

- The *General Interference* regime places no restrictions on the $c_v$ mode expansion coefficients, allowing the access waveguides to be located at any $x$ position on the multimode waveguide, except $x = 0$ and $x = \pm W_e/6$, where $W_e$ is the effective width of the MMI waveguide including Goos–Hähnchen shifts. $N$ identical images of the input field, each with power $P_{in}/N$, can be found at the intermediate lengths $z = \frac{\rho}{N} 3L_\pi$, where $\rho \geq 0$ and $N \geq 1$ are integers with no common divisor, and $L_\pi$ is the beat length of the two lowest order modes. A valid approximation for $L_\pi$ in a strongly guided InP-based MZM waveguide is $L_\pi = \pi/(\beta_0 - \beta_1) \simeq (4n_g W_e^2)/(3\lambda_0)$, where $n_g$ is the effective index method (EIM) calculated core refractive index, and $\lambda_0$ is the free-space wavelength.
- The *Restricted Paired* regime has $c_v = 0$ for $v = 2, 5, 8, 11 \ldots$ to reduce the length periodicity of the mode phase factor by a factor of three relative to the general interference regime. This selective excitation is possible by launching an even symmetric field $\Psi(x, 0)$ at $x = \pm W_e/6$ since at these positions the modes $v = 2, 5, 8, \ldots$ present odd symmetry against the even excitation and cause the overlap to vanish. The number of MMI access waveguides is limited to two. $N$ identical images of the input field $\Psi(x, 0)$ can be found at the intermediate lengths $z = \frac{\rho}{N} L_\pi$, where $\rho \geq 0$ and $N \geq 1$ are again two integers with no common divisor.
- The *Restricted Symmetric* regime excites only the even symmetric modes ($c_v = 0$ for $v = 1, 3, 5, 7 \ldots$) to reduce the length periodicity of the mode phase factor by a factor of four relative to the general interference regime. To achieve this excitation, the even symmetric field $\Psi(x, 0)$ is launched at the $x = 0$ midpoint in the multimode waveguide. The odd modes of the multimode waveguide are not excited by this launch due to a zero overlap caused by their zero crossing and odd symmetry at $x = 0$. This operating regime is obviously applicable only to $1 \times N$ MMI splitters. $N$ identical images of the input field $\Psi(x, 0)$ can be found at the intermediate lengths $L = \frac{\rho}{N}\left(\frac{3L_\pi}{4}\right)$, where $\rho \geq 0$ and $N \geq 1$ are once again two integers with no common divisor.

The fundamental mode field amplitude $\Psi_0(x, 0)$ for the MMI access waveguides can be approximated by a Gaussian beam with waist $\omega_0$. An intuitive analysis of MMI tolerances [83] considers each MMI output image, located at the self-imaging distance $z = L$, to be a focused Gaussian beam image of the input. The output coupling loss produced by a small $\delta L$ shift in the $z$ position of the output waveguide can be calculated by overlapping the defocused Gaussian beam with the Gaussian approximation to the output access waveguide fundamental mode. The length shift which produces a 0.5 dB output overlap loss penalty is approximately equal to the Rayleigh range. The Rayleigh range is defined in the region of a Gaussian beam focus, to be the axial distance from the point of minimum beam waist $\omega_0$ to the point where the beam diameter has increased to $\sqrt{\omega_0}$. Therefore, $\delta L \simeq \left(\pi n_g \omega_0^2\right)/(4\lambda_0)$. Note that $\delta L$ does not depend on the dimensions of the multimode waveguide. It is an absolute length tolerance that will depend only on the beam waist of the access waveguide mode for a given wavelength and material system. It is clear that the access waveguide width $W_a$, should be as large as possible to maximize the length tolerance of the MMI design.

The tolerance corresponding to the other design and fabrication parameters can be easily related to $\delta L$ as follows:

$$\frac{\delta L}{L} = 2\frac{\delta W_e}{W_e} \simeq \frac{|\delta\lambda_0|}{\lambda_0} \simeq \frac{\delta n_g}{n_g}. \tag{7.3}$$

Here, it is apparent that $L$ must be as short as possible to relax the tolerances of the other variables. The MMI designs of most interest therefore have $\rho = 1$. Since strongly-guided waveguides are used for the MMI access waveguides, the MMI width is not constrained by the possibility of coupling between access waveguides; however, there is a technological limitation in accurately defining ridge waveguides with sub micron spacing. Care must also be taken in fabrication to minimize the stress placed on the MMI ridge waveguide sidewalls since this can induce optical output split ratio imbalance in restricted paired MMI designs [84]. A tight tolerance on the MMI width is one of the most challenging fabrication issues for InP-based MZMs.

Imbalance is defined as the ratio of power in the cross-state output waveguide, relative to the power in the bar-state output waveguide. Hill et al. have theoretically shown that the maximum imbalance is typically much greater in the 2 × 2 general interference coupler than in the 2 × 2 paired interference coupler due to a $3\sqrt{3}$ increase in the modal phase errors at the output waveguides [85]. Even for the most promising 2 × 2 general interference coupler design, where the access waveguides are aligned to the outside edge of the MMI, they demonstrate inferior performance relative to the 2 × 2 paired interference coupler.

Care must be taken with the angles used on MMI end face walls perpendicular to the propagation direction in strongly-guided waveguides to minimize backreflections [23,86]. One approach that can be used to prevent backreflections from unused access waveguides on a 2 × 2 MMI is to terminate them abruptly into the semiconductor mesa, in an extreme limit of a strong-to-weak guiding waveguide transition. No backreflection occurs from a strong-to-weak guiding waveguide transition, since sufficient forward propagating radiation modes are available to collect the strongly-guided waveguide mode [87].

Besse et al. [88] introduced a class of MMI couplers that rely on a shaped interference section to achieve freely selectable splitting ratios. They replaced the standard rectangular geometry of the interference section with two linear transformed sections of equal length. The first section is linearly down-tapered, the second linearly up-tapered, in what they define as a butterfly geometrical configuration. The expression of MMI length $L$ in terms of $L_\pi$ remains unchanged in the butterfly MMI; however, $L_\pi$ must now be expressed as $L_\pi \cong (4n_g W_{e0} W_{e1})/(3\lambda_0)$, with $W_{e0}$ and $W_{e1}$ the effective widths of the MMI interference section at its input and its center, respectively. Butterfly MMI splitters and combiners are a convenient means of introducing the optical power imbalance in the modulator arms required to enable the chirped modulator designs introduced in Section 7.2.2.2 and shown in Figure 7.2b.

In [23], optimal 2 × 2 MMI designs for the strongly guided couplers in Figure 7.2a and 7.2b are proposed, using paired restricted interference with $W_{e0} = 10.5$ μm and $W_a = 2.4$ μm; an optimal 1 × 2 MMI splitter design is proposed using restricted symmetric interference with $W_e = 9$ μm and $W_a = 2.0$ μm. A 1 × 1 restricted symmetric MMI, used as an inline $TE_1$ mode filter, is also shown to be a tolerant MMI design that is effective at improving the $TE_0$ spectral purity in the output waveguides of various MMI splitter designs. Its value would be increased significantly if it doubled as a VOA, as discussed in Section 7.5.

### 7.3.3 Waveguide Bends

It is necessary to introduce s-bends, following the MMI splitter and prior to the MMI combiner, in order to increase the separation between the MZM waveguide arms. This is required to reduce the RF coupling between the high-speed phase shift electrodes present on each MZM arm waveguide. Care must be taken to avoid the s-bends exciting the higher order lateral optical waveguide modes, especially following the MZM MMI splitter. Beating between the $TE_0$ and $TE_1$ modes over the length of the interferometer modulation arms, can cause wavelength dependence in the output extinction ratio and insertion loss with a nonideal MMI combiner [23]. The selected separation for the MZM arms is 20 μm between waveguide centers in the sample MZM design shown in Figure 7.2. In a Cartesian MZM, the separation between upper and lower MZMs is also a compromise between minimizing the MZM RF crosstalk and minimizing the optical losses in longer bend waveguides [37]. Finally, there is value in 90° waveguide bends for on-chip optical routing to enable high-density integrated components such as a 10-channel polarization multiplexed return-to-zero differential-quadrature-phase-shift keying transmitter [40].

A simple framework for understanding bending loss was introduced in 1975 by Heiblum [89] using conformal transformations to derive an equivalent index profile for the waveguide bend. The first order approximation to the exact equivalent index distribution obtained with their conformal transformation showed that if $n(x)$ defines the refractive index of the waveguide, then a bent waveguide with radius $R$ behaves like a straight waveguide with a refractive index profile $n_{eq}^2(x) = n^2(x)\left(1 + \frac{2x}{R}\right)$, $x \ll R$. One consequence of this simple description of the waveguide refractive index profile is that every mode traveling around the bend must radiate energy. If the wavefront is to remain plane it must travel faster as it moves away from the center of curvature, and at some point it must exceed the speed of light. The locus of this point is known as the radiation caustic. Beyond the radiation caustic, the phase front must become curved and the field radiates away in the lateral direction. The radiation caustic moves closer to the waveguide as the radius of curvature diminishes. Bending loss is, therefore, the continuous radiation of mode power tangentially out of a curved waveguide as light travels around the bend. Given the index distribution in a bend, the mode profile will shift to the outer edge of the waveguide as the bending radius is decreased. The radiation caustic is located far from the edge of the waveguide mode in a strongly-guided structure; therefore, pure bending losses are not a very significant issue in strongly-guided waveguides.

When two waveguides of different radii of curvature and/or structure are connected to each other, the propagating mode supported by one waveguide transforms to every possible mode of the other waveguide at the junction. The radiation loss due to the lowest order mode field mismatch at the discontinuity is looked on as transition loss. The transmission coefficient of the power through the junction can be easily obtained from the overlap of the field profiles for the lowest order modes of the two connected waveguides. This transition loss may be reduced by changing the curvature in a gradual manner or, as proposed by Neumann [90], laterally offsetting the waveguide sections so as to reduce the overlap mode mismatch across the junction.

There are three simple waveguide configurations for an s-bend with a lateral offset $h$ and a length $l$:

1. An s-curve made of two circular arcs with a constant radius of curvature: $R = \pm\frac{L^2}{4h}\left(1 + \frac{h^2}{L^2}\right)$.
2. A cosine-based s-curve described by $x(z) = \frac{h}{2}\left[1 - \cos\left(\frac{h}{2}z\right)\right]$. Marcuse [91] introduced the use of the raised cosine as a transition function in 1978 and Ramaswamy [92] experimentally demonstrated low-loss for the first time with this function in 1981.
3. A sine-based s-curve where the center of the waveguide is defined by $x(z) = \frac{h}{L}z - \frac{h}{2\pi}\sin\left(\frac{2\pi}{L}z\right)$. The upper and lower waveguide boundaries should be defined such that the waveguide width remains constant at all transverse planes within the s-bend. Minford et al. [93] introduced this transition curve, having no discontinuities in the first and second spatial derivatives, to minimize losses due to curvature reversals and straight-to-curved waveguide transitions.

The total insertion loss for an s-bend is comprised of both the curvature loss and the transition loss at the straight-to-curved or curvature reversal transition. The cosine-based s-bend has discontinuous curvature at the entrance and exit ends of the bend, with the curvature varying monotonically towards the midpoint. Finite transition loss is, therefore, expected at the entrance and exit of the cosine-based s-bend. The sine-based s-bend, by comparison, has zero curvature at the entrance and exit ends, and at the midpoint, so lateral offsets are not required. However, the sine-based s-bend can potentially suffer from greater bending losses than the cosine-based s-bend due to its $\frac{4}{\pi}$ larger maximum instantaneous curvature for a given transition. The s-bend with two circular arcs will have the highest transition loss at the midpoint due to an abrupt curvature reversal. Lateral offsets are definitely required at the midpoint for low-loss circular arc transitions.

It is clear that for a strongly-guided waveguide, where the bending losses are negligible, a sine-based s-bend transition is the desirable choice since it minimizes both the transition losses and the excitation of the higher-order modes in the bend at the straight-to-curved transition. The greater maximum instantaneous curvature for the sine-based s-bend, in comparison with the cosine-based s-bend transition, is of no practical relevance. For a weakly-guided waveguide, the literature consensus is that the bending loss of the sine-based s-bend is more significant than the transition losses, so a cosine-based s-bend transition curve is recommended [94,95].

In [23], it is shown that a 100 μm to 120 μm long sine-based s-bend is sufficient to enable a 20 μm separation between MZM arm waveguide centers. It is also shown that the excess loss for a 90° bend, in a 2.0 μm wide waveguide, with a 50 μm bend radius, is as low as 0.3 dB for a 0.2 μm lateral waveguide offset at the straight to circular arc waveguide transitions. The excess loss for a 100 μm bend radius is as low as 0.05 dB with a 0.1 μm lateral waveguide offset.

### 7.3.4 Optical Spot-Size Converters

An optical mode spot-size converter (SSC) allows an adiabatic transition between an optical mode that is "optimal" for an active semiconductor device function and an optical mode that is "optimal" for interacting with the semiconductor device. In a strongly-guided InP-based MZM desirable characteristics from an integrated SSC include:

- Angular reduction and symmetrization of the MZM waveguide mode far field to enable efficient coupling to the device with micro-optic lenses. Improvements in coupling alignment tolerances are secondary to the reduction in coupling insertion loss.
- Wavelength independent optical mode conversion is required for full C-Band operation.
- The inclusion of the SSC must not impair high optical extinction ratios in the MZM. Extinction ratios as high as 40 dB do have practical applications.
- The SSC must be near single-mode. Light must be coupled from off-chip into the MZM so it is not practical to assume perfect modal launch conditions in the device. This constraint is not necessary for an output SSC on an MZM designed for integration with an optical source.
- Low polarization rotation in the SSC is required since the MZM is not usually designed to be polarization independent.
- The impact of the MZM InP doping profiles (especially p-doping) must be considered in the SSC design by either limiting the mode overlap to the doping through the use of a short SSC design, or removing the impact of the p-doping through He+ implantation or additional etch and regrowth processes.

An illustration of an SSC design [23,96] implemented on an InP-based MZM is given in Figure 7.2. Each SSC has a vertically tapered passive waveguide core within a laterally flared strongly-guided ridge. The SSC core is butt-joined to the MZM section by a precision dry etch and metal organic chemical vapor deposition selective area growth process. In total, three growth steps are required to form the full device with the overclad p-doped layers common between the MZM and the SSC. Care must be taken during fabrication to limit silicon contamination at the overgrowth interface between the MZM intrinsic core and the p-doped overclad layers [23]. The p-doping in the SSC contributes notable absorption losses that could be reduced further either through He+ implantation or additional selective area growth growth steps. Etch and regrowth intensive SSC designs are more problematic in InGaAlAs materials due to the rapid oxidation of exposed aluminum. There are no InGaAlAs-based MZM literature examples using SSCs to date; however, it is a viable area for future research given the significant development effort directed towards multigrowth InGaAlAs-based electroabsorption modulated Laser for 100 GbE applications [97].

Issues can arise when an $N^+$-doped InP substrate is used with strongly guided waveguides, since there exist higher-order "composite waveguide cavity" modes [98,99], guided laterally by the ridge etch and vertically by the positive index step between the $N^+$ substrate and the n-doped underclad InP layers. If this "composite waveguide cavity" mode is excited by misaligned input coupling, it can degrade the extinction ratio of the MZM. To remove this issue, the n-dopant induced refractive index change in InP can be used to add an antiguiding layer in the underclad to leak these "composite waveguide cavity" modes into the $N^+$ substrate [23]. These "composite waveguide cavity" modes are not present with the use of Fe-doped SI InP substrates because the Fe-doping has negligible impact on the real refractive index of InP.

The metal organic chemical vapor deposition selective area growth growth process enables the thickness of the SSC passive core to adiabatically taper over 350 μm, from 0.2 μm at the butt-joint, down to 0.05 μm at the facet. Simultaneously, the lateral ridge width flares exponentially over 300 μm, from 2.0 μm at the butt-joint, to 4.5 μm at the facet. The SSC transitions the modulator mode at the butt-joint to a facet mode with a 22° × 24° full-width half-maximum far field designed for low-loss coupling using microlenses. The MZM and SSC optical modes are shown in Figure 7.2. This SSC design is >2× longer than the required taper adiabatic limit and can easily be shortened [23].

InP-based MZMs that use this design of SSC have demonstrated insertion losses <6 dB (including coupling) and optical extinction ratios >30 dB over the full C-Band [30]. The back reflectivity of the SSC waveguide facet can be improved if desired by angling the waveguide approach to the facet. For an SSC waveguide angle of 6°, corresponding to an output beam angle of 19°, it is possible to have <−60 dB back reflection from the SSC output facet [23].

## 7.4 Radio Frequency Phase Shift Electrode Design for Indium Phosphide-Based Mach–Zehnder Modulator

The aim in traveling wave RF electrode design is to achieve a voltage wave that copropagates with the modulation envelope of the optical wave, at precisely the same velocity, and with low microwave loss over the length $L$ of the electrode. This permits the resulting optical phase modulation to accumulate monotonically along the length of the RF line irrespectively of frequency. Any velocity mismatch between the two waves causes phase walk-off that increases with either frequency or interaction length, and ultimately limits the $f_0$ electrical bandwidth of a lossless RF line to

$$f_0 = \frac{1.4 \cdot c}{\pi \left| n_o - n_\mu \right| L}, \tag{7.4}$$

where $n_o$ and $n_\mu$ are the optical and microwave effective refractive indices, respectively, and $c$ is the vacuum speed of light [100]. With InP-based MZM, a $V_\pi < 4$ V can be easily obtained for RF phase shifter lines <4 mm in length, due to the relatively high modulation efficiency of the waveguide structure and materials. From Eq. 7.4, a 4 mm long RF phase shift electrode on the waveguide structure in Figure 7.3b will have velocity mismatch limited bandwidth >30GHz, for a microwave effective index between 2.4 and 4.6. Velocity matching is therefore not a critical design criteria for InP-based MZMs in telecommunication applications.

In practice, the characteristic impedance of the traveling wave RF line must be matched to the typically 50 Ω system impedance to avoid large signal reflections; however, this impedance matching is generally difficult in InP-based MZMs due to the large capacitance per unit length that results from their high phase efficiency p-i-n waveguide structure (20 Ω–30 Ω typical.) It is for this reason that the early InP-based MZMs used simple, lumped-element, RF phase shift electrodes having an resistance-capacitance (RC) limited bandwidth [1]. These lumped-element RF phase shifter designs are, however, commercially significant at 10 Gb/s for negatively chirped (Section 7.2.2.2), or chip area sensitive applications.

Two broadly applicable approaches for the RF electrode design, shown in Figure 7.6, have been successfully demonstrated to date in InP-based MZMs. The first approach, with implementation examples

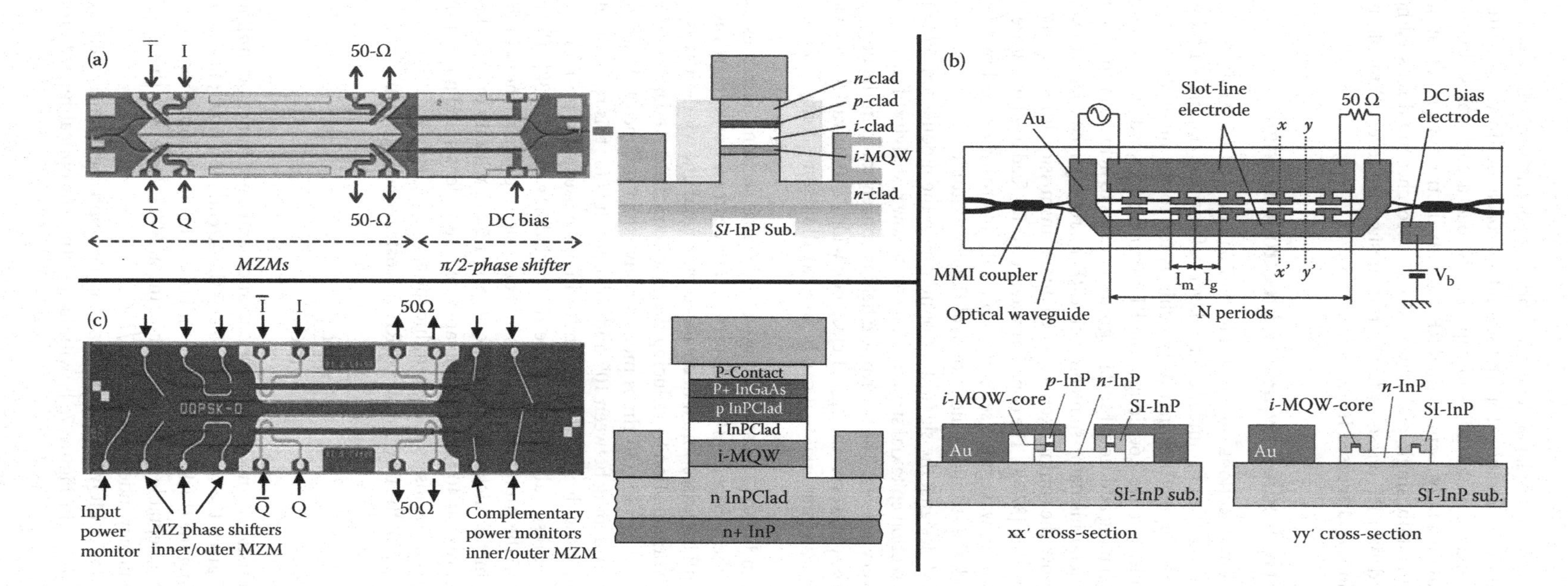


**FIGURE 7.6** (a) Coplanar microstrip RF electrode design for a DQPSK MZM using an SI substrate. (N. Kikuchi, et al., "80 Gb/s Low Driving-Voltage InP DQPSK Modulator with an n-p-i-n Structure," *IEEE Photonics Technol. Lett.*, Vol. 21, No. 12, pp. 787–789, © 2009 IEEE.) (b) Capacitively loaded coplanar stripline RF electrode design for an MZM. (S. Akiyama, H. Itoh, S. Sekiguchi, S. Hirose, T. Takeuchi, A. Kuramata, T. Yamamoto, "InP-Based Mach–Zehnder Modulator with Capacitively Loaded Traveling-Wave Electrodes," *J. Lightwave Technol.*, Vol. 26, No. 5, pp. 608–615, © 2008 IEEE.) (c) Coplanar microstrip RF electrode design for a DQPSK MZM using an N+ substrate. (With kind permission of Oclaro Inc.)

shown in Figure 7.6a and 7.6c, employs independent, conventional, coplanar RF microstrip electrodes on each interferometric arm of the modulator [19]. The second approach, outlined in Figure 7.6b, uses p-i-n epitaxy in a series push–pull configuration to capacitively load a coplanar stripline electrode [20–22]. The first approach fundamentally requires dual RF drive for electro-optic modulation, whereas the second approach is fundamentally single-end driven. Both approaches are discussed herein. The fundamentally independently driven RF phase shift electrode design has several limitations for zero chirp modulator implementations as discussed in Section 7.2.2.3.

### 7.4.1 Coplanar Microstrip Radio Frequency Electrode Design

The phase shifting waveguide cross sections shown in Figure 7.6a and 7.6c MZM device designs are composed of an inhomogeneous stack of semiconductor layers of variable doping with lossy metallic electrodes. In Figure 7.6a, the input RF signal to the chip is fed to the modulation section via a simple coplanar waveguide. In the Figure 7.6c design, the RF input signal is fed via a metal-insulator-semiconductor microstrip line having additional surface ground metals to help carry the return current and reduce the RF losses. The metal-insulator-semiconductor RF microstrip feed enables small bending radii, which can be used to remove the timing skew between dual transmission lines originating from a single side of the InP modulator chip [37,38]. The coplanar waveguide RF feed used in Figure 7.6a design is transferred, at the phase shifting waveguide section, into an RF microstrip signal by the n+ buffer layer (ground plane) underneath the strongly guided optical ridge waveguide.

In Figure 7.6a and 7.6c MZM designs, the signal to ground gap in the modulation section is determined by the waveguide cross-section epitaxial structure due to the low resistivity of the n+ buffer layer. Metal patterning tolerance is very relaxed. The microwave field is concentrated in the depletion region of both waveguide cross sections, however, the signal electrode width can be varied to produce stray field in the waveguide ridge cladding material that can be used to allow some control of the microwave refractive index.

The microwave attenuation in Figure 7.6a and 7.6c waveguide cross sections decreases as the ridge width decreases; it decreases as the p-doped InP cladding thickness decreases; it decreases as the p-doping concentration level in the InP increases; and it decreases as the signal line width increases. All these variations reduce the microwave attenuation by reducing the RF electric field overlap to resistive (lossy) dielectric materials. P-doped InP is ≈80–130 times more resistive than n-doped InP, when doped to give equivalent optical loss, for p-doping between $10^{17}$ and $10^{18}$ $cm^{-3}$. It is for this reason that the demonstrated microstrip line solutions with >30 GHz of bandwidth have used an n-p-i-n or n-i-n heterostructure to suppress the electrical signal loss due to the InP p-cladding layer [19,38]. The microwave attenuation is also dependent on the Au metal thickness and deposition process: The skin depth of Au is ≈ 385 nm at 40 GHz so the metal thickness should be at least 1.5 μm with up to 3 μm beneficial to minimize the impact of metal losses at 40 GHz [101]. Evaporated Au is denser and more uniform than plated Au and results in reduced metal losses at a given thickness.

The characteristic impedance $Z$ of the traveling wave RF microstrip line is given by

$$Z = \sqrt{\frac{R + i\omega L}{G + i\omega C}}, \tag{7.5}$$

where $R$, $L$, $G$, and $C$ are the resistance, inductance, conductance, and capacitance per unit length. It is difficult to change $R$, $L$, and $G$ given the requirement to keep the attenuation small; however, $C$ can be changed rather freely, by varying the thickness and/or width of the intrinsic doped region in the waveguide cross section to achieve acceptable impedance matching at 50 Ω. Note also that if $R$ and $G$ are sufficiently small, then $Z$ is frequency independent, $Z = \sqrt{L/C}$.

The thickness of the reference MQW structure cannot be appreciably increased beyond the 0.55 μm-thick 31-well design given in Figure 7.3b, without exciting an optical higher order vertical mode. To further reduce the waveguide capacitance, to achieve a 50 Ω impedance match, it is necessary to sacrifice modulation phase efficiency and increase the thickness of the intrinsic region into the MQW overclad InP layers. Alternatively, the modulation phase efficiency can be sacrificed longitudinally, by periodically implanting He+ ions into the InP p-doped overclad layers to change them to insulating material, in order to reduce the average capacitance per unit length [102].

Coplanar microstrip RF phase shift electrodes can be used for Cartesian MZM designs on either doped [37] or semiinsulating [38] InP substrates. The advantage of the p-i-n structure, used in [37] and shown in Figure 7.6c, is that it enables the integration of short, current driven, imbalance control electrodes, and multiple evanescently coupled InGaAs photodiodes, to allow monitoring and control of the DC optical phase in both the inner and outer MZMs of the Cartesian modulator. The n-p-i-n structure, used in [38] and shown in Figure 7.6a, requires longer, QCSE reverse biased imbalance control electrodes (1.5 mm long), with no easily integrated photodiode monitor feasible; however, ultimately higher bandwidth modulation designs are feasible with this low RF loss n-p-i-n waveguide cross section. In [37], it is shown that the reference MQW structure, on an N+ doped InP substrate, can enable RF phase shift electrode designs for full C-Band operation having a 3 dB EO electrical bandwidth >20 GHz, with return loss <10 dB, a $V_\pi$ <3.5 V, and with <0.5 dB voltage induced absorption loss. The n-p-i-n waveguide cross section in [38], with an intrinsic core thickness of 0.9 μm, a ridge width of 1.6 μm, and an electrode length of 3 mm, achieves a 3 dB EO electrical bandwidth of 30 GHz, with return loss <13.5 dB, for a $V_\pi$ of 2.8 V. This group has also achieved a 3 dB EO electrical bandwidth >40 GHz with an intrinsic core thickness of 1.3 μm, for a ridge width of 2.0 μm, over an electrode length of 3 mm, and with a metal thickness of 3 μm [101].

### 7.4.2 Series Push–Pull RF Phase Shifter Design

The segmented traveling-wave electrode design for an MZM has a coplanar stripline (CPS) geometry having two wide, low-loss conductors running parallel to the optical waveguides, but placed outside the MZM structure. An example is shown in Figure 7.6b. The dimensions of the CPS line are not critical provided appropriate conductor width to spacing is maintained. On an InP substrate, the CPS line has a microwave velocity faster than the optical group velocity, and a characteristic microwave impedance higher than 50 Ω.

Short lumped element electrode segments on the p-i-n optical waveguides are periodically connected, in shunt, to the CPS transmission line as shown in Figure 7.6b. This primarily capacitive loading of the CPS line lowers the microwave velocity and impedance of the line. The capacitance added per unit length is easily chosen during design through the selection of the gap length between electrode segments. The required isolation between the segments has been demonstrated using He+ implantation [102], proton implantation [103], or etch and regrowth of SI InP [20,104]. The loading electrode segments do not carry any of the axial currents in the transmission line so the propagation loss and the resistance per unit length of the loaded CPS line are primarily determined by the microwave properties of the unloaded CPS line.

This "loaded" or "slow-wave" transmission line structure also enables series push–pull MZM operation using only one electrical driver. In series push–pull operation, the two optical modes in the MZM arm waveguides are phase modulated *antiphase* by electrically connecting the two waveguides back-to-back in series to the CPS line, as shown in Figure 7.6b. Half the applied alternate current voltage will drop across each arm in opposite directions (antiphase), and the total loading capacitance will be only half of the loading capacitance in each arm due to the series connection. As can be seen in Figure 7.6b, the n-doped common base for the p-i-n phase shift electrode segments is partly extended to form an electrode which allows external DC reverse bias to be applied.

The slow-wave structure electrical design uses the loading capacitance of the phase shift electrode segments to lower the microwave velocity of the line to match the optical waveguide group velocity, and to lower the line microwave impedance to match the 50 Ω source impedance. The loading phase shift electrode segments will then act sequentially and in phase upon the propagating optical wave packet to produce a modulation length for the push–pull phase shift in the arms of the MZM. The microwave and optical designs are effectively decoupled. The slow-wave transmission line structure was first proposed and demonstrated for a GaAs-based MZM [100], and has been subsequently introduced to InP-based MZMs [20,22,103,104].

It is desirable to have the unloaded transmission line microwave index $n_\mu$ as low as possible to maximize the allowed capacitive loading. In a coplanar CPS line, the electric field travels equally in the air and the substrate, thus $n_\mu$ is derived from the mean permittivity $n_\mu = \sqrt{(1+\varepsilon_r)/2}$, where $\varepsilon_r$ is the dielectric constant of the InP substrate [100]. For $\varepsilon_r = 12.4$ in InP $n_\mu = 2.6$. Dielectrics can be placed between the CPS conductors and the InP substrate to further reduce $n_\mu$ down to the order of 2.2.

Using a simple LC circuit model the required loading capacitance is given by

$$C_L = (n_o^2 - n_\mu^2)/(cZ_o n_o), \tag{7.6}$$

where $n_o$ and $n_\mu$ are the loaded and unloaded microwave indices respectively, $c$ is the speed of light in a vacuum, and $Z_o$ is the characteristic impedance of the loaded line [100,105]. By this simple model, the loading capacitance per arm ($2 \times C_L$) can be on the order of 0.37 pF/mm assuming $Z_o$ is fixed by the 50 Ω source impedance, $n_\mu$ of 2.2 is feasible, and the loaded microwave index can be as high as 4.0 without practically limiting the bandwidth as discussed in Section 7.4. This 0.37 pF/mm of allowed loading capacitance enables a high longitudinal fill factor for electrode segments on the MZM arms when using the phase shifter designs given in Figure 7.3a (0.65 pF/mm) or Figure 7.3b (0.4 pF/mm.)

The frequency response of a segmented traveling wave MZM is also limited by the microwave loss, the microwave dispersion, and the electrical filter characteristic of the structure. The slow wave electrodes are inherent low pass filters with the upper end of the passband determined by the periodicity of the loading. To analyze the frequency response including all these additional effects, a microwave equivalent circuit model can be used [105].

The highest reported performance to date [22] using a slow wave structure demonstrates a 45 GHz packaged MZM 3 dB EO electrical bandwidth, using a 50% fill factor on a 4 mm long CPS line, with 62.5 μm-long active electrode segments. The $V_\pi$ was 2.6 V. The on-chip 3 dB EO bandwidth was extended to 63 GHz using a 40% fill factor on a 2 mm long CPS line, with 100 μm-long active electrode segments.

## 7.5 Value Added Functionality for Indium Phosphide-Based Mach–Zehnder Modulator Devices

The InP material system has a rich history as an integration platform. The complementary functions or features discussed here have been demonstrated either integrated with InP-based MZMs, or in epitaxial growth structures compatible with InP-based MZMs.

Deep helium implantation has been used to create point defects in zinc doped InP layers to remove carriers from participating in conduction, with minimal impact to the optical waveguide properties [106,107]. This helium implant process can be used to isolate electrodes and has been shown to remain stable under extensive thermal, electrical, and optical stress conditions.

Evanescently coupled waveguide InGaAs power detectors insensitive to the optical signal wavelength and input power have been monolithically integrated with the InP-based MZM presented in Section 7.2.2.3 and Section 7.2.2.2, as a simple extension of the spot-size converter fabrication process [29,30]. These detectors are placed on the complementary and/or in-line output waveguides to provide feedback for transmitter control (laser power control, control of imbalance control electrodes [68], etc.)

Leuthold and Joyner have proposed a method to actively tune the power splitting ratio in a 2 × 2 MMI [108] and in Section 2.2.7 of [23] the active tuning of the cross/bar MMI power splitting ratio between 1 and >1.7 is demonstrated. The tuning is achieved for <3 mA of applied current to helium implant isolated edge electrodes on a 10.3 μm 2 × 2 MMI and it produces <0.15 dB optical loss. This split ratio dynamic range, if applied to the 2 × 2 MMI combiner in the zero chirp modulator design presented in Section 7.2.2.3, produces sufficient optical power imbalance to move between zero chirp and the optimal negative chirp for maximum dispersion limited reach. A current tunable MMI has also been demonstrated using selective zinc diffusion [109].

An output power variable optical attenuator is a commonly required function in transmitters for practical optical communication systems. Early fixed wavelength MZM transmitters used integrated electroabsorption pads on the input of the InP-based MZM to provide this variable optical attenuator function [16,110]. The same processes used to implement the MMI tunable power splitter could be used to implement a wavelength independent variable optical attenuator on the output of an InP-based MZM, through selective asymmetric carrier injection in regions of a 1 × 1 MMI waveguide [111,112]. The operating principle for this low waveguide dependent variable optical attenuator is simple. The 1 × 1 MMI is a restricted symmetric interference device in which only even modes are excited. Therefore, by asymmetrically modifying the refractive index along a selected cross section within the MMI waveguide, such that a phase change of $\pi$ is induced, mode conversion of the even modes into odd modes is realized. The odd modes are rejected at the MMI output waveguide.

Semiconductor optical amplifiers have been integrated prior to MZMs that use InGaAsP/InP MQW cores, and lossless operation has been demonstrated in 10 Gb/s IMDD [113] and 40 Gbit/s DPSK [24] applications.

Future applications will benefit from exploration of a single InP chip for dual polarization Cartesian MZMs, through the monolithic integration of a TE to TM polarization converter and a polarization combining waveguide element. The demonstrated polarization manipulation functions in InP materials [114–116] have not used waveguide structures compatible with an MZM. This commercial application will hopefully spur further research in this area.

## 7.6 Monolithic Integration of an InP-Based MZM with a Tunable Laser Source

Stringent requirements on footprint, power dissipation, and cost are placed on widely tunable Transmitter Optical Subassemblies (TOSA) for use in the smallest DWDM transceiver form factors (such as the XFP.) The monolithic integration of an InP-based MZM with a tunable laser source, referred to herein as an ILMZ, is proving to be an attractive option to meet these TOSA requirements [25–27,117].

The current implementations of the ILMZ, with examples provided in Figure 7.7, are all formed through integration with waveguide core butt-coupling, which virtually eliminates the design

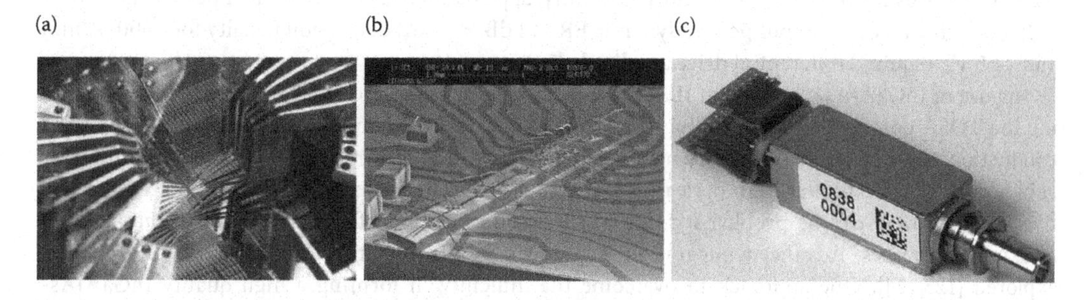


**FIGURE 7.7** (a) Integrated InP DSDBR tunable laser Mach–Zehnder modulator devices undergoing on-wafer test. (With kind permission of Oclaro Inc.) (b) Integrated InP DSDBR tunable laser Mach–Zehnder chip-on-carrier. (With kind permission of Oclaro Inc.) (c) ILMZ TOSA assembly for XFP transceivers. (With kind permission of JDS Uniphase Corporation.)

constraints on the MZM MQW core. Wavelength tuning, either thermally or through carrier injection, is performed within laser tuning sections external to the gain region, and a semiconductor optical amplifier is integrated to compensate for the MZM and laser cavity losses, as well as to provide a means for power leveling over wavelength, and to allow output blocking during channel switching. The MZMs in these devices use independent, lumped element phase shift electrodes, to minimize the device length required for differential drive amplitudes <3 $V_{pp}$. To justify the development effort (caused by growth and process complexity), the ILMZ has been initially targeted at the large and stable 10Gb/s IMDD market [118].

It is critical in the design of an ILMZ to minimize the impacts of optical, thermal and electrical cross-talk within the chip. Fixed phase/magnitude reflections back into the laser cavity exist at material interface butt-joints, at weak-to-strong guiding waveguide transitions, and at the MZM MMI optical splitter output. Variable phase/magnitude reflections exist at the MMI combiner output, and at the device facet. These optical reflections can be experimentally measured using optical low-coherence reflectometry [119]. An SSC on the end facet of the MZM interfacing to the fiber is generally not required in an ILMZ to reach to desired TOSA output power due to excellent coupling efficiency between the laser and MZM input; however, if no SSC is used, the facet waveguide must be weakly-guided and angled to guarantee sufficiently low back-reflections from the output facet [23,87].

A significant compromise is required in the selection of the waveguide orientation relative to the (100) InP substrate major flat. This compromise is between the process uniformity for the weakly-guided laser ridge width (single mode control), and the phase efficiency ($V_\pi$) of the MZM. In non-integrated devices, the MZM waveguide is oriented parallel to the major flat of the InP substrate such that the LEO phase shift adds to the QCSE induced phase shift, as discussed in Section 7.2.1; however, laser waveguides are typically oriented perpendicular to the major flat, such that a $HCl:H_3PO_4$ crystallographic wet clean-up etch [120,121], with high selectivity to InP, can be used to terminate the weakly-guided laser ridge without compromising control of the ridge width; hence, the requirement for a compromise. When the optical waveguide is oriented parallel to the major flat (MZM preferred direction), an $HCl:H_3PO_4$ wet clean-up etch makes the ridge width at the base of the laser waveguide rib directly proportional to the depth of the dry etch used to define the ridge, whereas for a waveguide oriented perpendicular to the major flat (laser preferred direction), the LEO effect acts to counteract the QCSE index change.

A second compromise is required in the selection of the p-doping concentration and thickness for the common overclad InP layers in the ILMZ. This compromise is between the tunable laser series resistance and the MZM/SSC passive optical losses.

A TOSA using an ILMZ for 10 Gb/s negative chirp applications has demonstrated performance over C-Band with >2.5 dBm output power, dynamic ER >10 dB, and <2.3 $V_{pp}$ required differential drive amplitude [25]. The TOSA power dissipation was <1.3 W at 75°C case temperature, enabled by an ILMZ operating temperature of 42°C. The ILMZ uses 750 μm-long lumped element phase shift electrodes. A second TOSA using an ILMZ for 10 Gb/s zero chirp applications has demonstrated performance over C-Band with >4.5 dBm output power, dynamic ER >12 dB, <1.5 dB dispersion penalty for ±800 ps/nm, and <3.5 $V_{pp}$ required differential drive amplitude [26].

The use of InGaAlAs materials in ILMZs is an active area of research [122] aimed at further reducing the TOSA power dissipation through semicooled operation at temperatures up to 55°C. The large conduction band offset in InGaAlAs materials provides strong electron confinement, enabling laser gain sections that can offer high optical output power at elevated temperatures [123]; however, selective area regrowth can be problematic due to the rapid oxidation of the aluminum in the material at exposed surfaces. Advanced integration techniques to form hetero-material interfaces are being explored [123,124]. One approach to overcome the difficulty in forming a high quality InGaAlAs-InGaAsP interface is using chlorine-based *in situ* cleaning of the InGaAlAs butt-joint interface in a metal organic chemical vapor deposition reactor, immediately prior to selective area growth of the InGaAsP materials. This has been shown to enable >97% optical coupling efficiency at the hetero-material butt-joint [124].

## 7.7 InP-Based MZM Packaging

An optical transmitter module using hybrid integration between an optical source and an MZM has been the dominant form for commercially available InP-based MZM devices since their introduction in 1995. Hybrid integration has offered several logistical advantages that have favored it to date. Prior to the general availability of tunable laser sources, managing DWDM wavelengths was simpler using separate lasers subcarriers; even today when a tunable laser source is used, the same laser subcarrier can be used within both tunable laser, and tunable transmitter modules. The ability to independently evolve and yield the laser and MZM subcomponents has also traditionally enabled superior performance from hybrid integrated transmitter modules, and enabled flexibility in targeting diverse application markets. In the future, as the tunable laser and MZM devices mature, and when stable performance expectations can be defined, that is, DWDM XFP transceivers), monolithic integration can be expected to play a larger commercial role.

Figure 7.8a shows a commercially successful, hybrid integrated, fixed wavelength, transmit optical train—used within various transmitter module incarnations between 1995 and the mid-2000s [16]. As in all high performance transmission sources, an optical isolator is needed to protect the laser from external feedback. Placing the isolator between the laser and modulator acts as a safeguard against reflections from the MZM output facet which, due to phase modulation, are particularly problematic. Within the package, aspheric lenses are used, first to collimate the laser output, and then following

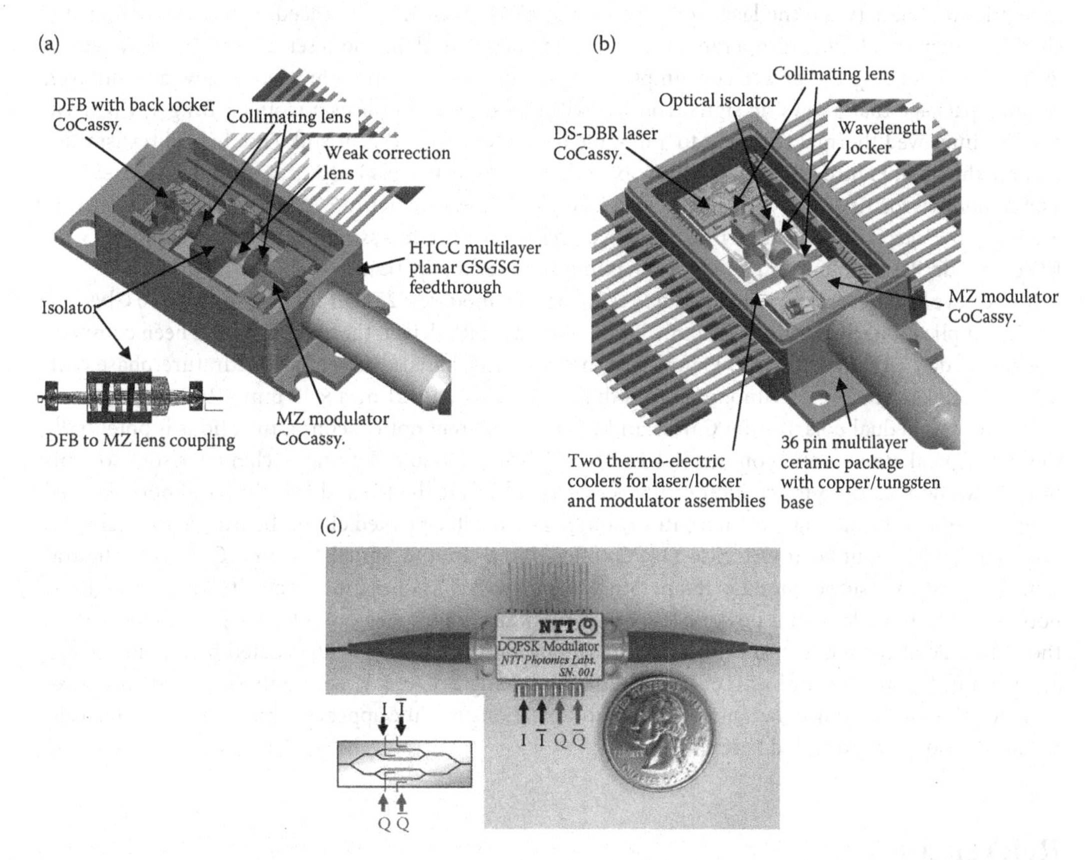


**FIGURE 7.8** (a) Copackaged DFB laser and InP-based MZM module, (b) Copackaged DS-DBR tunable laser and InP-Based MZM module. (c) 40 Gbit/s DQPSK MZM module. (N. Kikuchi et. al, "Full L-Band 40 Gbit/s Operation of Compact InP-DQPSK Modulator Module with Low Constant Driving Voltage of 3.5 Vpp," European Conference on Communications, 21–25 Sept., 2008, Mo.3.C4, © 2008 IEEE.)

the optical isolator to focus the laser output into the SSC on the MZM chip. Aspheric lenses offer the combination of low aberration, short focal length, high NA (up to 0.8), and small size, but they must be mounted in a stable configuration within around ±10 μm of their ideal positions, and with an angular tolerance of around ±1° [15]. A third relatively weak lens is included in the laser/MZM coupling train, mounted on the optical isolator, to steer the beam and optimize coupling without the need for ultra-fine adjustability. All the optics are fixed using UV cured epoxy. Coupling efficiency between the laser and MZM can be calculated with a Gaussian model and it is limited by the mismatch in laser/MZM mode ellipticity. All of the optics (including the laser and MZM subcarriers) mount on a single aluminum nitride ceramic frame, which in turn, mounts on a thermo-electric cooler. Since the MZM chip is small and dissipates little power, mounting it on the same TEC as the laser source is quite practical. Wavelength monitoring optics are fixed on the laser subcarrier and the RF termination components needed by the differential 10 Gb/s data signals are incorporated into the MZM subcarrier. All of the electrical connections, including the controlled impedance differential data inputs, are arranged along a single side of the 8.1 mm high hermetic Kovar package.

Figure 7.8b shows the evolution of the Figure 7.8a transmitter module to a widely tunable design. Many of the same concepts are reused; however, the fabrication is now facilitated by automated chip bonding, piece-part attachment, and optical alignment, to provide a consistently high degree of assembly control and yield [125]. The tunable laser is front locked to the International Telecommunications Union - Telecommunication Standardization Section (ITU-T DWDM) wavelength grid by the monitoring optics placed between the laser source and the MZM. The MZM is placed on a separate thermo-electric cooler enabling it to operate at a higher temperature than the laser source and wavelength locker, thus reducing the power consumption of the module. The module itself is now a multilayer, ceramic package that allows dc signal routing within its walls, and uses a Copper Tungsten (Cu-W) base for improved thermal coupling to a module backside mounted heat sink. The hybrid transmitter module shown in Figure 7.8b has also been extended to offer a 40 Gb/s DQPSK optical transmitter [66]. TOSA modules for 10 Gb/s DWDM XFP form factors, as shown in Figure 7.7c, must fit the tunable laser, wavelength locker, and the MZM into a smaller package while addressing the obvious challenges with I/O count, dc/RF isolation, and a reduced maximum power dissipation.

There have been several reported InP-based MZM only modules [22,65,126–128] for both 10 Gb/s and 40 Gb/s applications, including a design with an integrated RF driver, though none have been commercialized to date. As one example, Figure 7.8c shows an InP-based differential-quadrature-phase-shift keying modulator in a surface mountable module measuring only 21 mm × 13 mm × 10 mm [65].

An InP-based dual polarization Cartesian MZM for coherent optical communications is potentially the first logical stand-alone commercial InP-based MZM module. Desirable characteristics for this module would include: Surface mount type lead structures for the RF and DC electrical ports located at the bottom of the package, or alternatively offered on flexible printed circuit board (PCB) interfaces; single-ended RF input ports (XI, XQ, YI, YQ) offered in a Ground-Signal-Ground (G-S-G) configuration along only one side of module; temperature stabilized MZM operation, heat extracted through the bottom of the module with a power dissipation <500 mW; input and output fiber pigtails located on the same side of the module to minimize the overall footprint which is dominated by the fiber boots as shown in Figure 7.8c; and zero chirp modulation having $2\ V_\pi < 5\ V$, with only small variation over C-Band. This challenging coherent optical communications module appears achievable using the technology developments detailed herein.

## References

1. C. Rolland, R. S. Moore, F. Shepherd, and G. Hillier, "10 Gbit/s, 1.56 μm multiquantum well InP/InGaAsP Mach-Zender optical modulator," *Electron. Lett.*, vol. 29, no. 5, pp. 471–472, Jul. 1993.

2. J. Yu, C. Rolland, D. Yevick, A. Somani, and S. Bradshaw, "Phase-Engineered III-V MQW Mach–Zehnder Modulators," *IEEE Photonics Technol. Lett.*, vol. 8, no. 8, pp. 1018–1020, Aug. 1996.
3. W. Bardyszewski, D. Yevick, and C. Rolland, "Numerical studies of excitonic effects in the optical spectra of quantum well modulators," *SPIE Proc. Society For Optical Engineering Physics Simulation Optoelectronic Devices II*, iss. 2146, pp. 249–255, 1994.
4. W. Bardyszewski, D. Yevick, Y. Liu, C. Rolland, and S. Bradshaw, "Theoretical and experimental analysis of Mach–Zehnder quantum-well modulators," *J. Appl. Phys.*, vol. 80, no. 2, pp. 1136–1141, 1996.
5. D. A. B. Miller, D. S. Chemla, T. C. Damen, A. C. Gossard, W. Wiegmann, T. H. Wood, and C. A. Burrus, "Band-edge electroabsorption in quantum well structures—the quantum confined Stark-effect," *Phys. Rev. Lett.*, vol. 53, no. 22, pp. 2173–2176, 1984.
6. D. A. B. Miller, D. S. Chemla, T. C. Damen, A. C. Gossard, W. Wiegmann, T. H. Wood, and C. A. Burrus, "Electric field dependence of optical-absorption near the band-gap of quantum-well structures," *Phys. Rev.*, vol. 32, no. 2, pp. 1043–1060, 1985.
7. J. C. Cartledge, C. Rolland, S. Lemerie, and A. Solheim, "Theoretical performance of 10 Gb/s lightwave systems using a III-V semiconductor Mach–Zehnder modulator," *IEEE Photonics Technol. Lett.*, vol. 6, no. 2, pp. 282–284, 1994.
8. C. Rolland, M. S. O'Sullivan, H. B. Kim, R. S. Moore, and G. Hillier, "10 Gbit/s 120 km normal fiber transmission experiment using a 1.56 μm multiple quantum well InP/InGaAsP Mach–Zehnder modulator," in *Optical Fiber Communications*, San Diego, 1993, PD27.
9. H. Sano, "A multi-quantum well modulator for high bit rate transmission systems," in *4th OEC*, 1992, 17C3-2.
10. H. Sano, M. Aoki, M. Suzuki, M. Takahashi, T. Ido, T. Kawano, and A. Takai, "A high speed InGaAs/InAlAs MQW Mach–Zehnder type optical modulator," in *Conference on Optical Fiber Communications*, 1993.
11. D. M. Adams, C. Rolland, N. Puetz, R. S. Moore, F. R. Shepherd, H. B. Kim, and S. Bradshaw, "Mach–Zehnder modulator integrated with a gain-coupled DFB laser for 10 Gbit/s, 100 km NDSF transmission at 1.55 μm," *Electron. Lett.*, vol. 32, no. 5, pp. 485–486, 1996.
12. D. M. Adams, C. Rolland, A. Fekecs, D. McGhan, A. Somani, S. Bradshaw, M. Poirier, E. Dupont, E. Cremer, and K. Anderson, "1.55 μm transmission at 2.5 Gbit/s over 1102 km of NDSF using discrete and monolithically integrated InGaAsP/InP Mach–Zehnder modulator and DFB laser," *Electron. Lett.*, vol. 34, no. 8, pp. 771–772, 1998.
13. Ph. Delansay, D. Penninckx, S. Artigaud, J. G. Provost, J. P. Hebert, E. Bouchherez, J. Y. Emery, C. Fortin, and O. Le Gouuezigou, "10 Gbit/s transmission over 90–127 km in the wavelength range 1530–1560 nm using InP-based Mach–Zehnder modulator," *Electron. Lett.*, vol. 32, no. 19, pp. 1820–1821, Sep. 1996.
14. D. Penninckx, Ph. Delansay, E. Bouchherez, C. Fortin, and O. Le Gouuezigou, "InP/GaInAsP π-phase-shifted Mach–Zehnder modulator for wavelength independent (1530–1560 nm) propagation performance at 10 Gbit/s over standard dispersive fiber," *Electron. Lett.*, vol. 33, no. 8, pp. 697–698, Apr. 1997.
15. K. Anderson, "Design and manufacturability issues of a copackaged DFB/MZ module," in *Electronics Components and Technology Conference*, 1999, pp. 197–200.
16. K. Anderson and I. Betty, "Indium phoshide Mach–Zehnder chips are ideally suited to long reach metro," *Laser Focus World*, vol. 39, iss. 3, pp. 101–104, Mar. 2003.
17. http://www.300pinmsa.org.
18. http://www.xfpmsa.org.
19. K. Tsuzuki, T. Ishibashi, T. Ito, S. Oku, Y. Shibata, R. Iga, Y. Kondo, and Y. Tohmori, "40 Gb/s n-i-n InP Mach–Zehnder modulator with a π voltage of 2.2 V," *Electron. Lett.*, vol. 39, no. 20, pp. 1464–1466, 2003.

20. S. Akiyama, S. Hirose, T. Watanabe, M. Ueda, S. Sekiguchi, N. Morii, T. Yamamoto, A. Kuramata, and H. Soda, "Novel InP based Mach–Zehnder modulator for 40 Gbit/s integrated lightwave source," in *IEEE 18th Inter. Semiconductor Laser Conference*, Garmisch-Partenkirchen, Germany, 2002, p. 5758 (TuCl).
21. S. Akiyama, H. Itoh, S. Sekiguchi, S. Hirose, T. Takeuchi, A. Kuramata, and T. Yamamoto, "InP-Based Mach–Zehnder modulator with capacitively loaded traveling-wave electrodes," *J. Lightwave Tech.*, vol. 26, no. 5, pp. 608–615, Mar. 2008.
22. D. Hoffmann, S. Staroske, and K. O. Velthaus, "45 GHz bandwidth traveling wave electrode Mach–Zehnder modulator with integrated spot size converter," in *16th Int. Conf. on Indium Phosphide Related Materials*, Kagoshima, Japan, May–Jun. 2004, pp. 585–588 (ThA1-5).
23. I. Betty, "Strongly-guided $InP/In_{1-x}Ga_xAs_yP_{1-y}$ Mach–Zehnder modulator for optical communications," Ph.D. Thesis, University of Waterloo, Waterloo, ON, 2005.
24. N. Kikuchi, Y. Shibata, K. Tsuzuki, T. Yasui, M. Ishikawa, H. Ishii, M. Arai, T. Sato, Y. Kawaguchi, and F. Kano, "Full C-nand 40 Gbit/s DPSK modulation using lossless InP n-p-i-n Mach–Zehnder modulator monolithically integrated with SOA," in *Optical Fiber Communications Conference*, 22–26 Mar. 2009, OWQ3.
25. P. C. Koh, M. Ayliffe, E. R. Hegblom, M. Larson, G. T. Paloczi, A. Dahl, X. Huang, Y. Zhou, and Y. A. Akulova, "Ultralow power dissipation widely-tunable transmitter optical subassemblies for 10 Gb/s pluggable transceivers," in *Optical Fiber Communications Conference*, 22–26 Mar. 2009, OWJ6.
26. P.-J. Rigole, "Recent progress in the integration of MGY-based tunable lasers and Mach–Zehnder modulators," in *Optical Fiber Communication and Optoelectronic Exposition Conference*, 2007, SC2. S7.1.
27. M. Wale, "Photonic integration challenges for next-generation networks," in *European Conference on Communications*, Vienna, Austria, 20–24 Sep. 2009, 1.7.4.
28. R. A. Griffin, B. Pugh, J. Fraser, I. B. Betty, K. Anderson, G. Busico, C. Edge, and T. Simmons, "Compact, high power, MQW InP Mach–Zehnder transmitters with full-band tunability for 10 Gb/s DWDM," in *European Conference on Communications*, 2005, Th.2.6.2.
29. M. G. Boudreau, M. Scheer, I. Betty, L. Langley, and R. Longone, "An integrated waveguide detector for power control in an InP Mach–Zehnder modulator based 10 Gb/s transmitter," in *Int. Conf. on Indium Phosphide Related Materials*, 2006, pp. 352–355.
30. I. Betty, M. G. Boudreau, R. Longone, R. A. Griffin, L. Langley, A. Maestri, A. Pujol, and B. Pugh, "Zero chirp 10 Gb/s MQW InP Mach–Zehnder transmitter," in *Optical Fiber Communications Conference*, 25–29 Mar. 2007, OWH6.
31. R. A. Griffin, N. Swenson, D. Crivelli, H. Carrer, M. Hueda, P. Voois, O. Agazzi, and F. Donadio, "Combination of InP MZM transmitter and monolithic CMOS 8-state MLSE receiver for dispersion tolerant 10 Gb/s transmission," in *Optical Fiber Communications Conference*, 2008, OThO2.
32. P. J. Winzer, S. Chandrasekhar, C. R. Doerr, D. T. Neilson, A. Adamiecki, and R. A. Griffin, "42.7-Gb/s modulation with a compact InP Mach–Zehnder transmitter," in *European Conference on Communications*, 24–28 Sep. 2006, doi: 10.1109/ECOC.2006.4801156.
33. P. C. Koh, L. A. Johansson, Y. A. Akulova, G. A. Fish, G. T. Paloczi, M. Larson, M. Ayliffe, and L. A. Coldren, "Generation of 40Gbps duobinary signals using an integrated laser Mach–Zehnder modulator," in *Optical Fiber Communications Conference*, 22–26 Mar. 2009, OThN4.
34. R. A. Griffin, A. Tipper, and I. Betty, "Performance of MQW InP Mach–Zehnder modulators for advanced modulation formats," in *Optical Fiber Communications Conference*, 2005, paper OTuL5.
35. Y. Shibata, N. Kikuchi, K. Tsuzuki, W. Kobayashi, and H. Yasaka, "40 Gbit/s DPSK modulation using an InGaAsAs-InAlAs MQW n-i-n Mach–Zehnder modulator," *Electron. Lett.*, vol. 44, no. 21, pp. 1269–1271, Oct. 2008.

36. E. Yamada, A. Ohki, N. Kikuchi, Y. Shibata,T. Yasui, K. Watanabe, H. Ishii, R. Iga, and H. Oohashi, "Full C-band 40-Gbit/s DPSK tunable transmitter module developed by hybrid integration of tunable laser and InP n-p-i-n Mach–Zehnder modulator," in *Optical Fiber Communications Conference*, 2010, OWU4.
37. R. A. Griffin, and A. C. Carter, "Advances in InP optical modulators," in *Optical Fiber Communication and Optoelectronic Exposition Conference*, 2008. Oct.–Nov. 2008, SuF6.
38. N. Kikuchi, Y. Shibata, K. Tsuzuki, H. Sanjoh, T. Sato, E. Yamada, T. Ishibashi, and H. Yasaka, "80 Gb/s low driving-voltage InP DQPSK modulator with an n-p-i-n structure," *IEEE Photonics Technol. Lett.*, vol. 21, no. 12, pp. 787–789, Jun. 2009.
39. S. Corzine, P. Evans, M. Kato, G. He, M. Fisher, M. Raburn, A. Dentai, I. Lyubomirsky, R. Nagarajan, M. Missey, V. Lal, A. Chen, J. Thomson, W. Williams, P. Chavarkar, S. Nguyen, D. Lambert, T. Butrie, M. Reffle, R. Schneider, M. Ziari,C. Joyner, S. Grubb, F. Kish, and D. Welch, "10-channel × 40 Gb/s per channel DQPSK monolithically integrated InP-based transmitter PIC," in *Optical Fiber Communications Conference*, 2008, PDP18.
40. M. Kato, P. Evans, S. Corzine, J. Gheorma, M. Fisher, M. Raburn, A. Dentai, R. Salvatore, I. Lyubomirsky, A. Nilsson, J. Rahn, R. Nagarajan, C. Tsai, B. Behnia, J. Stewart, D. Christini, M. Missey, A. Spannagel, D. Lambert, S. Agashe, P. Liu, D. Pavinski, M. Reffle, R. Schneider, M. Ziari, C. Joyner, F. Kish, and D. Welch, "Transmitter PIC for 10-channel × 40 Gb/s per channel polarization-multiplexed RZ-DQPSK modulation," in *Optical Fiber Communications Conference*, 2009, OThN2.
41. R. A. Griffin, R. G. Walker, B. J. Buck, R. Powell, L. N. Langley, J. Hall, and A. C. Carter, "40Gb/s RZ GaAs transmitter with integrated wavefrom monitoring," in *European Conference on Communications*, 2002, 10.5.4.
42. R. A. Griffin, R. G. Walker, and R. I. Johnstone, "Integrated devices for advanced modulation formats," *IEEE/LEOS Workshop*, Jul., 2004, pp. 39–40 (FC1).
43. H. Sun, K.-T. Wu, and K. Roberts, "Real-time measurements of a 40 Gb/s coherent system," *Optics Express*, vol. 16, no. 2, Jan. 2008.
44. K. Roberts, M. O'Sullivan, K.-T. Wu, H. Sun, A. Awadalla, D. J. Krause, and C. Laperle, "Performance of dual-polarization QPSK for optical transport systems," *J. Lightwave Technol.*, vol. 27, no. 16, pp. 3546–3558, Aug. 2009.
45. P. Martin, "Accurate refractive-index measurements of doped and undoped InP by a grating coupling technique," *Appl. Phys. Lett.*, vol. 67, no. 7, pp. 881–883, 1995.
46. B. R. Bennett, R. A. Soref, and J. Del Alamo, "Carrier-induced change in refractive-index of InP, GaAs, and InGaAsP," *IEEE J. Quantum Electron.*, vol. 26, no. 1, pp. 113–122, Jan. 1990.
47. M. S. Whalen, and J. Stone, "Index of refraction of N-type InP at 0.633 μm and 1.15 μm wavelengths as a function of carrier concentration," *J. Appl. Phys.*, vol. 53, pp. 4340–4343, 1982.
48. L. Chusseau, P. Martin, C. Brasseur, C. Albert, P. Herve, P. Arguel, F. Lopez-Dupuy, and E. V. K. Rao, "Carrier induced change due to doping in refractive index of InP: Measurements at 1.3 and 1.5 μm," *Appl. Phys. Lett.*, vol. 69, p. 3054, 1996.
49. H. C. Casey, Jr. and P. L. Carter, "Variation of intervalence band absorption with hole concentration in p-type InP," *Appl. Phys. Lett.*, vol. 44, no. 1, pp. 82–83, 1994.
50. F. Fiedler and A. Schlachetzki, "Optical parameters of InP-based waveguides," *Solid-State Electron.*, vol. 30, no. 1, pp. 73–83, 1987.
51. G. M. Alman, L. A. Molter, H. Shen, and M. Dutta, "Refractive index approximations from linear perturbation theory for planar MQW waveguides," *IEEE J. Quantum Electron.*, vol. 28, no. 3, pp. 650–657, Mar. 1992.
52. S. Ohke, T. Umeda, and Y. Cho, "Optical waveguides using GaAs-$Al_xGa_{1-x}$ As multiple quantum well," *Optical Communications*, vol. 56, pp. 235–239, 1985.
53. W. Bardyszewski, C. Rolland, S. Bradshaw, and D. Yevick, "Resonant-level effects in absorption spectra of shallow quantum wells," in *Conference on Lasers and Electro-Optics*, 1998, CThX6.

54. I. Betty, M. G. Boudreau, R. A. Griffin, and A. Fekces, "An empirical model for high yield manufacturing of 10Gb/s negative chirp InP Mach–Zehnder modulators," in *Optical Fibre Communications Conference*, 2005, OWE5.
55. I. Betty, M. G. Boudreau, R. A. Griffin, and A. Fekces, "InP-based Mach–Zehnder modulators enable high-performance, compact, and full-band tunable transmitters for 10 Gb/s DWDM applications," in *Photonics North*, Quebec City, QC, 2007, 71-uNKS-254.
56. G. L. Li, and P. K. L. Yu, "Optical intensity modulator for digital and analog applications," *J. Lightwave Technol.*, vol. 21, no. 9, pp. 2010–2030, Sep. 2003.
57. N. Yoshimoto, T. Yamanaka, S. Kondo, Y. Noguchi, and K. Wakita, "Large field-induced refractive-index change on TM-polarized light in an InGaAlAsInAlAs MQW waveguide structure," *IEEE Photonics Technol. Lett.*, vol. 9, no. 2, pp. 200–202, Feb. 1997.
58. S. Nojima, and K. Wakita, "Optimization of quantum well materials and structures for excitonic electroabsorption effects," *Appl. Phys. Lett.*, vol. 53, no. 20, pp. 1958–1960, Nov. 1988.
59. M. R. Gokhale, P. V. Studenkov, J. Ueng-McHale, J. Thomson, J. Yao, and J. van Saders, "Uncooled 10 Gb/s 1310 nm electroabsorption modulated laser," in *Optical Fiber Communications Conference*, 2003, PD42-1.
60. F. Pockels, *Lehrbuch der Kristalloptik*, Leipzig: Teubner, 1906.
61. P. Maat, "InP-based integrated MZI switches for optical communication," Ph.D. Thesis, Department of Applied Physics, Delft University of Technology, Delft, The Netherlands, 2001.
62. P. Bravetti, G. Ghislotti, and S. Balsamo, "Chirp-inducing mechanisms in Mach–Zehnder modulators and their effect on 10 Gb/s NRZ transmission studied using tunable-chirp single drive devices," *J. Lightwave Technol.*, vol. 22, no. 2, pp. 605–611, Feb. 2004.
63. S. Balsamo, P. Bravetti, R. Brouard, and V. Rouffiange, "Effects of chirp mechanism on spectrum broadening of external modulators," in *Optical Fiber Communications Conference*, 2005, OME43.
64. S. Akiyama, H. Itoh, T. Takeuchi, A. Kuramata, and T. Yamamoto, "Wide-wavelength-band (30 nm) 10 Gb/s operation of InP-based Mach–Zehnder modulator with constant driving voltage of 2 Vpp," *Photonic Technol. Lett.*, vol. 17 no. 7 pp. 1408–1410, Jul. 2005.
65. N. Kikuchi, K. Tsuzuki, Y. Shibata, M. Ishikawa, T. Yasui, H. Ishii, H. Oohashi, T. Ishibashi, T. Akeyoshi, H. Yasaka, and F. Kano, "Full L-band 40 Gbit/s operation of compact InP-DQPSK modulator module with low constant driving voltage of 3.5 $V_{pp}$," in *European Conference on Communications*, 21–25 Sep. 2008, Mo.3.C4.
66. C. F. Clarke, R. A. Griffin, and T. C. Goodall, "Highly integrated DQPSK modules for 40 Gb/s transmission," in *Optical Fiber Communications Conference*, 22–26 Mar. 2009, NWD3.
67. D. McGhan, M. O'Sullivan, M. Sotoodeh, A. Savchenko, C. Bontu, M. Belanger, and K. Roberts, "Electronic dispersion compensation," in *Optical Fiber Communication Conference*, 2006, OWK1.
68. M. Wang, Y. Xu, and S. Gardner, "Analysis of Key Methods of MZ Setup in TTA," *Communications and Photonics Conference and Exhibition 2009 Asia*, 2009, ThW4.
69. Y. C. Zhu, F. H. Groen, D. H. P. Maat, Y. S. Oei, J. Romijn, and I. Moerman, "A compact PHASAR with low central channel loss," in *Proc. European Conference on Integrated Optics*, 1999, pp. 219–222.
70. Y. S. Oei, L. H. Spiekman, F. H. Groen, I. Moerman, E. G. Metaal, and J. W. Pedersen, "Novel RIE-process for high quality InP-based waveguide structures," in *Proc. 7th European Conference on Integrated Optics*, Delft, The Netherlands, 3–6 Apr. 1995, pp. 205–208.
71. U. Niggebrugge, M. Klug, and G. Garus, "A novel process for reactive ion etching on InP, using $CH_4$/$H_2$," *Inst. Phys. Conf. Ser.* 79, pp. 367, 1985.
72. D. L. Melville, S. M. Ohja, R. Moore, and F. R. Shepherd, "Application of $CH_4$/Ar and $CH_4/CO_2/H_2$ plasma chemistries in hydrocarbon RIE of InP-based optoelectronic devices," in *43rd AVS National Symposium*, Nov. 1996.
73. S. Vicknesh, and A. Ramam, "Etching characteristics of HBr-based chemistry on InP using the ICP technique," *J. Electrochem. Soc.*, vol. 151 no. 12, pp. C772–C780, 2004.

74. N. Kim, F. R. Shepherd, M. Boudreau, and I. Betty, "ICP etching of optical waveguides in InP-based materials using HBr and Cl2 chemistry," in *Eleventh Canadian Semiconductor Technology Conference*, Ottawa, 18–22 Aug., 2003, WP.15.
75. J. Lu, X. Meng, A. J. SpringThorpe, F. R. Shepherd, and M. Pokier, "Inductively coupled plasma etching of GaAs low loss waveguides for a traveling waveguide polarization converter, using chlorine chemistry," *J. Vac. Sci. Technol. A*, vol. 22, no. 3, pp. 1058–1061, May/Jun. 2004.
76. S. Oku, Y. Shibata, and K. Ochiai, "Controlled beam dry etching of InP using $Br_2$-$N_2$ gas," *J. Electron. Mater.*, vol. 25, pp. 585–591, 1995.
77. J. E. Haysom, R. Glew, C. Blaauw, R. Driad, D. MacQuistan, C. A. Hampel, J. E. Greenspan, and T. Bryskiewicz, "Improved P-doping profiles in lasers and modulators," in *Proc. 14th Int. Conf. on Indium Phosphide Related Materials*, 2002, pp. 627–630.
78. W. C. Dautremont-Smith, J. Lopata, and S. J. Pearton, "Hydrogen passivation of acceptors in P-InP," *J. Appl. Phys.*, vol. 66, no. 5, pp. 1993–1996, Sep. 1989.
79. R. Rousina-Webb, I. Betty, D. Sieniawski, F.R. Shepherd, and J.B. Webb, "The effect of process-induced stress in InP/InGaAsP weakly confined waveguides switch," *Optoelectronic Interconnects VII, Photonics Packaging and Integration II, SPEE*, vol. 3952, no. 1, pp. 168–177, 2000.
80. O. Bryngdahl, "Image formation using self-imaging techniques," *J. Optic. Soc. Amer.*, vol. 63, pp. 416–419, Apr. 1973.
81. R. Ulrich, "Image formation by phase coincidences in optical waveguides," *Optics Communications*, vol. 13, no. 3, pp. 259–264, 1975.
82. L. B. Soldano, and E. C. M. Pennings, "Optical multimode interference devices based on self-imaging: Principles and applications," *J. Lightwave Technol.*, vol. 13, pp. 615–27, Apr. 1995.
83. P. A. Besse, M. Bachmann, H. Melchior, L. B. Soldano, and M. K. Smit, "Optical bandwidth and fabrication tolerances of multimode interference couplers," *J. Lightwave Technol.*, vol. 12, no. 6, pp. 1004–1009, Jun. 1994.
84. L. Leick, J. H. Povlsen, and R. J. S. Pedersen, "Numerical and experimental investigation of 2 × 2 multimode interference couplers in silica-on-silicon," *Optical and Quantum Electron.*, vol. 33, pp. 387–398, 2001.
85. M. T. Hill, X. J. M. Leijtens, G. D. Khoe, and M. K. Smit, "Optimizing imbalance and loss in 2 × 2 3dB multimode interference couplers via access waveguide width," *J. Lightwave Technol.*, vol. 21, no. 10, pp. 2305–2313, Oct. 2003.
86. E. C. M. Pennings, R. van Roijen, M. J. N. van Stralen, P. J. de Waard, R. G. M. P. Koumans, and B. H. Verbeek, "Reflection properties of multimode interference devices," *IEEE Photonic Technol. Lett.*, vol. 6, no. 6, pp. 715–718, Jun. 1994.
87. I. Betty, "Experimental and theoretical determination of backreflections in a III-V semiconductor device," M.Sc. Thesis, Computer and Electrical Engineering, Queen's University, Kingston, ON, 1998.
88. P. A. Besse, E. Gini, M. Bachmann, and H. Melchoir, "New 1 × 2 multimode interference couplers with free selection of power splitting ratios," in *Proc. 20th European Conference on Communications*, Florence, 25–29 Sep. 1994, pp. 669–672.
89. M. Heiblum, "Analysis of curved optical waveguides by conformal transformation," *IEEE J. Quantum Electron.*, vol. QE-11, no. 2, pp. 75–83, 1975.
90. E.-G. Neumann, "Curved dielectric optical waveguides with reduced transition losses", *Microwaves, Optics and Antennas, IEE Proc. Pt. H*, vol. 129, no. 5, pp. 278–280, Oct. 1982.
91. D. Marcuse, "Length optimization of an S-shaped transition between offset optical waveguides," *Appl. Optics*, vol. 17, no. 5, pp. 763–768, 1978.
92. V. Ramaswamy, "Loss–loss bends for integrated optics," in *Proc. Conference on Lasers and Electro-Optics*, Washington, DC, 1981, THP1.
93. W. J. Minford, S. K. Korotky, and R. C. Alferness, "Low-loss Ti:$LiNbO_3$ waveguide bends at $\lambda$ = 1.3 μm," *IEEE J. Quantum Electron.*, vol. QE-18, no. 10, pp. 1802–1806, 1982.

94. A. Kumar, and S. Aditya, "Performance of S-Bends for integrated-optic waveguides," *Microwave and Optical Technol. Lett.*, vol. 19, no. 4, pp. 289–292, Nov. 1998.
95. K. T. Toai, and P.-L. Liu, "Modeling of Ti:LiNbO$_3$ waveguide devices: Part II-Shaped channel waveguide bends," *J. Lightwave Technol.*, vol. 7, no. 7, pp. 1016–1022, Jul. 1989.
96. K. Prosyk, R. Moore, I. Betty, R. Foster, J. Greenspan, P. Singh, S. O'Keefe, J. Oosterom, and P. Langlois, "Low loss, low chirp, low voltage, polarization independent 40 Gb/s bulk electro-absorption modulator module," in *Optical Fiber Communications Conference*, 2003, TuP3.
97. T. Fujisawa, M. Arai, N. Fujiwara, W. Kobayashi, T. Tadokoro, K. Tsuzuki, Y. Akage, R. Iga, T. Yamanaka, and F. Kano, "25 Gbit/s 1.3 μm InGaAlAs-based electroabsorption modulator integrated with DFB laser for metro-area (40 km) 100Gbit/s Ethernet system," *Electron. Lett.*, vol. 45, no. 17, pp. 900–902, Aug. 2009.
98. C. Rolland, G. Mak, W. Bardyszewski, and D. Yevick, "Improved extinction ratio of waveguide electro-absorption optical modulators induced by an InGaAs absorbing layer," *J. Lightwave Technol.*, vol. 10, no. 12, pp. 1907–1911, Dec. 1992.
99. G. Mak, C. Rolland, K. E. Fox, and C. Blaauw, "High speed bulk InGaAsP-InP electro-absorption modulators with bandwidth in excess of 20 GHz," *IEEE Photonics Technol. Lett.*, vol. 2, no. 10, pp. 730–733, Oct. 1990.
100. R. G. Walker, "High-speed III-V semiconductor intensity modulators," *IEEE J. Quantum Electron.*, vol. 27, no. 3, pp. 654–667, Mar. 1991.
101. K. Tsuzuki, T. Ishibashi, T. Ito, S. Oku, Y. Shibata, R. Iga, Y. Kondo, and Y. Tohmori, "A 40 Gbit/s InGaAlAs-InAlAs MQW n-i-n Mach–Zehnder modulator with a drive voltage of 2.3 V," *IEEE Photonics Technol. Lett.*, vol. 17, no. 1, pp. 46–48, Jan. 2005.
102. S.-W. Seo, J. Yan, J. H. Baek, F. M. Soares, R. Broeke, A. V. Pham, and S. J. Ben. Yoo, "Microwave velocity and impedance tuning of traveling-wave modulator using ion implantation for monolithic integrated photonic systems," *Microwave and Optical Technol. Lett.*, vol. 50, no. 8, pp. 2151–2155, Aug. 2008.
103. J. S. Barton, M. L. Masonovic, A. Tauke-Pedretti, E. J. Skogen, and L. A. Coldren, "Monolithically-integrated 40 Gbit/s widely-tunable transmitter using series push–pull Mach–Zehnder modulator SOA and sampled-grating DBR laser," in *Optical Fiber Communications*, 2005, OTuM3.
104. S. Akiyama, H. Itoh, T. Takeuchi, A. Kuramata, and T. Yamamoto, "Low-chirp 10 Gbit/s InP-based Mach–Zehnder modulator driven by 1.2 V single electrical signal," *Electron. Lett.*, vol. 41, no. 1, pp. 40–41, Jan. 2005.
105. G. L. Li, T. G. B. Mason, and P. K. L. Yu, "Analysis of segmented traveling-wave optical modulators," *J. Lightwave Technol.*, vol. 22, no. 7, pp. 1789–1796, Jul. 2004.
106. J. E. Haysom, I. Betty, K. Wong, M. Poirier, R. S. Moore, A. Ait-Ouali, and J. Lu, "Implant isolation applied to an InP Mach–Zehnder modulator," in *Eleventh Canadian Semiconductor Technology Conference*, Ottawa, Canada, 18–22 Aug. 2003.
107. St. J. Dixon-Warren, J. E. Haysom, I. Betty, J. Lu, and K. Hewitt, "Implant isolation in an indium phosphide optoelectronic device: A scanning spreading resistance microscopy study," *J. Vac. Sci. Technol. A*, vol. 22, no. 3, pp. 925–929, May/Jun. 2004.
108. J. Leuthold, and C. H. Joyner, "Multimode interference couplers with tunable power splitting ratios," *J. Lightwave Technol.*, vol. 19, no. 5, pp. 700–707, May 2001.
109. D. A. May-Arrioja, P. LiKamWa, C. Velasquez-Ordonez, and J. J. Sanchez-Mondragon, "Tunable multimode interference coupler," *Electron. Lett.*, vol. 43, no. 13, pp. 714–716, Jun. 2007.
110. M. Allard, R. A. Masut, and M. Boudreau, "Temperature determination in optoelectronic waveguide modulators," *J. Lightwave Technol.*, vol. 18, no. 6, pp. 813–818, 2000.
111. D. A. May-Arrioja, P. Likamwa, R. J. Selvas-Aguilar, and J. J. Sánchez-Mondragón, "Ultra-compact multimode interference InGaAsP multiple quantum well modulator," *Optical and Quantum Electron.*, vol. 36, pp. 1275–1281, 2004.

112. X. Jiang, X. Li, H. Zhou; J. Yang; M. Wang; Y. Wu, and S. Ishikawa, "Compact variable optical attenuator based on multimode interference coupler," *IEEE Photonics Technol. Lett.*, vol. 17, no. 11, pp. 2361–2363, Nov. 2005.
113. T. Yasui, Y. Shibata, K. Tsuzuki, N. Kikuchi, M. Ishikawa, Y. Kawaguchi, M. Arai, and H. Yasaka, "10-Gb/s 100-km SMF transmission using InP Mach–Zehnder modulator monolithically integrated with semiconductor optical amplifier," *IEEE Photonics Technol. Lett.*, vol. 20, no. 13, pp. 1178–1180, Jul. 2008.
114. H. El-Rafaei, "An InGaAsP/InP on-chip polarization bit interleaver for 40 Gb/s optical transmission systems," Ph.D. Thesis, Electrical and Computer Engineering, Queen's University, Kingston, ON, 2002.
115. M. M. Ragheb, H. H. El-Refaei, D. Khalil, and O. A. Omaret, "Design of compact integrated InGaAsP/ InP polarization controller over the C-band;" *J. Lightwave Technol.*, vol. 25, no. 9, pp. 2531–2538, Sep. 2007.
116. J. J. G. M. van der Tol, L. M. Augustin, A. A. M. Kok, U. Khalique, and M. K. Smit, "Use of polarization in InP-based integrated optics," in *Conference on Lasers and Electro-Optics*, 2008, CThM3.
117. J. S. Barton, E. J. Skogen, M. L. Masanovic, S. P. Denbaars, and L. A. Coldren, "A widely tunable high-speed transmitter using an integrated SGDBR laser-semiconductor optical amplifier and Mach-Zehnder modulator," *IEEE J. Sel. Topics. Quantum Electron.*, vol. 9, no. 5, pp. 1113–1117, Sep./Oct. 2003.
118. R. Clayton, A. Carter, I. Betty, and T. Simmons, "Cost-effective monolithic and hybrid integration for metro and long-haul applications," *Semiconductor Optoelectronic Devices for Lightwave Communication, SPIE*, vol. 5248, no. 1, pp. 67–79, 2003.
119. H. Chou, and W. V. Sorin, "High resolution and high sensitivity optical reflection measurements using white light interferometry," *Hewlett-Packard Journal*, vol. 44, no. 1, pp. 39–48, 1993.
120. L. A. Coldren, K. Furuya, and B. I. Miller, "On the formation of planar-etched facets in GaInAsP InP double heterostructures," *J. Electrochem. Soc.*, vol. 130, no. 9, pp. 1918–1926, 1983.
121. R. D. Dupuis, D. G. Deppe, C. J. Pinzone, N. D. Gerrard, S. Singh, G. J. Zydzik, J. P. van der Ziel, and C. A. Green, "InGaAs-InP heterostructures for vertical cavity surface emitting lasers at 1.65 μm wavelength," *J. Cryst. Growth*, vol. 107, no. 1–4, pp. 790–795, Jan. 1991.
122. D. J. Robbins, "Advances in monolithic integration of InP based optoelectronics," in *Conference on Lasers and Electro-Optics*, 2007, CTuH5.
123. N. D. Whitbread, A. J. Ward, B. de Largy, M. Q. Kearley, B. Asplin, P. J. Williams, and M. J. Wale, "AlGaInAs-InP C-band tunable DS-DBR laser for semi-cooled operation at 55°CT," in *European Conference on Communications*, 2008, We.3.C4.
124. K. Shinoda, S. Makino, T. Kitatani, T. Shiota, T. Fukamachi, and M. Aoki, "InGaAlAs-InGaAsP heteromaterial monolithic integration for advanced long-wavelength optoelectrics devices," *J. Quantum Electron.*, vol. 45, no. 9, pp. 1201–1209, Sep. 2009.
125. A. P. Janssen, "The transition from discrete optics to optical integration," in *Optical Fiber Communication and Optoelectronics Conference*, 17–19 Oct. 2007, JSYMP2.3.
126. K. Tsuzuki, H. Kikuchi, E. Yamada, H. Yasaka, and T. Ishibashi, "1.3-$V_{pp}$ push–pull drive InP Mach–Zehnder modulator module for 40 Gbit/s operation," in *Optical Fiber Communications Conference*, 2005, Th2.6.3.
127. K. Tsuzuki, K. Sano, N. Kikuchi, N. Kashio, E. Yamada, Y. Shibata, T. Ishibashi, M. Tokumitsu, and H. Yasaka, "0.3 $V_{pp}$ single-drive push–pull InP Mach–Zehnder modulator module for 43-Gbit/s systems," in *Optical Fiber Communications Conference*, 2006, OWC2.
128. K. Tsuzuki, N. Kikuchi, Y. Shibata, W. Kobayashi, and H. Yasaka, "Surface mountable 10-Gb/s InP Mach–Zehnder modulator module for SFF transponder," *IEEE Photonics Technol. Lett.*, vol. 20, no. 1, pp. 54–56, Jan. 2008.

# III

# Emerging Technologies

# 8

# Gallium Arsenide Modulator Technology




Robert G. Walker
*u2t Photonics UK Ltd*

John Heaton
*u2t Photonics UK Ltd*


## 8.1 Introduction

Photonic integrated circuits may be implemented in a variety of material systems. The high-speed optical modulator is a primary component, thus the material must have suitable electro-optical properties, limiting the practical choice to III-V semiconductors, ferroelectric ceramics (e.g., lithium niobate) or electro-optic polymers. Historically, gallium arsenide (GaAs) was the material of choice for mm-wave devices owing to high electron mobility and the availability of low-loss semiinsulating substrate. Foundry processes developed originally for the fabrication of GaAs monolithic microwave integrated circuits are ideally suited for extension into the optical domain and it is no accident that most of the early development of GaAs optical modulators was undertaken by institutions with established monolithic microwave integrated circuits expertise.

GaAs is environmentally stable, radiation-hard, and lacking undesired pyroelectric effects. Originally developed for aerospace and electronic warfare applications, these useful properties make GaAs high-speed modulator technology desirable for all system applications. GaAs-based devices are linear and well-suited for zero-chirp applications. Essentially electric field-operated, these are low-current devices which run cold and are stable, robust and long-lived. Compared with InP, GaAs is a low-cost technology, both for the cost of the raw material, and the process-route.

This chapter will focus on the differentiating electro-optic and high-speed aspects of design. Although essential and interesting, such ancillary structures as bends, Y-junctions and MMIs are common to other modulator types and will not be discussed here.

## 8.2 Electro-Optical Properties of GaAs

For the purposes of optical design, the primary characteristic of any semiconductor material is the fundamental band-edge wavelength ($\lambda_g$), whose photon energy corresponds to the valence-conduction band-gap and sharply demarks a longwave transparent-band from the shortwave opaque bands. The band-edge of GaAs, at 880 nm, is relatively remote from the standard telecoms wavelengths around 1550 nm and this determines its primary characteristics as an electro-optic material for these wavelengths: devices are optically broadband and relatively insensitive to wavelength. The relatively weak electro-optic (EO) effect is primarily linear; thus intensity-modulators must be interferometric and rather long if the drive voltage is to be low, hence the importance of *Traveling-Wave* designs for high-frequency.

The relative permittivity ($\varepsilon_r$) and refractive-index ($n = \sqrt{\varepsilon_r}$) of GaAs is high compared with most optical dielectrics, but is stable, varying little from zero to optical frequencies. For GaAs $n = 3.3769$ at 1550 nm and $\varepsilon_r = 12.9$ while for $LiNbO_3$ $n \cong 2.2$ and $\varepsilon_r \cong 36$.

### 8.2.1 Electro-Optic Effects

III-V compound-semiconductors are optically isotropic but, unlike elemental semiconductors, have a noncentrosymmetric crystal structure. This is polarizable, giving rise to a linear electrooptic (LEO) property (having the nature of an electric-field induced birefringence) and also a piezo-electric effect. The LEO is governed by tensor mathematics; thus, the sign and magnitude of the refractive index change depend on the relative orientation of the crystal, electric-field ($E$) and optical polarization [1]. The standard [100] cut material with $E$ aligned to the surface-vector yields a maximum LEO effect for polarization aligned to [011]—highly convenient as the [011] family of cleavage planes are used to create optical facets. GaAs modulators are always designed for TE polarization as there is no LEO effect for TM polarization using a surface-normal electric-field.

The refractive index change due to electric-field $E$ acting on electro-optic coefficient $r$ is: $\delta n = n^3 r\, E/2$. A high value for $n^3$, together with a low permittivity, does much to compensate GaAs for its weak electro-optic coefficient. An early survey of modulator technologies [2] showed that, in fact, GaAs is highly competitive with lithium niobate for high-speed modulation.

Quadratic and Free Carrier Effects. The field-induced broadening and red-shift of the fundamental absorption edge, known as the Franz–Keldysh effect, is used directly for electroabsorption of near band-edge wavelengths. The associated refractive index change, which varies as the square of the applied field, is termed the quadratic electro-optic effect. It is much greater than the LEO near $\lambda_g$ but rapidly diminishes with wavelength; in GaAs at 1550 nm the quadratic electro-optic effect nevertheless contributes a few percent. The QEO is not sensitive to polarization or crystal orientation and always yields a positive $\delta n$.

The presence of free carriers depresses the refractive-index, both by direct absorption, impacting both real and imaginary parts of the complex refractive index (the plasma effect) and by bandfilling [3] which causes an increase in the effective energy-gap when the conduction and/or valence bands are partially filled. Sweeping-out carriers by increasing the reverse-bias of a diode structure yields a positive $\delta n$, is a fast majority-carrier process and can contribute substantially to the total index change if the waveguide doping-profile is designed to make use of this effect. Note however, that "holes" in p-type material cause much more optical loss, but much less index-change compared with electrons in n-type materials.

All these effects combine constructively with suitable choice of waveguide orientation (in GaAs, perpendicular to a [011] facet) and produce a mesa profiled chemically-etched ridge. For convenience, the overall EO efficiency of the waveguide is summarized in the EO slope (= 360 · $\delta n/(\lambda \cdot \Delta V)$) in units of °.V$^{-1}$.mm$^{-1}$.

## 8.3 Waveguide Design

The key to efficient GaAs waveguide design is the GaAs/AlGaAs heterostructure. $Al_xGa_{1-x}As$ is closely lattice-matched to GaAs for all aluminum fractions. Increasing the aluminum-fraction (x) increases the band-gap and reduces the refractive-index—for example, to 3.225 for $x = 0.3$ [4].

The basic GaAs/AlGaAs multilayer hetero-structure confines light vertically by virtue of the refractive-index contrast between the two materials. GaAs has the higher index and so forms the waveguide core layer. The AlGaAs cladding layers must be sufficiently thick to eliminate optical leakage into the GaAs substrate below and absorption by any metallic electrode above.

Lateral confinement is imposed on the as-grown multilayer "slab" waveguide by etching the surface to create a rib or ridge waveguide. Modulation electrodes are furnished by in-grown doped, conductive sublayers below the light, and by ridge-top metallization above (see Figure 8.1). While the top electrode could be implemented as an ohmic-contact to an in-grown (or indiffused) $p^+$ capping layer, it is more usual to use a Schottky contact direct to the undoped surface. Excellent Schottky contacts are easily made to a GaAs surface, and their use simplifies the process considerably.

### 8.3.1 1D (Slab) Waveguide Optimization

Initial design of the optical waveguide involves adjusting the AlGaAs compositions and core thickness to achieve the desired slab-mode size and shape without incurring more than one higher-order mode; a single higher mode of odd parity is unlikely to cause any problems, but additional modes are undesirable. For given cladding compositions, a suitable core-layer thickness will generally be found at around the mode-1 cut-off value.

For highest EO efficiency, most of the confined optical profile should fall largely within the depletion-zone of the reverse-biased diode structure. Typically, the epitaxial material will be nominally undoped apart from n-type conductive-layers buried within the lower cladding to furnish the back-contact. An optimized waveguide design will be a compromise between high EO-overlap and high E-field intensity, while the capacitance should also be as low as possible. This generally pulls the design in the direction of high NA (small mode-size) and narrow waveguides while considerations of optical interfacing will pull in the other direction.

*Depletion Zone.* The depletion field may be calculated in one dimension from basic diode-law equations [5], in which the gradient of $E$ is proportional to the majority carrier concentration, and $E$ falls away from the rectifying junction to zero at the contacts. Knowing the doping-profile and applied voltage, the electric-field profile $E(y)$ can readily be derived and may be numerically overlapped with the calculated

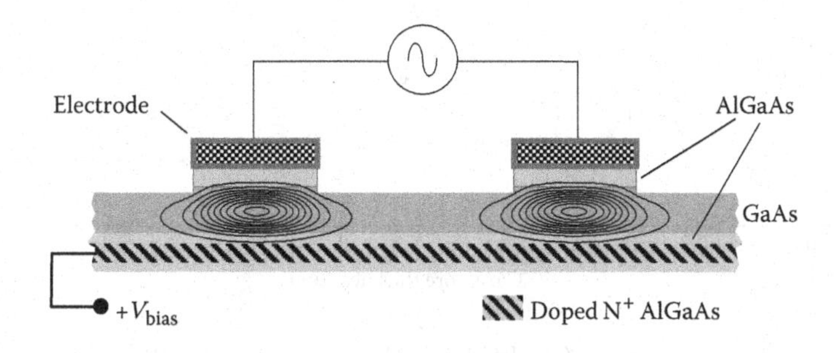


**FIGURE 8.1** Schematic GaAs/AlGaAs Mach–Zehnder cross-section in series push–pull configuration. Optical intensity contours are illustrated.

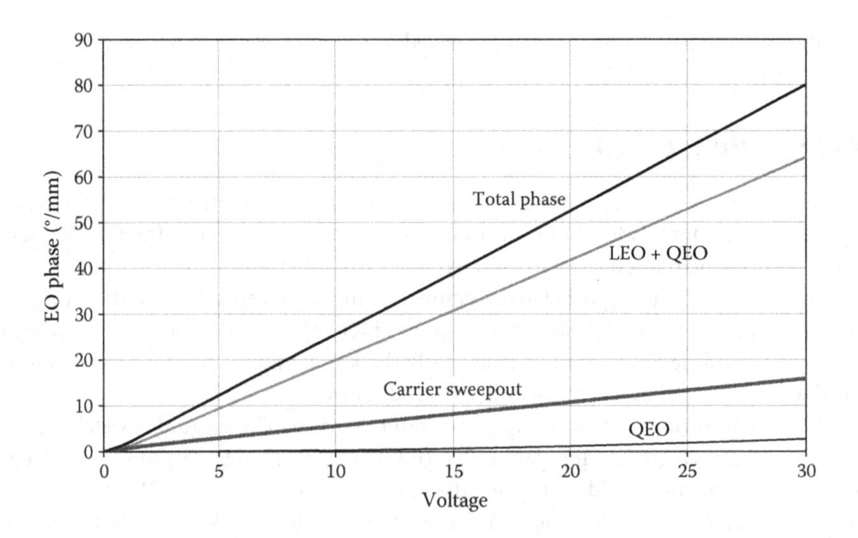


**FIGURE 8.2** Calculated electro-optical effects in a $GaAs/Al_{0.2}Ga_{0.8}As$ waveguide: core = 1.3 μm; depletion = 1.8 μm.

slab-waveguide mode. Quadratic EO and plasma sweep-out contributions to the index-change are also readily derived using, respectively, $E^2(y)$ and the change in depletion-depth between two voltage settings (see Figure 8.2). A *complex* mode-calculation yields optical losses due to doped and metallic layers while the capacitance may be estimated from a parallel-plate formula. With this information, a 1D first-pass optimization for high speed, low-loss, and low drive-voltage may be speedily performed. Figure 8.3 shows the result of such an optimization.

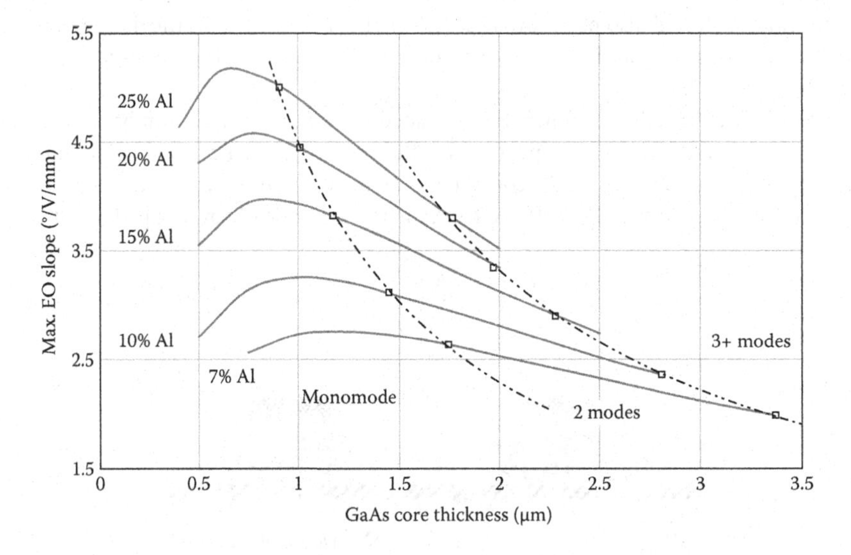


**FIGURE 8.3** Electro-optic design curves (1D slab) by cladding composition (%Al) and indicating modal transitions. Implicit variables are the upper cladding thickness (adjusted for constant low metal loss) and depletion-depth (adjusted for max. EO-slope).

### 8.3.2 $V_\pi$ and Figures of Merit

$V_\pi$ is defined as the applied voltage which achieves $\pi$ phase-shift in a given waveguide and is related inversely to the EO Slope. Consideration of the EO effects yields only point-values for the index-change, so the overall effect on the guided-mode strictly requires a full mode recalculation based on the perturbed index-profile. Fortunately, because this is only a small perturbation, an overlap-integral of optical intensity with $\delta n$ can be shown to be sufficiently accurate [2]. In one dimension, with the average electric-field $E = V/d$, $V_\pi$ can be written as

$$V_\pi = \frac{d.\lambda}{n^3 r.\xi.L},$$

where $L$ is the length, $\xi$ is the overlap-integral, and $d$ is depletion-depth.

$V_\pi$ alone is insufficient to qualify a high-speed modulator; we need also to consider speed-capability. Since the modulation-bandwidth of the electrode is inversely related to its capacitance ($C$), a useful figure of merit may be defined as the inverse product $(V_\pi C)^{-1}$ or, more generally, the ratio of EO slope to capacitance. In the parallel-plate approximation, $C = \varepsilon.L.W/d$, where $W$ is the width, the resultant figure of merit is independent of $L$ and $d$. 2D corrections to the 1D formulae [2] reintroduce a weak dependence on $d$ and modify the simple width sensitivity; nevertheless the above provides a useful guideline. As Figure 8.4 shows, the best overall figure of merit does not generally coincide with the highest EO slope.

### 8.3.3 Lateral Confinement—Rib and Ridge Waveguides

The light-wave is confined laterally by the depressed effective refractive-index in the outer etched regions. The end-points of the etch-depth continuum (see Figure 8.5) have quite different properties:

*Shallow Etch.* The outer slab waveguide is largely intact but has reduced modal effective-index due to the thinned upper cladding. The outer effective-index is just that of the residual slab-waveguide. Relatively weak confinement yields a waveguide which is monomode at undemanding widths and with a mode-profile extending laterally beyond the rib. Radiation (e.g., from a bend) is trapped by the outer

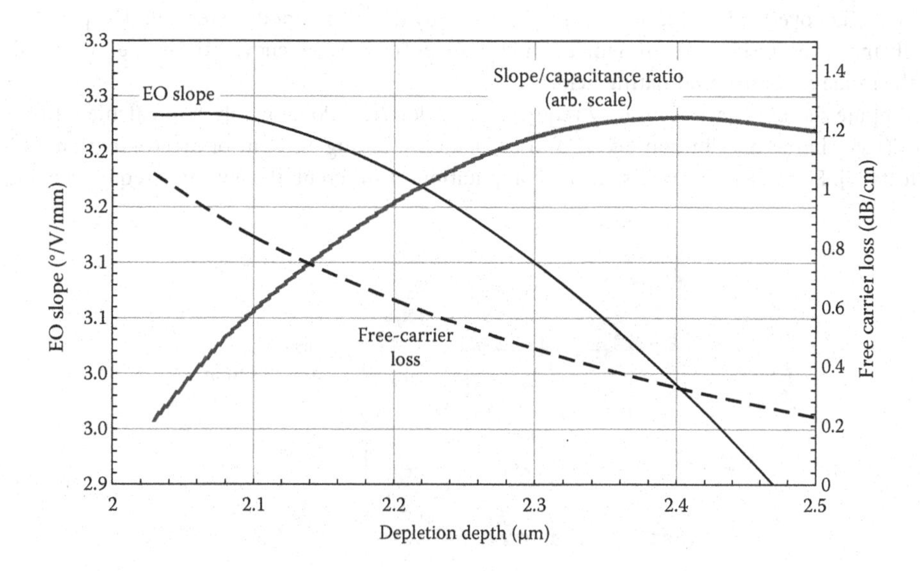


**FIGURE 8.4** Electro-optic slope, figure-of-merit and free-carrier loss, calculated for a fixed 1D slab-waveguide, varying the specified depth of the doped layers (i.e., the depletion-depth).

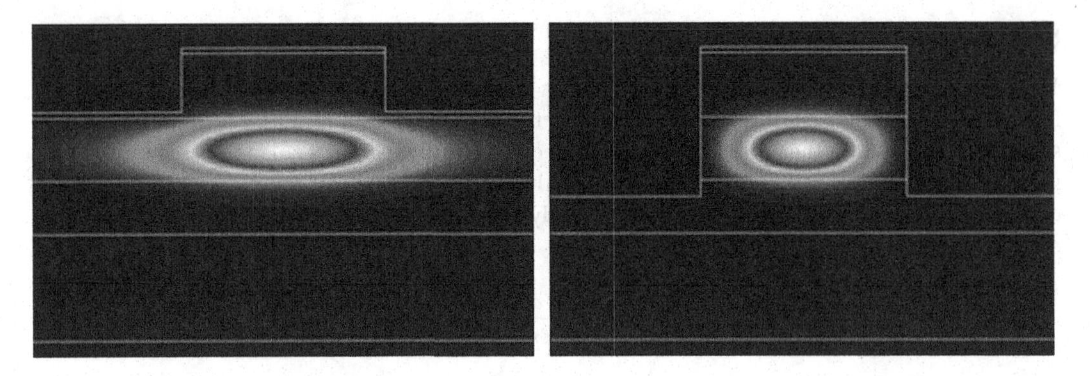

**FIGURE 8.5** Computed 2D fundamental-mode intensity cross-sections of dry-etched waveguides: shallow-etched, strip-loaded (left), deep-etched ridge (right).

slab-guides giving a 1D type of radiation behavior which is well represented by approximate (but rapid) 'Effective-Index' types of modeling [6].

*Fully Etched.* With the GaAs core fully etched away, the outer index is close to that of the cover material (usually air) making a very strongly confining waveguide. Apparently, heavily multimoded (unless very narrow), many of the modes are actually "leaky" because their effective-index is below that of the lower cladding. It has been reported that this geometry is susceptible to acousto-optic resonance arising from the piezoelectric and elasto-optic effects in GaAs [7].

An optimum for a given application may be sought between these extremes. With increasing etch-depth, the lateral mode-extension beyond the rib is rapidly attenuated (reducing $V_\pi$), bends can be made much smaller and unwanted directional-coupling between adjacent waveguides is diminished. However, the width constraint for monomode behavior becomes more demanding and scattering losses may increase.

### 8.3.4 2D Electro-Optic Waveguide Design

The effective EO overlap is reduced from the 1D value by the lateral mode extension. Deeper etch progressively increases the lateral confinement and improves the EO efficiency as the mode-tails are drawn under the influence of the modulating field.

To compute the EO effect accurately requires a 2D electrical Poisson-solver to calculate the vector E-field $E(x,y)$, charge-distribution $Q(x,y)$ and capacitance, taking account of heterojunction and surface effects [5]. Figure 8.6 illustrates such a computation. A differential solution is required using two

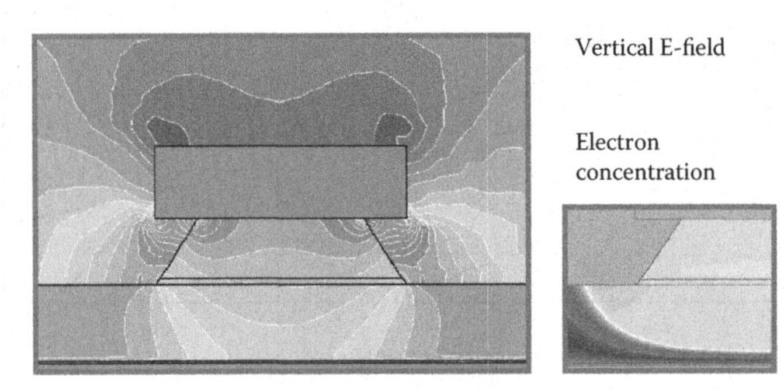


**FIGURE 8.6** Full 2D Poisson solutions of a wet-etched GaAs/AlGaAs electro-optic waveguide, showing contours of the vertical E-field component (left) and electron concentration (right).

voltage-values to avoid nonlinearity near the origin. Capacitance can be derived from $E$ by evaluating a surface-integral around the periphery of the electrode, in accordance with Gauss' Law.

The EO $\delta n$ is obtained from overlap-integrals of the $TE_{00}$ optical-intensity with: $E_y$ for the LEO (only the vertical-component of $E$ is effective); $|E^2|$ for the QEO; and $\Delta Q$ for plasma sweep-out. Figure 8.5 shows suitable $TE_{00}$ 2D mode-solutions. Comparison with 1D calculations show the effective lateral overlap to lie between 60% and 90% depending on the etch-depth.

## 8.4 Waveguide Fabrication

The multilayer source-material is generally grown by one of two epitaxial techniques: molecular beam epitaxy or metalorganic vapor phase epitaxy (also known as metalorganic chemical vapor deposition). Waveguide etching is either wet-chemical or dry plasma based, with the latter now most favored due to the vertical side-walls it makes possible.

With state-of-the-art projection aligners and micro-chip technology, it is possible to align and deposit narrow electrodes on top of the narrow ridges to the required precision. Such methods are versatile, allowing deep dry-etched waveguides and noncompromised (e.g., thick gold-plated) electrode systems to be used.

*Self-aligned* methods have also been used extensively [8]. Here, the electrode is itself used as the etch-mask to define the rib, with later removal of metal from passive waveguide sections. The choice of materials and etch-techniques is more limited: the metals must be robust against the etchant but must be removable subsequently without damage to the waveguide beneath. Aluminum has been used successfully with a wet chemical etch system; it is a good conductor, does not undercut excessively, and (unlike gold, for example) can be removed without damage to the GaAs using dilute hydrofluoric acid. Chemical etching yields a trapezoidal profile with undercut electrode [2,9]. This has benefits such as reduced capacitance and breakdown protection, but it also constrains the practical depth and width. Self-alignment is not usually suited to dry-etching due to sputtering and random redeposition of the metal.

### 8.4.1 Etch Stages

It is possible to design low-loss matched junctions between waveguides with different etch-depths [10], enabling various sections of a complex device to be optimally etched. A wider, deep-etched waveguide can, for example, be spot-matched to a narrower shallow-etched waveguide. Use of a durable etch-mask material (metal or dielectric) allows the etching to be repeated in multiple stages using temporary shielding of different regions. A finished device might combine regions with several different etch-depths depending on the desired local properties: (1) shallow etch, for unconditionally monomode operation (mode filtering); (2) intermediate etch for better lateral confinement and lower capacitance for efficient electro-optic waveguides; (3) deep etch to remove the GaAs core entirely, providing multimode interference (MMI) and spatial-filter functions; and (4) full epitaxial layer removal for electrical isolation.

### 8.4.2 Isolation and Air Bridge Techniques

Since the doped back-connection layers are present over the whole surface, large metallized areas for contacting or other purposes will, by default, contribute a similar capacitance per unit area as the electrodes. High-speed devices are grown on SI substrate to enable isolation of adjacent devices either by removal of the epitaxial material between by deep-etching [2,9] or by quelling its conductivity through ion-implantation [11]. By bombarding the surface with high-energy ion species such as $H^+$ (proton implantation), free carriers are compensated by either irradiation-induced damage or chemically related deep levels. Like deep-etching, ion-implantation has a secondary role of optical spatial-filtering as the implanted material is highly absorptive of light and so may be used to dispose of stray light trapped

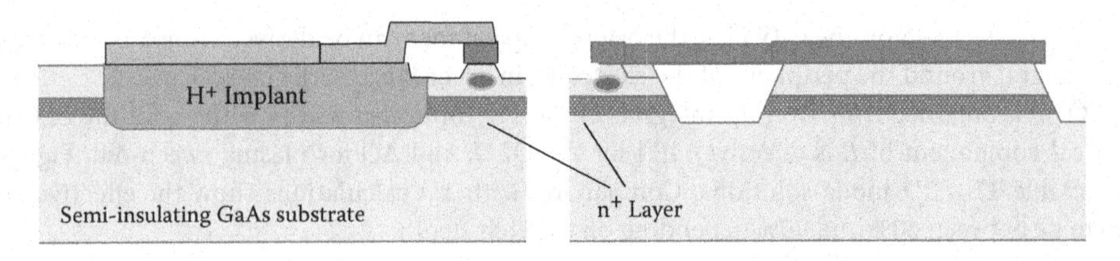


**FIGURE 8.7** Isolation and air-bridge methods: Implant-isolation with plated-over air-bridge (left), and trench-isolation with self-bridging using wet-etch undercut (right).

in the surface-slab; it has the additional advantages of surface planarity and low-reflectivity interfaces (Figure 8.7).

Air-bridges are needed to connect the RF modulation electrode out to a bond-pad or transmission-line without adding capacitance or perturbing the optical mode. With implantation-isolation, the nearly planar surface allows conventional MMIC air-bridge techniques to be used. With deep-trenching, pre-planarization will be necessary; alternatively a *self-bridging* technique can be used, in which narrow, possibly multistranded, air-bridges are created by undercut during a wet-chemical trench-etching [9] (Figure 8.7).

## 8.5 High Frequency Design

EO electrode structures present a largely capacitive load to the drive circuit; thus, provided the capacitance is known, the frequency-response is simple to calculate to a first approximation. However, for better accuracy, the long, narrow electrode configuration must be modeled as a transmission-line, whose RF characteristics may be computed using well-known transmission-line equations [12]. The complex propagation-constant and characteristic impedance of the EO waveguide are

$$\gamma = \alpha + j\beta = \sqrt{ZY} \quad \text{and} \quad Z_0 = \sqrt{Z/Y}\,, \tag{8.1}$$

where Z and Y are the distributed series-impedance and shunt-admittance:

$$Z = R + j\omega L \text{ and } Y = G + (C_d + C_f), \tag{8.2}$$

$\omega = 2\pi f$ is the angular frequency of the modulation signal and all parameters are per unit length. Inductance $L$ and direct fringing (or "coplanar") capacitance $C_f$ may be calculated numerically, or using one of the published analytic approximations [13–15]. Since III-V materials are not magnetic, a free-space solution of the conductor-geometry is adequate for inductance. $R$ is the electrode series resistance (line+return), obtained from the metal resistivity, taking account of high-frequency skin-effects. $C_d$ is the waveguide depletion capacitance discussed above, and $G$ is an effective shunt-conductance term derived from any additional resistances in the shunt-path—for example, resistance in the doped semiconductor layers.

The RF voltage-distribution along the electrode, and consequently the EO modulation function, may be calculated from these using standard transmission-line equations [12].

The EO function may be understood by considering a packet of photons traveling down the waveguide sampling the RF voltage. If the velocity of the photon-group matches that of the RF phase-fronts then the accumulation of EO phase-modulation will be linear and monotonic irrespective of frequency. If RF phase and optical group velocities are not matched, the phase of the modulation-envelope will progressively walk-off from that of the modulating RF voltage wave and frequency/length dependent cancellation will occur. The EO modulation function is calculated from the RF voltage distribution simply by multiplying by a RF/optical phase-difference term and integrating along the electrode.

The optical group-index $n_o$ of a GaAs/AlGaAs waveguide is derived from the group-velocity which defined as $\partial\omega/\partial\beta$. It is somewhat higher than the normal modal effective index $n_e$, and is due mainly to the bulk dispersion of GaAs. The corresponding RF-index ($n_\mu = \beta\lambda/2\pi$) is not, as will be seen, just a weighted-mean of the material-indices but is increased by slow-wave effects.

### 8.5.1 Push–Pull and Bias

A two-path interferometer, such as a Mach-Zehnder (MZ), with two EO waveguides back-connected by the doped contact layers will naturally operate in series push–pull mode [9,16] (see Figure 8.1). With equal capacitances series-connected, the effective capacitance is half that of each. The RF potential naturally divides equally between the two sides in antiphase yielding full intensity modulation upon recombination. Compared with single-sided drive, the Bandwidth/$V_\pi$ figure-of-merit is doubled—a major advantage of GaAs/AlGaAs MZ devices.

*Bias.* Electrically, the EO waveguides are diodes and must be kept under reverse-bias, and clear of the nonlinearity and high capacitance near zero voltage throughout the signal-cycle. This is arranged by maintaining the back-plane positive with respect to both top electrodes using a separate bias-contact, which can simply be a large-area Schottky pad, as it will be slightly forward-biased in use. For series push–pull, it is necessary that the doped back-plane immediately beneath the waveguides be allowed to float at RF frequencies; consequently, push–pull modulators are fabricated on SI substrate, with the back-plane decoupled from the bias circuit by means of low-pass filters. In effect, the diodes are parallel-connected at DC but series-connected at RF.

*Chirp.* When perfectly balanced and with an ideal recombiner, residual phase modulation is nulled and the chirp is zero. Intentional chirp can be engineered by unbalancing either the RF apportioning [17] and/or the optical split/recombination [18]. The residual phase-modulation has to be "bought" by degradation of some other parameter: by reduced bandwidth in the former case (net capacitance is higher) or by reduced modulation-depth in the latter case.

*Parallel Push–Pull.* An alternative arrangement, with the MZ arms cross-connected in parallel, is also possible [19]. Here, the advantage is one of compactness rather than bandwidth; with full drive-voltage across each arm, the length can be halved to achieve a given $V_\pi$, and capacitance. The fabrication-process with depletion-zone waveguides would be complex since the two backplanes must be isolated from each other and each connected to the top electrode of the other arm. However, where all-insulating materials are used, it is straightforward and is commonly used with lithium niobate. It has also been demonstrated with undoped GaAs [20].

### 8.5.2 Slow-Wave Effects

Semiconductor EO modulators of the depletion-zone type are inherently slow-wave in nature, as is apparent from Equations 8.1 and 8.2. The transverse diode-structures add capacitance $C_d$ without corresponding change to the series-impedance. Axial currents partition into the metal electrodes and the resistive back-plane material carries only the very short-scale shunt currents. The increase to the propagation-constant (and reduction in the characteristic-impedance) is described by the slow-wave factor:

$$\text{Slow-wave factor} = \sqrt{1+\frac{C_d}{C_f}}. \tag{8.3}$$

The slow-wave nature of the semiconductor waveguide modulator is of minor practical importance because the narrow electrodes incur too much RF loss for a very long modulator to be useful, while

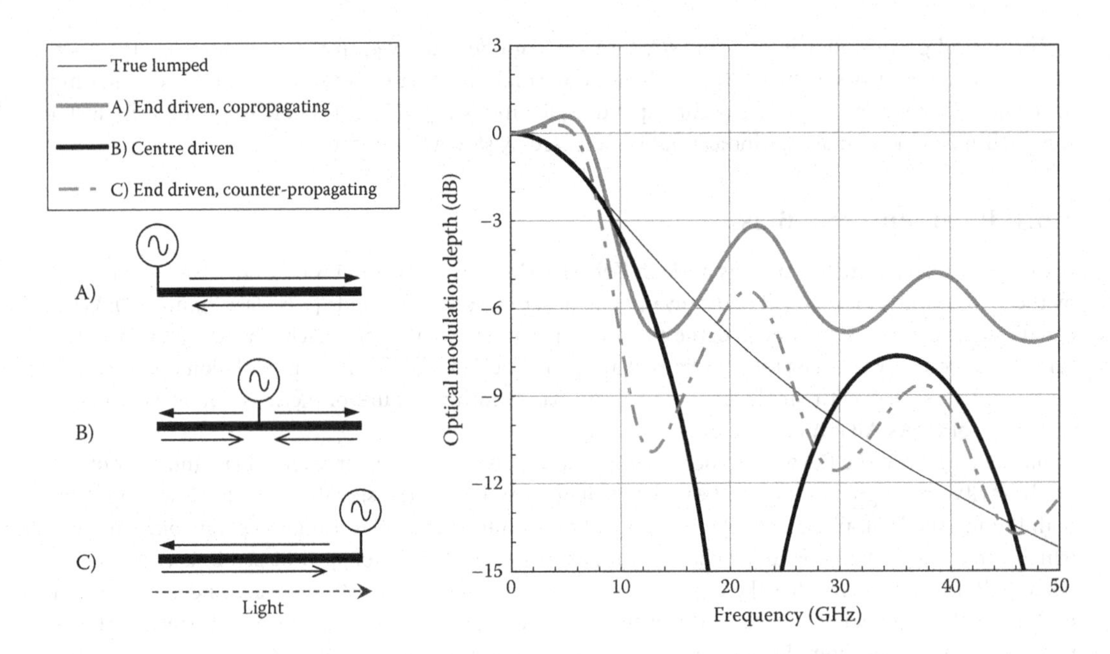


**FIGURE 8.8** Calculated EO frequency response of 2.5 mm "lumped" modulators, showing deviation from ideal lumped (dimensionless) behavior.

the velocity-match is unimportant in short structures (see Figure 8.8). It does, however, illustrate how the RF power-carrying structure might be dissociated from the electrode structures designed to accomplish the optical modulation, and each optimized for its distinct role.

## 8.6 The Segmented Traveling-Wave Modulator

The need for long EO sections (tens of mm) to achieve low $V_\pi$ has been noted (<3 V is considered desirable) and is due to the limited linear EO effects available in GaAs (see Figures 8.2 and 8.3). Traveling-wave design is necessary to achieve high modulation bandwidth in a long device. The segmented traveling-wave GaAs modulator, in which the modulation and RF transmission functions are largely separated into dedicated structures, was first devised and demonstrated by Walker [21]. Much of the subsequent work on GaAs modulators by others have used variations on this theme [20,22], and it has also been adapted for electro-absorption [23] and InP-based electro-optic modulators [24,25] (see also Chapter 7).

The two basic requirements for a high-bandwidth traveling-wave modulator are (1) that the high-frequency modulating signal should be able to propagate with low loss to the end of the modulating zone; (2) that the electrical phase-velocity should match the optical group-velocity; additionally, it is desirable that the response is not perturbed by electrical reflections due to load or source impedance mismatch.

Low RF loss is achieved by the use of a broad coplanar stripline (CPS) or coplanar waveguide (CPW), which is spatially distinct from the modulation electrodes. The latter are broken into short segments separated by passive spaces to ensure that the electrodes behave as reactive elements and not as a competing, lossy transmission-line. The optical waveguides will usually pass down the gap between the coplanar conductors while the electrodes sample the line-voltage periodically via air-bridge connections (see Figure 8.9). Two-conductor CPS is the more straightforward configuration for the series push–pull modulator type, while CPW coplanar waveguide would favor a parallel push–pull configuration (see Section 8.5.1) in which each of the two coplanar gaps carries just one of the MZ waveguides [20].

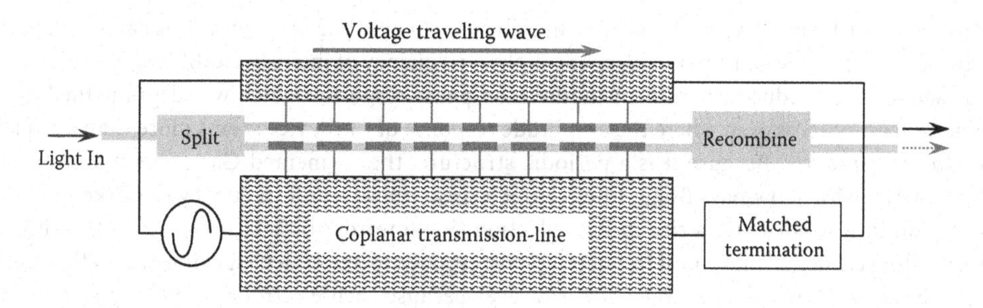


**FIGURE 8.9** Schematic of the segmented traveling-wave modulator configuration.

*Velocity Match.* While the optical mode is fully buried within the GaAs waveguide core, and has a group-index ($n_o$) around 3.55, the RF electrical wave on the coplanar line is only half-buried. To a good approximation, the effective-permittivity is a simple average of the GaAs and air-cover contributions, producing a much lower RF coplanar-index ($n_c$) around 2.65.

The electrode segments provide a periodic, reactive loading (capacitive at frequencies well below their resonance), creating a slow-wave structure. The slow-wave factor derived earlier (Equation 8.3) is applicable; this needs to be of value 3.55/2.65 = 1.339 to achieve velocity-match.

*Impedance Match.* The slow-wave factor also specifies a reduction in impedance from the coplanar value; thus, for 50 Ω (loaded), the unloaded coplanar impedance needs to be 67 Ω(= 1.339 × 50), which is readily achieved in a GaAs coplanar structure.

The CPS may be designed using a quasi-static, analytic approximation [13–15] or numerical microwave-design software, which is also able to calculate the attenuation and the variation in line parameters with frequency; that is, dispersion.

## 8.6.1 Loading Capacitance and Fill-Factor

The correct loading-capacitance ($C_L$) for velocity-match is readily derived from the basic transmission-line and slow-wave equations:

$$C_L = \frac{n_o^2 - n_c^2}{c \cdot Z_0 \cdot n_o}, \tag{8.4}$$

where $Z_0$ is the final loaded impedance (usually 50 Ω is desired) and $c$ is the vacuum light-speed, taken as $c = 0.3$ in order to work in convenient capacitance units of pF/mm. For the typical index values above, this evaluates to 0.105 pF/mm, which is very close to the absolute permittivity of GaAs ($\varepsilon_0 \varepsilon_r$ = 0.11377 pF/ mm). With width about twice the depletion-depth, the parallel-plate formula ($C = \varepsilon_0 \varepsilon_r W/d$) will be a fair approximation for the capacitance, so this waveguide aspect-ratio in a series push–pull configuration, will provide a reasonable velocity and impedance match in a segmented traveling-wave configuration.

If the raw loading capacitance is too high, it is simple to adjust the effective loading by reducing the *Fill Factor*—the fractional electrode length—of the segmented geometry.

## 8.6.2 Dispersion

Because different frequency-components of a complex signal on a transmission-line are dispersed by a frequency-dependent propagation velocity, the term *dispersion* has come to denote the frequency-dependence itself. RF dispersion in a traveling-wave modulator impacts the design in two ways: clearly, it means that the RF/optical velocity-match can only be correct at one frequency; this can be accommodated by designing for velocity-match near the highest frequency of interest. Additionally however, dispersion will tend to produce a nonlinear modulation-phase response, which will degrade the bit-error rate in long-haul systems even where the amplitude response of the device seems more than adequate.

*Structural Dispersion.* Because it is a periodic structure, the segmented GaAs modulator is inherently dispersive, with a low-pass Bragg-filter characteristic whose cutoff frequency is marked by a peak and inflexion in RF-index and a rise in RF reflection ($S_{11}$) towards 100% (see Figure 8.10). With ideal lumped loading elements, this cutoff frequency $f_c$ corresponds to an element spacing of half the effective wavelength, i.e., $\lambda_c/2$, where $\lambda_c = c/(f_c \cdot n_\mu)$. However because each electrode element is usually comparable in length to the spacing, (and is resonant at about twice this frequency—see Figure 8.8), the actual RF cutoff is rather lower, and the maximum modulation frequency lower still. In practice, high bandwidths are achieved simply by making the periodicity sufficiently fine. The cutoff is rather abrupt but structural-dispersion well below cutoff is generally small. Figure 8.10 illustrates the effect of segmentation fineness on the RF-dispersion and (consequently) on the modulation bandwidth (where the low-frequency velocity-match is close), while Figure 8.11 illustrates the effect of varying the fill-factor with fixed segmentation, showing the strong influence of the velocity-match.

*Coplanar Dispersion.* All transmission-lines are dispersive owing to changes in the transverse disposition of the RF current with frequency. The "skin effect"—concentration of RF current at the surface of a conductor—is well-known [12] and contributes to dispersion. Similarly, the analogous crowding of currents toward the edges of coplanar conductors [26] means that the effective-width of the conductor is also frequency-dependent; the wider the conductor, the more change is possible over frequency and the greater will be the dispersion.

*Higher RF Modes.* A third dispersive factor arises from the higher-order modes of the coplanar transmission-line. The number of eigenmodes is just the number of distinct conductors in the system in addition to a common ground-return (usually identified with the enclosing box or shielding); thus CPS has two modes (even and odd) while CPW has three. Continuous connection of one or more coplanar lines to the common-return will reduce the effective number of conductors and hence the possibility of higher modes. In practice, the increasing impedance of such connections (e.g., bond-wires) with

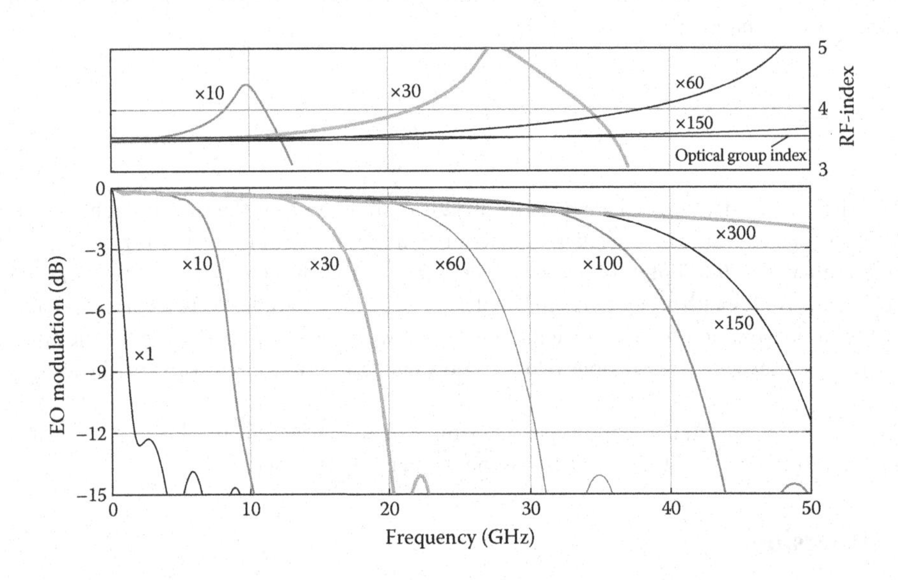


**FIGURE 8.10** Computed modulation responses and RF-Index (above) for a velocity-matched 49 Ω segmented traveling-wave modulator. Parameter ×*N* is the number of electrode segments over the fixed 30 mm length with a fill-factor of 95%. Quasi-static coplanar parameters were used.

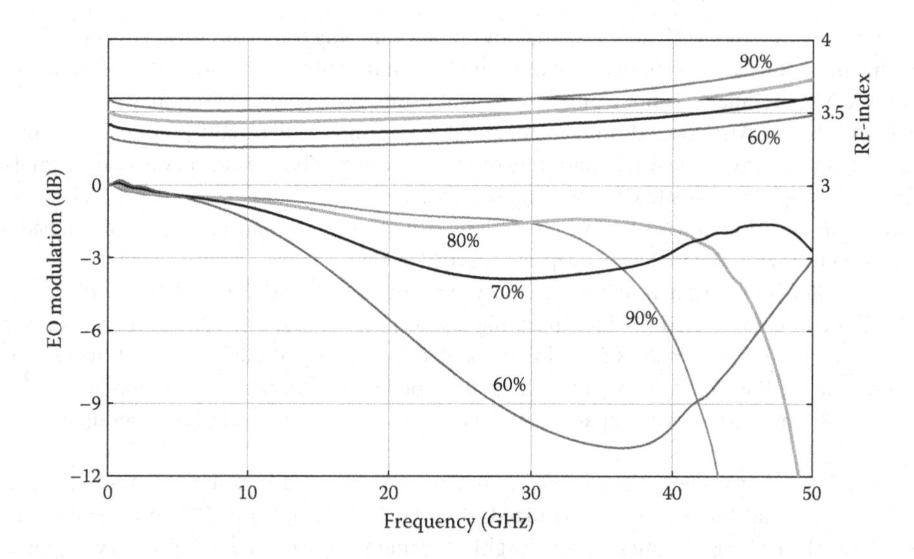


**FIGURE 8.11** As Figure 8.10 with the segmentation fixed at ×150 but varying the fill-factor (%) as indicated. The wider, more dispersive coplanar line here was characterized using *Sonnet*[em] (Sonnet Software inc.).

frequency allows the independence of coplanar conductors—hence higher-order modes—to reemerge at high frequencies.

At very high frequencies, even a single broad conductor can support higher modes as its two edges dissociate and are able to de-phase. Again, the solution to such problems is to keep dimensions as small as possible.

*Excitation of Higher RF Modes.* Any asymmetric disturbance or launch condition may excite higher modes. For example, tying one conductor of CPS to ground (0) at the RF input will launch a mix of even (++) and odd (+−) modes. The even-mode is incapable of EO modulation and is not slowed by the loaded-line but is unlikely to be correctly terminated. The reflected RF will be cross-coupled to the odd-mode by the input asymmetry at certain frequencies causing a "notchy", high-ripple modulation response.

### 8.6.3 Modeling of Segmented Traveling Wave Modulators

A complete model for the segmented traveling wave (TW) modulator was originally outlined by Walker [2,9] and a similar method more recently detailed by Li et al. [27]. Figures 8.10 and 8.11 were produced using a model of this type. The solution is built upon that of the single electrode (see Section 8.5). Briefly, a complex driving-point impedance is calculated for a representative electrode unit, usually (but not necessarily) assuming open-circuit ends and a central excitation-point. An effective electro-optic function encapsulates transit-time/velocity-match effects on the electrode and also accounts for RF voltage-drops across the resistive backplane material and inductive air-bridge connections. Transmission (or "chain") matrix formalism is used, first to build a segment-unit by chaining a short coplanar transmission-line section with one electrode driving-point; then to build the entire structure by chaining as many of these units as necessary. This yields RF S-parameters and a sequence of complex node voltages. The overall modulation is obtained by summing the electro-optic contributions using progressively phase-shifted versions of the electro-optic function to take account of the optical transit time.

## 8.7 Conclusions: High Functionality GaAs Modulators

The capability of GaAs modulator technology is demonstrated by a number of extended-functionality GaAs modulator variants [8]. These are mostly based on concatenations of lumped and segmented traveling-wave modulators with added spatial and modal filtering to ensure independence of the sections by shielding downstream inputs from the complementary outputs and other stray-light sources. If the input to a MZ modulator is not a pure fundamental-mode, its characteristic may be degraded and off/on voltages shifted. An odd mode input component, for example, will usually produce a fundamental-mode output, with an inverted switch-state. Since the modal-purity is likely to vary over the signal-cycle of the previous stage, this leads to coupling and instability.

Spatial filtration may comprise, for example, a long run of narrow mono-mode shallow-etched waveguide followed by a spot-size matched, deep-etch section [10].

10 Gb/s and 40 Gb/s nonreturn-to-zero (NRZ) and return-to-zero (RZ) modulators with $V_\pi$ less than 5 volts have been demonstrated [8], incorporating a second short MZ ($V_\pi \approx 28V$) for use as a variable optical attenuator. Similarly, in an RZ modulator, a second high-speed modulator section functions as a data-gate to create the RZ bit-stream from the output of a first nonreturn-to-zero modulator. A four-stage chirped RZ modulator with separate variable optical attenuator and phase-modulator has also been reported [28].

With modulators in parallel rather than concatenated in series, a differential quadrature phase shift Key (DQPSK) transmit function can be realized. Both GaAs [28] and InP [29] implementations have been reported. The DQPSK coding scheme is highly advantageous from a fiber-systems viewpoint [29].

The complexity of both the transmit and receive functions, with the need for compact bends and other high-level integrated features calls for high-functionality modulator devices with the capabilities of the GaAs/AlGaAs technology described in this chapter.

## References


1. J. F. Nye, *Physical Properties of Crystals*, New York: Oxford University Press, 1985.
2. R. G. Walker, "High speed electrooptic modulation in GaAs/GaAlAs waveguide devices," *J. Lightw. Technol.*, vol. 5, no. 10, pp. 1444–1453, Oct. 1987.
3. J. G. Mendoza-Alvarez, L. A. Coldren, A. Alping, R. H. Yan, T. Hausken, K. Lee, and K. Pedrotti, "Analysis of depletion edge translation lightwave modulators," *J. Lightw. Technol.*, vol. 6, no. 6, pp. 793–807, Jun. 1988.
4. M. A. Afromowitz, "Refractive index of $Ga_{1-x}Al_xAs$," *Solid State Commun.*, vol. 15, no. 1, pp. 59–63, Jul. 1974.
5. S. M. Sze and K. K. Ng, *Physics of Semiconductor Devices*, 3rd ed., New York: Wiley, 2007.
6. K. S. Chiang, "Analysis of the effective-index method for the vector modes of rectangular-core dielectric waveguide," *IEEE Trans. Microwave Theory Techniques*, vol. 44, no. 5, pp. 692–700, May 1996.
7. C. D. Watson, M. Poirier, J. M. Heaton, M. Lewis, and M. Boudreau, "Acousto-optic resonance in deep-etched GaAs–AlGaAs electro-optic modulators," *J. Lightw. Technol.*, vol. 22, no. 6, pp. 1598–1603, Jun. 2004.
8. R. G. Walker and R. A. Griffin, "GaAs photonic ICs for modulation and signal processing," in *IEEE Int. Microwave Symp., Phoenix, Arizona—Workshop: ICs 40 Gbit/s Data Rate Communications*, May 2001.
9. R. G. Walker, "High-speed III-V semiconductor intensity modulators," *IEEE J. Quantum. Electron.*, vol. 27, no. 3, pp. 654–667, Mar. 1991.
10. R. G. Walker, "Optical waveguide spatial filter," UK Patent: GB2 367 904, Apr. 17, 2002.
11. S. Ahmed, R. Gwilliam, and B. Sealy, "Proton implantation for isolation of n-type GaAs layers at different substrate temperatures," *Semicond. Sci. Tech.* vol. 16, no. 5, pp. 28–31, May 2001.
12. S. Ramo, J. Whinnery, and T. Van Duzer, *Fields & Waves in Communication Electronics*, New York: Wiley, 1965.
13. K. C. Gupta, R. Garg, I. Bahl, and P. Bharita, *Microstrip Lines and Slotlines*, Norwood, MA: Artech House, 1996.

14. S. Gevorgian, H. Berg, H. Jacobsson, and T. Lewin, "Basic parameters of coplanar-strip waveguides on multilayer dielectric/semiconductor substrates, part 1: High permittivity superstrates," *IEEE Microwave*, vol. 4, no. 2, pp. 60–70, Jun. 2003.
15. S. Gevorgian, H. Berg, H. Jacobsson, and T. Lewin, "Basic parameters of coplanar-strip waveguides on multilayer dielectric/semiconductor substrates, part 2: Low-permittivity superstrates," *IEEE Microwave*, vol. 4, no. 3, pp. 59–78, Sep. 2003.
16. R. G. Walker, "Broadband (6 GHz) GaAs/AlGaAs electro-optic modulator with low drive power," *Appl. Phys. Lett.*, vol. 54, no. 17, pp. 1613–1615, Apr. 1989.
17. R. G. Walker, "Optical modulator with pre-determined frequency chirp," UK Patent GB2 361 071A, Oct. 10, 2001.
18. P. Bravetti, G. Ghislotti, and S. Balsamo, "Chirp-inducing mechanisms in Mach-Zehnder modulators and their effect on 10 Gb/s NRZ transmission studied using tunable-chirp single drive devices," *J. Lightw. Technol.*, vol. 22, no. 2, pp. 605–611, Feb. 2004.
19. R. G. Walker and A. C. Carter, "Optical modulators operated in parallel push–pull mode," US Patent US7 082 237, Jul. 25, 2006.
20. R. Spickermann, S. R. Sakamoto, M. G. Peters, and N. Dagli, "GaAs/AlGaAs traveling wave electro-optic modulator with an electric bandwidth >40 GHz," *Electron. Lett.*, vol. 32, no. 12, pp. 1095–1096, Jun. 1996.
21. R. G. Walker, I Bennion, and A. C. Carter, "Low voltage, 50Ω GaAs/AlGaAs travelling-wave electro-optic modulator with bandwidth exceeding 25GHz," *Electron. Lett.*, vol. 25, no. 23, pp. 1549–1550, Nov. 1989.
22. N. Jager and Z. Lee, "Slow-wave electrode for use in compound semiconductor electrooptic modulators," *IEEE J. Quantum Electron.*, vol. 28, no. 8, pp. 1778–1784, Aug. 1992.
23. R. Lewen, S. Irmscher, U. Westergren, L. Thylen, and U. Eriksson, "Ultra high-speed segmented traveling-wave electroabsorption modulators," in *Optical Fiber Communication Conference 2003*, vol. 3, paper PD38.
24. J. Barton, M. Masonovic, A. Tauke-Pedretti, E. Skogen, and L. Coldren, "Monolithically-integrated 40 Gbit/s widely-tunable transmitter using series push–pull Mach-Zehnder modulator SOA and sampled-grating DBR laser," in *Optical Fiber Communications Conference 2005*, vol. 2, paper OTuM3.
25. S. Akiyama, H. Itoh, T. Takeuchi, A. Kuramata, and T. Yamamoto, "Low-chirp 10 Gbit/s InP-based Mach–Zehnder modulator driven by 1.2 V single electrical signal," *Electronics Letters*, vol. 41, no. 1, pp. 40–41, Jan. 2005.
26. C. L. Holloway and E. F. Kuester, "A quasi-closed form expression for the conductor loss of CPW lines, with an investigation of edge shape effects," *IEEE Trans. Microw. Theory Tech.*, vol. 43, no. 12, part 1, pp. 2695–2701, Dec. 1995.
27. G. L. Li, T. G. B. Mason, and P. K. L. Yu, "Analysis of segmented traveling-wave optical modulators," *J. Lightw. Technol.*, vol. 22, no. 7, pp. 1789–1796, Jul. 2004.
28. R. A. Griffin, R. Johnstone, R. G. Walker, S. Wadsworth, A. C. Carter, and M. J. Wale, "Integrated DQPSK transmitter for dispersion-tolerant and dispersion-managed DWDM transmission," in *Optical Fiber Communications Conference 2003*, pp. 770–771.
29. R. A. Griffin and A. C. Carter, "Advances in InP optical modulators," in *Optical Fiber Communication & Optoelectronic Exposition Conference (AOE)*, Shanghai, 2008, pp. 1–3.

# 9

# Electro-Optic Polymer Modulators

Larry R. Dalton
*University of Washington*

William H. Steier
*University of Southern California*



## 9.1 Overview of Polymer Materials and Devices

Organic electro-optic (OEO) materials and devices, particularly polymers, have been investigated and developed for a number of years. Compared to conventional crystalline materials, their advantages are the ability of chemists to molecularly engineer organic materials to achieve very high electro-optic (EO) coefficients and the ability of the device engineers to use spin cast fabrication technology and to therefore potentially integrate the EO devices with other opto-electronic and electronic devices and circuits. In the case of high-speed optical modulators, polymer materials have additional advantages in that their optical index of refraction is reasonably close to the microwave index of refraction, therefore, velocity matching is inherently possible, and that modulators with relatively long traveling electrodes can be made which allows one to simultaneously achieve low-drive voltage and broad bandwidth.

The questions raised concerning polymer devices have largely centered on their stability since organic materials have a reputation for instability over long times. The instability concerns are (1) the photostability after long-term exposure to laser radiation, (2) the long-term temperature stability of the molecular alignment and therefore the EO coefficient, and (3) the optical losses of optical waveguides fabricated from the materials. It has proven a difficult task to simultaneously solve all of these problems in one material. However, there has been significant progress on all of these issues and commercial polymer-based modulators are now beginning to appear and some very novel polymer-based EO devices, which are not possible using crystalline materials, have been demonstrated.

The question of where the polymer EO devices will likely find application is an important one and the answer can guide future research. It will be difficult for conventional polymer-based modulators to replace the well-developed lithium niobate ($LiNbO_3$) modulators that are used in most current long distance fiber communication networks since there must be a significant advantage before a new technology will be accepted into an established system. Polymer devices may be lower cost—although that remains to be proven—but in long distance fiber networks, a large number of modulators is not required and the cost per modulator is not a large concern. However, there are a number of applications, particularly in analog radio frequency (RF) fiber systems and computer back planes and interconnects, where polymer-based devices have a good future. In the case of computer back planes and interconnects, the cost per modulator and the possibility of integrating the modulator with silicon electronics will be the driving forces. In the case of RF analog systems, integration and the ability of polymer devices to be bent and to conform to a required surface shape are the advantages. Polymer modulators have the potential for very low half-wave voltage ($V_\pi$) since large EO coefficients have been demonstrated in some of the latest materials and Cox has pointed out that very low $V_\pi$ devices open up a number of possibilities in RF analog systems [1].

Molecules with large second order nonlinear optical coefficients have been synthesized and the chemistry of these molecules, as to maximizing the EO effect per molecule, is well-understood. When these molecules are incorporated into a host material and thin films are fabricated, the molecules are randomly oriented and some polar order must be introduced in order to establish the EO effect. This ordering is typically done by applying a large electric field at high temperature and then locking in the induced alignment by covalent bond formation (cross-linking or lattice hardening) chemistry. Understanding of the molecular alignment process and novel methods to lock in alignment have enabled recent advances in polymer EO materials; modification of the molecular shape (to reduce the electric dipole interaction between molecules) and the development of an understanding of the roles of various intermolecular interactions have been the most important developments. In addition, novel cross-linking (Diels-Alder/Retro-Diels-Alder) schemes that free the molecules to rotate at high temperature but then irreversibly lock in the alignment as the temperature is reduced, have been the advances that have provided good long-term stability of the poling.

The EO devices typically have a cladding-core-cladding waveguide structure and, because of the relatively high electrical conductivity of the EO polymer core relative to the cladding, it is difficult to get large poling electric fields into the core during device fabrication. The development of higher conductivity cladding materials has significantly improved the EO coefficients that can be achieved in devices.

In this chapter, we will review the state of the art and the significant recent advances in EO polymer materials and in EO polymer devices.

## 9.2 Types of Materials

The motivation for use of OEO materials is derived from the potential for ultrafast (femtosecond) responses to applied time-varying electric fields, very large EO activity (>400 pm/V), and versatile materials modification and processing options that facilitate integration with disparate materials [2–10]. An opportunity, but also a liability, associated with organic materials is the virtually endless potential for improvement of material performance by modification of material molecular composition and organization. Because of the seemingly endless array of materials that constitute OEO compounds, there is a wide variation of critical properties EO activity, optical loss, thermal stability, photochemical stability, processability, etc., and correspondingly, there is considerable confusion about the properties of OEO materials. For example, the thermal stability of organic materials can range from less than 100°C to greater than 400°C (e.g., for graphitic materials such as ladder polymers) based on variations of covalent bond density (including intermolecular bonding) and photochemical stability can vary by more than six orders of magnitude. No single OEO material has emerged as the clear choice for device fabrication and all necessary properties (EO activity, optical loss, thermal stability, photochemical stability, etc.) have yet to be optimized in a single material. OEO materials can be dipolar [2–5,11,12] or octupolar [13–16], although all device work to the present has focused on dipolar chromophores organized into acentric lattices (single crystals) [17–19], oriented films prepared by Langmuir–Blodgett or Merrifield sequential synthesis techniques [12,20–22], or partially oriented films prepared by electric field poling or a combination of electrical and optical poling [1–5,11,12,23].

## 9.2.1 Single Crystals

Dipolar OEO chromophores are normally prolate ellipsoidal in shape and as such tend to form centrosymmetric crystals driven by chromophore–chromophore dipolar interactions. To achieve EO-active (acentric) crystals requires that dipolar interactions be overcome by ionic forces. The chromophore 4'-dimethylamino-*N*-methyl-4-stilbazollium tosylate [19] is the prototypical example although many OEO crystals have been cataloged to date [18,24]. Most research focused on preparing single crystalline materials involves exploring different purification and crystal growth options. Control of structure is often limited to varying the size of the counterion associated with the charged chromophore. Because of the density of single crystal lattices and because of the strong forces holding the chromophores in place, thermal and photochemical stability are not normally problems with organic single crystal materials. However, using single crystalline materials in EO devices can prove challenging, particularly when there is a tendency for crystals to grow as long needles. Device production is particularly difficult for single crystalline materials when the melting temperature is above the decomposition temperature.

## 9.2.2 Self-Assembled Films

Self-assembled films are typically prepared by sequential (stepwise) synthesis techniques exploiting either ionic interactions (Langmuir–Blodgett technique [20]) or covalent bond coupling (Merrifield technique [12,21]). Chromophores are prepared for acentric assembly by exploiting interactions that permit only head-to-tail association of chromophores. A major problem with self-assembled films is that defects propagate and it is extremely difficult to avoid substantial disorder as the number of sequentially assembled layers exceeds 100; it typically requires 1000 layers or more to fabricate a 1-μm to 3-μm EO waveguide core. The propagation of defects and the increase of optical loss with the addition of subsequent layers are serious problems and have greatly limited the use of self-assembled films for development of prototype devices.

## 9.2.3 Electrically Poled Films

Electrically poled films are prepared by applying an electric field across a macromolecular film containing dipolar EO chromophores near the glass transition (melt) temperature of the film. The chromophore dipoles orient under the influence of the applied electric field and, in the absence of dipole–dipole interactions, the resulting acentric order parameter is $\langle \cos^3\theta \rangle = \mu F/5kT$ where $\theta$ is the angle between molecular *z*-axis and the poling field, $\mu$ is the chromophore dipole moment, $F$ is the effective electric field felt by the chromophore (the applied field, $E_{pol}$, screened by the dielectric medium), $kT$ is the thermal energy, and the factor 5 in the denominator appears for a three-dimensional or Langevin lattice. The poling acentric order (in the independent particle limit) is determined solely by the ratio of the poling to thermal energies. In real systems, a variety of spatially anisotropic interactions influence poling-induced order. Such interactions can both oppose and enhance poling-induced acentric order. Examples of typical OEO chromophores and host materials are given in Figure 9.1; these materials can be divided according to the following categories.

### 9.2.3.1 Chromophore-Polymer Composites

The most frequently studied OEO materials have consisted of chromophores dissolved into commercially available polymers to form composite materials (see Figure 9.1). The most studied polymers have included polymethylmethacrylate, polycarbonates including amorphous polycarbonate (APC), polyquinolines, and polyimides. Host polymers must have a sufficiently high glass transition temperature ($T_g$) that poling-induced order does not significantly relax at temperatures of interest (e.g., device operating temperatures). Host polymers must also exhibit adequate solubility in solvents used for spin casting

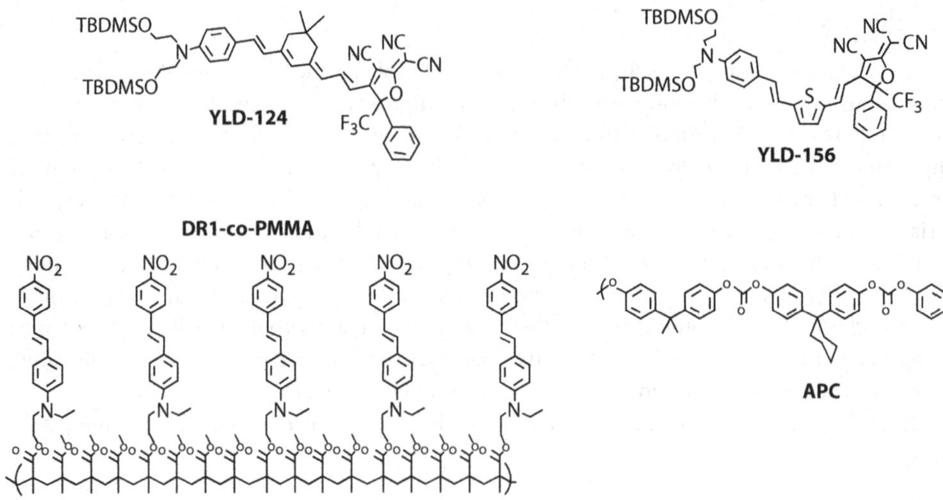

YLD-124 YLD-156 DR1-co-PMMA APC

**FIGURE 9.1** Typical chromophores and host materials are shown. The YLD-124 chromophore is a higher molecular first hyperpolarizability (β) variant of the familiar CLD chromophore (the $CF_3$ and phenyl substituents on the acceptor are replaced by methyl groups). The YLD-156 chromophore is a more active version of the familiar FTC chromophore. APC (amorphous polycarbonate) is one of the most commonly used polymer hosts in the preparation of EO composite materials. The DR1-co-PMMA is one of the most commonly used early EO materials; it is also used as a host material for the preparation of binary chromophore organic materials.

thin films. Because of a combination of reasonably high $T_g$ and good solubility in typical spin casting solvent, APC (see Figure 9.1) has become the most common choice for use as a polymer host. The great attraction for the use of composite materials is that these are straightforwardly prepared from readily available materials including commercially available host polymer materials. Moreover, different chromophores can be quickly evaluated using a common host such as APC.

For most composite materials, it can reasonably be assumed that interactions between chromophore guest and polymer host do not influence poling-induced acentric order. Two types of intermolecular (chromophore) interactions influence poling-induced order: (1) electronic dipole–dipole interactions among chromophores, and (2) shorter range nuclear repulsive interactions defined by chromophore shape [25–27]. There are two factors that influence the dipole–dipole interactions. The first factor, end-to-end approach, favors acentric order and the second factor, side-by-side approach, favors centrosymmetric order [25,26]. For prolate ellipsoidal shaped chromophores, side-by-side approach will dominate leading to centrosymmetric chromophore organization and a decrease of the acentric order parameter proportional to chromophore number density. The decrease in the order parameter with chromophore number density forces EO activity to exhibit a maximum as a function of chromophore number density. The design paradigm for improving EO activity of chromophore-polymer composite materials involves making chromophores more spherical through the addition of inert substituents to the chromophore structure [26]. This paradigm has permitted the preparation of EO composite materials that exhibit EO activity approaching 100 pm/V.

Composite materials do have some significant disadvantages. The requirement of high $T_g$ for the host polymer materials can lead to sublimation of chromophores during electric field poling. Chromophores can more easily phase separate from the polymer host than if the chromophores are covalently coupled to the host. Electrophoretic migration of chromophores can occur during electric field poling. Composite materials typically do not have thermal or photochemical stability that is as good as is obtained with more dense materials that are held together by covalent bonds (as we shall demonstrate later in this chapter).

#### 9.2.3.2 Covalently Coupled Chromophore Dendrimers and Polymers

While composite materials are convenient for the fabrication and testing of prototype devices for proof-of-principle demonstrations, they typically do not afford adequate thermal and photochemical stability through covalent anchoring of chromophores to their surrounding host materials. Composites also do not afford the opportunity of using covalent bonds to enhance and stabilize poling-induced acentric order. In this section, it is convenient to divide the discussion of covalently coupled materials into two parts: (1) materials that have their poling-induced order significantly influenced by intermolecular electrostatic interactions with nonchromophore components, and (2) materials that have their poling-induced order only slightly influenced or unaffected by intermolecular electrostatic interactions with nonchromophore components. The first class of materials is represented by the HDFD [12,28–30] and SBLD-1 [31,32] materials of Figure 9.2. In the case of HDFD, poling-induced order is enhanced by intermolecular electronic (quadrupolar) interactions among protonated and fluorinated phenyl (Frechet) dendrons. For SBLD-1, poling-induced acentric order is enhanced by coumarin–coumarin interactions. Force modulation scanning probe microscopy, intrinsic friction analysis, and dielectric relaxation spectroscopy permit quantitative measurement of the entropies (improvements in order) and correlation lengths (number of molecular units whose movements are correlated) of these interactions [29,30,33]. Given the nature of the interactions and the similarities of the structures of HDFD and SBLD-1, it is not surprising that both have similar $T\Delta S^*$ values (= 25 kcal/mol–30 kcal/mol) where $T$ is in kelvin and $\Delta S^*$ is the entropy change (enhancement of order) associated with phenyl-H/phenyl-F or coumarin–coumarin interactions. The effect of these interactions is also seen in measurements of the $\langle P_2 \rangle$ (= $0.5[3\langle \cos^2\theta \rangle - 1]$) order parameter by variable angle spectroscopic ellipsometry and variable angle polarization referenced absorption spectroscopy [11,12,32,34]. As seen in Figure 9.3, VASE permits measurement of the order parameters and relative orientations of coumarin and chromophore components of SBLD-1. Theory [33] permits correlation of $\langle \cos^2\theta \rangle$ (obtained

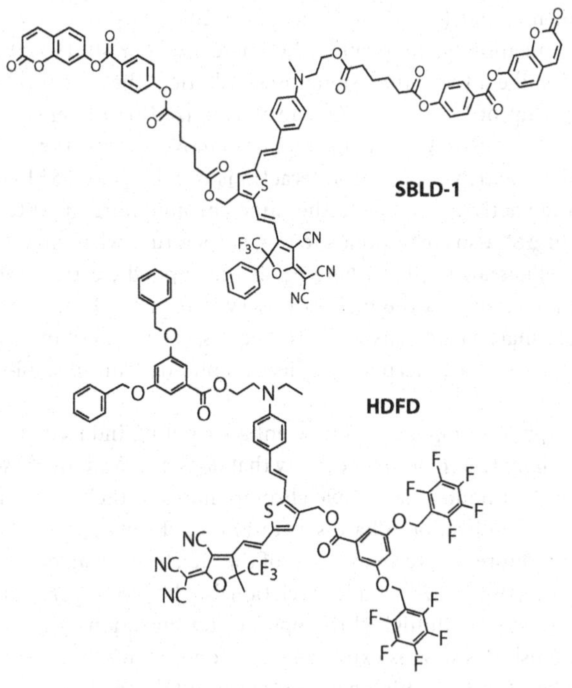

SBLD-1

HDFD

**FIGURE 9.2** Chemical structures of the SBLD-1 and HDFD EO dendrimers are shown.

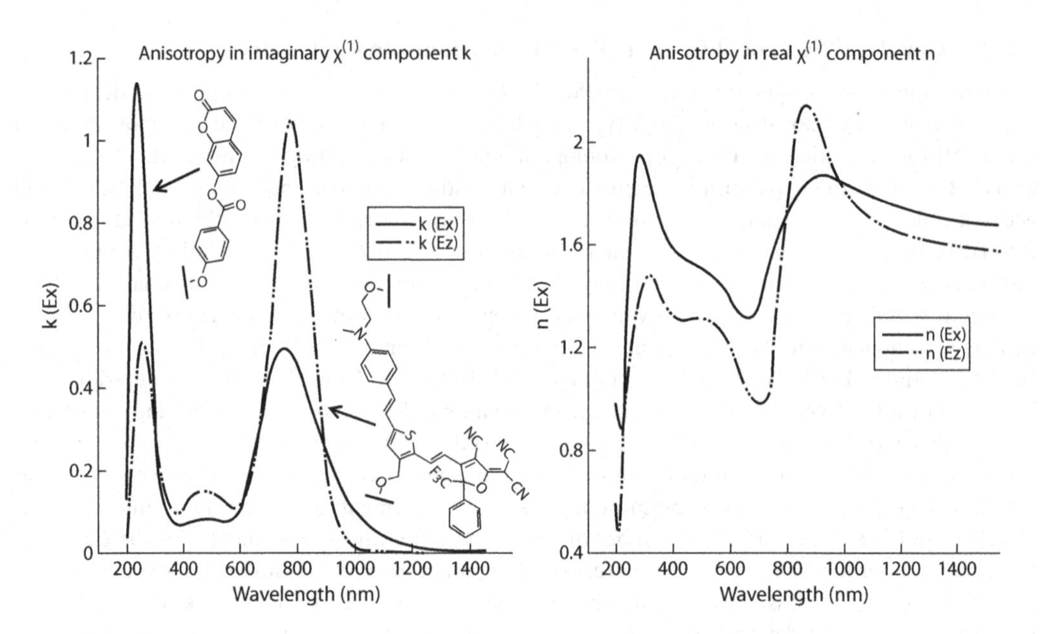


**FIGURE 9.3** Variable angle spectroscopic ellipsometry (VASE) of SBLD-1 is shown. These results illustrate that the chromophore and coumarin moieties transition moments are orthogonal. VASE and VAPRAS results permit determination of the $\langle P_2 \rangle$ order parameters of these moieties. The coumarin intermolecular interactions reduce the effective lattice symmetry of the EO dendrimer material to approximately 2.2.

from variable angle spectroscopic ellipsometry and variable angle polarization referenced absorption spectroscopy measurements) and $\langle \cos^3\theta \rangle$ (obtained from EO measurements) providing insight into the details of lattice organization (symmetry) for EO materials. For SBLD-1, this correlation suggests that the material behaves like a two-dimensional Bessel lattice (the actual "effective" lattice dimension is 2.2). The relationship of EO activity ($r_{33}$) to normalized poling energy (μF/kT) is known to depend on lattice symmetry (restriction of chromophore movement by the surrounding lattice). The phenyl-H/phenyl-F and coumarin-coumarin interactions of HDFD and SBLD-1 thus lead to factors of 2–3 enhancement of EO activity relative to the same chromophore in APC. The temperature corresponding to maximum ΔS* also corresponds to the temperature where maximum poling efficiency is observed. The poling efficiency ($r_{33}/E_{pol}$ where $E_{pol}$ is the applied electric poling field) observed for SBLD-1 (and similarly for HDFD) is the highest observed for any electrically poled material other than multi-chromophore materials discussed in the next section. These intermolecular electrostatic interactions also result in improved thermal stability and photochemical stability as will be discussed in subsequent sections.

Even if the nonchromophore components do not enhance poling-induced order, improved EO activity can be realized through increased number density that does not result in phase separation or attenuation of EO activity. Covalent bonds that couple chromophores to their surrounding host matrix can inhibit centrosymmetric crystallization (phase-separation) of chromophores. The multichromophore-containing dendrimers of Figure 9.4 are an example of the effects of covalent bond potentials. As noted in several communications [10,11,34,35], a linear relationship between $r_{33}/E_{pol}$ and chromophore number density, $N$, is observed even with high chromophore concentrations (e.g., $7 \times 10^{20}$ chromophores/cm$^3$). Such a linear relationship is what is expected for chromophores that behave as independent particles. The effects of covalent bond potentials have been well simulated by pseudoatomistic Monte Carlo/molecular dynamics theoretical approaches [11,12,35,36].

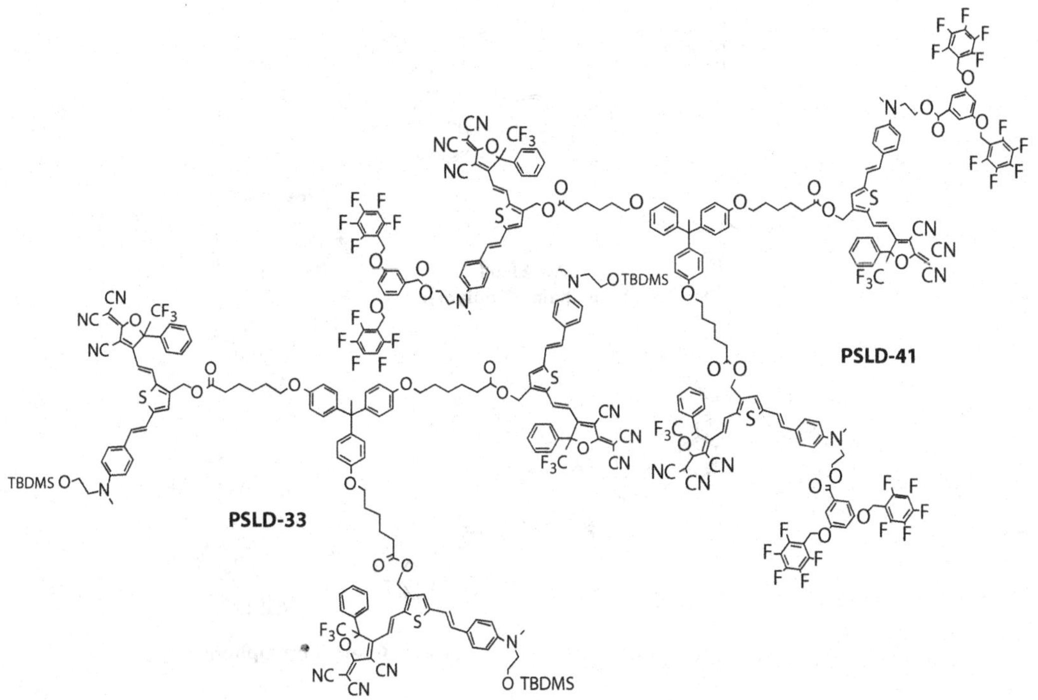

FIGURE 9.4 The chemical structures of the PLSD-33 and PSLD-41 dendrimers are shown.

### 9.2.3.3 Multichromophore Materials

Multichromophore materials refer to materials where both guest and host materials of composites contain chromophores [11,12,23,37–39]. An example is the YLD-124 chromophore of Figure 9.1 dissolved in the DR1-copolymethylmethacrylate material of Figure 9.1. The slope of the poling efficiency versus guest chromophore number density (an EO material's figure-of-merit) of this binary chromophore composite is $Nr_{33}/E_{pol}$ = 10 nm$^5$/molecules·V$^2$ compared to a value of $Nr_{33}/E_{pol}$ = 5 nm$^5$/molecules·V$^2$ for the same chromophore dissolved in APC. Another example is the YLD-124 chromophore of Figure 9.1 dissolved in the PSLD-44 dendrimer of Figure 9.4. Here, the poling efficiency increases to $Nr_{33}/E_{pol}$ = 15 nm$^5$/molecules·V$^2$, an increase of a factor of 3 over that observed with YLD-124 dissolved in APC. Again, theoretical modeling [38] is capable of simulating the results and demonstrates that the increase in poling efficiency can be attributed to specific interactions between guest and host chromophores. This is further demonstrated by laser-assisted poling experiments carried out on the YLD-124/DR1-copolymethylmethacrylate composite material [22]. It is well-known that irradiation of the charge transfer absorption of trans-DR1 with polarized light leads to optical poling of the DR1 chromophore. Photoinduced *trans-cis-trans* isomerization has the net effect of driving the DR1 chromophores into a plane (a 2-D or Bessel lattice). When this is done in the presence of an electric poling field, the acentric order of DR1 is increased. This increase in DR1 order produces a factor of 2.2 increase in the order of the guest chromophore through the interaction between guest and host chromophores. Poling efficiency $r_{33}/E_{pol}$ increases from 0.77 nm$^2$/V$^2$ to 1.7 nm$^2$/V$^2$ after the application of polarized light (optical poling).

Binary (and multi) chromophore-containing composite materials also exhibit improved optical properties. Because both the guest and host components are polar, solvatochromic shifts of the charge transfer resonances of both guest and host are not observed as a function of varying chromophore composition, which is in marked contrast to traditional composite materials where such shifts are large,

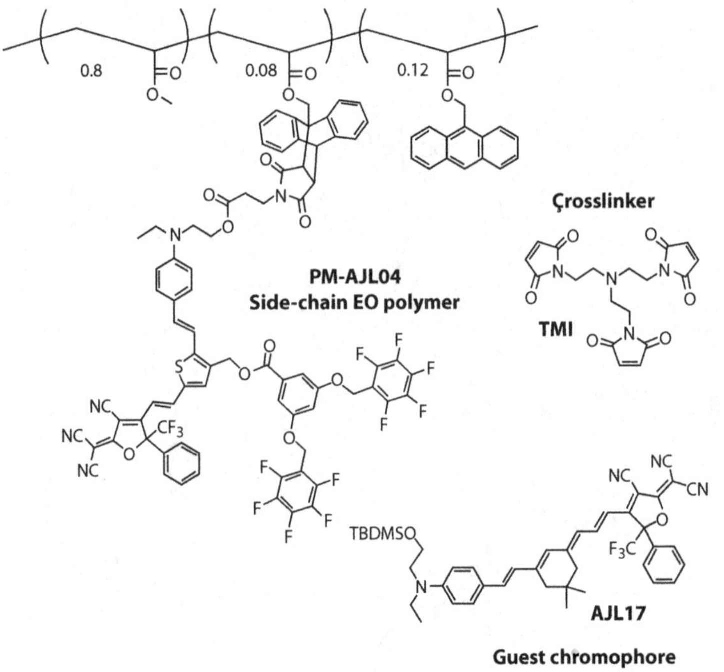

**FIGURE 9.5** The chemical structures of the components of the AJ404 EO material are shown. This material is hardened by Diels-Alder/Retro-Diels-Alder thermosetting reactions involving the crosslinker reacting with the antracene diene moieties.

reflecting the fact that chromophore addition makes the material more polar. Also, spectral line broadening is not observed for binary chromophore materials, reflecting the fact that the large free energy of mixing facilitates the achievement of high chromophore number densities without phase separation. These two observations contrast what is seen with traditional chromophore/polymer composite materials where phase separation and spectral line broadening associated with a distribution of dielectric environments is observed at much lower chromophore number densities. Currently, the largest poling-efficiencies ($r_{33}/E_{pol}$ = 4 – 5 nm$^2$/V$^2$) are observed for binary chromophore composite materials. The high chromophore number densities associated with these materials can result in unwanted conductivity near material glass transition temperatures with the result of limiting the poling field that can be applied (e.g., to values as low as 60 V/μm – 70 V/μm). This effect can be somewhat attenuated by depositing a nanoscopic (50 nm) layer of titanium dioxide between the EO materials and the poling electrode. These metal oxide layers may help to reduce charge injection and extraction. Binary chromophore composite materials, as with the other types of OEO materials discussed above, typically do not exhibit adequate thermal and photochemical stability without additional lattice hardening (i.e., introduction of covalent crosslinks). See Figure 9.5 for an example of a thermosetting binary chromophore composite material. Lattice hardening will be discussed later in this chapter.

## 9.2.4 Material Characteristics and Requirements

### 9.2.4.1 Optical Nonlinearity

Both large molecular, $\beta_{zzz}$, and macroscopic, $r_{33}$, optical nonlinearities are required for a successful EO material. In the simplest case, these two properties can be related for electrically poled materials by $r_{33} = N\langle\cos^3\theta\rangle\beta_{zzz}(\omega,\varepsilon)\ K(n,\varepsilon)$. Here, we emphasize that the molecular optical nonlinearity, $\beta_{zzz}$, depends upon the optical frequency, $\omega$, and the medium's dielectric permittivity, $\varepsilon$. The final term, K, is included above to indicate that there is another factor that depends on the index of refraction and the dielectric permittivity of the medium, but we will not consider it here. For some time, a variety of quantum mechanical computational methods have been successful at predicting trends for the variation of molecular optical nonlinearity with chromophore structure [39]. Recently, time-dependent density functional theory methods, where time-dependent fields are explicitly treated in the system Hamiltonian, have proven successful for the quantitative prediction of molecular optical nonlinearity including the dependence on $\omega$ and $\varepsilon$ [10,11,40,41] both of which can be significant. Currently, molecular optical nonlinearities are most commonly measured by wavelength-agile, femtosecond hyper Rayleigh scattering. These measurements are occasionally complemented by the measurement of the product of the chromophore dipole moment and the molecular optical nonlinearity by electric field induced second harmonic generation. The molecular optical nonlinearity measured by hyper Rayleigh scattering is an averaged value and theory must be used to relate this to the principle component of molecular first hyperpolarizability, $\beta_{zzz}(\omega,\varepsilon)$. Macroscopic EO activity is typically measured in simple thin films by Teng–Man [12,42,43] ellipsometry or by attenuated total reflection [11]. The Teng–Man method, which is very useful for *in situ* monitoring of the introduction of EO activity by electric field poling or the thermal relaxation of poling-induced activity, suffers from artifacts associated with thin OEO material layers and the use of indium tin oxide electrodes [42]. The attenuated total reflection method permits measurement of both tensor elements ($r_{33}$ and $r_{13}$) for poled materials and suffers less from measurement artifacts. EO activity can also be measured by other methods including by measuring the drive voltages required to operate EO devices.

Since molecular first hyperpolarizability and EO activity will depend on optical frequency and dielectric permittivity, it is not meaningful to state values for a particular chromophore or for an EO material unless details of material composition and processing are specified. In general, it is more meaningful to report values of $r_{33}/E_{pol}$ or $Nr_{33}/E_{pol}$ than simply reporting $r_{33}$ values. A linear dependence of $r_{33}$ on $E_{pol}$ is theoretically predicted and experimentally observed and thus reporting the ratio $r_{33}/E_{pol}$ (obtained by least square fitting of measurements at a number of poling field strengths) is statistically more reliable than a single $r_{33}$ value. For example, the SBLD-1 material deposited on an indium tin oxide (ITO) electrode yields $r_{33}/E_{pol}$ = 1.25 $nm^2/V^2$ compared to $r_{33}/E_{pol}$ = 0.43 $nm^2/V^2$ for the same chromophore (YLD-156) in polymethylmethacrylate at $1.7 \times 10^{20}$ molecules/$cm^3$ number density (near the value leading to maximum EO activity); this illustrates a factor of 3 improvement in poling efficiency. If SBLD-1 is deposited on a thin layer of titanium dioxide (on top of the ITO), then the poling efficiency of SBLD-1 increases to 1.92 $nm^2/V^2$. Comparing poling efficiencies yields quantitative insight into the effects of various intermolecular electrostatic interactions and conductivity effects. Comparing $Nr_{33}/E_{pol}$ values illustrates how these effects depend on chromophore number density. Fortunately, theoretical methods are now capable of providing quantitative prediction of all three parameters ($r_{33}$, $r_{33}/E_{pol}$, and $Nr_{33}/E_{pol}$). EO activities, $r_{33}$, approaching 500 pm/V (approximately 15 times that of $LiNbO_3$), have been achieved with thin film OEO materials. Because of problems associated with the relative conductivity of OEO materials and cladding materials, it has been difficult to achieve even 200 pm/V EO activity with the same materials in triple stack devices. The problem is even worse for OEO materials incorporated into silicon photonic circuitry (devices), where conductivity problems encountered during poling limit EO activity to values of approximately 30 pm/V–40 pm/V [8].

### 9.2.4.2 Optical Loss

Optical loss, expressed as waveguide loss in dB/cm, can arise either from absorption (intrinsic) or scattering (processing associated). The former can be dominated either by interband (charge transfer) electronic absorption or by vibrational overtone absorptions associated with hydrogen, whichever is larger for a particular material. For operation at telecommunication wavelengths, it is normally crucial to keep the wavelength of the maximum of the charge transfer absorption ($\lambda_{max}$) below 800 nm to avoid excessive (2 dB/cm or greater) absorption loss. For typical protonated EO materials, overtone vibrational absorption from hydrogen contributes about 0.7 dB/cm–1.2 dB/cm to absorption loss. Light scattering from various processing steps can contribute dramatically to propagation loss unless care is exercised to avoid surface roughness of waveguide walls and material inhomogeneity. Material inhomogeneity can be introduced during spin casting of thin films, poling of materials, and lattice hardening that are done to achieve adequate thermal and photochemical stability. If care is exercised in processing steps such as spin casting, reactive ion etching of waveguides, and deposition of cladding layers, then scattering losses can be kept to a few tenths of a dB/cm. Propagation loss is typically measured using either the cutoff or out-coupling methods [3,43]. In addition to propagation loss, total insertion loss also depends on coupling loss which is defined by the combination of mode size mismatch and index of refraction mismatch between silica fibers and OEO waveguides. Minimization of coupling loss has typically been achieved by use of some type of mode transformer. For OEO materials, the target has been to keep waveguide propagation losses to less than 2 dB/cm, which is important for achieving total device insertion loss of less than 6 dB.

#### 9.2.4.3 Thermal Characteristics

The critical parameter defining the thermal stability is the final $T_g$ of the material. $T_g$ is determined by the numbers and placement of covalent bonds coupling various components of the EO material. It is also influenced by the segmental flexibility of components. In general, one desires a modest $T_g$ for processing (e.g., electric field poling must be carried out below the decomposition temperature of components) but the highest possible glass transition temperature is desired after introduction of EO activity by poling. Numerous studies have shown that the operational stability of OEO materials is related to the final $T_g$ and, in general, it is desirable to have the material $T_g$ be at least 60°C above the operational temperature. More recent targets for thermal stability require a final $T_g$ of the EO material to be 200°C or greater.

It has been common practice to use lattice hardening (covalent bond cross-linking) chemistries to elevate material glass transition temperatures subsequent to poling. Throughout the 1990s, condensation reactions based on urethane or sol gel chemistries were popular; however, these reactions were difficult to control (e.g., sensitive to atmospheric moisture and prone to elimination products that disrupted the material lattices) [44,45]. Moreover, these chemistries seldom yielded material lattices with $T_g$ approaching 200°C. Since 2000, these earlier lattice hardening techniques have largely given way to cycloaddition reactions based either on reaction of fluorovinyl ether moieties to form cyclobutyl cross links [45–47] or Diels-Alder/Retro-Diels Alder chemistry [28,36,38,48] (see Figure 9.6). Both of these cycloaddition reactions avoid elimination products and yield final material glass transition temperatures of approximately 200°C. With Diels-Alder/Retro-Diels-Alder chemistry, the material $T_g$ can be defined by choice of dien and dienophile reactants; recently, EO materials with glass transition temperatures exceeding 250°C have been produced.

Glass transition temperature can be defined by many methods including differential scanning calorimetry and the onset of conductivity. However, perhaps the most relevant method for assessing the stability of poling-induced EO activity is *in-situ* measurement of EO activity employing Teng–Man ellipsometry [49] as a function of ramping temperature (at a rate of 5–10 degrees per minute). This measurement permits definition of the temperature at which EO activity begins to rapidly decay and can also provide insight into material inhomogeneity (if EO activity is lost in a stepwise or gradual fashion with temperature ramping).

Another method of assessing thermal stability is simply to monitor the drive voltage ($V_\pi$) of an operating device for long periods of time (e.g., thousands of hours), defining decay constants from normal kinetic analysis of the temporal data. Telcordia standards require 5000 h of stability at 85°C (and 85% humidity). It has been recently demonstrated that organic EO devices can meet Telcordia requirements [50].

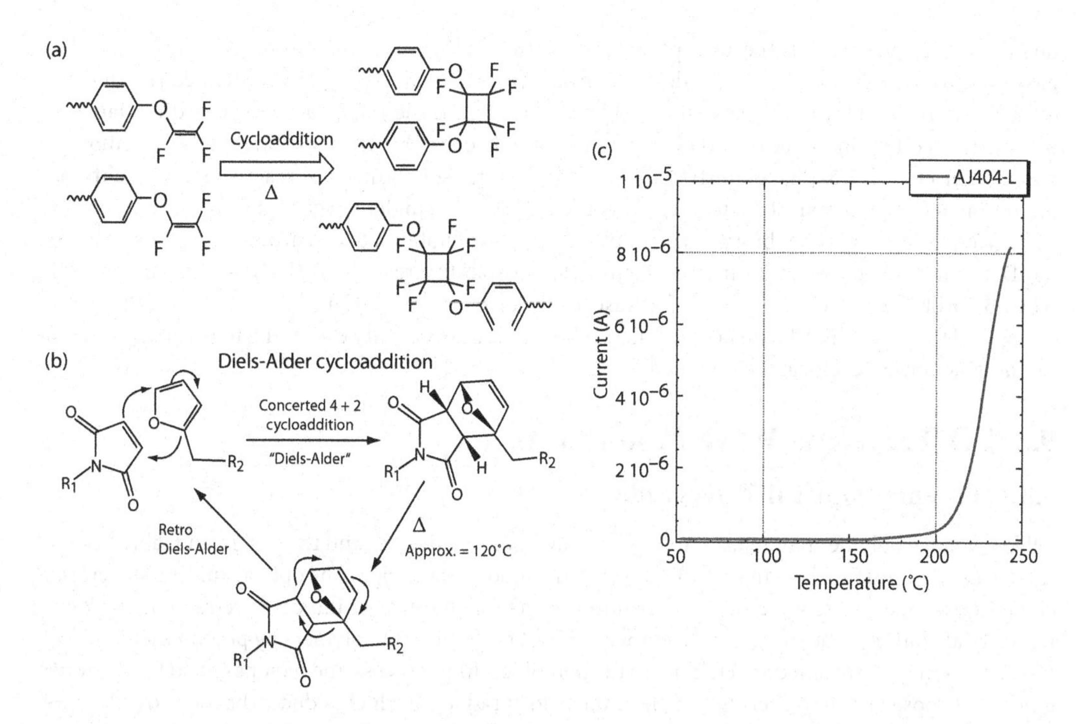


**FIGURE 9.6** Cycloaddition reactions capable of yielding material glass transition temperatures of 200°C or greater are shown. (a) The use of the fluorovinyl ether moiety to produce cyclobutyl crosslinks is shown. (b) Diels-Alder/Retro-Diels-Alder crosslinking is illustrated. (c) Measurement of material glass transition temperature (lattice softening) by recording the on set of conductivity with temperature ramping is shown. The material is the AJ404 material (see Figure 9.5) hardened by Diels-Alder/Retro-Diels-Alder chemistry.

### 9.2.4.4 Photochemical Stability

The photochemical stability of OEO materials (like that of most organic materials) is dominated by singlet oxygen chemistry. Except for a subset of organic chromophores that undergo photo-induced cyclization reactions, organic chromophores are quite robust in the absence of oxygen. Indeed, the photostability of organic materials has been shown to vary by as much as six orders of magnitude as a function of the inhibition of singlet oxygen chemistry.

Two general approaches have been adopted for the study of photochemical stability. The first is based on pump-probe spectroscopic experiments. With pump-probe experiments, the sample is irradiated with a strong optical field (the pump) at a wavelength of interest and the effect on the intensity of the interband charge transfer absorption of the EO chromophores is monitored with a weaker optical field (the probe). It is important that the probe light does not contribute to photochemical decay or artificially fast decay rates will be measured. Because singlet oxygen is activated when the chromophore absorbs light (the excitation is transferred to oxygen to produce singlet oxygen, which then attacks reactive sites on the chromophore), decay rates are a function of the proximity of the wavelength of the pump light to that of the charge transfer absorption maximum for the chromophore. Pump light at telecommunication wavelengths is far removed from the absorption maximum's wavelength so kinetic decay of the chromophores is very slow even for pump powers of 400 mW–1000 mW. Several thousand hours of constant pump irradiation are usually required to get adequate temporal kinetic decay data. To keep the probe light from influencing kinetics, the probe must be kept off except during the brief periods when the chromophore absorption is being sampled to record decay data. Stegeman et al. [12,51] developed a simple photostability figure-of-merit, $B/\sigma$, where $B^{-1}$ is the photodegradation quantum efficiency and $\sigma$ is the absorption cross section at the wavelength of interest. This analysis is based on the assumption of a single decay pathway. Günter et al. [12,52] have extended treatment to consider more complex decay pathways and have emphasized the importance of recording data over more extended temporal periods to obtain more realistic decay kinetics. Researchers at Corning [53] and Günter et al. [12,52] have illustrated the many factors (substitution of reactive sites on chromophores, material glass transition temperature, the presence of singlet oxygen quenchers, etc.) that can influence photochemical stability. As a simple example, we note that the chromophore/host mixture SBLD-1 exhibits a factor of 20 improvement in photostability relative to the same chromophore in APC. By protection of chromophore sites susceptible to attack by singlet oxygen [54], by lattice hardening, and by use of singlet oxygen quenchers, photochemical stability exceeding ten years at telecommunication power levels can be obtained.

# 9.3 EO Traveling Wave Modulators

## 9.3.1 Polymer Optical Waveguides

All low-voltage polymer modulators are optical wave-guiding devices and the design and fabrication of the optical waveguides plays an important role in the modulator design. The optical guiding structure is composed of an EO core surrounded by cladding layers that have lower indices of refraction to provide the vertical confinement. The lateral confinement is achieved in EO polymers by photobleaching, electric field poling, and reactive ion etching. In the photobleaching process, the regions outside of the waveguide are exposed to high-intensity visible or ultraviolet radiation which reduces the index of refraction of the core polymer material by photoinduced decomposition of the chromophores [55]. Typically, oxygen plays a role in the decomposition chemistry. Index of refraction changes on the order of 0.01 are possible. For poling-induced waveguides, the poling electrodes are patterned over the waveguide and therefore only the patterned waveguide core region is poled. The alignment of the choromphores (due to the poling) increases the index of refraction for the transverse magnetic (TM) polarization [56]. The TM polarization is along the poling E-field direction. In the reactive ion etching process, a ridge is etched in the core layer prior to spinning the upper cladding. The height of the ridge is designed to assure single mode operation at the desired wavelength [57].

Cladding polymers must satisfy several conditions and consequently a compromise is often required. The optical loss of the cladding material must be low and the cladding refractive index must be sufficiently lower than the index of the core to provide a reasonable index contrast. In addition, the three polymer materials (cladding/core/cladding) must be chemically compatible to allow spinning of one layer on top of the lower layer. Another important constraint that is sometimes overlooked is the direct current electrical conductivity of the claddings relative to the core at the poling temperature. The cladding conductivity must be lower than the core conductivity to assure that a significant portion of the poling voltage occurs across the core layer. Since the conductivity of the chromophore-containing core polymer is often relatively high, this electrical conductivity constraint makes it difficult to find good cladding polymers. Enami et al. [58] have reported that properly processed sol-gel can have high conductivity and they have achieved very good poling efficiency with a modulator using this material as cladding. Alternate approaches have been to corona pole the core layer prior to spinning the top cladding or to electrode pole the lower cladding/core stack, remove the poling electrode, and then spin the upper cladding.

The typical polymer modulator is based on reactive ion etching etched buried ridge waveguides (see Figure 9.7) since this provides a long-term stable waveguide structure. Buried ridge waveguides can be designed to be single mode by controlling the ridge height and core thickness. Even with a thick core layer that could support several modes in the vertical direction (slab modes), the ridge can be designed so that higher order slab modes are radiated horizontally out of the waveguide. This approach allows relatively thick core layers to still be single mode. The design is based on calculating the effective

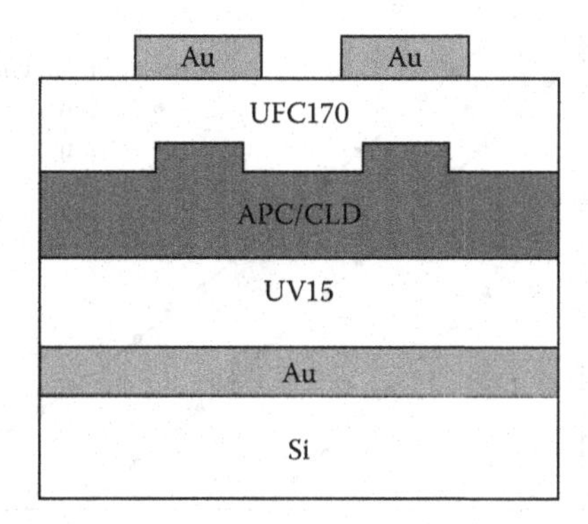


**FIGURE 9.7** Typical cross-section of a buried ridge waveguide showing the two arms of the Mach–Zehnder.

refractive index of the slab waveguide in the center (under the ridge) and the effective index of slab waveguide to the side (outside the ridge). This design procedure is discussed in the literature and Oh et al. [59] have calculated the effective refractive index for a set of specific polymer materials as shown in Figure 9.8.

In Figure 9.8 we see that if a 3 μm thick core layer is selected and the ridge height, $\Delta T$, is 0.4 μm, then the effective index of the lowest order slab mode in the center, $N_o$(center), is larger than the effective index of the lowest order slab mode outside the ridge, $N_o$(side), and therefore the lowest order mode is trapped under the ridge. However, the effective index of the next higher slab mode in the center, $N_1$(center), is smaller than the $N_o$(side) and therefore the higher order mode will radiate into the lowest order side mode. Any optical power that is initially coupled into the higher order slab mode will radiate out of the confined region under the ridge. There are obvious tradeoffs in the design. The core layer and the ridge width are selected to provide a reasonable coupling efficiency from a fiber into the core while the ridge height must be large enough and with reasonable tolerances to make etching of the ridge using typical reactive ion etching etch conditions possible.

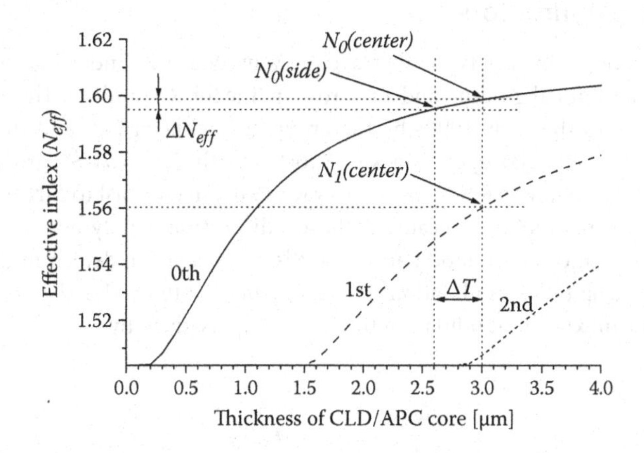


**FIGURE 9.8** Effective index of the planar waveguide as a function of the thickness of the core layer. The waveguide consists of UV15 lower cladding (1.504), CLD/APC core layer (1.612), and UFC170 upper cladding (1.488). In the rib waveguide structure, the waveguide side needs to be etched by $\Delta T$ to obtain the effective index contrast of $\Delta N_{eff}$ for the fundamental mode. (Reproduced from M.-C. Oh et al., *IEEE J. Sel. Topics Quant. Electr.*, 7, 5, 826–835, © 2001 IEEE.)

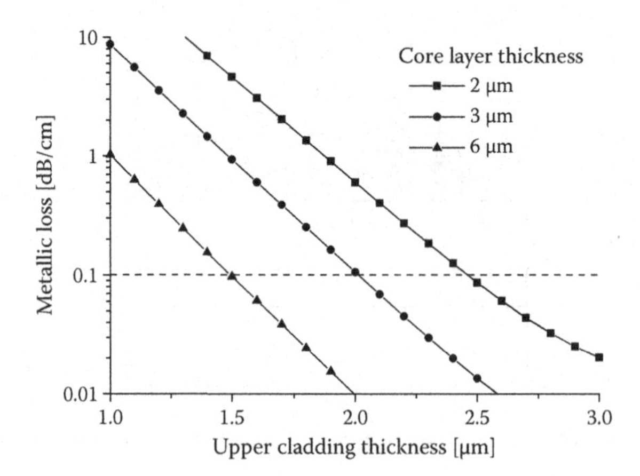


**FIGURE 9.9** Metallic loss of the planar waveguide structure including top and bottom electrode layers made of Cr and Au as a function of the cladding thickness for the three different core-thicknesses. The loss will be less than 0.1 dB/cm if the cladding thickness is greater than 2 mm for the 3-mm core waveguide. (Reproduced from M.-C. Oh et al., *IEEE J. Sel. Topics Quant. Electr.*, 7, 5, 826–835, © 2001 IEEE.)

The minimum cladding layer thickness is set to assure reasonably low optical loss due to the metal electrodes. For the example shown in Figure 9.9, Oh et al. has calculated the metal loss for the TM mode (optical E-field perpendicular to the metal surface) as a function of the cladding thickness for several core thicknesses. For the case of a 3-μm core, a cladding thickness of 2 μm keeps the metal loss less than 0.1 dB/cm. The example calculated by Oh et al. is a good design example that assures a single optical mode and low metal loss. The waveguide parameters are core thickness = 3 μm, ridge height = 0.4 μm, and cladding thickness = 2 μm. These numbers are for CLD/APC core (n = 1.612), UV15 lower cladding (n = 1.488), and UFC170 upper cladding (n = 1.488). Other core and cladding polymers can result in different designs but the numbers will not radically change.

## 9.3.2 High-Speed Modulators

High-speed polymer modulators are typically traveling wave Mach-Zehnder interferometers based on an optical waveguide for optical guiding and a microstrip line for RF guiding. The first demonstrations of this technology were in the early 1990s by the group at Lockheed [60] and the group at Hoechst Celanese [61] and the first 100 GHz operation was reported by the UCLA/USC groups [62]. These polymer devices use a microstrip line to guide the microwave signal in contrast to $LiNbO_3$ devices that typically use a coplanar transmission line. Because of the low dispersion of polymers, the velocity mismatch between the optical wave and the RF modulation wave is not a factor in determining the bandwidth of the polymer modulators up to RF modulation frequencies of over 100 GHz. The product of the interaction length, $L$, with the maximum modulation frequency, $f_{max}$, is given by

$$Lf_{max} = \frac{c}{2\left(n_{\mathrm{RF}} - n_{\mathrm{op}}\right)}, \tag{9.1}$$

where $n_{\mathrm{RF}}$ is the effective index of refraction of the modulation strip line and $n_{\mathrm{op}}$ is effective index of the optical waveguide. If we assume that the effective indices are the indices of the core and take as typical values for the EO polymer $n_{\mathrm{RF}} = 1.74$, $n_{\mathrm{op}} = 1.7$, we find $Lf_{max} = 300$ GHz cm. A 2 cm long device should operate at 150 GHz. Since a device design that corrects for a velocity mismatch is not required, polymer modulators are therefore of relatively simple design and have good overlap between the RF and the optical fields.

$V_\pi$ is often reported at a low frequency while the high-frequency response is given by the 3 $dB_e$ bandwidth. The 3 $dB_e$ bandwidth and $V_\pi$ both decrease with modulator length and there is therefore a trade-off between low $V_\pi$ and high bandwidth. Among the modulators with the lowest reported $V_\pi$ is a hybrid sol-gel device reported by Enami et al. [58] with $V_\pi = 0.65$ V and a 2.4 cm interaction length. The 3 $dB_e$ bandwidth was not reported. The first devices to achieve $V_\pi < 1$ V were demonstrated by Shi et al. [27].

The high-frequency response of polymer modulators is determined by the microwave loss of the microstrip line. The 3 $dB_e$ bandwidth is defined as the frequency at which the detected RF modulation power falls to half of the low frequency value or when the effective $V_\pi$ increases by a multiplicative factor of $\sqrt{2}$. Since the RF losses of most polymers are low [63], the RF loss of a microstrip line is dominated by the metal loss. Polymer modulators that have been fabricated with Au electrodes have an RF loss coefficient of ~0.7 dB $(cm)^{-1}(GHz)^{-1/2}$ [64]. A 2 cm interaction length has an RF bandwidth of ~17 GHz and a 1 cm device has an RF bandwidth of ~35 GHz. Figure 9.10 shows the measured RF response of the 1 cm and 2 cm devices. $V_\pi$ of the 2 cm device was 2.1 V. Lower loss coplanar lines have been reported [65] and therefore one should expect a coplanar design for a polymer modulator will have a larger bandwidth. Coplanar devices will require poling of the chromophores in the plane of the film [66].

Figure 9.11 shows the typical fabrication process of an all-polymer modulator. The polymer core in the two arms of the MZ interferometer is poled in opposite directions to achieve a push–pull operation which reduces $V_\pi$ by a factor of 2.

There is a bias instability observed in some polymer modulators. When a DC bias voltage is applied, the bias point continues to move and requires a continually higher bias voltage to stay at the quadrature point [67]. The presence of this instability depends on the polymer used and is due to charge movement that shields the core from the applied bias voltage. In some modulators, this drift can be stabilized and the bias point tracked by a bias control system [68]. Very effective and stable biasing has been reported using the thermo-optic effect [69]. One arm of the MZ interferometer was heated by an electrode and, because of the relatively large thermo-optic effect in polymers ($dn/dT \sim 10^{-4}$), only a small amount of electrical power was required for quadrature biasing. The thermal approach is stable and, essentially, drift free.

The long-term thermal and photo stabilities of polymer devices are serious system difficulties. While these problems have not been completely resolved, a good understanding of them has evolved in both materials and devices and thus, progress has been made on both issues.

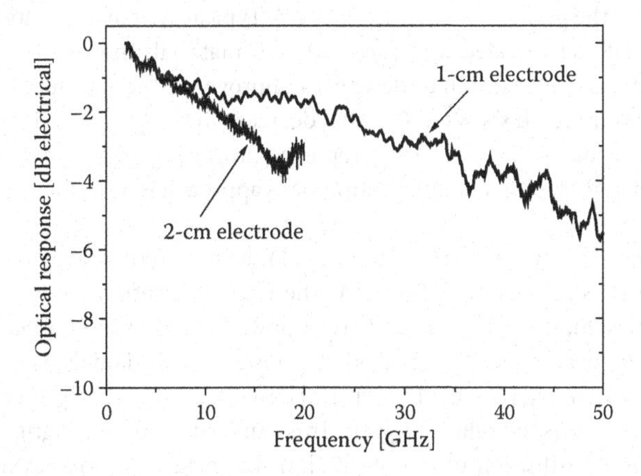


**FIGURE 9.10** The frequency response of the polymer MZ modulator with 1 cm and 2 cm long electrodes. The shorter device exhibits the 3-dB electrical bandwidth over 30 GHz to 2 GHz. (Reproduced from M.-C. Oh et al., *IEEE J. Sel. Topics Quant. Electr.*, 7, 5, 826–835, © 2001 IEEE.)

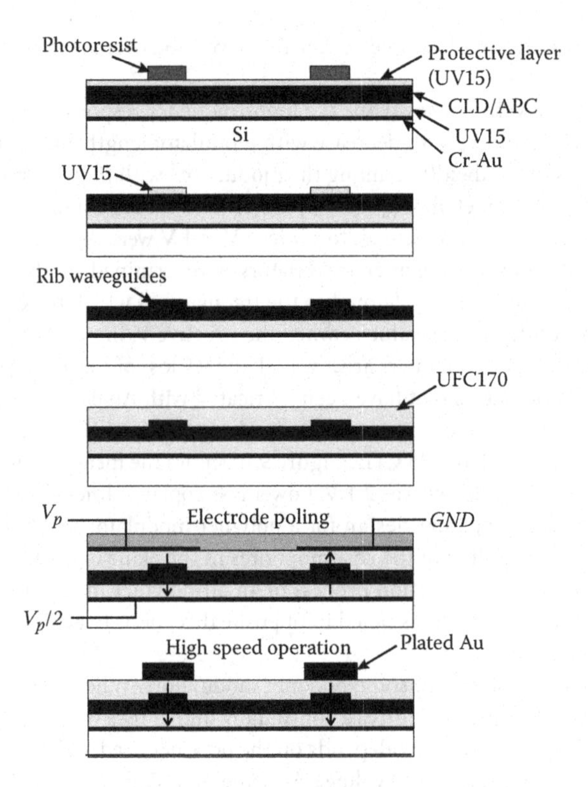


**FIGURE 9.11** Schematic diagram of the polymer modulator fabrication procedure. The polarity of the electric field for push–pull poling is denoted.

Long thermal stability of polymer EO modulators is related to the thermal stability of the poling of the polymer. To achieve the EO effect, the chromophores in the polymer must be aligned to some degree and this is conventionally done by applying a poling electric field with the sample held at a high temperature. If the modulator is operated at an elevated temperature over a length of time, this alignment will relax and the EO coefficient will decrease. The key to maintaining the alignment and improving the thermal stability is therefore to harden the polymer host either by using a high-$T_g$ material or by cross-linking the polymer after the poling is completed [70–72]. Typical system applications require long-term stability at 85°C. Modulators fabricated with the CLD/APC materials show a 25% increase in $V_\pi$ after 40 days at 60°C in air. There is now research underway to improve the thermal stability. An EO dendrimer material with cross-linking which shows only a 10% decrease in the EO coefficient after spending 100 h at 85°C in air has been reported [73]. A recently reported promising approach uses a Diels-Alder "click chemistry" to achieve a high-$T_g$ polymer after poling. This approach is described in the materials section of this chapter.

Photo stability is related to the relatively high optical intensity in the waveguides of a typical modulator (even at input powers of tens of milliwatts). The peak absorption of the chromophores occurs near 750 nm, but at these high intensities the optical power at 1300 nm or 1550 nm can damage and bleach the chromophores over a long time period. The result of this damage is a decrease in the index of refraction and a decrease in $r_{33}$ [74–76]. This optical damage requires the presence of oxygen and the key to preventing the damage is to exclude oxygen. This can be done by packaging, by using denser host polymers which inhibit the diffusion of oxygen [77], or by including moieties in the polymer which deactivate singlet oxygen. Figure 9.12 shows the change in insertion loss and $V_\pi$ of a modulator that is packaged in the inert gas Ar. The increase in the insertion loss when the modulator is operated in air is due to the decrease in the index of refraction of the core and the subsequent loss of waveguiding near the

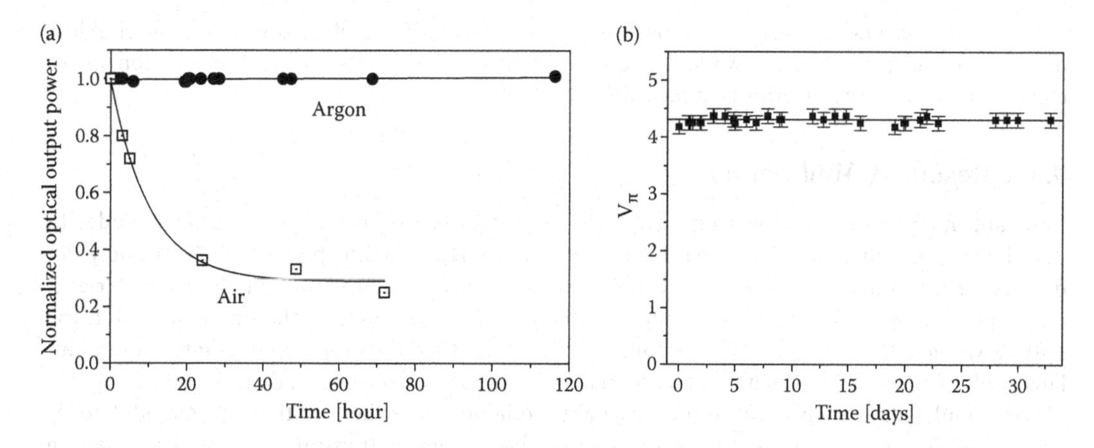


**FIGURE 9.12** (a) The change in waveguide transmission during the illumination with 10 mW at 1550-nm input to the waveguide for the waveguide sealed in Ar and another waveguide exposed to air. (b) Long-term measurement of the half wave voltage of the CLD/APC polymer MZ modulator in Ar with 20 mW at 1550-nm input. No considerable change of the EO coefficient was observed over 30 days. (Reproduced from M.-C. Oh et al., *IEEE J. Sel. Topics Quant. Electr.*, 7, 5, 826–835, © 2001 IEEE.)

input. After packaging, an inert gas is flowed through the package for ~30 min which allows the oxygen trapped in the polymer during processing to diffuse out. The inert gas is then sealed in the package. A recent publication [78] describes this packaging approach to demonstrate a stable polymer waveguide at 1550 nm with 100 mW coupled into the waveguide. Figure 9.13 shows stable transmission of an LPD-80 polymer [79] ridge waveguide for ~28 h at room temperature.

The optical loss in EO polymers has, up to now, not been as low as in $LiNbO_3$ and consequently the insertion loss of polymer modulators has been higher than that of $LiNbO_3$ devices. It is a very difficult chemical problem to synthesize EO polymers with simultaneously high $r_{33}$, low loss, and high stability.

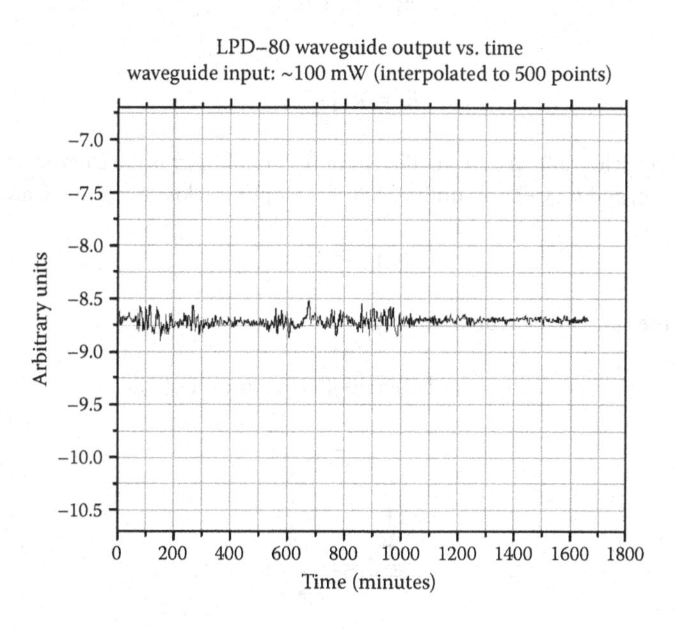


**FIGURE 9.13** Transmission of the LPD-80 waveguide packaged in $N_2$. The power coupled into the waveguide was 100 mW at 1550 nm. No photobleaching or change in transmission was observed over the ~28-h exposure.

EO polymer waveguides have been demonstrated with loss at 1550 nm of 1.2 dB/cm and the chip loss of a 3 cm modulator was ~5 dB, while the fiber to fiber loss was 9 dB–10 dB [59]. The insertion loss of commercial polymer modulators is in the 7 dB range [80].

## 9.3.3 Resonant Modulators

Resonant modulators based on ring waveguide designs or based on whispering gallery modes in discs have been demonstrated in several EO materials [81–84], including polymers [85]. The polymer devices are waveguide ring resonators coupled to a bus waveguide. The waveguide core material is an EO polymer and when a voltage is applied, the resonant wavelength of the ring is shifted. If the optical wavelength is offset from the resonant wavelength, then the output light is intensity modulated and if the optical wavelength is set to match the resonant wavelength, then the output light is phase modulated. The advantage of the resonant modulator over the traditional MZ modulator is an increase in sensitivity (lower $V_\pi$) because of the large slope in transmission versus wavelength (if the Q of the resonator is large). The downside of the resonant modulator is the reduced modulation bandwidth which is limited to approximately the full width at half maximum (FWHM) of the resonance [86]. All resonant modulators have an inherent trade-off between sensitivity and modulation bandwidth and the constant sensitivity-bandwidth product is largely determined by the EO coefficient of the polymer.

### 9.3.3.1 Overview of Ring Resonators

Prior to considering ring resonant modulators, we present a brief overview of ring resonators to define some of the pertinent parameters.

#### *9.3.3.1.1 Resonator Transfer Function*

A general picture of the microring resonator with a coupling waveguide is shown in Figure 9.14. Using the coupled mode theory [87], we can relate the electric fields in the coupling region as

$$E_3 = \tau E_1 + i\kappa E_2$$

$$E_4 = i\kappa E_1 - \tau E_2, \tag{9.2}$$

where $\kappa$ is the electric field transmission coefficient of the coupling, $\tau$ is the electric field reflection coefficient of the coupling, and $i = \sqrt{-1}$. Assuming that the coupling is lossless, $\tau$ and $\kappa$ are related by

$$|\kappa^2| - |\tau^2| = 1. \tag{9.3}$$

One round trip in the ring can be represented by

$$E_2 = \exp(-\alpha L/2)\exp(-j\beta L)E_4 = \alpha\ \exp(-i\beta L), \tag{9.4}$$

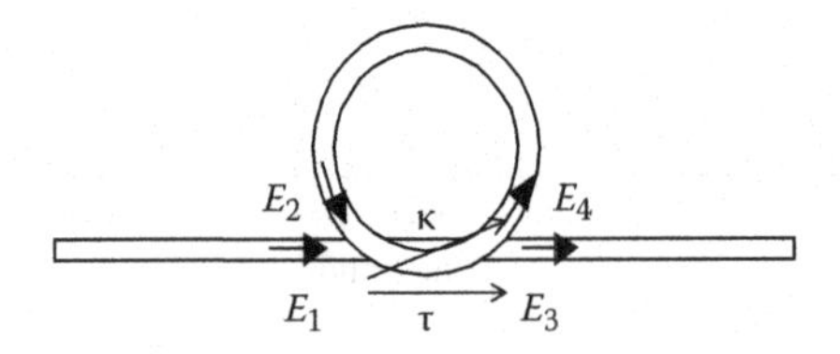


**FIGURE 9.14** Schematic of a resonant ring waveguide coupled to one bus waveguide.

where $\beta$ is the mode propagation constant in the ring given by

$$\beta = \frac{2\pi n_{\text{eff}}}{\lambda} \tag{9.5}$$

and $\alpha$ is the E-field transmission for one round trip in the ring

$$\alpha = \exp(-\alpha L/2). \tag{9.6}$$

In the expressions above,

$n_{\text{eff}}$ = effective index of refraction for the ring mode
$\alpha$ = intensity loss coefficient
$L$ = circumference of the ring.

By replacing $E_2$ from Equation 9.4 into Equation 9.2 and solving the equation for $E_4$ we obtain

$$\frac{E_4}{E_2} = \frac{i\alpha\kappa \exp(-i\theta)}{1-\alpha\tau \exp(-i\theta)}. \tag{9.7}$$

In the typical case, the coupling waveguide is phase matched to the ring waveguide and therefore $\kappa$ and $\tau$ are real numbers. In this case,

$$\frac{I_4}{I_2} = \frac{(1-\tau^2)\alpha^2}{1-2\alpha\tau\cos\theta-\tau^2\alpha^2}, \tag{9.8}$$

where

$$\theta = \beta L.$$

We can also obtain

$$\frac{E_3}{E_1} = \frac{\tau-\alpha s^{-i\theta}}{1-\alpha\tau s^{-i\theta}} \tag{9.9}$$

and if $\kappa$ and $\tau$ are real,

$$\frac{I_3}{I_1} = \frac{\alpha^2-2\alpha\tau\cos\theta-\tau^2}{1-2\alpha\tau\cos\theta+\alpha^2\tau^2}. \tag{9.10}$$

At resonance, $\theta = 2m\pi$ (where $m$ is an integer) and the throughput becomes

$$\frac{I_3}{I_1} = \frac{(\alpha-\tau)^2}{(1-\alpha\tau)^2}. \tag{9.11}$$

Critical coupling ($I_3 = 0$) occurs when $\alpha = \tau$.

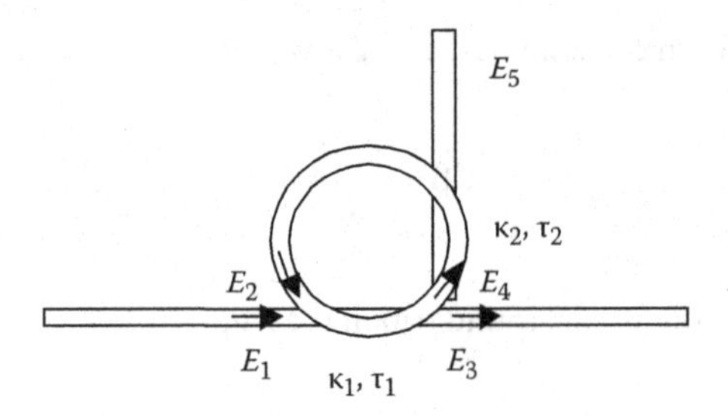


**FIGURE 9.15** Schematic of a resonant ring waveguide coupled to one input waveguide and one output waveguide.

If the microring has a second output waveguide as shown in Figure 9.15, the round trip expression (Equation 9.4) is modified by substituting $\alpha$ by $\alpha\tau_2$. The output intensity transmission ($I_3/I_1$) can then be calculated as

$$\frac{I_3}{I_1} = \frac{\tau_2^2\tau_1^2}{1 - 2\tau_1\tau_2\alpha\cos\theta - \alpha^2\tau_1^2\tau_2^2}, \tag{9.12}$$

where $\tau_1$ and $\kappa_1$ are the E-field reflection and transmission coefficients for input coupling, respectively, and $\tau_2$ and $\kappa_2$ are the E-field reflection and transmission coefficients for output coupling. The through-put intensity transmission ($I_3/I_1$) and output intensity transmission ($I_3/I_1$) at resonance ($\theta = 2m\pi$) can be calculated as

$$\frac{I_3}{I_1} = \frac{(\alpha\tau_2 - \tau_1)^2}{(1 - \tau_1\tau_2\alpha)^2} \tag{9.13}$$

and

$$\frac{I_3}{I_1} = \frac{\kappa_2^2\kappa_1^2}{(1 - \alpha\kappa_2\kappa_1)^2}. \tag{9.14}$$

Critical coupling ($I_3 = 0$) and the maximum of ($I_3/I_1$) occur when $\tau_1 = \tau_2\alpha$ and in this case

$$\left(\frac{I_3}{I_1}\right)_{\max} = \frac{1 - \tau_2^2}{1 - (\tau_2\alpha)^2}. \quad (I_3/I_1)_{\max} < 1 \tag{9.15}$$

If there is loss in the ring ($\alpha < 1$) then $(I_3/I_1)_{\max} < 1$ is always the case.

#### *9.3.3.1.2 Resonator Parameters*

There are a number of parameters that define the properties of microresonators. The first parameter is the Q-factor, which is defined by

$$Q = \frac{\lambda_0}{\Delta\lambda_{1/2}}. \tag{9.16}$$

where $\Delta\lambda_{1/2}$ is the (wavelength) full width half maximum of the dropped intensity or transmitted intensity and $\lambda_0$ is the resonant wavelength. One can derive $Q_L$, the loaded $Q$ (with coupling losses included):

$$Q_L = \frac{\pi L n_{\text{eff}} \sqrt{\alpha\tau_2\tau_1}}{(1\text{-}\alpha\tau_2\tau_1)\lambda_0}. \tag{9.17}$$

$Q_L$ for a single coupling waveguide can be obtained by setting $\tau_2 = 1$ and $Q_U$, the unloaded $Q$, can be obtained by setting $\tau_1 = \tau_2 = 1$.

The free spectral range (FSR) is the wavelength spacing between two resonances of the resonator (when m changes by one). If we neglect the dispersion of $n_{\text{eff}}$, the free spectral range is

$$FSR = \lambda_{m-1} - \lambda_m = \frac{\lambda^2}{n_{\text{eff}}(\lambda)2\pi R}. \tag{9.18}$$

The finesse, $F$, is given by

$$F = \frac{FSR}{\Delta\lambda_{1/2}} = \frac{\pi\sqrt{\alpha\tau_1\tau_2}}{(1-\tau_1\tau_2)}. \tag{9.19}$$

### 9.3.3.2 Sensitivity of Ring Resonant Modulators

The sensitivity of ring resonator-based modulators can be defined by comparing the half-wave voltage, $V_\pi$, of the resonant modulator to that of conventional MZ modulator. The schematic of a ring modulator is shown in Figure 9.16. The modulator consists of a ring resonator, made of an EO material, coupled to a bus waveguide. The amplitude of the output field $E_{\text{out}}$ is given by Equation 9.7:

$$E_{\text{out}} = H(\theta)E_{\text{in}} = \frac{\tau - \alpha s^{-i\theta}}{1 - \alpha\tau s^{-i\theta}} E_{\text{in}}. \tag{9.20}$$

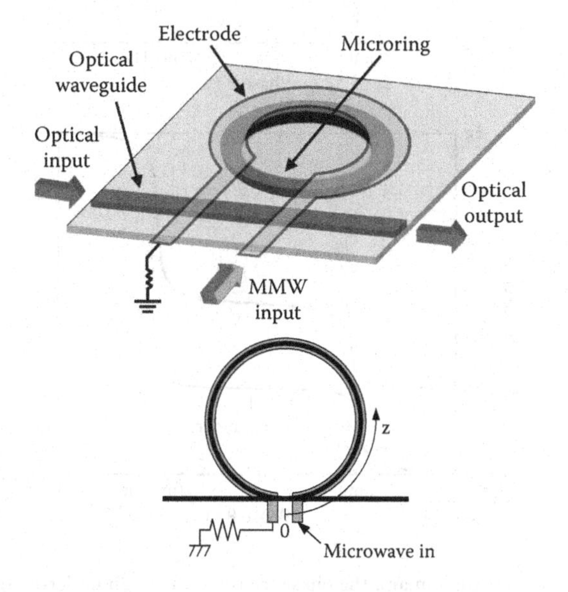


**FIGURE 9.16** Schematics of the traveling wave resonant ring modulator.

The transmission of a ring resonator is given by Equation 9.7:

$$T(\theta)=\left|H(\theta)\right|^2=1-\frac{(1-\alpha^2)(1-\tau^2)}{(1-\alpha\tau)^2+4\alpha\tau\sin^2(\theta/2)}=1-\frac{1}{1+\left(\frac{2F}{\pi}\right)^2\sin^2(\theta/2)}. \tag{9.21}$$

The second form of Equation 9.21 is valid when $F$ is large and $\theta$ is small. When $\alpha = \tau$, the transmission vanishes at the resonances ($\theta = 0$). In this case, the losses in ring are equal to the coupling losses, and this condition is called critical coupling. For $\alpha < \tau$, the resonator is said to be under-coupled and for $\alpha > \tau$ the resonator is said to be over-coupled. The phase of the transmitted light is given by the argument of Equation 9.20 as follows:

$$\Theta(\theta) = arg(H(\theta)). \tag{9.22}$$

Figure 9.17 shows the transmission and phase for over-coupled ($\alpha = 0.8$, $\tau = 0.7$) and under-coupled ($\alpha = 0.8$, $\tau = 0.9$) ring resonators. The transmission drops at a resonance and the phase undergoes a rapid variation with respect to the round-trip phase shift at the resonance. If the resonance of a ring resonator can be electro-optically tuned, ring resonators can be used for either intensity modulators using the

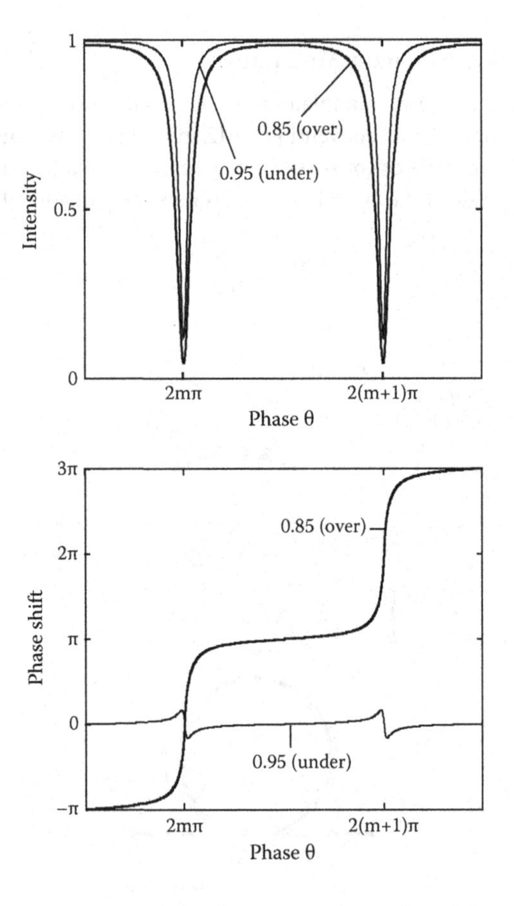


**FIGURE 9.17** The intensity transmission and the phase transmission of an under-coupled and an over-coupled resonator.

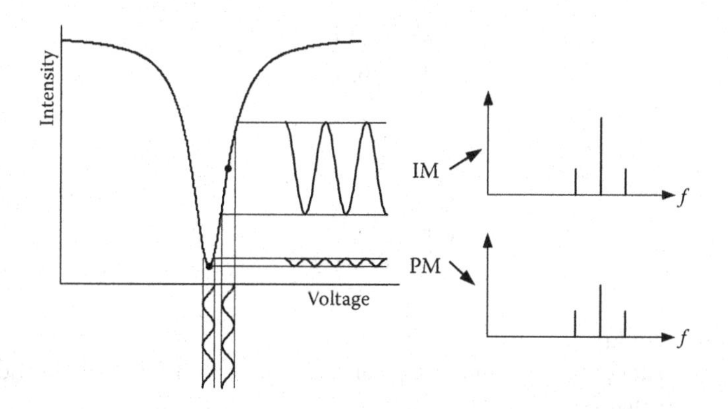


**FIGURE 9.18** Conceptual drawing of the intensity modulation (IM) and the phase modulation (PM) by an EO ring resonator and the frequency spectrum of the modulated light.

steep slope of the transmission near a resonance or phase modulators using the rapid variation of the phase at a resonance. The situation is illustrated in Figure 9.18. When the ring resonator is biased at the point of maximum slope of the transmission vs. modulation voltage curve, the output intensity of light will be strongly modulated by a small modulating voltage. In phase modulators operation, on the other hand, there is no intensity modulation at the modulation frequency, $\omega_m$, and a small amount at $2\omega_m$ due to the symmetry of $T(\theta)$ at an optical resonance. Also in phase modulators operation, the optical carrier is suppressed by optical resonance.

When a voltage, $V$, is applied to an EO ring, the phase shift, $\theta$, is expressed by

$$\theta = \theta_0 + \pi\frac{V}{V_\pi^0}, V_\pi^0 = \frac{\lambda g}{n_0^2 r_{33}\Gamma L}, \tag{9.23}$$

where $\theta_0$ is the bias phase, $L$ is the circumference of the ring, $n_0$ is the effective index of refraction of the ring waveguide, $\lambda$ is the free space optical wavelength, $r_{33}$ is the EO coefficient, $g$ is the electrode gap, and $\Gamma$ is the electrical-optical overlap integral. $V_\pi^0$ is the voltage that produces a $\pi$ phase shift in one round-trip in the resonator and is also the half-wave voltage of a MZ modulator with an interaction length of $L$. $V_\pi^0$ is given by material and device structure parameters.

#### 9.3.3.2.1 *Intensity Modulation*

For intensity modulator operation, the sensitivity of a ring modulator, $V_\pi^{\text{IM}}$ (an equivalent to $V_\pi^0$), can be defined by comparing the slope of the transmission, $|dT/dV|_{\max}$, of a ring modulator with that of a Mach–Zehnder (MZ) modulator with the same interaction length, as in [88,89]:

$$V_\pi^{\text{IM}} = \frac{\pi}{2}\left(\left|\frac{dT}{dV}\right|_{\max}\right)^{-1} = \frac{\pi}{2}\left(\left|\frac{dT}{d\theta}\frac{d\theta}{dV}\right|_{\max}\right)^{-1} = \frac{V_\pi^0}{2\left|\frac{dT}{dS}\right|_{\max}}. \tag{9.24}$$

The enhancement of the modulation sensitivity by optical resonance is the multiplicative factor $2|dT/d\theta|$.

*Biasing for Maximum Signal* The maximum slope of the transmission function, and therefore the lowest $V_\pi^{\text{eff}}$, for a single bus waveguide critically coupled microresonator modulator occurs at $T = 0.25$ with the optical frequency set at $\Delta f = \Delta f_{\text{FWHM}}/\left(2\sqrt{3}\right)$ off resonance where

$$\frac{dT}{d\theta} = \frac{\theta F}{\theta\pi\sqrt{3}} \tag{9.25}$$

and

$$V_\pi^{\text{eff}} = \frac{4\sqrt{3}\pi\lambda g}{\theta F n^3 r_{33} L\Gamma}. \tag{9.26}$$

Here, $F$ is the resonator finesse.

If we bias the modulator at the maximum slope of the transmission function, the modulated optical transmission can be written as

$$T = \frac{1}{4} + \frac{1}{2}\delta - c_3\delta^3 + c_4\delta^4 - c_6\delta^6 + \cdots, \tag{9.27}$$

where

$$\delta = \frac{\pi V_m}{V_\pi^{\text{eff}}}.$$

At this bias point there is no second-order distortion and $V_\pi^{\text{eff}}$ is a minimum.

For example, the condition of $\alpha = \tau = 0.8$ (critical coupling) gives $V_\pi^{\text{IM}} = 0.35V_\pi^0$ at $\theta_0 = 0.082\pi$ (the point of maximum slope). The higher finesse resonator gives a larger enhancement of the modulation.

*Biasing for Minimum Distortion* If the bias point is set to $f_{\text{FWHM}}/2$, where $T = 0.5$, the output power can be written as [90]

$$T = \frac{1}{2} + \frac{1}{2}\delta - \frac{3}{22}\delta^2 + \frac{3}{11}\delta^4 - \frac{10}{11}\delta^5 + \cdots, \tag{9.28}$$

where

$$\delta = \frac{\pi V_m}{V_\pi^{\text{eff}}}.$$

Under this biasing condition, there is no third-order term and

$$V_\pi^{\text{eff}} = \frac{\pi\lambda g}{F n^3 r_{33} L\Gamma}. \tag{9.29}$$

Hence, at this bias point, the third-order intermod is eliminated but $V_\pi^{\text{eff}}$ increases by 1.1 dB and the direct current optical power level increases by 3 dB.

#### *9.3.3.2.2 Phase Modulation*

In phase modulator operation, it is not straightforward to define an equivalent half-wave voltage [91], because the transmitted optical power depends on the parameters $a$ and $\tau$. At the condition of critical coupling, the slope of the phase $\Theta$, $|d\Theta/d\theta|_{\theta=0}$, is infinite and the transmittance, $T(0)$, goes to zero. An equivalent $V_\pi^{PM}$ can be defined which gives, in the small signal region, the same level of first-order sidebands as a conventional phase modulator with the same $V_\pi$ does:

$$V_\pi^{PM} \lim_{\theta\to 0} \pi\left(\left|\frac{d\Theta}{dV}\right| H(\theta)\right)^{-1} = \frac{V_\pi^0}{\lim_{S\to 0}\left|\frac{d\Theta}{dS}\right|\left|H(\theta)\right|}. \tag{9.30}$$

This equivalent $V_\pi^{PM}$ gives the intensity of the first modulation sideband as $\left|J_1\left(\pi V/V_\pi^{PM}\right)\right|^2$. It does not give the intensity ratio of the modulation sideband to the transmitted carrier. The transmitted carrier is given by $T(0)$. These intensities are consistent with the results calculated from the multiple round-trips approach [90]. Hence, one can use the equivalent $V_\pi^{PM}$ in the same way as $V_\pi$ of a conventional phase modulator, because the intensity of the first modulation sideband is the same. The resonant phase modulator produces a suppressed carrier phase modulation spectrum.

#### 9.3.3.3 Modulation Bandwidth and the Frequency Response of Resonant Modulators

As noted earlier, the modulation bandwidth of resonant modulators is always limited by the linewidth of the resonator. If the modulator is driven as a lumped circuit device, the bandwidth is also limited by the resistance-capacitance (RC) time constant of the device and by the drive circuitry. In many cases, the resonator is physically small and its capacitance is relatively small and, when driven by a 50 Ω source, the RC time constant is small and the bandwidth limiting parameter is the resonator's linewidth. However, resonant modulators can be used at multiples of the FSR and therefore used in a system with a high carrier frequency but with limited bandwidth around the carrier frequency. At multiples of the FSR, the RC time constant will become a limitation and these devices must be driven as RF traveling wave modulators. The traveling wave ring modulator is illustrated in Figure 9.16. A microstrip line electrode that is impedance-matched to the driving cable and termination covers an EO ring resonator. The traveling wave analysis will apply to operation at baseband and at high carrier frequencies.

When a modulation signal, $\sin(\omega_n t)$, is applied to an electrode in a ring resonator, the output amplitude, $E_{out}(t)$, using the multiple round-trip approach, is given by [88,89]

$$E_{out}(t) = \left\{\tau - (1-\tau^2)\sum_{n=1}^{\infty} \tau^{n-1}\alpha^n e^{-i\left[n\theta_0 + \delta_n \sin(\omega_n t - n\phi)\right]}\right\} E_{in}(t). \tag{9.31}$$

where $\phi = \omega_m/\text{FSR}$. The FSR is given by $c/n_0 L$. We assume that the group and effective indices of refraction of the waveguide are same. $\theta_0$ is the round-trip phase shift and $\delta_n$ is the modulation index of the nth round trip [88,89]. The dispersions of propagation constants, coupling coefficients, and losses are neglected. Assuming no microwave loss, we consider the drive signal

$$V(z,t) = V_0 \sin\left[\omega_m\left(t - \frac{n_m}{c}z\right)\right]. \tag{9.32}$$

where $n_m$ is the microwave effective index. The voltage seen at position $z$ along the ring resonator waveguide by photons that enter at $z = 0$ when $t = t$ is given by [91]

$$V(z,t) = V_0 \sin\left[\omega_m\left(t - \frac{\Delta n}{c}z\right)\right], \tag{9.33}$$

where $\Delta n = n_m - n_0$. The modulation index, $\delta_n$, is given by

$$\delta_n \sin(\omega_m t - n\phi) = \sum_{k=0}^{n-1} \int_0^L \Delta\beta \sin\left\{\omega_m \left[(t + kt_r) - \frac{\Delta n}{c} z\right] - n\phi\right\} dz$$

$$= \Delta\beta L \frac{\sin(\psi/2)}{\psi/2} \frac{\sin(n\phi/2)}{\sin(\phi/2)} \sin\left(\omega_m t - \frac{\psi}{2} - \frac{n+1}{2}\phi\right), \quad (9.34)$$

where $t_r = n_0 L/c$ is the optical round trip time and $\psi = \omega_m \Delta n L/c$ is the velocity matching factor.

The optical power of the transmitted signal can be given by substituting Equation 9.10 into Equation 9.7 and calculating the first component of the Fourier expansion of the output intensity:

$$I_{\omega_m} = \frac{2}{T} \int_0^T \left|E_{\text{out}}(t)\right|^2 e^{i\omega_m t}\, dt\,, \quad (9.35)$$

where $T = 2\pi/\omega_m$. The small signal frequency responses (in $dB_e$) of the modulated output from velocity matched ring modulators, optically biased to the maximum slope, with finesses of 10 and 30 are shown in Figure 9.19. The responses are normalized to the small signal response of an MZ modulator biased at quadrature with the same electrode length. The modulating frequency is normalized to the FSR. Velocity-matched traveling wave ring modulators provide high modulation efficiency at frequencies around all multiples of the FSR. The 3 $dB_e$ bandwidth for $F = 10$ is 0.174 and for $F = 30$ is 0.0516. Notice that this bandwidth is somewhat larger than the resonator line-width (0.1 for $F = 10$ and 0.033 for $F = 30$). The frequency scale around one of the optical resonances is expanded in Figure 9.20 to show the baseband response for $F = 10$.

Figure 9.21 provides a physical explanation of why the response is enhanced at frequency spacings of the FSR. At the baseband frequency, the optical source and the optical sidebands created by the modulation all fall within one optical linewidth. When the modulation frequency matches the FSR, the optical sidebands fall in the adjacent optical mode and therefore can resonate within the resonator.

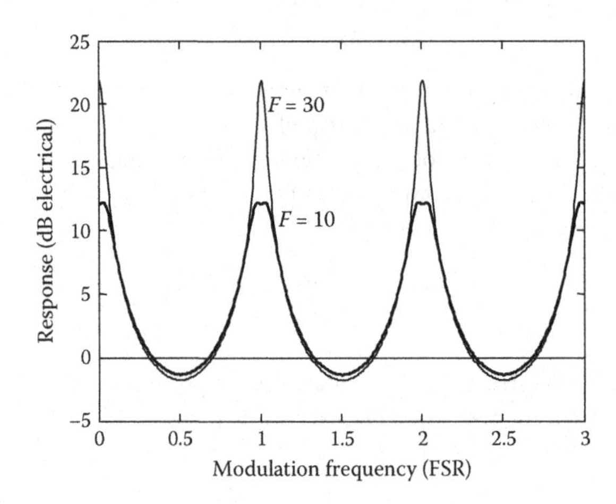


**FIGURE 9.19** Modulation frequency response of velocity matched traveling wave ring modulators with the finesse of 10 and 30. The responses are normalized by the response of an MZ modulator with the same electrode length.

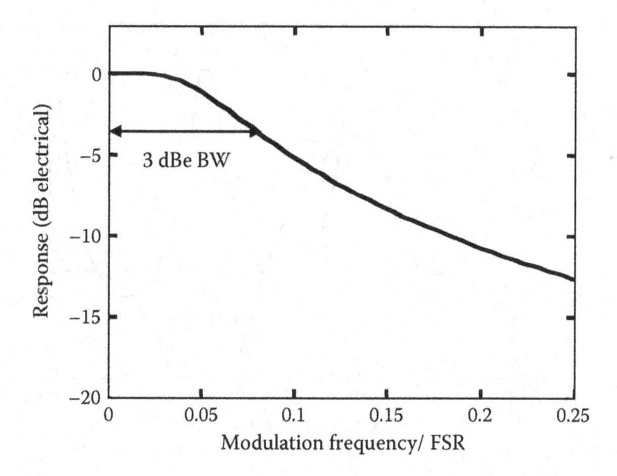


**FIGURE 9.20** Baseband modulation frequency response of resonant modulator. The 3 dBe bandwidth is larger than one half the FWHM of the optical resonance.

#### *9.3.3.3.1 Velocity Mismatch and Microwave Loss*

In addition to the velocity mismatch, another bandwidth limitation in traveling wave modulators is the loss of the microwave transmission line. For a given electrode dimension, the high-frequency microwave loss is determined by the skin depth and one expects a loss in dB/cm of $\alpha = \alpha_0 f^{1/2}$, where $a_0$ depends on electrode conductivity and geometry. Assuming no velocity mismatch, the effect of loss on the modulation index $\delta_n$ is [92]

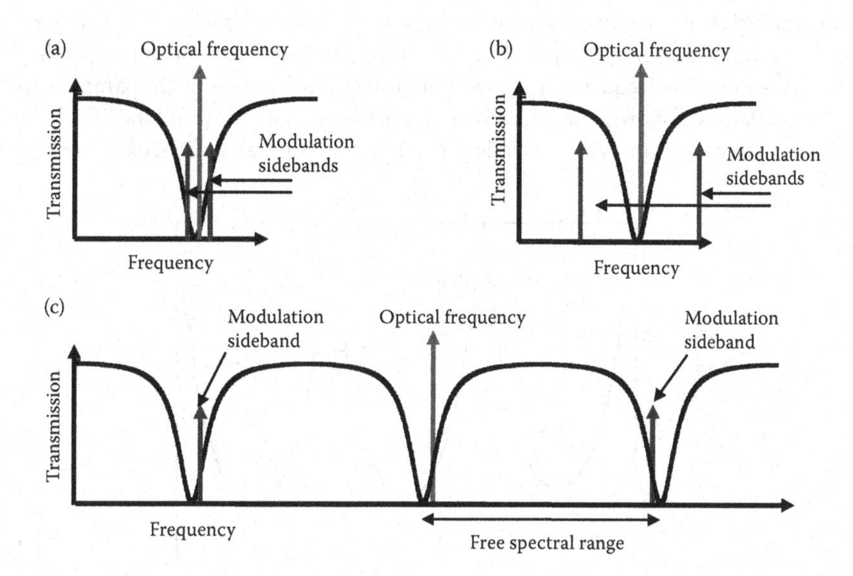


**FIGURE 9.21** Modulation frequency spectrum of the resonant modulator. Part a is the baseband case in which the optical carrier and the optical modulation sidebands all fall within the same optical resonance. Part b is the case in which the modulation frequency is greater than the FWHM of the resonator and the sidebands fall outside of the resonant response to cause the modulation efficiency to fall to near zero. In Part c the modulation frequency is increased to match the FSR of the resonator and the sidebands fall within the two adjacent modes of the resonator and the modulation efficiency is high.

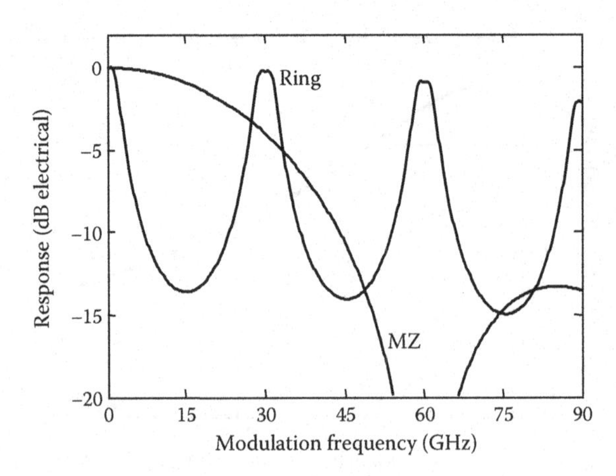


**FIGURE 9.22** Effect of velocity mismatch ($\Delta n = 0.2$, no microwave loss) in the modulation frequency response of the ring resonant modulator with FSR of 30 GHz and $F = 10$ compared to the equivalent broadband MZ modulator. The response is normalized by the low frequency response.

$$
\begin{aligned}
\delta_n \sin(\omega_m t - n\phi) &= \sum_{k=0}^{n-1} \int_0^L \Delta\beta e^{-\alpha_m z} \sin\left[\omega_m(t + kt_r) - n\phi\right] dz \\
&= \Delta\beta L \frac{1 - s^{\alpha_m L}}{\alpha_m L} \frac{\sin(n\phi/2)}{\sin(\phi/2)} \sin\left(\omega_m t - \frac{n+1}{2}\phi\right),
\end{aligned} \tag{9.36}
$$

where $\alpha_m = \alpha/8.7$ converts the power loss in dB/cm to an exponential amplitude loss coefficient.

To evaluate the effect of velocity mismatch and microwave loss in traveling wave ring modulators, we compare the small signal frequency response of the microring resonator output to that of a MZ modulator with the same low-frequency $V_\pi$. As a numerical example, we chose the parameters of the EO polymer ring modulator as follows. The refractive index of the polymer is 1.6 and the EO coefficient is 50 pm/V at a wavelength of 1.31 μm. The electrode gap is 10 μm. Assuming an FSR of 30 GHz, a finesse of

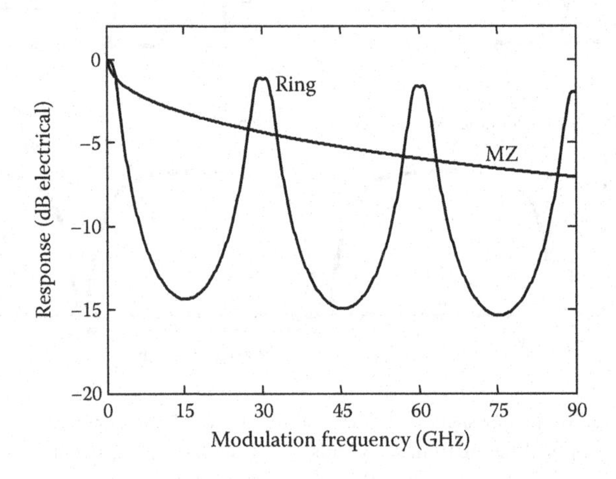


**FIGURE 9.23** Effect of microwave loss ($\alpha = 0.7$ dB/cm-GHz$^{1/2}$, $\Delta n = 0$) in the modulation frequency. Response of the ring resonant modulator with the FSR of 30 GHz ($L = 6.25$ mm) and $F = 10$ compared to the equivalent M-Z modulator ($L = 4.04 \times 6.25$ mm). The responses are normalized by the low frequency response.

10, and critical coupling, the circumference of the ring resonator is set to 6.25 mm and the $V_\pi^{eq}$ is 2.53 V. The electrode length of the equivalent MZ modulator would be 2.53 cm.

Figure 9.22 shows the effect of velocity mismatch on the modulation frequency response of the ring modulator and the equivalent MZ modulator for $\Delta n = 0.2$. The mismatch in $LiNbO_3$ is much larger ($\Delta n \sim 2$) but can be reduced by design while the mismatch is negligible in EO polymer modulators. We choose $\Delta n = 0.2$ as a compromise which clearly shows the effect. Figure 9.23 shows the effect of microwave loss on the modulation frequency response of the ring modulator and the MZ modulator with equal low-frequency $V_\pi$. The microwave power loss coefficient of the microstrip line is 0.7 dB/cm·GHz$^{1/2}$ which is typical for modulators [63]. No velocity mismatch is assumed. In both cases, the ring modulator shows higher modulation efficiency at frequencies around multiples of FSR than the equivalent MZ modulator does because of the shorter electrode length.

# References


1. C. H. Cox, III and E. I. Ackerman, "High electro-optic sensitivity ($r_{33}$) polymers: They are not just for low voltage modulators any more," *J. Physical Chem. B*, vol. 108, no. 25, pp. 8540–8542, Jun. 2004.
2. R. Lytel, G. F. Lipscomb, J. T. Kenney, and E. S. Binkley, "Large-scale integration of electro-optic polymer waveguides," in *Polymers for Lightwave and Integrated Optics*, L. A. Hornak, Ed., New York: Marcel Dekker, 1992, pp. 433–472.
3. T. A. Skotheim and J. R. Reynolds, Eds., *Handbook of Conducting Polymers, Third Edition: Conjugated Polymers, Theory, Synthesis, Properties, and Characterization*, Boca Raton: CRC Press, 2007.
4. L. R. Dalton, "Nonlinear optical polymeric materials: From chromophore design to commercial applications," *Adv. in Polym. Sci.*, vol. 158, pp. 1–86, 2001.
5. L. R. Dalton, P. A. Sullivan, D. Bale, B. Olbricht, J. Davies, S. Benight, I. Kosilkin, B. H. Robinson, B. E. Eichinger, and A. K.-Y. Jen, "Organic electro-optic materials: Understanding structure/function relationships critical to the optimization of electro-optic activity," in *Organic Thin Films for Photonics Applications*, W. Herman and S. Foulger, Eds., Washington, DC: ACS Symposium Series American Chemical Society, pp. 13–33, 2010.
6. Y. Huang, G. T. Paloczi, A. Yariv, C. Zhang, and L. R. Dalton, "Fabrication and replication of polymer integrated optical devices using electron-beam lithography and soft lithography," *J. Phys. Chem. B*, vol. 108, no. 25, pp. 8606–8613, Jun. 2004.
7. H.-C. Song, M.-C. Oh, S.-W. Ahn, W. H. Steier, H. R. Fetterman, and C. Zhang, "Flexible low-voltage electro-optic polymer modulators," *Appl. Phys. Lett.* vol. 82, no. 25, pp. 4432–4434, Jun. 2003.
8. Baehr-Jones, M. Hochberg, G. Wang, R. Lawson, Y. Liao, P. Sullivan, L. Dalton, A. Jen, and A. Scherer, "Optical modulation and detection in slotted silicon waveguides," *Optics Express*, vol. 13, no. 14, pp. 5216–5226, Jul. 2005.
9. T. Baehr-Jones, B. Penkov, J. Huang, P. Sullivan, J. Davies, J. Takayesu, J. Luo, et al., "Nonlinear polymer-clad silicon slot waveguide modulator with a half wave voltage of 0.25 V," *Appl. Phys. Lett.*, vol. 92, no. 16, pp. 163303-1-3, Apr. 2008.
10. J. Takayesu, M. Hochberg, T. Baehr-Jones, E. Chan, G. Wang, P. Sullivan, Y. Liao, J. Davies, L. Dalton, A. Scherer, and W. Krug, "A hybrid electro-optic microring resonator-based 1 × 4 × 1 ROADM for wafer scale optical interconnects," *J. Lightwave Technol.*, vol. 27, no. 4, pp. 440–448, Feb. 2008.
11. P. A. Sullivan and L. R. Dalton, "Theory-inspired development of organic electro-optic materials," *Acc. Chem. Res.*, vol. 43, no. 1, pp. 10–18, Jan. 2010.
12. L. R. Dalton, P. A. Sullivan, and D. H. Bale, "Electric field poled organic electro-optic materials: State of the art and future prospects," *Chem. Rev.*, vol. 110, no. 1, pp. 25–55, Jan. 2010.
13. H. M. Kim and B. R. Cho, "Second-order nonlinear optical properties of octupolar molecules structure–property relationship," *J. Mater. Chem.*, vol. 19, no. 40, pp. 7402–7409, 2009.

14. H. Akdas-Kilig, T. Roisnel, I. Ledoux, and H. Le Bozec "A new class of bipyrimidine-based octupolar chromophores: Synthesis, fluorescent and quadratic nonlinear optical properties," *New J. Chem.*, vol. 33, no. 7, pp. 1470–1473, 2009.
15. P. C. Ray and J. Leszcynski, "First hyperpolarizabilities of ionic octupolar molecules: structure-function relationships and solvent effects," *Chem. Phys. Lett.*, vol. 399, no. 1–3, pp. 162–166, Nov. 2004.
16. M. Blanchard-Desce, J.-B. Baudin, L. Jullien, R. Lorne, O. Ruel, S. Brasselet, and J. Zyss, "Towards highly efficient nonlinear optical chromophores: Molecular engineering of octupolar molecules," *Opt. Mat.*, vol. 12, no. 2–3, pp. 333–338, Jun. 1999.
17. S.-J. Kwon, O-P. Kwon, J.-I. Seo, M. Jazbinsek, L. Mutter, V. Gramlich, Y.-S. Lee, H. Yun, and P. Gunter, "Highly nonlinear optical configurationally locked triene crystals based on 3,5-dimethyl-2-cyclohexen-1-one," *J. Phys. Chem. C*, vol. 112, no. 21, pp. 7846–7852, May 2008.
18. Ch. Bosshard, K. Sutter, Ph. Pretre, J. Hulliger, M. Florsheimer, P. Kaatz, and P. Gunter, *Organic Nonlinear Optical Materials*, Basel, Switzerland: Gordon and Breach, 1995.
19. S. R. Marder, J. W. Perry, and W. P. Schaefer, "4-*N*-methylstilbazolium toluene-p-sulfonate salts with large second-order optical non-linearities," *J. Mater. Chem.*, vol. 2, no. 9, pp. 985–986, 1992.
20. M. Halter, Y. Liao, R. M. Plocinik, D. C. Coffey, S. Bhattacharjee, U. Mazur, G. J. Simpson, B. H. Robinson, and S. L. Keller "Molecular self-assembly of mixed beta zwitterionic and neutral ground-state NLO chromophores," *Chem. Mater.*, vol. 20, no. 5, pp. 1778–1787, Mar. 2008.
21. A. Facchetti, A. Abbotto, L. Beverina, M. E. van der Boom, P. Dutta, G. Evmenenko, G. A. Pagani, and T. J. Marks, "Layer-by-layer self-assembled pyrrole-based donor-acceptor chromophores as electro-optic materials," *Chem. Mater.*, vol. 15, no. 5, pp. 1064–1072, Mar. 2003.
22. A. Facchetti, E. Annoni, L. Beverina, M. Morone, P. Zhu, T. J. Marks, and G. A. Pagani, "Very large electro-optic responses in H-bonded heteroaromatic films grown by physical vapour deposition," *Nature Mater.*, vol. 3, no. 12, pp. 910–917, Nov. 2004.
23. B. C. Olbricht, P. A. Sullivan, G.-A. Wen, A. A. Mistry, J. A. Davies, T. R. Ewy, B. E. Eichinger, B. H. Robinson, P. J. Reid, and L. R. Dalton, "Laser-assisted poling of binary chromophore materials," *J. Phys. Chem. C*, vol. 112, no. 21, pp. 7983–7988, May 2008.
24. M. G. Kuzyk and C. W. Dirk, Eds., *Characterization Techniques and Tabulations for Organic Nonlinear Optical Materials*, New York: Marcel Dekker, 1998.
25. L. R. Dalton, B. H. Robinson, A. K. Y. Jen, W. H. Steier, and R. Nielsen, "Systematic development of high bandwidth, low drive voltage organic electro-optic devices and their applications," *Opt. Mater.*, vol. 21, no. 1–3, pp. 19–28, Jan. 2003.
26. B. H. Robinson and L. R. Dalton, "Monte Carlo statistical mechanical simulations of the competition of intermolecular electrostatic and poling field interactions in defining macroscopic electro-optic activity for organic chromophore/polymer materials," *J. Phys. Chem A*, vol. 104, no. 20, pp. 4785–4795, May 2000.
27. Y. Shi, C. Zhang, H. Zhang, J. H. Bechtel, L. R. Dalton, B. H. Robinson, and Wm. H. Steier, "Low (sub-1 volt) halfwave voltage polymeric electro-optic modulators achieved by control of chromophore shape," *Science*, vol. 288, no. 5463, pp. 119–122, Apr. 2000.
28. T.-D. Kim, J.-W. Kang, J. Luo, S.-H. Jang, J.-W. Ka, N. Tucker, J. B. Benedict, et al., "Ultralarge and thermally stable electro-optic activities from supramolecular self-assembled molecular glasses," *J. Am. Chem. Soc.*, vol. 129, no. 3, pp. 488–489, Jan. 2007.
29. T. Gray, T.-D. Kim, D. B. Knorr, Jr., J. Luo, A. K.-Y. Jen, and R. M. Overney, "Mesoscale dynamics and cooperativity of networking dendronized nonlinear optical molecular glasses," *Nano Lett.*, vol. 8, no. 2, pp. 754–759, Feb. 2008.
30. D. B. Knorr, Jr., X.-H. Zhou, Z. Shi, J. Luo, S.-H. Jang, A. K.-Y. Jen, and R. M. Overney, "Molecular mobility in self-assembled dendritic chromophore glasses," *J. Phys. Chem. B*, vol. 113, no. 43, pp. 14180–14188, Oct. 2009.

31. S. J. Benight, D. H. Bale, B. C. Olbricht, and L. R. Dalton "Organic electro-optics: understanding material structure/function relationships and device fabrication issues," *J. Mater. Chem.*, vol. 19, no. 40, pp. 7466–7475, Oct. 2009.
32. J. G. Grote, L. R. Dalton, P. Sullivan, B. H. Robinson, B. Eichinger, A. K.-Y. Jen, S. Benight, I. Kosilkin, and D. H. Bale, "Definition of critical structure/function relationships and integration issues for organic electro-optic materials," *Nonlinear Opt. Quant. Opt.*, vol. 40, no. 1–4, pp. 15–26, 2010.
33. D. B. Knorr, Jr., T. O. Gray, and R. M. Overney, "Intrinsic friction analysis—novel nanoscopic access to molecular mobility in constrainted organic systems," *Ultramicroscopy*, vol. 109, no. 8, pp. 991–1000, Jul. 2009.
34. S. J. Benight, L. E. Johnson, R. Barnes, B. C. Olbricht, D. H. Bale, P. J. Reid, B. E. Eichinger, L. R. Dalton, P. A. Sullivan, and B. H. Robinson, "Reduced dimensionality in organic electro-optic materials: Theory and defined order," *J. Phys. Chem. B*, vol. 114, no. 37, pp. 11949–11956, Sep. 2010.
35. P. A. Sullivan, H. L. Rommel, Y. Takimoto, S. R. Hammond, D. H. Bale, B. C. Olbricht, Y. Liao, et al., "Modeling the optical behavior of complex organic media: From molecules to materials," *J. Phys. Chem. B*, vol. 113, no. 47, pp. 15581–15588, Nov. 2009.
36. P. A. Sullivan, H. Rommel, Y. Liao, B. C. Olbricht, A. J. P. Akelaitis, K. A. Firestone, J.-W. Kang, et al., "Theory-guided design and synthesis of multi-chromophore dendrimers: An analysis of the electro-optic effect," *J. Amer. Chem. Soc.*, vol. 129, no. 24, pp. 7523–7530, Jun. 2007.
37. J. Luo, X.-H. Zhou, and A. K.-Y. Jen, "Rational molecular design and supramolecular assembly of highly efficient organic electro-optic materials," *J. Mater. Chem.*, vol. 19, no. 40, pp. 7410–7424, Oct. 2009.
38. T.-D. Kim, J. Luo, Y.-J. Cheng, Z. Shi, S. Hau, S.-H. Jang, X.-H. Zhou, et al., "Binary chromophore systems in nonlinear optical dendrimers and polymers for large electro-optic activities," *J. Phys. Chem. C*, vol. 112, no. 21, pp. 8091–8098, May 2008.
39. Y. V. Pereverzev, K. N. Gunnerson, O. V. Prezhdo, P. A. Sullivan, Y. Liao, B. C. Olbricht, A. J. P. Akelaitis, A. K.-Y. Jen, and L. R. Dalton, "Guest-host cooperativity in organic materials greatly enhances the nonlinear optical response," *J. Phys. Chem. C*, vol. 112, no. 11, pp. 4355–4363, Mar. 2008.
40. C. M. Isborn, A. Leclercq, F. D. Vila, L. R. Dalton, J. L. Brédas, B. E. Eichinger, and B. H. Robinson, "Comparison of static first hyperpolarizabilities calculated with various quantum mechanical methods," *J. Phys. Chem. A*, vol. 111, no. 7, pp. 1319–1327, Feb. 2007.
41. Y. Takimoto, F. D. Vila, and J. J. Rehr, "Real time time-dependent density functional theory approach for frequency-dependent non-linear optical response in photonic molecules," *J. Chem Phys.*, vol. 127, no. 15, pp. 154114-1–10, Oct. 2007.
42. C. C. Teng and H. C. Man, "Simple reflection technique for measuring the electro-optic coefficient of poled polymers," *Appl. Phys. Lett.*, vol. 56, no. 18, pp. 1734–1736, Apr. 1990.
43. P. D. H. Park, C. H. Lee, and W. N. Herman, "Analysis of multiple reflection effects in reflective measurements of electro-optic coefficients of poled polymers in multilayer structures," *Opt. Express*, vol. 14, no. 19, pp. 8866–8884, Sep. 2006.
44. L. R. Dalton, A. W. Harper, R. Ghosn, W. H. Steier, M. Ziari, H. Fetterman, Y. Shi, R. V. Mustacich, A. K.-Y. Jen, and K. J. Shea, "Synthesis and processing of improved second order nonlinear optical materials for applications in photonics," *Chem. Mater.*, vol. 7, no. 6, pp. 1060–1081, Jun. 1995.
45. S. S. H. Mao, Y. Ra, L. Guo, C. Zhang, L. R. Dalton, A. Chen, S. Garner, and Wm. H. Steier, "Progress towards device-quality second-order nonlinear optical materials: 1. Influence of composition and processing conditions on nonlinearity, temporal stability and optical loss," *Chem. Mater.*, vol. 10, no. 1, pp. 146–155, Jan. 1998.
46. S. Suresh, S. Chen, C. M. Topping, J. M. Ballato, and D. W. Smith, Jr., "Novel perfluorocyclo-butyl (PFCB) polymers containing isophorone derived chromophore for electro-optic [EO] applications," *Proc. SPIE*, 2003, vol. 4991, pp. 530–536, 2003.

47. S. T. Iacono, S. M. Budy, J. Jin, and D. W. Smith, Jr., "Science and technology of perfluorocyclobutyl aryl ether polymers," *J. Polym. Sci., Part 1: Polym. Chem.*, vol. 45, no. 21, pp. 5705–5721, Dec. 2007.
48. P. A. Sullivan, B. C. Olbricht, A. J. P. Akelaitis, A. A. Mistry, Y. Liao, and L. R. Dalton, "Tri-component Diels-Alder polymerized dendrimer glass exhibiting large, thermally stable, electro-optic activity," *J. Mater. Chem.*, vol. 17, no. 28, pp. 2899–2903, Jul. 2007.
49. P. A. Sullivan, A. J. P. Akelaitis, S. K. Lee, G. McGrew, S. K. Lee, D. H. Choi, and L. R. Dalton, "Novel dendritic chromophores for electro-optics: Influence of binding mode and attachment flexibility on electro-optic behavior" *Chem. Mater.*, vol. 18, no. 2, pp. 344–351, Jan. 2006.
50. R. Dinu, D. Jin, G. Yu, B. Chen, D. Huang, H. Chen, A. Barklund, E. Miller, C. Wei, and J. Vemagiri, "Environmental stress testing of electro-optic polymer modulators," *J. Lightwave Technol.*, vol. 27, no. 11, pp. 1527–1532, Jun. 2009.
51. A. Galvan-Gonzalez, M. Canva, G. I. Stegeman, R. Twieg, K. P. Chan, T. C. Kowalczyk, X. Q. Zhang, H. S. Lackritz, S. Marder, and S. Thayumanavan, "Systematic behavior of electro-optic chromophore photostability," *Opt. Lett.*, vol. 25, no. 5, pp. 332–334, Mar. 2000.
52. D. Rezzonico, M. Jazbinsek, P. Günter, C. Bosshard, D. H. Bale, Y. Liao, L. R. Dalton, and P. J. Reid "Photostability studies of π-conjugated chromophores with resonant and nonresonant light excitation for long-life polymeric telecommunication devices," *J. Opt. Soc. Amer. B*, vol. 24, no. 9, pp. 2199–2207, Sep. 2007.
53. M. E. DeRosa, M. He, J. S. Cites, S. M. Garner, and Y. R. Tang, "Photostability of high μβ electro-optic chromophores at 1550 nm," *J. Phys. Chem. B*, vol. 108, no. 25, pp. 8725–8730, Jun. 2004.
54. Y.-J. Cheng, J. Luo, S. Huang, X. Zhou, Z. Shi, T.-D. Kim, D. H. Bale, et al., "Donor-acceptor thiolated polyenic chromophores exhibiting large optical nonlinearity and excellent photostability," *Chem. Mater.*, vol. 20, no. 15, pp. 5047–5054, Aug. 2008.
55. M.-C. Oh, S.-S. Lee, and S.-Y. Shin, "Simulation of polarization converter formed by poling-induced polymer waveguides," *IEEE J. Quant. Elect.*, vol. 31, no. 9, pp. 1698–1704, Sep. 1995.
56. S. Kim, K. Geary, H. R. Fetterman, C. Zhang, C. Wang, and W. H. Steier, "Photo-bleaching induced electro-optic polymer modulators with dual driving electrodes operating at 1.55 mm wavelength," *Elect. Lett.*, vol. 39, no. 18, pp. 1321–1322, Sep. 2003.
57. R. A. Soref, J. Shidtchen, and K. Peterman, "Large single-mode rib waveguides in GeSi-Si and Si-on-$SiO_2$," *IEEE J. Quant. Electr.*, vol. 27, no. 8, pp. 1971–1973, Aug. 1991.
58. Y. Enami, C. T. DeRose, C. Loychik, D. Mathine, R. A. Norwood, J. Luo, A. K.-Y. Jen, and N. Peyghambarian "Low half-wave voltage and high electro-optic effect in hybrid polymer/sol-gel waveguide modulators," *Appl. Phys. Lett.*, vol. 89, no. 14, pp. 143505-1–3, Oct. 2006.
59. M.-C. Oh, H. Zhang, C. Zhang, H. Erlig, Y. Chang, B. Tsap, D. Chang, A. Szep, W. H. Steier, H. R. Fetterman, and L. R. Dalton,"Recent advances in electro-optic polymer modulators incorporating phenyltetracene bridged chromophores," *IEEE J. Sel. Topics Quant. Electr.*, vol. 7, no. 5, pp. 826–835, Sep. 2001.
60. D. G. Girton, S. L. Kwiatkowski, G. F. Lipscomb, and R. S. Lytel, "20 GHz electro-optic polymer Mach Zehnder modulator," *Appl. Phys. Lett.*, vol. 58, no. 16, pp. 1730–1732, Apr. 1991.
61. C. C. Teng, "Traveling-wave polymeric optical intensity modulator with more than 40 GHz of 3-dB electrical bandwidth," *Appl. Phys. Lett.*, vol. 60, no. 13, pp. 1538–1540, Mar. 1992.
62. D. Chen, H. R. Fetterman, A. Chen, W. H. Steier, L. R. Dalton, W. Wang, and Y. Shi, "Demonstration of 110 GHz electro-optic polymer modulators," *Appl. Phys. Lett.*, vol. 70, no. 25, pp. 3335–3337, Jun. 1997.
63. S. K. Mohapatra, C. V. Francis, K. Hahn, and D. W. Dolfi, "Microwave loss in nonlinear optical polymers," *J. Appl. Phys.*, vol. 73, no. 5, pp. 2569–2571, Mar. 1993.
64. H. Zhang, M.-C. Oh, A. Szep, W. H. Steier, C. Zhang, L. R. Dalton, H. Erlig, Y. Chang, D. H. Chang, and H. R. Fetterman "Push–pull electro-optic polymer modulators with low half-wave voltage and low loss at both 1310 nm and 1550 nm," *Appl. Phys. Lett.*, vol. 78, no. 20, pp. 3136–3138, May 2004.

65. J. Kondo, A. Kondo, K. Aoki, M. Imaeda, T. Mori, Y. Mizuno, S. Takatsuji, Y. Kozuka, O. Mitomi, and M. Minakata, "40-Gb/s X-Cut $LiNbO_3$ optical modulator with two-step back-slot structure," *J. Lightwave Tech.*, vol. 20, no. 12, pp. 2110, Dec. 2002.
66. A. Otomo, G. I. Stegeman, W. H. G. Horsthuis, and G. R. Mohlmann, "Strong field, in-plane poling for nonlinear optical devices in highly nonlinear side chain polymers," *Appl. Phys. Lett.*, vol. 65, no. 19, pp. 2389–2391, Nov. 1994.
67. H. Park, W.-Y. Hwang, and J.-J. Kim, "Origin of direct current drift in electro-optic polymer modulator," *Appl. Phys. Lett.*, vol. 70, no. 21, pp. 2796–2798, May 1997.
68. Y. Shi, W. Wang, J. H. Bechtel, A. Chen, S. Garner, S. Kalluri, W. H. Steier, D. Chen, H. R. Fetterman, L. R. Dalton, and L. Yu, "Fabrication and characterization of high speed polyurethane-disperse red 19 integrated electro-optic modulators for analog system applications," *IEEE J. Sel. Tropics in Quant. Electr.*, vol. 2, no. 2, pp. 289–299, Jun. 1996.
69. S. Park, J. J. Ju, J. Y. Do, S. K. Park, and M.-H. Lee, "Thermal bias operation in electro-optic polymer modulator," *Appl. Phys. Lett.*, vol. 83, no. 5, pp. 827–829, Aug. 2003.
70. T. C. Kowalczyk, T. Z. Kosc, K. D. Singer, A. J. Beuhler, D. A. Wargowski, P. A. Cahill, C. H. Seager, M. B. Meinhardt, and S. Ermer, "Crosslinked polyimide electro-optic materials," *J. Appl. Phys.*, vol. 78, no. 10, pp. 5876–5883, Nov. 1995.
71. H.-T. Man, and H. N. Yoon, "Long term stability of poled side-chain nonlinear optical polymer," *Appl. Phys. Lett.* vol. 72, no. 5, pp. 540–542, Feb. 1998.
72. W. Sotoyama, S. Tatsuura, and T. Yoshimura, "Electro-optic side-chain polyimide system with large optical nonlinearity and high thermal stability," *Appl. Phys. Lett.*, vol. 64, no. 17, pp. 2197–2199, Apr. 1994.
73. H. Ma, B. Chen, T. Sassa,, L. R. Dalton, and A. K.-Y. Jen, "Highly efficient and thermally stable nonlinear optical dendrimer for electrooptics," *J. Am. Chem. Soc.*, vol. 123, no. 5, pp. 986–987, Feb. 2001.
74. M. A. Mortazavi, H. N. Yoon, and C. C. Teng, "Optical power handling properties of polymeric nonlinear optical waveguides," *J. Appl. Phys.*, vol. 74, no. 8, pp. 4871–4876, Oct. 1993.
75. M. Mortazavi, K. Song, H. Yoon, and I. McCulloh, "Optical power handling of nonlinear polymers," *Polymer Reprints*, vol. 35, pp. 198–199, 1994.
76. A. Galvan-Gonzalez, M. Canva, G. I. Stegeman, R. Twieg, T. C. Kowalczyk, and H. S. Lackritz, "Effect of temperature and atmospheric environment on the photodegradation of some disperse red 1-type polymers," *Optics Lett.*, vol. 24, no. 23, pp. 1741–1743, Dec. 1999.
77. Y. Shi, W. Wang, W. Lin, D. J. Olson, and J. H. Bechtel, "Double-end crosslinked electro-optic polymer modulators with high optical power handling capacity," *Appl. Phys. Lett.*, vol. 70, no. 11, pp. 1342–1344, Mar. 1997.
78. S. Takahashi, B. Bhola, A. Yick, W. H. Steier, J. Luo, A. K.-Y. Jen, D. Jin, and R. Dinu, "Photo-stability measurement of electro-optic polymer waveguides with high intensity at 1550-nm wavelength," *J. Lightwave Technol.*, vol. 27, no. 8, pp. 1045–1050, Apr. 2009.
79. Material developed by Lumera, Inc., Bothell, WA.
80. Md. LX8400-GNL, Gigoptix, Palo Alto, CA.
81. D. A. Cohen, M. Hossein-Zadeh, and A. F. J. Levi,"High-Q microphotonic electro-optic modulator," *Solid-State Electronics*, vol. 45, no. 9, pp. 1577–1589, Sep. 2001.
82. T. Sadagopan, S. J. Choi, Sang Jun Choi, P. D. Dapkus, and A. E. Bond, "Optical modulators based on depletion width translation in semiconductor microdisk resonators," *IEEE Photon. Technol. Lett.*, vol. 17, no. 3, pp. 567–569, Mar. 2005.
83. Q. Xu, B. Schmidt, S. Pradhan, and M. Lipson, "Micrometre-scale silicon electro-optic modulator," *Nature*, vol. 435, pp. 325–327, May 2005.
84. V. S. Ilchenko, A. A. Savchenkov, A. B. Matsko, and L. Maleki, "Sub-micro watt photonic microwave receiver," *IEEE Photon. Technol. Lett.*, vol. 14, no. 11, pp. 1602–1604, Nov. 2002.
85. P. Rabiei, W. H. Steier, C. Zhang, and L. R. Dalton, "Polymer micro-ring filters and modulators," *J. Lightwave Technol.*, vol. 20, no. 11, pp. 1968–1975, Nov. 2002.

86. I.-L. Gheorma and R. M. Osgood, "Fundamental limitations of optical resonator based high-speed EO modulators," *IEEE Photon. Technol. Lett.*, vol. 14, no. 6, pp. 795–797, Jun. 2002.
87. R. Marz, *Integrated Optics: Design and Modeling*, Boston: Artech House, 1995.
88. H. Tazawa and W. H. Steier, "Analysis of ring resonator based traveling wave modulators," *IEEE Photon. Technol. Lett.*, vol. 18, no. 1, pp. 211–213, Jan. 2006.
89. H. Tazawa and W. H. Steier, "Erratum to 'Analysis of ring resonator-based traveling-wave modulations'," *IEEE Photon. Technol. Lett.*, vol. 18, no. 5, p. 727, Mar. 2006.
90. H. Tazawa and W. H. Steier, "Linearity of ring resonator based electro-optic polymer modulator," *Electronics Lett.*, vol. 41, no. 23, pp. 1297–1298, Nov. 2005.
91. H. Tazawa, Y.-H. Kuo, I. Dunayevskiy, J. Luo, A. K.-Y. Jen, H. R. Fetterman, and W. H. Steier, "Ring resonator-based electro-optic polymer traveling-wave modulator," *J. Lightwave Technol.*, vol. 24, no. 9, pp. 3514–3519, Sep. 2006.
92. B. Bortnik, Y.-C. Hung, H. Tazawa, B.-J. Seo, J. Luo, A. K.-Y. Jen, W. H. Steier, and H. R. Fetterman, "Electrooptic polymer ring resonator modulation up to 165 GHz," *IEEE J. Sel. Topics in Quant. Electr.*, vol. 13, no. 1, pp. 104–110, Jan. 2007.

# 10

# Silicon High-Speed Modulators


Ansheng Liu
*Intel Corporation*

Frederic Y. Gardes
*University of Surrey*

Ling Liao
*Intel Corporation*

Rebecca Schaevitz
*Intel Corporation*

Graham Reed
*University of Surrey*

Mario Paniccia
*Intel Corporation*




## 10.1 Introduction

Fiber optics is well established today due to the large capacity and reliability it provides, but it has a clear disadvantage in cost. Furthermore, components are typically large and expensive due to bulky fiber elements. This, linked to the time consuming assembly and packaging required for precision alignment of micro-optics with dimensions of the order of micron, has led to increasing interest in the possibility of building devices integrated on a single chip of transparent material and, consequently, benefitting from the much smaller footprint in integrated photonics. Ever since the earliest research on optical circuits, dating back to the 1970s, there have been visions of an optical superchip [1,2], containing a variety of integrated optical components to carry out light manipulation (generation, modulation, detection, switching, filtering, and amplification). In the early days of integrated photonics the work was associated with ferroelectric materials such as lithium niobate, and III-V semiconductors such as the gallium arsenide and indium phosphide based systems. This was mainly due to a large electro-optic effect enabling high modulation speed in the case of lithium niobate, and the possibility to integrate a source and other critical components in the case of III-V semiconductors.

However, although optical technologies offer unprecedented data rates over long distances, to date they have been designated primarily for high end applications. This is due to a number of factors, among them the technology used to fabricate mainstream optical components and until relatively recently, the lack of computing applications, where copper tracks currently still provide the necessary bandwidth, and the optical interconnect technology is still too expensive. Nowadays the mainstream electronic industry is based on the invention of the transistor at Bell laboratories in 1947 followed by the development of the electronic integrated circuit in 1959 and the first microprocessor in 1971, and is mainly based on the silicon substrate or silicon on insulator (SOI) technology. The dominance of silicon as the semiconductor of choice for microelectronics eventually led to the investigation of silicon photonic integrated circuits. This was primarily because of the potential attraction of integration of photonics with mainstream electronics and silicon technology, potentially enabling integrated optics to be fabricated in a cost effective manner.

To achieve photonic integration on a silicon platform, it is critical to develop various fundamental building blocks such as low loss silicon waveguides, silicon optical filters, silicon based lasers, silicon modulators, and silicon based photo-detectors. Although silicon is a transparent material in the wavelengths from ~1.1 μm to 1.55 μm, which is excellent for passive components, it is in fact difficult to consider silicon as an optical material for active components due to the two primary reasons. First, the indirect bandgap of silicon means that light emission from silicon is inefficient, and nontraditional techniques must be investigated if light emission is to be realized. Second, and more importantly for the subject of our concern in this text, the centrosymmetric crystal structure of silicon means that it does not exhibit a linear electro-optic (Pockels) effect. Since this is the traditional means of implementing an optical modulator in a waveguide based device, modulation of the refractive index of silicon must be carried out in another way. This chapter discusses silicon as an optical material for active photonic components. The progress, issues, and research associated with high-speed optical modulation in silicon are covered.

## 10.2 Electro-Optic Modulation in Silicon

Silicon (Si) is one of the most studied semiconductors in history, and literature is widely available. Due to its crystalline structural, silicon exhibits optical properties defined by the crystal symmetry and the indirect bandgap. These properties define the approach used to manipulate light in silicon via an electric signal.

### 10.2.1 Properties of Silicon

Si is a centrosymmetric, indirect bandgap semiconductor with a bandgap energy of 1.12 eV, which corresponds to optical absorption below a wavelength of 1.1 μm. Macfarlane et al. (1959) [3] measured the absorption edge with increasing temperature, and the results showed that even at relatively high temperatures (415 Kelvin ≡ 141.85 degrees Celsius), the absorption band shifts towards lower photon energy and approaches minimal absorption loss below approximately 0.95 eV, which corresponds to a wavelength of 1.3 μm. Therefore even in a high temperature environment, often found in high-speed electronic integrated circuits, silicon is suitable for the transmission of telecommunication wavelengths between 1.3 μm and 1.55 μm. In order to achieve active devices in silicon one must envision the use of doping to obtain p- and n-type regions. Varying the concentration of the doping will directly influence the electrical properties of the device. The question is mainly to what extent the doping will influence the material and hence the propagation of the light. The concentration of free carriers affects both the real and imaginary parts of the refractive index, and the theoretical change in absorption in semiconductors can be described by the well known Drude–Lorenz equation:

$$\Delta\alpha = \frac{e^3\lambda_0^2}{4\pi^2 c^3 \varepsilon_0 n}\left(\frac{N_e}{\mu_e (m_{ce}^*)^2} + \frac{N_h}{\mu_h (m_{ce}^*)^2}\right), \tag{10.1}$$

where $e$ is the electronic charge, $c$ is the velocity of light in vacuum, $\mu_e$ is the electron mobility, $\mu_h$ is the hole mobility, $m_{ce}^*$ is the effective mass of electrons, $m_{ch}^*$ is the effective mass of holes, $N_e$ is the free electron concentration, $N_h$ is the free hole concentration, $\varepsilon_0$ is the permittivity of free space, and $\lambda_0$ is the free space wavelength.

Soref and Lorenzo [4] evaluated Equation 10.1 for hole and electron concentrations varying between $10^{18}$ cm$^{-3}$ and $10^{20}$ cm$^{-3}$. The theoretical approach of the Drude–Lorenz equation indicates an absorption increase when the carrier concentration increases. By way of example, an injected hole and electron concentration of $10^{18}$ cm$^{-3}$ would correspond to a loss of around 10.86 dB/cm [5]. Spitzer and Fan (1957) [6] measured the free carrier absorption for varying n-type doping over a range of wavelengths from 1 μm to 50 μm.

The results from Spitzer and Fan [6] shows that increasing n type carrier concentration increases the absorption of the light at the wavelength of interest. If one compares the absorption coefficient of [4] and [6], one can see that for a wavelength of 1.3 μm, there is a discrepancy of the order of a factor of 2 for the absorption coefficient due to electrons, where the Drude–Lorentz model absorption coefficient is twice as small as the one measured by Spitzer and Fan. In 1987, Soref and Bennett [7], studied results in scientific literature and compared the theoretical results given by Equation 10.1 to experimental results.

The comparison shows the discrepancies between the Drude-Lorenz absorption equation and the experimental results. Based on the measured data, they produced expressions to evaluate the change in absorption due to a change in carrier concentration at wavelengths of 1.55 μm and 1.3 μm.

At λ= 1.55 μm this expression is

$$\Delta\alpha = \Delta\alpha_e + \Delta\alpha_h = 8.5 \times 10^{-18}\Delta N_e + 6 \times 10^{-18}\Delta N_h. \tag{10.2}$$

Similarly at λ= 1.3 μm:

$$\Delta\alpha = \Delta\alpha_e + \Delta\alpha_h = 6 \times 10^{-18}\Delta N_e + 4 \times 10^{-18}\Delta N_h. \tag{10.3}$$

where $\Delta N_e$ is the change in free electron concentration, $\Delta N_h$ is the change in free hole concentration, $\Delta\alpha_e$ is the change in absorption resulting from change in free electron carrier concentrations, $\Delta\alpha_h$ is the change in absorption resulting from change in free hole carrier concentrations. Applied to the measured absorption results from Spitzer and Fan (1957) [6], the equations provided by Soref and Bennett [7] give good agreement. This enables accurate modeling of devices where the losses due to free carriers may then be taken into account, at the wavelengths of 1.55 μm and 1.3 μm.

Now that the fundamentals of the silicon as a semiconductor and as a base optical material have been established, it is necessary to provide a summary of the possible ways to affect the light in silicon waveguides by the mean of an electric signal. In this section we follow the approach of Reed and Knights [5].

### 10.2.2 Electro-Optic Effects

Electro-optic effects are usually present in noncentrosymmetric crystals such as gallium arsenide and lithium niobate. Unstrained Si is centrosymmetric and does not exhibit any first order Pockels effect. Nevertheless in 2006, Jacobsen et al. [8] applied strain to Si by depositing a layer of silicon nitride. This had the effect of deforming the crystal structure, and hence introducing asymmetry to the silicon crystal. It resulted in a linear electro-optic effect of 15 pm·V. Previously, in 1987, Soref and Bennett [7] had studied electro-optic effects in silicon. Two electro-optic effects available to unstrained silicon were studied, the Kerr effect and the Franz–Keldysh effect (FKE). These two effects impact the light propagating in silicon in different ways. The Kerr effect is a second order electric field effect in which a change in real refractive index, $\Delta n$, is proportional to the square of the applied electric field where the change may be expressed as

$$\Delta n = s_{33} n_0 \frac{E^2}{2}, \tag{10.4}$$

where $s_{33}$ is the Kerr coefficient, $n_0$ is the unperturbed refractive index, and $E$ is the applied field. In this case, the sign of the refractive index change is not dependent on the direction within the crystal axis.

The theoretical quantification of the refractive index change due to the Kerr effect in Si at a wavelength of 1.3 μm is predicted to reach $10^{-4}$ at an applied field of $10^6$ V/cm (100 V/μm), which is above the breakdown field for lightly doped Si. This makes it impractical to use the Kerr effect for modulation in Si. Unlike the Pockels effect and the Kerr effect, the FKE gives rise to both electrorefraction and electroabsorption, although primarily the latter. The effect is due to distortion of the energy bands of the semiconductor upon application of an electric field. In effect, this shifts the bandgap energy, resulting in a change in the absorption properties of the crystal, particularly at wavelengths close to the bandgap, and hence a change in the complex refractive index. Soref and Bennett [7] also quantified the changes in refractive index due to the FKE. Whilst all their data was not presented graphically, they plotted the change in refractive index as a function of the applied electric field at a wavelength of 1.07 μm, the wavelength at which the effect is greatest. They also plotted data at 1.09 μm for comparison. It should be noted, however, that the absorption edge band is situated at around 1.1 μm. Thus the effect diminishes significantly at the telecommunications wavelengths of 1.31 μm and 1.55 μm.

The Franz–Keldysh effect in silicon induced index change as a function of applied electric field and the refractive index change reaches $10^{-4}$ at an applied field of $2 \times 10^5$ V/cm (20 V/μm), which is better than that evaluated for the Kerr effect. Nevertheless, the measured wavelengths are not in the 1.31 μm-1.55 μm telecommunication range, and this makes this effect impractical to use for modulation in Si unless the material is modified to increase the effect.

### 10.2.3 Thermo-Optic Effect

The thermo-optic effect is an effect for which the refractive index of Si is varied by applying heat to the material. The thermo-optic coefficient in silicon at the wavelength of 1.55 μm [9] is

$$\frac{dn}{dt} = 1.86 \times 10^{-4} \text{ per kelvin}. \tag{10.5}$$

The relation in Equation 10.5 quantifies the change in the refractive index per Kelvin increase in temperature. Therefore, if the Si can be raised in temperature by approximately 6°C in a controllable manner, the refractive index changes by $1.1 \times 10^{-3}$, hence, affecting the effective index of the mode propagating in the silicon waveguide. In 2000, Clark et al. [10], demonstrated that a 500-μm-long device using 10 mW of power could deliver a π phase shift in a waveguide thermally isolated from the substrate. A thermal change of approximately 7°C, was necessary to achieve this result which corresponds to a refractive index change of approximately $1.3 \times 10^{-3}$ over the length of the device. It is important to notice that although the thermal effect is relatively efficient in terms of refractive index change, it is also a slow modulation process and consequently high-speed modulation is unlikely to be realized using this effect.

### 10.2.4 Plasma Dispersion Effect

We observed previously that the concentration of free charges in silicon contributes to the loss via absorption. The imaginary part of the refractive index is determined by the absorption (or loss) coefficient. Therefore, it is clear that changing the concentration of free charges can change the refractive index of the material. We saw earlier that the Drude–Lorenz equation relates the concentration of electrons and holes to the change in absorption Equation 10.1. According to classical dispersion theory, free carriers will alter the real and imaginary parts of the Si dielectric constant. In optical terms, we say that the refraction and absorption indices are changed by amounts $\Delta n$ and $\Delta\alpha$, respectively. The analysis of Moss [11] and Lubberts [12], generalized for the two carrier types, showed that the corresponding equation relating the carrier concentration, $N$, to the change in refractive index, $\Delta n$, is

$$\Delta n = \frac{-e^2\lambda_0^2}{8\pi^2 c^2 \varepsilon_0 n}\left(\frac{N_e}{m_{ce}^*} + \frac{N_h}{m_{ch}^*}\right). \tag{10.6}$$

The changes in refractive index and absorption coefficient due to injected carriers in Si can be derived from the Kramers–Kronig relation between the real and the imaginary parts of the refractive index. The differential Kramers–Kronig dispersion relation is

$$\Delta n_r(\omega) = \frac{c}{\pi} P \int_0^{\infty} \frac{\Delta\alpha(\omega_1)}{\omega_1^2 - \omega^2} d\omega_1, \quad (10.7)$$

where $\Delta\alpha(\omega)=\alpha(\omega, \Delta N)-\alpha(\omega, 0)$, $\Delta N$ is the change in free-carrier concentration within crystalline silicon, c is the speed of light, $\omega$ the angular frequency.

Soref and Bennett [7] used experimental data from the absorption spectra available in the literature and Equation 10.7 to compute the integral and consequently the dependence of the change in the refractive index with carrier concentration. From these data, they produced expressions relating the changes in refractive index and absorption coefficients due to injection or depletion of carriers in silicon at wavelengths of both 1.3 μm and 1.55 μm. These expressions are now almost universally used for carrier concentration dependant evaluation of optical modulation in silicon. The expressions for absorption at both 1.3 μm and 1.55 μm are shown in Equations 10.2 and 10.3. The corresponding expressions for the change in refractive index are:

At $\lambda = 1.55$ μm:

$$\Delta n = \Delta n_e + \Delta n_h = -[8.8 \times 10^{-22} \Delta N_e + 8.5 \times 10^{-18} (\Delta N_h)^{0.8}]. \quad (10.8)$$

Similarly at $\lambda = 1.3$ μm:

$$\Delta n = \Delta n_e + \Delta n_h = -[6.2 \times 10^{-22} \Delta N_e + 6 \times 10^{-18} (\Delta N_h)^{0.8}], \quad (10.9)$$

where $\Delta n_e$ is the change in free electron concentration, $\Delta n_h$ is the change in free hole concentration, $\Delta n_e$ is the change in refractive index resulting from change in free electron carrier concentrations, $\Delta n_h$ is the change in refractive index resulting from change in free hole carrier concentrations.

It is interesting to note that for the plasma dispersion effect, the change in refractive index occurs over a wide range of wavelengths most notably at the communications wavelengths of 1.3 μm and 1.55 μm. Furthermore, it enables the fabrication of complementary metal-oxide-semiconductor compatible active devices in silicon, where a relatively low carrier concentration change, in the range of $1 \times 10^{17}$ to $1 \times 10^{18}$ cm$^{-3}$ enables a refractive index change between $1 \times 10^{-4}$ to $3 \times 10^{-3}$. We will show later that using the plasma dispersion effect in specific device structures, it is possible to achieve high-speed modulation in silicon.

## 10.3 Device Configurations

As mentioned above, modulation of the index of refraction in Si relies mainly on modulation of free carrier density. There are three basic device configurations that can be used to achieve carrier density modulation in silicon. The first and most extensively investigated device configuration is the forward biased p-i-n diode modulator [5]. A cross sectional view of a forward biased diode waveguide phase shifter is shown in Figure 10.1. It consists of an intrinsic silicon region in the center of the waveguide, a p-doped region, and an n-doped Si region on both sides of the rib. A metallic layer is deposited on top of the doped regions to form the anode and cathode for the p-i-n device. The doping concentration in both p-and n-regions is required to be high (~$10^{20}$ cm$^{-3}$) for Ohmic contacts between Si and metals with small contact resistance.

To obtain single-mode operation that is usually required for modulator performance, the p-i-n embedded waveguide dimension has to be properly chosen. Single-mode condition has been previously

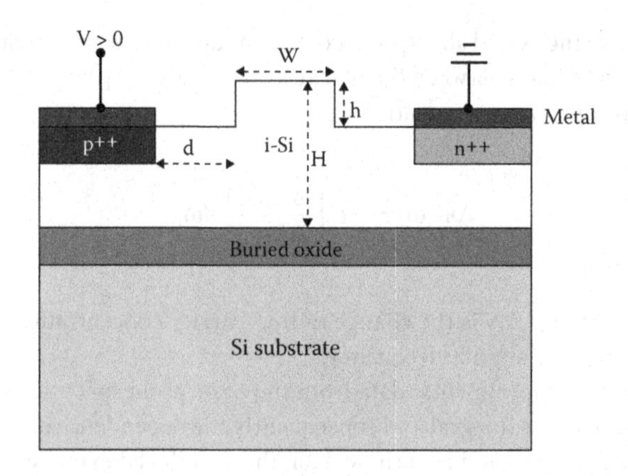


**FIGURE 10.1** Schematic of a forward biased p-i-n diode silicon modulator waveguide.

studied and relationship among the waveguide width (W), rib height (H), and etch depth (h) has been established. For large rib waveguides, the single-mode condition is given by [13,14]

$$\frac{W}{H} \leq \alpha + \frac{r}{\sqrt{1-r^2}}, \tag{10.10}$$

where $r$ is the ratio of the slab height to the overall rib height, that is, $(H-h)/H$, and $\alpha = 0.3$. For small rib waveguides, the single-mode condition is more difficult to describe by an empirical formula although it is possible over small ranges [15]. It can be, however, determined by numerical simulation using some commercially available modal solvers [16,17]. For example, for a waveguide having a rib width of 0.5 μm and a rib height of 0.5 μm, the modeling shows the waveguide etch depth of 0.25 μm still retains single mode operation.

An alternative device configuration is based on a metal-oxide-semiconductor (MOS) capacitor as shown in Figure 10.2 [18]. The MOS capacitor modulator waveguide is constructed with an n-type silicon slab and a p-type silicon rib separated with a thin oxide (gate oxide). Just like the forward biased p-i-n diode in Figure 10.1, heavily doped p++ and n++ regions are ion implanted before metal deposition to form Ohmic contacts. In common with more advanced p-i-n modulators [19], the MOS capacitor

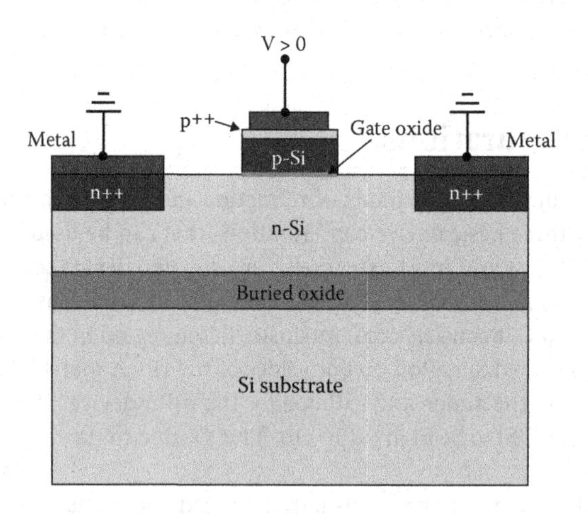


**FIGURE 10.2** Schematic of a MOS capacitor silicon modulator waveguide.

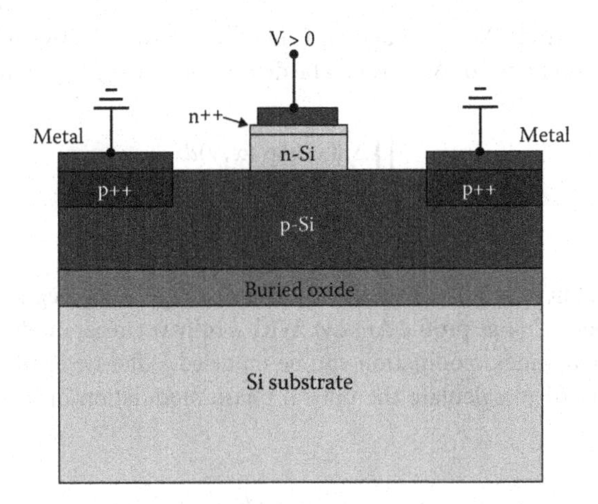


**FIGURE 10.3** Schematic of a reversed biased pn diode silicon modulator waveguide.

modulator has one metal contact on top of waveguide rib, while both contacts are on the waveguide slab for the forward biased p-i-n diode modulator in Figure 10.1. Therefore, it would be critical to design the metal contact for the MOS capacitor modulator to avoid or minimize the contact induced optical absorption loss. One efficient way to reduce the contact loss is to place the metal contact away from the waveguide rib center by adding a thin silicon "wing" on top of the rib [18].

When a forward bias is applied to the MOS capacitor, free carriers (electrons in n-type Si and holes in p-type Si) are accumulated on both sides of the gate oxide. The accumulated free carriers change the refractive index of silicon through the plasma dispersion effect, which in turn changes the optical phase of the propagating optical mode. As the accumulated charge layer thickness is usually very thin (on the order of nm's), the effective index change of the waveguide is strongly dependent on the overlap between the waveguide mode size and the charge layer thickness. Therefore, for an MOS capacitor based modulator, it is very important to optimize the device dimensions as well as the position of the gate oxide so that the optical mode interacts strongly with the space charge to ensure high phase modulation efficiency. Because the charge accumulation is based on majority carrier dynamics, the device speed can be very high, essentially governed by the resistance capacitance (RC) time constant.

A third device configuration option is based on a reverse biased pn junction, as shown in Figure 10.3. Such a device is very similar to the MOS capacitor except that there is no gate oxide in between the n-and p-type silicon regions. In a reverse biased pn diode, there is a depletion region in the vicinity of the junction in which no free carriers are present. The depletion width of the pn junction depends on the doping concentrations and applied reverse bias voltage. In response to the applied voltage, the depletion width change leads to a charge density change, which in turn changes the refractive index in the silicon waveguide. The performance of the reverse biased pn diode modulators also relies on electric-field induced majority carrier dynamics. Like the MOS capacitor, device bandwidth is primarily governed by the RC time constant and can be very high, above 30 GHz to obtain data rates from 10–40 Gb/s.

## 10.4 Phase Modulation in Silicon Modulator

As discussed above, index of refraction change in Si is related to the free carrier density change. Dependent upon the specific device configuration, the free carrier density change in response to the external drive voltage or current generally has spatial dependence across the waveguide mode. In turn, the refractive index modulation is also spatially dependent. When the index change is very small, as is usually the case for a Si modulator, we can use perturbation theory [20] to calculate the effective index change of the optical waveguide. With an optical field of $\varphi(x,y)$ the effective index modulation of the waveguide due to the index change of $\Delta n(x,y)$ can be determined by the following equation [20]:

$$\Delta n_{eff} = \frac{\iint \Delta n(x,y)\varphi^2(x,y)dx\,dy}{\iint \varphi^2(x,y)dx\,dy}. \tag{10.11}$$

Alternatively, it is convenient to simulate the effective index change of a waveguide by use of commercial modal solver with the index change profile $\Delta n(x,y)$. With a fully vectorial mode solver, the polarization dependence of the effective index modulation can be modeled. After determining the effective index change, it is straightforward to calculate the optical phase modulation of a waveguide containing a silicon phase modulator by

$$\Delta\phi = \frac{2\pi \Delta n_{eff} L}{\lambda}, \tag{10.12}$$

where $L$ is the phase modulator length and $\lambda$ the wavelength of light in vacuum. To characterize the phase modulation of the silicon modulator, we introduce a figure-of-merit of $V_\pi L_\pi$, where $V_\pi$ is the voltage required for $\pi$ radian phase shift for an $L_\pi$ long phase modulator. The smaller the $V_\pi L_\pi$, the better the phase modulation efficiency. In the following, we will discuss the phase modulation efficiency for the three Si modulator configurations considered above.

The free carrier density change in a forward biased p-i-n diode depends on the device structure, doping concentration, and carrier lifetimes. It can be modeled by commercially available software packages such as Silvaco [21]. Previous modeling shows that the free carrier density is relatively uniform cross the waveguide [5]. The maximum carrier density variation is ~10% for a rib waveguide having a width and a rib height of 3 μm–5 μm. The charge density change is slightly dependent on the etch depth as well as the sidewall angle. With a forward bias voltage of 0.9 V, the modeled average injected carrier density for such devices is $2.2 \times 10^{17}$ cm$^{-3}$. For the wavelength of 1.55 μm, the total index of refraction change due to electrons and holes is $-8.3 \times 10^{-4}$. The modulator length required for $\pi$ phase shift is estimated to be 934 μm. This leads to a $V_\pi L_\pi$ of 0.084 V·cm. The phase modulation efficiency of the forward biased p-i-n diode can be improved by shrinking the device size. For a rib waveguide having the width of 0.55 μm, rib height of 0.22 μm, and etch depth of 0.185 μm, a $V_\pi L_\pi$ of 0.036 V·cm has been experimentally demonstrated [22]. The results suggest that the forward biased p-i-n diode modulator has a relatively efficient phase modulation so that the device footprint can be small.

For an MOS capacitor modulator, the free carriers (electrons in n-type silicon and holes in p-type silicon) are accumulated in the vicinity of gate oxide under forward bias. The voltage-induced charge density change $\Delta N_e$ (for electrons) and $\Delta N_h$ (for holes) can be approximately described by [18]

$$\Delta N_e = \Delta N_h = \frac{\varepsilon_0 \varepsilon_r}{e t_{ox} t}[V - V_{FB}], \tag{10.13}$$

where $t_{ox}$ is the gate thickness, $t$ is the effective charge layer thickness, $V_{FB}$ is the flat band voltage of the MOS capacitor, and $V$ is the drive voltage.

Because the effective charge layer thickness is small (~10 nm), the effective index modulation of the MOS capacitor modulator is strongly dependent on the waveguide size, as is evident from Equation 10.11. For example, an MOS capacitor modulator with waveguide dimensions of 2.5 μm$^2$ × 2.3 μm$^2$ and 12 nm gate oxide has a measured $V_\pi L_\pi$ product of 7.8 V·cm [18]. This is two orders of magnitude larger than that for forward biased p-i-n diode modulator. By reducing the waveguide dimensions, however, one can significantly enhance the overlap between the charge density change and the optical mode. We

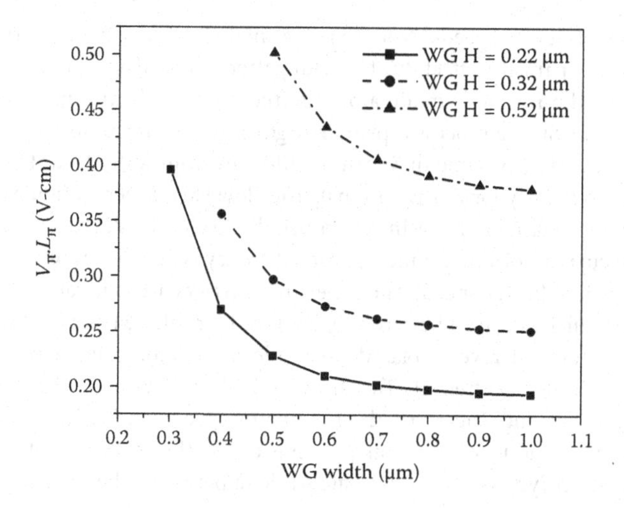


**FIGURE 10.4** Modeled phase efficiency ($V_{\pi}L_{\pi}$) of a MOS capacitor phase shifter as a function of the waveguide dimension, for a gate oxide thickness of 6 nm.

show in Figure 10.4 the simulated phase efficiency as a function of the waveguide size. In the simulation, a gate oxide thickness of 6 nm was used. From Figure 10.4 we see that the phase efficiency is increased by a factor of 2 by reducing the waveguide height from 0.5 μm to 0.2 μm. Comparing to the previously demonstrated MOS capacitor modulator with waveguide size of 2.3 μm × 2.5 μm [18], a 0.2 μm × 0.2-μm waveguide phase shift leads to modulator size reduction of a factor of 10. Comparing to the forward biased p-i-n diode modulator, the phase efficiency of the MOS capacitor modulator is still one order of magnitude smaller, (i.e., $V_{\pi}L_{\pi}$ is ~10× larger).

For the reverse biased pn junction, the depletion width depends on the bias voltage and doping concentrations. The charge density change associated with the depletion width change due to the bias voltage leads to the index of refraction change. Like the MOS capacitor modulator, the charge density change region for the reverse biased pn diode is usually small compared to the optical modal size. Therefore, we expect the phase efficiency of the reverse biased pn junction to be small as compared to the forward biased p-i-n diode modulator. In addition, because the depletion width is not linearly dependent on the drive voltage, we similarly expect the phase shift not to be linearly dependent on the voltage. This is quite different for the MOS capacitor modulator, which shows linear dependence of the phase shift vs. the drive voltage.

Comparing the three device configurations, it is clear that the forward biased p-i-n diode has the best phase modulation efficiency with a $V_{\pi}L_{\pi}$ of <0.04 V·cm. With a carefully designed waveguide, the MOS capacitor modulator can potentially lead to a $V_{\pi}L_{\pi}$ of 0.2 V·cm, one order of magnitude larger than that for the forward biased p-i-n diode modulator. For the reverse biased pn diode, the modeling suggests that it is feasible to achieve the phase efficiency of ~2 V·cm [23] for a submicrometer size waveguide at lower drive voltages. This is one order of magnitude larger than that for the MOS capacitor modulator. Although the forward biased p-i-n diode modulator has the best phase modulation efficiency, we will show that it has slow operation speed. It is therefore difficult to achieve high operation speed up to 40 Gb/s.

## 10.5 Speed Limitations in Silicon Modulators

The intrinsic modulation speed of a silicon modulator based on the free carrier dispersion effect is determined by how fast the free carriers are injected into and/or removed from the optical waveguide. Different modulator device configurations have different free carrier dynamics mechanisms. Therefore, we expect quite different device speeds. Note that the intrinsic device speed is referred to the device performance without any parasitic effect due to metal patterns and driver circuitry.

For the forward biased p-i-n diode modulator, the free carrier density change in silicon is achieved through current injection. Such a process is primarily governed by the minority carrier dynamics in Si. Although the minority carrier lifetime in Si can be different from wafer to wafer even from device to device, it is in general relatively long (1 μs-100 ns). The slow carrier generation and/or recombination processes for the forward biased current injection usually limits the device speed to the MHz range. We also note that the current injection (modulation) efficiency is dependent on the carrier lifetime. The shorter the carrier lifetime (faster speed), the fewer free carriers are generated. This tradeoff between speed and modulation efficiency must be carefully analyzed for different applications.

For the MOS capacitor and reverse biased pn diode modulators, the free carrier modulation is achieved through the electric-field effect. The device speed limitation can be simply considered due to the device RC time constant. Therefore, the modulator speed can be designed by properly choosing the device capacitance and resistance. Take as an example the MOS capacitor modulator having a p-type silicon layer and a n-type silicon with a gate oxide in between. The capacitance per unit length is approximately given by

$$C = \frac{\varepsilon_0 \varepsilon_r W}{t_{ox}}, \tag{10.14}$$

where $W$ is the gate oxide width. The total series resistance of the device is mainly due to the doped silicon regions because the metal contact has very small resistance assuming a good Ohmic contact between silicon and metal. Assuming the drive voltage is constant along the unit length of the MOS capacitor modulator, the total series resistance per unit length is given by

$$R = R_P + R_N = \rho_P \frac{W_P}{h} + \rho_N \frac{W_N}{h}, \tag{10.15}$$

where $\rho_P$ and $\rho_N$ are the resistivity of p-and n-doped silicon, respectively, $W_P$ and $W_N$ the p- and n-doped silicon width, $h$ the p- and n-doped silicon thickness. The 3-dB bandwidth due to the RC time limit is given by the well-known formula

$$f_{3dB} = \frac{1}{2\pi RC}. \tag{10.16}$$

We show in Figure 10.5 the simulated 3-dB bandwidth of a MOS capacitor modulator waveguide as a function of doping concentration. In the simulation, we assumed the p- and n-doping density is the same. The gate oxide thickness is $t_{ox} = 6$ nm. The gate oxide width is $W = 0.5$ μm. The doped silicon width is $W_P = W_N = 1$ μm. Three doped silicon heights, that is, $h = 0.1$, 0.15, and 0.2 μm were used. Note that the total waveguide height is equal to $2h$. We see from Figure 10.5 that the MOS capacitor modulator is clearly capable of high data rate operation. Of course, the waveguide loss is also dependent on the doping concentration. We should balance the device speed with the optical loss for the MOS capacitor modulator.

The speed of the reverse biased pn diode modulator is also determined by the RC time constant. The capacitance of the pn diode modulator is primarily determined by the depletion width. As the depletion width of the reverse biased pn diode is bias voltage dependent, the capacitance is also dependent on the drive voltage. In Figure 10.6, we plot the modeled direct current (DC) capacitance of a pn diode with $N_D = 2 \times 10^{17}$ cm$^{-3}$ (where $N_D$ is the concentration of donors) and $N_A = 5 \times 10^{17}$ cm$^{-3}$ (where $N_A$ is the concentration of acceptors) and waveguide dimension of W = 0.4, H = 0.4, and h = 0.22 μm. We see

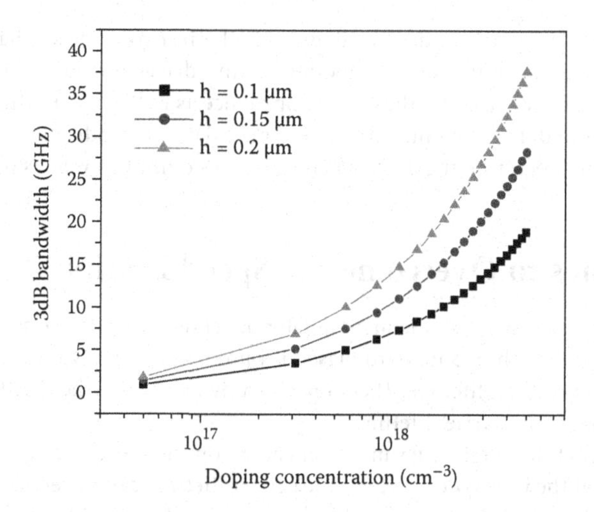


**FIGURE 10.5** Modeled 3-dB bandwidth of a MOS capacitor modulator as a function of doping concentration for three different silicon slab thicknesses, that is, h = 0.1, 0.15, and 0.2 μm. The same p-type and n-type doping concentrations were assumed. Gate oxide thickness is 6 nm.

that the capacitance decreases with an increase in the bias voltage, because the depletion width increases with an increase in the bias. Although there is a bias dependent DC capacitance change, the high-speed response of the pn diode modulator may be analyzed using the RC time constant with a capacitance at the DC bias voltage. For example, with an alternate current drive voltage between 0 V and 1 V, the DC bias voltage is 0.5 V. Therefore, we can use the capacitance value obtained with a bias voltage of 0.5 V to simulate the device speed.

Because the per-unit length capacitance of the reverse biased pn diode is much smaller than that for the MOS capacitor modulator, the required doping concentration can be much smaller for the same required bandwidth target. For example, one can achieve 10-GHz 3-dB bandwidth for a reverse biased pn diode with a doping level of ~$10^{17}$ cm$^{-3}$, whilst one needs ~$10^{18}$ cm$^{-3}$ for the MOS capacitor. Thus, the reverse biased pn diode is expected to have a much lower optical loss per unit length associated with the doping concentrations as compared to the MOS capacitor modulator.

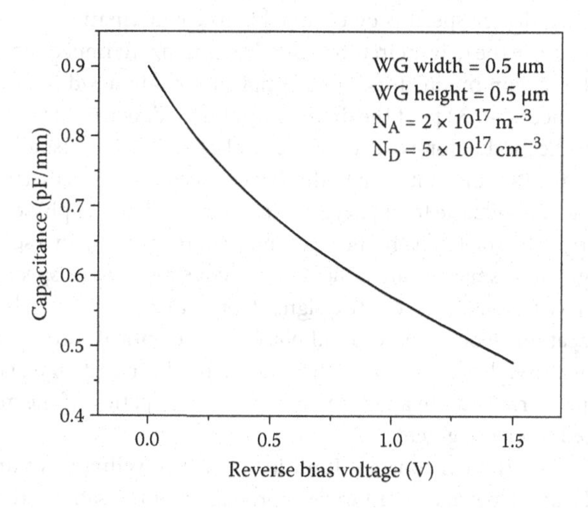


**FIGURE 10.6** Modeled capacitance of a reverse biased pn diode per unit length as a function of the DC reverse bias voltage. The p- and n-doping densities are $2 \times 10^{17}$ and $5 \times 10^{17}$ cm$^{-3}$, respectively.

We note that the device speed we described above is the intrinsic speed, that is, an ideal drive is assumed (drive voltage is constant along the modulator and driver impedance is zero). In real device operation, we have to take into account the driver impedance as well as the RF signal attenuation along the modulator. The speed of the combined driver and modulator depends on the drive schemes. In the following section, we will describe the drive scheme and speed limit as well as the techniques to overcome the speed limits.

## 10.6 Techniques to Overcome the Speed Limits

As mentioned above, the speed of the silicon modulator depends on the device configuration. For example, the modulation bandwidth of a forward biased p-i-n diode is usually limited to the MHz to GHz range because of the long carrier lifetime. Therefore, one way to increase the speed of the forward biased p-i-n diode is to minimize the carrier lifetime.

It has been shown that the carrier lifetime is dependent on the waveguide geometry. When the carrier scattering occurs at the waveguide surface, the carrier lifetime can be reduced due to the increased carrier surface recombination. Such a result has been reported in the literature [24]. In general, the smaller the waveguide dimension is, the shorter the carrier lifetime is. Another way to reduce the carrier lifetime is to introduce impurity in the silicon. For example, it has been shown that adding Au in Si leads to significant carrier lifetime reduction. Ion implantation is also an effective way to reduce the lifetime.

The above mentioned methods to overcome the speed also accompany a reduction in the free carrier generation rate, and in turn, the phase modulation efficiency. As the ultimate modulator performance depends on both the intrinsic modulator speed and drive circuitry, one can increase the modulator speed by adding preemphasis in the circuitry [22,24,25]. Namely, to compensate for the slow rise/fall time for the modulator, a spike of the drive voltage and/or current is designed in the drive circuit as proposed by Png et al. in 2004 [25]. As a result, the combined modulator and drive circuit gives a faster response. Such an approach has been demonstrated in the literature. 12.5 Gb/s and even 17.5 Gb/s have been reported with a silicon resonator based modulator [24]. 10 Gb/s operation in a silicon Mach–Zehnder modulator has also been demonstrated with a device size of few hundreds of micrometers [22]. The drawback of the preemphasis approach is that the required high voltage spike leads to high power dissipation. In addition, such a required high voltage may not be readily achievable for CMOS electronics.

Although the intrinsic speed of MOS capacitor and reverse biased pn diode modulators has been shown to be high, the real device speed is determined by the combination of the modulator response and drive electronics. Because the driver circuitry usually has nonzero impedance, the RC time constant is a key speed limiter for silicon modulators. For example, to obtain a 3-dB bandwidth of 10 GHz, the maximum total capacitance is ~0.32 pF if the driver output impedance is 50 Ω. However, to achieve the required phase shift for the modulator extinction, the total capacitance is usually larger than 1 pF.

One way to eliminate the RC constant limit to the device speed is to use a distributed drive scheme. In such an approach, one can divide the total phase shifter into a number of phase shifter segments. Each of the segments is electrically isolated from the next and each is driven with a separate driver. Once the capacitance of each segment is small enough, the RC time constant is not a speed limiter. To realize the distributed drive, we have to match the electric signal (correct timing) for each segment according to the optical wave propagation velocity, as the total phase for the optical wave is accumulated along the optical light propagating through each segment. Therefore, careful driver design and/or driver–modulator interface design must be carried out. In addition, total power dissipation of the modulator is the sum of all the drivers connected to each segment.

Another way to eliminate the RC constant limit is to use a traveling-wave drive scheme, in which both optical wave and radio frequency (RF) wave copropagate in the same direction. The RF signal is coupled into the modulator electrode input and the output is terminated with a load resistor. The traveling-wave scheme used for the Si modulator is the same as that used for $LiNbO_3$ and III-V semiconductor modulators. The two key design considerations for the traveling-wave drive are to ensure that the RF phase velocity matches the optical group velocity and that the electrodes have low RF attenuation. To minimize the RF reflection effect, the driver impedance, the traveling-wave electrode (transmission line) characteristic impedance, and the termination load must be matched. Because the MOS capacitor modulator has a very large capacitance per unit length, the traveling-wave electrode impedance is usually in the range of ~10 Ω. To drive such a low impedance transmission line, the power consumption is usually large. Also, it is not usually possible to obtain a commercial driver with such low impedance. For a reverse biased pn diode, the capacitance per unit length is usually much lower (~0.7 pF/mm), and it is possible to design a transmission line with an impedance of ~25 Ω. Thus, the traveling-wave drive is an efficient way to enhance the device speed for the reverse biased pn diode modulator.

## 10.7 Representative Modulator Performance

The first experimental demonstration of modulator devices based on high-speed majority carrier manipulation using structures such as an MOS capacitor was reported in 2004. Researchers from the Intel Corporation experimentally demonstrated a Si-based optical modulator with a bandwidth that exceeds 1 GHz [18] for the first time. This was a major milestone in silicon photonics for high-speed applications. Figure 10.7 shows a schematic of the reported device. This device operates by the free carrier effect and bears a close resemblance to a CMOS transistor. The device structure consists of n-type crystalline silicon with an upper "rib" of p-type polysilicon. The n-type and p-type regions are separated by a thin insulating oxide layer. Upon application of a positive voltage to the p-type polysilicon, charge carriers accumulate at the oxide interface, changing the refractive index distribution in the device. This in turn induces a phase shift in the optical wave propagating through the device.

The bandwidth of the device (a single 2.5 mm long phase modulator) was characterized in two ways in an integrated asymmetric Mach–Zehnder interferometer (MZI). The first technique was to drive the device with a 0.18-V rms sinusoidal source at the wavelength of 1.558 μm, using lensed fibers for coupling into and out of the device. Also, the normalized optical response of the MZI as a function of frequency (photoreceiver output voltage divided by on-chip drive voltage) clearly exceeds the 1 GHz 3 dB bandwidth. The second test was the application of a 3.5-V digital pulse pattern with a DC bias of 3 V. A 1-Gbit/s pseudorandom bit sequence was applied to the device and a high-bandwidth photoreceiver was used for detecting the transmitted optical signal. Here again, the optical signal faithfully reproduced the 1 Gbit/s electrical data stream. However, the on-chip loss for this device was rather high, at ~6.7 dB, and the device was also highly polarization dependent due to the horizontal gate oxide. Phase modulation

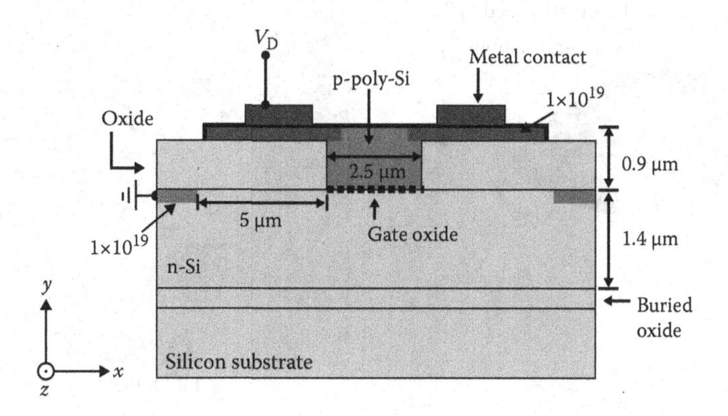


FIGURE 10.7 Schematic diagram of the silicon-based optical modulator demonstrated experimentally to exceed 1 GHz bandwidth fabricated using standard CMOS processing techniques. (Reprinted by permission from Macmillan Publishers Ltd., A. Liu et al., *Nature*, 427, 6975, 615–618, copyright 2010.)

efficiency for transverse electric (TE) polarization was larger than transverse magnetic (TM) polarization by a factor of 7.

The authors also suggested several methods that may improve the device performance even further. The first was replacing the p-type polysilicon with single crystal silicon, where the latter was expected to reduce on-chip loss by ~5 dB. Another suggestion was the reduction of the device dimensions as the capacitance is reduced through such shrinkage. They also suggested using a graded doping profile in the vertical direction such that higher doping densities exist in the areas close to the gate oxide and lower doping concentrations in the rest of the waveguide. Furthermore, the authors also reported that their modeling predicts that the device can be scaled to operate at 10 GHz, offering very significant improvement of silicon photonic devices.

A year later, Liao et al. [26], reported an improved version of the MOS optical modulator. This modulator waveguide size is smaller than the previous one, and comprises a 1.0-μm n-type doped crystalline Si region at the bottom and a 0.55-μm p-type doped crystalline Si-region on the top, with a 10.5-nm gate dielectric, sandwiched between them. In the previous version of the device, the waveguide cross-section was 2.5 μm × 2.3 μm and the top Si layer was polysilicon (poly-Si), which is significantly more lossy than crystalline Si due to defects and grain boundaries [26]. In the improved device the polysilicon was replaced by crystalline silicon via epitaxial lateral overgrowth, and the doping concentration was higher. In this smaller version of the phase shifter the mode-charge interaction is much stronger, which improved the $V_\pi L_\pi$ coefficient by 50%. A data rate of 10 Gb/s was reported for the modulator with an extinction ratio (ER) of 3.8 dB. Data transmission measurements suggested bandwidths ranging from 6 GHz (ER of 4.5 dB) to 10 GHz. The 6 GHz limitation is due to the driver design and wire bonding, which decreased the cut off frequency.

A further improvement to carrier accumulation based modulators has been reported recently by Lightwire Inc. [27]. The eye diagram displayed in Figure 10.8 indicates data transmission at 10 Gb/s with an ER of almost 9 dB, although no optical loss was reported. Their device is very compact (800 μm × 15 μm) which is possible due to a higher modulation efficiency with a submicrometer waveguide (cf. Figure 10.4). One of the possible drawbacks of this device and all reported devices based upon carrier accumulation, is the complexity of the fabrication process required to produce the MOS capacitor like structure as compared to the pn junction based modulator. Nevertheless, the advance in Si fabrication technology may simplify the MOS capacitor based modulator processing in the future.

In 2007, Liu et al. [28], experimentally demonstrated a pn diode carrier depletion based silicon optical modulator, similar to the depletion design reported in the literature by Gardes et al. [29]. Figure 10.9

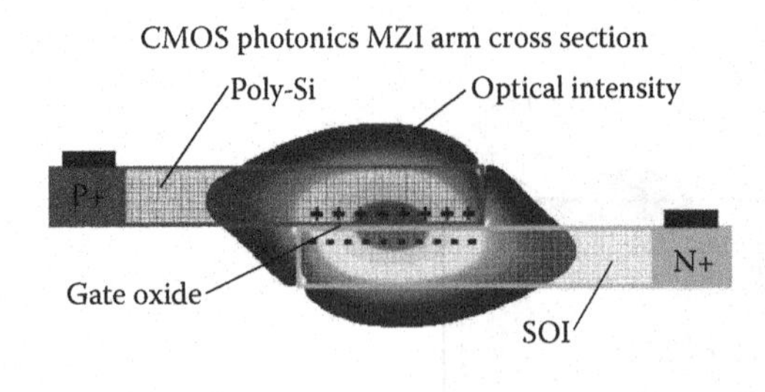


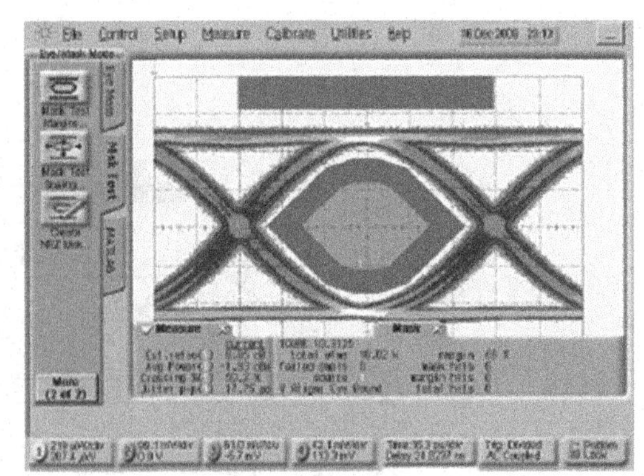


**FIGURE 10.8** Cross-sectional diagram of accumulation modulator (left) and 10-Gb/s eye diagram (right). (Reprinted from D. Andrea, "CMOS photonics today and tomorrow enabling technology," in OFC, 2009, San Diego, CA. With permission of Optical Society of America.)

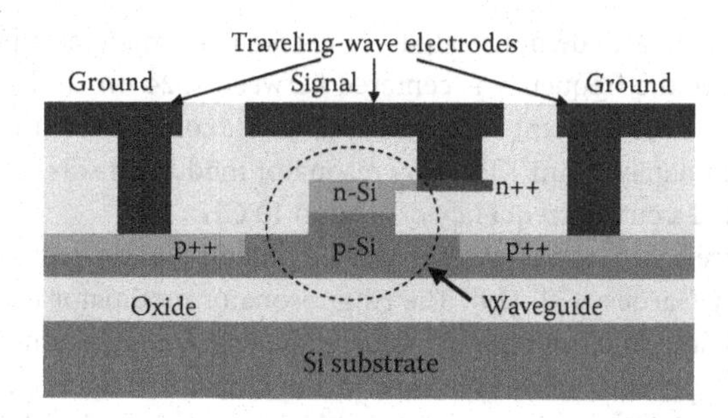


**FIGURE 10.9** Cross-sectional diagram of the reverse biased pn diode modulator. (Reprinted with permission from A. Liu et al., *Opt. Expression*, vol. 15, no. 2, pp. 660–668, Jan. 2007. © 2010 OSA.)

shows the schematic of the pn diode modulator. The modulation efficiency $V_\pi L_\pi$ of approximately 4 V·cm was reported.

Later in 2007, Liao et al. [30] presented further developments of the device, boasting an EO bandwidth of 30 GHz and data transmission at 40 Gb/s with an ER of about 1 dB. The frequency response and eye diagram are shown in Figure 10.10. To date, this is the fastest reported experimental optical modulator in silicon.

With the development of individual components, it is possible to integrate an array of these silicon modulators on a single chip to achieve even higher aggregate bandwidth. In 2008, Liu et al. [31], reported a silicon integrated photonic chip that contains eight silicon modulators with 1 × 8 demultiplexer and 8 × 1 multiplexer on a single SOI substrate. The multiplexer/demultiplexer is based on cascaded MZIs with thermo-optic phase tuning capabilities. Each of the modulators operated at 25 Gb/s and showed the possibility of transmitting data at an aggregate data rate of 200 Gb/s over a single optical fiber.

In the last few years, integration of photonic devices and CMOS electronics on the same chip has become a topic of interest in Si photonics. In early 2006, Gunn [32], demonstrated modulation in both MZI and a ring resonator with data rates up to 10 Gb/s. The authors stated that the modulator drivers were integrated on the chip, but did not provide any details about the electronics or the technology used to change the effective index of the mode in the waveguide. The information provided in [32] indicates that the waveguides are 500 nm wide and have a cross sectional area of 0.1 μm². The authors also stated that the typical ER was 5 dB when the modulator is driven at 2.5 V.

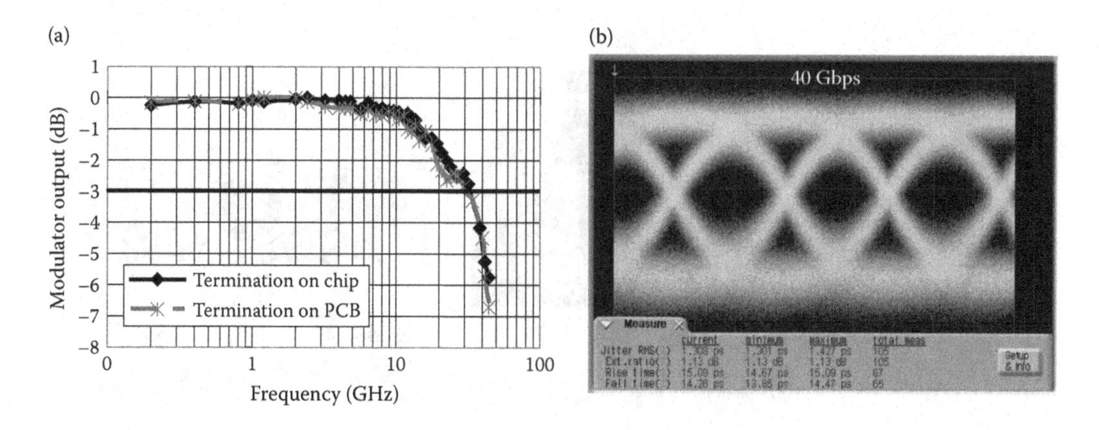


**FIGURE 10.10** (a) Optical response of the modulator as a function of the RF frequency for a 1-mm active area. (b) Optical eye diagram of the modulator with a 1-mm-long active area. (Reprinted with permission from L. Liao et al., *Electronics Lett.*, vol. 43, no. 22, doi: 10.1049/el: 20072253, Oct. 2007. © 2010 IET.)

As mentioned above, Gunn also demonstrated a ring resonator modulator [32]. The ring was used as a tunable notch filter where the frequency is centered between 1524 nm and 1525 nm. The ring radius was 30 μm and was a major improvement in terms of real estate compared to the proposed MZI modulators which occupy approximately 2 mm². The ring resonator modulator was operated at 10 Gb/s and the frequency response showed a cut off frequency of around 10 GHz.

In 2009, a similar device based on a ring resonator modulator and on depletion of a vertical pn junction was demonstrated by Gardes et al. [29]. The ring resonator modulator used a 300-nm-wide waveguide, 150 nm etch depth and 200 nm high rib waveguide, which enables single mode propagation. As shown in Figure 10.11, the pn junction is asymmetrical in size and in doping concentration in order to maximize the area of hole depletion that overlaps with the optical mode. The n-type region is 75 nm wide and the p type 225 nm wide, and the net doping concentration of this particular junction varies between $6 \times 10^{17}$ and $2 \times 10^{17}$ cm$^{-3}$, for n and p types, respectively. The modulator showed a 3-dB bandwidth of about 19 GHz for a ring radius of about 40 μm.

Whilst resonant structures are more compact than MZIs, they are very sensitive to temperature changes which can be problematic when the modulator is to be integrated, for example, with electronics, which tends to heat up during operation. As a result some DC tuning is normally required which complicates the design as well as increases power consumption.

A further development in the field of pn junction modulators has been demonstrated by Park et al. [33] who reported a carrier depletion MZI based modulator in 220 nm overlayer SOI using high p and n doping concentrations to achieve a high efficiency. A $V_\pi L_\pi$ efficiency of 2 V·cm and a loss of approximately 4 dB/mm were reported. A 3-dB electro-optic bandwidth of 7.1 GHz is reported and data transmission has been demonstrated at 12.5 Gb/s and 4 Gb/s with ERs of approximately 3 dB and 7 dB, respectively.

Modulation in an SOI based ring resonator structure has also been demonstrated with an electro-optic bandwidth about 30 GHz in a 1.7-mm circumference resonator [34] compared to 250 micrometer in [29]. The advantage of this device is the possibility to be able to actively tune the coupling area. The usually standard evanescent coupler is here replaced by a MZI enabling through the use of the thermo-optic effect to obtain critical coupling, hence, to maximize the extinction ration at resonance or to suppress the resonance by moving away from it.

Devices based on the carrier depletion of a pn junction or carrier accumulation in a photonic waveguide have been demonstrated to exceed 10 Gb/s and achieve up to 40 Gb/s. Nevertheless, increasing the speed of modulators typically shows a trade-off in terms of a reduction in extinction ratio and

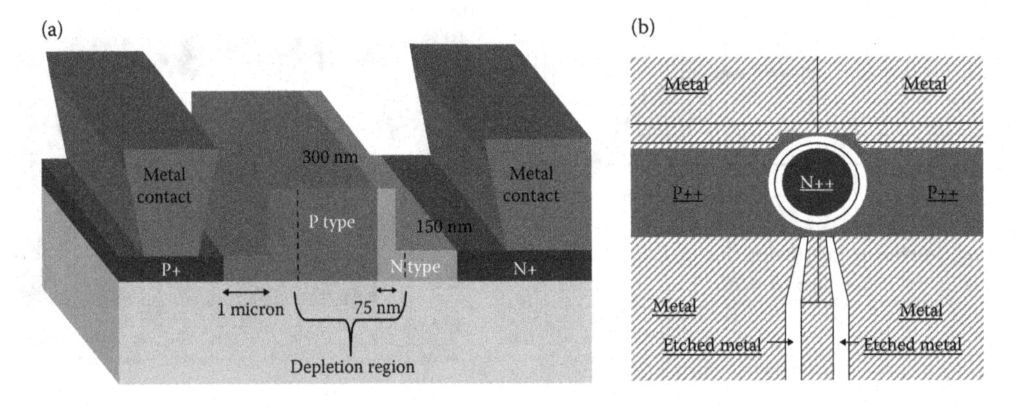


**FIGURE 10.11** (a) Cross section of a pn diode waveguide modulator; (b) top-down view of a ring resonator with metal contact. (Reprinted from F. Y. Gardes et al., *Opt. Express*, vol. 17, no. 24, pp. 21986–21991, Nov. 2009. With permission of Optical Society of America.)

this, together with a required increase in modulator efficiency, is an issue that needs to be addressed to achieve high density integration of active components in Si photonics.

## 10.8 Complementary Metal-Oxide-Semiconductor Compatible Silicon Modulator Techniques

In the earlier sections of this chapter, we reviewed the optical modulation mechanisms as well as the performance of silicon modulators having various device configurations. The most promising high-speed designs demonstrated to-date are those based on carrier-refraction in either an MZI or ring resonator implementation. MZI-based devices have the benefits of high speed, a broad wavelength range of operation, and temperature insensitivity, but they suffer from large footprint on the order of $mm^2$. Ring resonator based Si modulators, on the other hand, are one to two orders of magnitude smaller in size because the resonant effect is very sensitive to changes in index so even short device length is sufficient for modulation. This same sensitivity, however, means that the performance of these resonator-based devices is extremely susceptible to fabrication variations and changes in temperature and wavelength. As a result, device tuning is required to first set the operating point to compensate for fabrication errors, and dynamic tuning is needed to maintain this operating point over even small temperature variations or laser wavelength drifts.

While significant progress has been made using carrier-refraction for high-speed modulation in Si, the above-mentioned limitation of large device footprint or extreme temperature sensitivity prompted researchers to also investigate other physical mechanisms to achieve modulation in Si-compatible materials and structures. Devices that have shown the greatest promise are silicon-germanium (SiGe) electroabsorption modulators based on the FKE and proposed devices based on the quantum confined Stark effect (QCSE). Both effects are inherently fast and can theoretically achieve speeds greater than 100 GHz [35].

The FKE is a well-known phenomenon in bulk, direct bandgap semiconductors where the application of an electric field shifts the material's absorption edge to lower energy. This shift occurs because when an electric field is present, the electron and hole wavefunctions become Airy functions with "tails" that extend into the bandgap. The absorption spectrum therefore has oscillations above the bandgap and an exponential decay below it, allowing for absorption of lower energy light. While the FKE is weak in Si, it is relatively strong in tensile-strained germanium (Ge) because it has a band structure close to that of a direct bandgap material [36]. Using Ge-rich SiGe, epitaxially grown on a SOI platform, Ji-Feng Liu et al. 2007 experimentally demonstrated an FKE electroabsorption modulator that is only 50 μm long [36].

The QCSE is very similar to FKE in that it is also an electrical field induced change in optical absorption. The primary difference is that it is not a bulk material effect. It is observed in quantum well structures where the electron and hole confinement afforded by the barriers allows for exciton enhancement of the optical absorption. A comparison of QCSE and FKE absorption spectra will show the exciton effect present in the former as peaks leading to much sharper absorption edges. Consequently, QCSE is a stronger electroabsorption mechanism compared to FKE. In 2005, Kuo et al. demonstrated QCSE with pure Ge quantum wells and Ge-rich SiGe barriers on Si [37]. Similar structures exhibiting QCSE were later integrated into modulator devices, demonstrating proof-of-concept electroabsorption modulators [38,39], although no waveguide based device have yet been reported.

The remainder of this section will briefly describe models for FKE and QCSE and compare their applicability for optical modulation near 1310 nm and 1550 nm. SiGe FKE and QCSE modulator structures are very similar in many ways. First a thin buffer layer of Ge or SiGe is grown on Si and annealed to prevent defect propagation. Then a thicker layer of the same concentration SiGe is grown as either the active layer for the FKE or the virtual substrate for the QCSE. For the FKE modulators, a p-i-n diode is formed in this layer using the intrinsic material as the active region. The QCSE devices also have a p-i-n design, but the intrinsic region is filled with pairs of Ge well and Ge-rich SiGe barrier. Both material structures use the direct bandgaps of Ge (0.8 eV) and Si (4.0 eV) for optical modulation even though they are indirect bandgap materials [40]. The FKE devices have focused on materials with Ge concentrations >99% [35,36]. The QCSE devices use a wider range of material compositions: ~90% Ge for the virtual substrate, ~85% Ge for the barrier and 100% Ge for the well [36,38,39,41].

For both the FKE and the QCSE devices, material strain plays an important role in design and the resulting optical properties. The active layer for the FKE (and virtual substrate for the QCSE) is always grown beyond the critical thickness due to the large lattice mismatch between Si and the approximately 4% larger Ge. However, because material growth is done at elevated temperatures, the thermal coefficient mismatch between the Ge or Ge-rich epitaxial material and the Si wafer induces between 0.1% to 0.2% tensile strain in the epitaxy [42]. QCSE devices have additional strain due to the lattice mismatch between the barrier (tensile strain) and well (compressive strain) materials. This strain requires that the average concentration of the barrier and well pairs match that of the virtual substrate to prevent propagation of strain due to lattice mismatch.

For the FKE devices, the thermal tensile strain does not significantly change the indirect bandgap but does reduce the direct bandgap; as a result, it will push device operation beyond 1600 nm. To bring the operating range back into the C-band, the addition of a small amount of Si (<1%), with its very large bandgap, is sufficient to compensate for the strain-induced bandgap shrinkage [35,36]. For the QCSE, the quantum wells and barriers have lattice mismatch-induced strain in addition to the thermal strain. Since the barriers have a higher concentration of Si than the virtual substrate to compensate for the pure Ge wells, the barriers have tensile strain while the wells have compressive strain. Due to both the compressive strain and the quantum confinement, the device operation shifts toward shorter wavelengths between 1450 nm–1500 nm depending on the thickness of the well and the composition of the virtual substrate. Increasing well thickness, temperature and/or electric field can bring the operating wavelength range into the C-band.

To model the material potential for electroabsorption modulators, both thermal and lattice mismatch-induced strain must be considered. Not only does strain affect the overall bandgaps, it also affects the relevant masses and the shifting and splitting of the valence bands. Under tensile strain, the light hole band shifts down in energy, while the heavy hole shifts up. Under compressive strain, the opposite effect is observed. Indirect absorption is another very important parameter to model accurately for both FKE and QCSE-based devices, as it is an intrinsic source of insertion loss that is difficult to remove through engineering.

Calculations for the FKE are relatively simple and are well described by references [43–46]. The QCSE is more complex and has been modeled using a variety of methods including k.p, tight-binding, and tunneling resonance. The last method only determines the relative strength of absorption (due to the fraction of overlap of the electron and hole wavefunctions) and the energy of confinement of the electron and holes in the well material. The first two methods give more realistic models of the absorption spectra at different electric fields [47–53], with tight-binding being more precise but more complex and computationally intensive. For the purpose of comparing FKE and QCSE, k.p simulations are used. The k.p tool used in this study is provided by nextnano© and the code for the SiGe/Ge quantum well simulations is based on reference [53]. The results from k.p are primarily limited by an inaccurate prediction of the magnitude of exciton enhancement in absorption. Generally, the simulation gives values that are lower than found in experimental data, as shown in Figure 10.12.

In order to make the simulation more realistic when comparing possible ratios between high and low absorption at a given wavelength (or energy), indirect absorption is added to the k.p simulation for the QCSE as well as to the FKE model. When comparing modeled device performance, a figure of merit (FOM) of the change in absorption over absorption, or $\Delta\alpha/\alpha$ = ER/insertion loss), is used. This value assumes no insertion loss other than absorption due to the band edges. In other words, dopant loss, crystal defect loss, and device losses related to waveguides and coupling are not included. Consequently, if a device was fabricated with a given theoretical FOM, the expected FOM could be smaller.

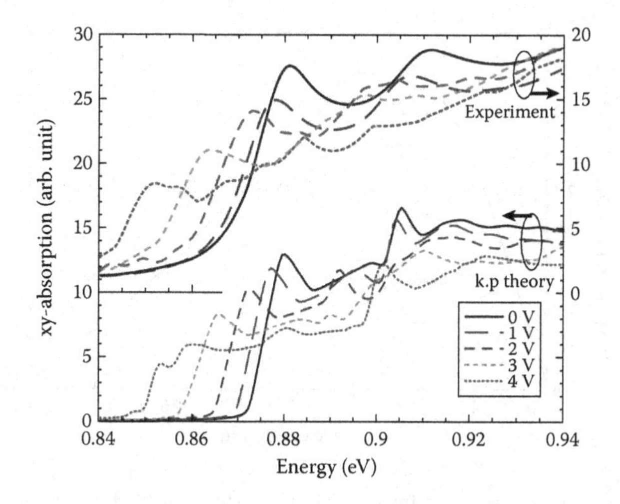


**FIGURE 10.12** Comparison of k.p (nextnano©) simulation results and experimental data.

Assuming the interested wavelengths of operation are around 1310 nm and 1550 nm, below is a series of figures that compare the FKE with the QCSE in SiGe material systems and for transverse electric (TE) mode. For the FKE, the optimal operating wavelength is ~1640 nm for pure Ge with 0.2% tensile strain. As Si is added to Ge to bring operation into the C-band, the FOM decreases rapidly as shown in Figure 10.13. While the value of ~3 is still reasonable for operation at 1550 nm, it is only 0.25 at 1310 nm. Assuming the best case strain scenario, without regard for whether it is physically achievable, the best FOM for 1310 nm operation is only ~0.65 (at 0.4% compressive strain). As a result, it is clear that the FKE in SiGe is unlikely to lead to practical modulators at 1310 nm. It is, however, a definite possibility at 1550 nm with a high FOM of ~3. As Figure 10.14 shows, it also has a possible wavelength bandwidth of >70 nm for an FOM>1.5. The results for 1550 nm compare well with previous experimental and theoretical work reported by Liu et al. [35,36].

In contrast to the FKE, the QCSE is a much stronger effect and offers the flexibility to shift the wavelength of operation by varying either the intrinsic material concentration or the well thickness. Figure 10.15 shows the possible FOM for quantum well structures designed for 1450 nm and 1300 nm wavelength ranges. Both sets of values are significantly better than the FOM modeled for the FKE.

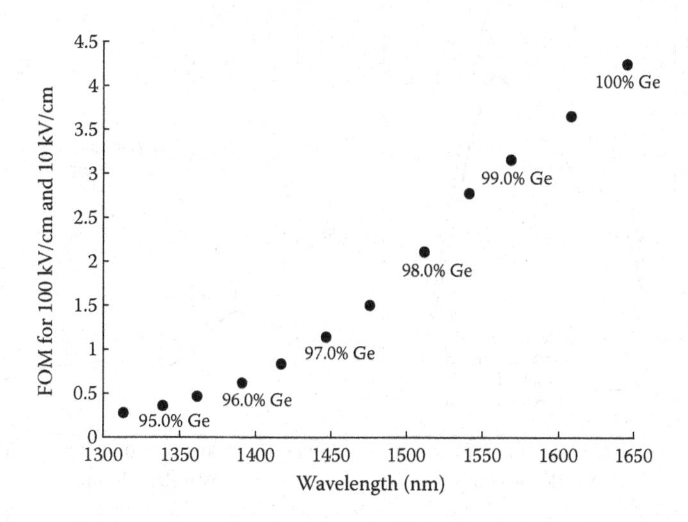


**FIGURE 10.13** FOM vs. wavelength for various epitaxial compositions and 0.2% tensile strain.

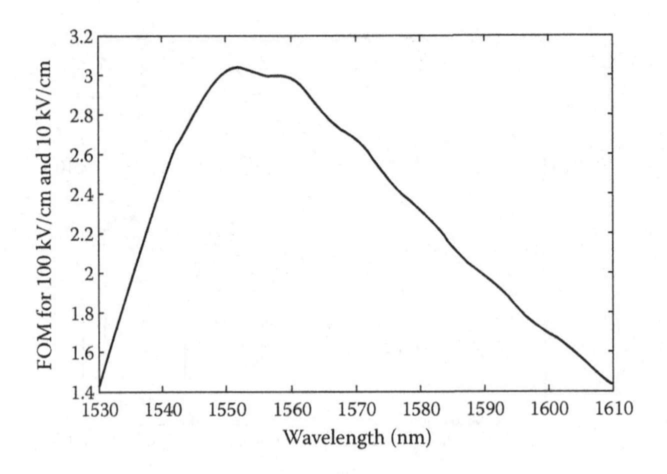


**FIGURE 10.14** FOM vs. wavelength for optimal composition (0.9925 Ge) and strain (~0.1%) at 1550 nm.

With the binary material system of Si and Ge to create the quantum well structures, the well material is always compressively strained. This means that the bandgap of the well is larger than that of bulk germanium's 0.8 eV, usually by ~20 meV-40 meV. In addition, quantum confinement pushes the bandgap even larger; the amount depends on the well width. Consequently, even relatively large well widths of 15 nm-25 nm produce peak FOM that is too small for practical applications for wavelengths above 1500 nm. Nevertheless, in 2008 Roth et al. [39] demonstrated that temperature tuning is an effective way of shifting the optimal wavelength of operation toward the C-band. The shift is due to the shrinkage of the Ge band gap with temperature and is approximately 0.7 nm/°C for this wavelength range.

While the SiGe QCSE promises better theoretical modulator performance compared to the SiGe FKE, material growth and subsequent processing are much more complicated with a quantum well-based device (QCSE) than with bulk material (FKE). Si and Ge interdiffuse relatively rapidly even at temperatures close to 400°C. While this diffusion does not affect device performance for bulk SiGe, as it would occur only at the interface to the Si substrate, it will change the effective well width quite

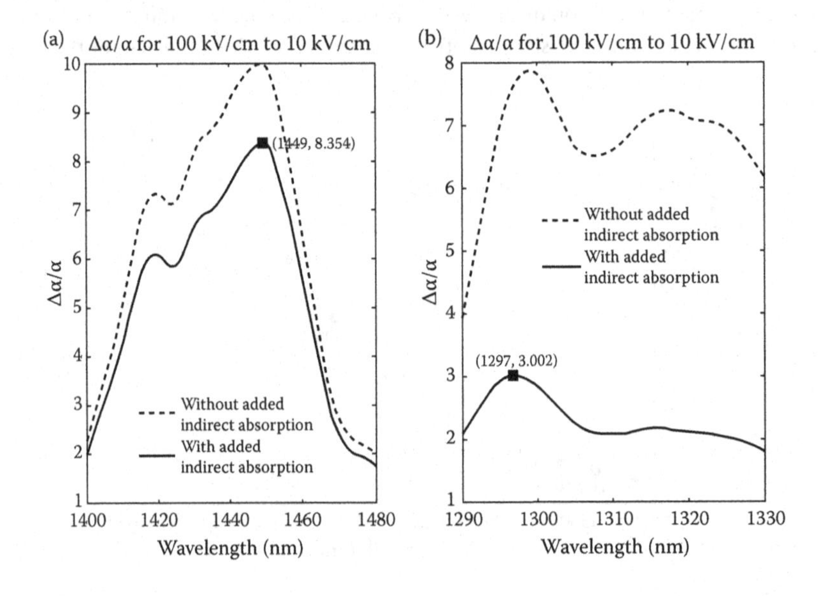


**FIGURE 10.15** FOM vs. wavelength for (a) 12.5-nm-wide 100% Ge well and (b) 10-nm-wide 96% Ge well.

significantly in quantum wells. Changing the well width drastically changes the wavelength of operation. Just a 2-nm shift in well width can result in a ~30 nm change in the optimal wavelength! Process integration with other optical devices and/or electronics could prove challenging as any thermal step, even rapid thermal anneals, could lead to unexpected and undesirable results. Furthermore, just growing quantum wells requires precise epitaxial calibration, ideally at the angstrom level, in order to meet specified wavelength targets. Of course, with the large wavelength bandwidth, a slight error in growth may not drastically affect device performance.

Both the FKE and the QCSE have been experimentally demonstrated in SiGe. A theoretical study explores their applicability for optical modulation near 1310 nm and 1550 nm by examining the FOM of $\Delta\alpha/\alpha$. Results of this study show that modulation at 1310 nm is only practical when employing QCSE. For 1550 nm, both effects are possible, but QCSE promises much higher FOMs. Its primary drawback is growth and processing sensitivities to temperature and layer thickness.

## 10.9 Concluding Remarks

Optical coupling between a single mode fiber and silicon modulator waveguide is one of the key issues for the device packaging. Because the submicrometer size silicon waveguide required for the modulator performance, significant modal size mismatch exists between optical fiber (core diameter ~9 μm) and silicon modulator chip. To overcome the modal size mismatch, a modal size converter or waveguide taper is needed for the Si modulator. A so-called inverted taper is one of the tapering schemes that attracted much attention in the last few years [54,55]. Such a taper not only has been proven to lead to a low coupling loss (~1 dB/facet), but also can be monolithically integrated with the Si modulator. The disadvantage of the inverted taper is the relatively large polarization dependent loss unless the taper tip size is very small, which is usually difficult to control due to the lithography and silicon etching. Alternatively, one may use a so-called double stage taper [56], which demonstrated ~1.5 dB/facet coupling loss. Such a taper has a much smaller polarization dependent loss. However, integration of the taper with silicon modulator is yet to be demonstrated.

Although it is critical for any applications, reliability of the Si modulator has not been an active research topic in the literature. Because the high-speed silicon modulator is either based on a reverse biased pn junction or a forward biased MOS capacitor, we expect the device reliability is not a big concern, as the corresponding electrical devices based on the diodes has been proven in the industry. However, the fully packaged modulator with optical coupling and RF connector is yet to be studied. Temperature dependence of the Si modulator depends on the device implementation schemes. For the MZI modulator with equal arm lengths, the device performance has almost no temperature dependence. However, for the resonance based modulator, one expects strong temperature dependence. For such devices, active control feedbacks have to be added to the devices, which not only add additional complexity but also power consumptions. Again, the temperature dependence of the fully packaged silicon modulator is to be studied.

Si photonics is based on one of the most studied semiconductors in history. Due to its structural properties, silicon exhibits optical properties that make it suitable for the transmission and the guiding of light in the near infrared. Further to this, the light at these wavelengths can be modulated using the plasma dispersion effect which, as discussed above, is the action of changing the index of refraction by changing the free carrier density. This effect can be controlled via an electrical potential which enables the fabrication of different optical modulator structures such as injection, accumulation or depletion based devices through the use of interferometers or resonant structures. In the last six years, silicon optical modulators have had their bandwidth increased by 2 orders of magnitude with high-speed modulation demonstrated at up to 40 Gb/s and researchers are now pushing towards the reduction of size and power consumption as well as large scale photonic integration. To do so as explained above, germanium can be added to the silicon such that effect such as the QCSE and the FKE can be used to modulate the light.

In light of the progress made in silicon optical modulation in the last few years, it is clear that optical modulators are moving towards highly integrated devices that can be used today for chip to chip or board to board interconnects, but in the near future, could be used for intrachip applications such as for core to core interconnects as well as a lot of other more conventional applications.

## References

1. R. A. Soref, "Silicon-based optoelectronics," *Proc. IEEE*, vol. 81, no. 5, pp. 1687–1706, Dec. 1993.
2. R. A. Soref, "Silicon-based photonic devices," in *1995 IEEE Int. Solid-State Circuits Conf.*, San Francisco, 1995, pp. 66–67.
3. G. G. Macfarlane, T. P. McLean, J. E. Quarrington, and V. Roberts, "Exciton and phonon effects in the absorption spectra of germanium and silicon," *J. Phys. Chem. Solids*, vol. 8, pp. 388–392, Jan. 1959.
4. R. A. Soref and J. P. Lorenzo, "All-silicon active and passive guided-wave components for $\lambda = 1.3$ and 1.6 μm," *IEEE J. of Quantum Elect.*, vol. 22, no. 6, pp. 873–879, Jun. 1986.
5. G. T. Reed and A. P. Knights, *Silicon Photonics: An Introduction*, West Sussex, England: John Wiley & Sons, Ltd., 2004.
6. W. Spitzer and H. Y. Fan, "Infrared absorption in n-type silicon," *Phys. Rev.*, vol. 108, no. 2, p. 268–271, Oct. 1957.
7. R. A. Soref and B. R. Bennett, "Electrooptical effects in silicon," *IEEE J. Quantum Elect.*, vol. 23, no. 1, pp. 123–129, Jan. 1987.
8. R. S. Jacobsen, K. N. Andersen, P. I. Borel, J. Fage-Pedersen, L. H. Frandsen, O. Hansen, M. Kristensen, et al., "Strained silicon as a new electro-optic material," *Nature*, vol. 441, no. 7090, pp. 199–202, May 2006.
9. G. Cocorullo and I. Rendina, "Thermo-optical modulation at 1.5 microns in silicon etalon," *Electronics Lett.*, vol. 28, no. 1, pp. 83–85, Jan. 1992.
10. S. A. Clark, B. Culshaw, E. J. C. Dawnay, and I. E. Day, "Thermo-optic phase modulators in SIMOX material," *Proceedings of SPIE*, vol. 3936, pp. 16–24, Jan. 2000.
11. T. S. Moss, G. J. Burrell, and B. Ellis, *Semiconductor Opto-Electronics*, London, England: Butterworth, 1973.
12. G. Lubberts, B. C. Burkey, F. Moser, and E. A. Trabka, "Optical properties of phosphorus-doped polycrystalline silicon layers," *J. Applied Physics*, vol. 52, no. 11, pp. 6870–6878, Nov. 1981.
13. R. A. Soref, J. Schmidtchen, and K. Petermann, "Large single-mode rib waveguides in GeSi-Si and Si-on-$SiO_2$," *IEEE J. Quantum Electronics*, vol. 27, no. 8, pp. 1971–1974, Aug. 1991.
14. O. Powell, "Single-mode condition for silicon rib waveguides," *J. Lightwave Technol.*, vol. 20, no. 10, pp. 1851–1855, Oct. 2002.
15. S. P. Chan, C. E. Png, S. T. Lim, G. T. Reed, and V. M. N. Passaro, "Single-mode and polarization-independent silicon-on-insulator waveguides with small cross section," *J. Lightwave Technol.*, vol. 23, no. 6, pp. 2103–2111, Jun. 2005.
16. Rsoft Design Group, Inc., 200 Executive Group Blvd., Ossining, NY 10562. Available: http://www.rsoftdesign.com.
17. Photon Design, Oxford, UK, Available: http://www.photond.com/index.htm.
18. A. Liu, R. Jones, L. Liao, D. Samara-Rubio, D. Rubin, O. Cohen, R. Nicolaescu, and M. Paniccia, "A high-speed silicon optical modulator based on a metal-oxide-semiconductor capacitor," *Nature*, vol. 427, no. 6975, pp. 615–618, Feb. 2004.
19. C. E. Png, S. P. Chan, S. T. Lim, and G. T. Reed, "Optical phase modulators for MHz and GHz modulation in silicon-on-insulator (SOI)," *J. Lightwave Technol.*, vol. 22, no. 6, pp. 1573–1582, Jun. 2004.
20. T. D. Visser, B. Demeulenaere, J. Haes, D. Lenstra, R. Baets, and H. Blok, "Confinement and modal gain in dielectric waveguides," *J. Lightwave Technol.*, vol. 14, no. 5, pp. 885–887, May 1996.
21. SILVACO International, Santa Clara, CA, USA.

22. W. M. J. Green, M. J. Rooks, L. Sekaric, and Y. A. Vlasov, "Ultra-compact, low RF power, 10 Gb//s silicon Mach-Zehnder modulator," *Opt. Express*, vol. 15, no. 25, pp. 17106–17113, Dec. 2007.
23. F. Y. Gardes, G. T. Reed, N. G. Emerson, and C. E. Png, "A sub-micron depletion-type photonic modulator in silicon on insulator," *Opt. Express*, vol. 13, no. 22, pp. 8845–8854, Oct. 2005.
24. S. Manipatruni, X. Qianfan, B. Schmidt, J. Shakya, and M. Lipson, "High speed carrier injection 18 Gb/s silicon micro-ring electro-optic modulator," in *20th Annu. Meeting IEEE Lasers and Electro-Optics Society, 2007*, Lake Buena Vista, FL, pp. 537–538.
25. C. E. Png, "Silicon-on-insulator phase modulators," Ph.D. dissertation, Dept. Elect. Eng., University Surrey, Guildford, England, 2004.
26. L. Liao, D. Samara-Rubio, M. Morse, A. Liu, D. Hodge, D. Rubin, U. D. Keil, and T. Franck, "High speed silicon Mach-Zehnder modulator," *Opt. Express*, vol. 13, no. 8, pp. 3129–3135, Apr. 2005.
27. D. Andrea, "CMOS photonics today and tomorrow enabling technology," in *OFC, 2009*, San Diego, CA.
28. A. Liu, L. Liao, D. Rubin, H. Nguyen, B. Ciftcioglu, Y. Chetrit, N. Izhaky, and M. Paniccia, "High-speed optical modulation based on carrier depletion in a silicon waveguide," *Opt. Express*, vol. 15, no. 2, pp. 660–668, Jan. 2007.
29. F. Y. Gardes, A. Brimont, P. Sanchis, G. Rasigade, D. Marris-Morini, L. O'Faolain, F. Dong, et al., "High-speed modulation of a compact silicon ring resonator based on a reverse-biased pn diode," *Opt. Express*, vol. 17, no. 24, pp. 21986–21991, Nov. 2009.
30. L. Liao, A. Liu, D. Rubin, J. Basak, Y. Chetrit, H. Nguyen, R. Cohen, N. Izhaky, and M. Paniccia, "40 Gbit/s silicon optical modulator for highspeed applications," *Electronics Lett.*, vol. 43, no. 22, doi: 10.1049/el:20072253, Oct. 2007.
31. L. Liao, A. Liu, J. Basak, H. Nguyen, M. Paniccia, Y. Chetrit, and D. Rubin, "Silicon photonic modulator and integration for high-speed applications," in *IEEE Int. Electron Devices Meeting, 2008*, San Francisco, pp. 1–4.
32. C. Gunn, "CMOS photonics for high-speed interconnects," *IEEE Micro*, vol. 26, no. 2, pp. 58–66, Mar. 2006.
33. J. W. Park, J.-B. You, I. G. Kim, and G. Kim, "High-modulation efficiency silicon Mach-Zehnder optical modulator based on carrier depletion in a PN diode," *Opt. Express*, vol. 17, no. 18, pp. 15520–15524, Aug. 2009.
34. D. M. Gill, M. Rasras, T. Kun-Yii, C. Young-Kai, A. E. White, S. S. Patel, D. Carothers, A. Pomerene, R. Kamocsai, C. Hill, and J. Beattie, "Internal bandwidth equalization in a CMOS-compatible Si-ring modulator," *IEEE Photonics Technol. Lett.*, vol. 21, no. 4, pp. 200–202, Feb. 2009.
35. J. F. Liu, D. Pan, S. Jongthammanurak, D. Ahn, C. Y. Hong, M. Beals, L. C. Kimerling, J. Michel, A. T. Pomerene, C. Hill, M. Jaso, K. Y. Tu, Y. K. Chen, S. Patel, M. Rasras, A. White, and D. M. Gill, "Waveguide-integrated Ge p-i-n photodetectors on SOI platform," in *3rd IEEE Int. Conf. Group IV Photonics, 2006*, Ottawa, Canada, pp. 173–175.
36. J. Liu, D. Pan, S. Jongthammanurak, K. Wada, L. C. Kimerling, and J. Michel, "Design of monolithically integrated GeSi electro-absorption modulators and photodetectors on a SOI platform," *Opt. Express*, vol. 15, no. 2, pp. 623–628, Jan. 2007.
37. Y. H. Kuo, Y. K. Lee, Y. Ge, S. Ren, J. E. Roth, T. I. Kamins, D. A. B. Miller, and J. S. Harris, "Strong quantum-confined Stark effect in germanium quantum-well structures on silicon," *Nature*, vol. 437, no. 7063, pp. 1334–1336, Oct. 2005.
38. J. E. Roth, O. Fidaner, R. K. Schaevitz, Y.-H. Kuo, T. I. Kamins, J. S. Harris, and D. A. B. Miller, "Optical modulator on silicon employing germanium quantum wells," *Opt. Express*, vol. 15, no. 9, pp. 5851–5859, Apr. 2007.
39. J. E. Roth, O. Fidaner, E. H. Edwards, R. K. Schaevitz, Y. H. Kuo, N. C. Herman, T. I. Kamins, J. S. Harris, and D. A. B. Miller, "C-band side-entry ge quantum-well electroabsorption modulator on SOIi operating at 1 V swing," *Electronics Lett.*, vol. 44, no. 1, pp. 49–50, Jan. 2008.

40. O. Madelung, Ed., I.-V. *Semiconductors: Intrinsic Properties of Group IV Elements and III-V, and I-VII Compounds, Landolt-Börnstein New Series Group III*, vol. 22, Berlin, Germany: Springer-Verlag, 1987.
41. R. K. Schaevitz, J. E. Roth, S. Ren, O. Fidaner, and D. A. B. Miller, "Material properties of Si-Ge/Ge quantum wells," *IEEE J. Selected Topics Quantum Electron.*, vol. 14, no. 4, pp. 1082–1089, Jul. 2008.
42. D. D. Cannon, J. Liu, Y. Ishikawa, K. Wada, D. T. Danielson, S. Jongthammanurak, J. Michel, and L. C. Kimerling, "Tensile strained epitaxial Ge films on Si(100) substrates with potential application in L-band telecommunications," *Applied Phys. Lett.*, vol. 84, no. 6, pp. 906–908, Feb. 2004.
43. J. Liu, D. D. Cannon, K. Wada, Y. Ishikawa, S. Jongthammanurak, D. T. Danielson, J. Michel, and L. C. Kimerling, "Silicidation-induced band gap shrinkage in Ge epitaxial films on Si," *Applied Phys. Lett.*, vol. 84, no. 5, pp. 660–662, Feb. 2004.
44. S. L. Chuang, *Physics of Optoelectronic Devices*, New York: Wiley, 1995.
45. H. Shen and F. H. Pollak, "Generalized Franz–Keldysh theory of electromodulation," *Phys. Rev. B*, vol. 42, no. 11, p. 7097–7102, Oct. 1990.
46. H. Shen and M. Dutta, "Franz–Keldysh oscillations in modulation spectroscopy," *J. Appl. Phys.*, vol. 78, no. 4, pp. 2151–2176, Aug. 1995.
47. D. J. Paul, "8-Band k.p. modeling of the quantum confined Stark effect in Ge quantum wells on Si substrates," *Phys. Rev. B*, vol. 77, no. 15, article 155323, Apr. 2008.
48. S. Richard, F. Aniel, and G. Fishman, "Band diagrams of Si and Ge quantum wells via the 30-band kbullp method," *Phys. Rev. B*, vol. 72, no. 24, article 245316, Dec. 2005.
49. M. Virgilio, M. Bonfanti, D. Chrastina, A. Neels, G. Isella, E. Grilli, M. Guzzi, G. Grosso, H. Sigg, and H. von Känel, "Polarization-dependent absorption in Ge/SiGe multiple quantum wells: Theory and experiment," *Phys. Rev. B*, vol. 79, no. 7, article 075323, Feb. 2009.
50. M. Virgilio and G. Grosso, "Valence and conduction intersubband transitions in SiGe, Ge-rich, quantum wells on [001] Si[sub 0.5] Ge[sub 0.5] substrates: A tight-binding approach," *J. Appl. Phys.*, vol. 100, no. 9, article 093506, Nov. 2006.
51. M. Virgilio and G. Grosso, "Conduction intersubband transitions at normal incidence in $Si_{1-x}Ge_x$ quantum well devices," *Nanotechnology*, vol. 18, no. 7, article 075402, Feb. 2007.
52. M. Virgilio and G. Grosso, "Type-I alignment and direct fundamental gap in SiGe based heterostructures," *J. Phys. Condensed Matter*, vol. 18, no. 3, pp. 1021–1031, Jan. 2006.
53. M. Virgilio and G. Grosso, "Quantum-confined Stark effect in Ge/SiGe quantum wells: A tight-binding description," *Phys. Rev. B*, vol. 77, no. 16, article 165315, Apr. 2008.
54. V. R. Almeida, R. R. Panepucci, and M. Lipson, "Nanotaper for compact mode conversion," *Opt. Lett.*, vol. 28, no. 15, pp. 1302–1304, Aug. 2003.
55. K. K. Lee, D. R. Lim, D. Pan, C. Hoepfner, W.-Y. Oh, K. Wada, L. Kimerling, K. P. Yap, and M. T. Doan, "Mode transformer for miniaturized optical circuits," *Opt. Lett.*, vol. 30, no. 5, pp. 498–500, Mar. 2005.
56. A. Barkai, A. Liu, D. Kim, R. Cohen, N. Elek, H.-H. Chang, B. H. Malik, R. Gabay, R. Jones, M. Paniccia, and N. Izhaky, "Double-stage taper for coupling between SOI waveguides and single-mode fiber," *J. Lightwave Technol.*, vol. 26, no. 24, pp. 3860–3865, Dec. 2008.

# 11

# Organic Electro-Optic Crystal Modulators

Mojca Jazbinsek
*ETH Zurich and Rainbow Photonics AG*

Peter Günter
*ETH Zurich and Rainbow Photonics AG*



## 11.1 Introduction

Organic crystals exhibiting second-order optical nonlinearities present an alternative to inorganic ferroelectric crystals, such as lithium niobate ($LiNbO_3$), potassium niobate ($KNbO_3$), etc., and III-V semiconductor materials, for linear electro-optic (EO) modulation employing the Pockels effect. Considering the fundamental material response, organic crystals offer advantages similar to those of poled-polymer systems, for example, a low dielectric dispersion and therefore the possibility to match the optical and radio frequency electric-field velocities in the material, which is promising for high-speed EO modulation with modulation bandwidths well beyond 100 GHz. On the other hand, fabrication and processing of organic crystals for optical waveguides needs different approaches than those used for polymers and inorganic materials.

Organic nonlinear optical (NLO) crystalline materials have been studied for more than two decades due to several advantages they offer compared to inorganic materials, for example, the figure-of-merit for second-harmonic generation can be two-orders of magnitude higher than that of $LiNbO_3$. Due to low dielectric dispersions and high optical nonlinearities, organic crystals are also very efficient generators of terahertz radiation by optical rectification or difference-frequency generation; they can be used to create a broad range of frequencies (>20 THz). Several bulk organic crystals with linear dimensions of more than 1 cm are now commercially available and are also integrated into commercial THz sources and spectrometers. More recently, organic crystals have been considered for integrated EO applications, due to their potential for low half-wave voltages and high-speed performance, as well as their superior temperature stability compared to poled-polymer systems.

In this chapter, we review the present status and future prospects of organic EO single crystals for integrated optical modulators. In Section 11.2 we introduce the origins of the EO response in organic crystals, compare their characteristic properties to those of inorganic materials and poled polymers, as well as present examples of the best organic EO crystals. In Section 11.3 we present the basics of organic material crystal growth using different techniques for bulk, thin-film, and one-dimensional wire growth. Section 11.4 introduces the basic design and challenges of EO modulators based on organic crystals, examples of organic crystalline waveguide fabrication techniques, and performance of the already demonstrated EO modulators. We conclude with a short summary and an outlook on future material and device-design development in Section 11.5.

## 11.2 Electro-Optic Effect in Organic Crystals

### 11.2.1 Organic Molecules for Electro-Optic Crystals

The basic design of organic nonlinear optical molecules for crystals is similar to NLO molecules for poled polymers and is commonly based on π-conjugated polar chromophores that have an asymmetric response to an external electric field [1–7]. π-conjugated structures are regions of delocalized electronic charge distribution resulting from the overlap of π-orbitals. The electron distribution can be distorted by substituents at both sides of the π-conjugated system, as illustrated in Figure 11.1. Under the influence of an external electric field, $E$, the total electric dipole moment of such a molecule can be expanded in a Taylor series as

$$p_i = \mu_i + \varepsilon_0 \alpha_{ij} E_j + \varepsilon_0 \beta_{ijk} E_j E_k + \cdots, \tag{11.1}$$

where μ is the permanent electric dipole moment; $\alpha$ is the linear polarizability, representing the linear electronic response to an applied field; and β is the first-order hyperpolarizability, which models the asymmetry of the electronic response with respect to the polarity of the applied field.

Optimization of the molecular nonlinear response by molecular engineering is only one part of designing nonlinear optical or electro-optic materials. To achieve a macroscopic second-order nonlinearity, the arrangement of the molecules plays a very important role. The chromophores in a material must be ordered in an acentric manner to achieve a macroscopic second-order nonlinear optical response. This is most often realized by incorporating the chromophores into a polymer matrix and poling the composite under the influence of a strong direct current electric field while the composite is held close to the glass transition temperature. Another option for obtaining an efficient macroscopic second-order active organic material is to order the chromophores in an acentric structure by crystallization. Such ordering offers the highest possible chromophore density and the best long-term orientational stability.

Extension of the π-conjugation and the strength of the donor/acceptor substituents can lead to molecules with a very high molecular nonlinearity, that is, high first-order hyperpolarizability, β, which is already well-understood [1–7]. However, such molecules are not always appropriate for incorporation into macroscopic materials. Controlled crystallization of large organic molecules with desired

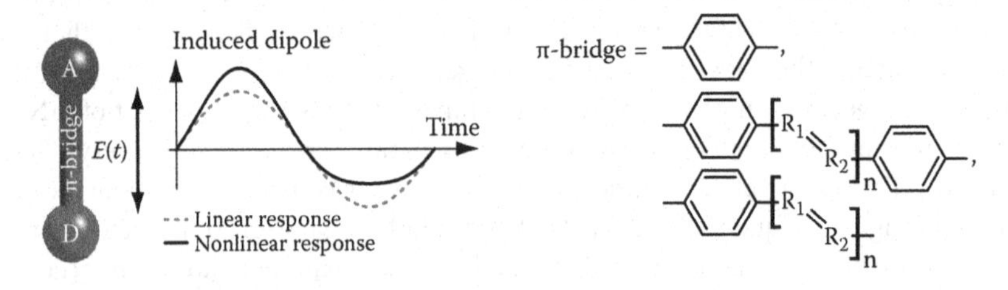


**FIGURE 11.1** Simple physical mechanism of the nonlinearity of the donor-acceptor substituted π-conjugated molecules. If we excite such molecules with an optical field, we induce an asymmetric polarization. The asymmetry is due to the nature of the substituents: the electron cloud (i.e., the electronic response) favors the acceptor over the donor.

optical properties is a challenging topic. A major problem with achieving this crystallization is that most organic molecules will crystallize in a centrosymmetric space group, which is usually attributed to dipolar interaction forces which favor antiparallel chromophore alignment. Prediction of the crystal structure based on the π-conjugated donor-acceptor chromophores is not yet possible. However, several crystal engineering approaches have been identified to obtain noncentrosymmetric nonlinear optical organic crystals. Most successful strategies include the use of molecular asymmetry, strong Coulomb interactions, non-rod-shaped π-conjugated cores, and a supramolecular synthetic approach. For details and examples see [5,8,9].

### 11.2.2 Macroscopic Electro-Optic Effect

The origin of the linear and nonlinear polarizability responses in organic materials is fundamentally different from that of inorganic ones. The electronic polarizability of molecular units presents the dominant contribution in organic materials. The polarizability of inorganic materials is based on strong bonding between the lattice components (ions), which are acting as additional polarizable elements. Contributions from ions or lattice vibrations are essential only for low (MHz to THz) frequencies of the applied electric field, since the dynamics of ions are much slower compared to that of electrons.

The free (unclamped or zero stress) EO coefficient $r^T$ in the low frequency region contains contributions from three different sources: acoustic lattice vibrations (acoustic phonons) $r^a$, optical lattice vibrations $r^o$, and electrons $r^e$

$$r^T = r^a + r^o + r^e = r^a + r^S, \tag{11.2}$$

where $r^S = r^o + r^e$ is the clamped (zero strain) EO coefficient. An analogous description is also valid for the linear response. At optical frequencies, far away from the electronic material resonances, the material response is purely of electronic origin. Therefore, the electronic contribution to the dielectric constant, $\varepsilon^e$, can be related to the refractive index, $n$, and the electronic contribution to the EO coefficient, $r^e$, can be related to the nonlinear optical coefficient for second-harmonic generation, $d$, as

$$\varepsilon^e = n^2 \text{ and } r^e = -\frac{4d}{n^4}. \tag{11.3}$$

Figure 11.2 schematically illustrates different lattice contributions and their frequency ranges for the linear response (dielectric constant, $\varepsilon$) and for the nonlinear response (EO coefficient, $r$).

Examples of the differences between the EO effects observed in inorganic and organic materials are given in Figure 11.3, where the time dependence of the refractive index change induced by a fast rectangular electric pulse is shown for $KNbO_3$ and 4-N, N-dimethylamino-4′-N′-methyl-stilbazolium tosylate (DAST) [10]. In $KNbO_3$, a large lattice contribution ($r^a$) and piezoelectric ringing are observed. Whereas in DAST, this contribution is very small and the response is mainly of electronic origin. From these measurements, the acoustic contributions to the EO effect $r^a = r^T - r^S$ can be determined and are listed in Table 11.1.

In Table 11.1, we compare the measured linear and nonlinear optical parameters and their frequency dependence in an organic crystal DAST and in inorganic crystals $LiNbO_3$ and $KNbO_3$. We see that in DAST the electronic contribution is dominant, whereas in inorganic crystals the greater part of the response comes from the acoustic and optical lattice vibrations. This difference is of essential importance for high-speed EO and terahertz wave applications.

There are advantages to building EO modulators with organic materials. In modulators made with organic materials, the modulating electric wave travels at about the same velocity as the optical wave due to the low dielectric constant in the low-frequency regime ($\varepsilon \approx n^2$). This is not the case for most inorganic EO materials where $\varepsilon >> n^2$. The velocity matching is important when building high-frequency EO modulators. The low dielectric constant of organic materials will also decrease the power requirement

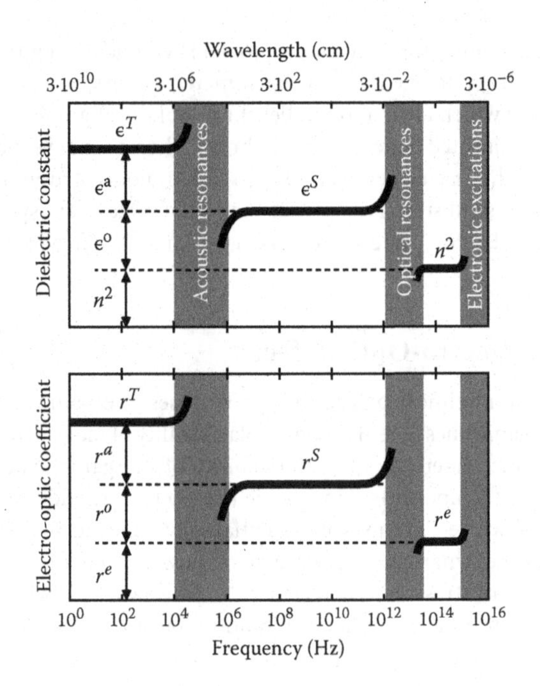


**FIGURE 11.2** Schematics of the material linear and nonlinear optical responses as a function of the frequency of the external electric field.

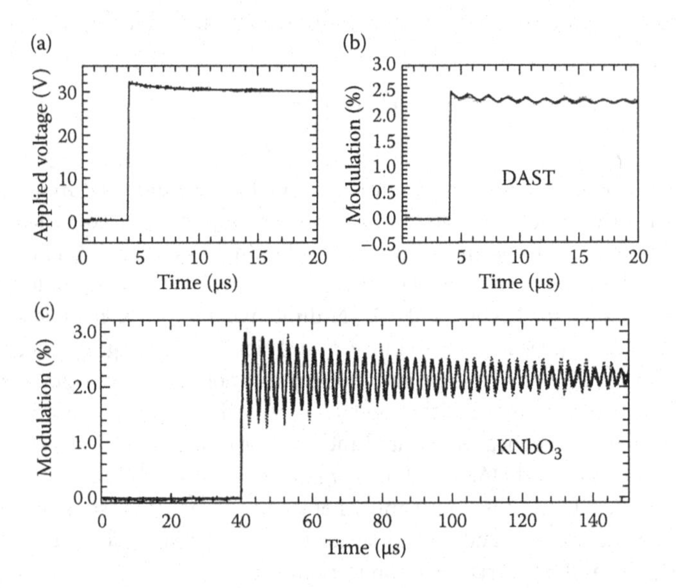


**FIGURE 11.3** (a) The applied step voltage and (b) the electro-optic response of $r_{111}$ in DAST at 1535 nm [10]. A small and negative contribution, resulting from acoustic excitation of the crystal, can be seen. For comparison, the electro-optic response, with a large contribution from acoustic phonons, in $KNbO_3$ is also shown (c). (Reproduced with permission from M. Jazbinsek, L. Mutter, and P. Gunter, *IEEE J. Sel. Top. Quantum Electron*, vol. 14, no. 5, pp. 1298–1311, © 2008 IEEE.)

**TABLE 11.1** Comparison of Inorganic Crystals $LiNbO_3$ and $KNbO_3$ to an Organic DAST Crystal with Respect to the Origins of the Linear and of the Nonlinear Optical Responses

| | $\varepsilon^T$ | $n^2$ | $r^T$ (pm/V) | $r^S$ (pm/V) | $r^e$ (pm/V) | $d$ (pm/V) | Reference |
|---|---|---|---|---|---|---|---|
| $LiNbO_3$ | 28 | 4.5 | 30 | 30 | 5.8 | 34 | [56] |
| $KNbO_3$ | 44 | 4.5 | 63 | 34 | 3.7 | 21 | [57] |
| DAST | 5.2 | 4.6 | 47 | 48 | 36 | 290 | [22] |

*Source:* M. Jazbinsek, L. Mutter, and P. Gunter, *IEEE J. Selected Topics Quantum Electron.*, vol. 14, no. 5, pp. 1298–1311, © 2008 IEEE. With permission

*Note:* Unclamped dielectric constant $\varepsilon^T$, refractive index $n$, unclamped $r^T$, clamped $r^S$, and electronic $r^e$ electrooptic coefficients, and nonlinear optical coefficient $d$. The values for $LiNbO_3$ and $KNbO_3$ are given at 1.06 μm, and for DAST at 1.535 μm.

of the EO modulators. Another advantage of organic over inorganic materials is the almost constant EO coefficient over an extremely wide frequency range. This property is essential when building broadband EO modulators and electric field sensors.

The interest in using organic crystals, rather than inorganic ones, as EO modulators stems from the above-mentioned advantages. In addition, the optical quality of molecular crystals as well as the long-term orientational and photochemical stabilities may be significantly superior to that of polymers [11].

### 11.2.3 Electro-Optic Effect in Terms of the Microscopic Nonlinearity

Using the oriented-gas model, the dominant electronic contribution to the EO effect in organic materials can be related to the molecular nonlinearity of the constituent molecules, that is, the first-order hyperpolarizability, β, and the orientational distribution of the molecules [12]. In a first approximation, the EO coefficients can be expressed by assuming that there is only one dominant tensor element of the dipolar molecules, $\beta_{zzz}$, which gives the following expression for the diagonal EO coefficients, $r_{kkk}$:

$$r_{kkk} = N f_{local} \langle \cos^3\theta_{kz} \rangle \beta_{zzz}, \tag{11.4}$$

where

$$f_{local} = \frac{2\left(f_k^{\omega}\right)^2 f_k^0}{n_k^4(\omega)},$$

and

$$f_i^{\omega} = \frac{n_i^2(\omega)+2}{3}.$$

Equation 11.4 shows the most important contributions to the linear EO effect in organic materials: the number density, $N$; the local-field factor, $f_{local}$; the molecular nonlinearity, β; and the orientational factor (the so-called order parameter $\langle \cos^3\theta_{kz} \rangle$, which is averaged over all molecules in a macroscopic system). Figure 11.4 schematically shows the molecules and projection angles, $\theta_{kz}$, between the polar axis, $k$, and the molecular axis, $z$. To maximize the diagonal EO coefficient, $r_{kkk}$, along the polar axis, the projection angles, $\theta_{kz}$, should be zero, that is, the charge transfer axes of the molecules should be parallel [12].

The relations between the micro- and the macroscopic nonlinearities (based on the oriented-gas model) give us a basic idea about the optimized packing of molecules. However, the relations still do not allow precise determination of the macroscopic EO coefficients from the known microscopic nonlinearities since intermolecular interactions may affect the molecular nonlinearities, β, in the solid state [13,14]. Nevertheless, for most practical considerations, including dispersion effects in organic materials and design strategies, the oriented-gas model describes the macroscopic second-order nonlinearities of various systems fairly well.

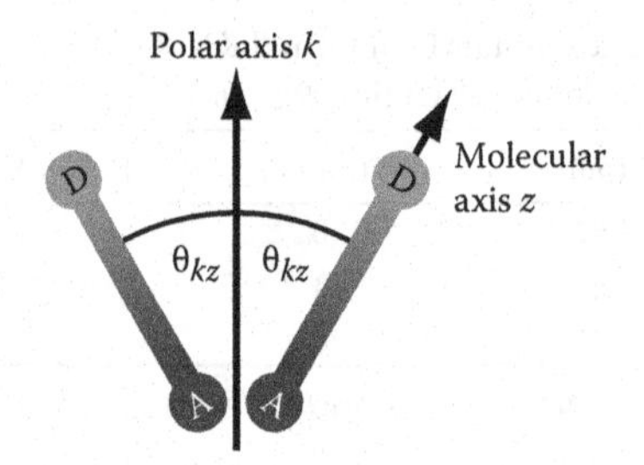


**FIGURE 11.4** Nonlinear optical molecules in the oriented-gas model. For optimizing the diagonal EO coefficient, the optimal $\theta_{kz} = 0°$.

## 11.2.4 DAST and Other Examples of Organic Electro-Optic Crystals

DAST is the most well-known and widely investigated organic EO crystal. DAST was first reported in 1989 by Marder et al. [15] and Nakanishi et al. [16] and is still recognized as the state-of-the-art organic nonlinear optical crystal. High optical quality and large size DAST crystals grown from methanol solution by the slow-cooling method [17,18] allow us to accurately determine the material's dielectric, linear, and nonlinear optical properties. Two reasons for the growing interest in obtaining high-quality DAST crystals are the high second-order NLO and EO coefficients, being respectively ten times and two times as large as those of the inorganic standard ($LiNbO_3$). DAST's high nonlinearities, in combination with a low dielectric constant, allow for high-speed EO applications and broadband THz wave generation.

The structural, dielectric, optical, EO, and nonlinear optical properties of DAST that are most relevant for photonic applications, including frequency conversion and THz wave generation have been recently reviewed [19]. Here, we summarize the material parameters most relevant for electro-optics.

DAST is an organic salt that consists of a positively charged stilbazolium cation and a negatively charged tosylate anion as shown in Figure 11.5a. The stilbazolium cation is one of the most efficient NLO active chromophores that pack in an acentric structure, whereas the counter ion tosylate is used to promote noncentrosymmetric crystallization [15,20]. The structure of DAST is shown in Figure 11.5b and 11.5c. The chromophores are packed with their main charge transfer axis oriented at about $\theta = 20°$ with respect to the polar axis, *a*. This packing results in a high order parameter of $\cos^3\theta = 0.83$, which is close to the optimum for EO applications.

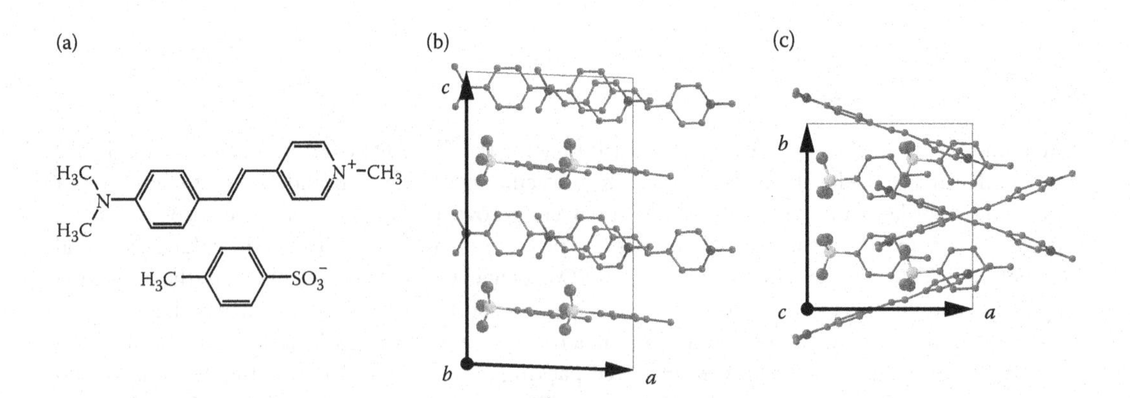


**FIGURE 11.5** (a) Molecular units of the ionic DAST crystal: the positively charged, NLO active chromophore methyl-stilbazolium and the negatively charged tosylate. (b–c) X-ray structure (projected along the crystallographic axes *b* and *c*) of the ionic DAST crystal with the point group symmetry *m* showing molecules from one unit cell. Hydrogen atoms have been omitted for clarity.

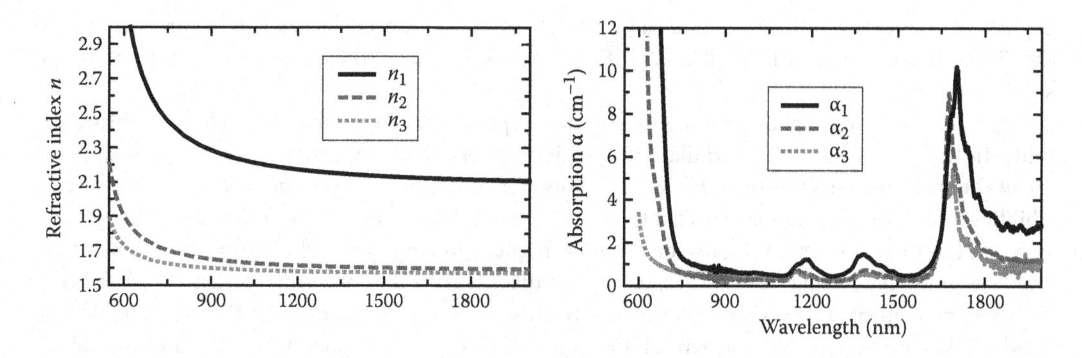


**FIGURE 11.6** Refractive indices $n_1$, $n_2$, $n_3$ and absorption coefficients $\alpha_1$, $\alpha_2$, $\alpha_3$ of DAST, indicated by full, dashed, and dotted curves, respectively. (Reproduced with permission from M. Jazbinsek, L. Mutter, and P. Gunter, *IEEE J. Sel. Top. Quantum Electron*, vol. 14, no. 5, pp. 1298–1311, © 2008 IEEE.)

Figure 11.6 shows the refractive indices and the absorption coefficients along the main optical axes as a function of wavelength [17,21]. DAST crystals are highly anisotropic with a refractive index difference $n_1 - n_2 > 0.5$ in the visible and infrared wavelength range. They show small absorption bands at 1700 nm, 1400 nm, and 1100 nm, which correspond to overtones of C-H stretching vibrational modes [22]. DAST crystals are well-suited for telecommunications applications, since they have a material absorption that is smaller than 1 cm$^{-1}$ at 1.3 μm and 1.55 μm wavelengths.

The dielectric constants of DAST in the low frequency range, below acoustic and optical lattice vibrations (see Figure 11.2), were measured to be $\varepsilon_1^T = 5.2 \pm 0.4$, $\varepsilon_2^T = 4.1 \pm 0.4$, and $\varepsilon_3^T = 3.0 \pm 0.3$ [21] and are considerably lower than those of inorganic EO materials (e.g., Table 11.1).

The low-frequency (unclamped) EO coefficients, $r_{ijk}^T$, of DAST are shown in Figure 11.7. The experimentally measured dispersion was modeled with the two-level model and is plotted with the solid curves in Figure 11.7 [21]; the deviation at shorter wavelengths stems from resonance effects when approaching the absorption edge.

DAST exhibits large EO coefficients and refractive indices, resulting in a high EO figure-of-merit $n^3r = 455 \pm 80$ pm/V at wavelength $\lambda = 1535$ nm. Therefore, the reduced half-wave voltage $V_\pi = \lambda/(n^3r)$ compares favorably to inorganic single crystals and poled EO polymers.

Theoretical evaluations show that the upper limits of second-order optical nonlinearities of organic crystals have, by far, not yet been reached [5]. Therefore, research and development of novel NLO organic

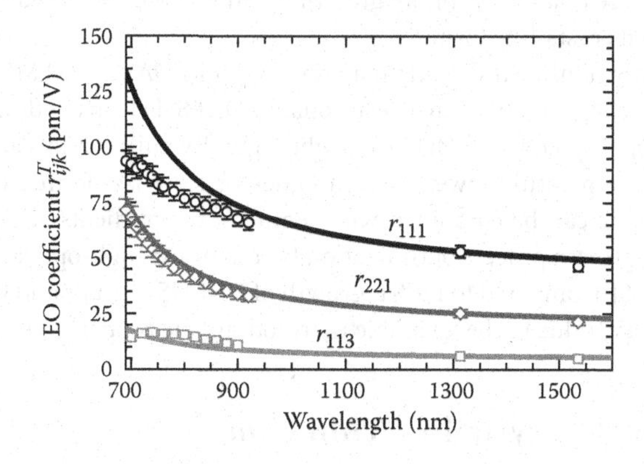


**FIGURE 11.7** Dispersion of the largest EO tensor coefficients of DAST: $r_{111}$, $r_{221}$, and $r_{113}$. (Reproduced with permission from M. Jazbinsek, L. Mutter, and P. Gunter, *IEEE J. Sel. Top. Quantum Electron*, vol. 14, no. 5, pp. 1298–1311, © 2008 IEEE.)

crystals continue. Several stilbazolium salts that have similar (or even superior) NLO properties than DAST has, have been identified [20,23–26]. These other salts may also have better crystal growth possibilities [27,28].

In general, the chromophores for highly nonlinear optical crystals exhibit a limited temperature stability. In the case of DAST (and similarly for its derivatives), the chromophores start to decompose at about the melting temperature of 256°C. Therefore, for most practical situations, the processing possibilities of stilbazolium salts are limited to solution-based techniques. On the other hand, melt growth is very attractive for several reasons, for example, higher growth rates, higher purity, and very attractive waveguide processing possibilities, as will be discussed in the following section. Until recently, the short $\pi$-conjugated chromophores with relatively low melting temperatures ($T_m < 150°C$), but also relatively low first-order hyperpolarizabilities, such as COANP (2-cyclooctylamino-5-nitropyridine), were the only organic nonlinear optical crystals obtained by melt growth techniques. Unfortunately, the EO figure-of-merit, $n^3r$, of these crystals may be one order of magnitude smaller than that of the best stilbazolium salts. Therefore, to design organic EO materials with a broad spectrum of processing possibilities, the challenge is to simultaneously achieve both high thermal stability and high nonlinearity in one compound. To avoid the tradeoff between nonlinearity and thermal stability of organic crystalline materials, different configurationally-locked polyene chromophores have been developed recently [29,30]. Several of these chromophores crystallize in a noncentrosymmetric structure and have a high powder second-harmonic generation efficiency in the same order of magnitude as that of DAST.

Configurationally-locked polyene crystals DAT2 (2-(3-(2-(4-dimethylaminophenyl)vinyl)-5,5-dimethylcyclohex-2-enylidene)malononitrile) and OH1 (2-(3-(4-hydroxystyryl)-5,5-dimethylcyclohex-2-enylidene)malononitrile) are particularly promising for integrated EO applications [31–34]. The big advantage of DAT2 is the excellent range of possibilities for thin-film processing from solution, vapor, and melt; although its EO properties are not as high as those of DAST due to a less optimal crystalline packing [29,31]. On the other hand, OH1 crystals show an EO effect as high as that of DAST crystals [35]. OH1 exhibits very good crystal processing possibilities from solution for both high optical quality bulk crystals [36] and large-area single crystalline thin films on substrates [37].

## 11.3 Crystal Growth Techniques

### 11.3.1 Solution, Melt, and Vapor Crystal Growth

Crystallization of organic materials is achieved by solution growth, melt growth, or vapor growth. Different, and in some cases rather complex, growth techniques are required to produce bulk, thin-film, or wire crystals. The choice of an appropriate growth technique depends on different material properties, as well as the desired crystalline form.

For many of the highly nonlinear organic materials, such as DAST, DSTMS (4′-dimethylamino-N-methyl-4-stilbazolium 2,4,6-trimethylbenzenesulfonate), DAPSH (trans-4-dimethylamino-N-phenyl-stilbazolium hexafluorophosphate) and OH1 the highest quality bulk crystals are grown from solution. Most commonly, slow temperature lowering techniques or slow isothermal evaporation techniques are used. These techniques can be combined with temperature gradients at the growth position. An example of the solution-growth process optimization for obtaining high-optical quality bulk crystals of DAST can be found in [18]. Compared to DAST, growth of DSTMS crystals and OH1 crystals from solution may be faster and easier due to the favorable thermodynamic properties of these materials [27,36].

### 11.3.2 Organic Single-Crystalline Thin Films

High quality, single crystalline thin films of highly NLO materials are essential for the fabrication of integrated photonic devices. If one starts with bulk crystals, then cutting, polishing and structuring procedures are required to fabricate waveguiding devices. These procedures are complicated, expensive, and time-consuming. Thin films may be much more conducive to building simpler and cheaper waveguiding structures for applications such as EO modulators. Various approaches have been investigated for the fabrication of single crystalline films using either solution, melt, or vapor growth techniques; examples are listed in Table 11.2. More details can be found in the respective references given in the table and an overview of different approaches for thin-film fabrication is presented in [38]. Several of the thin films presented in Table 11.2 are very compatible with integrated optical applications. Examples of successful demonstrations are discussed in Section 11.4.

### 11.3.3 Single-Crystalline Micro- and Nano-Wires

One of the very attractive solutions for integrated photonic devices is to directly grow micro- and nanostructures at the desired position. This can be done by structuring standard inorganic templates (made of glass, silicon, metals, and other materials) with void structures at positions where active organic crystalline materials are desired. This method was recently demonstrated by using melt-processable materials, namely the configurationally-locked polyene DAT2 [31] and the small chromophore COANP [39]. Both materials were chosen because of their favorable growth (from melt) characteristics as well as their tendencies for thin-film formation (see Table 11.2). With this method, several-mm long single crystalline wires with thicknesses from 30 nm to several μm have been obtained, as shown in Figure 11.8. More details on wire fabrication are given in Section 11.4.3.6.

## 11.4 Integrated Organic Crystalline Electro-Optic Modulators

### 11.4.1 Basic Modulator Design

With organic crystalline materials, several basic EO modulation schemes have recently been demonstrated, including both amplitude and phase modulation. Two basic amplitude modulation schemes considered here are illustrated in Figure 11.9.

The conventional scheme using a Mach-Zehnder (MZ) integrated optical modulator is based on the interference of optical waves traveling along the two arms of the interferometer, resulting in the following intensity transmission function:

$$T(t) = \cos^2\left(\frac{\Delta\phi(t)}{2}\right) = \cos^2\left(-\frac{1}{2}\frac{\pi n^3 r}{\lambda}V(t)\frac{L}{G}\right) = \cos^2\left(\frac{\pi}{2}\frac{V(t)}{V_\pi}\right), \tag{11.5}$$

where $\Delta\phi$ is the difference in phase for optical waves traveling along each arm of the interferometer due to the voltage applied along one arm of the interferometer, $L$ is the propagation length, and $G$ is the distance between the electrodes. In Equation 11.5, the half-wave voltage $V_\pi = (\lambda/n^3r)(G/L)$ determines the voltage needed to switch the transmittance from 0 to 1 and critically depends on the material EO figure-of-merit, $n^3r$, as well as the design of the waveguides by the geometrical factor $L/G$. If we operate the modulator at the point of the maximum slope of $dT/dV$ (here at $T(V) = 0.5$) by using an appropriate bias voltage, a linear intensity modulator is obtained (see Figure 11.9a).

Mach-Zehnder integrated-optic elements are compatible with complementary metal oxide semiconductor (CMOS) voltage levels and allow for high speed (>40 GHz) modulation in a traveling-wave configuration. Present devices based on $LiNbO_3$ are still relatively large in size (lengths in the order of 1 cm) and have relatively high power consumption (several watts). A promising structure for more compact and lower power devices is a microring resonator that allows for long electrical and optical interaction lengths with a very small size (on the order of 10 μm–100 μm).

The basic design of a microresonator is shown in Figure 11.9b. The device consists of one or two straight waveguides which are coupled to a microring waveguide. Light propagating in the input waveguide will be partially coupled into the microring. At resonator optical frequencies, the intensity of the

**TABLE 11.2** Growth Methods for Thin, Organic Single Crystalline Films That Have Second-Order Nonlinear Optical Activity

| Material | Thickness (μm) | Single Crystal Area | Reference |
|---|---|---|---|
| | **(a) Mechanical method** | | |
| *Cut and polish* | | | |
| DAST | 20–25 | ~20 mm² before polishing | [58] |
| *Etching* | | | |
| DAST | ~18 | 0.5–2 mm waveguide length | [45] |
| | **(b) Epitaxial growth method** | | |
| *Solution epitaxial growth* | | | |
| DAST | 10–35 | 4 mm² | [59] |
| OH1 | 0.1–4 | >2 cm² | [37] |
| *Organic molecular beam epitaxy* | | | |
| MNBA | 1–4 | 40 mm² | [60] |
| | **(c) Capillary method between two plates** | | |
| *Solution capillary method with shear* | | | |
| DAST | 0.1–18 | ~1 cm² | [61] |
| NPP | 3 | 1 cm² | [62] |
| *Solution capillary method without shear* | | | |
| DAST | 20 | 12 mm² | [59] |
| DSTMS | 30 | 30 mm² | [27] |
| DAT2 | 1–40 | 10 mm² | [63] |
| *Melt capillary method* | | | |
| MNA | 5–10 | 2–3 mm sides | [64] |
| MNA | 1 | 1 cm² | [65] |
| AANP | 6–14 μm | Not given | [66] |
| COANP | 5–20 μm | Few mm sides | [3,67] |
| NPP | 10 μm | 0.25 mm² | [68] |
| DAT2 | 0.025–1 | 7 mm waveguide length | [31] |
| | **(d) Planar solution growth method** | | |
| *Two-dimensional $\Delta T$ method* | | | |
| DAST | – | – | [59] |
| *Traveling cell method* | | | |
| DAST | 20–40 | 4 mm² | [59] |
| *Undercooled flow cell method* | | | |
| DAST | 40 | 11 mm² | [59] |
| | **(e) Vapor growth method** | | |
| DAT2 | 0.2–5 | 15 mm² | [69] |
| DAST | 1–5 | 0.2 mm diameter | [70] |

*Note:* DAST = 4'-dimethylamino-N-methyl-4-stilbazolium tosylate; OH1 = 2-(3-(4-hydroxystyryl)-5,5-dimethylcyclohex-2-enylidene)malononitrile; MNBA = 4'-nitrobenzylidene-3-acetamino-4-methoxyaniline; NPP = N-(4-nitrophenyl)-(L)-prolinol; DSTMS = 4'-dimethylamino-N-methyl-4-stilbazolium 2,4,6-trimethylbenzenesulfonate; DAT2 = 2-(3-(2-(4-dimethylamino-phenyl)vinyl)-5,5-dimethylcyclohex-2-enylidene)malononitrile; MNA = 2-methyl-4-nitroaniline; AANP = 2-adamantyl-amino-5-nitropyridine; COANP = 2-cyclooctylamino-5-nitropyridine.

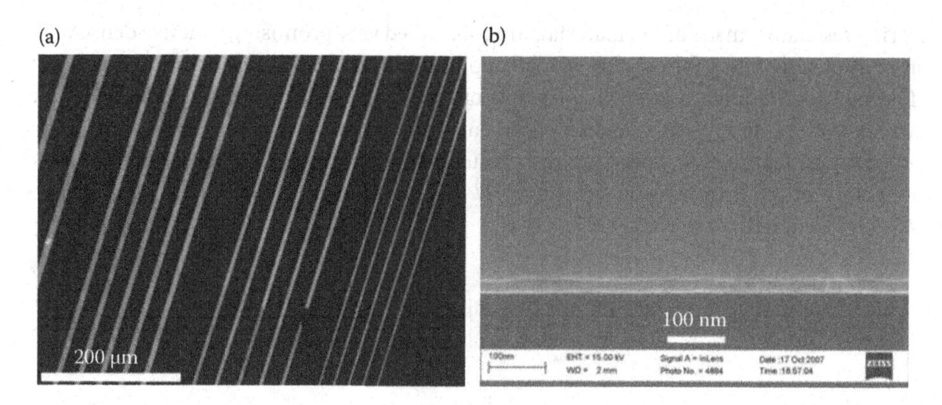


**FIGURE 11.8** (a) Transmission microscope image of approximately 25 nm thick, melt-grown, DAT2 crystalline stripes as seen between crossed polarizers. (b) Scanning electron micrograph of the corresponding end-facet. (Reprinted from H. Figi, M. Jazbinsek, C. Hunziker, M. Koechlin, and P. Gunter, *Opt. Express*, vol. 16, no. 15, pp. 11,310–11,327, July 2008. With permission of Optical Society of America.)

light in the ring will increase dramatically. The transmission function of a single waveguide coupled to a microring will depend on the finesse, $F$, of the microresonator and the circumference, $L$, as

$$T = 1 - \frac{T_0}{1 + \left(\frac{2F}{\pi}\right)^2 \sin^2\frac{\phi}{2}}, \tag{11.6}$$

where $\phi = 2\pi nL/\lambda$ is the round-trip phase change and $T_0$ depends on the coupling coefficient and the losses. For the so-called critical coupling $T_0 = 0$ (see e.g., [40]). If the ring is made of an EO material, we can tune its refractive index $n$ (or more precisely, the effective index of the waveguide mode) and therefore its resonant frequency by applying an external voltage.

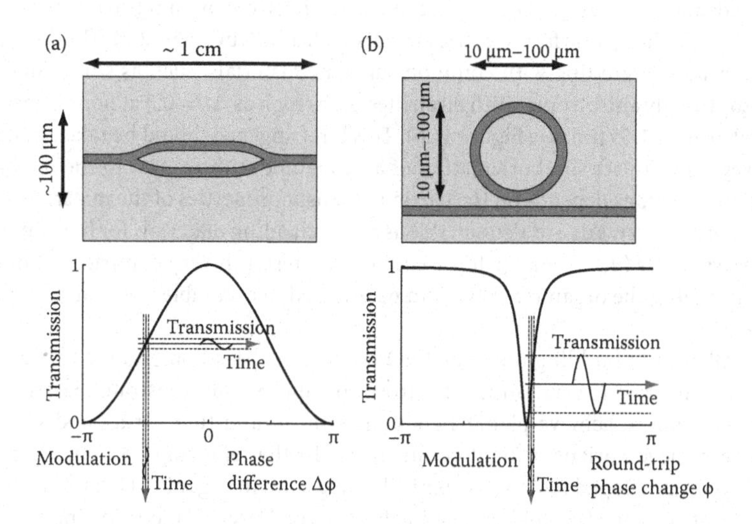


**FIGURE 11.9** Electro-optic modulation using a (a) Mach-Zehnder modulator and (b) microring resonator. The transmission function for the Mach-Zehnder modulator is according to Equation 11.5 and for the microring resonator according to Equation 11.6 using $F = 15$ and $T_0 = 1$.

Microring resonators made of EO materials are considered very promising as active dense wavelength division multiplexing filters for on-chip optical applications. Microring resonators are also very interesting for compact EO modulators. The principle of operation is illustrated in Figure 11.9. Compared to MZ devices, the same applied e-field amplitude may lead to much larger intensity modulations with microring devices. Figure 11.9 shows that the modulation sensitivity of the MZ device depends on the maximum slope of the transmission function: $dT/dV|_{max} = \pi/(2V_\pi)$. For microring resonators, we can define an equivalent half-wave voltage $V_\pi^{eq}$ [41] as

$$V_\pi^{eq} = \frac{\pi}{2}\left(\frac{dT}{dV}\bigg|_{max}\right)^{-1} = \frac{\pi}{2}\left(\frac{dT}{d\phi}\frac{dT}{dV}\bigg|_{max}\right)^{-1} \cong \frac{5}{2F}V_\pi, \tag{11.7}$$

considering $V_\pi$ as defined in Equation 11.5. From the above equation, we see that the voltage required to achieve the same intensity modulation is reduced by a factor of $2F/5$ in microresonators compared to MZ modulators, which may be substantial considering that $F$ is typically around 20 or more [41].

Microring modulators are therefore interesting because they may require lower driving voltages as compared to MZ modulators and they can have much smaller footprints (allowing high density integration). Using poled-polymer microring resonators and traveling-wave electrodes, modulation at 165 GHz has been demonstrated [42], which makes microring resonators highly promising for very compact and efficient EO modulators.

### 11.4.2 Overview of the Structuring Techniques for Organic Crystals

As for inorganic optical waveguides, we need to structure organic crystals with (sub-)micron precision so that a suitable refractive index contrast for optical waveguiding is achieved. For general optical waveguiding, a small refractive index contrast on the order of $\Delta n \sim 10^{-3}$ may be sufficient. However, for the small waveguides needed for large-scale integration, as well as for reducing the half-wave voltage of EO modulators, a larger index of refraction contrast is desired. For example, to fabricate microring resonators with a small radius (below 10 μm), the refractive index of the waveguide material should be larger than the cladding material by at least about $\Delta n \sim 0.5$ to avoid high losses. Compared to poled polymers, one advantage to using EO organic crystals is a relatively high refractive index, which can be as high as the values of inorganic ferroelectric crystals such as $LiNbO_3$ ($n \sim 2.2$). This high index contrast allows very efficient waveguiding with common substrate materials, such as silica ($SiO_2$). Organic EO crystals are also strongly anisotropic, with birefringence as high as $\Delta n = 0.5$ at nonresonant wavelengths (for example, at around 1.55 μm, see Figure 11.6). This birefringence should be taken into account when designing waveguides. The desired orientation of a waveguide with respect to the optical and electric field propagation directions depends on the particular tensor properties of the material (see examples in Section 11.4.3). Organic crystals are also suitable as active cladding materials for high-index silicon photonic passive waveguides ($n_{Si} \sim 3.5$). Such waveguides can result in very compact EO modulators with high figures-of-merit, if the organic crystals can be oriented in a suitable way. An example is discussed in Section 11.4.3.7.

The main challenges to building integrated EO modulators based on organic crystals are related to their processing. The organic crystal must be grown on appropriate substrate materials, grown in an appropriate orientation to achieve planar light confinement, and then structured with an appropriate technique to achieve horizontal light confinement. In the following section, we describe several techniques that were developed to fabricate optical waveguides in organic EO crystals including photolithography, photostructuring (photobleaching and femtosecond laser ablation), ion implantation, electron-beam irradiation, and direct deposition into prestructured inorganic templates. These techniques are schematically presented in Table 11.3 with some of their main features.

**TABLE 11.3** Structuring Methods Investigated for Organic, Single Crystalline Waveguides That Exhibit Second-Order Nonlinear Optical Activity

| Technique | Max. Index Contrast $\Delta n$ | Comments |
|---|---|---|
| **Photolithography** 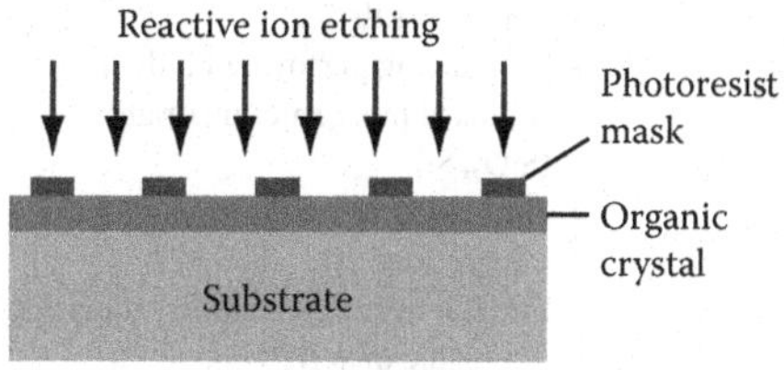<br>Reactive ion etching<br>Photoresist mask<br>Organic crystal<br>Substrate | 1.1 horizontal<br>0.6 vertical | • Thin films needed<br>• Side-wall quality depends critically on the optimization of reactive ion etching (RIE) |
| **Photobleaching** 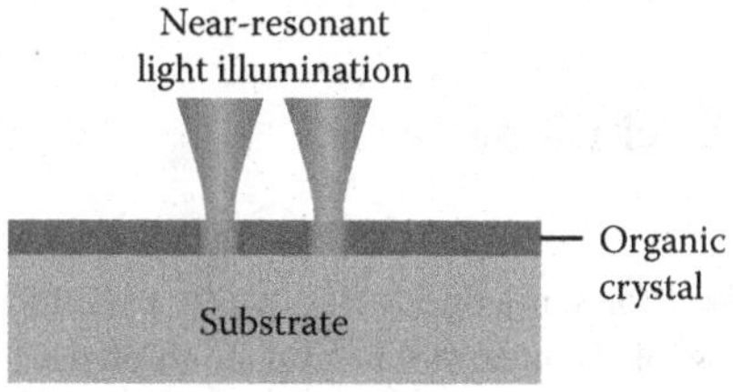<br>Near-resonant light illumination<br>Organic crystal<br>Substrate | 0.5 horizontal<br>0.6 vertical | • Thin films needed<br>• Smooth side walls |
| **Femtosecond laser ablation** 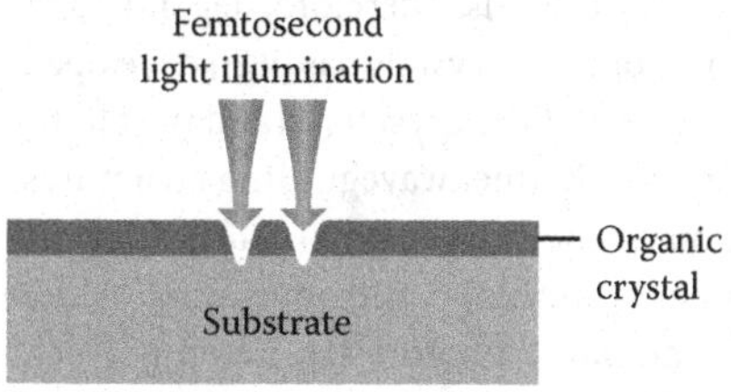<br>Femtosecond light illumination<br>Organic crystal<br>Substrate | 1.1 horizontal<br>0.6 vertical | • Thin films needed<br>• Side-wall quality depends critically on laser parameters |
| **Ion implantation** 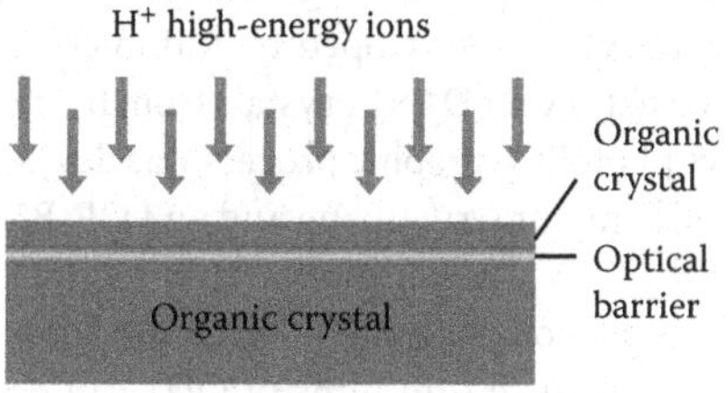<br>$H^+$ high-energy ions<br>Organic crystal<br>Optical barrier<br>Organic crystal | 0.1 vertical | • Provides vertical confinement if thin films are not available<br>• Smooth refractive-index gradients |
| **Electron-beam structuring** 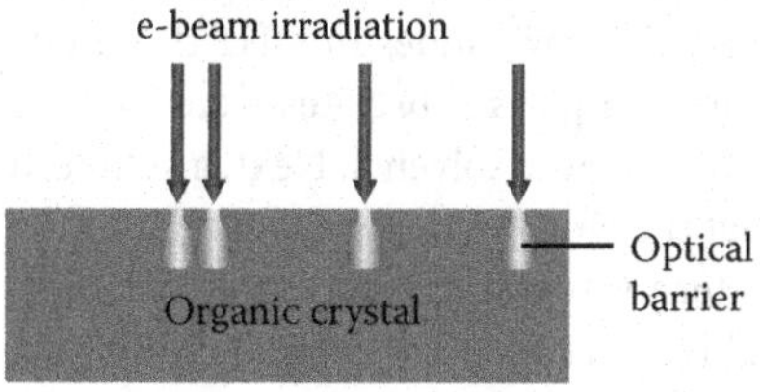<br>e-beam irradiation<br>Optical barrier<br>Organic crystal | 0.1 horizontal<br>0.1 vertical | • Thin films not needed<br>• Smooth side walls |

*(continued)*

**TABLE 11.3 (Continued)** Structuring Methods Investigated for Organic, Single Crystalline Waveguides That Exhibit Second-Order Nonlinear Optical Activity

| Technique | Max. Index Contrast $\Delta n$ | Comments |
|---|---|---|
| **Epitaxial Growth** | | |
| Epitaxially grown organic crystal<br>Other materials<br>Structured substrate | 0.6 horizontal<br>0.6 vertical | • Very versatile<br>• Structuring performed only in standard inorganic substrates ($SiO_2$, Si)<br>• Easy electrode or other material integration<br>• Limited by crystallization properties of organic material |

*Note:* More details are given in Section 11.4.3. $\Delta n$ is estimated assuming $SiO_2$ substrates (where applicable).

## 11.4.3 Examples of Modulators and Fabrication Techniques

### 11.4.3.1 Waveguide Fabrication by Photolithography

Standard photolithographic microfabrication techniques developed for semiconductors can be used for several inorganic nonlinear optical crystals and polymers, but most of them cannot be applied to organic crystals straightforwardly. This is because many organic crystals are incompatible with common photoresist solvents that will generally etch or even destroy the surfaces of the crystals. Therefore, depending on their specific physico-chemical properties, photolithography for some organic crystals has been developed.

A few organic crystals are insoluble in common organic solvents. For these crystals, standard photolithography and oxygen-reactive ion etching may be used to fabricate channel waveguiding structures, as demonstrated with MNBA (4′-nitrobenzylidene-3-acetamino-4-methoxyaniline) [43]. Channel waveguides were also produced in (-)MBANP ((-)-2-(α-methylbenzylamino)-5-nitropyridine), which is soluble in organic solvents, by using a special water-soluble inorganic photoresist: heteropolyacid [44]. However, both of these techniques cannot be used for highly nonlinear organic salts (like DAST) which are soluble in both water and the standard organic solvents used for photolithography. A special photolithographic technique to produce channel waveguides in DAST was developed by Kaino et al. [45,46], where a protecting polymethyl methacrylate layer was used to prevent DAST crystals from being dissolved in the photoresist solution and developer. An alternative photolithographic process was developed in our group [19] using the standard photoresist SU8 (MicroChem) for structuring and an LOR-B5 (MicroChem) lift-off resist as a protective cladding.

Photolithographic structuring of another promising EO crystal, phenolic polyene OH1, is less complex than for DAST crystals, since OH1 is not soluble in water. Also, since thin films of OH1 can be fabricated directly on glass substrates, OH1's structuring is analogous to structuring of inorganic materials. A photolithographic technique for the fabrication of wire optical waveguides in OH1 thin films is schematically depicted in Figure 11.10a–d. First, OH1 thin films on glass substrates were covered with a water-soluble polyvinyl alcohol layer to protect the OH1 from organic solvents. Next, a standard photoresist, SU8, was deposited on top and then structured so that the wires were perpendicular to the $x_3$ polar axis of the film (Figure 11.10c). Then, the waveguide pattern was transferred to the OH1 film by an etching process [33] which yielded OH1 wires as shown in Figure 11.10d and 11.10e. Finally, gold electrodes were vapor-deposited on the sides of the waveguides by using a simple shadow mask, resulting in a relatively large electrode separation of 50 μm, as shown in Figure 11.10f.

The resulting waveguides, with heights of 3.5 μm, exhibited propagation losses (at 980 nm) of 2 dB/cm, 9 dB/cm, and 17 dB/cm for waveguide widths of 7.6, 5.4, and 3.4 μm, respectively. The losses increase in narrower waveguides because of the very high core-cladding index contrast of $\Delta n = 1.23$. For phase-modulation measurements, a straight-waveguide sample with cross-sectional dimensions of $w \times h = 3.4\ \mu m \times 3.5\ \mu m^2$ was used (see Figure 11.10f). The measured half-wave voltage × interaction length

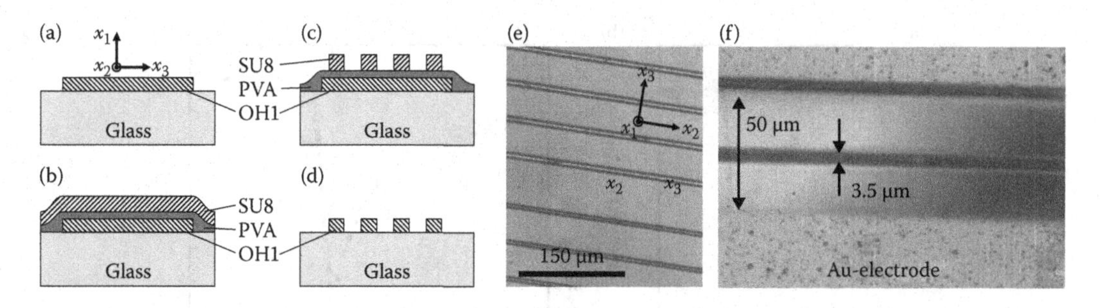


**FIGURE 11.10** (a)–(d) Processing steps for the fabrication of OH1 wire waveguides on glass substrates. (e) Microscope image of single crystalline OH1 wires on glass. (f) OH1 waveguide with deposited gold electrodes for EO modulation experiments. (Reprinted from C. Hunziker, S. J. Kwon, H. Figi, M. Jazbinsek, and P. Gunter, *Opt. Express*, vol. 16, no. 20, pp. 15 903–15 914, Sep. 2008. With permission of Optical Society of America.)

product, $V_\pi L$, resulting from the present configuration is 8.4 V·cm and 28 V·cm at 632.8 nm and 852 nm, respectively [33]. With an optimized electrode configuration of 1 μm spacing between the electrode and the waveguide, $V_\pi L$ is expected to be reduced to 0.3 V·cm. Therefore, this technique shows promise to achieve high-index-contrast, sub-1-V half-wave voltage, organic EO modulators with highly stable chromophore orientation.

#### 11.4.3.2 Photostructured Optical Waveguides

An alternative technique for waveguide patterning is photobleaching, which is often used for the structuring of polymer-based devices. Photobleaching refers to the change of chemical composition of molecules after high-intensity light exposure in the wavelength regime of high optical absorption. The decomposition leads to a decrease of the refractive index that can be used for confining light in the lateral direction. The refractive index of DAST crystals can be reduced by photobleaching using light that has wavelength within its absorption band (260 nm–700 nm) [47,48]. The depth range of photobleaching can be varied between 0.2 μm and 2.6 μm by selecting a suitable wavelength [48]. The refractive index change for light polarized along the polar axis, $x_1$, is relatively very high, about $\Delta n = 0.5$ at 1.55 μm, which is important for large-scale integration. Photobleaching was used to produce channel waveguides in thin DAST samples by Kaino et al. [46]. They used an ultraviolet UV resin as an undercladding and measured propagation losses of about 11 dB/cm.

Another possibility for direct-structuring of organic crystals with light is femtosecond laser ablation, which has been investigated for patterning of DAST surfaces [49]. Fluence ranges for optimal, almost damage free, ablation were determined at the wavelengths 775 nm, 600 nm, and 550 nm with a pulse width of about 170 fs. The groove profiles reveal that ablation of ridge waveguides in DAST using fs lasers is a promising alternative to photolithography for producing optical waveguides in organic crystals [49].

#### 11.4.3.3 Ion Implanted Waveguides

Since the growth of organic single crystalline thin films (with the thickness control needed for integrated optics) can be very challenging, ion-implantation was developed to produce planar waveguides in bulk organic crystals [50]. The refractive index of DAST decreases due to electronic excitations induced by ion irradiation, in contrast to inorganic materials, for which the major refractive index changes are due to ion-induced nuclear displacements.

In $H^+$ ion implanted DAST, the electronic loss curve has a peak at the end of the implantation depth range, which results in an optical barrier suitable for optical waveguiding. The measured profile of the refractive index, $n_1$, at $\lambda = 633$ nm for an implantation fluence of $\phi = 1.25 \times 10^{14}$ ions/cm$^2$ is shown in Figure 11.11a. Maximal peak refractive index changes of around $\Delta n = -0.2$ at 633 nm and $\Delta n = -0.1$ at 810 nm were measured [50].

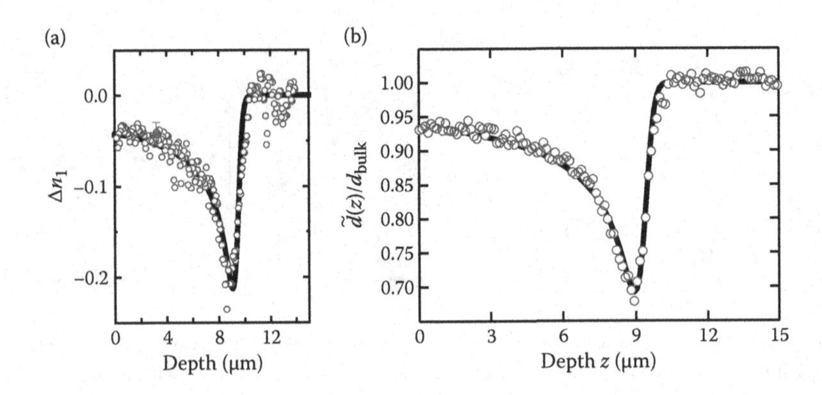


**FIGURE 11.11** (a) Measured refractive index change $\Delta n_1$ at $\lambda = 633$ nm as a function of the implantation depth. (Reprinted from L. Mutter, A. Guarino, M. Jazbinsek, M. Zgonik, P. Gunter, and M. Dobeli, *Opt. Express*, vol. 15, no. 2, pp. 629–638, Jan. 2007. With permission of Optical Society of America.) (b) Normalized susceptibility profile at $\lambda = 1176$ nm. The waveguides were produced in DAST by 1 MeV $H^+$ ion implantation with a fluence of $\phi = 1.25 \times 10^{14}$ ions/cm$^2$ at an angle of 60°. (Reprinted from L. Mutter, M. Jazbinsek, C. Herzog, and P. Gunter, *Opt. Express*, vol. 16, no. 2, pp. 731–739, Jan. 2008. With permission of Optical Society of America.)

To use ion-implanted waveguides for active integrated photonic devices, it is very important to maintain the high NLO and EO properties in the waveguide core region. After implantation, the nonlinear optical coefficient in the waveguide core region is preserved by more than 90%, as shown in Figure 11.11b [51].

Waveguiding in ion-implanted waveguides was demonstrated by using the conventional end-fire light coupling. The light was polarized parallel to $x_1$, which is the most interesting configuration for EO modulation in DAST. The estimated transmission losses are around 10 dB/cm at $\lambda = 1550$ nm [50]. Electro-optic phase modulation in ion-implanted waveguides was also demonstrated [51]. At $\lambda = 1550$ nm, the EO coefficient in $H^+$-implanted waveguides is $r_{111} = 42 \pm 10$ pm/V, which is about 10% lower than the bulk value of $47 \pm 8$ pm/V. This change agrees with the nonlinear susceptibility profile shown in Figure 11.11b. For DAST crystal modulators with an electrode distance less than 5 μm, the half-wave voltage-interaction length products for ion-implanted waveguides are expected to be below 1.7 V·cm.

#### 11.4.3.4 Electron Beam Induced Waveguides

Electron beam irradiation allows patterning EO channel waveguiding structures in bulk organic NLO crystals without the necessity for thin-film technology [52]. The electrons of the writing beam are scattered in the material and therefore the beam is widened in the target material. This circumstance can be exploited to directly write channel waveguides in bulk crystals by exposing two lines separated by the waveguide width as depicted in Figure 11.12a. The advantage of this configuration is that the waveguide core is mostly in virgin material, in which the nonlinear and EO properties are the same as in the bulk material.

DAST crystals were exposed to 30 keV electrons using an electron beam system. The refractive index decreased due to electron beam exposure. For a fluence of $\phi = 2.6$ mC/cm$^2$, a maximal refractive index reduction of $\Delta n = -0.3$ at $\lambda = 633$ nm was achieved [52]. The calculated two-dimensional refractive index cross-section after electron beam irradiation is shown in Figure 11.12b.

Channel waveguides and MZ modulators were produced by e-beam irradiation in DAST, using various fluences and waveguide dimensions. The $x_1x_3$ end-faces were subsequently polished and waveguiding was demonstrated by using standard end-fire coupling [52]. Electrodes were patterned after the e-beam exposure. The EO modulation measurements were performed in the experimental configuration shown in Figure 11.13b. The applied modulation voltage (amplitude of 10 V, lower curve) and the

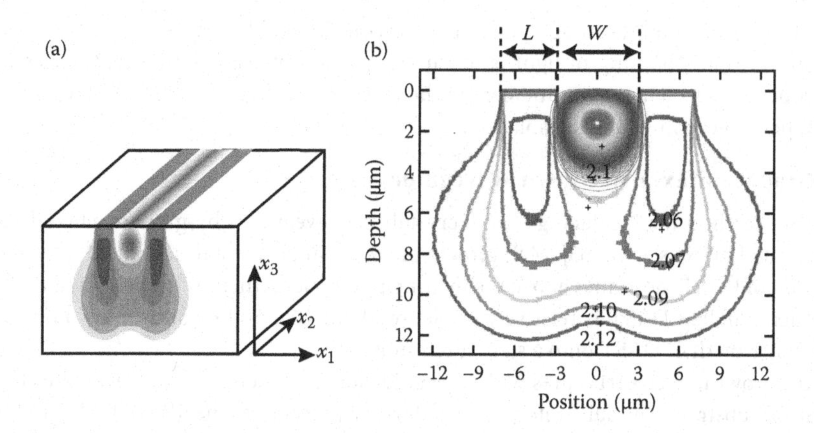


**FIGURE 11.12** (a) Concept of channel waveguide patterning in organic crystals: Two lines spaced by the waveguide core width are exposed to an electron-beam. An unexposed region (surrounded by an exposed area with lowered refractive index) is created, forming a waveguide between the two e-beam exposed lines. (b) Calculated 2D profile (contours) of the refractive index $n_1$ in DAST at a wavelength of $\lambda = 1.55$ μm. An electron fluence of $\phi = 2.6$ mC/cm$^2$ was used to create a line width of $L = 4$ μm and a waveguide core width of $W = 6$ μm. The corresponding intensity profile of the first-order guided mode with loss below 0.1 dB/cm is also depicted. (Reprinted from L. Mutter, M. Koechlin, M. Jazbinsek, and P. Gunter, *Opt. Express*, vol. 15, no. 25, pp. 16,828–16,838, Dec. 2007. With permission of Optical Society of America.)

measured modulation of $\lambda = 633$ nm light (upper curve) at the output of the MZ device are shown in Figure 11.13c for a waveguide width of $W = 4$ μm. The amplitude of the light modulation was about 20% of the average direct current level. At present, the half-wave voltage is still higher than 10 V, since the modulator dimensions and the electrode arrangement have not been optimized. The electrode spacing was relatively wide (20 μm), and the effective optical–electric interaction length was only $L_o = 0.85$ mm long.

Nevertheless, we find this technique very promising for structuring DAST as well as other organic crystals. E-beam patterning allows direct and single-exposure step-structuring of channel waveguides with particularly smooth side walls in bulk crystals. The depth and the lateral size can be precisely tuned in the range of 0 μm–12 μm needed for integrated optics. The refractive index contrast of about $\Delta n = 0.1$ at 1.55 μm allows curved structures with a radius of 100 μm and above. Compared to the ridge channel

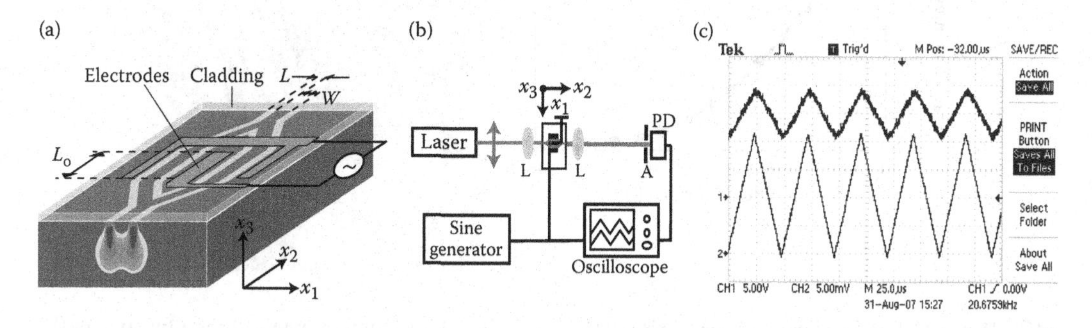


**FIGURE 11.13** (a) Mach-Zehnder modulator geometry in DAST with in-plane electrodes exploiting the EO coefficient $r_{111}$. (b) Set-up for the EO modulation experiment with a photodiode (PD), aperture (A) and lens (L). (c) The applied (10 V amplitude) modulation voltage (lower curve) and the detected photodiode signal (upper curve). (Reprinted from L. Mutter, M. Koechlin, M. Jazbinsek, and P. Gunter, *Opt. Express*, vol. 15, no. 25, pp. 16,828–16,838, Dec. 2007. With permission of Optical Society of America.)

waveguides fabricated by photolithography or femtosecond laser ablation, edge polishing e-beam waveguides for in- and out-coupling of light is much less demanding. Furthermore, there is no need for additional processing to achieve confinement in the vertical direction, as is required when using photolithography, photobleaching, or laser ablation.

#### 11.4.3.5 Graphoepitaxially Grown Waveguides

An interesting method for producing DAST crystalline waveguides by graphoepitaxial melt growth was demonstrated by Geis et al. [53]. Melt growth of DAST single crystals is limited due to the thermal decomposition of DAST molecules above the melting point. Nevertheless, it was shown that, in a nitrogen atmosphere, molten DAST is relatively stable for about 200–500 seconds. This duration was long enough for fast growth of DAST onto a structured substrate.

To seed the growth, microstructures with waveguide forms and crossed gratings were etched into an oxidized silicon substrate. The substrate was coated with a polycrystalline film of DAST and then heated above the melting point for a short time. The resulting crystals showed a reasonable orientation and a much higher quality than films grown without the grating-like microstructure. The *c*-axis was normal to the substrate to within approximately ±4°, as shown in Figure 11.14a. The graphoepitaxially grown waveguides with the desired orientation (*b*-axis parallel to the waveguide) exhibited optical losses below 10 dB/cm. Electro-optic modulation was also demonstrated in a MZ geometry using an ~100 nm titanium nitride (TiN) or chromium (Cr) conductive coating of the grating substrate prior to DAST growth [53].

#### 11.4.3.6 Melt Capillary Grown Waveguides inside Micro Channels

Another recently developed method for direct growth of organic waveguides from melt is based on single-crystalline growth inside microfluidic-like templates. For this technique, materials that are stable at the melting temperature were chosen. Therefore, the growth could be slower and highly controlled, which was beneficial for good quality and single crystallinity of the grown structures. The direction of the melt flow and temperature gradients determined the crystalline orientation, since most of the organic EO crystals tend to grow preferentially along one direction. Figure 11.15a illustrates the basic

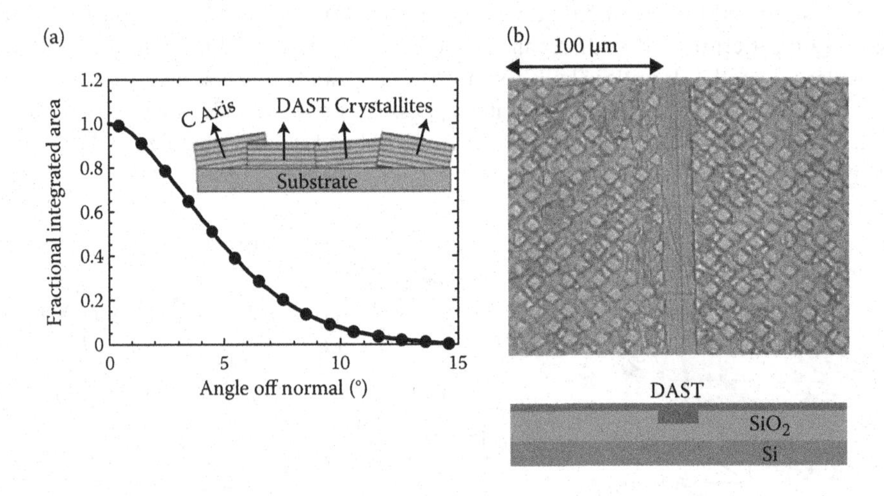


**FIGURE 11.14** (a) Distribution of the *c*-axis of DAST crystallites grown graphoepitaxially as a function of the angle from the structure normal. (b) Optical micrograph of a DAST waveguide 4 μm deep and 15 μm wide using polarized light (upper image). In-plane variation of 5° or more in the crystal axis is visible as a color variation when viewed through a microscope. Schematic cross-section (lower image). (Reprinted with permission from W. Geis, R. Sinta, W. Mowers, S. J. Deneault, M. F. Marchant, K. E. Krohn, S. J. Spector, D. R. Calawa, and T. M. Lyszczarz, *Appl. Phys. Lett.*, vol. 84, no. 19, pp. 3729–3731, May 2004. Copyright 2004, American Institute of Physics.)

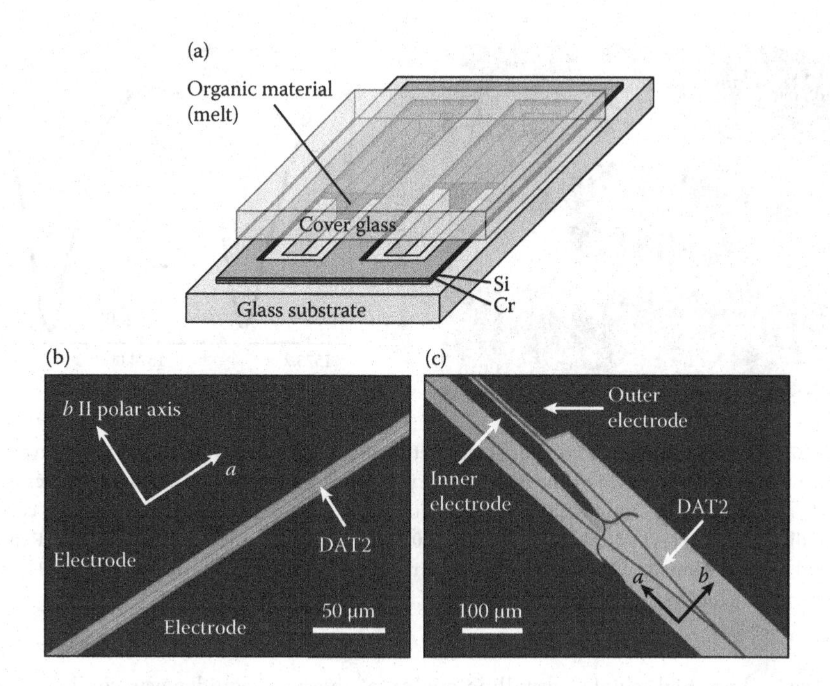


**FIGURE 11.15** (a) Schematic of melt capillary growth inside electrode-equipped, prestructured microchannels. Transmission microscope image between crossed polarizers of a DAT2-based (b) phase and (c) Mach-Zehnder EO modulator. (a and b, Reprinted from H. Figi, M. Jazbinsek, C. Hunziker, M. Koechlin, and P. Gunter, *Opt. Express*, vol. 16, no. 15, pp. 11,310–11,327, July 2008. With permission of Optical Society of America. c, Reprinted from H. Figi, M. Jazbinsek, C. Hunziker, M. Koechlin, and P. Gunter, *Proc. SPIE*, vol. 7599, pp. 75 991N–1–6, 2010. With permission. © 2010 SPIE.)

principle of this technique. First, borosilicate glass substrates were structured so that the waveguiding structures formed channels. Next, Cr electrodes were deposited on the sides of the EO material. Then, a thin silicon layer was added on top of the EO crystal, which was used for anodic bonding of the cover glass [31]. Finally, the organic material was placed at the edges of the cover glass and heated up to the melting temperature. The melt then started to flow into the channels by the capillary force. By cooling down in presence of a temperature gradient, single crystalline wires grew inside the channels. Using the melt-processable materials COANP and DAT2, several-mm long single crystalline wires have been obtained with this method (Figure 11.8). These wires had widths of several μm, and thicknesses of less than 30 nm to several μm.

By using this melt capillary growth technique, both phase and amplitude EO modulators were fabricated with DAT2 (see Figure 11.15b and 11.15c). These waveguides exhibited a high refractive index contrast between the DAT2 material and the surrounding glass of $\Delta n = 0.54$. EO modulation was measured for both transverse-electric (TE) and transverse-magnetic (TM) modes, which allowed estimation of the EO coefficients of DAT2: $r_{12} = 7.4 \pm 0.4$ pm/V and $r_{22} = 6.7 \pm 0.4$ pm/V at 1.55 μm. The half-wave voltage-length products of the fabricated structures were $V_\pi L = 78 \pm 2$ V·cm for TE-modes and $V_\pi L = 60 \pm 1$ V·cm for TM-modes at 1.55 μm, limited by the EO coefficients of the DAT2 material. Therefore, novel materials with state-of-the-art EO coefficients and the possibility for melt growth are greatly desired in order to improve the figures-of-merit of crystals grown with this technique.

By melt capillary growth, single-cyrstalline organic EO microring-resonator filters and modulators were demonstrated [39]. For these devices, organic COANP with very good melt-crystallization properties and a moderate EO coefficient ($r_{33} = 15 \pm 2$ pm/V at 633 nm) was employed. A top view, transmission microscope image of a COANP crystal grown in a microring resonator channel waveguide is shown

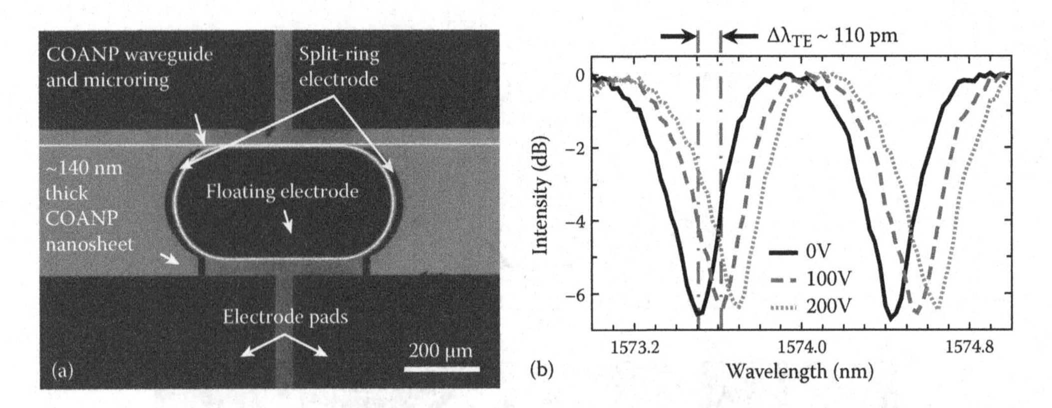


**FIGURE 11.16** (a) Transmission microscope image between crossed polarizers of a COANP waveguide with a racetrack microring resonator grown by the melt capillary method in prefabricated channels. (b) Resonance curve of a TE mode at a wavelength around 1.574 μm (solid line). The dashed and dotted lines are the corresponding electro-optically shifted curves by applying 100 V and 200 V, respectively, to the device electrodes. (Reprinted from H. Figi, M. Jazbinsek, C. Hunziker, M. Koechlin, and P. Gunter, *J. Opt. Soc. Am. B*, vol. 26, no. 5, pp. 1103–1110, May 2009. With permission of Optical Society of America.)

in Figure 11.16a. Very high single-crystalline quality of these waveguides was confirmed with optical waveguiding characterization. Typical devices fabricated showed almost perfectly symmetric high extinction ratio resonance peaks of about 10 dB, ring losses $\alpha = 12 \pm 0.3$ dB/cm, and a finesse $F = 6.2 \pm 0.2$. The measured TE spectrum of the racetrack resonator shown in Figure 11.16b showed a $\Delta\lambda = 110$ pm shift in response to an applied voltage of 100 V, corresponding to a frequency tunability of 0.11 GHz/V. Such tunability is comparable to what has been reported for ion-sliced $LiNbO_3$ microring resonators [54]. Here also, a great improvement in performance is expected if materials with state-of-the-art EO figures-of-merit ($n^3r$ of DAST or OH1 is more than one order of magnitude higher than for COANP) and higher index contrast (at 1.55 μm, $\Delta n$ with respect to borosilicate is about 0.15 for COANP and almost 0.7 for OH1 and DAST) can be used for melt growth.

#### 11.4.3.7 Crystalline Organic/Silicon Hybrid Modulators

In the last few years, silicon photonics has expanded as one of the most promising platforms for future on-chip, very-large-scale photonic integration. This is because of silicon's wide compatibility with present CMOS processing, as well as its high refractive index with a high index contrast with respect to $SiO_2$ substrates, allowing for very small waveguide cross sections of only about 200 nm$^2$ × 450 nm$^2$ [55]. Electro-optic modulation in silicon waveguides is possible based on charge injection, as discussed in Chapter 10. However, silicon does not exhibit an intrinsic Pockels effect, which is a preferred modulation process—in particular for high-speed modulation. Another possibility to achieve EO modulation in silicon waveguides is to integrate EO cladding materials. For most EO materials this is very challenging due to their limited processing possibilities within silicon structures. Very promising results with integrating EO cladding materials have been obtained by using poled polymers, as described in Chapter 9. Polymers can be relatively easily processed, but a high poling efficiency and long-term stability in submicron structures (required for silicon hybrid modulators) is difficult to achieve. Using organic EO crystals as cladding can avoid these issues. Next, we present recent first results on the hybrid integration of organic EO crystalline materials with silicon waveguides.

A cladding modulation scheme was chosen for the hybrid modulators. Here, the waveguide core is a high index-contrast, silicon-on-insulator waveguide and the cladding is an active EO material of a lower refractive index, as shown in Figure 11.17b.

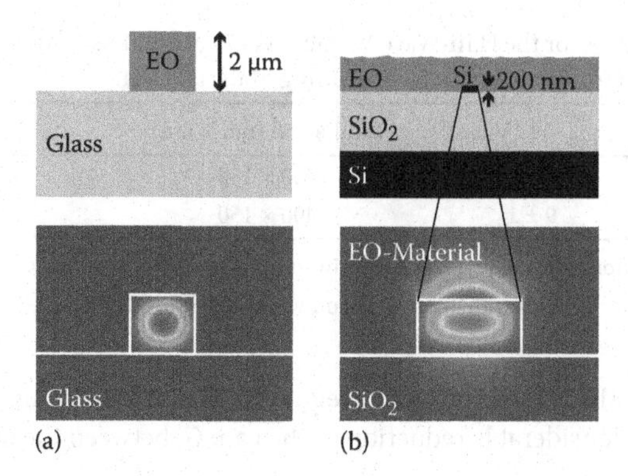


**FIGURE 11.17** Schematic of EO waveguides based on OH1: (a) OH1 wire waveguides discussed in Section 11.4.3.1, (b) hybrid OH1/silicon-photonics waveguides. In case (a), OH1 is a high-index material and is the core of the waveguide. While in (b), Si wire is the core waveguide and OH1 serves as an active cladding material.

An EO refractive index change in the cladding material results in a change of the effective index of the guided mode in the silicon-on-insulator core waveguide. OH1 was chosen as the active cladding material for MZ silicon-on-insulator modulators due to its favorable thin-film growth properties [37]. Since OH1 crystalline thin films grow with their $x_1$-axis perpendicular to the substrate, the optimal alignment is the one with the polar $x_3$-axis perpendicular to the electrodes on both sides of the waveguide. This alignment results in the employment of the EO coefficients $r_{113}$ and $r_{333}$ for TM- and TE-modes, respectively.

The processing steps of the fabrication of integrated MZ type modulators are depicted in Figure 11.18. Silicon waveguides, MZ-structures, and Cr electrodes were patterned using standard optical lithography. On top of these structures, large-domain, 2 μm–5 μm thick, single-crystalline OH1 was grown, as illustrated in Figure 11.18c. After thin film growth, the device was covered with a 2 μm thick polyvinyl alcohol protection layer. The structures were characterized by a standard end-fire coupling setup with single mode laser light in the 1530 nm–1610 nm wavelength range. The modulated signal, detected by the photodiode upon applying a sinusoidal voltage with 10 V amplitude across the electrodes, was visualized on an oscilloscope [34].

The cladding modulation efficiencies for these structures were determined numerically [34]. Table 11.4 contains the calculated modulation efficiencies and half-wave voltage-length products. The results are given for the experimental device configuration with dimensions $w \times h = 3$ μm × 200 nm and electrode spacing $G = 7$ μm (which agree well with the measured values), and an ideal configuration with single-mode waveguides of $w = 400$ nm, $h = 150$ nm or 180 nm, and $G = 600$ nm. MZ-modulators based on wire waveguides with OH1 as EO cladding can potentially lead to half-wave voltages-length products $V_{\pi,ideal} \cdot L$ below 1 V·cm. These results show that the cladding-modulation scheme is superior to that of

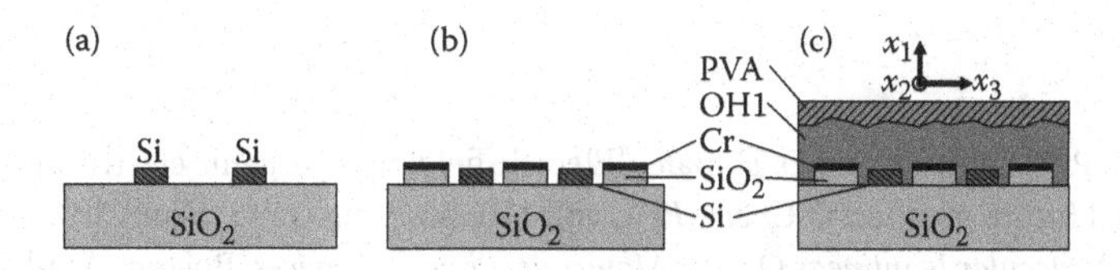


**FIGURE 11.18** Process steps for the fabrication of an electro-optically active MZ-interferometer based on the SOI waveguide technology. (a) Waveguide patterning using optical lithography. (b) Second lithography step for Cr electrode patterning. (c) Growth of the electro-optic single crystalline OH1 thin film, and spin-coating of the PVA protection layer. (Reprinted from M. Jazbinsek, C. Hunziker, S. J. Kwon, H. Figi, O. P. Kwon, and P. Gunter, *Proc. SPIE*, vol. 7599, pp. 75 990K–1–14, 2010 with permission. © 2010 SPIE.)

**TABLE 11.4** Calculated Values for the Half-Wave Voltage, $V_\pi \cdot L$, of SOI Based MZ-Modulators at Wavelengths around 1550 nm with the Electro-Optically Active Cladding Material OH1

| Mode | $\xi_{n_i,\text{exp}}$ (%) | $V_{\pi,\text{exp}} \times L$ (V·cm) | Ideal $w \times h$ (nm × nm) | $\xi_{n_i,\text{ideal}}$ (%) | $V_{\pi,\text{ideal}} \times L$ (V·cm) |
|---|---|---|---|---|---|
| TM | 60 | 31 ± 1 | 400 × 180 | 80 | 2.0 |
| TE | 13 | 9 ± 1 | 400 × 150 | 60 | 0.17 |

*Note:* The cladding modulation efficiency, $\xi_{n_i}$, is given by $\partial n_{eff}/\partial n_i|_{n_i}$, where $i = 1$ (TM mode) or 3 (TE mode). Exp: Experimentally demonstrated device with wide (3 μm) silicon waveguides. Ideal: Proposed MZ-structure with single-mode SOI wire waveguides [34].

conventional electro-optically active channel waveguides. The improvement is due to the stronger light confinement and possible considerable reduction of the gap, G, between the electrodes.

## 11.5 Summary and Outlook

Organic nonlinear optical crystalline materials are composed of highly polar chromophores with a highly asymmetric, ultrafast electronic response to external fields. They may exhibit high electro-optic figures-of-merit of more than $n^3r = 450$ pm/V in the 1.55-μm telecommunication window, low dielectric dispersion, and high thermal stability of polar order. Therefore, these materials show potential for high-speed and highly efficient electro-optic modulation.

In recent decades, there have been several promising organic EO crystalline materials developed for various applications. The basic molecular design is challenging mainly because calculating the crystalline packing of highly nonlinear optical molecules is not yet possible. Such calculations are desirable in order to predict highly favorable, noncentrosymmetric packing geometries, which are potentially useful for second-order nonlinear optical applications. Additionally, the crystal growth and processing of these materials is often very challenging. For the materials already identified, the main progress achieved in the last few years has been in the growth of high optical quality crystals and structuring of optical waveguides for EO applications.

For highly efficient integrated optics, planar, quasi-two-dimensional growth techniques are desired, since they are compatible with the already established waveguide technology and therefore greatly reduce the material cost and processing time. Several thin-film growth techniques (including solution, melt, and vapor) are still being investigated to achieve single crystals on various structures. In the last few years, several promising lateral structuring techniques were also developed, including photolithography, direct photo-, electron-beam, and proton structuring, as well as direct growth of waveguides on top of or inside of prestructured inorganic substrates. Based on this recent progress, integrated EO modulators fabricated by different techniques have been demonstrated—including phase, amplitude, and microring resonator modulators. These first demonstrations highlight exciting areas for future development: from novel material development, to optimizing the structuring processes and the electro-optic efficiency, to integration of photonic circuits (including hybrid integration) with silicon nanophotonic waveguides.

## References

1. L. R. Dalton, P. A. Sullivan, and D. H. Bale, "Electric field poled organic electro-optic materials: State of the art and future prospects," *Chem. Rev.*, vol. 110, no. 1, pp. 25–55, Jan. 2010.
2. J. Zyss, Ed., *Molecular Nonlinear Optics: Materials, Physics, Devices*. Boston: Academic Press, 1994.
3. C. Bosshard, K. Sutter, P. Prêtre, J. Hulliger, M. Flörsheimer, P. Kaatz, and P. Günter, *Organic Nonlinear Optical Materials*. Amsterdam, The Netherlands: Gordon and Breach Science Publishers, 1995.
4. C. Bosshard and P. Gunter, "Electro-optic effects in organic molecules and polymers," in *Nonlinear Optics of Organic Molecules and Polymers*, H. S. Nalwa and S. Miyata, Eds. Boca Raton, FL: CRC Press, 1997, p. 391.

5. C. Bosshard, M. Bösch, I. Liakatas, M. Jäger, and P. Günter, "Second-order nonlinear optical organic materials: Recent developments," in *Nonlinear Optical Effects and Materials*, P. Günter, Ed. Berlin: Springer Series in Optical Science, 2000, vol. 72, pp. 163–300.
6. M. G. Kuzyk, "Physical limits on electronic nonlinear molecular susceptibilities," *Phys. Rev. Lett.*, vol. 85, no. 6, pp. 1218–1221, Aug. 2000.
7. K. Clays and B. J. Coe, "Design strategies versus limiting theory for engineering large second-order nonlinear optical polarizabilities in charged organic molecules," *Chem. Mater.*, vol. 15, no. 3, pp. 642–648, 2003.
8. H. S. Nalwa and S. Miyata, Eds., *Nonlinear Optics of Organic Molecules and Polymers.* Boca Raton, FL: CRC Press, 1997.
9. M. Jazbinsek and P. Günter, "Organic molecular nonlinear optical materials and devices," in *Introduction to Organic Electronic and Optoelectronic Materials and Devices*, ser. Optical Science and Engineering, S. Sun and L. Dalton, Eds. Boca Raton, FL: CRC Press, 2008, pp. 421–466.
10. R. Spreiter, C. Bosshard, F. Pan, and P. Gunter, "High-frequency response and acoustic phonon contribution of the linear electro-optic effect in DAST," *Opt. Lett.*, vol. 22, no. 8, pp. 564–566, Apr. 1997.
11. D. Rezzonico, S. Kwon, H. Figi, O. Kwon, M. Jazbinsek, and P. Gunter, "Photochemical stability of nonlinear optical chromophores in polymeric and crystalline materials," *J. Chem. Phys.*, vol. 128, no. 12, pp. 124 713-1–6, Mar. 2008.
12. J. Zyss and J. L. Oudar, "Relations between microscopic and macroscopic lowest-order optical nonlinearities of molecular-crystals with one-dimensional or two-dimensional units," *Phys. Rev. A*, vol. 26, no. 4, pp. 2028–2048, Oct. 1982.
13. C. Bosshard, R. Spreiter, and P. Gunter, "Microscopic nonlinearities of two-component organic crystals," *J. Opt. Soc. Am. B*, vol. 18, no. 11, pp. 1620–1626, Nov. 2001.
14. I. Liakatas, C. Cai, M. Bosch, M. Jager, C. Bosshard, P. Gunter, C. Zhang, and L. R. Dalton, "Importance of intermolecular interactions in the nonlinear optical properties of poled polymers," *Appl. Phys. Lett.*, vol. 76, no. 11, pp. 1368–1370, Mar. 2000.
15. S. R. Marder, J. W. Perry, and W. P. Schaefer, "Synthesis of organic salts with large 2nd-order optical nonlinearities," *Science*, vol. 245, no. 4918, pp. 626–628, Aug. 1989.
16. H. Nakanishi, H. Matsuda, S. Okada, and M. Kato, "Organic and polymeric ion-complexes for nonlinear optics," in *Materials Research Society International Meeting on Advanced Materials*, M. Doyama, S. Somiya, and R. P. H. Chang, Eds., Pittsburgh, PA: Materials Research Society, 1989, pp. 97–104.
17. F. Pan, M. S. Wong, C. Bosshard, and P. Gunter, "Crystal growth and characterization of the organic salt 4-N,N-dimethylamino-4'-N'-methyl-stilbazolium tosylate (DAST)," *Adv. Mater.*, vol. 8, no. 7, pp. 592–595, Jul. 1996.
18. B. Ruiz, M. Jazbinsek, and P. Gunter, "Crystal growth of DAST," *Cryst. Growth Des.*, vol. 8, no. 11, pp. 4173–4184, Nov. 2008.
19. M. Jazbinsek, L. Mutter, and P. Gunter, "Photonic applications with the organic nonlinear optical crystal DAST," *IEEE J. Sel. Top. Quantum Electron.*, vol. 14, no. 5, pp. 1298–1311, 2008.
20. S. R. Marder, J. W. Perry, and C. P. Yakymyshyn, "Organic salts with large 2nd-order optical nonlinearities," *Chem. Mater.*, vol. 6, no. 8, pp. 1137–1147, Aug. 1994.
21. F. Pan, G. Knopfle, C. Bosshard, S. Follonier, R. Spreiter, M. S. Wong, and P. Gunter, "Electro-optic properties of the organic salt 4-N,N-dimethylamino-4'-N'-methyl-stilbazolium tosylate," *Appl. Phys. Lett.*, vol. 69, no. 1, pp. 13–15, Jul. 1996.
22. C. Bosshard, R. Spreiter, L. Degiorgi, and P. Gunter, "Infrared and Raman spectroscopy of the organic crystal DAST: Polarization dependence and contribution of molecular vibrations to the linear electro-optic effect," *Phys. Rev. B*, vol. 66, no. 20, p. 205107, Nov. 2002.
23. Z. Glavcheva, H. Umezawa, Y. Mineno, T. Odani, S. Okada, S. Ikeda, T. Taniuchi, and H. Nakanishi, "Synthesis and properties of 1-methyl-4-2-[4-(dimethylamino)phenyl]ethenylpyridinium p-toluenesulfonate derivatives with isomorphous crystal structure," *Jpn. J. Appl. Phys. Part 1*, vol. 44, no. 7A, pp. 5231–5235, Jul. 2005.

24. Z. Yang, M. Jazbinsek, B. Ruiz, S. Aravazhi, V. Gramlich, and P. Gunter, "Molecular engineering of stilbazolium derivatives for second-order nonlinear optics," *Chem. Mater.*, vol. 19, no. 14, pp. 3512–3518, Jul. 2007.
25. B. J. Coe, J. A. Harris, I. Asselberghs, K. Clays, G. Olbrechts, A. Persoons, J. T. Hupp, R. C. Johnson, S. J. Coles, M. B. Hursthouse, and K. Nakatani, "Quadratic nonlinear optical properties of N-aryl stilbazolium dyes," *Adv. Funct. Mater.*, vol. 12, no. 2, pp. 110–116, Feb. 2002.
26. H. Figi, L. Mutter, C. Hunziker, M. Jazbinsek, P. Gunter, and B. J. Coe, "Extremely large nonresonant second-order nonlinear optical response in crystals of the stilbazolium salt DAPSH," *J. Opt. Soc. Am. B*, vol. 25, no. 11, pp. 1786–1793, Nov. 2008.
27. Z. Yang, L. Mutter, M. Stillhart, B. Ruiz, S. Aravazhi, M. Jazbinsek, A. Schneider, V. Gramlich, and P. Gunter, "Large-size bulk and thin-film stilbazolium-salt single crystals for nonlinear optics and THz generation," *Adv. Funct. Mater.*, vol. 17, no. 13, pp. 2018–2023, Sep. 2007.
28. J. Ogawa, S. Okada, Z. Glavcheva, and H. Nakanishi, "Preparation, properties and structures of 1-methyl-4-2-[4-(dimethylamino)phenyl]ethenylpyridinium crystals with various counter anions," *J. Cryst. Growth*, vol. 310, no. 4, pp. 836–842, Feb. 2008.
29. O. P. Kwon, B. Ruiz, A. Choubey, L. Mutter, A. Schneider, M. Jazbinsek, V. Gramlich, and P. Gunter, "Organic nonlinear optical crystals based on configurationally locked polyene for melt growth," *Chem. Mater.*, vol. 18, no. 17, pp. 4049–4054, Aug. 2006.
30. O. P. Kwon, S. J. Kwon, M. Jazbinsek, F. D. J. Brunner, J. I. Seo, C. Hunziker, A. Schneider, H. Yun, Y. S. Lee, and P. Gunter, "Organic phenolic configurationally locked polyene single crystals for electro-optic and terahertz wave applications," *Adv. Funct. Mater.*, vol. 18, no. 20, pp. 3242–3250, Oct. 2008.
31. H. Figi, M. Jazbinsek, C. Hunziker, M. Koechlin, and P. Gunter, "Electro-optic single-crystalline organic waveguides and nanowires grown from the melt," *Opt. Express*, vol. 16, no. 15, pp. 11 310–11 327, July 2008.
32. H. Figi, M. Jazbinsek, C. Hunziker, M. Koechlin, and P. Gunter, "Integrated electro-optic devices of melt-processable single-crystalline organic films," *Proc. SPIE*, vol. 7599, pp. 75 991N-1–6, 2010.
33. C. Hunziker, S. J. Kwon, H. Figi, M. Jazbinsek, and P. Gunter, "Fabrication and phase modulation in organic single-crystalline configurationally locked, phenolic polyene OH1 waveguides," *Opt. Express*, vol. 16, no. 20, pp. 15 903–15 914, Sep. 2008.
34. M. Jazbinsek, C. Hunziker, S. J. Kwon, H. Figi, O. P. Kwon, and P. Gunter, "Hybrid organic crystal/silicon-on-insulator integrated electro-optic modulators," *Proc. SPIE*, vol. 7599, pp. 75 990K-1–14, 2010.
35. C. Hunziker, S. J. Kwon, H. Figi, F. Juvalta, O. P. Kwon, M. Jazbinsek, and P. Gunter, "Configurationally locked, phenolic polyene organic crystal 2-3-(4-hydroxystyryl)-5,5-dimethylcyclohex-2-enylidene-malononitrile: linear and nonlinear optical properties," *J. Opt. Soc. Am. B*, vol. 25, no. 10, pp. 1678–1683, Oct. 2008.
36. S. J. Kwon, M. Jazbinsek, O. P. Kwon, and P. Gunter, "Crystal growth and morphology control of OH1 organic electro-optic crystals," *Cryst. Growth Des.*, vol. 10, no. 4, pp. 1552–1558, Apr. 2010.
37. S. J. Kwon, C. Hunziker, O. P. Kwon, M. Jazbinsek, and P. Gunter, "Large-area organic electro-optic single crystalline thin films grown by evaporation-induced local supersaturation with surface interactions," *Cryst. Growth Des.*, vol. 9, no. 5, pp. 2512–2516, Mar. 2009.
38. M. Jazbinsek, O.-P. Kwon, C. Bosshard, and P. Gunter, "Organic nonlinear optical crystals and single crystalline thin films," in *Handbook of Organic Electronics and Photonics*, H. S. Nalwa, Ed., Stevenson Ranch, CA: American Scientific Publishers, 2008.
39. H. Figi, M. Jazbinsek, C. Hunziker, M. Koechlin, and P. Gunter, "Electro-optic tuning and modulation of single-crystalline organic microring resonators," *J. Opt. Soc. Am. B*, vol. 26, no. 5, pp. 1103–1110, May 2009.
40. D. Rezzonico, M. Jazbinsek, A. Guarino, O. P. Kwon, and P. Gunter, "Electro-optic Charon polymeric microring modulators," *Opt. Express*, vol. 16, no. 2, pp. 613–627, Jan. 2008.

41. I. L. Gheorma and R. M. Osgood, "Fundamental limitations of optical resonator based high-speed EO modulators," *IEEE Photonics Technol. Lett.*, vol. 14, no. 6, pp. 795–797, Jun. 2002.
42. B. Bortnik, Y. C. Hung, H. Tazawa, B. J. Seo, J. D. Luo, A. K. Y. Jen, W. H. Steier, and H. R. Fetterman, "Electro-optic polymer ring resonator modulation up to 165 GHz," *IEEE J. Sel. Top. Quantum Electron.*, vol. 13, no. 1, pp. 104–110, Jan. 2007.
43. S. Fukuda and T. Gotoh, *Laser Kenkyu (Rev. Laser Eng.)*, vol. 21, p. 1134, 1993 [in Japanese].
44. K. Tsuda, T. Kondo, F. Saito, T. Kudo, and R. Ito, "New fabrication method of channel optical waveguides of organic-crystal using an inorganic photoresist," *Jpn. J. Appl. Phys. Part 2*, vol. 31, no. 2A, pp. L134–L135, Feb. 1992.
45. K. Takayama, K. Komatsu, and T. Kaino, "Serially grafted waveguide fabrication of organic crystal and transparent polymer," *Jpn. J. Appl. Phys. Part 1*, vol. 40, no. 8, pp. 5149–5150, Aug. 2001.
46. T. Kaino, B. Cai, and K. Takayama, "Fabrication of DAST channel optical waveguides," *Adv. Funct. Mater.*, vol. 12, no. 9, pp. 599–603, Sep. 2002.
47. B. Cai, K. Komatsu, and T. Kaino, "Refractive index control and waveguide fabrication of DAST crystals by photobleaching technique," *Opt. Mater.*, vol. 21, no. 1–3, pp. 525–529, Jan. 2003.
48. L. Mutter, M. Jazbinsek, M. Zgonik, U. Meier, C. Bosshard, and P. Gunter, "Photobleaching and optical properties of organic crystal 4-N, N-dimethylamino-4'-N'-methyl stilbazolium tosylate," *J. Appl. Phys.*, vol. 94, no. 3, pp. 1356–1361, Aug. 2003.
49. P. Dittrich, R. Bartlome, G. Montemezzani, and P. Gunter, "Femtosecond laser ablation of DAST," *Appl. Surf. Sci.*, vol. 220, no. 1–4, pp. 88–95, Dec. 2003.
50. L. Mutter, A. Guarino, M. Jazbinsek, M. Zgonik, P. Gunter, and M. Dobeli, "Ion implanted optical waveguides in nonlinear optical organic crystal," *Opt. Express*, vol. 15, no. 2, pp. 629–638, Jan. 2007.
51. L. Mutter, M. Jazbinsek, C. Herzog, and P. Gunter, "Electro-optic and nonlinear optical properties of ion implanted waveguides in organic crystals," *Opt. Express*, vol. 16, no. 2, pp. 731–739, Jan. 2008.
52. L. Mutter, M. Koechlin, M. Jazbinsek, and P. Gunter, "Direct electron beam writing of channel waveguides in nonlinear optical organic crystals," *Opt. Express*, vol. 15, no. 25, pp. 16 828–16 838, Dec. 2007.
53. W. Geis, R. Sinta, W. Mowers, S. J. Deneault, M. F. Marchant, K. E. Krohn, S. J. Spector, D. R. Calawa, and T. M. Lyszczarz, "Fabrication of crystalline organic waveguides with an exceptionally large electro-optic coefficient," *Appl. Phys. Lett.*, vol. 84, no. 19, pp. 3729–3731, May 2004.
54. A. Guarino, G. Poberaj, D. Rezzonico, R. Degl'Innocenti, and P. Gunter, "Electro-optically tunable microring resonators in lithium niobate," *Nat. Photonics*, vol. 1, no. 7, pp. 407–410, 2007.
55. P. Dumon, G. Priem, L. R. Nunes, W. Bogaerts, D. Van Thourhout, P. Bienstman, T. K. Liang, et al., "Linear and nonlinear nanophotonic devices based on silicon-on-insulator wire waveguides," *Jpn. J. Appl. Phys. Part 1*, vol. 45, no. 8B, pp. 6589–6602, 2006.
56. M. Jazbinsek and M. Zgonik, "Material tensor parameters of $LiNbO_3$ relevant for electro- and elasto-optics," *Appl. Phys. B*, vol. 74, no. 4–5, pp. 407–414, Apr. 2002.
57. M. Zgonik, R. Schlesser, I. Biaggio, E. Voit, J. Tscherry, and P. Gunter, "Materials constants of $KNbO_3$ relevant for electro-optics and acoustooptics," *J. Appl. Phys.*, vol. 74, no. 2, pp. 1287–1297, Jul. 1993.
58. F. Pan, K. McCallion, and M. Chiappetta, "Waveguide fabrication and high-speed in-line intensity modulation in 4-N,N-4'-dimethylamino-4'-N'-methyl-stilbazolium tosylate," *Appl. Phys. Lett.*, vol. 74, no. 4, pp. 492–494, Jan. 1999.
59. S. Manetta, M. Ehrensperger, C. Bosshard, and P. Gunter, "Organic thin film crystal growth for nonlinear optics: present methods and exploratory developments," *C. R. Phys.*, vol. 3, no. 4, pp. 449–462, May 2002.
60. T. Yamashiki and K. Tsuda, "Low-loss waveguides of benzylidene-aniline derivatives by organic molecular beam heteroepitaxy," *Opt. Lett.*, vol. 28, no. 5, pp. 316–318, Mar. 2003.
61. M. Thakur, J. J. Xu, A. Bhowmik, and L. G. Zhou, "Single-pass thin-film electro-optic modulator based on an organic molecular salt," *Appl. Phys. Lett.*, vol. 74, no. 5, pp. 635–637, Feb. 1999.

62. L. G. Zhou and M. Thakur, "Molecular orientation in single crystal thin films of N-(4-nitrophenyl)-(l)-prolinol," *J. Mater. Res.*, vol. 13, no. 1, pp. 131–134, Jan. 1998.
63. O. P. Kwon, S. J. Kwon, H. Figi, M. Jazbinsek, and P. Gunter, "Organic electro-optic single-crystalline thin films grown directly on modified amorphous substrates," *Adv. Mater.*, vol. 20, no. 3, pp. 543–545, Feb. 2008.
64. Y. Kubota and T. Yoshimura, "Endothermic reaction aided crystal-growth of 2-methyl-4-nitroaniline and their electro-optic properties," *Appl. Phys. Lett.*, vol. 53, no. 26, pp. 2579–2581, Dec. 1988.
65. S. Gauvin and J. Zyss, "Growth of organic crystalline thin films, their optical characterization and application to non-linear optics," *J. Cryst. Growth*, vol. 166, no. 1–4, pp. 507–527, Sep. 1996.
66. P. M. Ushasree, K. Komatsu, T. Kaino, and T. Taima, "Growth of organic crystal thin film by a new method-rectangular heater heated pedestal growth method," *Mol. Cryst. Liq. Cryst.*, vol. 406, pp. 119–127, Jan. 2003.
67. A. Leyderman and Y. L. Cui, "Electro-optical characterization of a 2-cyclo-octylamino-5-nitropyridine thin organic crystal film," *Opt. Lett.*, vol. 23, no. 12, pp. 909–911, Jun. 1998.
68. Z. F. Liu, S. S. Sarkisov, M. J. Curley, A. Leyderman, and C. Lee, "Thin film electro-optic modulator based on single crystal of N-(4-nitrophenyl)-(l)-prolinol (NPP) grown from melt by the modified bridgman method," *Opt. Eng.*, vol. 42, no. 3, pp. 803–812, Mar. 2003.
69. A. Choubey, O. P. Kwon, M. Jazbinsek, and P. Gunter, "High-quality organic single crystalline thin films for nonlinear optical applications by vapor growth," *Cryst. Growth Des.*, vol. 7, no. 2, pp. 402–405, 2007.
70. M. Baldo, M. Deutsch, P. Burrows, H. Gossenberger, M. Gerstenberg, V. Ban, and S. Forrest, "Organic vapor phase deposition," *Adv. Mater.*, vol. 10, no. 18, pp. 1505–1514, Dec. 1998.

# 12

# Complex Oxide Electro-Optic Modulators



Yalin Lu
*U.S. Air Force Academy*

## 12.1 PLZT Transparent Ceramics

One of the major achievements in the field of optical ceramics was the development of transparent ferroelectric oxide $Pb_{1-x}La_x(Zr_{1-y}Ti_y)_{1-x/4}O_3$ (PLZT). PLZT with a composition of $Pb_{0.92}La_{0.08}(Zr_{0.65}Ti_{0.35})_{0.98}O_3$ was first produced by Haertling and Land in 1969 at the Sandia National Laboratories [1]. Incorporation of lanthanum (La) ions into the lattice enhances the densification rate of PZT ceramics, leading to pore-free homogeneous ceramic microstructures. Since its inception, a great deal of research has been carried out on PLZT and a far deeper understanding of many interrelated properties and phenomena associated with it has been achieved [2]. Advantages of PLZT include high optical transparency from visible to mid-infrared wavelengths, large electro-optic (EO) coefficients, very fast EO response, its solid state nature, the ability to make it into large crystals, its isotropic optical properties, and the potential, as thin films, to integrate it with many other optical materials and semiconductors. There have been many successful application demonstrations of PLZT such as displays, ultrafast optical shutters, fast EO modulators, optical tunable filters, laser beam deflectors, variable optical attenuators, and polarization compensators [3].

### 12.1.1 Composition and Phase Diagram of PLZT

When $Ti^{4+}$ in $PbTiO_3$ is partially replaced by $Zr^{4+}$ with a molar ratio of $x$, a solid solution $x$PbZrO$_3$-(1-$x$)PbTiO$_3$ binary system is formed. This solid solution is called lead zirconate titanate (PZT) and its chemical formula is $Pb(Zr_xTi_{1-x})O_3$. PZT has a perovskite structure with $Ti^{4+}$ and $Zr^{4+}$ ions randomly occupying the "A-sites" of the structure. Figure 12.1 shows the temperature–molar ratio phase diagram of PZT. The $T_c$ line separates cubic paraelectric and ferroelectric phases. A morphology phase boundary (MPB) is defined as the division between two ferroelectric phases; tetragonal structures are on the Ti-rich side

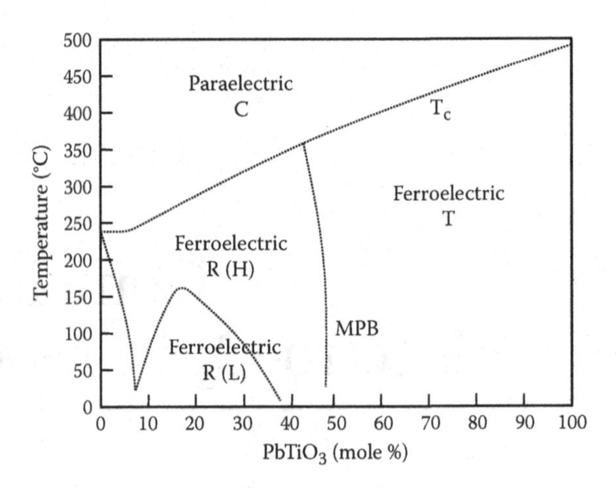


**FIGURE 12.1** $PbTiO_3$–$PbZrO_3$ subsolidus phase diagram.

and rhombohedral structures are on the Zr-rich side. At room temperature, the MPB is at the point Zr/Ti = 53/47. In the region where Zr/Ti lies between 100/0 and 94/6, the solid solution is in an antiferroelectric orthorhombic phase.

Compositions of transparent EO ceramics are mainly PZT with additional chemical modifiers such as La, bismuth (Bi), and barium (Ba) ions. Among them, La ions were found to affect PZT the most. Incorporating La ions into the PZT structure reduces light scattering from grain boundaries by reducing the oxygen octahedral unit cell distortion, which minimizes the index of refraction difference at grain boundaries. Adding La ions also produces a homogeneous solid solution due to the high solubility of La ions in PZT (up to 35% on the $PbTiO_3$-rich side). La ions also are found to be able to enhance the densification process and control grain growth by promoting a highly uniform microstructure. The chemical formula of PLZT can be generally written as

$$Pb_{1-x}La_x(Zr_{1-y}Ti_y)_{1-0.25x}V_{0.25x}O_3, \tag{12.1}$$

where La ions partially replace Pb ions on the "A-sites" and the charge is balanced by the creation of "B-site" lattice vacancies (represented by $V$ in the formula). The concentration of $La_x$ (atomic %), may vary from 2 to 30, and the ratio of $y$ = Zr/(Zr + Ti) may vary from 0 to 1. PLZT composition can be represented with the notation "$x/y/(1 - y)$." For example, PLZT 8/65/35 represents a composition of $Pb_{0.92}La_{0.08}(Zr_{0.65}Ti_{0.35})_{0.98}O_3$.

The phase diagram of PLZT at room temperature is shown in Figure 12.2, from which we can surmise a few facts: (1) actual solubility of La in PZT depends on the PZT composition and it can change from a few percent near $PbZrO_3$ to ~35% in pure $PbTiO_3$; (2) excess La doping results in mixed phases, typically $La_2Zr_2O_7$ and $La_2Ti_2O_7$, which further reduces the transparency; (3) in the homogeneous region, adding La reduces the Curie temperature nearly linearly at a rate of about 37°C/at.% of La; and (4) for the Zr/Ti ratio of 65/35, 9% of La will reduce the temperature of the stable rhombohedral phase to below room temperature. This temperature drop can cause enormous dielectric, EO, and piezoelectric effects. These effects have been coined "MPB behavior" [4]. A lot of research has been conducted on PLZT compositions in a region of 6~9/65/35 for EO applications. The selection of PLZT composition should be considered carefully based on the requirements of a specific application because there is a trade-off between maximizing the EO effect and maintaining operation reliability.

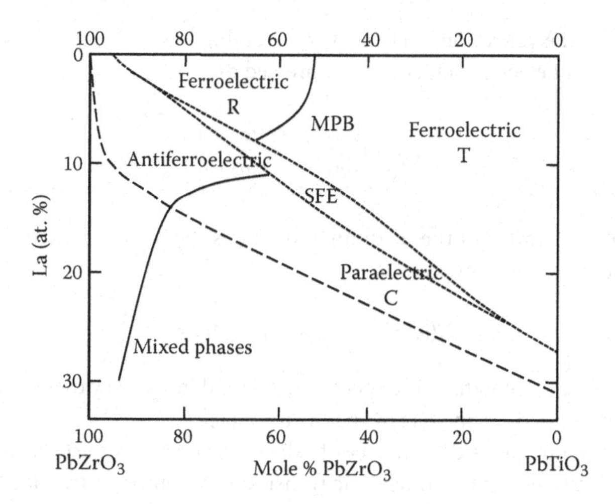


**FIGURE 12.2** PLZT phase diagram at room temperature. SFE: ferroelectric phase of a slim loop.

## 12.1.2 Dielectric, Optical, and EO Effect in PLZT

### 12.1.2.1 Dielectric Constant and Dielectric Spectrum

The dielectric constant of a lossy dielectric can be expressed as

$$\varepsilon(\omega) = \varepsilon'(\omega) - i\varepsilon''(\omega), \tag{12.2}$$

where $\varepsilon'(\omega)$ is the real part of dielectric constant and $\varepsilon''(\omega)$ is the imaginary. Instead of $\varepsilon''$, $\tan\delta$ (tangent of dielectric loss angle) is most frequently used to express the dielectric loss and $\tan\delta = \varepsilon''/\varepsilon'$. The variation of $\varepsilon(\omega)$ as a function of the frequency of the applied electromagnetic field is usually called the dielectric spectrum. Both the real and imaginary parts are not independent of each other and are correlated by the well-known Kramers–Kronig relations [5]:

$$\varepsilon'(\omega) - \varepsilon_\infty = \left(\frac{2}{\pi}\right)\int_0^\infty \varepsilon''(\omega')\omega'/\left(\omega'^2 - \omega^2\right)d\omega' \tag{12.3a}$$

$$\varepsilon''(\omega) = \left(\frac{2}{\pi}\right)\int_0^\infty \left[\varepsilon'(\omega') - \varepsilon_\infty\right]\omega/(\omega^2 - \omega'^2)d\omega', \tag{12.3b}$$

where $\omega'$ is an integration variable with units of angular frequency. The dielectric spectrum provides information about the polarizability, polarization mechanisms, and lattice vibrations of the material. In different frequency ranges, relaxation-type responses are determined by space-charge polarization and dipole orientation polarization. On the other hand, resonance-type responses are determined by atomic (or ionic) polarization, valence electronic polarization, and inner electronic polarization. For EO modulators, dielectric properties of the material in both microwave frequency and optical frequency ranges are important. Microwave frequencies correspond to operation frequencies of most driving electrical fields and the optical frequencies to the optical waves being modulated.

At normal incidence, the reflection coefficient, $r(\omega)$, defined as the amplitude ratio of the reflected electric field to the incident electric field, can be expressed as

$$r(\omega) = \frac{n(\omega) + i\kappa(\omega) - 1}{n(\omega) + i\kappa(\omega) + 1}, \tag{12.4}$$

where $n(\omega)$ is the refractive index of the medium and $\kappa(\omega)$ is the distinction coefficient. Therefore, the complex refractive index can be expressed as

$$N(\omega) = n(\omega) + i\kappa(\omega) = [\varepsilon(\omega)]^{1/2} \tag{12.5}$$

and $\varepsilon' = n^2 - \kappa^2$ and $\varepsilon'' = 2n\kappa$. Therefore, the spectra of real and imaginary dielectric constants at optical frequencies can be determined completely by measuring $r(\omega)$.

Measurement of the temperature dependence of a dielectric is also important. The real dielectric constant, $\varepsilon'$, peaks at the ferroelectric to paraelectric transition. At temperatures above this transition, the low frequency dielectric constant varies following the Curie-Weiss law

$$\varepsilon' = \frac{C}{T - T_c}, \tag{12.6}$$

where $C$ is the Curie-Weiss constant and $T_c$ is the Curie temperature. Normally, $T_c$ and $C$ can be determined by the peak position from the $T$-$\varepsilon'$ curve and the slope of the curve $(\varepsilon')^{-1}$ versus $T$, respectively.

Figure 12.3 shows the influence of La dopant on dielectric constants of the $x$/65/35 PLZT family. A 2.0% La doping increases the real part of the dielectric constant from 27,299 of PZT to 28,955 [6]. Further increasing the La concentration reduces the peak dielectric constant, displaces the peak to lower temperature, and broadens the peak of the dielectric spectrum. The peak tangent loss overlaps with the peak dielectric permittivity and increases when La concentration increases. Such characteristics are typical of relaxor ferroelectric materials exhibiting diffusive phase transitions.

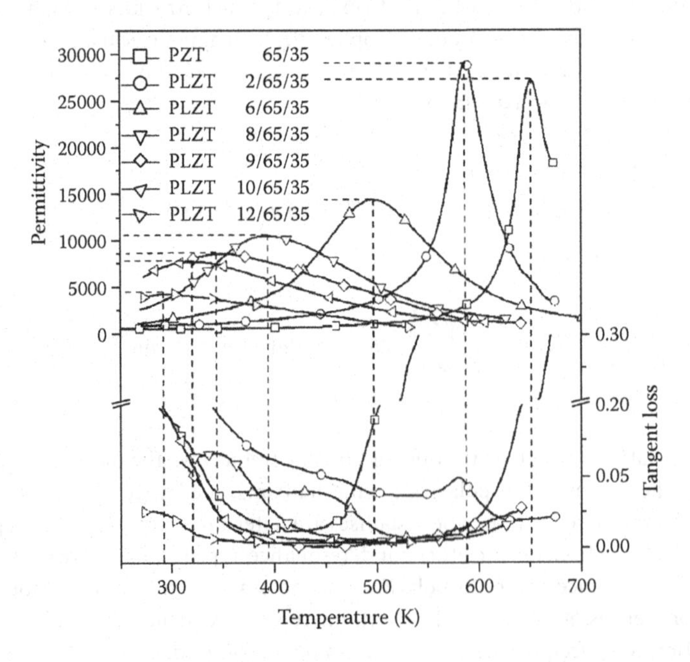


**FIGURE 12.3** Dielectric constants of PLZT $x$/65/35 when $x$ varied from 0% to 12%.

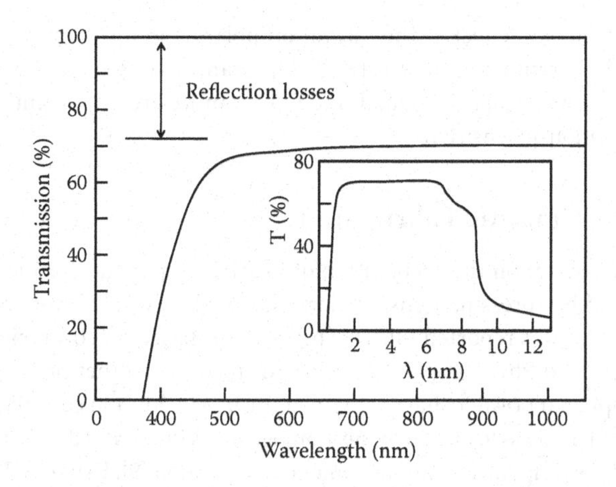


**FIGURE 12.4** Transmittance of PLZT 9/65/35 ceramics.

### 12.1.2.2 PLZT Optical Transparency

Figure 12.4 shows a typical transmission curve from visible to mid-infrared wavelengths for PLZT 9/65/35. The transparency is a function of La concentration as well as the Zr/Ti ratio. Normally, the maximum transparency can be achieved using compositions along the $T_c$ phase boundary. For example, the Zr/Ti 65/35 ratio composition will be the most transparent when the La concentration ranges from 8 at.% to 16 at.% and the 10/90 ratio will be the most transparent when La concentration is between 22% and 28%. This result will be instructive for developing new transparent EO ceramics.

### 12.1.2.3 EO Effect in PLZT

After electrical poling, a transparent EO ceramic usually behaves like a uniaxial negative crystal ($n_e - n_o = \Delta n < 0$) with the remanent polarization coinciding with the optical axis. The $\Delta n$ changing with an applied electric field, $E$, is called the EO effect and it follows linear ($\Delta n = -1/2n^3rE$) and quadratic ($\Delta n = -1/2n^3RE^2$) rules. Here, $r$ and $R$ are linear and quadratic EO coefficients, respectively. Figure 12.5 shows various regions in the PLZT composition diagram possessing different EO characteristics: A (memory), B (linear EO effect), and C (quadratic EO effect). Compositions in Region A have low coercive fields, a square hysteresis loop, and large EO coefficients. Region B corresponds to compositions that have high

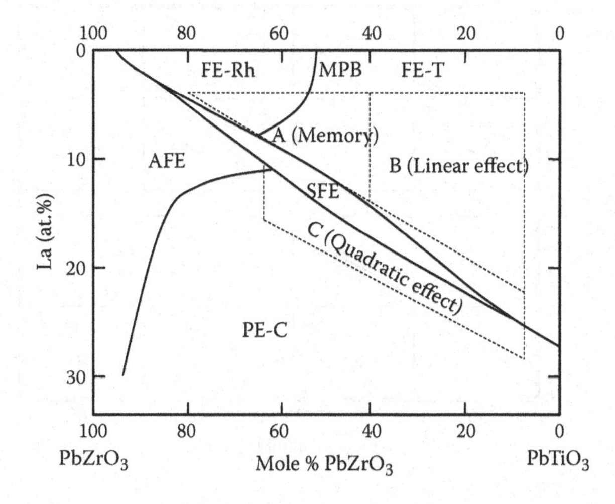


**FIGURE 12.5** Various regions in PLZT possessing different EO characteristics.

coercive fields and a linear EO effect at the saturation polarization. Ceramics in Region C have very small coercive fields. After removing the electric field, ceramics in Region C become optically isotropic without field-induced birefringence. Quadratic EO ceramics are important for optical modulators because of their polarization-insensitivity.

### 12.1.3 Bulk PLZT Ceramic Fabrication

In 1971, a detailed manuscript on the preparation of PLZT transparent ceramics, obtained by conventional mixed oxides and hot-pressing densification, was published [7]. Only one year later, a powder process by chemical routes was reported [8]. This process was based on co-precipitation of alkoxides in the presence of proportioned PbO. Despite the many advances with other processes, hot-pressed PLZT ceramics from coprecipitated powders have remained the most common ceramics for EO devices.

A two-stage calcination process to obtain highly homogeneous PLZT was developed, in which all constituents are dissolved and coprecipitated from the same source solution [9]. Using PLZT 10/65/35 as an example, the powders will be produced by the conventional mixed-oxide method, being subjected to their first calcination at 1173 K for three hours. Then, the calcined powder will be dissolved in a solution of nitric acid under controlled pH at 343 K. After that, precipitation will be promoted by adding $HNO_3$ to the solution to make the pH between 9 and 10. The slurry, after rinsing, filtering and vacuum drying, will be subjected to a second stage of calcination. Through this process, cold-pressed pellets can be prepared from ball-milled powders. Prior to milling, 2 wt.% excess of PbO was added to compensate losses from volatilization.

Sol-gel processing of ceramics has also attracted great interest. Such processing confers a high degree of chemical homogeneity and allows superb control of stoichiometry to the material [10]. In addition, the powders obtained by this method are finely divided and are characterized by improved reactivity and sinterability compared to powders prepared by conventional processing. In this process, lead acetate [$Pb(OOCCH_3)_2.3H_20$], lanthanum acetate hydrate [$La(OOCCH_3)_3.xH_20$], zirconium n-propoxide [$Zr(OC_3H_7)_4$], and titanium n-butoxide [$Ti(OC_4H_9)$] can be employed as precursors. The process begins by dissolving La, Zr, and Ti precursors in absolute ethanol, followed by mixing and heating to 78°C. Appropriate amounts of lead acetate and water are slowly added to the mixture and stirred to form a clear solution of hydroxide. The hydroxide will then be heated to 78°C for 40 hours to yield a completely white amorphous powder. After an additional 10 wt.% of PbO is added to the powder, it is pressed into pellets.

Figure 12.6 shows the temperature dependence of some hysteresis loops for PLZT 10/65/35 [11]. When the temperature is increased, the coercive field and the remanent polarization gradually decrease. At

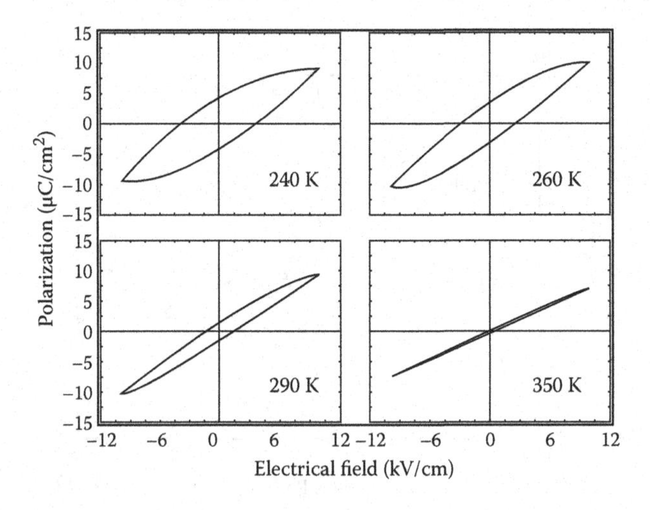


**FIGURE 12.6** Temperature dependence of hysteresis loops for PLZT 10/65/35.

350 K, the hysteretic behavior disappears completely and the curve of polarization versus electric field becomes a "slim loop," as commonly observed for ferroelectric relaxor materials. The coercive field and remanent polarization peak at 240 K ($E_c$ = 3.5 kV/cm, $P_r$ = 4.4 μC/cm$^2$). The slight rounded shape in the hysteresis loops is attributed to point-defect conduction mechanisms.

### 12.1.4 PLZT Thin Films

#### 12.1.4.1 PLZT Thin Film Fabrication

PLZT thin films have attracted much attention in recent years for their use in optical devices such as waveguide EO modulators, ultrafast EO shutters, and laser displays. Normally, fabrication of PLZT thin films may follow either vacuum or nonvacuum deposition techniques. The vacuum deposition methods include RF-magnetron sputtering, chemical vapor deposition, and pulsed laser deposition. The nonvacuum deposition methods include sol-gel, aerosol, and "hybrid" metal-organic chemical liquid deposition.

For EO applications, epitaxial PLZT thin films are preferred due to their excellent optical clarity and EO properties. However, optical isotropy of the thin films should be considered when designing EO devices. Thin film epitaxy can be classified into the three following types: (1) highly oriented polycrystalline films consisting of a complete *z*-axis orientation and a random orientation in the *x-y* plane; (2) crystalline films with a single crystal-like texture consisting of a complete *z*-axis orientation and low angle grain boundaries in the *x-y* plane; and (3) truly single crystalline films, which exhibit no grain boundaries except for other defects such as twins, stacking faults, and dislocations to relax the stress between the film and the substrate.

Vacuum deposition methods are very useful for growing epitaxial PLZT thin films. Epitaxial PLZT thin films have been grown on many nearly lattice-matched substrates including $LaAlO_3$, $(La_{0.29},Sr_{0.71})(Al_{0.65},Ta_{0.35})O_3$, $SrTiO_3$, *r*-cut sapphire, MgO, and a few semiconductors (such as silicon, Si) with buffer layers. Vacuum deposition methods have serious limitations because of the processing complexity of multicomponent complex oxide systems, and are normally inconvenient and expensive in practice. Nonvacuum deposition methods can provide low temperature synthesis, precision composition control, and high purity, but have problems related to the actual open-system operation and difficulties in eliminating organics from the thin films.

The epitaxial growth mechanisms for both vacuum and nonvacuum depositions are also different. For sputtering, the growth begins with the condensation of a vapor of atoms and ions onto the surface of the substrate, followed by nucleation initiated on the lattice-matched substrate surface, and then a subsequent epitaxial growth. In the sol-gel process for example, the growth is related to the crystallization of the solid amorphous film prepared by spin-coating and subsequent drying. The sol-gel process is generally more complicated and difficult than the sputtering process due to the fact that in the sol-gel process nuclei will form not only on the interface, but also on film surface and/or on impurity particles. The digitized metal-organic chemical liquid deposition method actually uses much diluted, metal-organic sources as precursors. A digitized coating/drying/crystallization process then allows atomic level growth without a vacuum. This can be considered a "hybrid" technique and is therefore efficient in growing exptaxial PLZT thin films. Figure 12.7 shows a schematic of such system [12].

#### 12.1.4.2 PLZT Thin Film on Sapphire, $SrTiO_3$, and Si

For ferroelectric PLZT *x*/65/35 compositions, their pseudocubic lattice constants are $a = b$ ~4.05 Å. Sapphire has a hexagonal corundum structure. For *r*-cut sapphire, it has $a$ = ~4.76 Å and $c$ = ~13.0 Å, with lattice mismatches to the *a*-axis and the *b*-axis of PLZT of ~15% and ~1.2%, respectively (when overlapping three PLZT *b*-axis unit cells with one sapphire *c*-axis unit cell). Here, the negative sign indicates the compressive strain applied to the PLZT layer. Despite the large lattice mismatch between PLZT and *r*-cut sapphire, it actually allows for the growth of a highly *c*-oriented PLZT thin film [13].

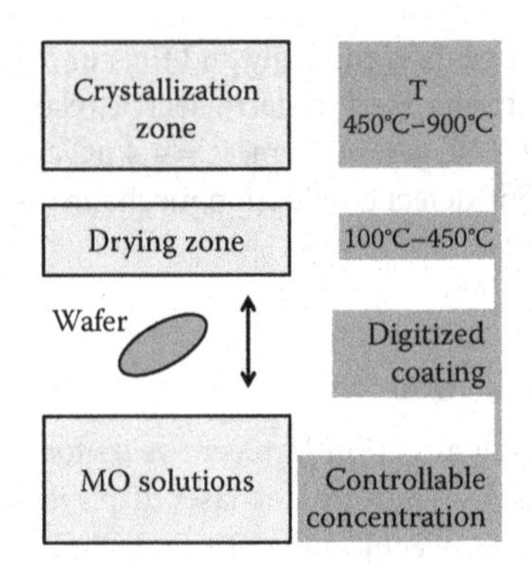


**FIGURE 12.7** Schematic of the MOCLD nonvacuum deposition system.

$SrTiO_3$ has the same perovskite structure as PLZT and a lattice constant of 3.905 Å at room temperature. Lattice mismatch between the two is –2.3% [14]. (001)-cut silicon has $a = b = 5.431$ Å, with a large lattice mismatch to PLZT of ~25%. Direct epitaxial growth of PLZT on Si is difficult, which necessitates using intermediate buffer layers.

Anisotropic thermal expansion coefficients along different crystalline axes in a single crystal play a key role in epitaxial growth. Large thermal expansion differences between the thin film and the substrate will cause a gradual build-up of strong anisotropic internal strains, until the epitaxial growth is lost. The sapphire crystal's thermal expansion coefficients along the *a*- and *c*-axes are about $4.5 \times 10^{-6}$ and $5.3 \times 10^{-6}$ (1/K), respectively [15]. They are close to the coefficient, $5.7 \times 10^{-6}$ (1/K), of PLZT. Cubic $SrTiO_3$ and Si have thermal expansion coefficients of about $10.8 \times 10^{-6}$ and $2.61 \times 10^{-6}$ (1/K), respectively. They are quite different from the thermal expansion coefficients of PLZT. Sapphire and Si have attracted research interest because they can be integrated together through the Si-on-Sapphire (SOS) technique [16]. For this process, sapphire is ($1\bar{1}02$)-cut to nearly match (001)-cut Si, as shown in Figure 12.8.

It should be pointed out that epitaxial growth of oxide thin films on Si has fundamental problems related to Si's low reactivity. Surface oxidation takes place easily to form amorphous Si oxide and oxide cations diffuse into Si at a relatively high rate. Buffer layers have to be used to address these issues [17]. Yittrium-stabilized zirconia (YSZ) grows epitaxially on Si. The typical epitaxial relationship between YSZ and Si is (001) of YSZ || (001) of Si and (100) of YSZ || (100) of Si (cubic-to-cubic alignment). If using the YSZ as a buffer layer, PLZT can epitaxially grow on YSZ/Si. In this case, the epitaxial relationship is (011) of PLZT || (001) of YSZ. This indicates a 45° degree rotation of the PLZT polarization with respect to the substrate, which is smaller than the polarization rotation in (001)-oriented PLZT thin films. Ceria ($CeO_2$) also grows epitaxially on YSZ with cubic-to-cubic alignment. Recently, $CeO_2$/YSZ buffer layers

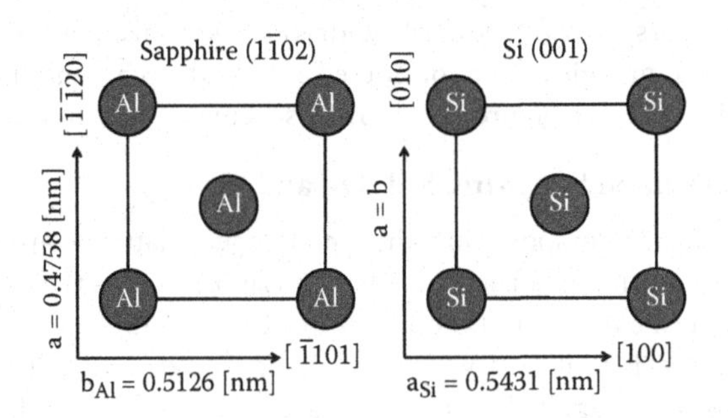


**FIGURE 12.8** Schematic of ($1\bar{1}02$)-cut sapphire and (001)-cut Si.

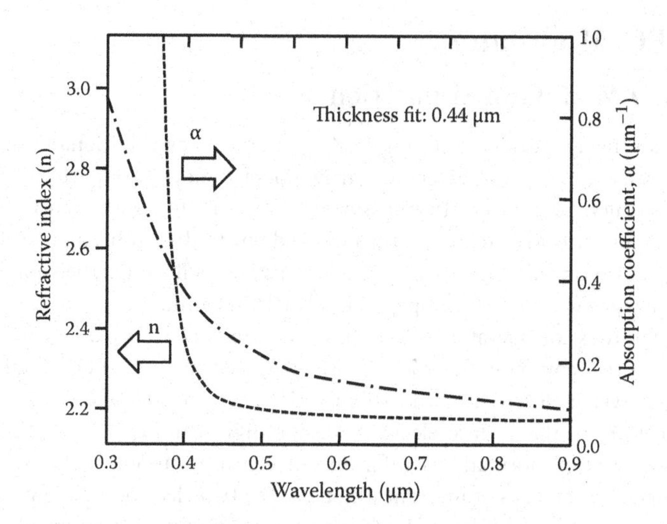


**FIGURE 12.9** Refractive index and absorption vs. wavelength of a PLZT thin film on sapphire.

have attracted much interest, due to the possibility of growing (001)-oriented PLZT thin films [18]. An additional benefit of using buffer layers is the possibility to epitaxially grow bottom electrodes using semiconductor oxides such as strontium ruthenium oxide ($SrRuO_3$, SRO) and niobium-doped $SrTiO_3$. For example, 0.1~0.7 wt.% Nb doping of $SrTiO_3$ can make the material metallic. It is therefore useful either as a substrate or as a buffer layer because of its lattice-matching and conducting functions [19]. For example, epitaxial Nb-doped $SrTiO_3$ (001) thin films can be integrated on (001) Si with a TiN template layer by RF-sputtering in an Ar environment at a substrate temperature of 540°C. Here, the orientation relationship is simply cube-on-cube as in Nb-$SrTiO_3$ (001)[110]||TiN(001) [110]||Si(001)[110] [20].

Figure 12.9 shows the dispersion curves of the refractive indices as well as the absorption versus wavelength for a 0.44 μm thick, PLZT 9/65/35 film grown on sapphire [13]. These curves were from the best fits to the optical transmission data of the sample, assuming that both the refractive indices and the absorption follow a single-term Sellmeier equation. The absorption edge of the PLZT film is approximately 330 nm, which is slightly lower than the value of 370 nm for bulk PLZT 9/65/35 ceramics.

EO effects in epitaxial PLZT thin films depend strongly on the orientation of the film. Understanding the anisotropy of EO effects allows for optimal design of the electrodes in EO modulators. Table 12.1 lists EO coefficients of $r_{13}$, $r_{33}$, and $r_c$ in PLZT 8/65/35 thin films having orientations of (100), (101), and (111), respectively. These films were grown on $SrTiO_3$ substrates cut along the (100), (101), and (111) directions, respectively. Here $r_c = r_{33} - (n_{13}/n_{33})^3 r_{13}$. $n_{13}$ and $n_{33}$ are ordinary ($n_o$) and extraordinary ($n_e$) refractive indices, respectively. (100)-oriented, thin-film PLZT shows a minimum birefringence. The large anisotropic EO effect in PLZT thin films can be explained by a rotation of the refractive indices under the applied electrical field that is also accompanied by spontaneous polar cluster polarization switching [21].

**TABLE 12.1** EO Coefficients for Different PLZT Thin Film Orientations

| Orientation | $r_{13}$ (pm/V) | $r_{33}$ (pm/V) | $r_c$ (pm/V) |
|---|---|---|---|
| (100) | 27 | 27 | 0 |
| (101) | 24 | 68 | 44 |
| (111) | −9 | 90 | 99 |

*Source:* Reprinted with permission from K. Sato et al., *Appl. Phys. Lett.*, 87, 25, 251927 (1–3), Copyright 2005, American Institute of Physics.

## 12.2 PLZT EO Modulators

### 12.2.1 PLZT EO Modulator Simulation

EO phase modulation using weakly scattering PLZT materials was traditionally achieved by surface electrodes on thin wafers [22]. Past research primarily modeled modulator performance using a simple Jones calculus, assuming that optical scattering is negligible. In reality, and when a large phase modulation is required, optical scattering can be greatly increased due to the strong applied electric and optical fields. This complex scattering will also cause the depolarization effect. Furthermore, saturation of the EO-induced birefringence with increased applied electric field cannot be ignored. Therefore, modeling a PLZT modulator requires consideration of the above-mentioned effects [23].

Figure 12.10 shows two common-electrode alignments used in PLZT EO modulators: (1) surface-electrode and (2) transverse-electrode. On PLZT materials, evaporation of CrAu is a standard process for electrode fabrication. In the surface-electrode design, light can be incident either on the surface (along *y*-axis, a configuration normally used for bulk devices) or on one of the polished edges either between the two surface electrodes or underneath one of the two electrodes (along *z*-axis, a configuration normally used for waveguide devices). In the transverse alignment, edge-incidence is typical. The two alignments can become the basis for large-scale devices including interdigital surface electrodes that are commonly used in transverse modulators.

In order to take optical scattering, depolarization, and birefringence saturation into full consideration, an experimental determination of the phase variation versus the applied electric field is necessary. Such an experiment can be performed using a "sandwich" structure similar to that in Figure 12.10b. A homogeneous electric field, $E = V/d$, can be generated inside the PLZT plate, assuming that the two electrodes are large, parallel, and flat. In a typical measurement setup, a laser beam, linearly polarized at 45° with respect to the electric field, passes through the sandwich and then through a linear polarizer whose transmission axis is at −45° with respect to the electric field. The normalized optical transmittance, *T*, of the setup varies as a function of the relative phase according to the relation [24]

$$T = \frac{1}{2}a^2 + \frac{1}{2}b^2 - ab\sin(\Phi), \tag{12.7}$$

where $\Phi$ is the relative phase induced by the electric field and $a^2$ and $b^2$ are transmittances for the two orthogonal components of the light. Dependence of $a^2$, $b^2$, and $T$ on the applied electric field and on the incident light intensity can be measured by setting the second polarizer at different angles relative to the electric field. By solving Equation 12.7, the relationship between the relative phase and either $E$ or the incident light intensity can be determined. In general, the resulting relationship is not linear due to

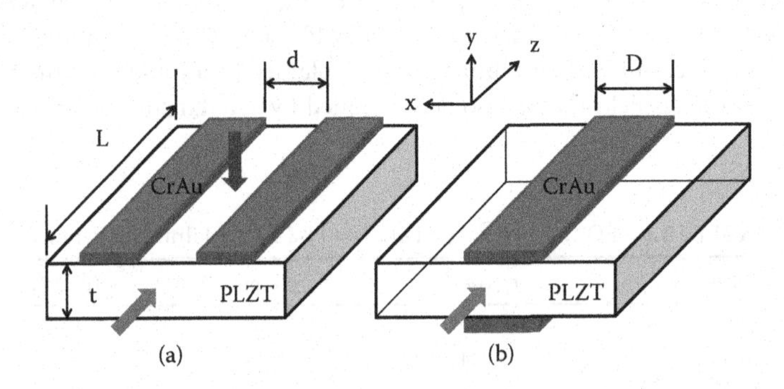


**FIGURE 12.10** Two common electrode alignments for EO modulators: (a) surface-electrode and (b) transverse-electrode.

the depolarization effect. By defining the fractions of depolarized light corresponding to vertically and horizontally polarized incident components as $c_a$ and $c_b$, respectively, Equation 12.7 becomes

$$T = \frac{1}{2}\left(a^2 + c_a^2\right) + \frac{1}{2}\left(b^2 + c_b^2\right) - ab\sin(\Phi). \quad (12.8)$$

Further, if we let $A^2 = a^2 + c_a^2$ and $B^2 = b^2 + c_b^2$, and assume $c_a = c_b = c$, then Equation 12.8 becomes

$$T = \frac{1}{2}(A^2 + B^2) - \sqrt{A^2 - c^2}\sqrt{B^2 - c^2}\,\sin(\Phi)\,. \quad (12.9)$$

Figure 12.11a shows typical experimental results from a PLZT sample of $t = 2.01$ mm and $L = 300$ μm (Figure 12.10b). $T$ was measured when the second polarizer was set at 45° with respect to the electric field. For $A^2$ and $B^2$, the second polarizer was set vertically and horizontally, respectively. The observed intensity decrease when $E > {\sim}7 \times 10^5$ (V/m) is due to increased scattering. The relative phase as a function of the electric field is shown in Figure 12.11 b. Saturation appears when $E > 1.2 \times 10^6$ (V/m), and reaches

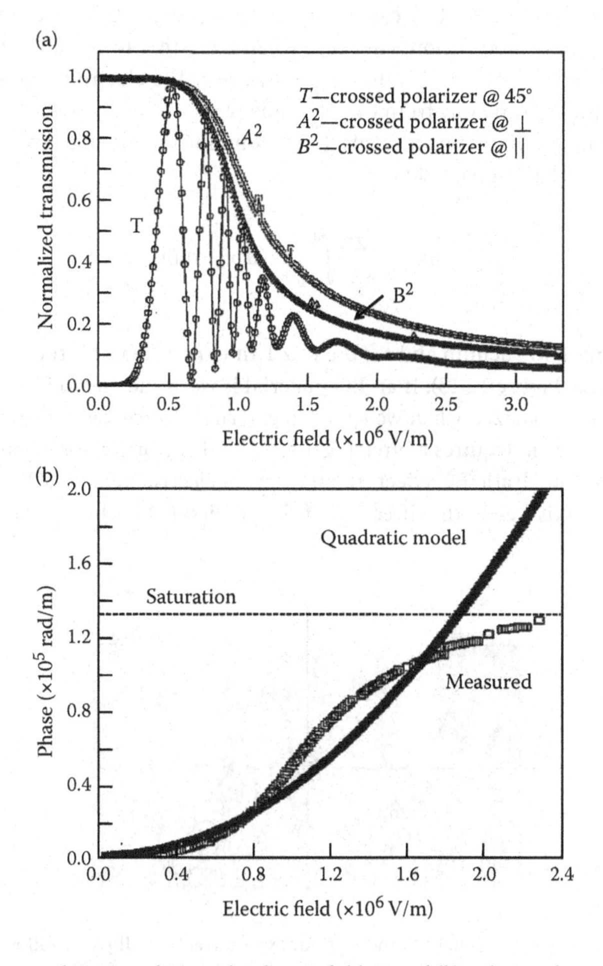


**FIGURE 12.11** (a) Measured $T$, $A^2$, and $B^2$ vs. the electric field, $E$, and (b): relative phase versus $E$.

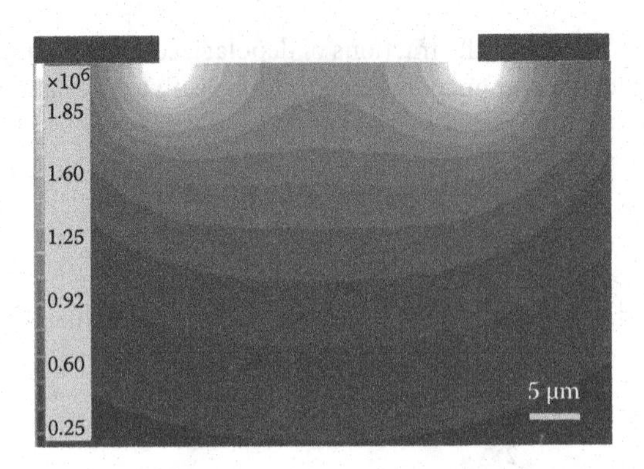


**FIGURE 12.12** Simulated electric field contours of an electrode using the FEA method.

a maximum value when $E > 2.2 \times 10^6$ (V/m). The best-fit curve, using a quadratic EO coefficient of $R = 6 \times 10^{-16}$ $(V/m)^2$, is also shown [25].

The second step in modulator modeling is the determination of the applied field distribution, which can be calculated using commercial Finite Element Analysis software. Figure 12.12 shows a typical plot of electric field contours between surface electrodes (250 μm wide) that have a gap of 50 μm. According to Figure 12.11b, little phase modulation will occur when $E$ is either too small ($<0.2 \times 10^6$ (V/m)) or too large ($>2.2 \times 10^6$ (V/m)). In Figure 12.12, the bright area near the electrodes is saturated. Modulation diminishes farther than 100 μm away from the electrode [25].

The last step in modulator modeling is to effectively combine the previous two results, using the relationship describing the relative phase, $\Phi(x)$:

$$\Phi(x) = \frac{2\pi}{\lambda} \int_{y=0}^{t} \Delta n(\theta(x,y))dy, \tag{12.10}$$

where $\lambda$ is the wavelength in vacuum and $t$ is the PLZT thickness. $\theta(x,y)$ is the electric field vector with respect to the $x$-axis (see Figure 12.13). If an EO material is isotropic due to its cubic structure at room temperature, it becomes polarized when we apply an external electric field, demonstrating anisotropic optical characteristics (i.e., it acquires a birefringence). The third-order nonlinear optical properties of the material lead to the quadratic EO effect. In an external electric field, PLZT behaves like a uniaxial crystal and the optical axis lies in the direction of the applied field. The induced ordinary and extra-

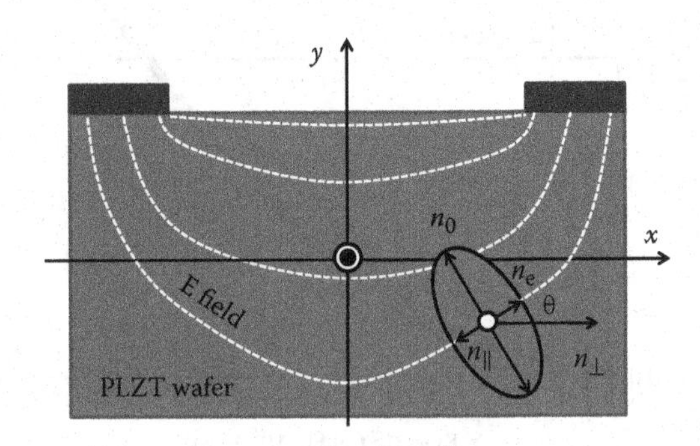


**FIGURE 12.13** Schematic showing that the induced index of refraction ellipsoid follows the tangent of electric field lines.

ordinary indices of refraction thus align according to the electric field contour shown in Figure 12.12 and can be expressed as

$$n_0 = n - \frac{1}{2}n^3 R_{12} E^2 \text{ and } n_e = n - \frac{1}{2}n^3 R_{33} E^2, \tag{12.11}$$

$$\Delta n(E) = n_0 - n_e = \frac{1}{2}n^3 R E^2, \tag{12.12}$$

where $R = R_{12} - R_{33}$, $R_{12}$, and $R_{33}$ are the quadratic EO coefficients, $n$ is the refractive index of PLZT in zero external field, and $\Delta n(E)$ is the induced optical birefringence.

In Figure 12.13, the refractive indices parallel and perpendicular to the surface electrodes can be expressed approximately as

$$n_{\parallel} \approx n_0 \text{ and } n_{\perp} \approx \left[ \frac{\cos^2(\theta(x,y))}{n_e^2} + \frac{\sin^2(\theta(x,y))}{n_0^2} \right]^{1/2}, \tag{12.13}$$

where

$$\theta(x,y) = \tan^{-1}\left( \frac{E_y(x,y)}{E_x(x,y)} \right). \tag{12.14}$$

In Equation 12.9, the relative index change $\Delta n(\theta(x,y))$ is then defined as $n_{\perp} - n_{\parallel}$. When designing a PLZT EO modulator, following the three procedures given above will allow one to optimize the three most important modulator parameters: the electrode width, the electrode gap, and $V_{\pi}$.

### 12.2.2 PLZT Mach-Zehnder Modulators (MZMs)

In this section, two PLZT EO waveguide MZM designs are introduced: ridge-type and reversed ridge-type. In both cases, Y-branch waveguides are normally used to make Mach-Zehnder modulators. Figure 12.14 shows a typical ridge-type cross-section and an actual Y-branch image [26]. In this particular design, the substrate is Nb:$SrTiO_3$, the buffer layer is PLZT 9/65/35 (2.6 μm thick), and the waveguide layer is PLZT 2.5/52/48 (2.2 μm thick). The ridge's height and width are 0.5 μm and 5.0 μm, respectively. This size is required to maintain single-mode operation. An indium tin oxide layer is used as the top electrode. Fabrication of such a structure involves epitaxial growth of the two PLZT thin films followed

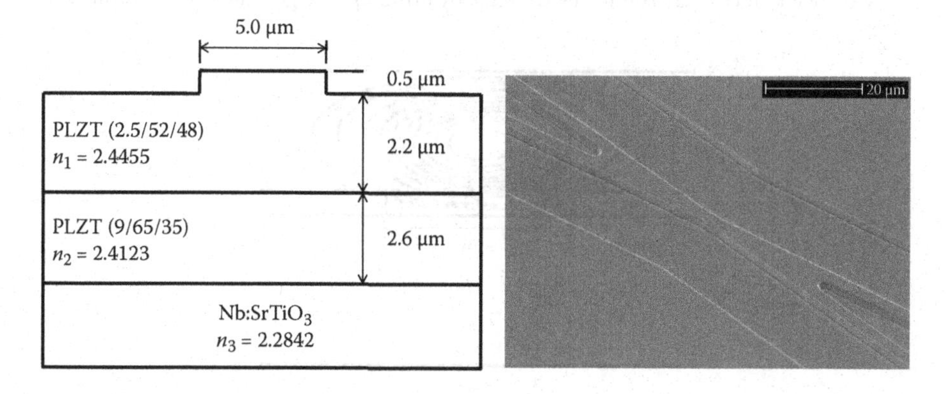


**FIGURE 12.14** Ridge-type MZM's waveguide cross-section.

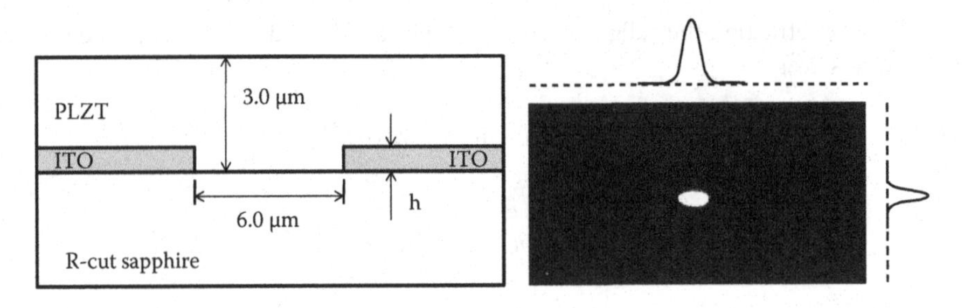


**FIGURE 12.15** Reversed ridge-type waveguide's cross-section.

by a dry etching using $CHF_3$ mixed with $H_2$ to create the ridges. The calculated insertion loss for this waveguide is about 13.2 dB and the actual measured insertion loss is about 16.0 dB. The measured PLZT quadratic EO coefficient is about two orders of magnitude smaller than that of the bulk material and a large -drift (~40 mV/hr) has also been observed.

The large insertion loss is mainly attributed to the small size of the waveguide mode (estimated to be ~9.7 dB). Large differences between the estimated and experimental insertion loss can be explained by volume and ridge interface scattering. Dry etching may also be one of the major reasons for deteriorated EO performance in PLZT thin films. To overcome such problems, a reversed ridge-type waveguide was proposed, which can be much simpler to fabricate (see Figure 12.15) [27]. This device can be made directly on an *r*-cut sapphire substrate. The process starts with a coating of indium tin oxide as both a spacer and potentially a bottom electrode, followed by etching the indium tin oxide to create reversed ridge patterns, and then by a subsequent epitaxial growth of PLZT on the structure. In a demonstration device, the reversed ridge width was 6.0 μm, and the PLZT 8.5/65/35 film thickness was 3.0 μm. The indium tin oxide spacer thickness, *h*, was varied from 0.4 μm to 2 μm, in order to study the dependence of the waveguide mode on *h*.

When *h* was reduced from 2.0 to 0.4 μm, the waveguide changed from a multimode guide (supporting the 4 lowest-order modes at $h = 2.0$ μm) to a single-mode one ($h = 0.4$ μm, Figure 12.16). The single-mode profile's area was estimated to be ~24 μm², which is about two times larger than that shown above in Figure 12.14. The measured quadratic EO coefficient for the PLZT 8.5/65/35 thin film is about 0.25 $\times 10^{-16}$ $(m/V)^2$. This value is smaller than that in the bulk crystal, but is about an order of magnitude greater than that shown in Figure 12.14. This improvement is a result of direct epitaxial growth of PLZT film on the *r*-cut sapphire. The total loss in a 10 mm long waveguide was measured to be about 7.1 dB. Of that total loss, about 2.7 dB was attributed to propagation loss for the 1.55 μm light, about 1.8 dB to reflection losses at the interfaces, and approximately 2.6 dB to the coupling loss. Losses have been significantly reduced, in comparison to the designs, by utilizing a large single-mode profile, improved

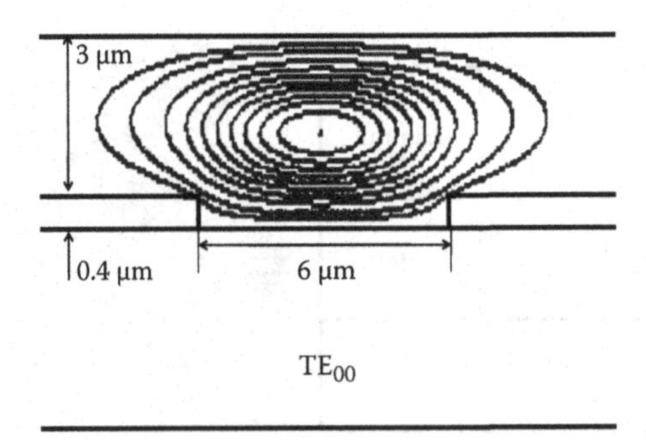


**FIGURE 12.16** Optical mode profile in a 0.4 μm high reversed ridge-type single-mode waveguide.

PLZT film quality (reduced volume scattering), and reduced edge and interface scattering. This modulator delivers an extinction ratio of ~14.4 dB and a modulation depth of 0.93 at an operation frequency of 1 MHz.

### 12.2.3 PLZT Modulators Integrated with Si

A new trend in developing EO modulators is to make them compatible with the complementary-metal-oxide-semiconductor processes of the semiconductor industry. As the clock rate of modern semiconductor chips continues to increase, problems related to conventional electrical interconnects have become critical. Using optical interconnects instead of the electrical interconnects will likely solve those problems. Advantages to using PLZT EO modulators on Si, in comparison with semiconductor-based modulators (for example, modulators based on electro-absorption and microring resonators), include smaller size, lower voltage, and wider operation wavelength range.

Figure 12.17a illustrates a simple reversed-ridge structure Mach Zehnswe PLZT EO modulator on Si [28]. It consists of an $SiO_2$ insulator on an Si substrate, a lower electrode layer, a lower cladding layer, a PLZT core, an upper cladding layer, and an upper electrode. The reversed-ridge waveguide design is implemented into the design to significantly reduce the influence of side-wall roughness caused by etching processes. By using a smaller core and a perpendicularly applied electric field, this modulator can easily reach gigahertz frequency operation. Figure 12.17b shows one interconnect scheme in a partially etched ridge-type PLZT EO modulator. The above two device performances can be further improved by using the above-mentioned buffer layers to achieve epitaxial growth of PLZT on Si.

### 12.2.4 PLZT Spatial Light Modulators

A high-performance spatial light modulator (SLM) is the key component for many applications such as image processing and optical computing. The SLM, typically a two-dimensional array of light modulators, spatially encodes data bits onto a light beam as an image [29]. An SLM is required to be low-voltage, low power, and fast, in order to attain a high frame rate. Direct integration of the light-modulating layer with a silicon integrated circuit is therefore useful for utilizing functional elements of the circuit to apply the required driving voltages to the modulation layer. PLZT-on-Si has been demonstrated successfully in the past for making 1D and 2D SLMs [30]. The major challenges are related to the integration, pixel density, and the cost of the devices. EO-based SLM is still an interesting area of technology development, as the PLZT-on-Si technology is continually improving and the demand for such SLMs continues to increase.

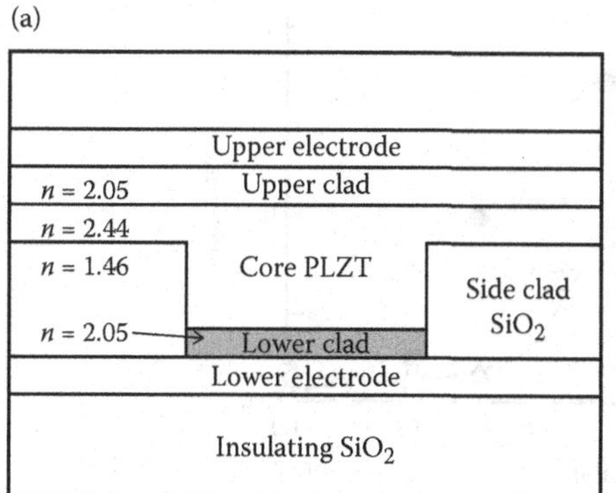


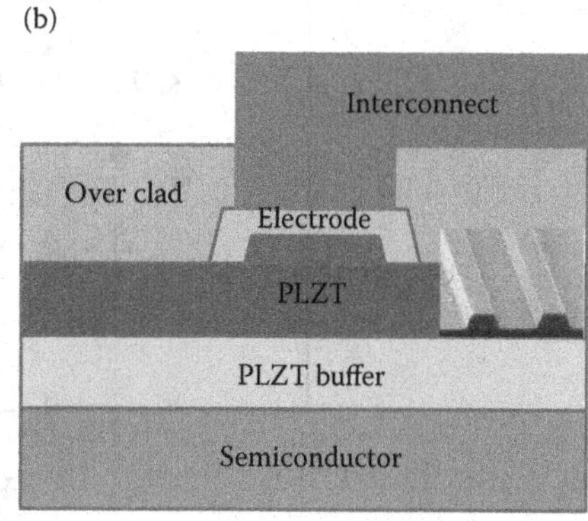


**FIGURE 12.17** (a) A reversed-ridge PLZT modulator on Si. (b) A partially etched ridge-type PLZT waveguide on semiconductor (inset is the actual image of the ridge waveguide).

## 12.3 Other EO Materials/Modulators

### 12.3.1 $Ba_{1-x}Sr_xTiO_3$ (BST)

$BaTiO_3$ is a tetragonal crystal at room temperature. The tetragonal to cubic phase transition occurs at temperatures higher than 120°C. This phase transition can be described as a distortion of the tetragonal structure. This distortion occurs when the titanium atom within the oxygen octahedron is shifted from the octahedral corner toward the center. By doping the crystal with foreign cations, the phase transition temperature ($T_c$) can be varied. When $Ba^{2+}$ ions in $BaTiO_3$ are partially replaced by $Sr^{2+}$ with a molar ratio $0 \leq x \leq 1$, the electronic and optical properties of solid solution $Ba_{1-x}Sr_xTiO_3$ (BST) can be tailored over a broad range. Composition dependence of thin film BST's dielectric constant and its linear and quadratic EO coefficients are shown in Figure 12.18a and 12.18b, respectively [31]. Table 12.2 lists the measured values of both the linear and quadratic EO coefficients for two BST compositions with molar ratios of $x = 0.3$ and $x = 0.5$. These values are much larger than those for PLZT films listed in Table 12.1.

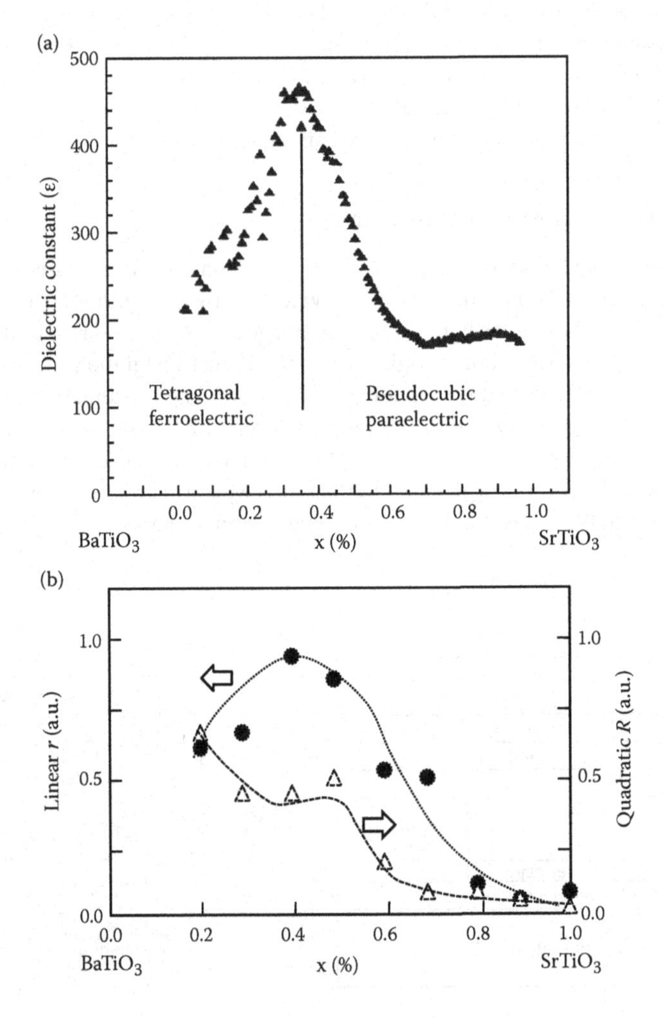


**FIGURE 12.18** (a) Composition dependence of thin film BST's dielectric constant. (b) Linear and quadratic EO coefficients vs. the composition.

**TABLE 12.2** EO Coefficients of BST Thin Films

| Films | $r$ (× $10^{-12}$ m/V) $E$ Increasing | $r$ (× $10^{-12}$ m/V) $E$ Decreasing | $R$ (× $10^{-16}$ $m^2/V^2$) |
|---|---|---|---|
| $x = 0.5$ | 70 | 115 | 1.3 |
| $x = 0.3$ | 110 | 230 | 0.55 |

## 12.3.2 PMN-PT EO Ceramics, Thin Films, and Superlattices

The recent invention of La-modified lead magnesium niobate-lead titanate (PMN-PT) transparent EO ceramics represents a major development in the optical ceramics field [32]. Such materials can deliver better EO performance than PLZT (larger EO coefficients and greater reliability) and have been successfully used in commercial photonic components for optical communications. In PMN-PT phase diagrams, an almost vertical MPB separates rhombohedral and tetragonal regions (temperature vs. $x$). The typical MPB behaviors, such as enhanced piezoelectric effects and maximized dielectric constants, are very pronounced in the material. Similar to PMN-PT, lead zinc niobate-lead titanate materials have also gained much attention in recent years.

Large EO effects in PMN–PT thin films have been demonstrated recently. Quadratic EO coefficients in PMN–PT films as large as $1.38 \times 10^{-16}$ $(m/V)^2$ were observed at the material's MPB that corresponds to a PMN to PT ratio of around 67/33 [12]. This MPB composition has a phase transition around 175°C, which is significantly higher than room temperature. PMN–PT films also show very good optical clarity and low propagation loss (<3 dB/cm) [33]. PMN–PT is more durable than the members of the PZT family, which is required for handling high laser powers. The EO effects are relatively small along the main axis [34], but about an order of magnitude enhancement of EO effects has been demonstrated by using a structural coupling between two crystalline states (tetragonal and rhombohedral) with nearly equal energies [35]. Figure 12.19 shows the (001) direction EO enhancement and its dependence on the superlattice periodicity [36].

## 12.3.3 Periodically Poled $LiNbO_3$ (PPLN)

In a PPLN crystal, its intrinsic polarization is periodically reversed, allowing very efficient quasiphase-matched nonlinear processes to occur [37]. In a PPLN crystal, the nonlinear optical coefficients periodically change sign due to the periodic domains. As a result, an excited parametric wave will have a π phase shift when passing through the domain boundary. If each domain thickness is equal to the coherence length, the excited parametric wave from each domain will interfere constructively [38]. PPLN

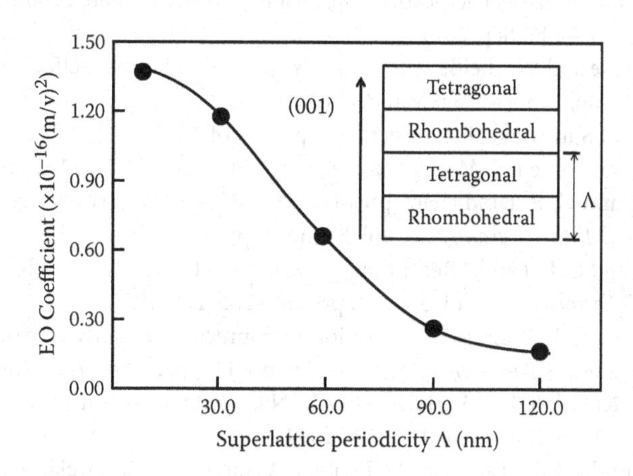


**FIGURE 12.19** (001) direction EO enhancement and the dependence on the superlattice's periodicity.

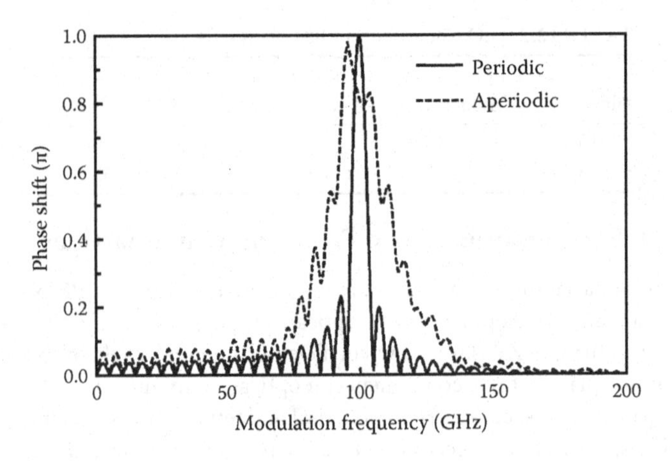


**FIGURE 12.20** Phase shift vs. frequency for both periodically and aperiodically poled LN crystals.

has also been widely used in piezoelectric and EO processes. The piezoelectric and EO coefficients also change sign periodically in PPLN and these changes produce interesting effects.

EO PPLN can be used as either an intensity modulator or a phase modulator. The former application uses PPLN as a dynamic grating driven by an applied field that diffracts wavelengths satisfying the Bragg condition. The phase modulator achieves broadband phase modulation by matching the velocity of the driving microwave field and the propagating optical wave (quasivelocity matching). Figure 12.20 shows the phase shift as a function of modulation frequency of a modulator made out of PPLN (that has a poling period of 1.44 mm) and aperiodically poled $LiNbO_3$ [39]. The latter material has its laminar thickness randomly varied with an average periodicity of 1.44 mm.

# References


1. L. M. Levinson, Ed., *Electronic Ceramics*, New York: Marcel Dekker Inc., 1998.
2. Y. H. Xu, *Ferroelectric Materials and Their Applications*, Amsterdam: North Holland Publishing Co., 1991.
3. J. T. Cutchen, J. O. Harris, Jr., and G. R. Laguna, "PLZT electro-optic shutters: Applications," *Appl. Opt.*, vol. 14, no. 8, pp. 1866–1873, Aug. 1975.
4. A. Garg and D. C. Agrawal, "Effect of rare earth (Er, Gd, Eu, Nd and La) and bismuth additives on the mechanical and piezoelectric properties of lead zirconate titanate ceramics," *Mater. Sci. Eng. B*, vol. 86, no. 2, pp. 134–143, Sep. 2001.
5. E. Gorges, P. Grosse and W. Theiß, "The Kramers–Kronig-Relation of effective dielectric functions," *Z. Phys. B*, vol. 97, no. 1, pp. 49–54, Mar. 1994.
6. M. Płońska and Z. Surowiak, "Piezoelecric properties of *X*/65/35 PLZT ceramics depended of the lanthanum (*X*) ions contents," *Mole. Quant. Acous.*, vol. 27, pp. 207–211, 2006.
7. G. H. Haertling and C. E. Land, "Hot-pressed $(Pb,La)(Zr,Ti)O_3$ ferroelectric ceramics for electro-optic applications," *J. Am. Ceram. Soc.*, vol. 54, no. 1, pp. 1–11, Jan. 1971.
8. G. H. Haertling and C. E. Land, "Recent improvements in the optical and electro-optic properties of PLZT ceramics," *Ferroelectrics*, vol. 3, no. 1, pp. 269–280, Feb. 1972.
9. B. A. Menegazzo and J. A. Eiras, "Preparation of coprecipitated ferroelectric ceramic powders by two-stage calcination," *J. Am. Ceram. Soc.*, vol. 76, no. 11, pp. 2734–2736, Nov. 1993.
10. D. S. Yoon, C. J. Kim, J. S. Lee, W. J. Lee, and K. No, "Epitaxial growth of sol-gel PLZT thin films," *J. Mater. Res.*, vol. 9, no. 2, pp. 420–425, Feb. 1994.
11. I. A. Santos, C. Endo, A. L. Zanin, M. H. Lente, J. A. Eiras, and D. Garcia, "Hot-pressed transparent PLZT ceramics from low cost chemical processing," *Mater. Res.*, vol. 4, no. 4, pp. 291–295, Oct. 2001.

12. Y. Lu, G.-H. Jin, M. Cronin-Golomb, S.-W. Liu, H. Jiang, F.-L. Wang, J. Zhao, S.-Q. Wang, and A. J. Drehman, "Fabrication and optical characterization of $Pb(Mg_{1/3}Nb_{2/3})O_3$-$PbTiO_3$ planar thin film optical waveguides," *Appl. Phys. Lett.*, vol. 72, no. 23, pp. 2927–2929, Jun. 1998.
13. B. Tunaboylu, P. Harvey, and S. C. Esener, "Characterization of dielectric and electro-optic properties of PLZT 9/65/35 films on sapphire for electro-optic applications," *IEEE Trans. Ultrason. Ferrelectr. Freq. Contr.*, vol. 45, no. 4, pp. 1105–1112, Jul. 1998.
14. O. Nordseth, T. Tybell, J. K. Grepstad, and A. Royset, "Sputter-deposited $(Pb,La)(Zr,Ti)O_3$ thin films: Effect of substrate and optical properties," *J. Vac. Sci. Technol. A*, vol. 27, no. 3, pp. 548–553, May 2009.
15. M. Lucht, M. Lerche, H.-C. Wille, Y. V. Shvyd'ko, H. D. Rüter, E. Gerdau, and P. Becker, "Precise measurement of the lattice parameters of $\alpha$-$Al_2O_3$ in the temperature range 4.5–250 K using the Mössbauer wavelength standard," *J. Appl. Cryst.*, vol. 36, part 4, pp. 1075–1081, Aug. 2003.
16. T. Nakamura, H. Matsuhashi, and Y. Nagatomo, "Silicon on sapphire (SOS) device technology," *Oki Technical Rev.*, vol. 71, no. 4, pp. 66–69, Oct. 2004.
17. M. Kondo, K. Maruyama, and K. Kurihara, "Epitaxial ferroelectric thin films on silicon substrates for future electronic devices," *FUJITSU Sci. Tech. J.*, vol. 38, no. 1, pp. 46–53, Jun. 2002.
18. O. Nordseth, T. Tybell, and J. K. Grepstad, "Epitaxial $(Pb,La)(Zr,Ti)O_3$ thin films on buffered Si(100) by on-axis radio frequency magnetron sputtering," *Thin Solid Films*, vol. 517, no. 8, pp. 2623–2626, Feb. 2009.
19. M. A. Saifi and L. E. Cross, "Dielectric properties of strontium titanate at low temperature," *Phys. Rev. B*, vol. 2, no. 3, pp. 677–684, Aug. 1970.
20. C. Wang, "Preparation and characterization of epitaxial conductive Nb-$SrTiO_3$ thin films on Si substrates," *J. Electron. Mater.*, vol. 39, no. 2, pp. 187–190, Feb. 2010.
21. K. Sato, M. Ishii, K. Kurihara, and M. Kondo, "Crystal orientation dependence of the electro-optic effect in epitaxial lanthanum-modified lead zirconate titanate films," *Appl. Phys. Lett.*, vol. 87, no. 25, pp. 251927 (1–3), Dec. 2005.
22. P. E. Shames, P. C. Sun, and Y. Fainman. "Modeling of scattering and depolarizing electro-optic devices. I. Characterization of lanthanum-modified lead zirconate titanate," *Appl. Opt.*, vol. 37, no. 17, pp. 3717–3725, Jun. 1998.
23. P. E. Shames, P. C. Sun, and Y. Fainman, "Modeling of scattering and depolarizing electro-optic devices. II. Device simulation," *Appl. Opt.*, vol. 37, no. 17, pp. 3726–3734, Jun. 1998.
24. A. Yariv and P. Yeh, *Optical Waves in Crystals,* New York: Wiley-Interscience, 1993, ch. 5.
25. P. Shames, P.-C. Sun, and Y. Fainman, "Modeling and optimization of electro-optic phase modulator," in *Physics and Simulation of Optoelectronics Devices IV, Proc. SPIE*, vol. 2693, Bellingham, WA: SPIE, 1996, pp. 787–796.
26. R. Thapliya, Y. Okano, and S. Nakamura, "Electro-optic characteristics of thin-film PLZT waveguide using ridge-type Mach–Zehnder modulator," *J. Lightwave Technol.*, vol. 21, no. 8, pp. 1820–1827, Aug. 2003.
27. G. H. Jin, Y. K. Zou, V. Fuflyigin, S. W. Liu, Y. L. Lu, J. Zhao, and M. Cronin-Golomb, "PLZT film waveguide Mach–Zehnder electrooptic modulator," *J. Lightwave Technol.*, vol. 18, no. 6, pp. 807–812, Jun. 2000.
28. T. Shimizu, M. Nakada, H. Tsuda, H. Miyazaki, J. Akedo, and K. Ohashi, "Gigahertz-rate optical modulation on Mach–Zehnder PLZT electro-optic modulators formed on silicon substrates by aerosol deposition," *IEICE Electrics Express*, vol. 6, no. 23, pp. 1669–1675, Dec. 2009.
29. M. Roy, C. J. R. Sheppard, and P. Hariharan, "Low-coherence interference microscopy using a ferroelectric liquid crystal phase-modulator," *Opt. Express,* vol. 12, no. 11, pp. 2512–2516, May 2004.
30. T.-H. Lin, A. Ersen, J. H. Wang, S. Dasgupta, S. Esener, and S. H. Lee, "Two-dimensional spatial light modulators fabricated in Si/PLZT," *Appl. Opt.*, vol. 29, no. 11, pp. 1595–1603, Apr. 1990.
31. J. Li, F. Duewer, C. Gao, H. Chang, X.-D. Xiang, and Y. Lu, "Electro-optic measurements of the ferroelectric-paraelectric boundary in $Ba_{1-x}Sr_xTiO_3$ materials chips," *Appl. Phys. Lett.*, vol. 76, no. 6, pp. 769–771, Feb. 2000.

32. K. K. Li, Y. Lu, and Q. W. Wang, "Electro-optic ceramic material and device," U.S. Patent 6 890 874, May 10, 2005.
33. Y. Lu, B. Gaynor, C. Hsu, G. Jin, M. Cronin-Golomb, F. Wang, J. Zhao, S.-Q. Wang, P. Yip, and A. J. Drehman,"Structural and electro-optic properties in lead magnesium niobate titanate thin films," *Appl. Phys. Lett.*, vol. 74, no. 20, pp. 3038–3040, May 1999.
34. Y. Lu, J. Zheng, M. Croning Golomb, F. Wang, H. Jiang, and J. Zhao, "In-plane electro-optic anisotropy of $(1-x)Pb(Mg_{1/3}Nb_{2/3})O_3-xPbTiO_3$ thin films grown on (100)-cut $LaAlO_3$," *Appl. Phys. Letts.*, vol. 74, no. 25, pp. 3764–3766, Jun. 1999.
35. Y. Lu, 2004. "Dielectric and ferroelectric behaviors in $Pb(Mg_{1/3}Nb_{2/3})O_3-PbTiO_3$ rhombohedral/tetragonal superlattices," *Appl. Phys. Lett.*, vol. 85, no. 6, pp. 979–981, Aug. 2004.
36. Y. Lu, W. J. Mandeville, and R. J. Knize, "Electro-optic effect in relaxor ferroelectric films and superlattices," *Integrated Ferroelectrics*, vol. 80, no. 1, pp. 29–37, Nov. 2006.
37. Y.-L. Lu, L. Mao, S.-D. Cheng, N.-B. Ming, and Y.-T. Lu, "Second-harmonic generation of blue light in $LiNbO_3$ crystal with periodic ferroelectric domain structures," *Appl. Phys. Lett.*, vol. 59, no. 5, pp. 516–518, Jul. 1991.
38. Y.-L. Lu, L. Mao, and N. B. Ming, "Blue-light generation by frequency doubling of an 810-nm cw GaAlAs diode laser in a quasi-phase-matched $LiNbO_3$ crystal," *Opt. Lett.*, vol. 19, no. 14, pp. 1037–1039, Jul. 1994.
39. Y.-Q. Lu, J.-J. Zheng, Y.-L. Lu, N.-B. Ming, and Z.-Y. Xu, "Frequency tuning of optical parametric generator in periodically poled optical superlattice $LiNbO_3$ by electro-optic effect," *Appl. Phys. Lett.*, vol. 74, no. 1, pp. 123–125, Jan. 1999.

# IV

# Practical Issues in Modulator Operation

# 13

# Modulator Drivers




John J. DeAndrea
*Finisar Corporation*


## 13.1 Introduction

Electrical drivers for optical modulation must bridge the gap between electronic digital multiplexing circuits and amplify these digital levels up to voltages of the modulator technology chosen. The amplification by the driver should provide minimal degradation of the multiplexed signal characteristics, with minimal amplitude or bit transition jitter of the signal applied to the modulator. The driver characteristics are affected by the topology used in the amplifier driver, the integrated circuit (IC) technologies used in the design, the radio frequency (RF) broadband coupling of the design from the electrical multiplexer to the optical modulator, interstage coupling of cascaded amplifier stages, and the packaging technologies used for building these solutions. The following sections will detail each of the items and explain the practical implementations of these solutions used in the present market.

## 13.2 Driver Topologies and Characteristics

There are various driver topologies that can be designed and implemented to meet the various modulator requirements: single-ended drive amplifiers for single-ended input modulators, dual drive amplifiers

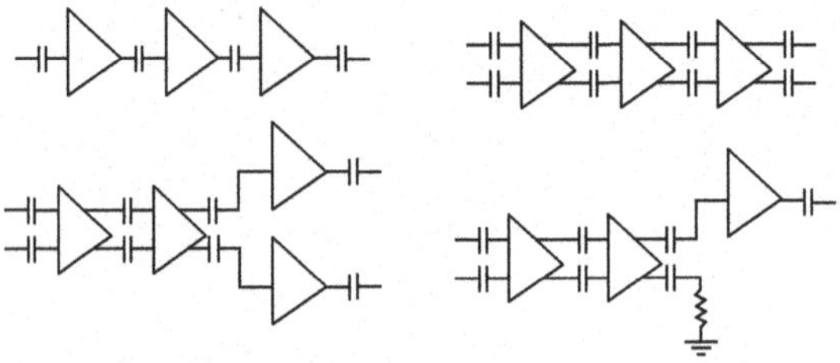

**FIGURE 13.1** Four types of driver topologies.

for differential input modulators, a combination of a differential stage amplifier followed by two single-ended drivers, or a differential stage gain section followed by a single stage amplifier for a single-ended modulator. These topologies can be designed in a linear or limiting fashion, each having its own pros and cons in implementation, packaging, associated costs, and volume capability. The following sections will outline these options and investigate the differences. Figure 13.1 shows the basic block diagrams of the four types of topologies to be examined.

### 13.2.1 Single-Ended Driver Topology

The single-ended driver topology is the simplest approach for amplifying digital signals and coupling them to the optical modulator. These consist of single-ended RF transistors biased and matched for input and output impedance of the transmission lines and cascaded to allow for amplifying the digital multiplexer signals to the higher voltage required by the modulator. Capacitive coupling between each amplification stage is used to allow for proper direct current (DC) biasing of each transistor stage. The bias point of the transistors in each stage affects the RF input return loss, output return loss, and gain of the transistor. The transistor amplifiers are typically built using hybrid packaging techniques, where transistor die are mounted on a substrate, wire-bonded, and additional discrete components added for biasing and matching this type of driver. Transistor types used for cascaded amplifiers can be bipolar junction transistor (BJT) technology as well as field effect transistor (FET) topologies. Multiple stages are required to be cascaded to meet the requirement of amplifying multiplexer output voltage to the required modulator input voltage. The quantity of stages used is related to the required gain of the driver, the transistor type used, losses in the connections, matching networks, and the control of the amplifier. Figure 13.2 shows a typical cascaded transistor amplifier.

For the cascaded transistor approach, each transistor stage can be designed for fast rise and fall times using high $f_t$, smaller geometry devices for the initial stages, then larger geometry devices for the high voltage swing required at the final output stage. The higher output swing comes at the cost of lower rise time, but this can be simulated, designed, and tuned for the application and frequency of operation needed.

In addition to transistors, the traveling wave amplifier (TWA) IC can be used, and cascaded to amplify the signals. These stages are also coupled with capacitors to allow for proper biasing of the TWA

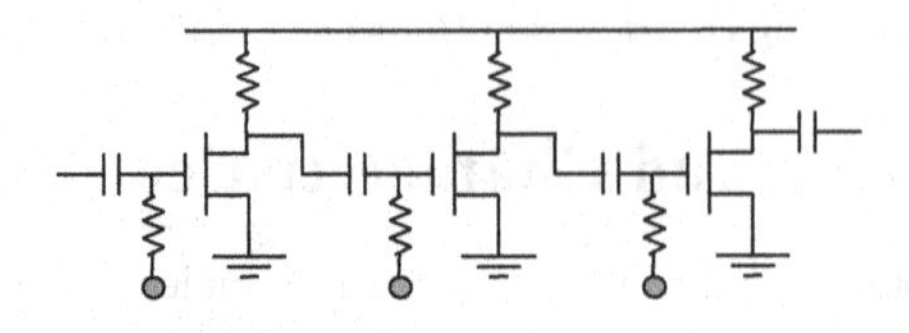

**FIGURE 13.2** Cascaded transistor amplifier.

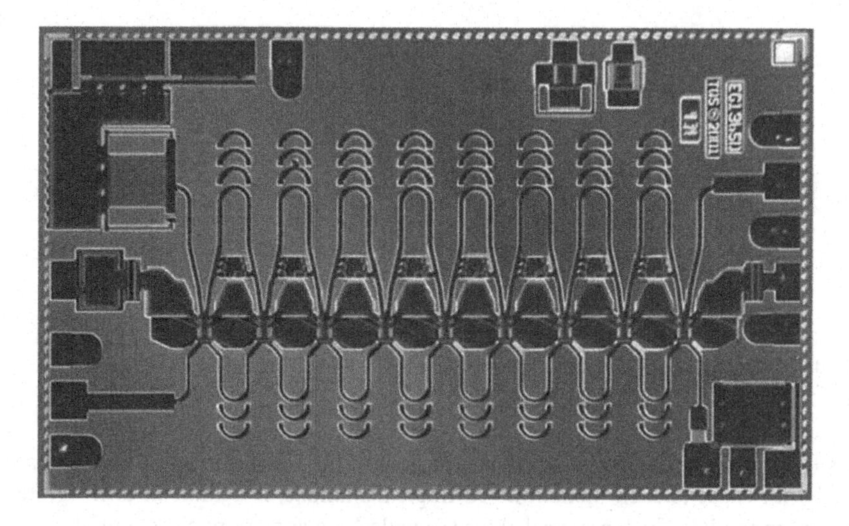

**FIGURE 13.3** GaAs TWA. (Reprinted with permission from TriQuint Corporation. © 2008 TriQuint Corporation.)

ICs. These monolithic devices are single-ended input/output and can be produced in large quantities in standard semiconductor manufacturing processes and packaged in either surface mount technologies or supplied in die form for use in hybrid packaging designs. These types of TWAs are available today from multiple vendors and foundries. Figure 13.3 shows a single gallium arsenide (GaAs) die form of a typical TWA used in both the RF industry and in the optical driver solutions available commercially, from TriQuint Semiconductor [1].

Depending on the amplitude requirements for the optical modulator, cascading single-ended TWAs may be required to bring the multiplexer voltages to suitable levels for the modulator. These devices, however, do require additional circuitry to bias and control, as well as the need for packaging and connecting the multiplexer to the modulator.

## 13.2.2 Differential Driver Topology

The differential driver allows for differential coupling of digital multiplexer outputs to either differential modulators or single-ended modulators. One of the advantages of these differential drivers is the use of differential transistor pairs and the ability to couple these to differential lines from the multiplexer. The use of differential lines in communications helps in preventing degradation of signal quality due to ground plane noise and/or coupled noise to these lines. Differential lines coupled to differential amplifiers reduce the potential of electromagnetic interference (EMI), have better power supply rejection, and allow for simpler control of amplitude and duty cycle of the output driver signal. Differential drivers are advantageous as they are suited for monolithic production technologies and can be produced at high volumes. They do require high breakdown voltage for the transistors used, as in the case of high-frequency processes; these breakdown voltages tend to go lower as speeds increase. Therefore, the differential monolithic driver will tend to limit the type of modulator that can be driven and will limit its application as a single IC. It can, however, be used to drive low $V_\pi$ modulators or as a preamplification stage in cascaded designs. Figure 13.4 shows a simplified differential transistor driver. Both BJT topologies as well FET topologies are shown.

These differential stages typically have matched input and output impedances and a current source for each stage. The current source is designed to limit the peak-to-peak swing to a predetermined value or can be adjusted to change the peak-to-peak swing. The input resistances are biased to a voltage

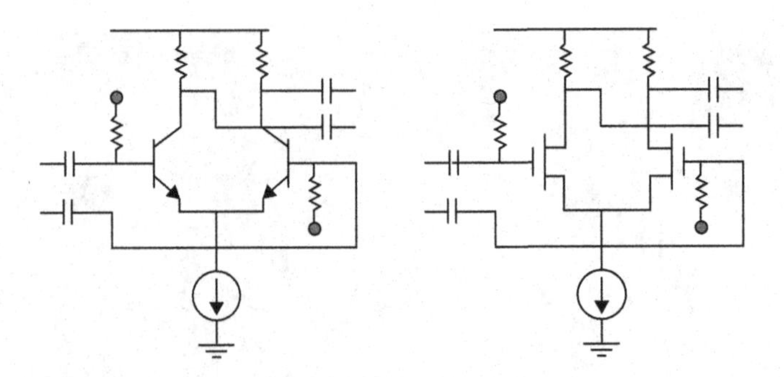

**FIGURE 13.4** Differential transistor amplifier topology.

that allows for optimal frequency response and/or gain of the transistors used. The stages shown are DC blocked with capacitors. In monolithic IC implementations, it is difficult to integrate high values capacitors on the IC, which are needed for the wide-frequency response required for digital telecom systems. A solution for broadband connection of interstage amplifiers is to use diodes (or base/emitter junctions of the transistors in the process used for the design) to DC couple the prior stage to the next stage and realize a higher gain driver. Figure 13.5 shows a bipolar cascaded DC-coupled topology.

The dotted line section shows a simplified monolithic circuit design with two stages of differential amplification and DC coupling between stages. External broadband capacitors would be used for interconnecting the amplifier/driver with the multiplexer and externally to a monolithic driver.

## 13.2.3 Linear Driver

Linear drivers have advantage of near duplication of the incoming waveforms and transmission to the modulator of interest. Various optical modulation techniques typically use standard nonreturn to zero (NRZ) electrical outputs and with proper modulator biasing create alternate modulation formats in the optical domain. A simple "on" (1) or "off" (0) signal of the multiplexer is amplified and applied to the modulator. An exception to the typical on/off digital driver output is the optical duo-binary modulation, where the driver amplifies the signal to twice the $V_\pi$ of the modulator (still 2 levels of amplitude) but is followed by a low pass filter whose 3 db bandwidth is on the order of the bit rate. Using a precoder, the resulting electrical domain signal has 3 discrete levels, with proper biasing of the modulator, results in a binary optical signal equivalent to the source electrical bit stream. Figure 13.6 shows the block diagram of the driver circuit with the associated stages and output filter.

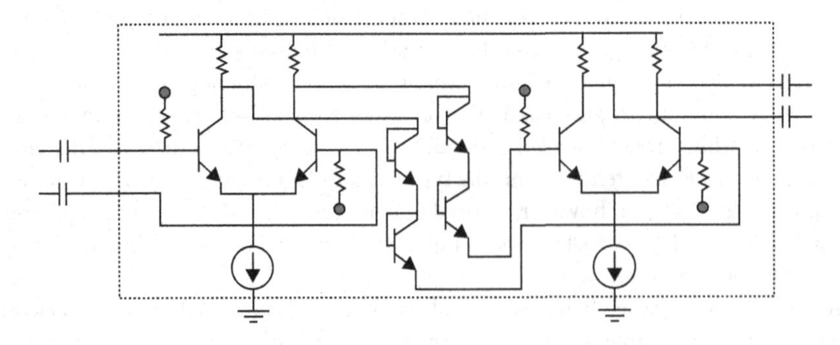

**FIGURE 13.5** DC-coupled differential multistage amplifier.

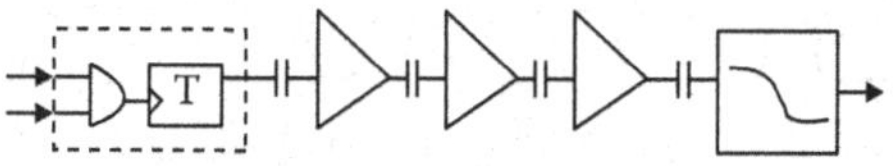

FIGURE 13.6 Duo-binary driver diagram.

The driver is designed to be linear in its output, allowing for proper adjustment of the amplitude to properly drive the modulator and generate the duo-binary waveform, and in this example uses the low pass filter to generate a three-level waveform. Figure 13.7 shows the NRZ data stream and resulting voltage waveform from the driver system that is used to drive the modulator.

The prior example shows an implementation of three-level electrical signals, also known as "PAM3" (pulse amplitude modulation) in the electrical domain. Using a linear driver will allow for alternate modulation formats, as an example, "PAM4" and 4 levels of drive voltage. The multilevel output in the amplitude domain can therefore be directly applied to a modulator and direct multilevel modulation in the optical domain. A method for such an encoding and the resulting waveforms are shown in Figure 13.8.

Multilevel transmission coding is one way of increasing the total bit rate throughput of the system, while reducing the required bandwidth of each component in the system, the digital multiplexer components, the driver electronics, and the optical modulator. The encoded multilevel signal can be driven into a linear driver and applied to the modulator of interest. Each level within a bit time would correspond to 2 bits of data, doubling the bit rate of the system. This does, however, pose challenges in the receiver/decoder and would not be tolerant of typical fiber nonlinearity and channel characteristics. However, for short distance and higher bit rate applications, the technique does enable higher throughput with existing technologies.

## 13.2.4 Limiting Driver

Limiting drivers take advantage of the fact that power supplies applied to the circuits are at a fixed voltage. Two techniques will be discussed: the first being the use of a differential amplifier stage and the second is the single-ended stage and associated biasing and supply voltages.

The results of the technique are the following:

1. Stable peak-to-peak voltage independent of run length (quantity of successive binary "1" values)
2. Stability of output voltage independent of gain variations of the output stage
3. Stability of the output independent of the supply voltage variations
4. Stability of the output voltage independent of input voltage variations
5. Ability to control duty cycle/crossing point of the output voltage waveform

In a differential pair transistor stage, typically, a current bias source is adjusted to a value that biases each of the transistors such that the resistive drop will support the peak-to-peak voltage before saturation. Assume the stages are biased by a current source at 120 mA current, then with no input voltage,

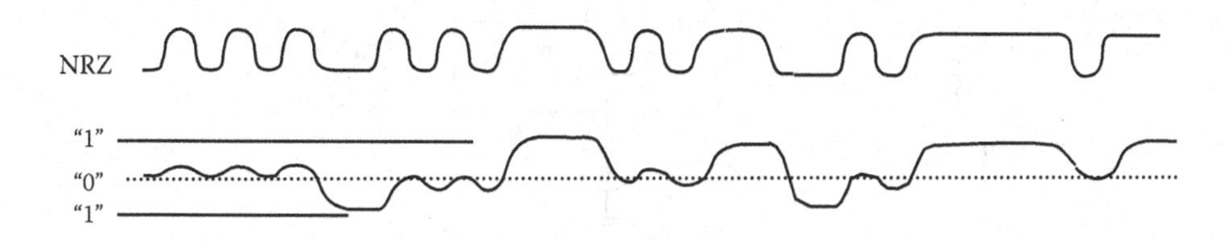


FIGURE 13.7 Input and output waveform of duo-binary driver.

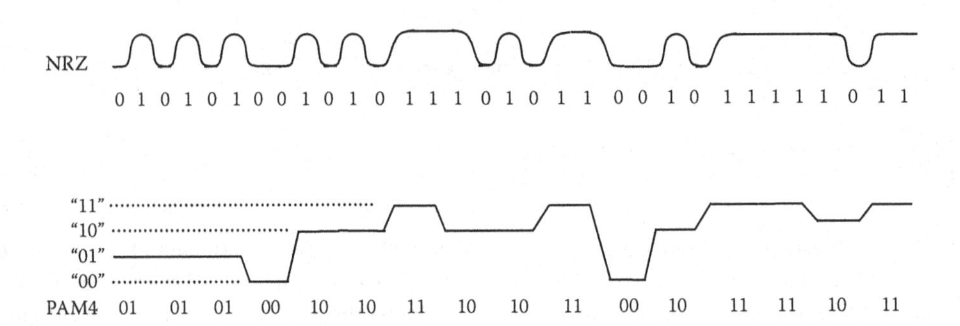


**FIGURE 13.8** Input and output waveform PAM4.

each stage will simply bias at ½ $I_{bias} \times R_{value}$, or 60 mA × 50 Ω = 300 mV DC drop. For a given supply (assuming 3.3 V for most digital systems), then the DC value of the output of this stage is 3.00 V. Assume a cascade of two differential stages, each with a voltage gain of 4, and an input of 400 mV peak-to-peak input. Differential stage 1 will result in a differential output voltage of 1.60 V peak to peak, or one side at 800 mV peak to peak. The current required for this peak-to-peak voltage swing is therefore 800 mV/50 Ω = 16 mA peak to peak. The resultant bias current at each resistor in the differential pair due to the input voltage is 52 mA to 68 mA and remains linear and unsaturated. At stage 2, the input peak-to-peak voltage of 1.6 V would result in a peak-to-peak output voltage of 6.4 V and a single-ended swing of 3.2 V; therefore, the peak-to-peak current is 3.2 V/50 Ω = 64 mA. The bias of the current source for each resistor of the differential pair, 60 mA, would not support a peak-to-peak output current swing of 64 mA at linear voltage gain of the stage. Stage 2 is therefore "limiting" based on the current source value chosen, and the output swing would remain a constant peak-to-peak voltage of 3.0 V. The limiting in this case is the current available from the current source (60 mA vs. the 64 mA required for linear gain). This voltage would remain constant for changes and input peak-to-peak voltage swing and also constant based on the 120-mA current source. The advantage of this will be discussed later for crossover/duty cycle control of the drive voltage into the optical modulator. Figure 13.9 shows an implementation of the cascaded differential saturated output stages and associated waveforms.

In a single-ended amplifier stage, either single transistors or TWAs can be used for a limiting stage. The technique relates to using the power supply of the amplifier and the associated gain with the driver stage. Single-ended limiting can be used to optimize the total driver design, for power dissipation, duty cycle control, and output peak-to-peak swing. Assume a single-ended transistor amplifier or TWA is biased at 50 mA, and the resistance used is 50 Ω for the stage, a DC drop of 50 mA × 50 Ω = 2.5 V results for this stage. From a typical DC supply for a digital system of 5 V, the resultant DC value of 2.5 V will

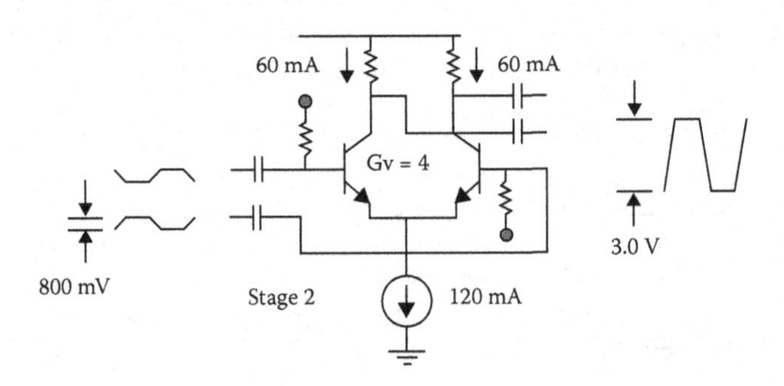


**FIGURE 13.9** Differential saturating bipolar amplifier.

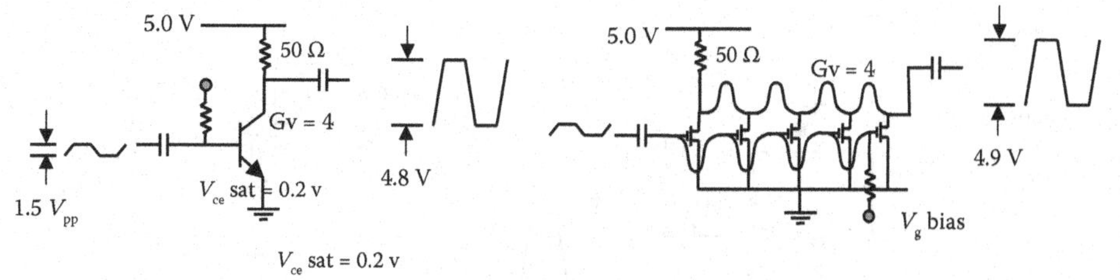

FIGURE 13.10 Single-ended saturating amplifiers.

appear on the output with no input voltage peak-to-peak applied voltage. Assuming the voltage gain of the TWA is 4, then at an input voltage peak to peak of 1.5 V and the resultant output voltage would be 6 V peak to peak. The power supply, however, would only allow a maximum swing of 0 V to 5 V DC, minus the $V_{ce}$ saturation of 0.2 V for the transistor then a peak-to-peak output voltage of 4.8 V from this stage. For the single-ended TWA stage in saturation, the peak-to-peak voltage would be 5.0 V minus the $V_{ds}$ saturation voltage of the transistors used in the traveling wave topology. In both cases, the stages are "limiting" based on the bias current, output resistance, and gain of the amplifier. Figure 13.10 shows an implementation of the single-ended saturated output stage and associated waveforms.

## 13.2.5 Hybrid Topologies

A mix of the various types of drivers, single-ended, differential, linear, or limiting can be assembled to meet the requirements for various types of modulators, voltage requirements, and bandwidths. These topologies each have their advantages and strengths, based on applications and requirements. First, by having a differential input, the output multiplexer can take advantage of coupled lines in transmission from the multiplexer outputs to the driver input. This improves power supply noise rejection and noise immunity, decreasing the likelihood of signal degradation. Additionally, using a single-ended output stage in saturation, the duty cycle can be adjusted and applied to the modulator, resulting in a balanced duty cycle in the optical domain. Figure 13.11 shows a differential input stage followed by a single-ended output stage.

Using this hybrid topology approach, the signal quality of the differential coupled lines from the multiplexer can be preserved and controlled by the differential stage. The single-ended stage, which is connected to one side of the differential pair, is amplified to a high peak-to-peak level and used to drive a single-ended modulator. Using this hybrid topology, the power dissipation is reduced but still can generate high peak-to-peak driving signals with duty cycle control and noise immunity.

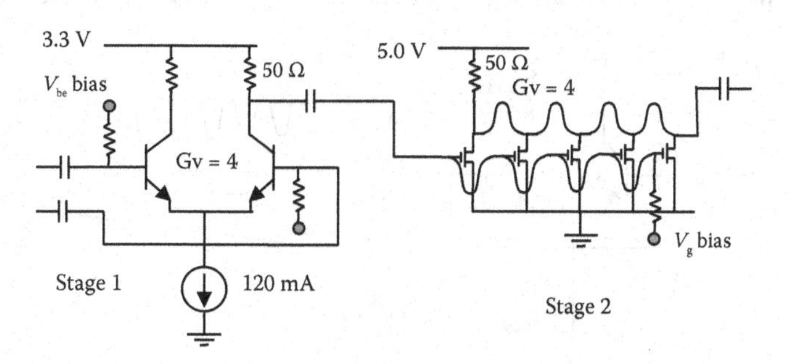


FIGURE 13.11 Hybrid amplifier stage.

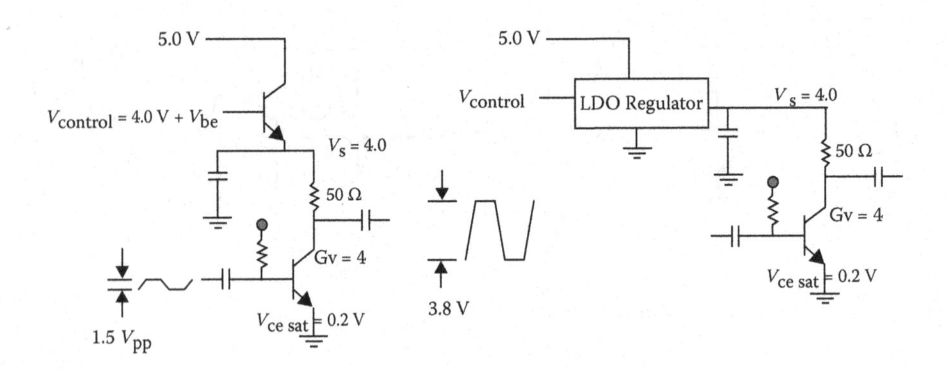


**FIGURE 13.12** Amplifier output voltage control.

## 13.3 Driver Output Voltage and Tuning Capability

The ability to both amplify multiplexer output voltages and control the characteristics of the amplified waveform into the modulator is critical to system performance. The adjustment of the amplified signal is important to match the modulator $V_\pi$ for keeping the extinction ratio to a maximum value for the modulator chosen. Adjusting duty cycle and/or crossover is also critical for systems and necessary for high quality optical signals with balanced duty cycles. The following sections will investigate the techniques and tradeoffs of controlling both amplitude of the signals and the output characteristics of the waveform generated for the modulator inputs.

### 13.3.1 Amplitude Control

Amplitude control can be accomplished in various ways within each type of topology. In the case of a single-ended saturating driver, the power supply voltage can be increased for the stage in saturation, and the resultant peak-to-peak voltage varied. This holds true for both transistor and traveling wave single-ended topologies. In typical digital systems, the power supplies available will be at some nominal voltage, with a tolerance above and below this typical value. Due to the variations in system power supplies, regulating the supply to a lower DC value and using the last stage in the driver in saturation will result in the output voltage tuned to a required value. Simply adding a bipolar transistor to the power supply can provide a voltage control method for each of the single-ended saturated drivers discussed previously. Additionally, the use of a low dropout regulator IC can also control the supply voltage to the saturated stage and result in stable control of the output peak-to-peak swing of the driver. Figure 13.12 shows how control of the supply voltage results in the change in peak-to-peak voltage for the single-ended driver.

For the differential driver stage, the amplitude can be controlled by adjustment of the current source when the gain of the stage is in saturation, as discussed in Section 13.2.4 and allows for adjustment of the peak-to-peak swing to optimally drive the modulator. In Figure 13.13, the current control is ramped

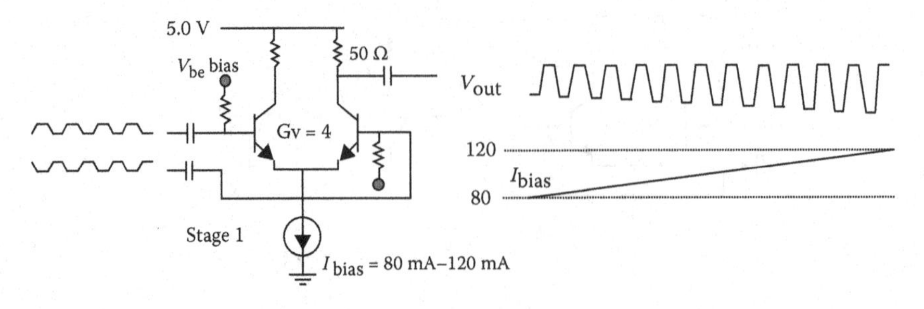


**FIGURE 13.13** Differential amplifier voltage control.

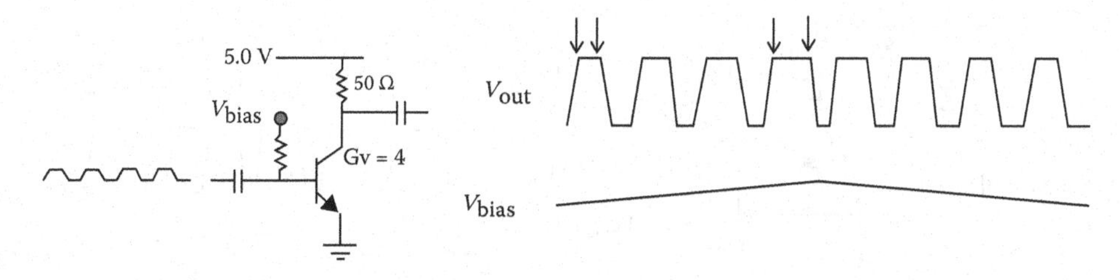


FIGURE 13.14 Amplifier pulse width control.

and shows how the peak-to-peak swing changes based on the control signal for the differential pair driver stage.

Amplitude control is required to optimize the optical performance and must be tailored for the modulator used in the design. The voltage control will typically be adjusted for the modulator $V_\pi$ and maximize the optical extinction ratio. The higher extinction ratio will minimize the bit error rate for longer lengths of fiber, allowing for a maximum peak-to-peak swing and decision point at the receiver, and will decrease the amplified spontaneous emission (ASE) buildup of the "0" level for a cascade of optical amplifiers found in long distance transmission links.

## 13.3.2 Crossover/Duty Cycle Control

The ability to control the duty cycle of the driver output is important for various reasons. For electro-absorption modulators, the crossover of the driver can correct the crossover offset of the modulator. In the case of lithium niobate modulators, similar corrections of the optical crossover point are required due to fabrication, and technology variations. Some techniques for controlling the crossover/duty cycle of the electrical driver will now been explained.

Using the technique of gain compression for a single-ended stage is one way of controlling the crossover point in the electrical waveform. This method also uses the fact that the driver has a finite bandwidth and, as a result, has an associated rise and fall time from the pulsed response of the driver output. Assuming we have the output stage of the driver in compression, then by applying a DC input offset signal to the bias of the stage, the output waveform can be controlled and the width of the pulse on the output of the driver stage. The technique is illustrated in Figure 13.14.

The increase in bias voltage causes an increase in bias current in transistor Q1, and the result is the 0 level is now a wider pulse. In the same way, a TWA stage can also be controlled when in compression and with finite output rise and fall times. Figure 13.15 shows the equivalent pulse width control for this type of driver stage.

In the differential pair, a similar effect can be accomplished by offsetting the bias voltage between the two inputs of the transistor bases in the differential pair. In this control method, the finite bandwidth

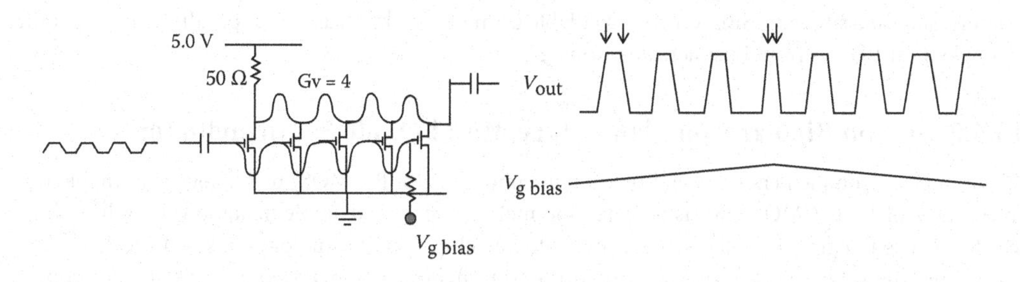


FIGURE 13.15 TWA pulse width control.

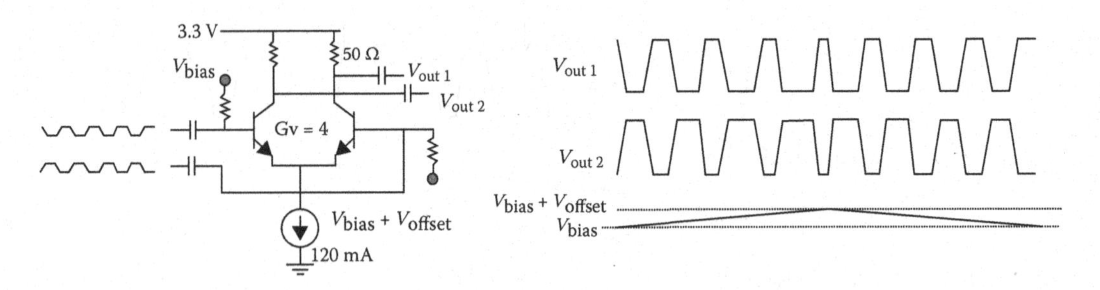


**FIGURE 13.16** Differential amplifier pulse width control.

and rise/fall time of the driver can be used to generate a variable pulse width output. In Figure 13.16, the differential pair outputs with offset bias controlled will result in the controlled variable width output, and complementary signals result.

The ability to control the pulse width/duty cycle of the electrical signal is required to optimize the optical signal for the modulator technology chosen in a given application. For a given modulator whose transfer function is not symmetric in the optical domain, the crossover point can be compensated with the electrical input crossover of the driver.

## 13.4 Major IC Technologies Used in Modulator Drivers

Various semiconductor technologies are available for the implementation of a driver system. RF technology helped in the development of wideband, high frequency transistors for discrete designs. As speeds of transistor technologies increased, the ability to monolithically integrate various topologies and functions have become available for solutions. The speed and breakdown voltage can be optimized for specific driver requirements. The following will list technologies available for designing driver systems.

### 13.4.1 Silicon Complementary Metal Oxide Semiconductor

Silicon complementary metal oxide semiconductor (Si CMOS) technology has been the leader in increasing of transistor density per unit area of the available silicon wafer. This scaling down of transistor size reduces the power dissipation as well as increasing the speed of the transistors. As speeds increase and transistor size decrease, the breakdown voltages decrease as well as the supply voltages for the ICs. Although optimal for microprocessors and digital logic and used widely in the telecommunication and data communication areas for the logic multiplexer and demultiplexer circuits, the scaling for higher-frequency and bit rate applications are counter to the requirement for high-voltage, high-bandwidth driver requirements. The technology, however, is suitable for lower voltage drive conditions and has demonstrated use in driving lower voltage vertical cavity surface emitting laser (VCSEL) and distributed feedback lasers (DFB) lasers in a single IC form. These type of circuits are available from multiple vendors, such as a single VCSEL IC driver available from Phyworks [2] as well as parallel drive capability in single IC at 10 Gb/s from Ensphere Solutions [3].

### 13.4.2 Silicon Bipolar Complementary Metal Oxide Semiconductor

Silicon bipolar complementary metal oxide semiconductor (Si BiCMOS) technology has enabled the integration of both CMOS transistor gates for monolithic digital implementation as well as bipolar transistors integrated on the same semiconductor chip. This technology makes it ideal in building monolithic solutions for driver ICs, incorporating hybrid topologies on a single silicon chip. As in the standard CMOS technology, scaling has occurred in BiCMOS technology, higher frequency of operation by decreasing transistor geometry, and decreasing supply voltages, which decrease total IC power. Again, the technology has developed and evolved counter to the requirements of high-voltage, high-frequency drivers and speeds needed for 10 Gb/s and 40-Gb/s transmission. Scaling of the technology to smaller geometry devices results in higher speed switching, but at lower voltage swings and lower breakdown voltages.

### 13.4.3 Silicon Germanium Bipolar Complementary Metal Oxide Semiconductor

Silicon germanium bipolar complementary metal oxide semiconductor (SiGe BiCMOS) technology was a result of research in the epitaxial growth layers of silicon transistor technology. Instead of scaling transistors to smaller geometries, the use of germanium in the epitaxial layer of the transistors increased the mobility and switching speed of the transistors fabricated with the technology [4]. The technology allows standard high-volume and low-cost manufacturing of volume ICs with the added benefit of higher speed. The higher speed of the transistors allows larger geometry devices to be designed into the monolithic IC and therefore can be used for higher voltage driver designs. A differential topology with 3.8-V output swing and 10-Gb/s data rates has been demonstrated using this technology [5]. The breakdown voltage is still a limiting factor in developing high voltage single-ended designs, but high-bandwidth, moderate-drive voltages make this a good choice for drivers.

### 13.4.4 Gallium Arsenide Heterojunction Bipolar Transistor

GaAs as a substrate has been widely used since the late 1960s for microwave amplification with discrete devices developed such as metal semiconductor field effect transistors (MESFET), Gunn, and impact ionization avalanche transit time (IMPATT) diodes [6]. GaAs heterojunction bipolar transistor technology (HBT) can be used in the design of single-ended and differential designs for drivers and have high $f_t$ transistor frequency response with moderate breakdown voltages. GaAs HBT have over twice the breakdown voltage of SiGe and is therefore advantageous in designing the various topologies in high speed and voltage ICs [7]. GaAs HBT can be used in implementing high-power broadband TWAs, realizing gains of 8 db and 3-db bandwidths of 26 GHz [8]. The TWAs can be cascaded or built as a differential + TWA topology in a practical design. This technology offers a sweet spot for implementing hybrid topologies monolithically, achieving high speed and moderate voltages.

### 13.4.5 Gallium Arsenide Pseudomorphic High Electron Mobility Transistor

Gallium arsenide pseudomorphic high-electron mobility transistor (PHEMPT) technology has been the key semiconductor choice for RF, broadband high power applications. The high breakdown voltage (>10 V), high $f_t$ (>90 GHz) and high-power transistors provide the most prevalent technology used for highest-speed and high-voltage drivers. Successful ICs have been designed using differential pairs and capable of driving 40 Gb/s at 2.9 V [9]. A single TWA IC is capable of 9 db gain out to bandwidths up to 70 GHz [10]. When optimized for speed and power, the TWA built with PHEMTs can obtain flat, broadband gain response (30 kHz to 40 GHZ) with digital output voltages in the 7-V range at 40-Gb/s rates [11]. The high-volume monolithic fabrication is also key in its prevalence in the driver market allowing for optimal tradeoff of costs and technology.

### 13.4.6 Indium Phosphide Heterojunction Bipolar Transistors

Indium phosphide heterojunction bipolar transistors (InP HBT) technology has some of the highest recorded speeds for transistors ever fabricated, with $f_t$ reported in excess of 600 GHz [12] InP HBT technology has lower breakdown voltage than GaAs PHEMT and cannot generate some of the higher voltages required to drive the higher speed modulator technologies. This aspect as well has higher fabrication costs make this less attractive for the volume high-speed driver market.

## 13.5 Driver RF Broadband Phase and Amplitude Response

The driver system must have broadband high-quality phase and amplitude response to have minimal degradation on the digital input cover the modulation bandwidths and run lengths found in the data communication and telecommunication markets. For 40-Gb/s communications, a 30-kHz to 40-GHz frequency range creates some challenges in preserving high-quality digital signals. This requires careful design techniques, not only in the technology chosen for the semiconductor portion, but also in the packaging and connection techniques from driver to the multiplexer and driver to modulator interface. The broadband RF performance requirement is needed to ensure clean low jitter transitions for the high speed digital data patterns. Any mismatches in characteristic impedance from transitions of the multiplexer to circuit board, circuit board to driver, and then driver to modulator will contribute to degradation of the signals. These design techniques will now be discussed for drivers.

### 13.5.1 Scattering Parameters

When designing a driver system, an understanding of transmission lines, two port networks and impedance matching techniques is required to optimize the various stages and interconnections in the system. Scattering (*S*) parameters are used to characterize two port networks and indicate gain and phase of input and output reflected power, forward transfer gain, and reverse isolation of the amplifier [13]. By utilizing *S* parameter measurements available on network analyzers, characterization and optimization of the driver can be done during the design process. The characteristic impedance of connections and active elements in the design can be analyzed and optimized. Figure 13.17 is an example of measured *S* parameters for a typical amplifier showing the input match (S11) output match (S22) and the gain (S21) of the part.

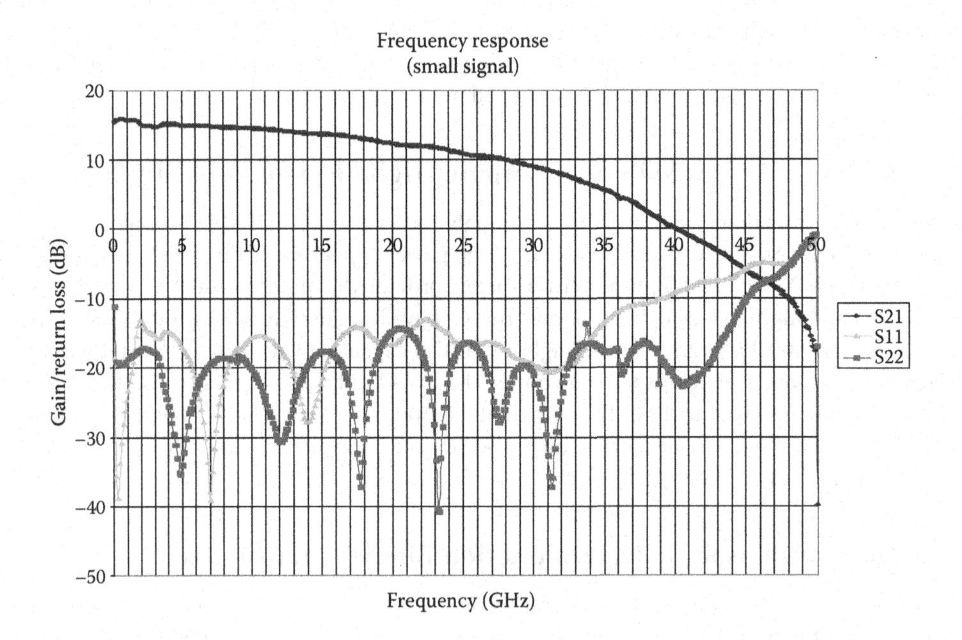


**FIGURE 13.17** Typical amplifier *S* parameters.

Using the network analyzer to characterize each component in the driver line up, the composite response of the driver system can be optimized for low jitter operation. One of the most important aspects for building the driver is to have broadband and flat frequency response. Gain variation of even 1 dB in the S21 frequency response can show up as pattern-dependent amplitude jitter in the time domain. Inter stage and coupling mismatch also introduces distortion of the digital patterns. Interconnections of transistor and IC amplifiers are critical, and care must be taken in designing these transitions.

### 13.5.2 RF Strip Line, Micro Strip, and Coplanar Lines

In both IC and circuit board layouts, RF connections are used to couple the active and passive elements together, and various types of transmission lines are used. Figure 13.18 shows the cross-section geometry of four types of interconnections.

The strip line connection is most often used in circuit board interconnections of high speed lines. They have an advantage in that electromagnetic emissions are minimal due to top and bottom ground planes and prevent cross talk from other signals within the circuit design. A disadvantage for strip line transmission lines is difficulty in controlling the transition from the internal trace to the top side of the circuit, which leads to signal degradation.

The micro strip line is one of the most common interconnects used for inter stage amplifier connections as well as circuit board to driver from the multiplexer. A single transmission line is used for connections on single-ended amplifiers to the driver. The coplanar line is often used in circuit boards for connections of multiplexer to connector or driver input as well as within an IC for transistor or TWA interconnections. Coplanar lines can be designed with or without a ground plane on the bottom side. A nongrounded coplanar line must be suspended and therefore is less prevalent and used mostly for suspended substrate hybrid designs, where the circuit is mounted internally with a space under the substrate. The grounded coplanar line allows for mounting of the IC or substrate to a metal housing. The coplanar line has the advantage of ground connections on the top side, which allows wire bond connections for IC to transmission lines for both ground and signal.

### 13.5.3 Connectors and Mismatches

One of the challenges in building optical systems with different technologies is interconnecting the parts with different technologies and associated packages together in a system. Transitioning from an IC to a package lead or package to a circuit board can cause impedance mismatch as shown by the network analyzer *S* parameters and show up as jitter on an oscilloscope. Connectors are often used for package coupling between PCB multiplexer and driver as well as the driver to modulator. There are various connectors used in the industry and each has their own advantages. The inner diameter dimension is often used as a description of the connector.

The subminiature version A (SMA) is one of the most prevalent connector types and has been widely used in the RF industry for interconnections. The usable bandwidth is in the DC to 18-GHz range, it is

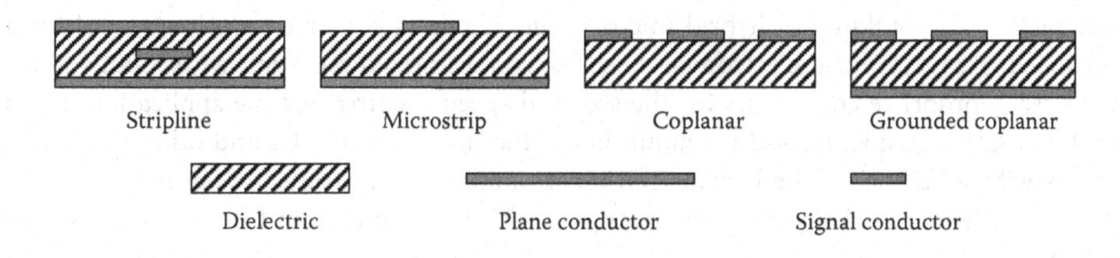


**FIGURE 13.18** Four types of transmission lines.

moderate in size, and it is being used for many hybrid module to circuit board interconnections in the 10-Gb/s market. The connection requires threaded male and female versions for connections and has 3.5 mm outer diameter, with a Teflon dielectric. Because of its larger size and lower bandwidth, alternatives were needed to reduce size and increase the usable frequency.

The subminiature version B (SMB) was developed as a push–pull version of the SMA, requiring no threading for connection. The parts simply require a push–pull mating, but the bandwidth for this type are only usable to about 4 GHz, below the requirements of even the 10-Gb/s market. Both 2.6 mm and 2 mm outer diameter are available in the SMB connector.

In addition to the SMA and SMB versions, precision types were developed that use air as the dielectric to improve the transition and impedance between the male and female halves of the connectors. The 3.5-, 2.9-, 2.4-, and 1.85-mm versions are available and cover the frequency range from 18 GHz to 65 GHz. These all require threading to make or break connection between the cables and drivers or modulators.

As a result of the market requirement for a smaller, higher-frequency connection system, Corning Gilbert [14] developed the GPO® series of connectors, with a smaller size and push–pull connection technique. The GPO connector is usable from DC to 26 GHz. Its small size (circular diameter of 0.120 inch) makes it usable for the 10-gigabit driver and modulator market. The GPPO series with a diameter of 0.092 inch extends the frequency to 40 GHz and can be used for applications to 40 gigabits. Finally, the G3PO series (0.085-inch diameter) enable connections to 60 GHz.

## 13.6 Packaging Technologies

The packaging technology used for driver designs has evolved and changed based on various factors. RF hybrid packaging has been available for many years in the point-to-point radio and antennae markets as well as the satellite industry. Surface mount technology (SMT) has enabled high-volume, low-cost manufacture and has evolved by careful design methods for high-frequency operation. The packaging of the driver influences both the transistor and IC technology used in the design and the performance that can be obtained.

### 13.6.1 Historical Advancement in Packaging Technologies

For high-speed electronics, the engineer is faced with choosing various technologies brought together in a module. These modules, using a mix of discrete and monolithic electronic components, are known as "hybrids." Hybrid packaging has been used in both space applications and RF amplifiers due to its flexibility to incorporate mixed technologies and simple connector inputs and outputs. Hybrids will sustain a market to the point where volume would warrant a single IC to be developed for the intended gain, power, and application. The integration step leads to alternate packaging for smaller size, typically eliminating connectors. The conversion from a hybrid driver to surface mount occurred in the 2.5-Gb/s driver market and repeated in the 10-Gb/s market.

### 13.6.2 Hybrid Packaging

The construction and building of hybrid microcircuits is one of the first techniques used for building high speed circuits. Hybrid electronics consist of connectorized modules, with input and outputs designed with appropriate connectors for the size and speed required for the application. Internally, a substrate, usually ceramic, is used for mounting of the transistors or ICs and other associated passive devices with wire bonds. The hybrids can incorporate different transmission lines for the interconnections of the input to output signals. A typical 40-Gb/s hybrid from Centellax [15] is shown in Figure 13.19.

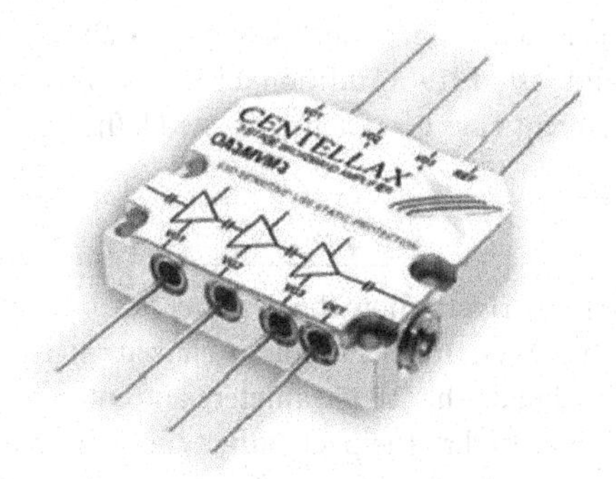


**FIGURE 13.19** Hybrid driver. (Reprinted with permission from Centellax. © 2010 Centellax.)

### 13.6.3 Surface Mount Packaging

Surface mount technology has allowed for high-volume manufacturing of semiconductors. A surface mount package is constructed with copper, copper alloys, or ferrous composite alloys [16]. The lead frames are manufactured in long strips by either chemical etching or metal stamping. These long lead frames allow for automated manufacturing and are contained in high-volume IC manufacturing lines, where IC die mounting, wire bonding, and transfer molding take place. The transfer molding process is an epoxy resin that encapsulates the lead frame and IC and is then cut from the lead frame to individual packaged ICs for use in surface mount designs.

In addition to lead frame transfer molded packages, a similar surface mount package can be built with ceramic carrier and ceramic cover. Although higher in cost, this type of packaging keeps the internal IC in air as compared to the higher dielectric constant of transfer molded resin. The air enables higher bandwidth of operation, eliminating the higher capacitance of the transfer molded type of design. Figure 13.20 shows several views of typical surface mount IC in a leadless package configuration from GigOptix [17].

## 13.7 Paths to 100-Gb/s Transmission

As speeds are pushed from 40 gigabits to 100 gigabits, a serial NRZ 100-Gb/s driver in a surface mount package would create many challenges in both packaging and IC fabrication technology at these speeds. A direct 100-gigabit driver would require the latest in connector technology and would most likely be built in a hybrid package design. Direct serial 100-Gb/s multiplexer and demultiplexers have been realized with the SiGe process, but the drivers for this serial rate are not yet available [18]. Optical serial rates

**FIGURE 13.20** Surface mount drivers. (Reprinted with permission from Gigoptix. © 2010 Gigoptix.)

of 107 Gb/s NRZ differential quadrature phase shift keying (DQPSK) with two serial 53.5 Gb/s serial data streams have been built and tested [19]. Additional 53.5-Gb/s serial rates in a NRZ vestigial side band (VSB) transmission experiment have been demonstrated [20].

## 13.8 Conclusions

Various topologies used in building drivers allow the designer flexibility in optimizing the design based on the type of modulator used in the system. The topology chosen can be used to control the voltage and duty cycle to optimize the drive signal. The RF parameters and interconnections influence the type of technology chosen for the design as well as the packaging type. Volume requirements and cost targets for the market have resulted in the transition of hybrid connectorized driver designs to surface mount in the 2.5 Gb/s and 10-Gb/s range. The trend for smaller and lower-cost drivers will continue even at 40-Gb/s rates as volumes continue to increase.

## References


1. TriQuint Semiconductor, Product Data Sheet, DC-18GHz, TGA1328-SCC.
2. Phyworks Corporation, Product Brief, PHY1075.
3. Ensphere Solutions, ESI 1014 Data Sheet, 4-Port 10Gbit/s Laser Diode Driver.
4. D. Harame and B. Meyerson, "The early history of IBM's SiGe mixed signal technology," *IEEE Trans. Electronic Devices*, vol. 48, pp. 2555–2565, Nov. 2001.
5. S. Mandegaran and A. Hajimiri, "A breakdown voltage doubler for high voltage swing drivers," *Proc. IEEE 2004 Custom Integrated Circuits Conf.*, Orlando, FL, 2004, p. 106.
6. S. Long and S. Butner, "Fundamentals of GaAs Devices," in *Gallium Arsenide Digital Integrated Circuit Design*, 1st ed., New York: McGraw-Hill, 1990, p. 11.
7. M. Wilson, "GaAs and SiGeC BiCMOS cost comparison—is SiGeC always cheaper?" in *2003 Int. Conf. Compound Semiconductor Mfg.*, Scottsdale, AZ, 2003.
8. C. Meliani, M. Rudolph, and W. Heinrich, "On-chip GaAs-HBT broadband-coupled high-bitrate modulator driver TWAs," in *European Gallium Arsenide Other Semiconductor Application Symp., 2005*, Paris, pp. 661–664.
9. Z. Lao, A. Thiede, U. Nowotny, H. Lienhart, V. Hurm, M. Schlechtweg, J. Hornung, et al., "40 Gb/s high-power modulator driver IC for lightwave communication systems," *IEEE J. Solid-State Circuits*, vol. 33, no. 10, pp. 1520–1526, Oct. 1998.
10. M. Sato, T. Hirose, and Y. Watanabe, "A 70-GHz bandwidth and 9-dB travelling wave amplifier using 0.15-μm gate InGaP/InGaAs HEMPTS with coplanar transmission line technology," *IEICE Technical Report, Electron Devices*, 100(549), pp. 9–14, 2001.
11. Avago Technologies, *AMMC-5024, 30KHz-40 GHz TWA Operational Guide*, Application Note 5359, AV02-0704EN, Oct. 10, 2007.
12. L. W. Hafez and M. Feng, "Experimental demonstration of pseudomorphic heterojunction bipolar transistors with cutoff frequencies above 600 GHz," *Appl. Phys. Lett.*, vol. 86, no. 15, paper 152101, Apr. 2005.
13. R. Carson, "Transistor parameters and stability," in *High-Frequency Amplifiers*, 2nd ed., New York: John Wiley and Sons, 1982.
14. Corning Gilbert Inc. Microwave Technical Library, Microwave Catalog [Online], available: http://www.corning.com/gilbert/microwave_products/technical_library/index.aspx.
15. Centallax Coproration, Modulator Drivers, OA4MVM3 smd-00049 rev. D.
16. *Standard Specification for Integrated Circuit Lead Frame Material*, ASTM Standard F375-89, 2005.
17. GigOptix, GX6120 and GX6155 data sheets [Online], available: http://products.gigoptix.com.
18. P. Winzer and G. Raybon, "100G Ethernet—a review of serial transport options," in *IEEE/LEOS Summer Topical Meetings*, Portland, OR, 2007, paper TuE4.1, pp. 7–8.

19. G. Raybon, P. J. Winzer, A. H. Gnauck, A. Adamiecki, D. A. Fishman, N. M. Denkin, Y.-H. Kao, et al., "107-Gb/s transmission over 700 km and one intermediate ROADM using LambdaXtreme® transport system," in *Conf. Optical Fiber Communications*, San Diego, CA, 2008, pp. 1–3.
20. K. Schuh, E. Lach, B. Junginger, and B. Franz, "53.5 Gbit/s NRZ-VSB modulation applying a single Mach–Zehnder modulator and transmission over 21 km SSMF with electronic dispersion compensation," in *Conf. Optical Fiber Communications*, San Diego, CA, 2009, pp. 1–3.

# 14

# Reliability of Lithium Niobate Modulators




Hirotoshi Nagata
*JDSU*


## 14.1 Introduction

All devices deployed in optical communication systems are required to operate reliably for a minimum of 20 years under field conditions. The reliability of lithium niobate (LN) modulators is primarily linked to device performance under temperature changes and under the applied DC bias voltage. In this chapter, we describe the underlying physics and material science associated with reliable performance.

## 14.2 The Need for DC Biasing

In theory, a symmetric Mach–Zehnder optical interferometer (MZ modulator) will have identical phase delays in each arm and will constructively interfere, producing a maximum intensity at the output (see Figure 14.1). In general application, a DC bias is applied to the device such that with no applied modulation (AC) signal, the device is shifted to a quadrature point. This quadrature point is sometime called as an operation point and is chosen on either negative or positive slope of the modulation curve depending on required chirp in the prechirped modulator. The AC modulation signal is applied to drive the output between minimum and maximum output powers. For proper control, the MZ modulator must be maintained at the quadrature point throughout the device operation. In practice, wafer processing variations result in asymmetric phase delays in the two arms. Thus, a DC bias must be applied to compensate for processing variations.

In addition to process variations, modulator designs may also introduce phase asymmetry. Ideally, the modulator is manufactured with identical strain or stress in each of its MZ waveguide arms. In this case, any temperature dependent shift of the operation point is very small. However, design requirements for high-speed modulators and the manufacturing assembly process inevitably introduce a large strain (or stress) asymmetry to MZ waveguides. The magnitude of the generated mechanical asymmetries vary with temperature changes even within the normal operation temperature range of 0°C–70°C.

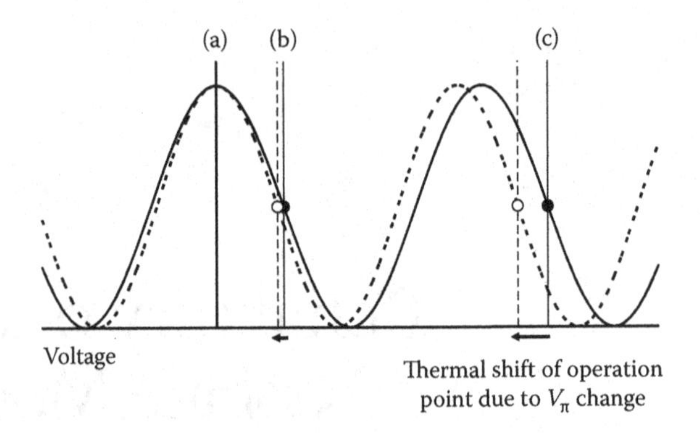


**FIGURE 14.1** Schematic illustration showing the impact of phase retardation and $V_\pi$ change on shifts of the operating point. Operating points on the negative slope of the modulation curve are shown by black or white circles. (a) Ideally prepared symmetric MZ LN modulator outputs a maximum optical intensity at 0 V, and this point is not affected by $V_\pi$ shrinkage or expansion. (b) Example of the DC bias setting at the closest operating point of the ideally prepared symmetric LN modulator; DC bias = $V_\pi/2$. $V_\pi$ change causes a slight shift of the operating point. (c) If the modulation output includes $2\pi$ retardation due to strain asymmetry of the MZ waveguides, the closest operating point is set at $2V_\pi$ away from (b) although the actual applied DC bias is $V_\pi/2$. In this case the same change in $V_\pi$ would cause a larger shift of the operating point than (b).

In *z*-cut LN modulator, a pyroelectric effect along the *z*-axis of the $LiNbO_3$ crystal generates another temperature dependent fluctuation of the operation point.

From the above reasons, DC biasing, and its feedback control, is a mandatory operation scheme in nearly all high-speed LN modulators. In the following sections, the underlying mechanisms will be described and current practice and performance will be illustrated. Section 14.4 contains a brief description of a 2.5-Gb/s LN modulator in which special design and manufacturing technologies are applied to minimize mechanical strains in the MZ waveguides to allow for bias-free operation.

## 14.2.1 Operation Point Shift Due to Mechanical Stress/Strain

Congruent (i.e., nonstoichiometric) LN single-crystal wafers have an *R3c* space system and exhibit ferroelectric properties [1]. As described in Chapter 6, LN is the material of choice for optical modulators because it has a large electro-optic coefficient. Hence, an applied electrical field can efficiently change dielectric constants, i.e., optical indexes, of the crystal through a distortion of ionic field in the LN crystal. Besides having a large electro-optic coefficient, LN also has large piezo-optic coefficient. Thus, application of mechanical stress/strain also distorts the location of ions and ionic field and changes the optical indexes. This latter effect or the associated piezo-electric effect of LN crystal expands the application area of LN to surface acoustic wave (SAW) devices, but at a same time, this phenomenon negatively affects stability and controllability of LN optical modulators.

Typically, the AC modulation voltage (half wave voltage, or $V_\pi$) of 10-Gb/s LN MZs is required to be smaller than 7 V to be compatible with commercially available high-speed RF drivers. Higher-speed modulators require even lower voltages. To meet this voltage requirement, MZ waveguides are designed to be 40 mm or longer due to electro-optic coefficient of LN and the typical overlap of the RF field between the electrodes and optical waveguides. As a result, LN modulator die have a very high dimensional aspect ratio—the die length is 10–20 times larger than the width. This is a fundamental reason that mechanical strains are inevitably introduced into MZ waveguides through the various process and assembly steps of the LN modulator.

A typical wafer fabrication process for LN modulator includes (1) Ti in-diffused waveguide formation at ~1000°C, (2) ~1-μm-thick dielectric (e.g., silicon oxide) buffer layer deposition and succeeding oxidation annealing at ~600°C, (3) charge dissipation layer deposition (e.g., amorphous Si; see Section

14.2.2), and (4) >10-μm-thick electrode plating. All of these process steps are sources of asymmetric strain generation in MZ waveguides. Ti ions substitute at cation sites of LN crystal resulting in wafer warping through volume expansion of the wafer surface. Deposition of the buffer and charge dissipation layers onto whole surface of the wafer causes additional wafer warping due to the large thermal expansion mismatch between LN ($1.5 \times 10^{-5}$) and the upper layers ($5.0 \times 10^{-7}$ for $SiO_2$). The mechanical strains generated during wafer fabrication are transferred to the LN modulator die as mechanical distortion in the MZ interferometer, and they generate phase retardation of the optical output through interaction with the piezo-optic property of LN. In the case of *z*-cut LN modulator die, the generation of mechanical strain in the waveguides on the order of $1 \times 10^{-5}$ can cause optical phase retardation by $2\pi$ or more [2]. This means that process induced mechanical deformation would cause optical phase shift comparable to application of $>2V_\pi$ voltages.

Figure 14.1 illustrates the influence of the strain induced phase retardation to thermal shift of the operation point. Because the optical index is a function of temperature and wavelength, $V_\pi$ of LN modulator also changes with temperature and wavelength (smaller $V_\pi$ at higher temperature or shorter wavelength). Intuitively, the temperature dependent change of $V_\pi$ induces a relatively larger thermal shift of the operation point in a modulator having larger internal mechanical strain. However, this is not the main contributor to thermal instability of LN modulator optical output.

## 14.2.2 Operation Point Shift Due to Temperature Change

There are two other contributors that cause a thermally dependent shift of the operation point of an LN modulator. First is the thermally dependent change of mechanical strains in MZ waveguides and second is the pyroelectric effect of LN crystal, which is another intrinsic property of a ferroelectric crystal [1].

The thermal expansion mismatch between the LN waveguide substrate and other constituent materials is the source of thermally dependent mechanical strains. The bonding of a long LN modulator die in a metal package, which is usually done by using electrically conductive adhesive material, can cause thermal strains through the cure process of the adhesive material at elevated temperature and through the thermal expansion mismatch between package material and LN crystal (thermal expansion coefficient $\alpha = 1.54 \times 10^{-5}$/K perpendicular to the *z*-axis). However, when using a stainless steel ($\alpha = 1.60 \times 10^{-5}$) package to minimize the thermal expansion mismatch, one can ignore the influence of packaging on the thermal shift of the operation point. Isolating the die from the package by suspending it in space with two gold wires wire-bonded to electrode pads was demonstrated to see impact of the packaging process, in which the free-standing LN modulator die still showed a similar magnitude of thermal shift of the operation point with that of the packaged modulator [3].

Correlation studies between mechanical strains induced at each of the wafer process steps and the magnitude of thermal shift of LN modulator optical output suggests that the electrode formation process is the largest contributor of fabrication induced strains [4]. To achieve high-speed RF modulation exceeding 10 Gb/s, coplanar Au electrodes with a thickness >10 μm are deposited on the MZ modulator surface as schematically shown in Figure 14.2. In the typical design of *z*-cut LN modulator, a narrow center electrode is placed over one of the MZ waveguides, while wide planar grounds are placed at both sides of the center electrode. The ground electrode extends to cover the other MZ waveguide and become an origin of large and temperature-dependent asymmetry of mechanical strain between the MZ waveguide pair. Therefore, minimization of mechanical asymmetry between center and ground electrodes while maintaining RF performance is the key to the electrode design for high-speed LN modulators. A symmetric *x*-cut LN modulator design, which outputs zero-chirp modulation, introduces much less thermal shift than *z*-cut LN modulator. Introduction of intentional asymmetry in the electrode geometry or design on an *x*-cut LN waveguides causes large thermal shifts similar to *z*-cut performance [5].

In addition to the effects described above, pyroelectric charges can influence the thermal stability for an LN modulator. In fact, they can be the largest source of instability and they present a significant technical challenge. LN crystals have a pyroelectric property in which the magnitude of the spontaneous

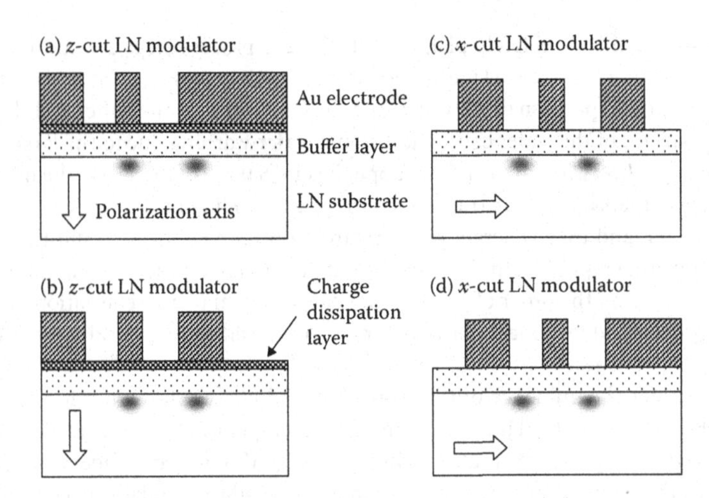


**FIGURE 14.2** Schematic illustration of the cross section of (a and b) *z*-cut LN modulator and (c and d) *x*-cut LN modulator die. Optical waveguides for *z*-cut LN modulator are usually prepared on the –*z*-face of the LN substrate and covered by dielectric buffer layer and charge dissipation layer. The *x*-cut LN modulator does not need the charge dissipation layer because the spontaneous polarization axis is parallel to the modulator surface and pyroelectric charges do not appear on the surface. Mechanical strain asymmetry of MZ waveguides pair is largest in design (a) and smallest in design (c). In *z*-cut LN modulator, design (b) could decrease mechanical asymmetry and reduce the thermal shift of the operating point relative to design (a). Even in *x*-cut LN modulator, introduction of mechanical asymmetry as in design (d) deteriorates the thermal shift of the operating point.

polarization along the *z*-axis (*c*-axis in crystallographic notation) changes with temperature variation [1]. The change of the spontaneous polarization is translated to a change of the corresponding electrical field and it generates electrical charges. Because of the electrically insulating nature of LN at relatively low temperatures (device operation temperature), those charges are accumulated on the *z*-cut faces of the LN modulator die and behave as an uncontrolled DC bias source. Under gradual temperature change, the accumulated pyroelectric charges can give an operation point shift much larger than that caused by mechanical strain fluctuation. The magnitude of the thermal shift due to pyroelectric charges is hard to estimate, but it tends to become larger in modulator designs with larger exposed area of *z*-cut surface or closer distance between exposed *z*-cut surface and a center electrode. A faster change of the ambient temperature causes apparently larger shifts of the operation point due to less time for charge relaxation. If the crystal has a sufficiently high electric conductivity, the generated pyroelectric charges are immediately neutralized. For SAW device application, so-called black LN wafers are commercially available. These are designed to have >1000 times larger DC conductivity than normal congruent LN wafer by chemical reduction technique [6]. However, for optical modulator applications, this technique cannot be used because the increased electrical conductivity leads an increase of optical absorption and increases optical propagation loss along the waveguides.

Minimization of the pyroelectric charge effect had been a key research subject for realization of stable LN modulators. By proper processing, stable *z*-cut LN modulators have been achieved with a reliability estimation of over 15 years of operation life [7]. This modulator has a thin semiconductive layer deposited on top of the buffer layer. The additional layer acts as a continuous surface electrode to dissipate surface charges and to remove the optical output instability due to pyroelectric charges. For process simplicity, the charge dissipation layer is formed by depositing an amorphous and high-resistivity film of Si or doped Si on the buffer layer by a sputtering or vacuum evaporation deposition technique. The fundamental structure of a *z*-cut LN modulator with the charge dissipation layer is illustrated in

Figure 14.2. Because of the engineering importance of the charge dissipation, various designs of charge dissipation layers have been proposed. For example, Minford and Sneh [8] propose $Si_2Ti_xN_{8/3-x}$ as the charge dissipation layer material to cover at a least a surface of the $SiO_2$ buffer layer and also the backside of the LN modulator die. Burns et al. [9,10] propose the use of silicon titanium oxynitride layer as a surface charge dissipation layer and another layer to cover the other three faces of the LN modulator die and to electrically connect both ends of the surface charge dissipation layer.

In an *x*-cut LN modulator, *z*-cut surfaces are exposed on side faces of the die and, since they are far from the MZ region, there is no direct impact of pyroelectric charges on optical output stability. For *x*-cut devices, no charge dissipation layer is needed on the buffer layer surface. However, application of the charge dissipation layer on *x*-cut face could further stabilize the output signal. This may be due to leaky fields onto the *x*-cut surface from charges pyroelectrically induced along *z*-axis (perpendicular direction to *x*-cut waveguides). The *z*-cut surfaces exposing along the side of the modulator die is usually coated with an electrically conductive film and grounded, to avoid the occurrence of arcing of pyroelectric charges. Arcing to the surface electrodes would cause sudden shift or hopping of the operation point and disturb the DC bias control loop.

## 14.3 Contributors to DC Drift

The technical breakthroughs described above for the suppression of the thermal instability of the LN modulator, particularly the suppression of the intrinsic pyroelectric effect, improved LN modulator performance. Despite these improvements, the changes in bias conditions can be large enough (~1-V shift of the operation point or more under the range of normal operation temperatures) that control loops are needed to assure device performance over its lifetime. In addition to these mechanisms, LN modulators do also show drift of the operating point under DC bias. Fortunately, this can also be controlled by feedback loops as described in Chapter 15. In this section, the current understanding of "DC drift" will be described. Due to the dielectric nature of the LN substrate and buffer layer, a supplied DC bias induces electro-optic phase changes in the waveguides through a dielectric relaxation process and with a certain delay determined by relaxation time constant *RC*. This mechanism causes a monotonic increase of the required DC bias and appears as "DC drift" in the bias voltage control loop.

The physics of the DC drift of LN modulators has not been quantitatively studied yet, and only phenomenological or empirical explanations have been reported in studies of long-term reliability of the LN modulator. Nonetheless, there is a large body of data that allows the estimation of the DC bias growth over the system life (20–25 years at 30°C–50°C ambient temperature). Simulation of the LN modulator die cross section by an equivalent circuit including *RC* relaxation components is a common way to analyze and predict the DC bias drift behavior over a time. Such circuit analyses were done in the very early R&D stage of LN modulators by several groups and were applied to the search for buffer layer materials to reduce the DC drift phenomenon [11–13]. An *RC* circuit analysis has been used to understand the properties of the buffer layer that delay and suppress the DC bias drift [14]. Figure 14.3 shows the *RC* circuit used in this study [14], in which the LN substrate and dielectric buffer layer are considered as two major contributors to DC bias drift phenomenon. Later, another group applied a more complete *RC* circuit model to simulate these data. In this later study, the network included every single boundary of constituent layers and attempted a more detailed and quantitative analysis of the DC drift behavior [15].

Empirically, the LN substrate and buffer layer can be shown as a key contributor to the DC bias drift. It is clear that the DC drift behavior of LN modulators having buffer layers is largely different from that of unbuffered LN modulators. It is also clear that the buffer layer chemistry greatly affects the magnitude of DC drift. For larger applied DC bias voltages (i.e., high electrical stress condition), the contribution of a boundary between the buffer layer and the charge dissipation layer to the DC drift can also be observed (see Section 14.3.4).

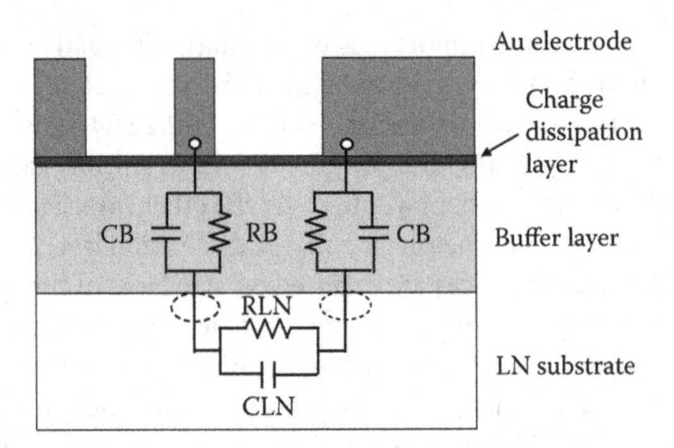


**FIGURE 14.3** Schematic diagram of *z*-cut LN modulator cross section and the corresponding *RC* equivalent circuitry. C and R represent electrical capacitance and resistance of the dielectric buffer layer (CB or RB) or LN substrate (CLN or RLN), respectively.

### 14.3.1 Contribution of Buffer Layer to DC Drift

In Figure 14.3, CB and RB are the electrical capacitance and resistance of the dielectric buffer layer, respectively, and CLN and RLN are those for the LN substrate [14]. When the DC bias voltage $V$ is applied to the electrodes of such a circuit, a magnitude of the voltage actually applied to the LN waveguides VLN is affected by these $C$ and $R$ parameters. The proposed *RC* circuit model indicates that the spontaneous VLN(0) at time $t = 0$ is mainly determined by the capacitances as CB/(CB + 2CLN), while the long-term VLN(∞) at $t$ = infinity is determined by the resistances as RLN/(RLN + 2RB). A relation of VLN(0) = VLN(∞) means no change of the applied voltage throughout an operation time and absence of the DC bias drift. When VLN(0) > VLN(∞), the actual voltage at the waveguides will decrease with time, and additional DC bias voltage needs to be supplied to compensate for the VLN decrease. This is referred to as positive DC drift. Similarly, VLN(0) < VLN(∞) corresponds to negative DC drift. Because LN crystal has a partial ionic nature rather than complete covalent bond structure [1], RLN is not much larger than the RB of a common dielectric buffer layer material such as $SiO_2$. Therefore, RLN/(RLN + 2RB) tends to be small and results in relatively smaller VLN(∞) and thus VLN(0) > VLN(∞), causing an undesirable positive DC drift. To reduce the magnitude of the positive DC drift, reference [16] proposes an idea to reduce the buffer layer resistance RB by chemical doping to the silicon oxide film. However, a dopant causing ionic conductivity such as $Na^+$ will generate electrical leakage through the buffer layer and deteriorate the DC drift. Well-controlled material and process design in the selection of dopants and adjustment of their concentration are necessary to obtain a lower DC drift LN modulator.

As another approach to suppress the positive DC drift, improvement of LN substrate resistance RLN by minimization of $OH^-$ through dry atmosphere waveguide diffusion process has been reported [17]. $OH^-$ fragments are contained on the O–O plane of the LN crystal structure and migrate under application of DC field [18]. Hence, minimization of $OH^-$ content in the crystal can reduce drift.

### 14.3.2 Drift Behavior for Buffered and Unbuffered Modulators

As shown in Figure 14.2a, the *z*-cut LN modulator design requires a buffer layer covering the MZ waveguides to prevent optical absorption by contact of metal electrodes to the waveguides. In *x*-cut design, since the electrodes are placed adjacent to the waveguides (Figure 14.2c), DC bias electrodes can be placed before or after the RF electrodes along MZ waveguides, and the buffer layer can be removed from that DC bias electrodes section. For high-speed *x*-cut LN modulator, its RF section must have the buffer layer to achieve RF impedance matching between electrodes and LN waveguides. The unbuffered DC bias section may show a different DC drift behavior from that of the buffered LN modulator because there is competition of *RC* relaxation processes between the buffer layer and the LN substrate. This section presents various accelerated aging test results displaying different DC drift behaviors observed on the buffered *z*-cut and the unbuffered *x*-cut LN modulators.

Since DC drift is an inherent phenomenon in LN modulators, the magnitude of the DC bias must stay within voltage limits of the automatic bias control circuit over the operational life (20–25 years) and across operation temperatures. To assure device performance over life, accelerated aging test is performed under bias as a part of reliability qualification tests of LN modulators. Usually, temperature acceleration and voltage acceleration mechanism are considered in the design of the test condition, and the samples are operated at various constant temperatures and with a range of initial bias voltages. For test purpose, the initial DC bias voltage can be chosen independent on any specific position on the modulation curve, and the DC bias voltage can be feedback controlled to lock the operation point at the initial bias position. Accelerated testing has shown different drift acceleration mechanisms between buffered and unbuffered LN modulators. For the unbuffered *x*-cut LN modulator, the initial DC bias ($V_s$) was found to be an almost linear acceleration factor. This result means that applied bias voltages can be simply normalized by the corresponding $V_s$, and DC drift reliability analysis becomes very simple, i.e., only temperature is an acceleration factor of the mechanism [19]. On the buffered *z*-cut LN modulators, extended test results suggest a weak nonlinear contribution of $V_s$.

Figure 14.4a and 14.4b show typical drift curves obtained at 85°C and 135°C, respectively. As is clearly seen in Figure 14.4b, the unbuffered *x*-cut LN modulator exhibits two bias peaks, and the drift profile eventually reaches a maximum and then decreases. The second broad peak grows higher than the first peak and thus determines the drift reliability of the unbuffered *x*-cut LN [20]. On the other hand, neither peaking nor saturation of the applied bias voltages is seen for the buffered *z*-cut LN modulators, except for a small step-like change of the bias voltage, e.g., at ~40 hours on the Z-LN profile of Figure 14.4b. Drift tests with extended temperature and $V_s$ conditions indicated that the time to this bias-step has a temperature dependency. This monotonically increasing DC bias drift is a characteristic of the buffered LN modulator. Indeed, an *x*-cut LN modulator having the buffer layer under the bias electrodes also shows DC bias growth similar to the buffered *z*-cut LN modulator [21].

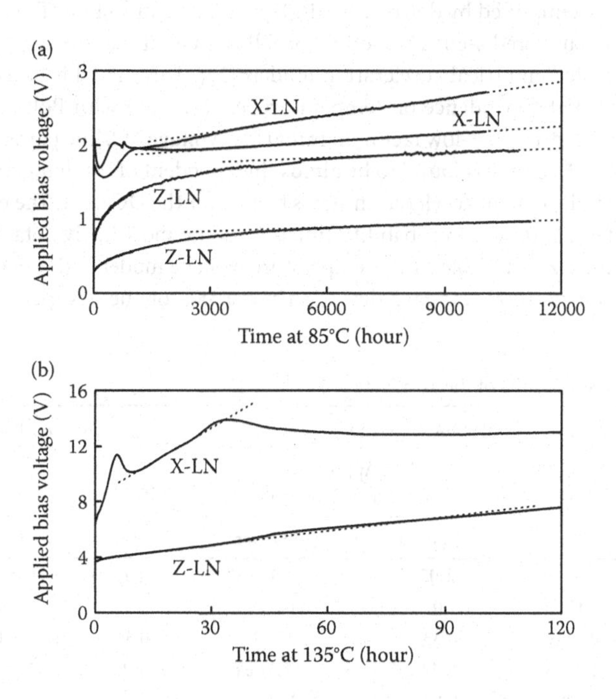


**FIGURE 14.4** Typical DC bias drift curves for buffered *z*-cut and unbuffered *x*-cut LN modulators tested at (a) 85°C or (b) 135°C, using a constant phase bias feedback control technique.

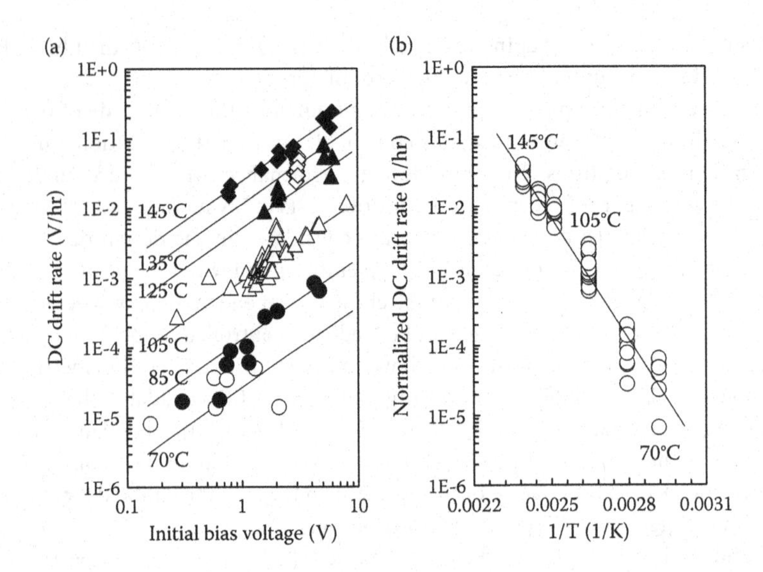


**FIGURE 14.5** (a) Temperature, $T$, and initial bias voltage, $V_s$, dependencies of DC drift rate measured on the buffered $z$-cut LN modulator (before bias-step). (b) Arrhenius acceleration models of the DC drift rates of the buffered $z$-cut LN modulators (before bias-step). For simplicity, a linearly fitted drift rate was used in this data analysis (see the dotted lines in Figure 14.4a and 14.4b).

Figure 14.5a shows a relationship between the first drift rates of the buffered $z$-cut LN modulators (before the bias-step) and the corresponding initial bias voltages ($V_s$), which were measured at various test temperatures. For simplicity, a linearly fitted drift rate was used in this data analysis, and a definition of the drift rate is exemplified by dotted lines in Figure 14.4a and 14.4b. The plots suggest that the DC drift behavior of the buffered $z$-cut LN modulator follows a temperature–nonthermal model that is one of commonly accepted empirical acceleration models [22]. This model is based on a simple multiplication of an exponential dependence on temperature, $\exp(-E_a/kT)$ with Boltzmann constant $k$ and absolute temperature $T$ and a power-low factor on initial bias voltage, $V_s^n$. The parameter $n$ is determined by the slope of Figure 14.5a and was found to be almost independent of test temperatures. It is seen that $T$ and $V_s$ are mutually independent acceleration stress factors for the DC drift. The maximum likelihood (ML) analysis indicates a lognormal probability function as the most likely data distribution, and the fitting results are summarized in Table 14.1 ("temperature–voltage model"). $E_a$ = 1.12 eV and $n$ = 1.15 eV were estimated with a lognormal standard deviation $\sigma$ = 0.384 for the first part of the buffered $z$-cut

**TABLE 14.1** ML Analyses Results of the Test Data

| | | Temperature–Voltage Model | | | Arrhenius Model | | |
|---|---|---|---|---|---|---|---|
| | | | 95% Approximate Confidence Intervals | | | 95% Approximate Confidence Intervals | |
| | Parameter | Mean | Lower | Upper | Mean | Lower | Upper |
| Z-LN | $E_a$ | 1.12 eV | 1.07 eV | 1.17 eV | 1.15 eV | 1.11 eV | 1.20 eV |
| | $n$ | 1.15 | 1.02 | 1.28 | – | – | – |
| | $\sigma$ | 0.384 | 0.436 | 0.339 | 0.393 | 0.446 | 0.346 |
| X-LN | $E_a$ | 1.22 eV | 1.11 eV | 1.34 eV | 1.25 eV | 1.15 eV | 1.35 eV |
| | $n$ | 0.988 | 0.892 | 1.08 | – | – | – |
| | $\sigma$ | 0.882 | 0.981 | 0.794 | 0.878 | 0.974 | 0.791 |

LN modulators DC drift. Given that the contribution of the above obtained nonlinear acceleration factor $V_s^{1.15}$ is very small compared to the exponential temperature factor one can model the performance assuming a linear dependence on $V_s$. Figure 14.5b shows a simple Arrhenius plot of data of Figure 14.4a, where a linear contribution of $V_s$ (i.e., $n = 1$) is assumed and the DC drift rates were normalized by the corresponding $V_s$. The ML analysis results are shown in Table 14.1 ("Arrhenius model"). As expected due to the small nonlinearity of the $V_s$ factor, the values of $E_a = 1.15$ eV and $\sigma = 0.393$ are very close to the results of the "temperature–voltage model."

Similar ML analysis of the unbuffered *x*-cut LN modulator data results in values of $E_a = 1.22$ eV and $n = 0.988$ with $\sigma = 0.882$, as shown in Table 14.1. A remarkable thing is that the unbuffered *x*-cut LN modulator exhibited an almost complete linear dependency of the DC bias drift on the initial bias voltage $V_s$. Analysis of Arrhenius plots of the normalized drift rates (assuming $n = 1$) provided $E_a = 1.25$ eV with 95% confidence intervals limits of 1.15 eV and 1.35 eV.

In the case of the unbuffered LN modulator, the temperature dependency of its DC drift is expected to be very similar to the temperature dependency of electrical resistance of the LN substrate. DC resistances of commercially available *z*- and *x*-cut LN wafers were measured around 100°C (close to the drift test temperatures) and showed that the resistance along *z*-axis has two $E_a$'s, $E_a = 1.33$ eV beyond ~90°C while 0.57 eV below 90°C [23]. On the other hand, no bending of the Arrhenius plot was reported for DC resistance along *x*-axis and $E_a = 1.40$ eV between 40°C and 150°C. Figure 14.6a replots these data as resistivity, calculated using the reported sample dimensions. In the unbuffered *x*-cut LN modulator, the applied DC field passes through along both *x*- and *z*-axes, as schematically shown in Figure 6b. Under such a simple series resistive circuit, $E_a = 1.4$ eV along the more resistive *x*-axis path is the primary contributor to the total circuit $E_a$, as shown in Figure 14.6b. In the calculation, a ratio of a path-length along *x*-axis ($L_x$) to a length along *z*-axis ($L_z$) was assumed to be 4/20, 6/20, or 8/20, taking into consideration plausible DC penetration depths and common gap distances between electrode pairs. In this model, the kink around 90°C becomes very small, and $E_a = 1.2$ eV–1.3 eV was calculated over the range 40°C to 150°C. This is very consistent with $E_a = 1.25$ eV obtained for the unbuffered *x*-cut LN modulator DC drift.

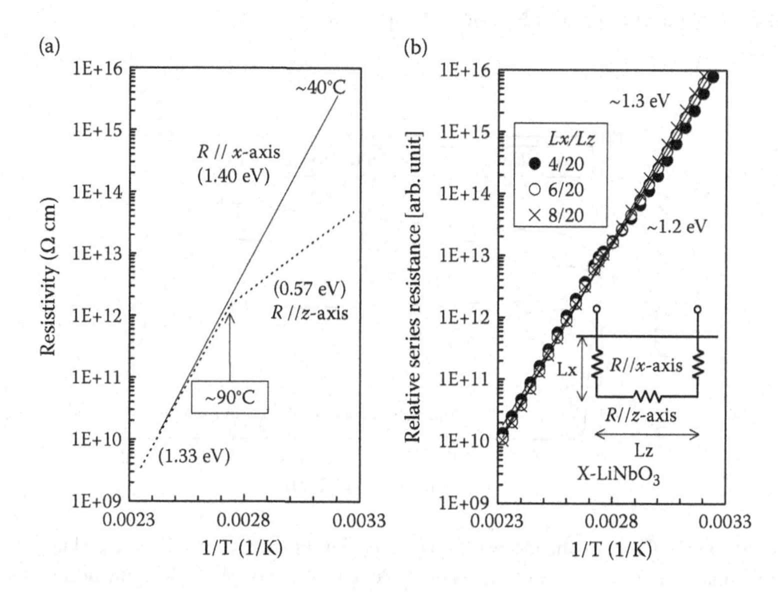


**FIGURE 14.6** The temperature dependency of (a) resistivity measured on LN wafers and (b) relative series resistance calculated for *x*-cut LN modulator design. In (a), the solid line and dotted line show the results measured on commercially available *x*- and *z*-cut wafers, respectively.

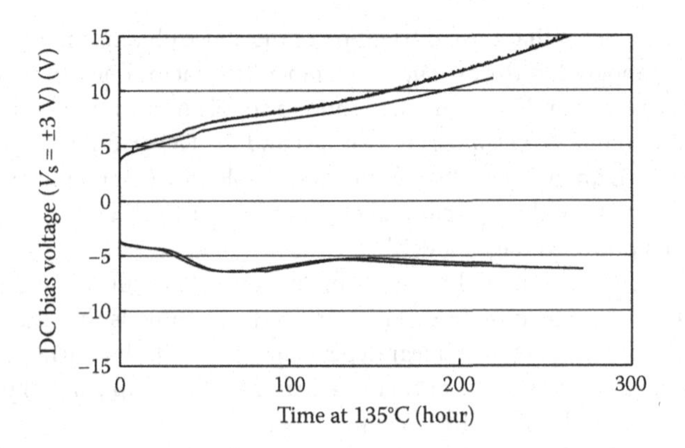


**FIGURE 14.7** DC bias drift of the buffered *z*-cut LN modulators tested at 135°C with the initial DC bias at 3 V or −3 V.

## 14.3.3 Effect of Bias Polarity

Extended DC drift tests with varying initial DC bias voltages demonstrate almost no bias polarity dependency for the unbuffered *x*-cut LN modulators [19]. However, the buffered *z*-cut LN modulators show a DC bias polarity dependency in their DC drift behavior as shown in Figure 14.7. An obvious feature of the bias polarity dependency of the buffered *z*-cut LN modulators is a negative peaking (or bumping) of the DC bias voltage, which appears when the initial bias has a negative polarity. As shown in Figures 14.8 and 14.9, the behavior of the bump depends on the chemistry of the buffer layer and the magnitude of the initial bias voltage. In the experiments for Figure 14.8, the concentration of the $In_2O_3$ dopant in the silicon oxide based buffer layer was intentionally changed. The highly doped buffer layer exhibits a larger bump on the drift curve tested with the negative bias voltages. Since an occurrence of too large bump may make the DC bias control loop unstable, the buffer layer chemistry must be one of the important control parameters in LN modulator process design.

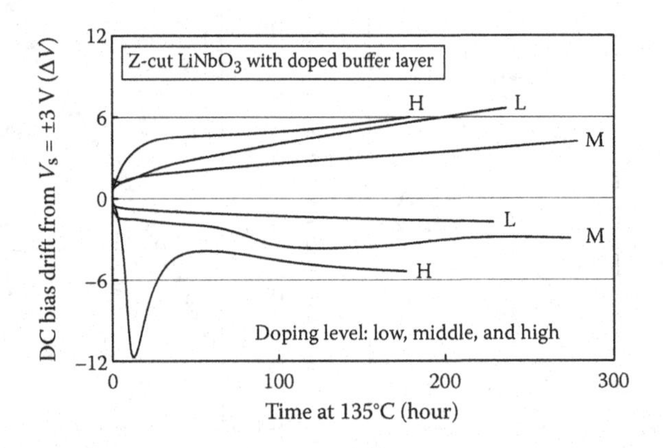


**FIGURE 14.8** Effect of buffer layer chemistry on variation of DC bias drift behavior and on bias polarity. The buffer layer of the modulator samples is made of silicon oxide doped with multiple dopants including $In_2O_3$. In this test, doping level (%) of indium oxide is intentionally changed from low (<5 mol%) to middle and to high (>10 mol%). Results of Figure 14.7 correspond to the test modulator with the middle-level doped buffer layer. The curves show the bias voltage growth from the initially applied DC bias $V_s$ = 3 V or −3 V. Test temperature is 135°C.

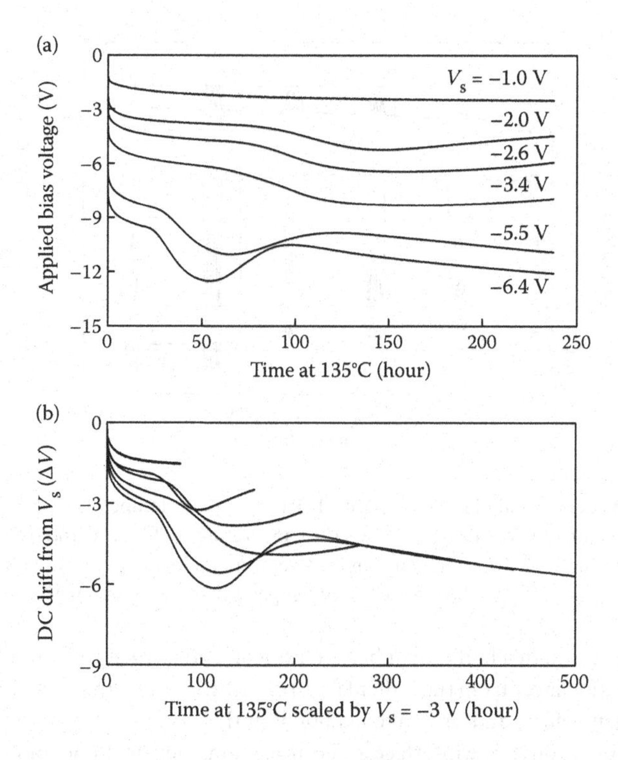


**FIGURE 14.9** (a) Initial bias voltage ($V_s$) dependency of the DC drift behavior observed when a negative bias is applied to the buffered *z*-cut LN modulator. In (b), the horizontal time axis of (a) under various $V_s$ is linearly scaled by $V_s = -3$ V, showing a likely linear dependence of the time for the start of the inflection point on $V_s$.

## 14.3.4 Contribution of the Boundary between Material Layers to DC Drift

For the buffered *z*-cut LN modulators, in addition to the above bias polarity dependency of DC drift behavior, highly stressed test results suggest an increase of the temperature dependent shift of the bias point voltage [24]. Figure 14.10 shows the biased aging test results of two buffered *z*-cut LN modulators, which are programmed to alternate temperature cycles between 25°C and 85°C, with 40-hour periods of aging at 135°C, after the initial bias application at 135°C ($V_s = \pm 3$ V). The extent of the bias voltage change over temperature grows with increase of elapsed time under continuous DC biasing. The experiments also suggested a tendency of larger deterioration of the temperature dependent bias voltage change with negative DC bias application [25].

Nagata [26] discusses those findings in DC bias drift behavior of buffered *z*-cut LN modulators and proposes that the boundary chemistry between the doped buffer layer and the charge dissipation layer contributes to the phenomena. Experiments using *x*-cut LN modulators only with the similar doped buffer layer (no charge dissipation layer; Figure 14.2c) show neither the bias bump on negatively biased drift profile nor deterioration of the temperature dependent bias voltage change. Although the actual mechanism of this somewhat anomalous drift phenomenon has yet to be understood, the following hypothesis is considered as a likely contributor. Metallic dopants such as indium may substitute silicon and bind with oxygen, but their bonding strength is usually weaker than that between silicon and oxygen. Thus, the doped silicon oxide would be considered as a chemically unstable material. On the other hand, the charge dissipation layer is usually made of Si-based material such as amorphous silicon to obtain a certain electrical conductivity. When the Si-based charge dissipation layer is deposited on the doped silicon oxide buffer layer, the charge dissipation layer may chemically reduce the buffer layer at the interface due to the high oxygen affinity of silicon atom (ion). For example, because the bonding strength between

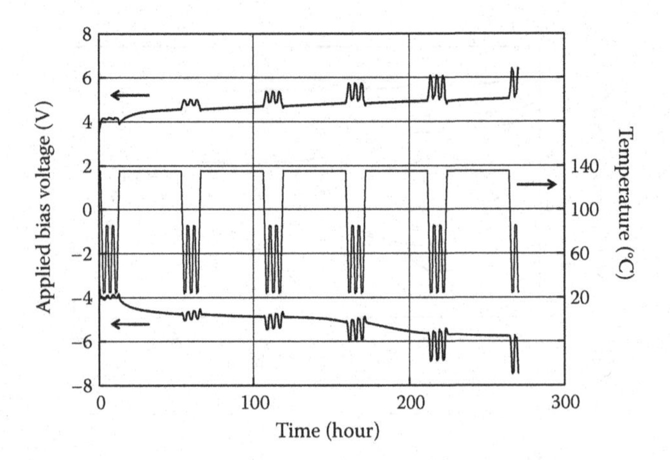


**FIGURE 14.10** Results of a repeated cycle of biased 135°C aging and temperature cycles (25°C to 85°C) for a buffered *z*-cut LN modulator. The middle curve shows the temperature profile. The initial DC voltage was set at 3 V or −3 V at 135°C. Dwell time at 135°C is 40 hours, and temperature cycle ramp rate is ±1 °K/min. The amount of a temperature dependent voltage change during the biased temperature cycles grows with increased time under bias.

indium and oxygen is less than a half of that between silicon and oxygen [27], some oxygen may diffuse into the charge dissipation layer from the buffer layer, resulting in the generation of oxygen defects at the interface. Those oxygen defects and/or weakened bonds at the interface would enhance the occurrence of electrical breakdown through the interface and cause anomalous drift phenomena.

### 14.3.5 Activation Energy and Reliability

The unbuffered *x*-cut LN modulators shows roll-over type DC drift (Figure 14.4), and their long-term reliability is to be determined by the relative magnitude of the peak bias voltage against the bias voltage limit of the system [20]. However, the buffered LN modulators perform a growing type DC drift independent of the LN orientation, and acceleration factors of the DC drift become a key aspect of their long-term reliability estimation. For practical reliability estimation purpose, the temperature can be considered the dominant acceleration factor. DC drift activation energies $E_a$ has been reported by various LN modulator suppliers and are summarized in Table 14.2. Reported $E_a$ varies between 1.0 eV and 1.4 eV. This range of activation energies means a broad range of acceleration factors. For instance, when extrapolating from DC drift data at 100°C to normal operation temperature, this $E_a$ difference gives ~7 times difference in the test time required to demonstrate reliable long-term behavior; i.e., for tests

**TABLE 14.2** List of DC Drift Activation Energy $E_a$ Reported from Various LN Modulator Suppliers

| Modulator Type | $E_a$ (eV) | Source | Year | Reference |
|---|---|---|---|---|
| Unbuffered *x*-cut LN | 1.23 | JDSU | 2004 | [21] |
| | ~1.4 | JDSU | 2002 | [31] |
| | 1.4–1.7 | Sumitomo | 2000 | [32] |
| Buffered *x*-cut LN | 1.4 | Sumitomo | 2000 | [33] |
| Buffered *z*-cut LN | 1.0–1.2 | JDSU | 2004 | [34] |
| | 1.35 | Agere | 2001 | [35] |
| | 1.1 | Lucent | 1998 | [36] |
| | 1.4–1.6 | Fujitsu | 1995 | [14] |
| | 1.0 | Sumitomo | 1995 | [37] |
| | 1.0 | Fujitsu | 1992 | [7] |

at 100°C we require 1780 test hours when using $E_a$ = 1.0 eV to demonstrate reliable performance for 25 years at 50°C but only 259 hours when assuming $E_a$ = 1.4 eV. Similarly, we require 123 hours (1.0 eV) or 6.2 hours (1.4 eV) when using an aging temperature of 135°C. Clearly, the values of activation energies are dependent on the specifics of the device design and fabrication.

In reliability estimation, constant temperature operation at a typical module temperature is usually assumed for simplicity in design and analysis of the stressed bias aging tests. When we need to consider the impact of operation temperature changes to the bias drift reliability, the phenomenon discussed at Figure 14.10 also needs to be considered. Due to the growth of the temperature dependent bias point change over DC biased duration, the effective range between maximum and minimum bias voltage limits of the system will shrink with operation time [25].

As the result of the engineering efforts to overcome the above discussed bias stability concerns, high reliability of LN modulators has been proven by several modulator suppliers. For example, 10-Gb/s *z*-cut LN modulators with only one field failure in over 740,000,000 in-service device hours has been reported. This corresponds to a failure rate of 6.4 FITs with a 95% confidence interval.

## 14.4 Bias-Free Modulator

Due to the DC drift phenomena of LN modulators, users must deploy auto bias control circuitry to maintain the operation point over time. A typical control circuit taps off a small portion (typically a few percent) of the optical output and monitors it with either an external photodiode through a fiber coupler spliced to a modulator output fiber or an internal monitor photodiode that was integrated inside the modulator package. One design detects the signal that is radiated from the Y-branch of the MZ by photodiodes placed after the LN die [28].

To avoid the added of bias control circuits, some systems demand bias-free LN modulators. Because the application of any small DC voltage can cause DC drift, the operation point of such a modulator must be adjusted by a nonelectrical mean and must be stable across the temperature. The latter requirement is an especially critical-temperature-dependent shift of the operation point must be within a range that does not deteriorate mask margin of the optical output. This performance requirement imposes requirements of symmetry and minimal internal stresses on the modulator design. The factors of asymmetric *z*-cut LN modulator die design with high dimensional aspect ratio, waveguide formation at high temperature, buffer layer deposition, and formation of thick electrodes pattern described above are obstacles for the design of a bias-free LN modulator. Those design limitations cannot be achieved in higher speed LN modulators, but a 2.5-Gb/s zero-chirp bias-free *x*-cut LN modulator is commercially available and has field-proven high reliability. This device is designed with annealed proton-exchange (APE) waveguides that are prepared at much lower process temperature than Ti in-diffused waveguides, no buffer layer, and a thin symmetric electrodes pattern [29]. The optical output from these low stress symmetric modulators is adjusted near the quadrature point of the optical transfer curve (position b of Figure 14.1) by a nonelectrical method and hence eliminates the need for external DC biasing over operational temperatures and wavelength. Long-term reliability of these bias-free APE modulators is demonstrated by the stability of 2.5-Gb/s PRBS extinction ration under unbiased high temperature aging [30].

# References

1. A. M. Prokhorov and Y. S. Kuz'minov, *Physics and Chemistry of Crystalline Lithium Niobate,* Bristol: Adam Hilger, 1990.
2. H. Nagata, K. Kiuchi, and T. Sugamata, "Refractive index fluctuations in deformed Ti:$LiNbO_3$ waveguides due to $SiO_2$ overlayer deposition," *Appl. Phys. Lett.*, vol. 63, no. 9, pp. 1176–1178, Aug. 1993.
3. H. Nagata, K. Kiuchi, and T. Saito, "Studies of thermal drift as a source of output instabilities in Ti:$LiNbO_3$ optical modulators," *J. Appl. Phys.*, vol. 75, no. 9, pp. 4762–4764, May 1994.

4. H. Nagata, S. Oikawa, and M. Yamada, "Comments on fabrication parameters for reducing thermal drift on $LiNbO_3$ optical modulators," *Opt. Eng.*, vol. 36, no. 1, pp. 283–286, Jan. 1997.
5. K. Higuma, Y. Hashimoto, M. Yatsuki, and H. Nagata, "Electrode design to suppress thermal drift in lithium niobate modulators," *Electron. Lett.*, vol. 36, no. 24, pp. 2013–2014, Nov. 2000.
6. P. F. Bordui, D. H. Jundt, E. M. Standifer, R. G. Norwood, R. L. Sawin, and J. D. Galipeau, "Chemically reduced lithium niobate single crystals: Processing, properties and improved surface acoustic wave device fabrication and performance," *J. Appl. Phys.*, vol. 85, no. 7, pp. 3766–3769, Apr. 1999.
7. M. Seino, T. Nakazawa, Y. Kubota, M. Doi, T. Yamane, and H. Hakogi, "A low dc-drift modulator assured over 15 years," in *OSA Optical Fiber Communications Conf., 1992*, pp. 325–328.
8. W. J. Minford and O. Sneh, "Apparatus and method for dissipating charge from lithium niobate devices," U.S. Patent 5 949 944, Sep. 7, 1999.
9. W. K. Burns, L. A. Hess, and V. Agarwal, "Buffer layer structures for stabilization of a lithium niobate device," U.S. Patent 6 654 512, Nov. 25, 2003.
10. W. K. Burns, L. A. Hess, and V. Agarwal, "Buffer layer structures for stabilization of a lithium niobate device," U.S. Patent 6 661 934, Dec. 9, 2003.
11. S. Yamada and M. Minakata, "Dc-drift in $LiNbO_3$ optical waveguide devices," *Jpn. J. Appl. Phys.*, vol. 20, no. 4, pp. 733–737, Apr. 1981.
12. C. M. Gee, G. D. Thurmond, H. Blauvelt, and H. W. Yen, "Minimizing dc drift in $LiNbO_3$ waveguide device," *Appl. Phys. Lett.*, vol. 47, no. 3, pp. 211–213, Aug. 1985.
13. R. A. Becker, "Circuit effect in $LiNbO_3$ channel-waveguide modulators," *Opt. Lett.*, vol. 10, no. 8, pp. 417–419, Aug. 1985.
14. M. Seino, T. Nakazawa, S. Taniguchi, and M. Doi, "Improvement of dc drift characteristics in Ti:$LiNbO_3$ modulators," *Tech. Rep. IEICE*, vol. OCS95-66, pp. 55–60, 1995.
15. S. K. Korotky and J. J. Veselka, "An RC network analysis of long term Ti:$LiNbO_3$ bias stability," *J. Lightwave Technol.*, vol. 14, no. 12, pp. 2687–2697, Dec. 1996.
16. M. Seino, T. Nakazawa, T. Yamane, Y. Kubota, M. Doi, K. Sugeta, and T. Kurahashi, "Optical waveguide device," U.S. Patent 5 404 412, Apr. 4, 1995.
17. H. Nagata, J. Ichikawa, M. Kobayashi, J. Hidaka, H. Honda, K. Kiuchi, and T. Sugamata, "Possibility of dc drift reduction of Ti:$LiNbO_3$ modulators via dry $O_2$ annealing process," *Appl. Phys. Lett.*, vol. 64, no. 10, pp. 1180–1182, Mar. 1994.
18. R. G. Smith, D. B. Fraser, R. T. Denton, and T. C. Rich, "Correlation of reduction in optically induced refractive-index inhomogeneity with OH content in $LiTaO_3$ and $LiNbO_3$," *J. Appl. Phys*, vol. 39, no. 10, pp. 4600–4602, Sep. 1968.
19. H. Nagata, N. Papasavvas, and D. R. Maack, "Bias stability of OC48 x-cut lithium niobate optical modulators: four years of biased aging test results," *IEEE Photonics Technol. Lett.*, vol. 15, no. 1, pp. 42–44, Jan. 2003.
20. H. Nagata, Y. Li, D. R. Maack, and W. R. Bosenberg, "Reliability estimation from zero-failure $LiNbO_3$ modulator bias drift data," *IEEE Photon. Technol. Lett.*, vol. 16, no. 6, pp. 1477–1479, Jun. 2004.
21. H. Nagata, Y. Li, W. R. Bosenberg, and G. L. Reiff, "DC drift of x-cut $LiNbO_3$ modulators," *IEEE Photon. Technol. Lett.*, vol. 16, no. 10, pp. 2233–2235, Oct. 2004.
22. P. A. Tobias and D. C. Trindade, *Applied Reliability,* 2nd ed., Boca Raton: Chapman & Hall/CRC, 1995.
23. V. Atuchin, T. I. Grigoria, H. Nagata, and J. Ichikawa, "Puncture of $LiNbO_3$ under strong dc electrical fields," in *Asia-Pacific Conf. Fundamental Problems Opto-Microelectronics*, 2001, pp. 91–96.
24. H. Nagata, N. F. O'Brien, W. R. Bosenberg, G. L. Reiff, and K. R. Voisine, "Dc voltage induced thermal shift of bias point in $LiNbO_3$ optical modulators," *IEEE Photon. Technol. Lett.*, vol. 16, no. 11, pp. 2460–2462, Nov. 2004.
25. H. Nagata, Y. Li, A. Finch, and K. R. Voisine, "Bias point thermal shift growth in z-cut $LiNbO_3$ modulators," *IEEE Photon. Technol. Lett.*, vol. 17, no. 6, pp. 1184–1186, Jun. 2005.

26. H. Nagata, "Electro-optic waveguide device capable of suppressing bias point DC drift and thermal bias point shift," US Patent 7 231 101, Jun. 12, 2007.
27. D. R. Lide, Ed., *CRC Handbook of Chemistry and Physics*, 83rd ed., Boca Raton: CRC Press, 2002–2003.
28. N. Miyazaki, K. Ooizumi, T. Hara, M. Yamada, H. Nagata, and T. Sakane, "$LiNbO_3$ optical intensity modulator packaged with monitor photodiode," *IEEE Photon. Technol. Lett.*, vol. 13, no. 5, pp. 442–444, May 2001.
29. E. L. Wooten, K. M. Kissa, A. Yi-Yan, E. J. Murphy, D. A. Lafaw, P. F. Hallemeier, D. R. Maack, D. V. Attanasio, D. J. Fritz, G. J. McBrien, and D. E. Bossi, "A review of lithium niobate modulators for fiber-optic communications systems," *IEEE J. Select. Topics Quantum Electron.*, vol. 6, no. 1, pp. 69–82, Jan. 2000.
30. H. Nagata, Y. Li, K. R. Voisine, and W. R. Bosenberg, "Reliability of non-hermetic bias-free $LiNbO_3$ modulators," *IEEE Photon. Technol. Lett.*, vol. 16, no. 11, pp. 2457–2459, Nov. 2004.
31. H. Nagata, Y. Li, I. Croston, D. R. Maack, and A. Appleyard, "Dc drift activation energy of $LiNbO_3$ optical modulators based on thousands of hours of active accelerated aging tests," *IEEE Photon. Technol. Lett.*, vol. 14, no. 8, pp. 1076–1078, Aug. 2002.
32. H. Nagata, Y. Ishizuka, and K. Akizuki, "Temperature dependency of x-cut $LiNbO_3$ modulator dc drift," *Electron. Lett.*, vol. 36, no. 23, pp. 1952–1953, Nov. 2000.
33. H. Nagata, "Activation energy of dc-drift of x-cut $LiNbO_3$ optical intensity modulators," *IEEE Photonics Technol. Lett.*, vol. 12, no. 4, pp. 386–388, Apr. 2000.
34. H. Nagata, G. D. Feke, Y. Li, and W. R. Bosenberg, "Dc drift of z-cut $LiNbO_3$ modulators," *IEEE Photonics Technol. Lett.*, vol. 16, no. 7, pp. 1655–1657, Jul. 2004.
35. W. J. Minford, "Advances in the reliability of $LiNbO_3$ integrated optic devices," in ITCOM 2001, *paper 4532-09*.
36. R. S. Moyer, R. Grencavich, F. F. Judd, R. C. Kershner, W. J. Minford, and R. W. Smith, "Design and qualification of hermetically packaged lithium niobate optical modulator," *IEEE Trans. Comp., Packag. Manufact. Technol. B*, vol. 21, no. 2, pp. 130–135, May 1998.
37. H. Nagata, H. Takahashi, H. Takai, and T. Kougo, "Impurity evaluation of $SiO_2$ films formed on $LiNbO_3$ substrates," *Jpn. J. Appl. Phys.*, vol. 34, no. 2A, pp. 606–609, Feb. 1995.

# 15

# Bias Control Techniques



Pak S. Cho
*University of Maryland*

## 15.1 Modulator Bias Control

Long-haul fiber-optic telecommunication systems operating in the S-, C-, and/or L-band of silica single-mode fiber typically employ external high-speed modulators to produce optical digital modulation formats for transmission data rates at and beyond 10 Gb/s. These optical modulation formats include conventional on–off keying (OOK) and the widely accepted phase-shift keying (PSK) capable of higher spectral efficiency. An example of a high-speed optical modulator device for OOK or PSK generation is an integrated lithium niobate (LN) waveguide electro-optic traveling-wave Mach–Zehnder modulator (MZM). LN MZMs are also used to generate other modulation formats such as optical single side-band [1], pulse-position modulation [2], and quadrature amplitude modulation [3]. LN MZMs are also extensively used in analog RF/microwave photonics (see Chapter 3) and have also been considered in unguided free-space atmospheric line-of-sight laser communications [4,5].

To produce optical OOK an electrical nonreturn-to-zero (NRZ) data signal and a DC bias voltage is applied to the MZM. The bias voltage sets the MZM at an optimum operating point for best performance. The bias point of the MZM, however, drifts due to various intrinsic and extrinsic factors described in Chapter 14. Bias tracking is therefore required to maintain the optimum bias point of the MZM for long-term stable operation. In this chapter, optimum biases, bias-error-induced penalties, and examples of bias control circuits for MZM-based devices for generation of optical OOK and PSK (specifically quadrature PSK or QPSK) are discussed. The subject of modulator bias for an emerging complex modulation format, optical orthogonal frequency-division multiplexing, will be briefly described.

### 15.1.1 Optimum Bias for Optical OOK

For an optical format such as NRZ-OOK using conventional square-law or direct-detection receiver, i.e., linear conversion of optical intensity to electrical voltage, an optical transmitter that converts the applied electrical NRZ waveform linearly to the optical intensity provides the best signal fidelity. Consequently, the optimum bias point of an MZM for optical NRZ-OOK generation is the quadrature (quad) or half-power transmission point where the most linear region of the sinusoidal transfer response (optical intensity vs. applied electrical voltage) is located as shown in Figure 15.1a.

Consider an ideal (perfect RF frequency response, perfect Y-branch, perfect symmetry, lossless waveguides, etc.) lossless $x$-cut single-drive push–pull type MZM with an input CW optical field $E_i = A_i\exp(j2\pi f_o t)$ at a carrier frequency $f_o$, the modulated output optical field can be expressed as

$$E_o = E_i\cos\left[\left(\phi_s + \phi_B\right)/2\right] = E_i\cos\left[(\pi/2)\left(V_s(t)/V_\pi^{RF} + V_B/V_\pi^{DC}\right)\right], \tag{15.1}$$

where $V_s(t)$ is the NRZ drive signal, $V_B$ is the DC bias voltage applied to the MZM, and $V_\pi^{RF}$ and $V_\pi^{DC}$ are the RF and DC half-wave or switching voltages of the MZM, respectively. Note that $V_B$ also varies with time in response to the bias drift during bias tracking but at a much slower rate (<100 Hz [6]) than $V_s(t)$, thus $V_B$ can be considered as a near-static quantity. The output optical intensity of the MZM with an input intensity $I_i \propto |E_i|^2$ is given by

$$I_o = I_i\cos^2\left[\left(\phi_s + \phi_B\right)/2\right] = (I_i/2)\left\{1 + \cos\left[\pi\left(V_s(t)/V_\pi^{RF} + V_B/V_\pi^{DC}\right)\right]\right\}. \tag{15.2}$$

As shown in Figure 15.1a, the optimum bias (±quad) closest to zero is $V_B = \mp V_\pi^{DC}/2$ or $\phi_B = \phi_{opt} = \mp\pi/2$ for optical NRZ-OOK with the same ($\phi_B = -\pi/2$) or opposite ($\phi_B = +\pi/2$) polarity as $V_s(t)$, where $I_o(\phi_{opt}) = (I_i/2)\left[1 \pm \sin\left(\pi V_s(t)/V_\pi^{RF}\right)\right]$. Applying an ideal NRZ drive signal (noiseless perfect square pulse) with a peak-to-peak voltage of $V_{pp} = V_\pi^{RF}$ produces $I_o = I_i$ or 0 when $V_s(t) = \pm V_\pi^{RF}/2$ giving an ideal zero extinction ratio of $r_e = 0$. The extinction ratio as a function of the bias error, $\phi_e \equiv \phi_B - \phi_{opt}$, is defined here as the linear ratio of the OFF state ($I_{OFF}$) to the ON state ($I_{ON}$) output optical intensity, i.e., $0 \le r_e(\phi_e) \equiv \left(I_{OFF}/I_{ON}\right) < 1$. A drive voltage swing of $V_{pp} = V_\pi^{RF}$ also minimizes transfer of nonideal

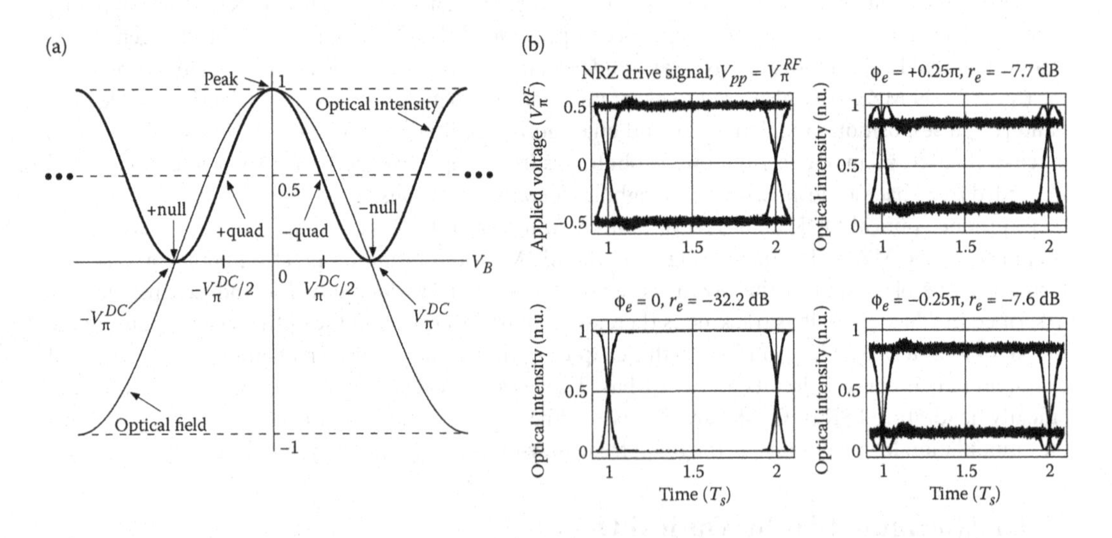


**FIGURE 15.1** (a) Optical field and intensity (heavy line) transmission response of an ideal MZM vs. applied bias voltage. ±quad: quadrature or half-power point. Peak and ±null are the maximum and minimum optical intensity transmission points. (b) Simulated eye waveforms of NRZ (PRBS: $2^9 - 1$) drive signal at $V_{pp} = V_\pi^{RF}$ with additive white Gaussian noise (top left), normalized optical output for $\phi_e = 0$ (bottom left), $\phi_e = +0.25\pi$ (top right), and $\phi_e = -0.25\pi$ (bottom right). $T_s$ is the symbol period (inverse of the clock rate) of the NRZ signal.

NRZ electrical noise (e.g., overshoots and ringing near the transitions) to the optical output thanks to the nonlinear compression response near the peak and null points as can be seen in Figure 15.1a. This effect can be seen on the left column of Figure 15.1b for the simulated eye waveforms of the nonideal NRZ drive signal with additive white Gaussian noise and the normalized output optical intensity at the optimum bias, $I_o(\phi_e = 0)/I_i$.

### 15.1.2 Nonoptimum Bias Penalty for Optical OOK

Deviation of the bias from the optimum point (+quad or $\phi_{opt} = -\pi/2$), i.e., $\phi_e \neq 0$, gives rise to a distorted optical NRZ-OOK signal where $I_{ON}$ decreases and $I_{OFF}$ increases degrading the extinction ratio even with $V_{pp} = V_\pi^{RF}$. This is illustrated in Figure 15.1b (right column), where simulated eye waveforms of $I_o(\phi_e)/I_i$ produced by the noise-added nonideal NRZ drive signal (top left of Figure 15.1b) are shown for $\phi_e = \pm 0.25\pi$. Also shown are the computed $r_e(\phi_e)$ values in decibels. Note that noise of the NRZ drive signal appears prominently on the distorted optical NRZ-OOK signals for $\phi_e = \pm 0.25\pi$, while the noise are significantly suppressed for $I_o(\phi_e = 0)/I_i$ (bottom left of Figure 15.1b). The impact of extinction ratio degradation caused by bias drift as well as $V_{pp} \neq V_\pi^{RF}$ on the receiver sensitivity can be quantified by the power penalty in decibel given approximately by [7]

$$p_{ext}(\phi_e) \approx 10\log_{10}\left[\left(1+r_e(\phi_e)\right)/\left(1-r_e(\phi_e)\right)\right]. \tag{15.3}$$

The penalty is zero for the ideal case of $r_e(0) = 0$. In practice, however, $r_e(0) << 1$ is not zero due to factors such as nonideal MZM and/or imperfect NRZ drive waveform. The dependency of the penalty on these nonideal factors not related to the bias error can be removed by defining a relative penalty: $P_{ext}(\phi_e) \equiv p_{r_e}(\phi_e) - p_{r_e}(0)$. The penalty can also be obtained from the bit-error rate (BER) measurement of the detected optical NRZ-OOK output followed by estimation of the $Q$-factor from the BER. The $Q$-factor is related to the BER via

$$\mathrm{BER}(Q) = (1/2)\mathrm{erfc}\left(Q/\sqrt{2}\right). \tag{15.4}$$

A reverse operation [8] based on the above equation can be applied to obtain an estimated $Q$-factor from the BER. A relative penalty in decibel can be obtained from:

$$P_{BER}(\phi_e) \equiv 10\log_{10}\left\{\left[Q(\phi_e = 0)\right]/\left[Q(\phi_e \neq 0)\right]\right\}. \tag{15.5}$$

Figure 15.2 shows computed $P_{ext}(\phi_e)$ and $P_{BER}(\phi_e)$ in decibel vs. $\phi_e$ for $V_{pp} = V_\pi^{RF}$ (left column) and vs. $V_{pp}$ for $\phi_e = 0$ (right column). Optical and thermal noise limited cases are shown for signal-to-noise ratio (SNR) corresponds to a BER of $10^{-3}$ for the ideal case, i.e., $Q(\phi_e = 0) \approx 3.09$ at $V_{pp} = V_\pi^{RF}$. A relative penalty of less than 0.1 dB can be achieved with $|\phi_e| < 0.05\pi$ (9°) for $V_{pp} = V_\pi^{RF}$. For $\phi_e = 0$, a ±10% deviation of $V_{pp}$ from $V_\pi^{RF}$ can be tolerated for a penalty within 0.1 dB.

### 15.1.3 Modulator Bias Stability Control Techniques

The bias point of an MZM drifts as a result of relative optical phase change caused by environmental perturbations, e.g., temperature, humidity, stress, aging. In addition, there are device-to-device variations in the bias and RF switching voltage due to manufacturing tolerances/offset. To ensure long-term stability of the transmitter, bias of the MZM must be maintained at the optimum point dynamically.

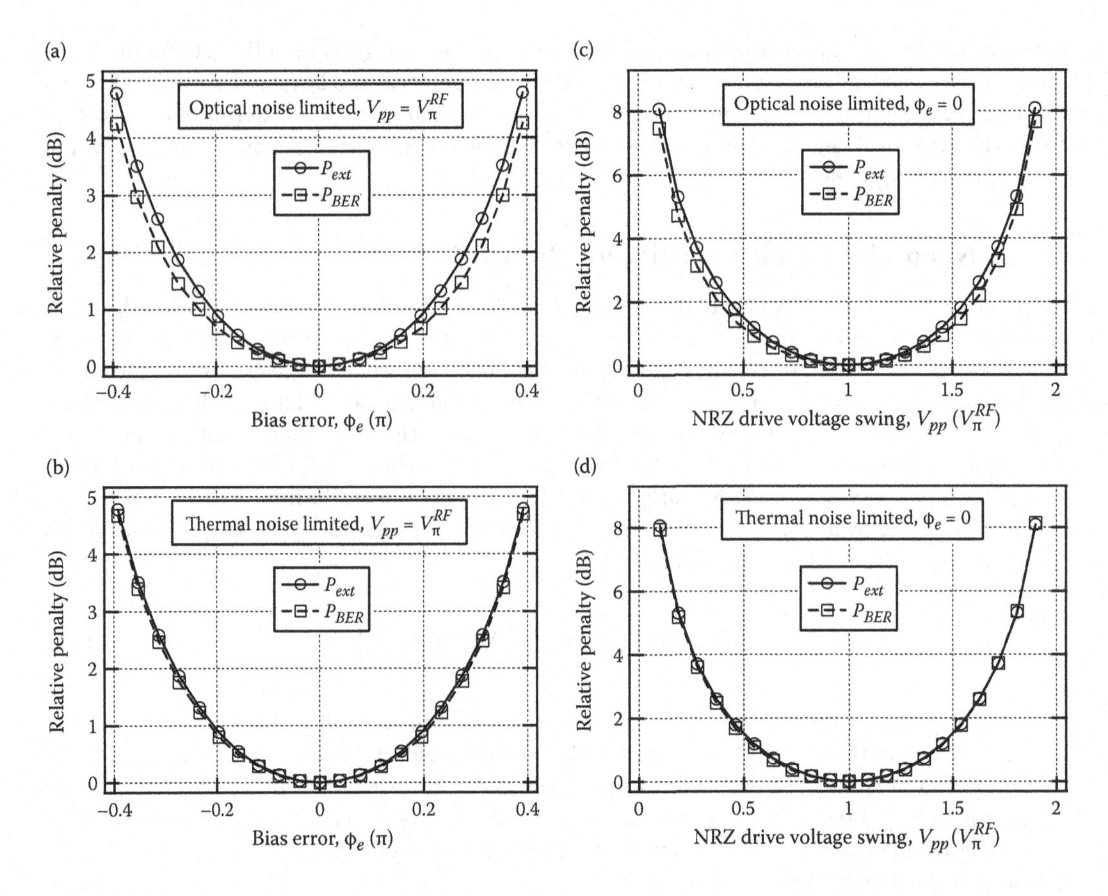


**FIGURE 15.2** Computed $P_{ext}(\phi_e)$ and $P_{BER}(\phi_e)$ of optical NRZ-OOK vs. $\phi_e$ (left column: a and b) and vs. $V_{pp}$ (right column: c and d) for optical (top row: a and c) and thermal (bottom row: b and d) noise limited cases where the SNR corresponds to $10^{-3}$ BER for $\phi_e = 0$ and $V_{pp} = V_\pi^{RF}$.

This is accomplished via a bias control circuit that performs automatic bias acquisition, tracking, and drift compensation.

Modulator bias stability control techniques for optical OOK generation are well-established. Commercial bias control systems for MZM are readily available from laboratory instruments to integrated board-level solutions. Many U.S. patents [9] related to optical modulator bias stability monitoring and control circuit have been issued since as early as 1973 [10] for applications ranging from digital/analog communications to electric field sensing [11]. Almost all modulator bias control circuits use a servo-loop or closed-loop feedback control configuration as depicted in Figure 15.3a. A fraction (few percent) of the MZM output power is tapped off via a fiber-optic tap coupler followed by a photodetector (e.g., a PIN photodiode). The photocurrent monitor signal is directed to a bias controller where the monitor signal is processed (e.g., filtered and amplified) to produce one or more feedback signals that contain information (magnitude and direction) on bias drift of the MZM, if any, from the optimum bias point. The feedback signal is usually compared with a reference signal producing an error signal with a polarity opposite to the drift. The nonzero error signal can then be used to tune a DC voltage source applied to the MZM bias electrode if available to compensate for the drift. Optimum bias is reached when the error signal vanishes at which point bias tuning terminates until bias error occurs producing a nonzero error signal and the process repeats.

Closed-loop bias control circuit can be divided into two general categories: active and passive. Active or dither-based control circuit perturbs the output of the MZM using an externally applied reference

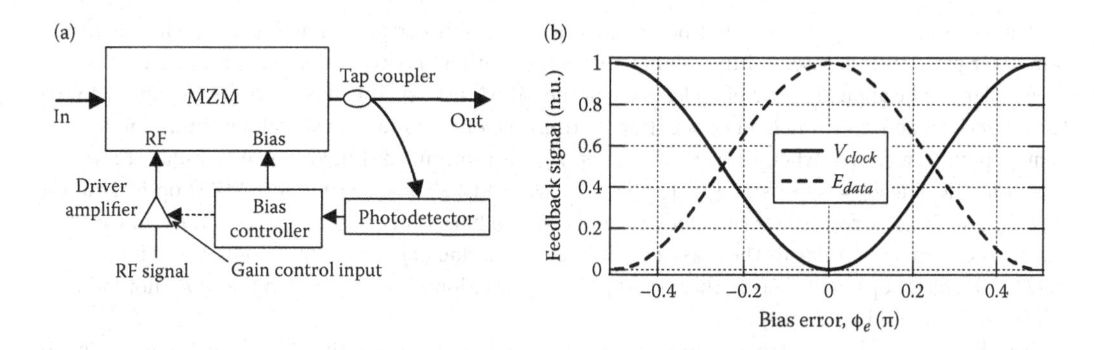


**FIGURE 15.3** (a) Schematic of a typical MZM closed-loop bias control circuit. (b) $V_{clock}$ and $E_{data}$ vs. $\phi_e$ for optical NRZ-OOK with $V_{pp} = V_\pi^{RF}$ computed using a $2^9 - 1$ long PRBS. Quad point: $\phi_e = 0$. RF spectral integration bandwidth for $E_{data}$: 4% of symbol rate.

or pilot signal. The pilot signal is typically a low frequency, low amplitude sinusoidal waveform (square wave or bipolar pulse are also used). The upper limit of the pilot dither frequency ($f_d$) is chosen to be well below the lowest frequency component of the information signal to minimize direct interference. Most widely used $f_d$ range from 0.5 kHz to 10 kHz. Modulation strength of the pilot signal is usually small to minimize impact on the information signal but large enough to maintain adequate SNR for robust bias tracking. Modulation strength of the pilot signal from less than 1% to ~5% of the MZM switching voltage is typical.

A common commercial active MZM bias control solution for NRZ-OOK uses a dither pilot signal to modulate the amplitude of the NRZ drive signal [10,12–16]. This can be achieved by applying the dither pilot signal with a frequency $f_d$ to the gain control input of the data driver amplifier as depicted in Figure 15.3a. The dither signal varies the amplitude of the input signal producing an amplitude-modulated NRZ data signal at the driver amplifier output. It can be shown that the RF power at $f_d$ denoted by $V(f_d)$ extracted from the monitor signal, e.g., using a bandpass filter centered at $f_d$, is minimized when quad bias is achieved for a NRZ drive voltage swing of $V_\pi^{RF}$. Using Equation 15.2, one can write for $\phi_s = \phi_{NRZ}[1 + \delta\sin(\omega_d t)]$ the output optical intensity of the MZM as follows

$$I_o(\phi_B) = (I_i/2)\Big[1 + \cos\big(\phi_{NRZ}\big[1+\delta\sin(\omega_d t)\big]\big)\cos(\phi_B) - \sin\big(\phi_{NRZ}\big[1+\delta\sin(\omega_d t)\big]\big)\sin(\phi_B)\Big], \quad (15.6)$$

where $\omega_d = 2\pi f_d$, $\delta$ is the pilot signal modulation index and $\phi_{NRZ}$ is the optical phase shift produced by the NRZ drive signal for $\delta = 0$. For a nominal NRZ drive voltage swing of $V_\pi^{RF}$ giving $\phi_{NRZ} = \pm\pi/2$, $I_o$ at the optimum bias (+quad), i.e., $\phi_B = \phi_{opt} = -\pi/2$, is

$$I_o(\phi_{opt}) = (I_i/2)\{1 \pm \cos[(\pi/2)\delta\sin(\omega_d t)]\}, \text{ for } \phi_{NRZ} = \pm\pi/2. \quad (15.7)$$

Due to the term $\cos[(\pi/2)\delta\sin(\omega_d t)]$ above, only even harmonics of the pilot signal are present in $I_o$ while all odd harmonics including $f_d$ vanish, i.e., $V(f_d) = 0$. Minimizing $V(f_d)$ therefore ensures optimum bias at the +quad point is achieved. Using this approach, an error signal is produced in the bias controller by multiplying $V(f_d)$ and the pilot signal at $f_d$ followed by a low-pass filter and/or an integrator. The polarity of the error signal is determined by whether $V(f_d)$ and the pilot signal is in or out of phase. A phase comparator can also be used, which provides an error signal proportional to the magnitude of its input signals, i.e., $V(f_d)$, and the pilot signal, as well as to the cosine function of the instantaneous phase difference between these input signals. An example of this dithering technique using amplitude-modulated NRZ that does not use $V(f_d)$ to produce an error signal can be found in [17].

For some applications, it may not be practical to apply a dither pilot signal to modulate the driver amplifier gain. In this case, a dither signal combining with a bias control voltage produced by a bias controller can be applied to the bias electrode of the MZM instead. The bias controller is similar to the one described above in which $V(f_d)$ is extracted from the optical monitor signal. Optimum bias at the +quad point is achieved when the error signal or $V(f_d)$ is minimized [18]. Following similar analysis as above and letting $\phi_B = \phi_b + \delta\sin(\omega_d t)$, $\phi_b = \phi_{opt} = -\pi/2$, and $\phi_s = \phi_{NRZ} = \pm\pi/2$, the MZM optical output is $I_o(\phi_{opt}) = (I_i/2)\{1 \pm \cos[\delta\sin(\omega_d t)]\}$. Once again, only even harmonics of $f_d$ are present in $I_o$ while all odd harmonics vanish. Similar to the bias control circuit described above an error signal that minimizes $V(f_d)$ will ensure optimum bias at the +quad point. A variation of this approach that does not monitor $V(f_d)$ can be found in, e.g., [19].

The dither-based bias control techniques described above require that the NRZ drive voltage swing to be very near, if not exactly, at the RF switching voltage or $V_\pi^{RF}$. Drive voltage swing deviation from $V_\pi^{RF}$ not only degrades the precision and the robustness of the bias control circuit but also affect the extinction ratio even if the bias is optimum. The deviation can be attributed to device-to-device variation of $V_\pi^{RF}$, drift of the driver amplifier gain and of $V_\pi^{RF}$ due to temperature and aging. Tracking and maintaining a stable $V_\pi^{RF}$ at the output of the driver can be achieved using closed-loop amplifier gain control together with an MZM bias control circuit as described in [20–23], for example.

Passive or dither-less bias control, on the other hand, does not employ perturbation or dither pilot signal. For example, Morin and Giraud [24] describe a microprocessor-based bias controller using the fact that the RF power of the frequency component at the symbol or clock rate of the detected optical NRZ-OOK signal, $V_{clock}$, is minimal when the MZM bias reaches the quad point ($\phi_e = 0$) as shown in Figure 15.3b. A decision on whether to increase or decrease the applied bias voltage by a small amount can be determined by choosing the bias voltage that produces a smaller $V_{clock}$ extracted from the optical output. Another dither-less bias control technique suggested by Cox and Ackerman [18] uses a microprocessor to monitor the energy of the low-frequency spectral component of the detected optical output of the MZM as the bias increases and decreases by a small amount. This low-frequency spectral energy denoted by $E_{data}$, induced by the NRZ binary data pattern, reaches a maximum value when the bias is optimum at the quad point as shown in Figure 15.3b. For null or peak point operation, the same technique still applies except that the bias is adjusted in a manner that minimizes $E_{data}$. The bias can be continuously adjusted in response to the feedback signal (e.g., $E_{data}$ or $V_{clock}$) or adjusted only if needed. Examples of other dither-less bias control techniques can be found in [23,25,26].

In addition to NRZ-OOK, MZM bias control techniques have been reported for other OOK formats such as duobinary generated by a null-biased MZM [27] and return-to-zero OOK (RZ-OOK). For optical RZ-OOK generation, an MZM bias control technique using a dither pilot signal at $f_d$ combined with an electrical RZ-OOK driving waveform generating an error signal that minimizes $V(f_d)$ has been proposed [28]. A dither-less bias control circuit for an MZM driven by an electrical RZ-OOK waveform has also been suggested [25,26]. For long-haul high date rate transmission, an external pulse carver MZM driven by a sinusoidal waveform and a second MZM for NRZ data modulation is usually preferred for high-speed optical RZ-OOK generation. This configuration provides the flexibility to produce optical pulses with useful duty cycles (ratio of the optical pulse FWHM to the pulse period) such as 0.33, 0.5, and 0.67 (carrier suppressed RZ) corresponding to bias points at the peak, quad, and null, respectively [29]. The pulse carver approach does not require electronics to produce the high-speed electrical RZ-OOK waveforms or bipolar RZ data pulses but correct phase alignment between the optical pulse train from the pulse carver and the NRZ drive signal at the data MZM is necessary. Bias control techniques using dithering described earlier can also be applied to the pulse carving MZM [15]. Separate bias control circuits for the data MZM and for the pulse carver including phase alignment control have been suggested [30,31].

For nondigital applications such as analog RF/microwave photonics, modulator bias control is critical to achieve maximum dynamic range of the optical communication link. As a result of the MZM nonlinear response, second-order harmonic and intermodulation distortion increase rapidly as the bias deviates from the quad point. A typical bias control technique [32,33] for analog applications uses a dither pilot signal at $f_d$ combined with the bias voltage before applying it to the bias electrode of the MZM. For small drive signal, quad bias is achieved when all even harmonics of the pilot signal vanish with only odd harmonics present in the optical output. The second harmonic or $V(2f_d)$ (RF power at $2f_d$) is extracted from the optical output followed by synchronous demodulation producing an error signal that adjusts the bias voltage in a manner that minimizes $V(2f_d)$, i.e., zero error signal $\Rightarrow V(2f_d) = 0$. Since $V(2f_d)$ is significantly weaker than $V(f_d)$, a lower SNR compared with detecting $V(f_d)$ is obtained, which reduces the control loop robustness. A higher modulation index of the pilot signal can be used to increase $V(2f_d)$ but at a cost of signal distortion. The SNR can be improved without increasing the modulation by using two phase-synchronized pilot frequencies, $f_{d1}$ and $f_{d2}$, in which $V(|f_{d1} - f_{d2}|)$ is extracted and minimized [34]. The advantage is that $V(|f_{d1} - f_{d2}|)$ is much stronger than $V(2f_d)$ giving a better SNR. A different bias control approach [35] that does not monitor the pilot signal harmonics has been proposed in which a bipolar square-wave dither pilot signal together with a bias voltage controlled by a microprocessor is applied to the MZM. Bias control using dithering can be detrimental to the analog photonic link performance by reducing its intermodulation-free dynamic range [36]. Examples of dither-less MZM bias control techniques developed for microwave photonics and radio-over-fiber links can be found in [37,38].

The control voltage of commercial bias controller is usually available between ±10 V and ±18 V. If the bias point drift beyond these limits bias reset [13,35] becomes necessary, which might be unacceptable for some applications. Fortunately, fabrication technology of LN MZM as well as bias drift has improved significantly over the years (see Chapter 14). Commercial LN MZMs are available that can be continuously operated for more than 20 years at 50°C without requiring bias reset [39].

## 15.2 Modulator Bias Control for Optical QPSK

Optical quadrature PSK (QPSK) is a two bits per symbol modulation format, which has been receiving wide attention for high capacity transmission [29,40–42] as demonstrated by recent 100 Gb/s and 200-Gb/s experiments [43,44]. Optical QPSK along with polarization-division multiplexing (PDM) is being considered by the telecom industry as the standard signal format for 100-Gb/s ultralong haul dense wavelength division multiplexing (DWDM) networks as of 2010 [45]. A modulator for QPSK is generally more complex than OOK modulator although a single dual-drive MZM has been proposed for optical QPSK [46]. Best performance and robustness are usually achieved using an I/Q or quadrature modulator (QM) [29,47] as depicted in Figure 15.4a. In this section, optimum bias, bias point tracking, and bias-error-induced penalty of a QM for optical QPSK generation are described. Note that QPSK and differential QPSK (DQPSK) transmitters are similar except that the binary information is usually differentially precoded for DQPSK. Optimum biases of the QM for both formats are therefore identical. Further details on (D)QPSK can be found in Chapter 2.

### 15.2.1 Optimum Bias and Analysis of Bias Control for Optical QPSK

The QM consists of two parallel single-drive push–pull MZMs nested in an MZ interferometer (MZI) as depicted in Figure 15.4a. Optical QPSK at the QM output can be produced by combining two independent optical binary PSK (BPSK) signals in quadrature with $\phi_{IQ} = \pi/2$. The BPSK signals are produced by two single-drive push–pull MZMs driven by NRZ signals. Combining the BPSK signals in quadrature with $\phi_{IQ} = \pi/2$ minimizes crosstalk induced by interference. Optimum bias for each MZM of the QM is the ±null point or $V_B = \mp V_\pi^{DC}$ ($\phi_B = \phi_{opt} = \mp\pi$) as shown in Figure 15.1a. This produces optical BPSK with an almost exact $\pi$ phase shift for any NRZ drive voltage swing between zero and $2V_\pi^{RF}$ [29]. This can be seen by substituting $V_B = -V_\pi^{DC}$ into Equation 15.1 giving

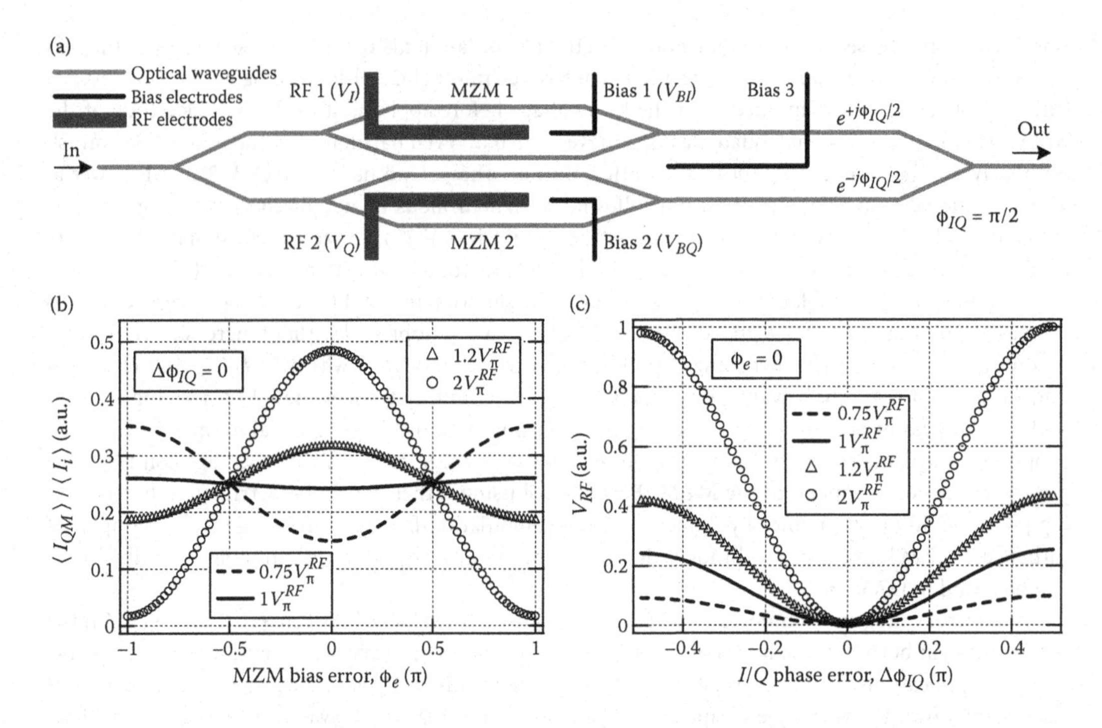


**FIGURE 15.4** (a) Schematic of a quadrature modulator with two parallel single-drive push–pull MZMs nested in an MZI. Only the RF and bias electrodes are shown. (b) Simulated $\langle I_{QM}\rangle/\langle I_i\rangle$ vs. $\phi_e$ with $\Delta\phi_{IQ} = 0$ and (c) $V_{RF}$ vs. $\Delta\phi_{IQ}$ with $\phi_e = 0$ for $V_{pp} = 0.75, 1, 1.2,$ and $2V_\pi^{RF}$. $V_I$ and $V_Q$: $2^{11} - 1$ and $2^9 - 1$ long PRBS. (Reproduced with permission from P. S. Cho, J. B. Khurgin, and I. Shpantzer, *IEEE Photonics Technol. Lett.*, vol. 18, no. 21, pp. 2209–2211, © 2006 IEEE.)

$$E_o = E_i \cos\left[(\pi/2)\left(V_s(t)/V_\pi^{RF} + V_B/V_\pi^{DC}\right)\right] = E_i \sin\left[(\pi/2)\left(V_s(t)/V_\pi^{RF}\right)\right]. \tag{15.8}$$

The sign of $E_o$ is identical to the sign of $V_s(t)$, giving an optical phase shift of $E_o$ that toggles exactly between 0 and $\pi$ for $V_s(t) = \pm V_{pp}/2$, where $0 < V_{pp} \leq 2V_\pi^{RF}$, producing a near perfect optical BPSK with minimal amplitude modulation (except at the NRZ transitions). Minimum optical modulation loss (ratio of the output to input average optical power neglecting intrinsic loss of the modulator) is achieved with $V_s(t) = \pm V_\pi^{RF}$ or $V_{pp} = 2V_\pi^{RF}$ since $E_o = \pm E_i$ (except at the NRZ transitions) producing an output optical intensity equal to the input intensity. Using $V_{pp} = 2V_\pi^{RF}$ also minimizes transfer of NRZ electrical drive imperfections and noise to the optical output due to the nonlinear compression response near the peak point. Consider now the output optical field of the QM given by

$$E_{QM} = (E_i/2)\left[\cos(\phi_I/2)e^{j\phi_{IQ}/2} + \cos(\phi_Q/2)e^{-j\phi_{IQ}/2}\right], \tag{15.9}$$

where $\phi_{I,Q} = \pi\left(V_{I,Q}/V_\pi^{RF} + V_{BI,BQ}/V_\pi^{DC}\right)$, $V_I$ and $V_Q$ are the NRZ drive signals applied to the two MZMs biased at $V_{BI}$ and $V_{BQ}$, respectively. The output optical intensity of the QM with an input intensity $I_i \propto |E_i|^2$ is $I_{QM} \propto |E_{QM}|^2$

$$I_{QM} = (I_i/8)[2 + \cos(\phi_I) + \cos(\phi_Q) + 4\cos(\phi_I/2)\cos(\phi_Q/2)\cos(\phi_{IQ})]. \tag{15.10}$$

Without loss of generality, consider the average output optical power of the QM, $\langle I_{QM}\rangle$, over a period of time $T$ for the case of $V_I = V_Q = V_s(t)$, $V_{BI} = V_{BQ} = V_B$, and $\phi_{IQ} = \pi/2$

$$\langle I_{QM}\rangle = \left(\langle I_i\rangle/4\right)\left\{1 + T^{-1}\int_0^T \cos\left[\pi\left(V_s(t)/V_\pi^{RF} + V_B/V_\pi^{DC}\right)\right]dt\right\}. \tag{15.11}$$

For ±null biasing where $V_B = \mp V_\pi^{DC}$, the conditions for maximum and minimum of $\langle I_{QM}\rangle$ with respect to $V_B$ are

$$\begin{aligned}
\langle I_{QM}\rangle_{\max} &\Rightarrow \left.\frac{\partial^2\langle I_{QM}\rangle}{\partial V_B^2}\right|_{V_B=\mp V_\pi^{DC}} < 0 \Rightarrow \cos\left[\pi\frac{V_s(t)}{V_\pi^{RF}}\right] < 0 \Rightarrow V_\pi^{RF} < V_{pp} \le 2V_\pi^{RF}, \\
\langle I_{QM}\rangle_{\min} &\Rightarrow \left.\frac{\partial^2\langle I_{QM}\rangle}{\partial V_B^2}\right|_{V_B=\mp V_\pi^{DC}} > 0 \Rightarrow \cos\left[\pi\frac{V_s(t)}{V_\pi^{RF}}\right] > 0 \Rightarrow 0 \le V_{pp} < V_\pi^{RF}.
\end{aligned} \tag{15.12}$$

Therefore, null bias of the MZM of the QM can be achieved by maximizing (minimizing) $\langle I_{QM}\rangle$ for $V_\pi^{RF} < V_{pp} \le 2V_\pi^{RF}$ ($0 \le V_{pp} < V_\pi^{RF}$). Figure 15.4b shows simulated optical modulation loss or $\langle I_{QM}\rangle/\langle I_i\rangle$ versus the MZM bias error ($\phi_{e1} = \phi_{e2} = \phi_e$) for $V_{pp}$ = 0.75, 1, 1.2, and $2V_\pi^{RF}$. The simulation uses 12.5 Gb/s NRZ pseudorandom binary sequence (PRBS) signals with a word length of $2^{11} - 1$ and $2^9 - 1$ applied to the two MZMs of the QM with $\phi_{IQ} = \pi/2$ or zero I/Q phase error ($\Delta\phi_{IQ} \equiv \phi_{IQ} - \pi/2 = 0$). From Equation 15.10, the output optical intensity of the QM with +null biases for the nested MZMs, i.e., $V_{BI} = V_{BQ} = -V_\pi^{DC}$, can be written as

$$\begin{aligned}
I_{QM} = (I_i/8)\Big\{&2 - \cos\left(\pi V_I/V_\pi^{RF}\right) - \cos\left(\pi V_Q/V_\pi^{RF}\right) \\
&+ 4\sin\left[\pi V_I/\left(2V_\pi^{RF}\right)\right]\sin\left[\pi V_Q/\left(2V_\pi^{RF}\right)\right]\cos(\phi_{IQ})\Big\}.
\end{aligned} \tag{15.13}$$

Using an optimum drive voltage swing of $V_{pp} = 2V_\pi^{RF}$ for the two independent NRZ signals $V_I$ and $V_Q$ with binary values of $\pm V_\pi^{RF}$, $I_{QM}$ becomes (ignoring the NRZ transitions)

$$I_{QM} = (I_i/2)[1 \pm \cos(\phi_{IQ})], \text{ for } V_I = \pm V_Q. \tag{15.14}$$

It is clear that data-like binary pattern as a result of interference will appear in $I_{QM}$ if $\phi_{IQ} \neq \pi/2$ generating low-frequency components in the RF spectrum of $I_{QM}$. Therefore, a minimum integrated RF spectral power of $I_{QM}$ in the low-frequency region (e.g., fraction of the clock rate) is an indication that $\phi_{IQ}$ is close to $\pi/2$. Figure 15.4c shows the simulated low-frequency RF spectral energy of $I_{QM}$ (denoted by $V_{RF}$) vs. $\Delta\phi_{IQ}$ for different $V_{pp}$ using similar NRZ drive signals with additive white Gaussian noise as in Figure 15.4b. Bias errors for both null-biased MZMs are assumed to be zero in this case, i.e., $\phi_{e1} = \phi_{e2} = \phi_e = 0$. Note that except for a scaling factor due to the optical modulation loss (see Figure 15.4b) the dependence of $V_{RF}$ on $\Delta\phi_{IQ}$ is not affected by $V_{pp}$. Figures 15.5a and 15.5b show simulated RF spectrum of $I_{QM}$ for $\Delta\phi_{IQ} = 0$ and $\Delta\phi_{IQ} = -\pi/2$, respectively.

Based on the analysis above and simulation results shown in Figures 15.4b,c and 15.5a,b, a dither-less bias control algorithm for optical QPSK was developed that searches for the optimum bias points of the QM [47]. Figures 15.5c to 15.5e show typical simulation results of the QM bias control loop. A 12.5-GSym/s optical QPSK transmitter with the QM dither-less bias control approach described above was constructed and tested experimentally [47]. BER measurement using a one-symbol-delay

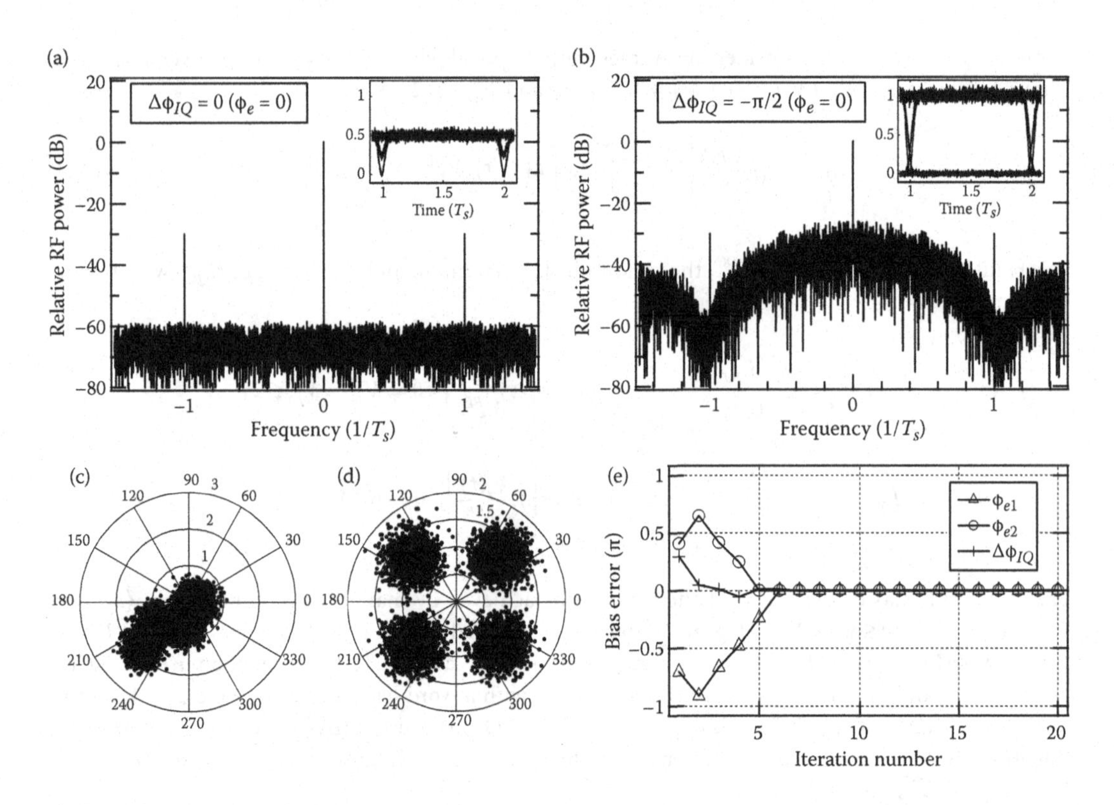


**FIGURE 15.5** RF spectrum of $I_{QM}$ with $\phi_e = 0$ and $V_{pp} = 2V_\pi^{RF}$ for (a) $\Delta\phi_{IQ} = 0$ ($\phi_{IQ} = \pi/2$) and (b) $\Delta\phi_{IQ} = -\pi/2$ ($\phi_{IQ} = 0$). Insets show corresponding waveforms of $I_{QM}$. Bottom row: QM bias control loop simulation result for $V_{pp} = 1.2V_\pi^{RF}$. Constellation plots of the QM optical field output at (c) startup and (d) after 20 iterations of the control loop. (e) Convergence plot of bias errors for $V_{pp} = 1.2V_\pi^{RF}$. Bias errors converged in seven iterations to $\phi_{e1} \approx 0.0042\pi$, $\phi_{e2} \approx 0.0028\pi$, and $\Delta\phi_{IQ} \approx -0.0023\pi$. (Reproduced with permission from P. S. Cho, J. B. Khurgin, and I. Shpantzer, *IEEE Photonics Technol. Lett.*, vol. 18, no. 21, pp. 2209–2211, © 2006 IEEE.)

interferometer receiver reveals a 1 dB power penalty at $10^{-9}$ BER compared with manual bias adjustment. This is attributed to the limited drive voltage swing of only $V_{pp} \sim 1.2V_\pi^{RF}$ where $\langle I_{QM}\rangle/\langle I_i\rangle$ is relatively flat and insensitive to $\phi_e$ for $V_{pp}$ close to $V_\pi^{RF}$ as can be seen in Figure 15.4b. Reduced penalty is expected using a QM with a lower $V_\pi^{RF}$ and/or driving the QM with a higher $V_{pp}$. In addition to the dither-less bias control circuit for optical QPSK described above, dither-based bias controllers for the QM have also been suggested [48,49]. Another technique uses integrated monitor waveguides in a $LiNbO_3$ QM to facilitate bias monitoring has been reported [50]. Commercial dither-based bias controllers for QM are also available [51].

## 15.2.2 Nonoptimum Bias Penalty for Optical QPSK

The impact of QM bias error and nonoptimum NRZ drive voltage swing, i.e., $\phi_{e1,2} \neq 0$, $\Delta\phi_{IQ} \neq 0$, and $V_{pp} < 2V_\pi^{RF}$, on the optical QPSK signal was quantified via BER computation. The $Q^2$-factor was obtained from the BER via Equation 15.4 and a reverse operation described in Section 15.1.2. Figure 15.6 shows the simulated $Q^2$ penalty vs. $\phi_{e1,2}$ and $\Delta\phi_{IQ}$. The penalty is referenced to the $Q^2$-factor for the ideal case of $\phi_{e1,2} = \Delta\phi_{IQ} = 0$ and $V_{pp} = 2V_\pi^{RF}$ with the SNR corresponds to a BER of $10^{-3}$ or $Q^2 \approx 9.8$ dB achieved by adding a fixed amount of optical additive white Gaussian noise. Ideal coherent detection of the QPSK signal was used in the simulation with no receiver impairments, e.g., no phase noise and no frequency

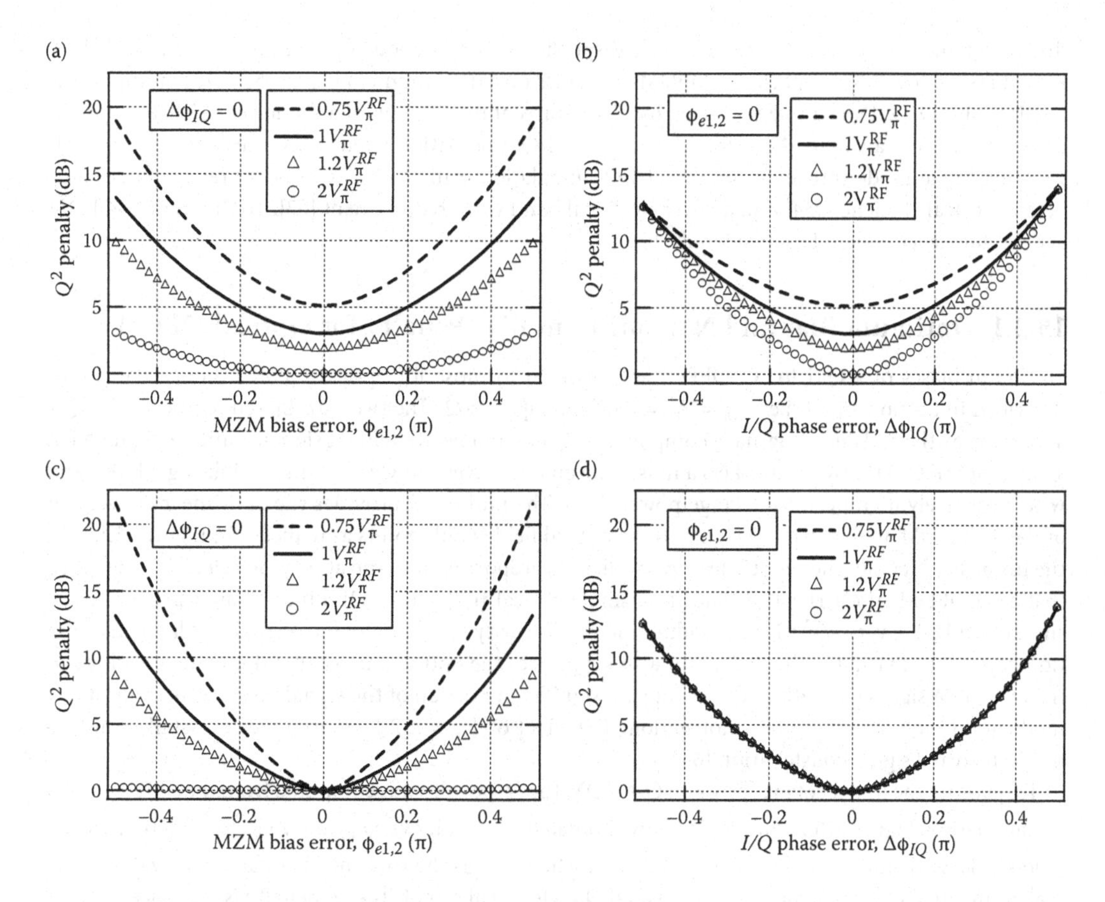


**FIGURE 15.6** Simulated $Q^2$ penalty of optical QPSK vs. $\phi_{e1,2}$ (left column) and vs. $\Delta\phi_{IQ}$ (right column) for $V_{pp}$ = 0.75, 1, 1.2, and $2V_\pi^{RF}$. Top and bottom rows represent results for uncompensated and compensated optical modulation loss, respectively. $V_I$ and $V_Q$: $2^{11} - 1$ and $2^9 - 1$ long PRBS.

offset of the transmitting and local lasers. The effect of optical modulation loss or reduction of output optical power (see top row of Figure 15.6) due to nonoptimum biases and nonideal drive voltages giving rise to increasing penalty is evident and consistent with the results shown in Figure 15.4b. The bottom row of Figure 15.6 shows the $Q^2$ penalty in which the optical modulation loss has been removed by normalizing the output optical power of the QM such that the optical SNR is constant. Note the insensitivity of the penalty to $\phi_{e1,2}$ for $V_{pp} = 2V_\pi^{RF}$, as shown in Figure 15.6c, where the optical modulation loss is compensated giving a $\phi_{e1,2} \approx \pm 0.34\pi$ ($\pm 61.2°$) for a penalty within 0.1 dB for ideal coherent detection. The penalty is more sensitive to $\Delta\phi_{IQ}$ but does not depend on $V_{pp}$ giving a $\Delta\phi_{IQ} \approx \pm 0.02\pi$ ($\pm 3.6°$) for 0.1 dB penalty as shown in Figure 15.6d. Different bias error tolerances are expected if receiver impairments are introduced.

## 15.3 Modulator for Optical OFDM Generation

Optical orthogonal frequency-division multiplexing (O-OFDM) is an emerging multicarrier modulation format capable of high spectral efficiency [52]. It is analogous to the RF version of OFDM employed in wireless communications standards (e.g., WiFi, WiMAX) to mitigate microwave frequency-selective multipath fading [53]. O-OFDM uses many subcarriers modulated at a much lower data rate for transmission [54]. Long symbol length and the use of cyclic prefix make O-OFDM tolerant to group velocity dispersion and robust to polarization mode dispersion in single-mode optical fiber transmission [55]. Recent research on high-capacity O-OFDM transmission can be found in [52,56–58]. Depending on the applications, O-OFDM can be produced by direct current modulation of a diode laser [59], by external modulation using a single MZM [60], or by direct I/Q modulation via a quadrature modulator (QM) [61]. For single-mode transmission, direct I/Q modulation using a QM is usually preferred due to its better spectral efficiency [54] and a lower electrical bandwidth requirement [62]. To this end, O-OFDM generation via a QM will be considered.

### 15.3.1 Optimum Bias and Nonoptimum Bias Penalty for Optical OFDM

Optimum biases of a QM for O-OFDM are identical to those for optical QPSK produced by a QM described in Section 15.2.1, i.e., $V_{BI} = V_{BQ} = -V_\pi^{DC}$ and $\phi_{IQ} = \pi/2$. The two null-biased MZMs of the QM are driven by the real and imaginary components of the complex RF OFDM signal. Unlike a digital NRZ signal, the RF OFDM signal displays a noise-like analog waveform with no discernible logic levels and with a relatively high peak-to-average power ratio. The null point provides the best linear conversion of the RF OFDM analog driving signal to the optical field (±null points as depicted in Figure 15.1a). In the linear field region, the modulated optical field is proportional to the applied RF electrical signal so that each OFDM subcarrier translates to a single optical frequency [63]. Linear conversion is critical since O-OFDM is very sensitive to modulation nonlinearity as a result of bias errors and large modulation depth [61]. Biasing at the null point also suppresses the optical carrier that does not contribute to the O-OFDM signal strength and it occupies a significant portion of the signal power, especially at low modulation index, carrying no information. The effect of $\phi_{IQ} \neq \pi/2$ is interference crosstalk resulting into a distorted signal constellation [64].

To quantify the impact of bias error on O-OFDM the $Q^2$ penalty was computed from the BER obtained via simulation. Figures 15.7a and 15.7b show simulated $Q^2$ penalty vs. $\phi_{e1,2}$ and $\Delta\phi_{IQ}$ for different modulation indexes defined as $M \equiv \pi\sqrt{2}\,V_{rms}/V_\pi^{RF}$ [65] where $V_{rms}$ is the root-mean-square (rms) value of the real or imaginary component of the applied OFDM electrical signal. The $Q^2$ penalty is referenced to a $Q^2 \approx 9.8$ dB or $10^{-3}$ BER for $\phi_{e1,2} = \Delta\phi_{IQ} = 0$ and for $M \ll 1$ with optical additive white Gaussian noise. To remove the penalty incurred by the optical modulation loss (ratio of the output to input average optical power neglecting intrinsic loss of the QM) or $L_M$, the amount of optical noise was adjusted according to the QM output power to keep the optical SNR constant. The simulation uses 256 subcarriers with QPSK modulation where the two tributaries are $2^{11} - 1$ and $2^9 - 1$ long PRBS. Ideal coherent detection of the O-OFDM signal was used in the simulation with no receiver impairments, e.g., no phase noise and no frequency offset of the transmitting and local lasers. As can be seen, O-OFDM is very susceptible to $\phi_{e1,2}$ for small $M$ but not as severe for $M \geq 1$. Small $M$ values also yield large optical losses, for example, with optimum biasing $L_M(M = 0.15) \approx -29$ dB vs. $L_M(M = 1) \approx -13$ dB. On the other hand, as can been in Figure 15.7c and 15.7d large modulation is detrimental with higher penalty. This is attributed to clipping and the nonlinearity of the MZM transfer response as more intense intermodulation components are generated producing more distortion to the O-OFDM signal at higher $M$ values. From Figure 15.7a and 15.7b, a bias error within ±0.03π (±5.4°) is required for both $\phi_{e1,2}$ and $\Delta\phi_{IQ}$ at $M = 1$, i.e., $V_{rms} = V_\pi^{RF}/\left(\pi\sqrt{2}\right)$, to keep the penalty below 0.1 dB for ideal coherent detection. Different bias error tolerances are expected in the presence of receiver impairments and for modulation other than QPSK.

### 15.3.2 Bias Control for Optical OFDM

Conventional bias control technique such as dithering using pilot-tone for null biasing of the MZMs of the QM, e.g., detect and minimizes $V(f_d)$, might be applicable to O-OFDM. An option is to use an unoccupied subcarrier as a dither pilot signal for bias control as demonstrated in a subcarrier-multiplexed

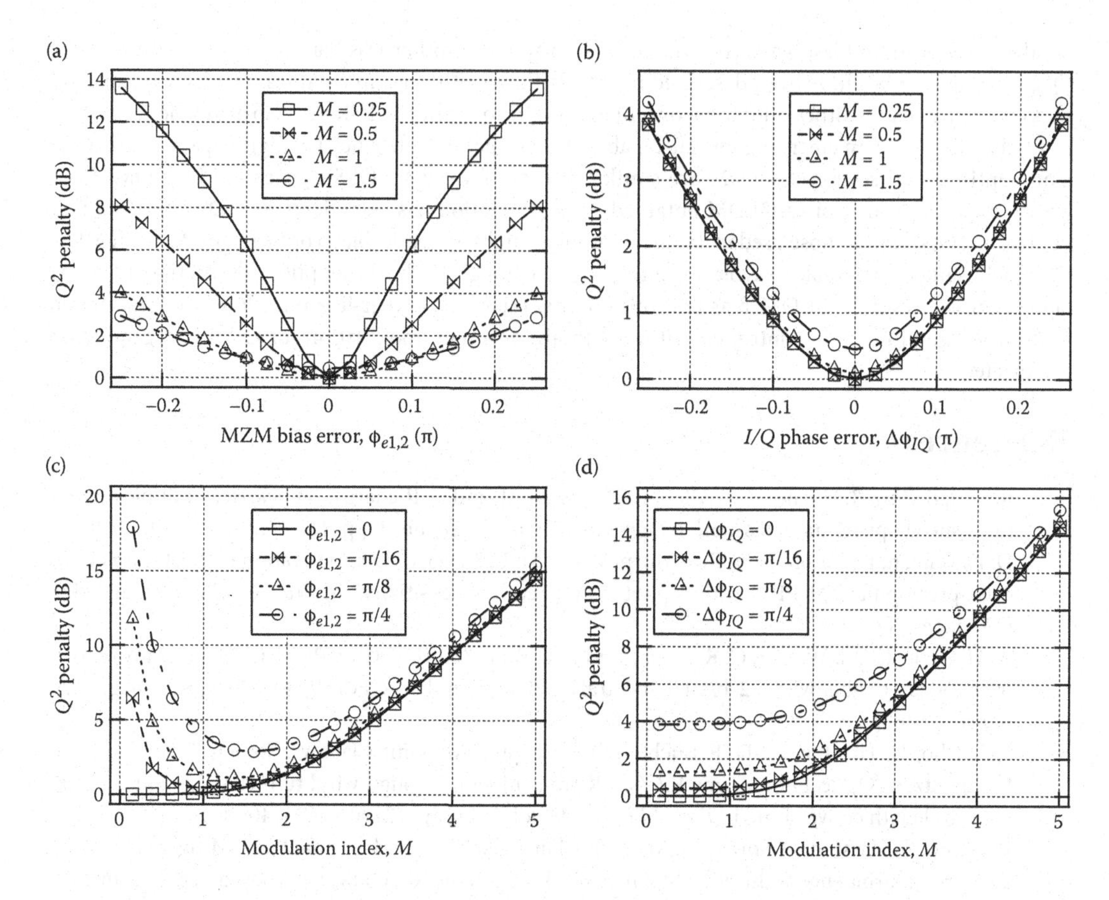


**FIGURE 15.7** Simulated $Q^2$ penalty for O-OFDM vs. (a) $\phi_{e1,2}$ ($\Delta\phi_{IQ} = 0$) and vs. (b) $\Delta\phi_{IQ}$ ($\phi_{e1,2} = 0$) for different $M$ values. $Q^2$ penalty vs. $M$ for different $\phi_{e1,2}$ and $\Delta\phi_{IQ}$ are shown in (c) and (d), respectively. QPSK modulation ($2^{11} - 1$ and $2^9 - 1$ long PRBS) was used with 256 subcarriers.

system [66]. Since O-OFDM performance depends on the bias as well as the modulation index, a bias control circuit combined with a driver amplifier gain control circuit to maintain an optimum modulation index is preferable for robust tracking performance. Dither-based bias control can be detrimental to analog photonics link performance by reducing its intermodulation-free dynamic range [36] when a dither pilot-tone is applied to an MZM. As a result of the analog-like OFDM baseband waveform applied to the QM, degradation of O-OFDM performance using dither-based bias control might be expected. Dither-less bias control of the QM for O-OFDM generation is likely preferred.

## 15.4 Summary

Dynamic tracking of optical modulator bias point at the transmitter is critical to maintain long-term stable operation for applications from communications to sensing. Optimum bias of an MZM for OOK generation and bias-error–induced sensitivity penalty has been discussed. Examples of dither-based and dither-less bias control circuits for OOK and analog photonics applications were introduced. Commercial bias controllers for ±quad, peak, and null point tracking with low bias error are available [67].

Optical QPSK with PDM is gaining wide acceptance as the standard format for 100 Gb/s transmission. Optical QPSK is typically produced using an I/Q modulator or QM with three independent biases.

Analysis of optimum bias, bias-error–induced penalty and a dither-less bias control algorithm of the QM for optical QPSK have been described in details along with experimental test results. Commercial bias controllers using dither pilot tone for QM applicable for optical QPSK are also available [51].

Optical OFDM is an emerging format capable of high spectral efficiency for high-capacity transmission. Optimum biases for O-OFDM are similar to those for optical QPSK generated by a QM. Bias-error–induced penalty for O-OFDM obtained from simulation has been described. Dither-based bias control circuits similar those used in analog photonics might be applicable to optical OFDM. Dither-less approach is likely preferable to avoid potential negative impact of the dither pilot tone on the O-OFDM signal. Future research in QM bias control circuits, especially dither-less ones [68], in conjunction with driver amplifier gain control circuit to maintain optimum modulation for O-OFDM generation is expected.

## References

1. R. Hui, B. Zhu, R. Huang, C. T. Allen, K. R. Demarest, and D. Richards, "Subcarrier multiplexing for high-speed optical transmission," *J. Lightwave Tech.*, vol. 20, no. 3, pp. 417–427, Mar. 2002.
2. D. O. Caplan, B. S. Robinson, R. J. Murphy, and M. L. Stevens, "Demonstration of 2.5-Gslot/s optically-preamplified M-PPM with 4 photons/bit receiver sensitivity," in *Proc. OFC 2005*, Anaheim, CA, paper PDP32.
3. A. H. Gnauck, P. J. Winzer, C. R. Doerr, and L. L. Buhl, "10×112-Gb/s PDM 16-QAM transmission over 630 km of fiber with 6.2-b/s/Hz spectral efficiency," in *Proc. OFC/NFOEC 2009*, San Diego, CA, paper PDPB8.
4. P. S. Cho, G. Harston, K.-D. F. Büchter, D. Soreide, J. M. Saint Clair, W. Sohler, Y. Achiam, and I. Shpantzer, "Optical homodyne RZ-QPSK transmission through wind tunnel at 3.8 and 1.55 μm via wavelength conversion," in *Proc. SPIE*, 2009, vol. 7324, pp. 73240A1–73240A12.
5. P. S. Cho, G. Harston, D. Soreide, J. M. Saint Clair, Y. Achiam, and I. Shpantzer, "Mitigation of weak and strong turbulence-induced fading in optical homodyne RZ-QPSK via delay-diversity transmission," in *Proc. SPIE* 2009, vol. 7324, pp. 73240M1–73240M9.
6. S. K. Korotky and J. J. Veselka, "An RC network analysis of long term Ti:LiNbO$_3$ bias stability," *J. Lightw. Technol.*, vol. 14, no. 12, pp. 2687–2697, Dec. 1996.
7. D. A. Fishman and B. S. Jackson, "Transmitter and receiver design for amplified lightwave systems," in *Optical Fiber Telecommunications IIIB*, I. P. Kaminow and T. L. Koch, Eds., San Diego, CA: Academic Press, 1997, p. 76.
8. N. S. Bergano, "Wavelength division multiplexing in long-haul transoceanic transmission systems," *J. Lightw. Technol.*, vol. 23, no. 12, pp. 4125–4139, Dec. 2005.
9. Issued patents and patent applications on or relating to modulator bias control are available at the web site of U.S. Patent and Trademark Office: http://www.uspto.gov.
10. A. Waksberg and J. I. Wood, "Automatic optical bias control for light modulators," U.S. Patent 3 780 296, Dec. 18, 1973.
11. N. Kato, H. Ito, T. Ichikawa, T. Motohiro, and T. Hioki, "Stabilizing device for optical modulator," U.S. Patent 5 907 426, May 25, 1999.
12. N. Kuwata and H. Nishimoto, "Optical transmitter," U.S. Patent 5 170 274, Dec. 8, 1992.
13. A. Noda, "Optical modulation device having bias reset means," U.S. Patent 5 742 268, Apr. 21, 1998.
14. Y. Nagakubo, S. Ibukuro, A. Hayashi, and T. Tsuda, "Light transmitter having an automatic bias control circuit," U.S. Patent 5 900 621, May 4, 1999.
15. S. Gronbach, "Method and apparatus for controlling a bias voltage of a Mach–Zehnder modulator," U.S. Patent 7 075 695 B2, Jul. 11, 2006.
16. G. Nahapetian, C.-H. Chen, S. Shang, and C. Schulz, "Optical modulator control system," U.S. Patent 7 106 486 B1, Sep. 12, 2006.
17. M.-L. Kao and Y.-K. Park, "Modulator-based lightwave transmitter," U.S. Patent 5 208 817, May 4, 1993.

18. C. H. Cox and E. I. Ackerman, "Modulator bias control," U.S. Patent 7 369 290 B1, May 6, 2008.
19. T. Tajima, "DC bias controller for optical modulator," U.S. Patent 5 726 794, Mar. 10, 1998.
20. J. A. Wilkerson, Jr., J. V. Wernlund, A. Fejzuli, and S. P. Reddy, "Reference frequency quadrature phase-based control of drive level and DC bias of laser modulator," U.S. Patent 6 539 038 B1, Mar. 25, 2003.
21. D. S. Olesen, "Method and an apparatus for modulating light," U.S. Patent 6 810 159 B2, Oct. 26, 2004.
22. J. K. Sikora, "Method and system for first-order RF amplitude and bias control of a modulator," U.S. Patent 6 687 451 B1, Feb. 3, 2004.
23. D. Schneider, J. A. Wilkerson, Jr., and J. V. Wernlund, "Mach–Zehnder modulator bias and driver gain control mechanism," U.S. Patent 6 700 907 B2, Mar. 2, 2004.
24. S. Morin and F. Giraud, "Apparatus for servo-controlling the bias voltage of a light source," U.S. Patent 5 440 113, Aug. 8, 1995.
25. I. Watanabe and S. Komurasaki, "Optical modulation device," U.S. Patent 4 306 142, Dec. 15, 1981.
26. H. Satoh, K. Tanaka, and Y. Ozeki, "Driving circuit for an optical signal modulator," U.S. Patent 5 805 328, Sep. 8, 1998.
27. C. R. Stook and J. B. Wood, "Device for Mach–Zehnder modulator bias control for duobinary optical transmission and associated system and method," U.S. Patent 7 155 071 B2, Dec. 26, 2006.
28. M. Henry, "Apparatus and method for adjusting the control signal of an electro-optical modulator," U.S. Patent 6 587 249 B2, Jul. 1, 2003.
29. A. H. Gnauck and P. J. Winzer, "Optical phase-shift–keyed transmission," *J. Lightw. Technol.*, vol. 23, no. 1, pp. 115–130, Jan. 2005.
30. R. C. Fuller, Y.-H. Kao, and F. J. Peragine, "Method and apparatus for transmitting a modulated optical signal," U.S. Patent 6 671 079 B2, Dec. 30, 2003.
31. J. K. Sikora, "System and method for generating return-to-zero (RZ) optical data in a digital lightwave communications system," U.S. Patent 6 952 534 B1, Oct. 4, 2005.
32. D. J. Allie and J. D. Farina, "Electro-optic modulator having gated-dither bias control," U.S. Patent 5 400 417, Mar. 21, 1995.
33. F. Mussino, G. Ravasio, and C. Zammarchi, "Bias system in an optical CATV modulator," U.S. Patent 5 812 297, Sep. 22, 1998.
34. A. H. Kou, T. K. Yee, and N. L. Swenson, "Automatic bias control for electro-optic modulators," U.S. Patent 6 046 838, Apr. 4, 2000.
35. W. H. Terbrack and M. G. Lee, "Automatic bias controller for electro-optic modulator," U.S. Patent 5 003 624, Mar. 26, 1991.
36. E. I. Ackerman and C. H. Cox III, "Effect of pilot tone-based modulator bias control on external modulation link performance," in *Int. Topical Meeting Microwave Photonics*, Oxford, England, 2000, paper TU4.6.
37. B. Onillon, P. Danès, B. Bénazet, and O. Llopis, "An optical link for microwave clock distribution using optical carrier suppression and DC drift compensation," *Microwave Optical Technol. Lett.*, vol. 49, no. 7, pp. 1634–1637, Jul. 2007.
38. D. Chanda and A. Sesay, "Wireless signal-preamble assisted Mach–Zehnder modulator bias stabilization in wireless signal transmission over optical fibre," *European Trans. on Telecommun.*, vol. 19, no. 6, pp. 669–679, Oct. 2008.
39. F. Heismann, S. K. Korotky, and J. J. Veselka, "Lithium niobate integrated optics: selected contemporary devices and system applications," in *Optical Fiber Telecommunications IIIB*, I. P. Kaminow and T. L. Koch, Eds., San Diego, CA: Academic Press, 1997, p. 413.
40. R. A. Griffin, R. L. Johnstone, R. G. Walker, J. Hall, S. D. Wadsworth, K. Berry, A. C. Carter, M. J. Wale, J. Hughes, P. A. Jerram, and N. J. Parsons, "10 Gb/s optical differential quadrature phase shift key (DQPSK) transmission using GaAs/AlGaAs integration," in *Proc. OFC 2002*, Anaheim, CA, paper FD6–1.

41. P. S. Cho, V. S. Grigoryan, Y. A. Godin, A. Salamon, and Y. Achiam, "Transmission of 25-Gb/s RZ-DQPSK signals with 25-GHz channel spacing over 1000 km of SMF-28 fiber," *IEEE Photonics Technol. Lett.*, vol. 15, no. 3, pp. 473–475, Mar. 2003.
42. P. S. Cho, G. Harston, C. J. Kerr, A. S. Greenblatt, A. Kaplan, Y. Achiam, G. L.-Yurista, M. Margalit, Y. Gross, and J. B. Khurgin, "Investigation of 2-bit/s/Hz 40-Gb/s DWDM transmission over 4 × 100-km SMF-28 fiber using RZ-DQPSK and polarization multiplexing," *IEEE Photonics Technol. Lett.*, vol. 16, no. 2, pp. 656–658, Feb. 2004.
43. M. Salsi, H. Mardoyan, P. Tran, C. Koebele, E. Dutisseuil, G. Charlet, and S. Bigo, "155×100Gbit/s coherent PDM-QPSK transmission over 7,200km," in *Proc. ECOC 2009*, Vienna, Austria, paper PD2.5.
44. P. J. Winzer, A. H. Gnauck, G. Raybon, M. Schnecker, and P. J. Pupalaikis, "56-Gbaud PDM-QPSK: coherent detection and 2,500-km transmission," in *Proc. ECOC 2009*, Vienna, Austria, paper PD2.7.
45. *100G Ultra Long Haul DWDM Framework Document from the Optical Internetworking Forum*, [Online] Available: http://www.oiforum.com/public/documents/OIF-FD-100G-DWDM-01.0.pdf.
46. K.-P. Ho and H.-W. Cuei, "Generation of arbitrary quadrature signals using one dual-drive modulator," *J. Lightw. Technol.*, vol. 23, no. 2, pp. 764–770, Feb. 2005.
47. P. S. Cho, J. B. Khurgin, and I. Shpantzer, "Closed-loop bias control of optical quadrature modulator," *IEEE Photonics Technol. Lett.*, vol. 18, no. 21, pp. 2209–2211, Nov. 2006.
48. R. Griffin, "Modulation control," U.S. Patent 7 116 460 B2, Oct. 3, 2006.
49. T. Terahara, T. Hoshida, K. Nakamura, Y. Akiyama, H. Ooi, J. C. Rasmussen, and A. Miura, "Quadrature phase-shift keying modulator and phase shift amount controlling method for the same," U.S. Patent 7 603 007 B2, Oct. 13, 2009.
50. T. Kawanishi, T. Sakamoto, T. Miyazaki, M. Izutsu, T. Fujita, K. Higuma, S. Mori, and J. Ichikawa, "Integrated $LiNbO_3$ modulator for high-speed optical quadrature phase shift keying," in *Proc. COTA 2006*, Whistler, BC, Canada, paper CWC2.
51. See, for example, http://www.yylabs.com.
52. H. Takahashi, A. Al Amin, S. L. Jansen, I. Morita, and H. Tanaka, "DWDM transmission with 7.0-bit/s/Hz spectral efficiency using 8×65.1-Gbit/s coherent PDM-OFDM signals," in *Proc. OFC/NFOEC 2009*, San Diego, CA, paper PDPB7.
53. S. Hara and R. Prasad, *Multicarrier Techniques for 4G Mobile Communications*. Boston, MA: Artech House, 2003.
54. J. Armstrong, "OFDM for optical communications," *J. Lightw. Technol.*, vol. 27, no. 3, pp. 189–204, Feb. 2009.
55. W. Shieh, W. Chen, and R. S. Tucker, "Polarization mode dispersion mitigation in coherent optical orthogonal frequency division multiplexed systems," *Electron. Lett.*, vol. 42, no. 17, pp. 996–997, Aug. 2006.
56. M. Nazarathy, J. Khurgin, R. Weidenfeld, Y. Meiman, P. Cho, R. Noe, I. Shpantzer, and V. Karagodsky, "Phased-array cancellation of nonlinear FWM in coherent OFDM dispersive multi-span links," *Opt. Express*, vol. 16, no. 20, pp. 15777–15810, Sep. 2008.
57. Y. Ma, Q. Yang, Y. Tang, S. Chen, and W. Shieh, "1-Tb/s per channel coherent optical OFDM transmission with subwavelength bandwidth access," in *Proc. OFC/NFOEC 2009*, San Diego, CA, paper PDPC1.
58. H. Masuda, E. Yamazaki, A. Sano, T. Yoshimatsu, T. Kobayashi, E. Yoshida, Y. Miyamoto, et al., "13.5-Tb/s (135×111-Gb/s/ch) no-guard-interval coherent OFDM transmission over 6,248 km using SNR maximized second-order DRA in the extended L-band," in *Proc. OFC/NFOEC 2009*, San Diego, CA, paper PDPB5.
59. D. Qian, J. Yu, J. Hu, P. N. Ji, and T. Wang, "11.5-Gb/s OFDM transmission over 640 km SSMF using directly modulated laser," in *Proc. ECOC 2008*, Brussels, Belgium, paper Mo.3.E.4.
60. S. L. Jansen, I. Morita, T. C. W. Schnek, N. Takeda, and H. Tanaka, "Coherent optical 25.8-Gb/s OFDM transmission over 4160-km SSMF," *J. Lightw. Technol.*, vol. 26, no. 1, pp. 6–15, Jan. 2008.

61. Y. Tang, W. Shieh, X. Yi, and R. Evans, "Optimum design for RF-to-optical up-converter in coherent optical OFDM systems," *IEEE Photonics Technol. Lett.*, vol. 19, no. 7, pp. 483–485, Apr. 2007.
62. W. Shieh, H. Bao, and Y. Tang, "Coherent optical OFDM: theory and design," *Opt. Express*, vol. 16, no. 2, pp. 841–859, Jan. 2008.
63. B. J. C. Schmidt, A. J. Lowery, and J. Armstrong, "Experimental demonstration of 20 Gbit/s direct-detection optical OFDM and 12 Gbit/s with a colorless transmitter," in *Proc. OFC 2007*, Anaheim, CA, paper PDP18.
64. W.-R. Peng, B. Zhang, X. Wu, K.-M. Feng, A. E. Wilner, and Sien Chi, "Compensation for I/Q imbalances and bias deviation of the Mach–Zehnder modulators in direct-detected optical OFDM systems," *IEEE Photonics Technol. Lett.*, vol. 21, no. 2, pp. 103–105, Jan. 2009.
65. Y. Tang, K.-P. Ho, and W. Shieh, "Coherent optical OFDM transmitter design employing predistortion," *IEEE Photonics Technol. Lett.*, vol. 20, no. 11, pp. 954–956, Jun. 2008.
66. Q. Jiang and M. Kavehrad, "A subcarrier-multiplexed coherent FSK system using a Mach–Zehnder modulator with automatic bias control," *IEEE Photonics Technol. Lett.*, vol. 5, no. 8, pp. 941–943, Aug. 1993.
67. See, for example, http://www.photonicsinc.com, http://www.yylabs.com, and http://www.photoline.com.
68. P. S. Cho and M. Nazarathy, "Bias control for optical OFDM transmitters," *IEEE Photonics Technol. Lett.*, vol. 22, no. 14, pp. 1030–1032, Jul. 2010.

# V

# Applications beyond the Digital Telecom System

# 16

# Broadband Electric Field Sensors

Vittorio M. N. Passaro
*Politecnico di Bari*



## 16.1 Introduction

During the last years, interest in electromagnetic field (EMF) sensors has widely increased. In fact, EMF measurements have a critical part in various scientific and technical areas, such as process control, electric field monitoring in medical apparatuses, ballistic control, electromagnetic compatibility, and microwave integrated circuit testing.

Moreover, in the twentieth century, the environmental exposure to artificial sources of EMFs was increased and the potential health effects of EMFs have received particular attention [1,2]. In particular, effects of electric and magnetic fields at frequencies less than 100 kHz on physiology, behavior, reproduction, and cancer development have been investigated and consequences of heating produced by high frequency EMFs have been widely studied. The presence of effects on human health from medium- and high-intensity EMFs and the precautionary principle induced many states to impose maximum permissible EMF exposure levels [3]. The Institute of Electrical and Electronics Engineers (IEEE), the International Radiation Protection Association (IRPA), the World Health Organization (WHO), and the European Union also spread recommendations about safety levels with respect to human exposure to EMFs [4–7]. To verify compliance with legal exposure limits, it is obviously necessary to develop high sensitivity EMF sensors, working in either the near- or far-field region [8].

The near-field region is well-known as the region in proximity to an antenna or other radiating structure, in which electric and magnetic fields do not have a substantially plane-wave character, but they vary considerably from point to point. The near-field region is further subdivided into the reactive near-field region, closest to the radiating structure and containing most or nearly all of the stored energy, and the radiating near-field region, where the radiation field dominates over the reactive field, but it does not have a plane-wave character. In the near-field region, which extends out to a distance of $d^2/\lambda$ from the EMF source (where $d$ is the largest dimension of the effective aperture of the source and $\lambda$ is the electromagnetic radiation wavelength), it is necessary to separately estimate the electric and magnetic field intensities.

The far-field region, also called the free space region, is the region of the field of an antenna where the angular field distribution is essentially independent of the distance from the antenna and the field has a predominantly plane-wave character. In the far-field region, one usually measures the electric field intensity and, from this value, the magnetic field intensity can also be estimated.

Conventional EMF measurement systems currently use active metallic probes, which can disturb the measured EMF and make sensor very sensitive to electromagnetic noise. Photonic EMF sensors exhibit great advantages with respect to the conventional ones: a very good galvanic insulation, high sensitivity, and very wide bandwidth. For these reasons, in the last twenty years a great research effort has been devoted to integrated optics and optical fiber EMF sensors.

In this chapter, different configurations proposed for electric field optical sensing are shown. Several optical fiber sensors for electric field measurement are examined, including those realized by jacketing the fiber with electrochromic, thermo-optic, and polymeric materials and those measuring induced strain by electrostrictive and piezoelectric transducers. Optical fiber sensors exploiting electro-optic and Joule effects are also presented.

Moreover, integrated optical electric field sensors are reviewed, in which a microwave signal provided by an antenna modulates the optical signal generated by an optical source, including those sensors based on interferometric or coupled-cavity schemes, on the electro-optic or electro-absorption effect, or on lithium niobate or polymer materials. Finally, other approaches to designing electric field sensors, including active sensors and those based on the electrostriction effect, are briefly introduced.

## 16.2 Electric Field Optical Fiber Sensors

Electric field optical fiber sensors are often realized by coating an optical fiber with materials whose optical properties change proportionally to an applied electric field. Polymer-dispersed liquid crystals (PDLCs) are mixtures of liquid crystals (LCs) and one polymer where micron-size droplets of LCs are formed in the polymer matrix using phase separation techniques [9–11]. In the absence of an applied electric field, LC directors inside the droplets do not have any preferred orientation, being randomly distributed. Light is scattered due to the mismatch of refractive indices between the LC droplets and the polymer matrix. Usually, the average size of these droplets is much smaller than the film thickness and the light is scattered many times inside the film. As a result, there is little light transmitted through the film. When an electric field is applied, the LC directors align themselves parallel to the externally applied electric field. In this case, the effective refractive index seen by light with normal incidence becomes equal to the ordinary refractive index of the LC droplets ($n_o$) and, because the refractive index of the selected polymer matrix ($n_p$) is chosen to be equal to $n_o$, the film appears transparent, as in Figure 16.1. Thus, an exposed-core multimode optical fiber coated with PDLCs (Figure 16.2) can be used to detect an external electric field [12].

In presence of an external electric field, the evanescent field propagating in the fiber cladding is altered by changes of PDLC transmittance. Thus, the change in the PDLC optical properties (due to the application of an electric field) results in a change of optical power at the output of optical fiber. Thus the electric field strength is detected by monitoring the intensity of light exiting from the fiber. However, this kind of electric field sensor presents a low sensitivity due to the weak interaction between the evanescent field and LC droplets. Moreover, they show a long response time (around 3 min) which is determined by the electrical time constant of the device.

Other materials exhibiting optical property modifications produced by external electric fields are electrochromic materials. These polymers have polyconjugated structures, and they are insulators in the pure state, becoming conductive after electrochemical oxidation or reduction [13]. An electrochromic polymer, $Fe_4[Fe(CN)_6]_3$, dissolved in pure water has been used to realize an electric field optical fiber sensor [14]. It is constituted by a tapered optical fiber (obtained from a single mode commercial optical

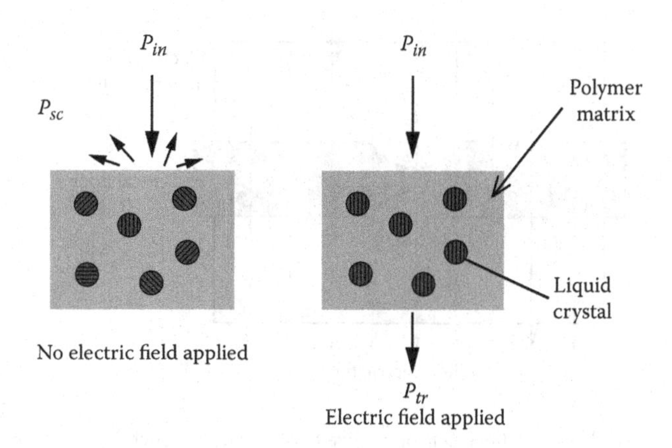


**FIGURE 16.1** Schematic representation of polymer-dispersed liquid crystal electro-optic properties. (Reprinted from *Progress in Quantum Electronics*, 30, V. M. N. Passaro et al., 45–73, Copyright 2006, with permission from Elsevier.)

fiber) surrounded by 0.2 ml of the solution obtained by dissolving 10 mg of soluble $Fe_4[Fe(CN)_6]_3$ in 3 ml of pure water, as in Figure 16.3. This kind of electric field sensor presents a response time of the order of hundreds of seconds, depending on the measured electric field intensity. Both of the previously described electric field optical fiber sensors can be used only for low frequency measurement because of their long response times.

Using an optical fiber thermo-optic (TO) probe, it is possible to estimate the average squared electric field in a wide frequency range (above 30 GHz). By measuring the temperature rise of a resistive element in an electromagnetic field with an optically sensing thermometer, an electromagnetic field sensor can be obtained with a good spatial resolution, and a flat response over a wide frequency band.

The optical fiber thermometer is based on the properties of certain fluorescent phosphors that present temperature-dependent lifetimes of their excited states [15]. Therefore, by exciting such a phosphor and then measuring its decay time, one can measure the temperature of the phosphor and consequently of anything being in thermal equilibrium with it. In this temperature probe, a small quantity (about 40 μg) of a phosphor is attached at the optical fiber end by a clear bonding agent. The phosphor is excited by a pulse of light carried by the fiber, and the light emitted by the decaying phosphor is carried back down the same fiber to determine the decay time. To turn this temperature sensor into an electric-field probe, one needs to introduce a resistive element that will undergo Joule heating due to the current induced in the element when it is placed in an electromagnetic field. Different configurations of TO effect-based fiber sensors have been proposed [16], as sketched in Figure 16.4. With this sensor, high field levels can be measured but sensor sensitivity is not enough for some applications (minimum detectable electric field of 60 V/m).

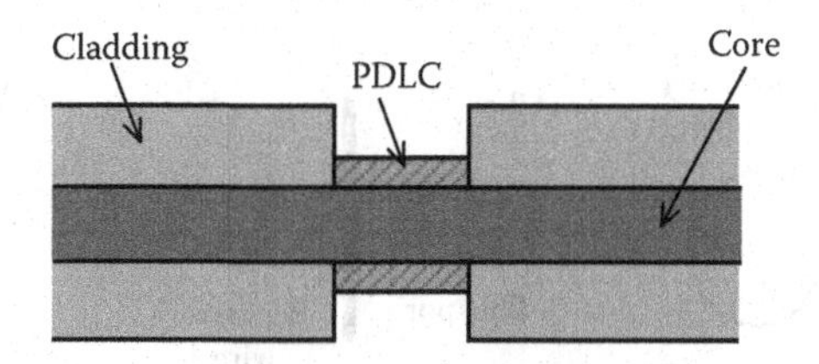


**FIGURE 16.2** Electric field sensor realized by an exposed-core multimode fiber optic coated with polymer-dispersed liquid crystal. (Reprinted from *Progress in Quantum Electronics*, 30, V. M. N. Passaro et al., 45–73, Copyright 2006, with permission from Elsevier.)

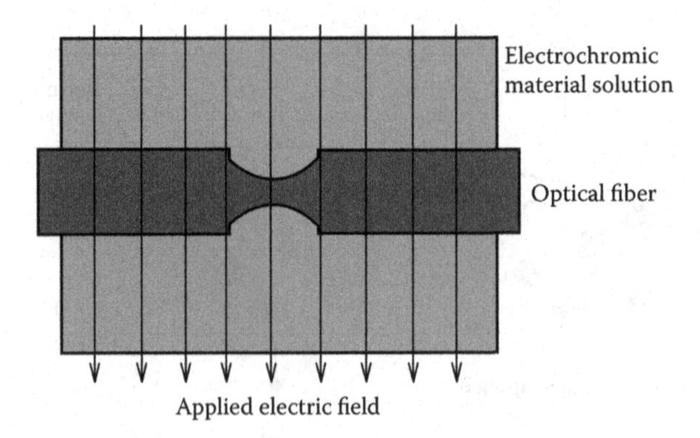


**FIGURE 16.3** Electric field optical fiber sensor realized using an electrochromic polymer. (Reprinted from *Progress in Quantum Electronics*, 30, V. M. N. Passaro et al., 45–73, Copyright 2006, with permission from Elsevier.)

Other electric field optical fiber sensors use piezoelectric or electrostrictive transducers to generate a phase shift in the optical signal propagating in the optical fiber. This phase shift is due to the strain produced by the interaction between the applied electric field and piezoelectric or electrostrictive materials. Piezoelectric materials exhibit a linear relationship between induced strain and applied electric field whereas in electrostrictive materials this relationship is quadratic. Electrostrictive materials are centrosymmetric and typically show negligible induced strain. Recently, however, new composites have been synthesized that exhibit very high induced strain [17]. All these sensors use an optical fiber Mach–Zehnder interferometer to measure the phase shift produced by an external electric field.

In [18], a polyvinylidene fluoride ($PVF_2$) piezoelectric transducer has been proposed. In this sensor, the fiber constituting the sensing arm of the interferometer is stripped of its plastic jacket and is then epoxied onto a 25-μm-thick $PVF_2$ strip, obtaining a sensitivity of 90 $V/m\sqrt{Hz}$ at 10 Hz. Moreover, a $(Pb_{0.73}Ba_{0.27})_{0.97}Bi_{0.02}(Zr_{0.7}Ti)_{0.3}O_3$ (Ba:PZT) electrostrictive transducer has been used, improving the sensitivity down to 10 $mV/m\sqrt{Hz}$ at 1 Hz [19].

For microwave-integrated circuit testing, electric field optical fiber sensors have been developed, in which an electro-optic crystal is attached to the fiber end face. In the sensor proposed in [20], a gradient

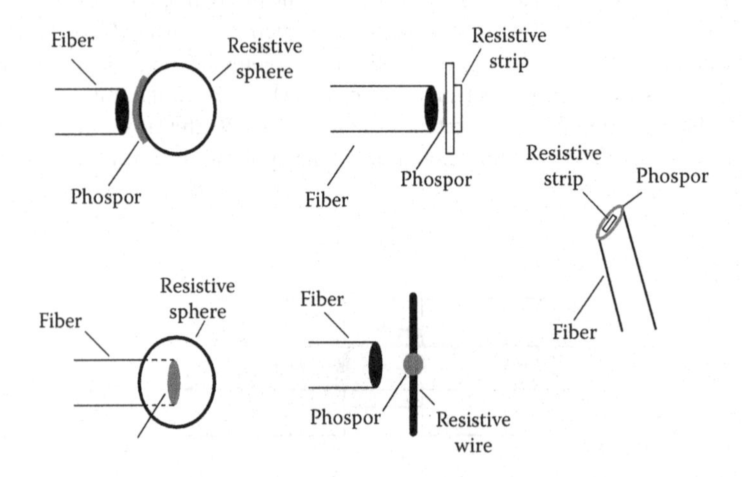


**FIGURE 16.4** Different configurations of thermooptic effect based fiber sensors. (Reprinted from *Progress in Quantum Electronics*, 30, V. M. N. Passaro et al., 45–73, Copyright 2006, with permission from Elsevier.)

index (GRIN) lens with diameter of 1.0 mm and length of 5.0 mm is mounted at the end face of the fiber to focus the beam onto the surface of a micromachined GaAs crystal, which is attached to the GRIN lens using transparent cement. The optical beam reflected from the probe is phase modulated by the RF electric field. This signal is sent to a photodiode by a beam splitter that also converts the phase shift into an intensity modulation.

An electric field sensor based on a ZnTe electro-optic crystal has been studied in [21], proving a 20-GHz bandwidth. A microcavity electro-optic fiber probe has been recently demonstrated using layers of different crystals ($LiTaO_3$, ZnSe, $MgF_2$), for minimally invasive high frequency field sensing. Full three-dimensional reconstruction of vectorial electric field is also possible in X- and Ka-band radiation measurements [22]. A recent demonstration of a broadband electro-optic electric field sensor was used to obtain a minimum detectable electric field intensity of 100 mV/m and a bandwidth of more than 500 MHz [23]. The device is based on a new approach: the fabrication on a side-polished fiber of an electro-optic polymer ring resonator, whose resonance frequencies are changed by the electric field under measurement, through the electro-optic effect. Sensitivity as low as 1 mV/m has been predicted.

# 16.3 Integrated Optical Electric Field Sensors

## 16.3.1 Electric Field Sensors Based on Electro-Optic Modulators

Integrated optical electric field sensors were proposed twenty-five years ago for the first time. They were based on the use of electro-optic Mach–Zehnder modulators and exhibited a minimal detectable electric field of 1 V/m and a bandwidth of 50 MHz [24]. In these devices, a short dipole antenna was connected to the modulator electrodes and the modulator was realized by titanium thermal diffusion in $LiNbO_3$ (see Figure 16.5).

A Mach–Zehnder-based E-field sensor exploiting a discone antenna (Figure 16.6) instead of dipole antenna has been also proposed [25]. This device is optimized to operate from 2 GHz to 18 GHz and electric fields as low as 0.1 V/m can be measured with a good linearity.

The devices described above are optimized for small intensity electric field measurements. For high intensity electric field measurements, two approaches have been proposed. In the first, a Mach–Zehnder modulator is used in conjunction with a capacitive voltage divider. The capacitive divider is used to divide out a small voltage, which can then be applied directly to the Mach–Zehnder modulator [26]. In the second, a Mach–Zehnder modulator is directly exposed to the external electric field and an asymmetry between the two interferometer arms is produced by coating the diffused waveguides with loading strips having different thicknesses. This asymmetry induces a fixed phase difference between optical signals propagating in the interferometer arms [27].

In $LiNbO_3$ electro-optic modulators, an elastic wave at acoustic frequency can appear in the substrate [28,29], when a voltage is applied to the electrode. This wave changes the optical waveguide refractive

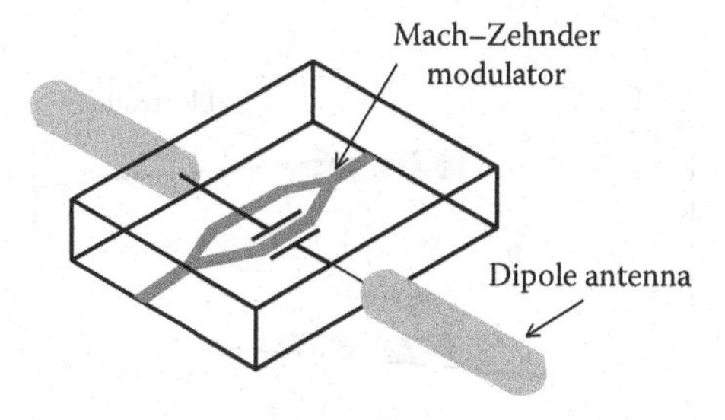


**FIGURE 16.5** Integrated optical electric field sensor based on use of electro-optic $LiNbO_3$ Mach–Zehnder modulator. (Reprinted from *Progress in Quantum Electronics*, 30, V. M. N. Passaro et al., 45–73, Copyright 2006, with permission from Elsevier.)

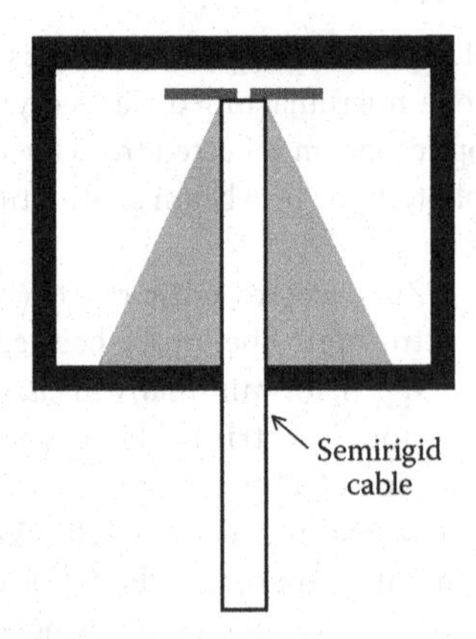


**FIGURE 16.6** Discone antenna for E-field sensors operating from 2 GHz to 18 GHz. (Reprinted from *Progress in Quantum Electronics*, 30, V. M. N. Passaro et al., 45–73, Copyright 2006, with permission from Elsevier.)

index due to the piezoelectric effect. In E-field sensors exploiting a $LiNbO_3$ modulator, this refractive index change affects the device sensitivity and makes the sensitivity not uniform over the operating frequency range. To overcome this problem, it has been proposed [30] to change the substrate width along the propagation direction (Figure 16.7). This way, it is possible to largely decrease (from 4 dB to 1 dB) the electric field device sensitivity deviation in the frequency operating range.

As demonstrated in [31], the output voltage $V_{ph}$ of the photodetector that receives the optical signal from a Mach–Zehnder interferometer modulator is given by

$$V_{ph} = \frac{\eta P_{in}}{2}\left[1+\cos\left(\pi\frac{V_c}{V_\pi}+\phi\right)\right], \tag{16.1}$$

where $P_{in}$ is the input optical power of the modulator, $\eta$ is a conversion factor that includes both photodetector efficiency, insertion losses, and optical fiber propagation losses, $\phi$ is a bias phase shift caused by the intrinsic phase difference between the interferometer arms, $V_\pi$ is the half-wave voltage of the optical modulator, and $V_c$ is the voltage applied to the optical modulator. From Equation 16.1, it is clear that a powerful optical source and a low half-wave voltage will improve the sensitivity. This approach has been followed in sensor designed in [32]. In this sensor, a Nd:YAG laser has been used as optical source to obtain an optical power of 11 dBm at the input of the modulator. Moreover, $V_\pi$ for the $LiNbO_3$ modulator is equal to 4 V (Figure 16.8). By this way a minimum detectable electric field of 0.079 mV/m (around 750 MHz) and a bandwidth of 1 GHz have been obtained.

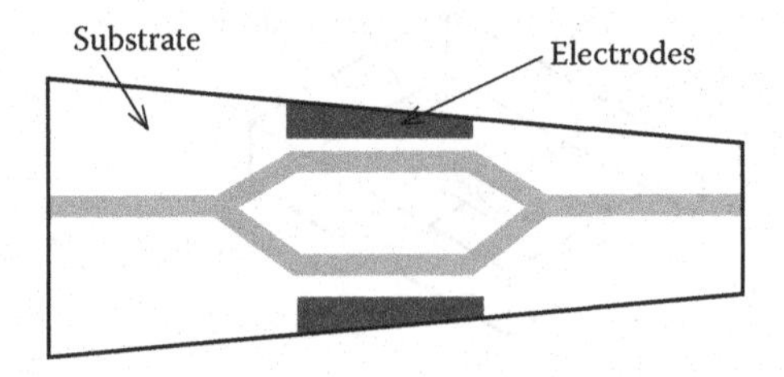


**FIGURE 16.7** Mach–Zehnder electro-optical modulator in which substrate width changes along the propagating direction. (Reprinted from *Progress in Quantum Electronics*, 30, V. M. N. Passaro et al., 45–73, Copyright 2006, with permission from Elsevier.)

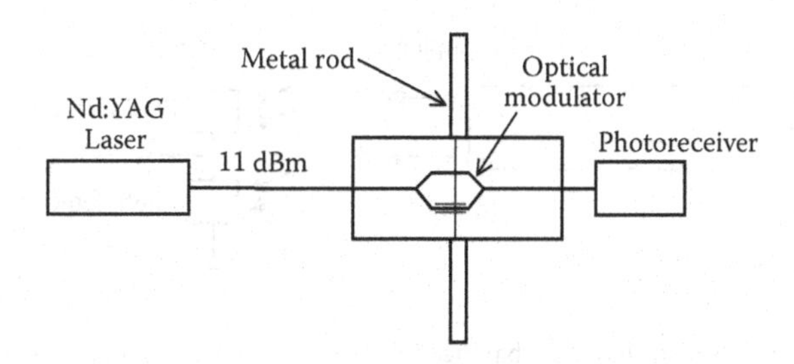


**FIGURE 16.8** Optical waveguide electric field sensor with high optical power input to the modulator and low half-wave voltage $V_\pi$. (Reprinted from *Progress in Quantum Electronics*, 30, V. M. N. Passaro et al., "Electromagnetic field photonic sensors," 45–73, Copyright 2006, with permission from Elsevier.)

Because the electrode capacitance increases linearly with length, long modulator electrodes yielding a low $V_\pi$ can cause a considerable impedance mismatch between the dipole antenna and the same modulator electrodes. To solve this problem, that affects both sensor sensitivity and bandwidth, a segmented electrode configuration has been developed [33], as sketched in Figure 16.9. Using this configuration, an E-field sensor with a bandwidth of 3 GHz and a minimum detectable field of 0.04 mV/m (around 1 Hz) has been demonstrated [34]. Moreover, a segmented-type microminiaturized device with flat frequency response from 750 MHz to 6 GHz and minimum detectable electric field of about 10 mV/m has been recently presented [35].

Electric field sensors with this kind of configuration have been used to monitor EMF generated by the annular phased-array applicator SIGMA-60 of the BSD-2000 system adopted for cancer treatment by hyperthermia [36]. Finally, an electric field microsensor based on a Mach–Zehnder modulator, having a microannular antenna directly integrated with the traveling-wave electrodes on the same $LiNbO_3$ substrate, has been proved in the frequency range from 50 MHz to 2 GHz, with a minimum detectable electric field intensity of 3.9 mV/m [37].

To achieve a further improvement of sensitivity, a balanced detection scheme may be implemented [38]. This choice requires a two output port sensor and a photodetection system constituted by two photoreceivers (Figure 16.10). One benefit of this detection scheme is the rejection of common-mode noise, due to instability of optical source intensity. For a high level of relative intensity noise in the emitted light, a sensitivity improvement of about 34 dB could be obtained with this configuration.

In integrated optical electric field sensor design, great attention to the stability with respect to temperature changes is needed. In particular, when a sensor is used outdoors, for example, for monitoring voltages in transmission lines, the characteristics of the sensor must be stable over a wide temperature range as much as possible. To reach this goal, in the sensor proposed in [34], electrodes have been covered by both an indium tin oxide (ITO) overlay and an additionally evaporated layer of amorphous Si. An alternative approach to further improve the sensor stability with respect to any temperature drift is the use of a polarization modulator [39] instead of an intensity modulator.

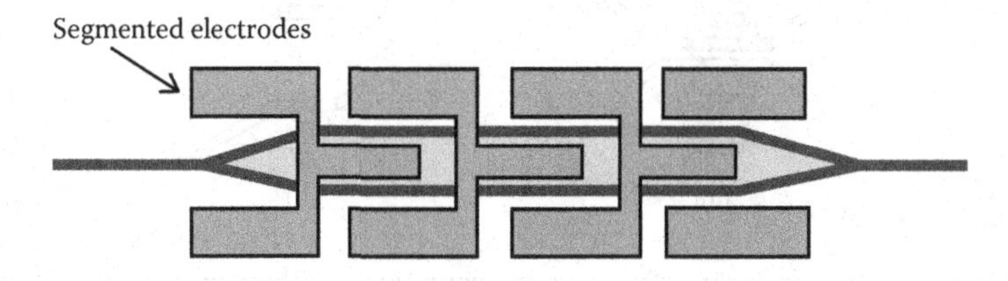


**FIGURE 16.9** E-field sensor structure with segmented electrodes. (Reprinted from *Progress in Quantum Electronics*, 30, V. M. N. Passaro et al., 45–73, Copyright 2006, with permission from Elsevier.)

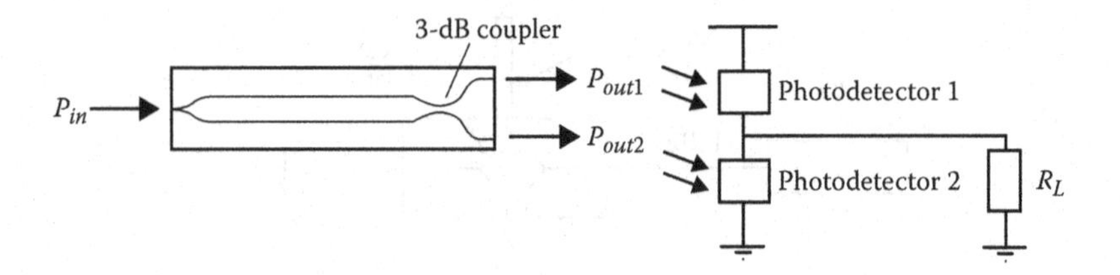


**FIGURE 16.10** Electric field sensor having a balanced detection scheme. (Reprinted from *Progress in Quantum Electronics*, 30, V. M. N. Passaro et al., 45–73, Copyright 2006, with permission from Elsevier.)

Changes in sensitivity $\gamma_E$ or bias $\Phi_B$ due to any temperature drift will degrade the measurement accuracy. To estimate the sensor stability with respect to temperature drift, following parameters may be calculated: $1/\gamma_E \partial\gamma_E/\partial T$, $\partial\phi_B/\partial T$, where $T$ is the absolute temperature of sensor surrounding environment. In the modulator designed in [39], values of $1.3 \times 10^{-5}$ K$^{-1}$ and $2.5 \times 10^{-4}$ rad/K have been achieved, respectively. Thus, this sensor exhibits a measurement error of ±7% in a broad temperature range, from −30°C to 90°C.

The Mach–Zehnder configuration is the most largely used to realize integrated optical E-field sensors based on electro-optic effect. It has been also used to measure high-voltage electric fields, larger than 250 kV/m [40]. However, other interferometer architectures are also possible. In [41], four- and three-port coupler interferometer configurations (see Figure 16.11) have been proposed to realize electric field sensors with a wide linearity range and a good thermal stability.

The four-port configuration allows good sensitivity and complementary output channels, but the need for an active bias voltage has been a significant deterrent for adopting this kind of modulator for E-field measurements. On the other hand, the three-port coupler design can be passively biased and offers a good thermal stability. Using a three-port coupler modulator, an electric field sensor having a bandwidth of 3 GHz and a dynamic range of 65 dB has been obtained.

Another interferometer configuration used to design broadband electric field sensors is based on the use of a mirror that enables construction of an interferometer using only one coupler (Figure 16.12) [42]. With this sensor, a bandwidth of about 3.5 GHz and a sensitivity of 4.5 V/m has been obtained.

Lithium niobate is surely the most popular electro-optic material to design and realize integrated optical E-field sensors. However, using this ferroelectric material, it is not easy to obtain modulators operating up to about 100 GHz [43], due to $LiNbO_3$'s high permittivity at microwave frequencies. Using polymeric electro-optic materials, it may be possible to overcome this limit. Recently, polymer-based technology has permitted the demonstration of ultrafast electro-optic Mach–Zehnder modulators, having wideband frequency response to over 100 GHz [44–46], and electro-optic microring resonators for sensing [47]. In these modulators, guest–host polymers, consisting of a highly nonlinear chromophore and amorphous polycarbonate (APC) or polymethylmethacrylate (PMMA), are used as active

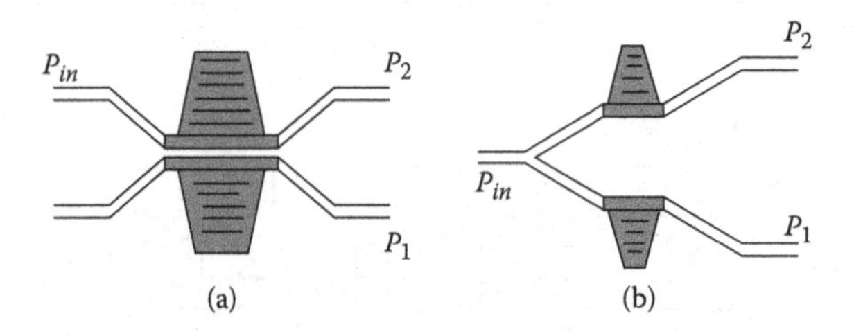


**FIGURE 16.11** (a) Four-port couplers interferometer configuration. (b) Three-port couplers interferometer configuration. (Reprinted from *Progress in Quantum Electronics*, 30, V. M. N. Passaro et al., 45–73, Copyright 2006, with permission from Elsevier.)

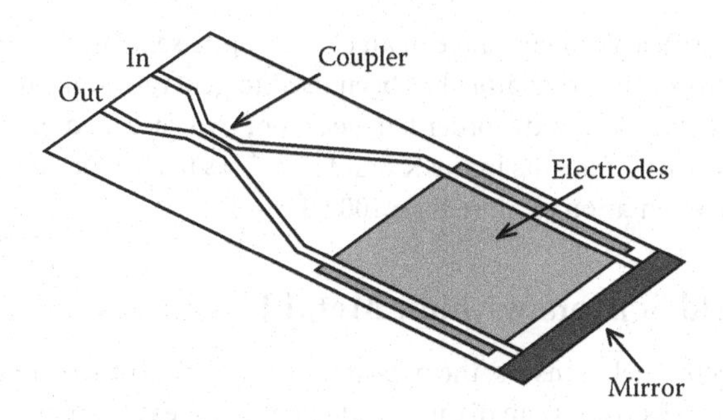


**FIGURE 16.12** Interferometer modulator exploiting a mirror. (Reprinted from *Progress in Quantum Electronics*, 30, V. M. N. Passaro et al., 45–73, Copyright 2006, with permission from Elsevier.)

electro-optic materials. The most important properties of these materials are their very large electro-optic coefficients and their low dielectric constants at microwave frequencies that permit very good velocity matching between the optical field propagating in modulator waveguide and the microwave signal propagating in the modulator electrodes. Moreover, since polymer modulators are limited only by the microstrip electrode losses, a very wide bandwidth becomes obtainable, larger than 300 GHz within at least 16-mm interaction length [48].

In electric field measurements, the polarization of electromagnetic waves under measurement is not always uniform, so sometimes it is necessary to use isotropic sensors [49,50]. In these sensors, the electric signal provided by three orthogonal dipole antennas arrives to three Mach–Zehnder optical modulators. Each of these modulators produces an optical signal whose intensity is proportional to applied voltage. The three optical signals are sent to three photodetectors providing electric signals that, once correctly processed, permit to calculate the total electric field strength as $E_{total} = \sqrt{E_x^2 + E_y^2 + E_z^2}$, where $E_x$, $E_y$, and $E_z$ are the electric field components picked up by the three dipole antennas in the $x$-, $y$-, and $z$-directions, respectively.

Electric field sensors using waveguide optical modulators were originally developed to measure high-frequency electric fields, and the probe antenna was fabricated small enough to avoid any perturbation of high frequency electric field under measurement. For measuring low frequency electric fields, sizes of the probe antenna do not have to be limited, and so wide effective area antennas are used in these applications. In [51], an electric field sensor optimized for low frequency measurements has been proposed (Figure 16.13). This sensor was fabricated using a plate-type probe antenna and tested for measuring electric fields in the range of 0.1 V/cm to 60 V/cm, at frequencies between 60 Hz and 100 kHz.

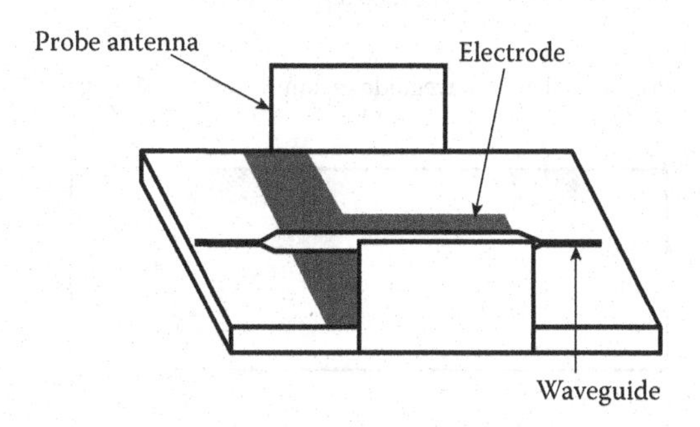


**FIGURE 16.13** Electric field sensor optimized for low frequency measurements. (Reprinted from *Progress in Quantum Electronics*, 30, V. M. N. Passaro et al., 45–73, Copyright 2006, with permission from Elsevier.)

To further enhance the measurement range of an electro-optic sensing system, a technique based on cascading multiple electro-optic modulators has been used to generate high frequency sidebands using low drive frequencies. Fourth- and sixth-order harmonics of two cascaded modulators have been used in near-field electric field measurements in the K-band [52]. This technique could extend the features of microwave K-band instrumentation to more than 100 GHz.

### 16.3.2 Electric Field Sensors without Metal Electrodes

In integrated optical electric field sensors, the presence of metal electrodes can create several problems and limitations. The metal electrodes can disturb the field under measurement and can degrade the sensor bandwidth due to associated capacitance. Electric field sensors operating without metal electrodes have been reported [53]. In this approach, the Mach–Zehnder interferometer sensor is placed directly in the electric field being measured, and this field modulates the incident optical signal. To this aim, reverse poling of $LiNbO_3$ crystal [54] must be used to reverse the sign of the $r_{33}$ coefficient in the region corresponding to one interferometer arm, as in Figure 16.14.

By this way, optical signals propagating in the interferometer arms acquire opposite phase shifts during their propagation. The interference of these two optical signals produces an amplitude-modulated optical signal arriving to the photodetector. In this electric field sensor without electrodes, the reverse-poled region is produced by titanium diffusion and the optical waveguide is formed by a proton exchange process. With the sensor just described, a sensitivity of 0.22 V/m $\sqrt{\text{Hz}}$ and a bandwidth of 1 GHz have been achieved. Moreover, an all-dielectric polymer waveguide electric field sensor realized in a Mach–Zehnder interferometer geometry has demonstrated flat response up to 12 GHz, good sensitivity (2 V/m), and very high dynamic range (>100dB) [55].

### 16.3.3 Electric Field Sensors Based on Electroabsorption Modulators

To improve sensitivity and bandwidth of integrated optical electric field sensors, electroabsorption (EA) modulators instead of $LiNbO_3$ electro-optic ones can be used. EA modulators exploit the electroabsorption phenomenon (change of absorption coefficient of an optical material when a voltage is applied across it) to modulate the intensity of an optical signal [56–58]. To enhance the magnitude of the electroabsorption effect, multiple quantum well (MQW) layers realized with III–V semiconductor

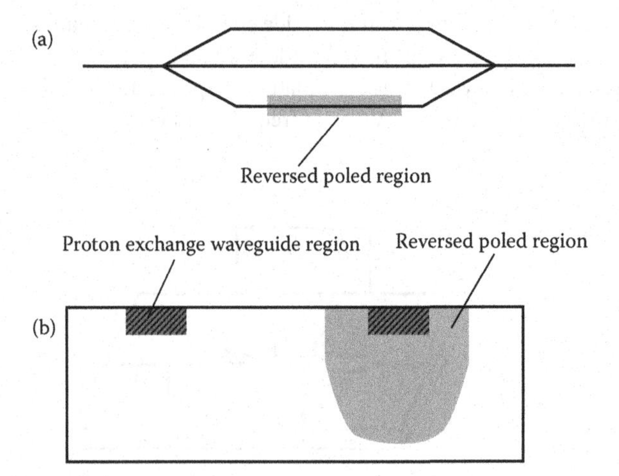


**FIGURE 16.14** Electric field sensor with no metal electrodes. (a) Interferometer structure. (b) Cross section through interferometer arms. (Reprinted from *Progress in Quantum Electronics*, 30, V. M. N. Passaro et al., 45–73, Copyright 2006, with permission from Elsevier.)

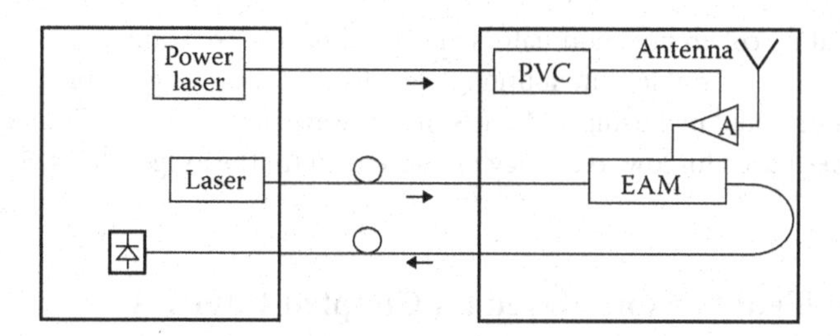


**FIGURE 16.15** Electric field sensor based on electroabsorption modulator. (Reprinted from *Progress in Quantum Electronics*, 30, V. M. N. Passaro et al., 45–73, Copyright 2006, with permission from Elsevier.)

materials are used to form the modulator guiding region. By adopting a traveling-wave configuration for electrodes, EA modulators with bandwidths larger than 50 GHz have been demonstrated [59].

Another electric field sensor using an EA modulator has also been proposed [60]. In this sensor, to keep interference with the measured field at a minimum level, the sensor system has been split into a sensor head and a remote read-out unit, which are connected by optical fibers. The read-out unit contains the laser (operating at 1.55 μm) that provides the optical signal to be modulated by EA modulator; a second laser (at 0.84 μm) is used for remote powering of the transimpedance amplifier in the sensor head; a photodiode connected to the EA modulator output. The sensor head is remotely powered via an integrated photovoltaic cell (PVC) array. This choice permits control of the performance of the sensor by adjusting the bias point of the transimpedance amplifier that links the modulator with the short planar monopole antenna, as shown in Figure 16.15. With this sensor, a minimum detectable electric field of 0.1 V/m and a bandwidth of 6 GHz have been obtained.

### 16.3.4 Electric Field Sensors Based on Coherence Modulators

Coherence modulation of light from a low-coherence optical source has been proposed as an alternative modulation technique for transmitting information in optical fiber links. In these systems, electrical signals are imprinted on light as a sequence of optical delays greater than the coherence time of the optical source. In this way it is possible to design a multichannel E-field optical sensor capable of measuring different applied electric fields [61].

To introduce optical delays, electro-optic $LiNbO_3$ coherence modulators have been used [62]. These modulators are based on the velocity mismatch introduced by an electro-optic $LiNbO_3$ crystal between quasi-TE and quasi-TM modes, as sketched in Figure 16.16.

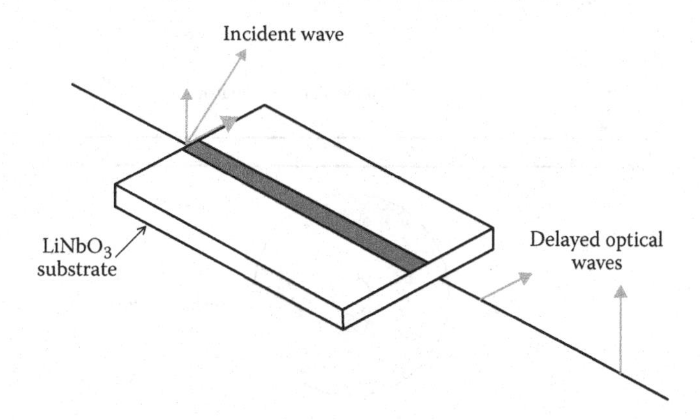


**FIGURE 16.16** Coherence modulator for E-field sensors. (Reprinted from *Progress in Quantum Electronics*, 30, V. M. N. Passaro et al., 45–73, Copyright 2006, with permission from Elsevier.)

Delays introduced by coherence modulators are the sum of a constant contribution depending on the electrode length and a variable contribution depending on the voltage applied to the electrodes. At the receiver, light is demodulated using a Michelson two-wave interferometer. This kind of electric field optical sensor has been used for low frequency measurement (in the range 0 KHz–20 KHz) and exhibits good linearity [63].

### 16.3.5 Electric Field Sensors Based on Coupled Cavities

The resonance frequency shift of two coupled cavities [64,65] can be used as alternative approach to measure electric field strength. In the sensor proposed in [66], electric fields picked up by an antenna produce variations in the resonance frequency of the system constituted by a Fabry–Perot cavity and a disk resonator (see Figure 16.17). Then, it is possible to estimate the external electric field by either monitoring the shift of the coupled cavities' resonance frequency or measuring the change of light intensity at the system output at a given wavelength.

The sensor has been designed to be built using silicon-on-insulator (SOI) technology, and simulations of its behavior show a minimum detectable electric field at the dipole antenna applied to the sensor electrodes of 15 V/m, with a bandwidth around 500 MHz. Improvements in sensitivity up to three to four times than allowed by a simple Fabry–Perot cavity are expected.

### 16.3.6 Active Electric Field Sensors

In active electric field optical sensors, the electric signal picked up from antenna directly modulates the intensity of optical signal generated by laser. This modulated signal is sent to a remote unit in which a photodetector produces a photocurrent proportional to the intensity of received light. The optical source requires a bias current, and so a power supply is needed, which can be electrical or optical. The latter approach is usually preferred because an optical power supply does not disturb the electric field being measured.

In the sensor proposed in [67], an optical signal generated by a vertical-cavity surface-emitting laser (VCSEL) is modulated by the electric signal that arrives from a dipole antenna. The VCSEL bias current is provided by an array of eight silicon solar cells. Light arrives to the solar cell array from an arc lamp by a glass fiber bundle. The modulated optical signal generated by the VCSEL is sent to the photodiode by an optical fiber, as in Figure 16.18. This sensor exhibits a sensitivity of 50 $\mu V/m \sqrt{Hz}$ (at 1 Hz) and a bandwidth of 1 GHz.

A similar E-field sensor has been presented in [68]. In this sensor, a modulated optical signal is generated by an MQW-ridge waveguide laser, biased utilizing a 2 V solar cell. This sensor exhibits a minimum

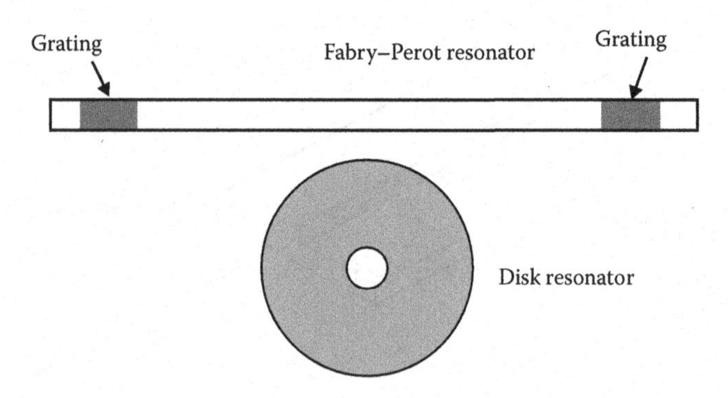


**FIGURE 16.17** Electric field sensor based on two coupled cavities. (Reprinted from *Progress in Quantum Electronics*, 30, V. M. N. Passaro et al., 45–73, Copyright 2006, with permission from Elsevier.)

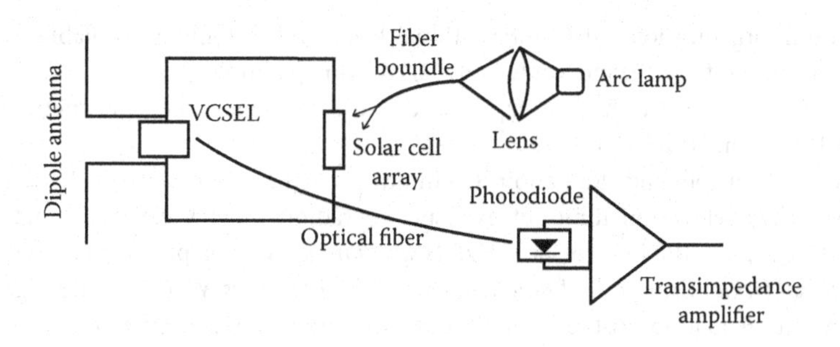


**FIGURE 16.18** VCSEL-based active electric field sensor, with light amplitude modulated by the electric signal arriving from dipole antenna. (Reprinted from *Progress in Quantum Electronics*, 30, V. M. N. Passaro et al., 45–73, Copyright 2006, with permission from Elsevier.)

detectable electric field of 10 mV/m in the frequency range of 0.5 GHz to 12 GHz and a signal-to-noise ratio of 10 dB (at 10 Hz).

### 16.3.7 Other Electric Field Sensors

A different approach of electric field optical sensor is based on the electrostriction effect on the whispering gallery modes (WGM) of polymeric microspheres. This effect creates an elastic deformation of the dielectric material due to the applied electrostatic field, thus inducing a refractive index change in the microsphere. This change can be measured as a WGM resonance frequency shift, allowing electric field measurements as small as 500 V/m [69].

Electric field sensor arrays have also been investigated. These arrays are constituted by an electro-optic crystal in close proximity to the core of an optical fiber and rely on resonant mode coupling. By coupling D-fibers with multiple electro-optic crystals (for example, lithium niobate, potassium titanyl phosphate), an array of sensing points in the same fiber platform can be achieved for electromagnetic pulse measurements with good sensitivity, as low as 100 V/m [70].

## 16.4 Conclusions

This section has briefly reviewed the wide variety of proposed configurations for photonic sensing of electromagnetic fields. Among several approaches for optical fiber sensors, thermo-optic probes show good performance in terms of bandwidth (around 30 GHz), while electrostrictive-based fiber sensors have very good sensitivity, 10 mV/m. A recent approach based on electro-optic polymer ring resonators fabricated on side-polished fibers has improved the sensitivity down to 1 mV/m.

Integrated optical electric field sensors have also been widely studied during the last 20 years and high sensitivities, good noise immunity, and very wide bandwidths have been obtained. In particular, Mach–Zehnder interferometer approaches for sensing allow, in general, very good sensitivity over a very large range from 0.079 mV/m to more than 250 kV/m. Substantial improvements in bandwidth can be obtained using electro-optic polymers, with values much larger than 100 GHz. To realize more compact devices, active electric field sensors that should have 10-mV/m sensitivity and 12-GHz bandwith have been proposed.

## References

1. J. Gasos and P. Stavroulakis, *Biological Effects of Electromagnetic Fields*, Berlin: Springer, 2003.
2. C. Polk, *Handbook of Biological Effects of Electromagnetic Fields*, Boca Raton, FL: CRC Press, 1995.

3. World Health Organization. 2010, International EMF Project Online. Available: http://www.who.int/docstore/pehemf/EMFStandards/who0102/Worldmap5.htm.
4. Safety levels with respect to human exposure to radio frequency electromagnetic fields, 3 kHz to 300 GHz, IEEE Standard C95.1-1999.
5. International Non-Ionizing Radiation Committee of the International Radiation Protection Association, "Guidelines on limits of exposure to radiofrequency electromagnetic fields in the frequency range from 100 kHz to 300 GHz," *Health Physics*, vol. 54, pp. 115–123, Jan. 1988.
6. *Environmental Health Criteria 16: Radiofrequency and Microwaves*, World Health Organization, 1981.
7. Council recommendation on the limitation of exposure of the general public to electromagnetic fields (0 Hz to 300 GHz), *Official Journal of the European Communities (OJEC)*, 1999.
8. Recommended practice for measurements and computations of radio frequency electromagnetic fields with respect to human exposure to such fields, 100 kHz–300 GHz, IEEE Standard C95.3-2002.
9. P. N. Prasad, *Nanophotonics*, New York: Wiley, 2004.
10. F. Simoni, *Nonlinear Optical Properties of Liquid Crystals and Polymer Dispersed Liquid Crystals*, Singapore: World Scientific, 2004.
11. J. Mark, K. Ngai, W. W. Graessley, L. Mandelkern, E. Samulski, J. Koenig, and G. Wignall, *Physical Properties of Polymers*, Cambridge: Cambridge University Press, 2004.
12. M. Tabib-Azar, B. Sutapun, T. Srikhirin, J. Lando, and G. Adamovsky, "Fiber optic electric field sensors using polymer-dispersed liquid crystal coatings and evanescent field interactions," *Sensors Actuators A*, vol. 84, no. 1–2, pp. 134–139, Aug. 2000.
13. P. S. Neelakanta, *Handbook of Electromagnetic Materials*, Boca Raton, FL: CRC Press, 1995.
14. C. Fernandez-Valdivielso, I. R. Matias, M. Gorraiz, F. J. Arrequi, C. Bariain, and M. Lopez-Amo, "Low cost electric field optical fiber detector," in *Opt. Fiber Sensors Conf. Tech. Dig.*, 2002, vol. 1, pp. 499–502.
15. M. Fox, *Optical Properties of Solids*, Oxford: Oxford University Press, 2002.
16. J. Randa, M. Kanda, and R. D. Orr, "Thermo-optic designs for electromagnetic-field probes for microwaves and millimeter waves," *IEEE Trans. Electromagn. Compatibility*, vol. 33, no. 3, pp. 205–214, Aug. 1991.
17. D. W. Richerson, *Modern Ceramic Engineering*, New York: Marcel Dekker, 1992.
18. K. P. Koo and G. H. Sigel Jr., "An electric field sensor utilizing a piezoelectric polyvinylidene fluoride ($PVF_2$) film in a single-mode fiber interferometer," *IEEE J. Quantum Electron.*, vol. 18, no. 4, pp. 670–675, Apr. 1982.
19. S. T. Vohra, F. Bucholtz, and A. D. Kersey, "A fiber optic DC and low frequency electric field sensor," in *Opt. Fiber Sensors Conf. Proc.*, pp. 418–421, 1992.
20. K. Yang, L. P. B. Katehi, and J. F. Whitaker, "Electric field mapping system using an optical-fiber-based electrooptic probe," *IEEE Microwave Wireless Components Lett.*, vol. 11, no. 4, pp. 164–166, Apr. 2001.
21. S. Wakana, T. Ohara, M. Abe, E. Yamazaki, M. Kishi, and M. Tsuchiya, "Novel electromagnetic field probe using electro-flash magneto-optical crystals mounted on optical-fiber facets for microwave circuit diagnostics," in *IEEE MTT-S International Microwave Symposium Digest*, Piscataway, NJ: IEEE Press, 2000, pp. 1615–1618.
22. D.-J. Lee and J. F. Whitaker, "An optical-fiber-scale electro-optic probe for minimally invasive high frequency field sensing," *Opt. Express*, vol. 16, no. 26, pp. 21587–21597, Dec. 2008.
23. H. Sun, A. Pyajt, J. Luo, Z. Shi, S. Hau, A. K.-Y. Jen, L. R. Dalton, and A. Chen, "All-dielectric electro-optic sensor based on a polymer microresonator coupled side-polished optical fiber," *IEEE Sensors J.*, vol. 7, no. 4, pp. 515–524, Apr. 2007.
24. H. Bassen and G. S. Smith, "Electric field probes—A review," *IEEE Trans. Antennas Propagation*, vol. 31, no. 5, pp. 710–718, Sep. 1983.
25. P. Young, "Electro-optic E-field sensors for shielding measurements up to 18 GHz," in *IEEE Int. Symp. Electromagn. Compatibility Proc.*, pp. 87–91, Aug. 1995.

26. N. A. F. Jaeger and L. Young, "High-voltage sensor employing an integrated optics Mach–Zehnder interferometer in conjunction with a capacitive divider," *J. Lightwave Technol.*, vol. 7, no. 2, pp. 229–235, Feb. 1989.
27. N. A. F. Jaeger and L. Young, "Asymmetric slab and strip-loaded integrated optic devices for the measurement of large electric fields," *J. Lightwave Technol.*, vol. 5, no. 6, pp. 745–750, Jun. 1987.
28. K. K. Wong, *Properties of Lithium Niobate*, London: INSPEC, 2002.
29. P. Basséras, R. J. D. Miller, and S. M. Gracewski, "Theoretical analysis of acoustic transients in lithium niobate electro-optic modulators," *J. Appl. Phys.*, vol. 69, no. 11, pp. 7774–7781, Jun. 1991.
30. N. Kuwabara and R. Kobayashi, "Frequency response improvement of electric field sensor using optical modulator," in *IEEE Instrum. Measure. Technol. Conf.*, 1994, vol. 1, pp. 21–24, May 1994.
31. K. Kawano, T. Kitoh, O. Mitomi, T. Nozawa, and H. Jumonji, "A wideband and low-driving power phase modulator employing $LiNbO_3$ optical waveguide at 1.5 μm wavelength," *IEEE Photonics Technol. Lett.*, vol. 1, no. 2, pp. 33–34, Feb. 1989.
32. N. Kuwabara, K. Tajima, R. Kobayashi, and F. Amemiya, "Development and analysis of electric field sensor using $LiNbO_3$ optical modulator," *IEEE Trans. Electromagn. Compatibility*, vol. 34, no. 4, pp. 391–396, Nov. 1992.
33. Z. Fuwen, C. Fushen, and Q. Kun, "An integrated electro-optic E-field sensor with segmented electrodes," *Microwave Opt. Technol. Lett.*, vol. 40, no. 4, pp. 302–305, Feb. 2004.
34. T. Meier, C. Kostrzewa, K. Petermann, and B. Schueppert, "Integrated optical E-field probes with segmented modulator electrodes," *J. Lightwave Technol.*, vol. 12, no. 8, pp. 1497–1503, Aug. 1994.
35. H.-Y. Lee, T.-H. Lee, W.-T. Shay, and C.-T. Lee, "Reflective type segmented electrooptical electric field sensor," *Sensors Actuators A*, vol. 148, no. 2, pp. 355–358, Dec. 2008.
36. P. Wust, J. Berger, H. Fähling, J. Nadobny, J. Gellermann, W. Tilly, B. Rau, K. Petermann, and R. Felix, "Scanning E-field sensor device for online measurements in annular phased-array systems," *Int. J. Radiat. Oncol. Biol. Phys.*, vol. 43, no. 4, pp. 927–937, Mar. 1999.
37. T.-H. Lee, W.-T. Shay, and C.-T. Lee. "Integrated electrooptical electromagnetic field sensor with Mach–Zehnder waveguide modulator with annular antenna," *Microwave Opt. Technol. Lett.*, vol. 50, no. 12, pp. 3125–3128, Dec. 2008.
38. M. Schwerdt, J. Berger, B. Schüppert, and K. Petermann, "Integrated optical E-field sensors with a balanced detection scheme," *IEEE Trans. Electromagn. Compatibility*, vol. 39, no. 4, pp. 386–390, Nov. 1997.
39. O. Ogawa, T. Sowa, and S. Ichizono. "A guided-wave optical electric field sensor with improved temperature stability," *J. Lightwave Technol.*, vol. 17, no. 5, pp. 823–830, May 1999.
40. R. Zeng, Y. Zhang, W. Chen, and B. Zhang, "Measurement of electric field distribution along composite insulators by integrated optical electric field sensor," *IEEE Trans. Dielectrics Electrical Insulation*, vol. 15, no. 1, pp. 302–310, Feb. 2008.
41. M. Kanda and K. D. Masterson, "Optically sensed em-field probes for pulsed fields," *Proc. IEEE*, vol. 80, no. 1, pp. 209–215, Jan. 1992.
42. E. Heyman, J. Shiloh, and B. Mandelbaum, *Ultra-Wideband, Short-Pulse Electromagnetics 4*, New York: Plenum Press, 1999.
43. K. Noguchi, O. Mitomi, and H. Miyazawa, "Millimeter-wave $Ti:LiNbO_3$ optical modulators," *J. Lightwave Technol.*, vol. 16, no. 4, pp. 615–619, Apr. 1998.
44. D. Chen, H. R. Fetterman, A. Chen, W. H. Steier, L. R. Dalton, W. Wang, and Y. Shi, "Demonstration of 110 GHz electro-optic polymer modulators," *Appl. Phys. Lett.*, vol. 70, no. 25, pp. 3335–3337, Jun. 1997.
45. M.-C. Oh, H. Zhang, C. Zhang, H. Erlig, Y. Chang, B. Tsap, D. Chang, A. Szep, W. H. Steier, H. R. Fetterman, and L. R. Dalton, "Recent advances in electrooptic polymer modulators incorporating highly nonlinear chromophore," *IEEE J. Selected Topics Quantum Electron.*, vol. 7, no. 5, pp. 826–835, Sep. 2001.
46. M. Lee, H. E. Katz, C. Erben, D. M. Gill, P. Gopalan, J. D. Heber, and D. J. McGee, "Broadband modulation of light by using an electro-optic polymer," *Science*, vol. 298, no. 5597, pp. 1401–1403, Nov. 2002.

47. A. Chen, H. Sun, A. Pyayt, L. R. Dalton, J. Luo, and A. K.-Y. Jen, "Microring resonators made in poled and unpoled chromophore-containing polymers for optical communication and sensors," *IEEE J. Selected Topics Quantum Electron.*, vol. 14, no. 5, pp. 1281–1288, Sep. 2008.
48. F. Dell'Olio, V. M. N. Passaro, and F. De Leonardis, "Simulation of a high speed interferometer optical modulator in polymer materials," *J. Computational Electronics*, vol. 6, no. 1, pp. 297–300, Sep. 2007.
49. K. Tajima, R. Kobayashi, and N. Kuwabara, "Experimental evaluation of broadband isotropic electric field sensor using three Mach–Zehnder interferometers," *Electron. Lett.*, vol. 34, no. 11, pp. 1130–1132, May 1998.
50. S. Diba and H. Trzaska, "Isotropic receive pattern of an optical electromagnetic field probe based upon Mach–Zehnder interferometer," *IEEE Trans. Electromagn. Compatibility*, vol. 39, no. 1, pp. 61–63, Feb. 1997.
51. Y. K. Choi, M. Sanagi, and M. Nakajima, "Measurement of low frequency electric field using Ti:$LiNbO_3$ optical modulator," *IEE Proc. J. Optoelectron.*, vol. 140, no. 2, pp. 137–140, Apr. 1993.
52. D.-J. Lee and J. F. Whitaker, "Bandwidth enhancement of electro-optic sensing using high-even-order harmonic sidebands," *Opt. Express*, vol. 17, no. 17, pp. 14909–14917, Aug. 2009.
53. D. H. Naghski, J. T. Boyd, H. E. Jackson, S. Sriram, S. A. Kingsley, and J. Latess, "An integrated photonic Mach–Zehnder interferometer with no electrodes for sensing electric fields," *J. Lightwave Technol.*, vol. 12, no. 6, pp. 1092–1098, Jun. 1994.
54. T. Suhara and M. Fujimura, *Waveguide Nonlinear-optic Devices*, Berlin: Springer, 2003.
55. R. Forber, W. C. Wang, D.-Y. Zang, S. Schultz, and R. Selfridge, "Dielectric EM field probes for HPM test and evaluation," in *Annual ITEA Technology Review*, Cambridge (UK): International Test and Evaluation Association, 2007.
56. G. P. Agrawal, *Lightwave Technology: Components and Devices*, New York: Wiley, 2004.
57. C. Pollock and M. Lipson, *Integrated Photonics*, Norwell: Kluwer Academic Publishers, 2003.
58. K. Wakita, *Semiconductor Optical Modulators*, Berlin: Springer, 1998.
59. K. Kawano, M. Kohtoku, M. Ueki, T. Ito, S. Kondoh, Y. Noguchi, and Y. Hasumi, "Polarisation insensitive travelling-wave electrode electroabsorption (TW-EA) modulator with bandwidth over 50 GHz and driving voltage less than 2V," *Electron. Lett.*, vol. 33, no. 18, pp. 1580–1581, Aug. 1997.
60. R. Heinzelmann, A. Stohr, D. Kalinowski, and D. Jaeger, "Miniaturized fiber coupled RF E-field sensor with high sensitivity," in *IEEE Proc. 13th Annu. Meeting Lasers Electro-Opt. Soc.*, 2000, vol. 2, pp. 525–526.
61. C. Gutiérrez-Martínez, G. Trinidad-García, and J. Rodríguez-Asomoza, "Electric field sensing system using coherence modulation of light," *IEEE Trans. Instrumentation Measurement*, vol. 51, no. 5, pp. 985–989, Oct. 2002.
62. H. Porte, J.-P. Goedgebuer, and R. Ferriere, "An $LiNbO_3$ integrated coherence modulator," *J. Lightwave Technol.*, vol. 10, no. 6, pp. 760–766, Jun. 1992.
63. C. Gutiérrez-Martínez and J. Santos Aguilar, "Electric field sensing scheme based on matched $LiNbO_3$ electro-optic retarders," *IEEE Trans. Instrumentation Measurement*, vol. 57, no. 7, pp. 1362–1368, Jul. 2008.
64. K. Vahala, *Optical Microcavities*, Hackensack: World Scientific, 2004.
65. C. Manolatou and H. A. Haus, *Passive Components for Dense Optical Integration*, Berlin: Springer, 2002.
66. V. M. N. Passaro and F. De Leonardis, "Modeling and design of a novel high-sensitivity electric field silicon-on-insulator sensor based on a whispering-gallery-mode resonator," *IEEE J. Selected Topics Quantum Electron.*, vol. 12, no. 1, pp. 124–133, Jan. 2006.
67. W. Mann, and K. Petermann, "VCSEL-based miniaturised E-field probe with high sensitivity and optical power supply," *Electron. Lett.*, vol. 38, no. 10, pp. 455–456, May 2002.
68. J. Svedin and J. Önnegren, "A wide-band electro-optical E-field probe powered by a solar cell," in *1996 Int. Topical Meeting Microwave Photonics Tech. Dig.*, pp. 21–24 suppl.
69. T. Ioppolo, U. K. Ayaz, and M. V. Otugen, "Whispering gallery mode-based micro-optical sensor for electromagnetic field detection," in *AIAA Aerospace Conference*, 2009, pp. 1814.
70. R. Gibson, R. Selfridge, and S. Shultz, "Electric field sensor array from cavity resonance between optical D-fiber and multiple slab waveguides," *Appl. Opt.*, vol. 48, no. 19, pp. 3695–3701, Jul. 2009.

# 17

# Passive Millimeter-Wave Imaging Using Optical Modulators

Yao Peng
*University of Delaware*

Christopher Schuetz
*University of Delaware*

Dennis W. Prather
*University of Delaware*



## 17.1 Overview

Human vision lies within the range of electromagnetic radiation most effectively emitted by the sun, approximately between the wavelengths of 400 nm and 700 nm. Other ranges of the electromagnetic spectrum, such as X-ray, infrared (IR), terahertz radiation, radio-frequency (RF), etc., have been used for applications where they provide a significant advantage over visual observation systems. For example, radar systems provide significantly larger range and resistance to atmospheric disturbances for remote tracking of aircraft. IR systems are very effective at detecting and imaging scenes of varying temperatures or emissivities, enabling the creation of thermal imaging systems. However, the millimeter-wave (MMW) regime remains relatively unexplored despite of the unique ability of MMW radiation to penetrate fog, clouds, smoke, and thin dielectrics while maintaining high resolution. Many potential applications exist for remote aerial imaging of obscured ground scenes and for security scanning. The lack of development in these areas is mainly attributed to the complex nature of MMW sources and detectors. For detection of passive emissions at terrestrial temperatures, very high detection sensitivity is required for an imager since the native blackbody radiation at millimeter wavelengths decreases by a factor of ~$10^8$ from its peak value. To date, most imaging technologies have been limited by the cost of low-noise amplifiers (LNAs) and the physical limitations imposed by the diffraction limit, which require bulky imagers to achieve even modest resolutions.

Microwave detectors are typically based on semiconductor techniques that rely on nonlinearities in the response of these devices as mixers for downconversion or direct rectification of the detected signal. IR band detectors have achieved the most success using calorimetric detectors, such as microbolometers, which use the thermally dependent properties of materials to detect radiated energy. Although efforts have been made to extend both diode-based [1–5] and microbolometer [6–8] techniques to the MMW portion of the spectrum, both of these techniques have significant shortcomings in this frequency range. Calorimetric detector technologies that work well at infrared wavelengths have insufficient sensitivity for passive imaging in the MMW regime. Detectors such as superheterodyne receivers and amplified direct detection receivers, while capable of the desired sensitivities, have only recently been applied to parts of the MMW regime due to difficulties in device fabrication and packaging for high operational frequencies. Parasitic device capacitance and carrier transit times become significant barriers to efficient direct detection or downconversion at frequencies approaching 100 GHz. For remote imaging applications, atmospheric transmission windows are available at frequencies of nominally 35, 89, 140, and 220 GHz. Since a higher frequency of operation generally allows higher spatial resolution from an imager for a fixed aperture size, there is a consistent push for access to the high frequency windows, which are currently exceedingly difficult to access with current electronic technologies. In addition, such microwave-style devices require a high level of integration or scanning mechanisms to achieve desired pixel densities for imaging applications.

In this chapter, recently developed techniques for detection of passive MMW radiation based on optical upconversion techniques are detailed. These techniques use high-frequency optical modulators to convert received MMW radiation to the optical domain where they can be detected using traditional optical photodetectors, which offer near ideal square-law detection with low noise. This detection technique provides several potential advantages over more traditional MMW detection technologies including: detector sensitivities sufficient for detecting passive radiation without the use of high-frequency amplification or cryogenic cooling, wide collection bandwidths, and the ability to readily create distributed aperture imagers.

The underlying technique for optoelectronic MMW detection is shown schematically in Figure 17.1. Native MMW radiation of objects is first collected by a broadband horn antenna. The MMW signal is then fed via transmission lines to an electro-optic modulator. In the modulator, the optical source is modulated by the MMW signal through an induced change in the modulator material properties. While, in principle, any of the modulator types discussed in this text would be an acceptable means of achieving this modulation, the requirement for the modulator to respond at MMW frequencies with the maximum possible modulation efficiency significantly narrows the range of modulation devices

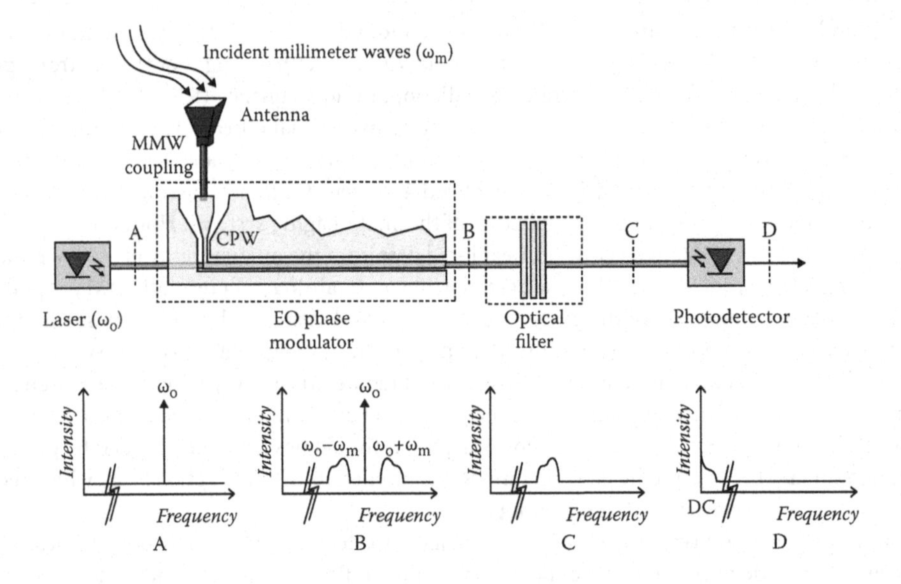


**FIGURE 17.1** The schematic of mechanism of optical up-conversion MMW detector. (Reproduced from C. J. Huang et al., $LiNbO_3$ optical modulator for MMW sensing and imaging, *Proc. SPIE*, vol. 6548, pp. 65480I1-7. © 2007 SPIE. With permission.)

that are effective in this application. At the time of the publication of this text, the lithium niobate (LN), traveling-wave modulator remains the most effective modulator technology for this application. While both phase-only and Mach–Zehnder amplitude modulators offer their own distinct advantages for this application, phase-only modulators in general provide the highest level of conversion efficiency of MMW energy into optical sidebands.

In such modulators, a refractive index change is induced by the MMW field applied by coplanar electrodes tuned to match propagation velocity with the optical carrier. The induced phase change will transfer the MMW power to the sidebands of the carrier in the optical frequency domain. The imposed sidebands are located at a frequency offset from the carrier equal to the incoming MMW frequency. The modulated optical carrier is then suppressed by a bandpass filter. The remaining sidebands are finally converted to a DC photocurrent by a low frequency photodetector. The resultant photocurrent, which is directly proportional to the MMW power fed to the EO modulator, is used to form a MMW image.

The above optics-based architecture possesses advantages for MMW imaging over traditional radio frequency approaches. For example, a wider collection bandwidth is available in our method since optical modulators can be made with relatively wideband operation of several octaves or more. Effectively, this increases the available passive radiation for detection and thus results in better temperature sensitivity of the detector. In addition, amplification is not necessary for modest sensitivities due to the better signal-to-noise figure inherently provided by the optical isolation of the detection electronics from the antenna element. Moreover, the relatively small size of the optical components opens the door for the dense integration of pixels necessary for high-resolution imaging. Finally, the optical upconversion process preserves the phase of the MMW energy, thereby enabling techniques for imaging based on coherent detection. Since such imagers can be implemented in a planar fashion, this technique could enable high resolution imagers to be fabricated without the volumetric scaling typically encountered in traditional focal plane arrays. A prototype system based on distributed aperture techniques with optical processing of the upconverted energy has been demonstrated [9].

## 17.2 MMW Detection Using Optical Techniques

Broadband, high–dynamic range relaying of microwave and MMW signals using optical techniques has been well demonstrated in the literature [10,11] and in previous chapters of this text. However, such techniques have not typically been used for applications requiring high sensitivities due to the conversion losses associated with electro-optic modulation and excess noise derived from sources such as laser relative intensity noise (RIN) and photodetector shot noise. Primarily, research in this area has focused on using high-bandwidth, RF-matched photodetectors to preserve spectral and phase information of the modulation signal. Such high-bandwidth photodetectors place significant limits on the link design by imposing an output impedance requirement that is typically 50 Ω to match RF transmission lines. This relatively low output impedance of the detector serves to severely limit the sensitivity of the link and increase the effect of thermal noise in the detection process. However, for radiometric applications, where preserving spectral and phase information of the detected energy is not necessary, the need for such high-speed, low-impedance photodetectors may be avoided by optical filtering and carrier suppression techniques. The few studies that have been performed regarding the use of optical upconversion for passive imaging [12–14] have been primary focused on the use of optical processing and beam-forming techniques and have used high levels of low noise amplification on the front end to set the noise floor of the system or have used bulk electro-optic crystals to image actively illuminated sources [15]. However, such detection may be utilized for high-sensitivity detectors without low-noise amplifiers when implemented with carrier suppression and low noise photodetectors.

The key components of this process are detailed in Figure 17.1. As shown, an optical source is modulated via incident MMW radiation fed via transmission lines to an optical modulator, which yields sidebands on the optical carrier that are directly proportional to both the collected MMW and the incident optical carrier power. The efficiency of this process is determined by the properties of the modulator, which will be discussed in detail in the next section. After upconversion, the optical signal is filtered to remove the optical carrier and fed to an optical photodetector. The photodetector output current is then directly proportional to the energy in the sidebands and, consequently, the incident MMW radiation. Since the photocurrent is simply responding to the total energy in the sidebands, the photodetector need not operate at MMW frequencies and rather need only have enough bandwidth to respond to the desired scan or slew rate of the imager. In turn, this low-bandwidth requirement enables the use of extremely high-gain transimpedance amplifiers (TIAs), which are commercially available to gains of $10^{13}$ Ω [16]. These high gain TIAs serve the dual purpose of both increasing the overall sensitivity of the MMW detector in V/W and lowering the contribution of Johnson/thermal noise at the output of the photodetector, which is, in most instances, the limiting noise floor of this detection technique. Since modern optical components enable conversion efficiencies from MMW to optical sidebands that approach unity, such detectors should be capable of achieving noise equivalent powers comparable to those of the optical photodetectors, which are on the order of tens of $\mathrm{fW}/\sqrt{\mathrm{Hz}}$.

As mentioned previously, it is desirable for such detectors to operate over the maximum possible range of bandwidths. In this process, the bandwidth of operation is limited, in principle, on the low end by the ability of the optical filter to reject the optical carrier and, on the high end, by roll-off of the modulator response. For this reason, to achieve detectors that operate in MMW transmission bands and achieve higher resolution imagery, it is desirable to design optical modulators that operate at frequencies up to 220 GHz. Later in this chapter, work optimized to deliver optical modulators operating in the 89-GHz transmission band is presented.

## 17.3 Modulator Conversion Efficiency

In the optical detection scheme, obtaining sufficient conversion efficiency of MMW energy into the sidebands is critical to achieving the desired levels of detector sensitivity. Although many factors such as system optical losses, detector responsivity, and postdetection amplification contribute to the system conversion efficiency, the efficiency with which the modulator transfers energy into the carrier's sidebands is one of the prime factors.

For a phase modulator under sinusoidal modulation of frequency $\omega_m$ the output optical field, $E_0$, can be expressed as

$$E_0(t) = E_i e^{i\omega_0 t} e^{im\cos(\omega_m t + \varphi_m)}, \tag{17.1}$$

where $m$ is the modulation amplitude expressed as $m = \pi V_m/V_\pi$ and $E_i e^{j\omega_0 t}$ is the incident optical field on the modulator. Using the Jacobi–Anger expansion, Equation 17.1 can be expressed as

$$E_0(t) = E_i e^{j\omega_0 t} \sum_{n=-\infty}^{\infty} i^n J_n(m) e^{in(\omega_m t + \varphi_m)}, \tag{17.2}$$

where $J_n$ is the $n$th-order Bessel function. Expressed in frequency space, the output field can be reduced to a series of frequency components

$$E_0(\omega_0 \pm k\omega_m) = E_i J_n(m) e^{in\varphi_m}, \tag{17.3}$$

or converted to the respective intensities of each frequency component,

$$I(\omega_0 \pm k\omega_m) = I_i J_k^2(m). \tag{17.4}$$

For the purposes of this discussion we are primarily interested in modulation amplitudes $m<<1$. Using the small argument Bessel function approximation, $J_0(x) \cong 1-x^2/4$ and $J_1(x) \cong x/2$, and neglecting high-order terms, Equation 17.4 then becomes

$$I(\omega_0 \pm k\omega_m) \cong I_i \begin{cases} 1-\dfrac{m^2}{2} & k=0 \\ \dfrac{m^2}{4} & k=\pm 1 \end{cases}. \tag{17.5}$$

Substituting $m = \pi V_m/V_\pi$ to Equation 17.5, we obtain

$$I(\omega_0 \pm k\omega_m) \cong I_i \begin{cases} 1-\dfrac{\pi^2}{2V_\pi^2}V_m^2 & k=0 \\ \dfrac{\pi^2}{4V_\pi^2}V_m^2 & k=\pm 1 \end{cases}, \tag{17.6}$$

where $V_m$ is the applied modulation voltage and $V_\pi$ is the intensity modulation half-wave voltage. Since the input power to the modulator is proportional to the square of the applied voltage, and using the relation $I_m = V_m^2/2Z_e$, we can express the optical power in the first-order sidebands, $I_{FSB}$, as

$$I_{FSB} = \left(\frac{\pi^2 Z_e}{2V_\pi^2}\right) I_i I_m \equiv \eta_{mod,p} I_i I_m, \tag{17.7}$$

where $Z_e$ is the electrode impedance and $\eta_{mod}$ is defined as the modulation efficiency, with units of inverse watts. In the following, we use the modulation efficiency, $\eta_{mod}$, as the figure of merit for the modulator design and optimization, which is not a common method since most $LiNbO_3$ modulators are characterized by the half-wave voltage, $V_\pi$. Using $\eta_{mod}$ is, however, more convenient from the standpoint of MMW imaging because that modulation efficiency is directly proportional to the sideband power according to Equation 17.7. This relationship also allows for relatively simple testing of the modulator performance at these high frequencies by simply characterizing the optical sidebands apparent on an optical spectrum analyzer as described by Shi et al. [17].

## 17.4 Traveling-Wave Modulator Design and Optimization

In a traveling-wave modulator, the electrode is designed as a transmission line overlaying the optical waveguide. The modulating electrical signal on the electrode travels in the same direction as the modulated optical signal. At the end of the interaction region, the phase change induced by the electrical signal is accumulated along the length of the electrode. $LiNbO_3$ is used mainly due to its large EO coefficient, $r_{33} \approx 30\ pm/V$. For $z$-cut $LiNbO_3$ wafers, $r_{33}$ is perpendicular to the wafer surface. Usually, optical and RF transmission lines are along the $y$-axis. In this configuration, a TM-polarized optical mode will be modulated most efficiently. Using the transmission line theory, the $V_\pi$ for a traveling-wave phase modulator can be expressed as [18]

$$V_\pi = \frac{2\pi G\Gamma_E}{k_o n_o^3 rL} \frac{Z_m + Z_g}{Z_m} \left| \frac{e^{-\gamma_m L}}{1-\Gamma_l \Gamma_g e^{-2\gamma_m L}} \left( \frac{1-e^{-u_-}}{u_-} + \Gamma_l \frac{1-e^{-u_+}}{u_+} \right) \right|^{-1}, \tag{17.8}$$

with

$$\gamma_m = \alpha_m + j\beta_m = \left(\alpha_c\sqrt{f} + \alpha_d f\right) + j\frac{2\pi f}{c}n_m,$$

$$u_\pm = \pm\left(\alpha_c\sqrt{f} + \alpha_d f\right)L + j\frac{2\pi f}{c}(n_o \pm n_m)L,$$

$$\Gamma_l = \frac{Z_l - Z_m}{Z_l + Z_m},$$

$$\Gamma_g = \frac{Z_g - Z_m}{Z_g + Z_m},$$

where $Z_g$, $Z_l$, and $Z_m$ are the impedances of the source, load, and transmission line, respectively; $k_o$ and $n_o$ are the vacuum wave-vector and the mode index; $r$ is the related material EO coefficient; $L$ is the length of the interaction region; $G$ is the width of the gap between the signal and ground electrodes of the transmission line; $f$, $n_m$, $\alpha_c$, and $\alpha_d$ are the frequency, the refractive index, the conductor loss, and the dielectric loss of the MMW signal propagating in transmission line, respectively. $\Gamma_E$ is the correction factor related to the overlap of the electrical field and the optical mode.

A primary concern of design and optimization of a modulator structure for MMW frequencies arises in the large dispersion of the dielectric index going from MMW to optical frequencies. For example, the relative dielectric constants of $z$-cut $LiNbO_3$ are $\varepsilon_{ry} = 43$ and $\varepsilon_{rz} = 28$ in the parallel and perpendicular directions, respectively. While in a Ti in-diffused $LiNbO_3$ optical waveguide, the optical effective index is about 2.19 ($\varepsilon_r = 4$). As a result, the modulating MMW signal travels significantly slower than the optical carrier. Therefore, the electrode and waveguide structures have to be designed and optimized for velocity matching between the RF mode and the optical mode. There are a variety of methods that can increase the RF propagation velocity. A general guideline can be obtained by studying the propagation velocity, $v_{RF}$, and the characteristic impedance, $Z_0$, of a transmission line:

$$v_{RF} = \frac{1}{\sqrt{LC}} \quad Z_0 = \sqrt{\frac{L}{C}}, \tag{17.9}$$

where $L$ and $C$ are inductance and capacitance per unit length of the transmission line, respectively. Under a quasistatic approximation, replacing dielectrics with air and keeping the line geometry exactly the same, we can find that $L$ of the air line is the same as $L$ of the original one. However, the new capacitance per unit length, $C_a$, is different. We also know that a wave propagates in the air line at the speed of light, $c_l$. Hence,

$$c_l = \frac{1}{\sqrt{LC_a}}. \tag{17.10}$$

Combining Equations (17.9) and (17.10), we obtain

$$v_{RF} = c_l\sqrt{\frac{C_a}{C}} \quad Z_0 = \frac{1}{c_l\sqrt{CC_a}}. \tag{17.11}$$

Equation 17.11 shows that the RF propagation velocity can be increased by decreasing the capacitance of the transmission line or by increasing the capacitance of the air line. For example, velocity matching can be improved by adding a thin layer of silicon oxide under the electrodes. Since silicon oxide has a

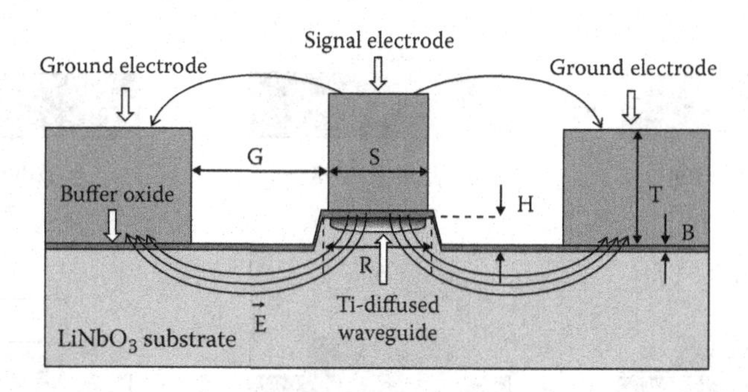


**FIGURE 17.2** Cross section of the phase modulator design.

dielectric constant of about 3.9, an oxide buffer layer partially fills the transmission line with a lower index medium and thus decreases *C*. However, a very thick buffer layer is not desirable for two reasons: first, a thick buffer layer reduces the E-field intensity across the optical waveguide that is under the buffer layer and thus increases $V_\pi$; second, a thick buffer layer introduces more dielectric loss, especially at high frequencies. Transmission line capacitance can also be effectively reduced using a ridge structure, where the $LiNbO_3$ substrate between the electrodes is etched down. As a result, high dielectric constant $LiNbO_3$ is replaced by air in the E-field that is close to the center electrode. In addition to increasing RF propagation velocity, a high dielectric constant ridge also helps to confine the E-field under the electrode in the vertical direction, which improves the overlap of the optical and RF modes, as well as the effective EO coefficient by aligning the E-field to the $r_{33}$ direction. An additional velocity increase can be obtained with thicker electrodes. Increasing electrode thickness lifts more E-field into the slot between the electrodes, which subsequently increases $C_a$ and thus $v_{RF}$. Thick electrodes also lower the device's conduction loss.

In practice, all aforementioned methods are employed for simultaneous matching of both propagation velocity and the impedance. Figure 17.2 shows the cross-section of the phase modulator design, where *S* represents the width of the traveling wave electrode, *G* represents the gap between the electrode and the ground, *T* the thickness of the electrodes, *R* the width of the ridge, *H* the height of this ridge, and *B* the buffer layer thickness. The cross-section parameters are simulated using a finite-element model. The same model is also used for design optimization. As a result, the dimensions of the modulator are the following: $S = 8$ μm, $G = 25$ μm, $T = 30$ μm, $R = 9.5$ μm, $H = 3.5$ μm, and $B = 0.9$ μm.

## 17.5 Modulator Fabrication and Characterization

Figure 17.3 shows the fabrication procedure for the designed modulator. It starts with a *z*-cut $LiNbO_3$ substrate. After lithography and liftoff, thin Ti lines are formed on the $LiNbO_3$ substrate surface. These Ti lines serve as the source for Ti in-diffusion to form optical waveguides. The diffusion is conducted in a furnace at a temperature of about 1000°C. Ridge structures are formed over the optical waveguides by lithography and high-power inductive coupled plasma (ICP) etching. After cleaning and proper surface preparation, a thin oxide buffer layer and several metal seed layers are deposited on the substrate surface. This is followed by spin-coating and exposure of a thick resist layer to form the desired high-aspect ratio resist template. Uniform gold electrodes are subsequently formed by electroplating with strong agitation. Finally, the resist and seed layers are removed.

The basic fabrication process needs to be engineered in many aspects for better device performance and processing repeatability. For example, the Ti in-diffusion process has been systematically studied. We found that Ti lines were easily damaged by surface discharge from the $LiNbO_3$ wafer during the initial heating of the furnace before titanium is oxidized. The problem was solved using a better-designed furnace that had a more uniform temperature profile and a slower heating ramp in the diffusion process.

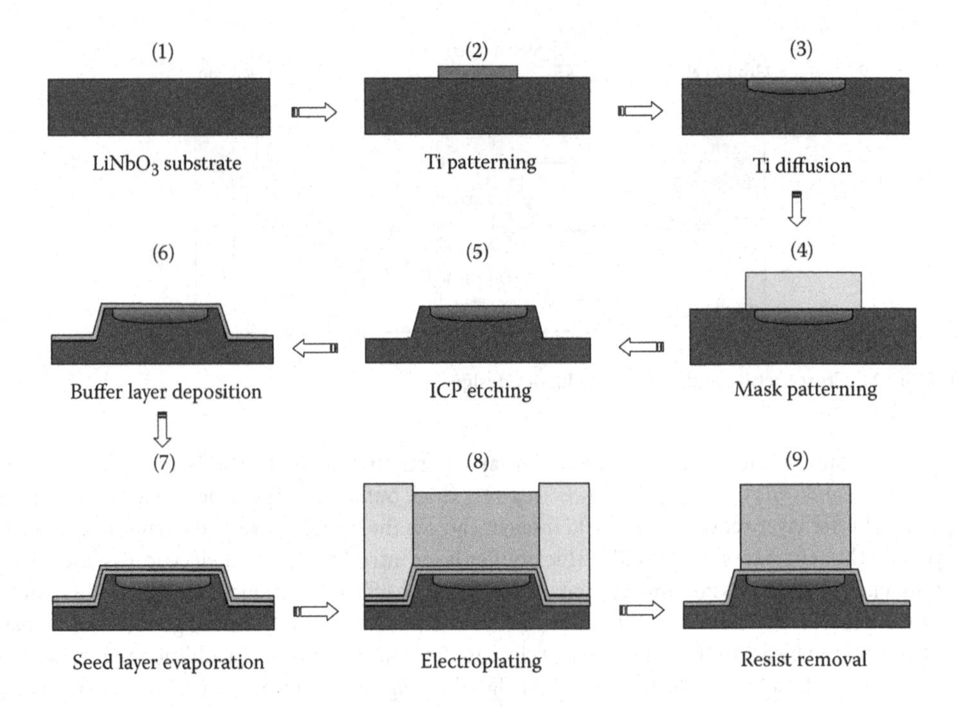


**FIGURE 17.3** Fabrication process of the $LiNbO_3$ phase modulator. (Reproduced from P. Yao et al., Design, fabrication and characterization of $LiNbO_3$ optical modulator for high sensitivity MMW imaging system, *Proc. SPIE*, vol. 6948, pp. 694808-1-8. © 2008 SPIE. With permission.)

We also compared a dry diffusion process and a wet diffusion process and concluded that the dry process is better in terms of pattern uniformity and process repeatability.

A $LiNbO_3$ wafer is dry etched using a mixture of Ar and CF4 in an ICP etcher. Generally speaking, very high ICP power and bias voltage are required for an acceptable etching rate. Due to the physical nature of the etching process, "grass" formed by material redeposition is a common problem. We found that grass formed in the etching process was related to the thermal dissipation path of the wafer. When air bubbles are present between the etched wafer and loading tray, grass occurs in the corresponding areas. We have developed a wafer bonding process under vacuum that creates a uniform backside

**FIGURE 17.4** Comparison of etching results (a) with "grasses" and (b) without grasses. (Reproduced from P. Yao et al., Design, fabrication and characterization of $LiNbO_3$ optical modulator for high sensitivity mmW imaging system, *Proc. SPIE*, vol. 6948, pp. 694808-1-8. © 2008 SPIE. With permission.)

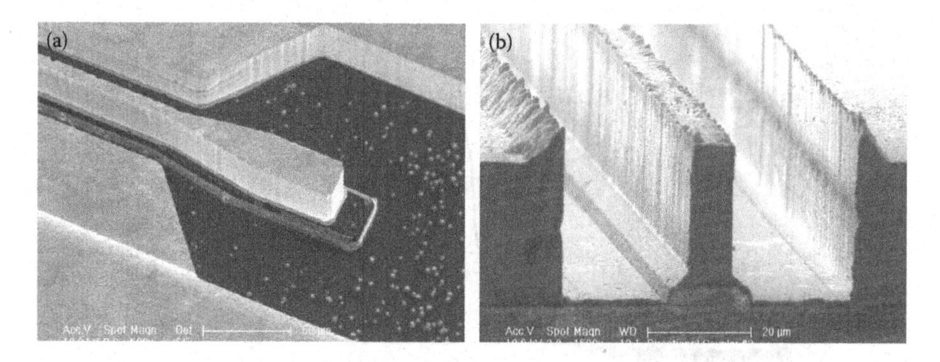


**FIGURE 17.5** Micrographs of fabricated modulator structures. (a) SEMs of the coupling end of the CPW. (b) SEM of cross section of a fabricated modulator. The structure was diced and polished at the end face. Seed layer was not stripped for imaging purpose.

thermal path. As a result, a grass-free clean etch has been obtained. A comparison of etch results with and without grass is shown in Figure 17.4.

In addition, a thick-resist lithography process has been studied in terms of spin uniformity, exposure absorption, and internal stress to serve as a high–aspect ratio template for the subsequent electroplating step. A thick resist layer was spin-coated in an environment with a controlled solvent vapor pressure, which substantially improved thickness repeatability and surface uniformity. To get electrodes with straight sidewalls, we used a long-pass filter during the exposure for more uniform dose distribution in the vertical direction. Resist internal stress may crack patterns at the corners after development. We solved this problem by modifying the structure layout and relaxing the crosslinked resist film before final development. With the refinements of the thick resist lithography process, we were able to fabricate coplanar waveguide electrodes of uniform high aspect ratio. Figure 17.5 shows the coupling end and the polished cross section of a fabricated modulator. The device has a 3.5-µm-high ridge structure and 30-µm-high electrodes. The signal electrode is 8 µm wide and 25 µm away from the ground surfaces. There is a 0.9-µm silicon oxide buffer layer and a 200-nm Ti–Au seed layer on the $LiNbO_3$ surface. The seed layer was not stripped for imaging purposes.

The fabricated modulator chip is integrated with polarization maintaining (PM) fiber using V-groove holders. Insertion losses less than 4.5 dB are typical. Most of the losses come from the end coupling loss between $LiNbO_3$ and fiber due to the index and mode mismatches. The modulator is then characterized using a basic setup consisting of a laser source at 1550 nm, a vector network analyzer (VNA) and an optical spectrum analyzer (OSA), as shown in Figure 17.6.

Electrical properties of a modulator can be calculated based on measured S-parameters, which are usually measured in the 0 GHz to 50-GHz range, using the VNA. Two standard 50-Ω ground-signal-ground (GSG) probes connected to coaxial cables are brought in contact with the CPW electrodes at both ends of the modulator. The electrical properties are then extracted from the S-parameters by calculating the ABCD matrix using transmission theory. Typical results for the dielectric loss $\alpha_d$, the

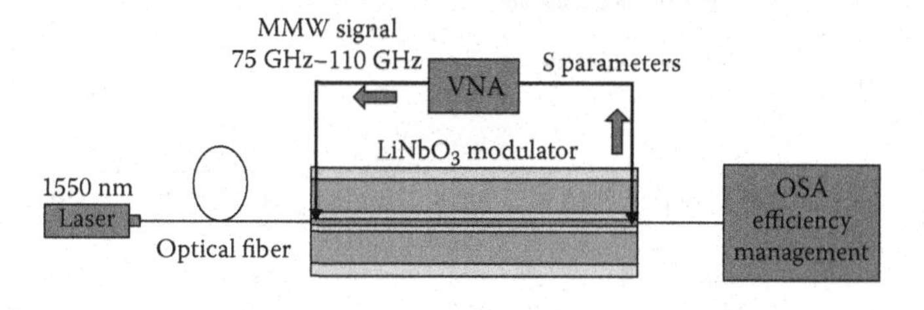


**FIGURE 17.6** Experiment setup for modulator characterization.

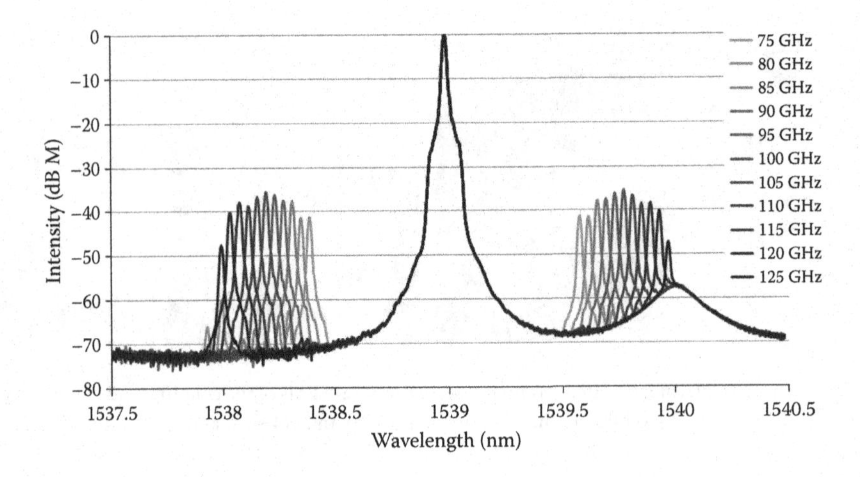


**FIGURE 17.7** Sidebands of the optical signal after modulation.

conductor loss $\alpha_m$, the impedance $Z$, and the effective index $n$ are 0.023 dB/(cm GHz), 0.24 dB/(cm GHz$^{1/2}$), 41.5 Ω, and 2.21, respectively.

Figure 17.7 shows the sidebands of the optical signal after modulation. Due to good velocity matching between the optical and RF modes, the modulator has response over a broadband up to 125 GHz. As the frequency approaches 125 GHz, the sidebands start to diminish, mainly due to cutoff of the WR10 waveguides that are used to feed the RF signals. Modulation efficiency can be obtained by comparing the power in the lower sidebands to the optical reference power and the RF input power.

Figure 17.8 shows the modulation efficiencies of eight modulators with the same design parameters and the same fabrication processes. Four of the eight modulators, shown in broken lines, have a substrate thickness of 500 μm; substrates of the other four modulators are lapped down from the backside to about 100 μm. As indicated by the results, thinning the modulator substrate improves the modulation efficiency. This can be explained by the induced substrate modes for thick substrates. For a substrate thicker than 125 μm, substrate modes may be excited at the coupling point or through defects

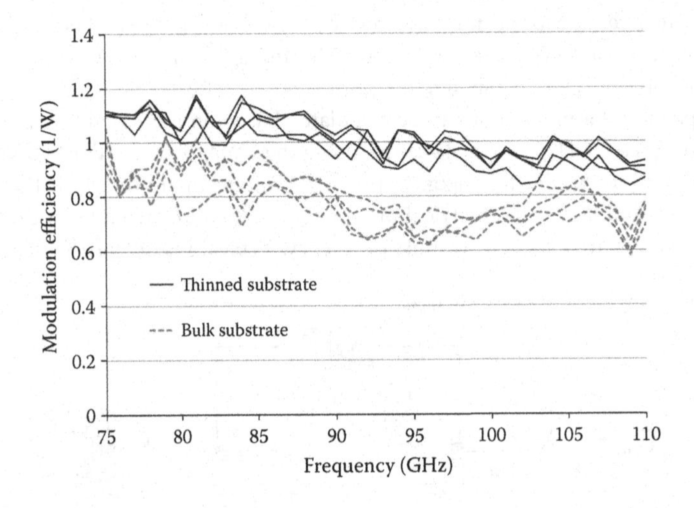


**FIGURE 17.8** Comparison of modulation efficiencies of devices from a common fabrication run where a portion of the devices were postprocessed to thin the substrates.

along the transmission line. The presence of substrate modes introduces peaks and valleys in the CPW transmission curve depending on the working frequency. Corresponding dips are also observed in the modulation efficiency curves since substrate modes contain energy leaking from the CPW. When the leaky modes are suppressed using a thinned substrate, better modulation efficiencies are achieved, as indicated by solid lines in Figure 17.8. Using these processes, modulation efficiencies as high as 1 $W^{-1}$ at 95 GHz have been achieved.

## 17.6 MMW Detector Experimental Results

The modulator technology described in the previous section has been used to demonstrate passive sensitivity levels for the optical detection technique without the use of low noise amplifiers or cryogenic cooling. As shown in Figure 17.9, the chip-scale modulator was fed with an optical source of ~200 mW generated by an amplified and filtered distributed feedback laser operating at a nominal wavelength of 1546.6 nm. The modulator chip used for these experiments was from a previous generation with a nominal conversion efficiency of 0.7 $W^{-1}$. A WR-10 horn antenna was coupled to the modulator electrodes by means of a GSG probe. The output of the modulator was filtered using a series pair of commercially available optical add/drop multiplexers centered nominally 0.8 nm (100 GHz) off of the carrier prefilters. The resultant passed sideband was detected using a fiber-coupled InGaAs photodiode whose output was amplified using a commercial TIA with a nominal gain of 10 GΩ. The output of the TIA was fed to a lock-in amplifier synchronized to an RF absorber-covered chopper wheel passed in front of the WR-10 horn. To characterize the system responsivity and noise performance, the output of the horn antenna was presented with a section of Cummings AN-72 absorber material soaked in liquid nitrogen, which presented a temperature contrast between chop states of approximately 213 K. Based on this setup, a temperature responsivity of 32.1 μV/K, a noise floor of 20.1 $\mu V/\sqrt{Hz}$ and a noise equivalent temperature difference (NETD) of 1.2 $K/\sqrt{Hz}$ were measured. A MMW detection bandwidth larger than the entire W band (>35 GHz) was demonstrated by testing the system with a swept MMW source. Further details on this setup may be found in [19].

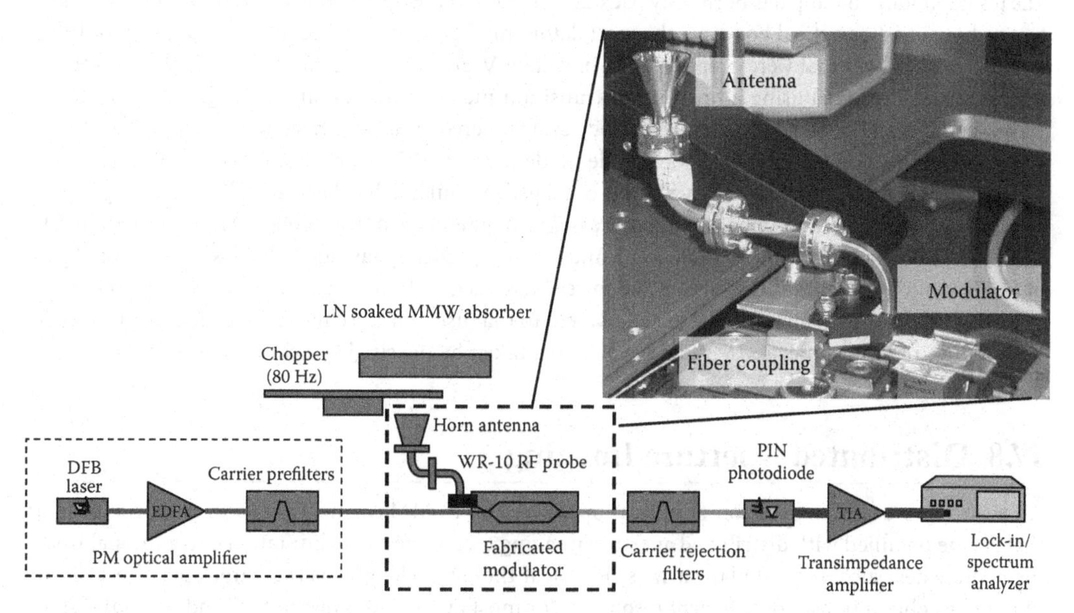


**FIGURE 17.9** Test setup used for characterization of passive performance of optical upconversion, MMW detection technique. (Reproduced from C. J. Huang et al., $LiNbO_3$ optical modulator for MMW sensing and imaging, *Proc. SPIE*, vol. 6548, pp. 65480I1-7. © 2007 SPIE. With permission.)

(a) (b)

**FIGURE 17.10** Real MMW image of a boat demonstrates the successful development of LN phase modulator for W-band. (a) MMW images of a boat. (b) Corresponding photograph of the same boat.

## 17.7 Imaging Results Obtained by a Single-Pixel Scanner

The tested modulator chip has been fully packaged including the optical fiber, WR-10 coupling and protective housing. For optical bonding, the modulator chip was first polished at both end faces and then bonded to PM fibers that were prepositioned in silicon V-grooves. Coupling to the WR-10 rectangular waveguide was realized using a fin-coupled transition mechanism. The fin protrudes down from the waveguide top and touches the signal electrode at its tapered end. The fin structure allows the electric field to smoothly transfer from the waveguide mode to the CPW mode. As a result, −2.8 dB electrical insertion loss at 94 GHz was achieved. The packaged modulator has been installed in a single pixel scanning imager, where a gimbal-mounted Cassegrain antenna is used to collect MMW radiation. The collected MMW signal is fed to a PIN switch and then amplified by cascaded LNAs with a nominal gain of 19 dB to allow an improved scan rate and lower image acquisition time. The enhanced signal is fed to our LN modulator with an optical setup equivalent to that used for the chip-scale testing described previously. Figure 17.10 shows a MMW image of a boat taken by the single pixel imager using the $LiNbO_3$ phase modulator.

## 17.8 Distributed Aperture Imaging

An even more useful application of optical upconversion to MMW imagers comes from the benefits that can be obtained with distributed aperture imaging techniques. One limitation of traditional imaging approaches, such as focal plane arrays (FPAs), at these wavelengths is that large aperture sizes are required to obtain images of sufficient resolution for most applications due to the fundamental restrictions on resolution imposed by the diffraction limit. With traditional FPAs imagers, increased resolution also requires a subsequent scaling in the depth of the imager, resulting in a volumetric increase in imager size with increased resolution. Distributed aperture techniques could open up a wider range of applications by mitigating the volume requirements of high resolution/large-aperture imagers. Unfortunately, traditional distributed aperture techniques encounter significant complications due to significant detection, routing, and processing requirements of such distributed arrays. These problems can be potentially addressed using optical techniques. Using electro-optic modulators to upconvert received MMW fields onto an optical carrier, such fields can be readily captured, routed, and processed using optical techniques, thereby providing significant advantages over more traditional heterodyne imagers. This imaging technology could potentially provide high-resolution, passive imagery without the volumetric scaling with aperture size imposed by traditional FPA designs.

Optical processing of distributed aperture data has been demonstrated in the seminal work by Blanchard et al. at microwave frequencies [12] and extended to a 1-D system in the MMW regime [13]. This technique relies on using the spatial Fourier transform properties of an optical lens to perform the numerous correlations required to regenerate the image from the sampled $u$–$v$ plane. In fact, digital correlation algorithms are essentially methods for performing discrete spatial Fourier transforms and require increasingly numerous correlations (~$n^2/2$ for $n$ nodes) as the number of antenna nodes grows. Using smaller optical wavelengths, Fourier transform operations may be carried out using a simple small optical lens and a photodetector array. As such, the sampled image is generated in real time without the use of complicated correlation engines.

Conceptually, as shown in Figure 17.11, this approach may be thought of as a technique for (1) using antennas to discretely sample the complex amplitude of the MMW signal at various locations in a distributed array, (2) converting the captured complex amplitude to optical wavelengths using electro-optic modulators, (3) routing upconverted signals to a central processor fiber array that mimics the MMW array layout, (4) performing a continuous spatial Fourier transform of the discretely sampled aperture using simple optics at shorter wavelengths, and (5) capturing regenerated imagery in real time using a standard optical camera.

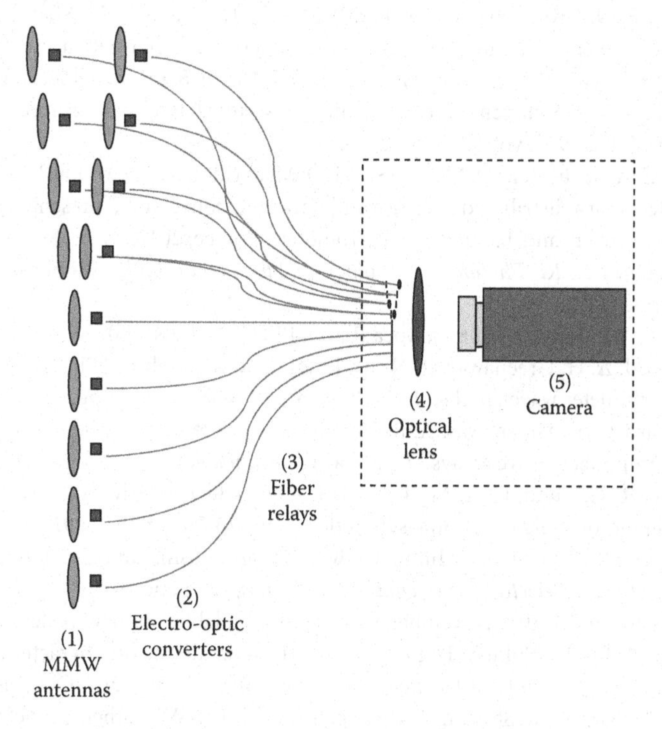


**FIGURE 17.11** Schematic representation of a distributed aperture, optical upconversion imager. (Reproduced from C. A. Schuetz et al., Sparse aperture millimeter-wave imaging using optical detection and correlation techniques, *Proc. SPIE*, vol. 6548, pp. 65480B1-7. © 2007 SPIE. With permission.)

This method promises to provide a viable way to achieve large effective apertures without the volumetric scaling implied by traditional imaging techniques and does so in a manner that does not require complex electronic processing. Such techniques may well open a range of applications for MMW imaging that were previously impractical due to the large size and weight demands of more traditional imaging techniques.

## References

1. H. Kazemi, J. D. Zimmerman, E. R. Brown, A. C. Gossard, G. D. Boreman, J. B. Hacker, B. Lail, and C. Middleton, "First MMW characterization of ErAs/InAlGaAs/InP semimetal–semiconductor–Schottky diode ($S^3$) detectors for passive millimeter-wave and infrared imaging," in *Proc. SPIE*, 2005, vol. 5789, pp. 80–83.
2. H. P. Moyer, R. L. Bowen, J. N. Schulman, D. H. Chow, S. Thomas III, T. Y. Hsu, J. J. Lynch, and K. S. Holabird, "Sb-heterostructure diode detector W-band NEP and NEDT optimization," in *Proc. SPIE*, 2006, vol. 6211, p. 62110J.
3. H. P. Moyer, T. Y. Hsu, R. L. Bowen, Y. K. Boegeman, P. W. Deelman, S. Thomas III, A. T. Hunter, J. N. Schulman, A. Luukanen, and E. N. Grossman, "Low noise Sb-heterostructure diode detectors for W-band imaging arrays without RF amplification," in *Proc. SPIE*, 2005, vol. 5789, pp. 84–92.
4. J. N. Schulman, D. H. Chow, and D. M. Jang, "InGaAs zero bias backward diodes for millimeter wave direct detection," *IEEE Electron Device Lett.*, vol. 22, no. 5, pp. 200–202, May 2001.
5. J. N. Schulman, V. Kolinko, M. Morgan, C. Martin, J. Lovberg, S. Thomas III, J. Zinck, and Y. K. Boegeman, "W-band direct detection circuit performance with Sb-heterostructure diodes," *IEEE Microwave Wireless Components Lett.*, vol. 14, no. 7, pp. 316–318, Jul. 2004.
6. F. J. Gonzalez, M. A. Gritz, C. Fumeaux, and G. D. Boreman, "Two dimensional array of antenna-coupled microbolometers," *Int. J. Infrared Millimeter Waves*, vol. 23, no. 5, pp. 785–797, May 2002.
7. A. Luukanen, E. N. Grossman, A. J. Miller, P. Helisto, J. S. Penttila, H. Sipola, and H. Seppa, "An ultra-low noise superconducting antenna-coupled microbolometer with a room-temperature read-out," *IEEE Microwave Wireless Components Lett.*, vol. 16, no. 8, pp. 464–466, Aug. 2006.
8. A. Luukanen and V.-P. Viitanen, "Terahertz imaging system based on antenna coupled microbolometers," in *Proc. SPIE*, 1998, vol. 3378, pp. 34–44.
9. T. D. Dillon, C. A. Schuetz, R. D. Martin, S. Shi, D. Mackrides, and D. W. Prather, "Passive millimeter wave imaging using a distributed aperture and optical upconversion," presented at Millimetre Wave and Terahertz Sensors and Technology III, Toulouse, France, 2010.
10. W. S. C. Chang, Ed., *RF Photonic Technology in Fiber Optic Links*, Cambridge, UK: Cambridge University Press, 2002.
11. C. H. Lee, Ed., *Microwave Photonics*, Boca Raton, FL: CRC Press, 2007.
12. P. M. Blanchard, A. H. Greenaway, R. N. Anderton, and R. Appleby, "Phase calibration of arrays at optical and millimeter wavelengths," *J. Opt. Soc. Am. A*, vol. 13, no. 7, pp. 1593–1600, Jul. 1996.
13. P. M. Blanchard, A. H. Greenaway, A. R. Harvey, and K. Webster, "Coherent optical beam forming with passive millimeter-wave arrays," *J. Lightwave Technol.* vol. 17, no. 3, pp. 418–425, Mar. 1999.
14. J. A. Lovberg, R.-C. Chou, C. A. Martin, and J. A. J. Galliano, "Advances in real-time millimeter-wave imaging radiometers for avionic synthetic vision," in *Proc. SPIE*, 1995, vol. 2463, pp. 20–27.
15. A.-I. Sasaki and T. Nagatsuma, "Millimeter-wave imaging using an electrooptic detector as a harmonic mixer," *IEEE J. Selected Topics Quant. Electronics*, vol. 6, no. 5, pp. 735–740, Sep. 2000.
16. FEMTO Messtechnik. (2010, November 12) GmbH, Available: http://www.femto.de.
17. Y. Shi, L. Yan, and A. E. Willner, "High-speed electrooptic modulator characterization using optical spectrum analysis," *J. Lightwave Technol.*, vol. 21, no. 10, pp. 2358–2367, Oct. 2003.
18. C. J. Huang, C. Schuetz, R. Shireen, T. Hwang, S. Shi, and D. W. Prather, "Development of photonic devices for MMW sensing and imaging," in *Proc. SPIE*, 2006, vol. 6232, p. 62320Q.
19. C. Schuetz, "Optical techniques for millimeter-wave detection and imaging," Ph.D. dissertation, Dept. Elec. Comp. Eng., Univ. Delaware, Newark, DE, 2007.

# 18

# Optical Beam Forming for Microwave and Millimeter Waves

Harold R. Fetterman
*University of California at Los Angeles*

Seongku Kim
*University of California at Los Angeles*

Matthew R. Fetterman
*University of California at Los Angeles*



## 18.1 Introduction

Recent years have seen significant advances in optical technologies, including new types of high-frequency optical modulators, broadband optical fibers, and high-speed optical detectors. These advances in optics have enabled exciting new developments in microwave and millimeter-wave technologies.

The ability to place and transmit microwave and millimeter-wave (MMW) signals on optical beams is a relatively new capability and is currently limited to relatively low powers. A new direction [1], which involves high-power optical signals, is currently under intense research. However, even with low optical power levels it is possible to transmit the optical signals, with their RF modulation, long distances with excellent signal-to-noise ratio [2]. Links with spurious-free dynamic range (SFDR) of 130 dB have been demonstrated, and with the introduction of linearization techniques, should increase by as much as 15 dB [3].

A critical technology to optical beam forming for microwave and millimeter-wave applications is a high-performance electro-optic modulator. For many applications, commercially available $LiNbO_3$ modulators offer sufficient performance, with <3-dB optical loss, <2-V switching voltage $V_\pi$, and a modulation bandwidth exceeding 20 GHz. Research continues into developing new modulator technologies, with high linearity, low $V_\pi$, ultralow-power consumption, and extremely low optical loss for cascading multiple modulators. Low cost and capability for integration are also a consideration in systems that may require tens or hundreds of devices.

Using transmission over significant distances, it is possible to control the time of arrival of RF signals by transit delays on optical fibers or optical guiding structures. This capability permits the development of new types of optically controlled phase array antennas, which will have extremely broad bandwidths. Essentially, it is possible to make arrays where any frequency will have the correct phase to point in a given direction. This is because the phase at the antenna is determined by a time delay that does not depend on frequency. This "true time delay" (TTD) approach is particularly suitable for short pulses [4]. Such TTD phased arrays are difficult to construct without optics, especially at higher RF frequencies.

Optics offers additional advantages to phased array radars. In a CW or long-pulse system, we can use phase control to point in different directions with a look-up table for each frequency of interest. This type of system also has advantages using optical components and can be easily made to point in a continuous range of angles. The use of optical control also has significant advantages in getting the signals to the radiating element in a 3D radar phase control system without huge waveguides or heavy coaxial cables.

The control of phase has applications far beyond forming parallel microwave and millimeter-wave beams. Beams can be focused and shaped for long range transmission. Additionally, beams can be formed for imaging arrays and object identification.

A particularly interesting version of these phased arrays is to make a phase conjugating array using the capability of optical fiber millimeter-wave interconnections [5]. This concept in many ways duplicates the use of nonlinear crystals at visible wavelengths. However, in this microwave and millimeter-wave application, we can have additional gain. Essentially, mixers and amplifiers can be introduced into the system to increase the signal levels. This permits the development of autotracking systems, high-power beam forming, bistatic radars, and even conformal smart antennas for airplanes.

Advantages in optics can also be applied to RF technology. Optical signal processing (OSP) permits the introduction of logic elements and novel processing tools based upon the control of delays and amplitude in a timed configuration. This will open up entirely new areas of development for both beam forming and logic elements. An example of OSP applications is TTD radar with continuous tunability and constant amplitude.

Finally, we will look at applications of this technology to airport security and imaging systems. A holographic version [6], related to our phase conjugation, seems to have exciting potential. There are a number of approaches to holographic technology that are supported by optical interconnections. In one approach, the use of new optical digital conversion techniques becomes vital.

## 18.2 Optical Phased Array Radar

The phased array radar offers significant advantages over the conventional dish radar, which has a single radiating element. One advantage of the phased array radar is that it can steer beams at a much higher speed than a conventional dish radar because the beam movement is driven by the electronics and not by mechanical movement. Another advantage is that large phased arrays can be constructed without the mechanical and weight issues of large dishes. Phased arrays are used in a variety of applications, from large ground-based missile warning radars to ship-based fire control radars.

The phased array transmitter must control the relative phase of each transmitting element, and the phased array receiver must be able to both combine the incident radar signals and add the correct phase to each signal so that the beam points in the desired direction. A phased array system that delays each emitter by the same time delay is referred to as a TTD system.

A significant challenge to building electronic phased array radars has been developing a method to adjust the RF phase at each transmitter/receiver. Electronic RF phase shifters are heavy and consume significant power and become significantly expensive as the RF frequency increases.

In this section, we describe phased array radars that utilize optical techniques to control the RF phase. The optical system can have advantages in size, weight, power, and speed over the electronic method of adjusting the RF phase. A secondary benefit of optical phase control is that the optics allow for antenna remoting, so that the control optics and electronics can be placed at the base station. Fiber optics transmit the signals to the antenna, so that the antenna can be lightweight and portable.

Researchers have developed different designs for optical phased array radars. There are serial and parallel architectures. There are wavelength division multiplexing (WDM) architectures and direct modulation architectures. We will review optical phased arrays with TDD, as well as optical phased arrays that are based on shifting the phase of the RF signal. It is important to note that in the development of these systems it was originally very important to reduce the number of lasers involved. At that time, lasers were expensive and had short lifetimes, in contrast to the low-cost, high reliability lasers that are readily available today.

In Figure 18.1, we show the architecture for an optical phased array TTD based on WDM technology. This design uses a single tunable laser and high-frequency modulator in conjunction with a chirped grating that was developed with a serial feed [7]. Essentially, fiber optic delay lines are loaded in a serial manner, tuning the wavelength rapidly to get the appropriate time delays. When all the lines are ready, the output is switched on and the antenna radiates in the desired direction. This configuration shows the inherent power of using optical fibers where significant time delays can be obtained with little or no loss. It also shows the available resources such as optical Bragg reflectors fabricated in the low loss optical fibers. It was also possible, given the advances in planar optical guiding structures, to fabricate many of these components in a monolithic and integrated form.

As in all these radar systems, the problem of receiving is more complex and virtually all of these systems, both serial and parallel, require an optical modulator for each element. Essentially, for the TTD, there must be a device to convert the RF signal to an optical one at each antenna element.

A typical receiver system [8] that avoids the need for a modulator for every antenna element is shown in Figure 18.2. In this case, the receiver system is not TTD but uses mixers. Some of the timing information at the edge of a very short pulse is then lost in this approach, but it preserves many of the best features and uses optical delay lines effectively.

Moving to current versions of TTD system in Figure 18.3, we show the elements of new parallel systems [9] using WDM structures to switch in the required optical delays. Essentially, in this WDM approach there are relatively inexpensive lasers with a different wavelength for each antenna element. These lasers are tuned a relatively small amount, 0.17 nm about their center wavelength, to choose the appropriate delays. Using a multichannel chirped fiber grating (MCFG), it is then possible to get continuously variable delays from every laser. This is potentially an exciting configuration since, in addition to scanning a plane wavefront, it is possible to use such systems to form focusing beams for imaging applications.

The time delay can be set by using other approaches rather than Bragg gratings. Methods include a dispersive medium and even different paths determined by special holograms in the delay materials [10]. Because of the need for multiple modulators in the receiver circuit, TTD systems tend to be rather costly

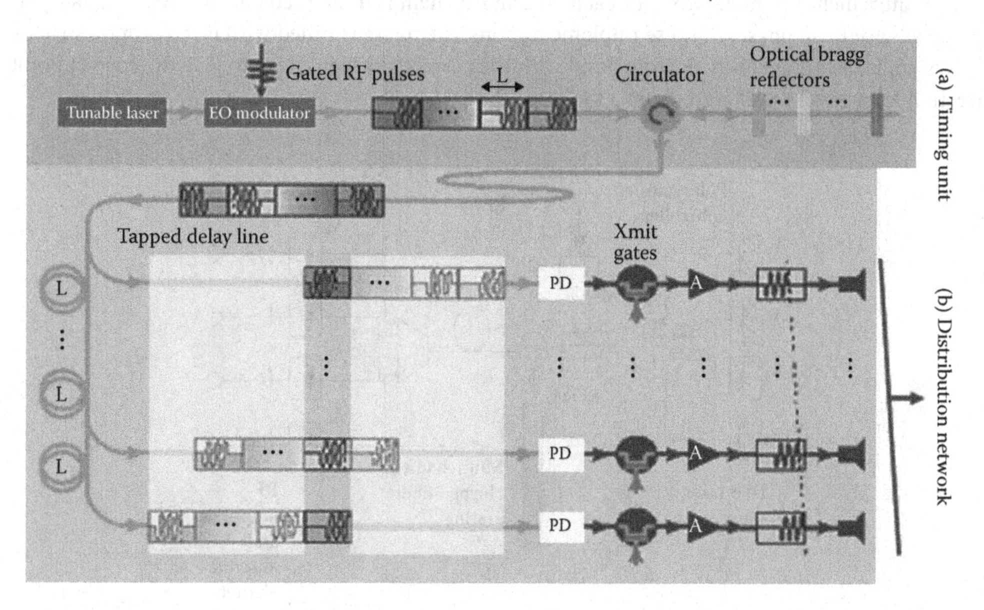


**FIGURE 18.1** Basic design of the optically controlled, serially fed phased array transmitter. The tunable laser selects the delay on the fiber grating relative to the gate. Then the tapped delay line converts the serial signal into a parallel feed. (Reproduced with permission from Y. Chang, B. Tsap, H. R. Fetterman, D. A. Cohen, A. F. J. Levi, and I. Newberg, *IEEE Microwave Guided Wave Lett.*, vol. 7, no. 3, pp. 69–71, © 1997 IEEE.)

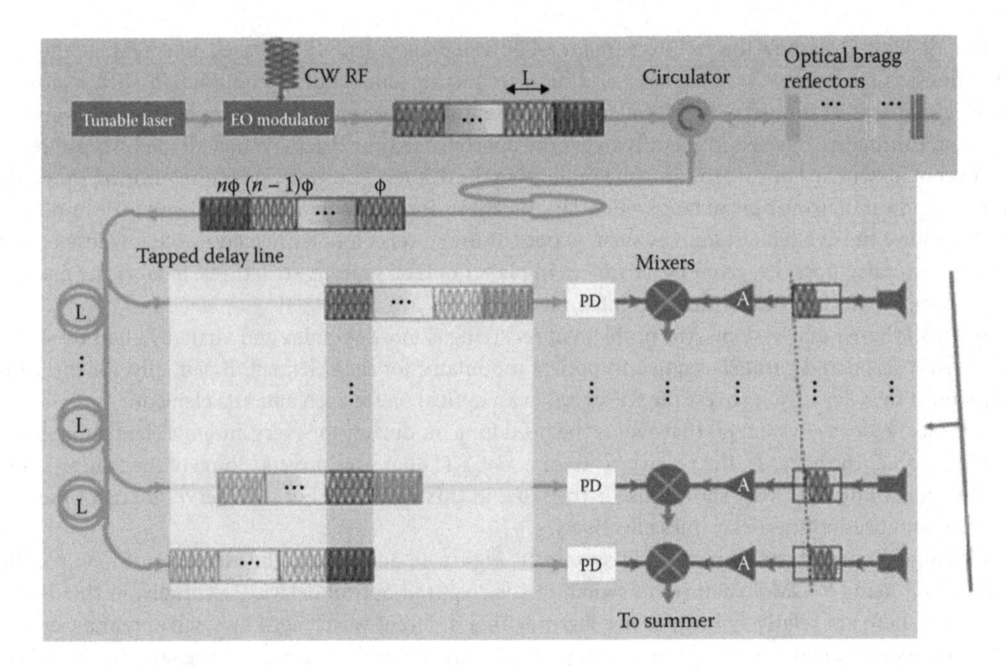


**FIGURE 18.2** Receiver system using a serial approach and mixers for a phase array system. (Reproduced with permission from B. Tsap, Y. Chang, H. R. Fetterman, A. F. J. Levi, D. A. Cohen, and I. Newberg, *IEEE Photonics Technol. Lett.*, vol. 10, no. 2, pp. 267–269, © 1998 IEEE.)

and complex. It is in general usually required to make the conversion from the microwave to an optical carrier to achieve wide instantaneous bandwidth such as required in using short pulses.

The requirement for modulators for each antenna element is in marked contrast to the array systems that depend only on phase. In these implementations, the phase is tuned with electro-optic structures. The system is instantaneously narrow-band with the correct phase inserted to give the correct pointing direction for a given wavelength from a look-up table.

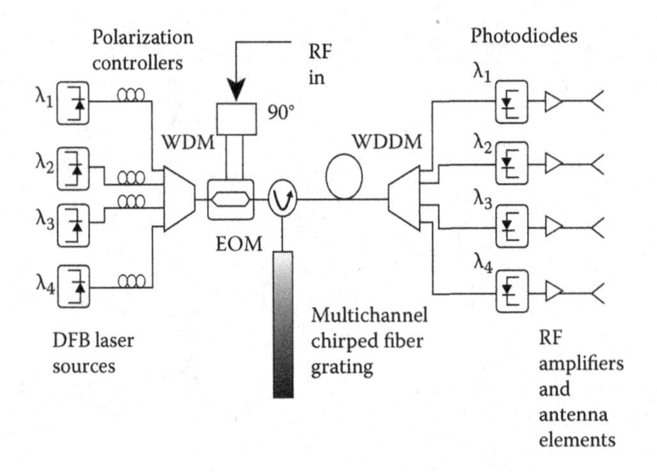


**FIGURE 18.3** A modern parallel transmit system using WDM and a multichannel chirped fiber grating for delays. (Reproduced with permission from B. Hunter, M. E. Parker, and J. L. Dexter, *IEEE Trans. Microwave Theory Techniques*, vol. 54, no. 2, pp. 861–867, © 2006 IEEE.)

An interesting integrated approach uses optical mixing to define the phase of the millimeter-wave system rather than the time delay. A single sided modulator is used to generate the microwave modulated optical signal. An electro-optic phase shifter is then used to select the desired phase for the particular element. These systems, of course, have many of the advantages of weight and remoting common to TTD systems we have examined.

Figure 18.4 shows a four element phased array driver with continuous tuning of the microwave signal from the mixers at the antenna element. A four-element chip, in this case using nonlinear polymers, is shown which works very well with a relatively simple receiver based upon mixing and heterodyne detection. The problem, of course, is that the pointing direction is a strong function of microwave frequency so that this does not function extremely well with instantaneous broadband or short pulses. However, the system is extremely lightweight, continuous in directivity and can be formed in a monolithic fashion using newer materials such as electro-optic polymers.

The calculated RF phase and power characteristics revealed that the phase of the RF signal can be controlled by changes in the control voltage $V_{Cont}$ and varied almost linearly up to 140°. In practice, these systems exhibited a lack of linearity as the control voltage is tuned over $2\pi$ that was caused by the presence of the carrier signal from the single side band (SSB) modulator. A simple solution has been developed to solve this problem under small signal operation. If the carrier signal is fully suppressed in the SSB modulator, these unwanted effects could be significantly eliminated. Figure 18.4 represents a balanced photonic RF phase shifter, in which an additional arm in the inner MZ (Mach–Zehnder) interferometer is inserted that is intended to suppress the carrier signal in the SSB modulator so as to balance the system. If the input optical signal with unit magnitude at a frequency of $\Omega$ is $E_{in}(t) = e^{i\Omega t}$, the output optical field and the resulting intensity at the modulation frequency $\omega_{RF}$. can be expressed as

$$E(t)=\frac{1}{4(1+\alpha)}e^{i\Omega t}\left[e^{i\Delta\sin(\omega_{RF})t}+e^{i\Delta\cos(\omega_{RF})t+i\frac{\pi}{2}}+2\alpha e^{i\phi_{Bal}}+2(1+\alpha)e^{i\phi_{Cont}}\right], \tag{18.1}$$

$$I_{\omega_{RF}}(t)=\frac{1}{4(1+\alpha)^2}A_{RF}J_1(\Delta)\cos(\omega_{RF}+\varphi_{RF}), \tag{18.2}$$

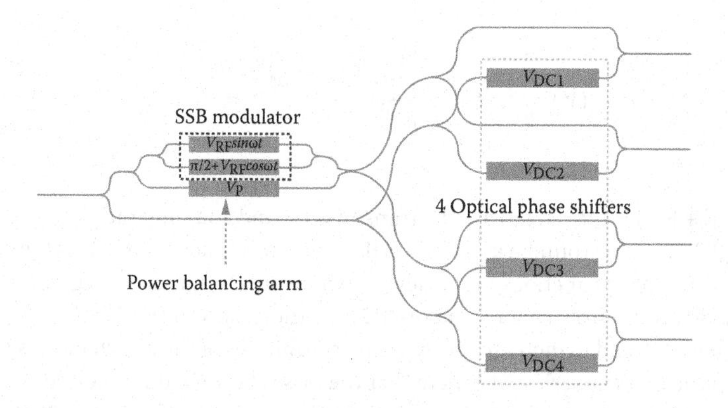


**FIGURE 18.4** Schematic diagram representing the four-element RF phase shifter array with the balanced design. It consists of single sideband modulator on upper arm of a Mach–Zehnder modulator and optical phase modulators on the four DC control arms. The SSB modulator with two input RF signals generates a carrier at $\Omega$ and a sideband at $\Omega + \omega_{RF}$. The control DC bias $V_{Cont}$ applied to the four optical phase modulators induces a phase-shifted optical carrier at $\Omega$. The mixing of these signals in a photodiode gives rise to the RF signal at $\omega_{RF}$ with a variable phase controlled by $V_{Cont}$. (Reproduced with permission from J. Han, B. Seo, S. Kim, H. Zhang, and H. Fetterman, *J. Lightwave Technol.*, vol. 21, no. 12, pp. 3257–3261, © 2003 IEEE.)

where

$$A_{RF} = \left\{ \left[ J_0(\Delta) + 2\alpha\cos\phi_{Bal} + 2(1+\alpha)\cos\phi_{Cont} \right]^2 + \left[ J_0(\Delta) + 2\alpha\sin\phi_{Bal} + 2(1+\alpha)\sin\phi_{Cont} \right]^2 \right\}^{\frac{1}{2}}, \quad (18.3)$$

$$\varphi_{RF} = \tan^{-1}\left[ \frac{J_0(\Delta) + 2\alpha\sin\phi_{Bal} + 2(1+\alpha)\sin\phi_{Cont}}{J_0(\Delta) + 2\alpha\cos\phi_{Bal} + 2(1+\alpha)\cos\phi_{Cont}} \right]. \quad (18.4)$$

Here $V_\pi$ is the half-wave voltage, $\Delta = \pi V_{RF}/V_\pi$ is the modulation depth, $\phi_{Cont} = \pi(V_{Cont}/V_\pi)$ is the optical phase shift by the control DC bias, $\phi_{Bal} = \pi(V_{Bal}/V_\pi)$ is the optical phase shift by the balancing DC bias, and $\alpha^2$ is the optical power splitting ratio at the balancing arm. For the choice of $\alpha = J_0(\Delta)/\sqrt{2}$ with $\phi_{Bal} = 5\pi/4$, the undesirable terms $J_0(\Delta)$ completely disappear meaning that the carrier signal is fully suppressed in the SSB modulator and the ideal characteristics for the RF phase and power can be obtained such that

$$A_{RF}^2 = const., \varphi_{RF} = \tan^{-1}\left[ \frac{2\sin\phi_{Const}}{2\cos\phi_{Cont}} \right] = \phi_{Cont}. \quad (18.5)$$

This indicates that the RF power will not vary and the RF phase shift will be highly linear with respect to the control DC voltage. The modified design makes these devices very suitable for optically controlled phase array antenna systems with precise steering.

In recent years, there have been significant advances in the field of optical signal processing. These systems use combinations of delays and amplitude control to achieve the same capabilities as digital signal processing electronics. In the case of OSP, rings are used to obtain delays. Since the polymers materials were relatively high in loss, one would expect that the signal amplitude would be a function of delay. However, using the configurations possible with the OSP approach, these losses could be made constant as a function of delay in these TTD systems. In Figure 18.5, we show a continuous variable TTD system that has a relatively flat output. The effective transfer function of the OSP element is found as

$$H_{TTD}(V) = t_1 t_3 \frac{t_2 - \alpha e^{j\left(\pi\frac{V}{V_\pi} - \tau\omega\right)}}{1 - \alpha t_2 e^{j\left(\pi\frac{V}{V_\pi} - \tau\omega\right)}} - c_1 c_3 e^{j\pi\frac{V_P}{V_\pi}} = H_{P1}(V) - H_{P2}, \quad (18.6)$$

where $t_i$ and $c_i(|t_i|^2 + |c_i|^2) = 1, i = 1,2,3)$ are the transmission and coupling ratio of the $C_i$ coupler, $\alpha$ is the round trip loss (RTL), $\tau$ is the round trip time, $\omega$ is the angular frequency, and $V_\pi$ is the half-wave voltage. $H_{P1}$ and $H_{P2}$ are the transfer functions of the device, which pulses in each waveguide arm experience.

The OSP TTD approach uses ring resonators [12] for the delays in this basic unit configuration. To make this more selective and to increase the dynamic capabilities of these systems it is necessary to gang several of these basic units in series. It is here that the losses become dominant. It will be necessary to reduce the unit loss below 3 dB for this technology to become useful for real systems. Ultimately, this approach can also be used to perform other logic elements including linearized modulation and arbitrary wavefunction generation [13].

The use of this approach becomes very important as the delays used in larger arrays become longer. Then, the losses become more significant and the ability to drive the array is effectively reduced since all the signals have to have the same amplitude. Because the optical signal processing approach can address this major problem, it will be more important in future systems where size will be a significant factor.

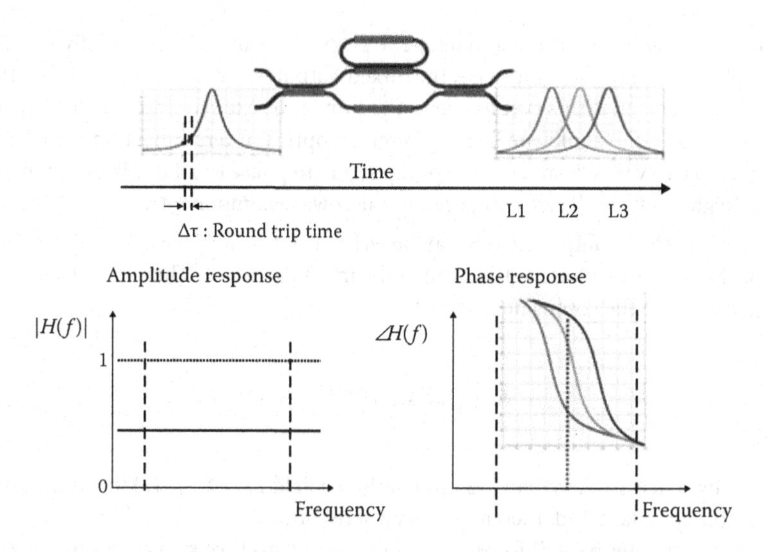


**FIGURE 18.5** This shows the basic unit cell used in the optical signal processing approach on top. It consists of a conventional ring structure and a bypass waveguide. The lower right shows the amplitude of the system with the proper bias voltages and the left shows the phase responses as a function of voltage.

## 18.3 Optical Phase Conjugation

Following the idea of using optics for beam control, it would be extremely useful if one could extend this technology to systems that use nonlinear material technologies. One of the most useful of this class of technology is the concept of phase conjugation. The basic idea to permit the extension of optical beam forming technology to phase conjugation is to use an array of antenna elements to sample the millimeter-wave beam at different positions of the incident wavefront. Then microwave mixers are used to generate phase conjugated currents that are then radiated. If there is a dense enough sampling (Nyquist), the radiated beam will be the phase-conjugate wave of the incident beam. Systems based upon this technology will have remarkable applications to communications and radar [14].

In Figure 18.6, we show that by mixing an incoming signal with twice the frequency, the conjugate wave can be effectively generated in the mixer. The signal is then amplified and reradiated either using

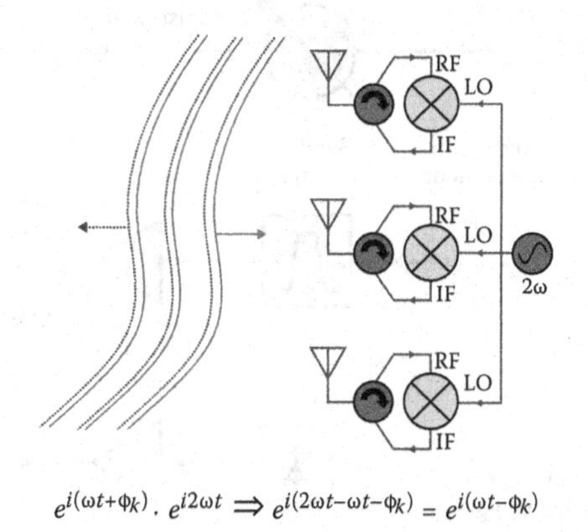


$$e^{i(\omega t+\phi_k)} \cdot e^{i2\omega t} \Rightarrow e^{i(2\omega t-\omega t-\phi_k)} = e^{i(\omega t-\phi_k)}$$

**FIGURE 18.6** Phase conjugation of a millimeter-wave signal using an antenna surface with mixing with twice the frequency. (Reproduced with permission from Y. Chang, H. R. Fetterman, I. L. Newberg, and S. K. Panaretos, *IEEE Trans. Microwave Theory Techniques*, vol. 46, no. 11, pp. 1910–1919, © 1998 IEEE.)

a circulator or a separate transmitting antenna. The 2ω pump signal has to be delivered to all elements at the same amplitude and phase; otherwise, the mixed output will contain a term other than the signal that depends on *j*th element. If this ever happens, the sum of the excited field at each element will be distorted and will not form the conjugate beam. Therefore, optical interconnection is the crucial technology implemented to carry this 2ω microwave pump signal in phase to all mixing elements because of its low loss, light weight, and small size compared to microwave components. Using difference frequency generation in a mixer, the IF output current can be written as $I_C \propto e^{i2\omega t} \times e^{-i(\omega t - \varphi_j)} = e^{i(\omega t + \varphi_j)}$. This current component has the conjugate phase $+\varphi_j$ instead of the input phase $-\varphi_j$. Therefore, when it is delivered to the antenna, it will excite the conjugate field at $r_j$:

$$E_{C_j}(r_j) \propto A(r_j)e^{i(\omega t + \varphi_j)} + c.c. \tag{18.7}$$

When the sampling spacing is less than λ/2, the combined field $E_C = \Sigma_j E_{C_j}(r)$ forms the phase-conjugate wave on the sampling surface and therefore everywhere [14].

The signal from the antenna will form a beam that will travel back to the origin, correcting for all distortions as it retraces its path. Of course, the problem in this system is to get twice the frequency to every mixer with the same phase. Here is where the importance of the optical feed comes into play as shown in Figure 18.7.

In some cases, this can be used in auto-tracking systems such as in satellite communications. This approach would have the satellite send down a pilot beam, which would then be autotracked by the conjugated beam in a secure link. Essentially the conjugated beam would retrace its path back to the satellite with the information signal attached. All of the distortion in the path would be automatically corrected by this approach.

Two-dimensional free-space phase conjugation at 10.24 GHz has been demonstrated, with diffraction-limited results, utilizing an optically injected artificial nonlinear microarray. By extending this 1D array into a 2D surface as shown in Figure 18.8, complete three-dimensional wave-front reconstruction can be realized at MMW frequencies. In Figure 18.9, we show the actual implementation with an 8 × 16

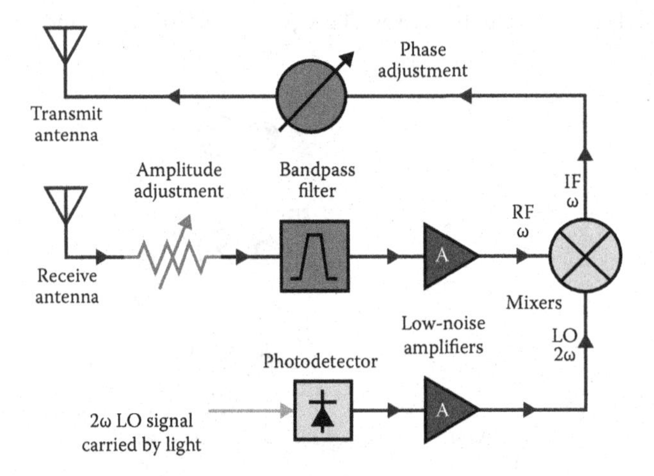


**FIGURE 18.7** An optical feed is used to bring twice the frequency of the conjugated signal with the same phase to each mixer in the array. (Reprinted with permission from Y. Chang, H. R. Fetterman, I. L. Newberg, and S. K. Panaretos, *Appl. Phys. Lett.*, vol. 72, no. 6, pp. 745–747, Feb. 1998, Copyright 1998, American Institute of Physics.)

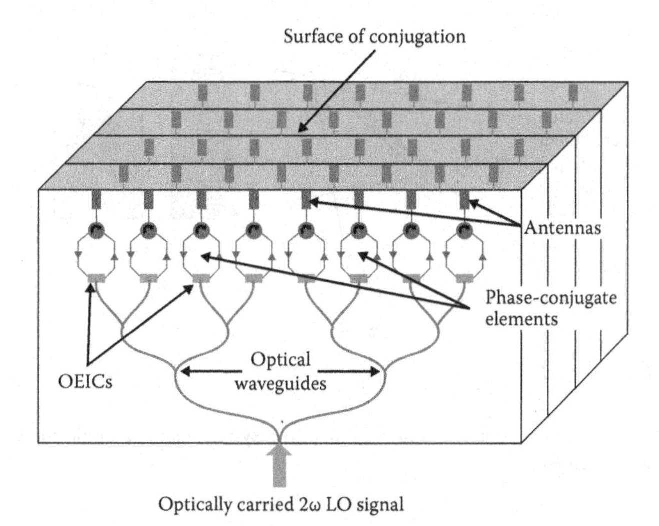


**FIGURE 18.8** Monolithic one-dimensional arrays forming a two-dimensional artificial nonlinear microwave surface for generating three-dimensional microwave phase conjugation. It shows the basic configuration using endfire antennas, circulators, and the optical feeds in a monolithic design. (Reproduced with permission from Y. Chang, H. R. Fetterman, I. L. Newberg, and S. K. Panaretos, *IEEE Trans. Microwave Theory Techniques*, vol. 46, no. 11, pp. 1910–1919, © 1998 IEEE.)

array. The extra elements in the vertical direction provide for transmit antenna elements without use of circulators.

Other applications involve using this technology to generate high-power single-mode beams. This particular application is shown in Figure 18.10. Typically, it is very hard to combine beams from many sources into a well-defined single-mode source. As shown in Figure 18.10, a beam splitter is used to illuminate, using a low-power single-mode source, a phase conjugating system with gain. In this way, a high-power single-mode signal can be generated. In another configuration, the beam can be used on one side of the phase conjugator and radiated from the other. It is important to note that the surface on the plane does not need to have any particular shape. Since it is phased by the incoming beam, it automatically achieves the correct phase relationship to represent an idea antenna.

**FIGURE 18.9** An actual picture of the fabricated phase conjugation array. It has an 8 × 16 configuration with endfire antennas and uses separate antennas for receive and transmit. Therefore, it has 16 elements in one dimension eliminating the circulator.

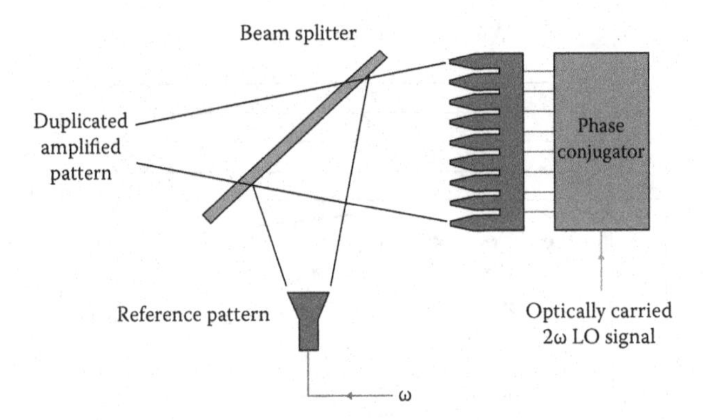


**FIGURE 18.10** A high-power single-mode phase conjugating system using a single-mode laser to illuminate a phase conjugating system with gain.

Another interesting application is to make a system called a bistatic radar for difficult radar targets. This bistable configuration is radar in which the source and the receiver operate from different locations. Such systems have particular application in situations that involve low-level observables such as stealth airplanes. These typically have shapes that do not reflect back to the radiation source. However, the use of a phase conjugated system in a second location, near the transmitter, will reflect the signal from the target back in exactly the right way to go back to the sender. Since it can have gain, it can make an almost invisible target much brighter.

Finally, another application of this type of automatic beam forming is the use of conformal antenna on airplanes [16]. It also performs the function of autotracking a pilot beam from a source such as a satellite. It will return a pilot beam from the surface as if it were a perfect parabolic surface and will even correct for any distortion in the path (Figure 18.11).

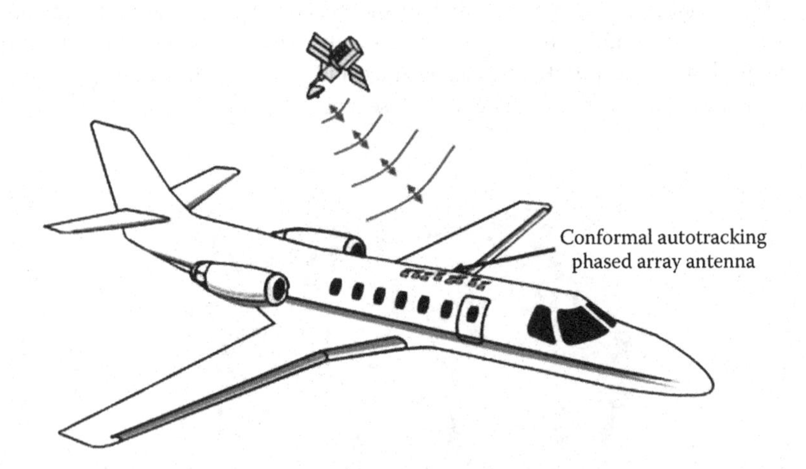


**FIGURE 18.11** A conformal phase conjugating system can communicate with a satellite as if it were an ideal antenna. It also performs the function of autotracking a pilot beam. A conformal antenna, using optical interconnections, on the side of an airplane can effectively function as a perfect antenna and can be used for autotracking to other planes or to satellites.

## 18.4 Imaging and Holography

Conventional imaging devices that use scanning antenna elements are often large and relatively expensive if they are confined to be able to process signals at millimeter-wave frequencies. An optical interconnect approach to the imaging array becomes much simpler and compact to fabricate since no coaxial cables or waveguides are required. It is desirable to work at extremely low powers and is necessary to minimize the physical array size and still obtain the highest resolution. The optically fed phased array has a rapid scanning speed and a flat form factor.

The photonic phased array system offers many advantages in millimeter-wave (95-GHz) imaging. Active imaging in particular offers significant improvement over passive, and at millimeter-wave frequencies, represents no known health problems at modest power levels. In the active mode of imaging, one basically looks at reflections and many objects will only be visible if the surface angle is such that it directly reflects from the transmitter to the detector. Therefore, active systems for security screening will often require the subject to rotate relative to the array to see all angles.

An advantage of active imaging is that the receiver sensitivity required is much less than in a passive system. As we have noted in making this active system, it is particularly useful to have a photonic approach. Microwave striplines and cables are extremely lossy at the higher frequencies (95 GHz and up). Also, the photonic approach enables remoting, so that instead of placing complex electronic components at the imaging site, they can be remoted at a base station. Furthermore, using optical interconnects permits the systems to be made without electronic amplifiers and still have acceptable signal-to-noise ratios.

A major recent application is in airports where images of people might be taken to look for explosives and contraband. The frequency of interest is often at millimeter waves (100 GHz or so) to obtain the necessary penetration through clothing with sufficient resolution. The problem is that there are many directions that must be examined to form useful 3D images.

One innovative way to solve this problem is to use a holographic approach [5] where the phase and amplitude are recorded for each antenna element. In many ways, this technology is very similar to phase conjugation in that the phase and amplitude are determined for each antenna element. Then having both the phase and the amplitude for a reflected beam, it is possible to calculate a computer hologram of the subject. The phase can be determined by a simple measurement of the in-phase and out-of-phase elements at millimeter-wave frequencies. A computer can perform the transform to then actually form a 3D image. The phase and amplitude measurement is usually accomplished by a heterodyne circuit such as shown in Figure 8.12. In this transceiver, the millimeter-wave signal is generated using a millimeter-wave oscillator, typically a Gunn diode oscillator. The transmitted signal is launched using a small wide-beamwidth antenna, often a small pyramidal waveguide horn. The transmitted signal is reflected from the target and then received by the receive antenna. The received signal is then mixed with in-phase (0° phase shift) and quadrature (90° phase shift) signals coupled from the millimeter-wave oscillator.

Since the system also becomes relatively heavy and difficult to implement in large arrays, optical interconnections represents an important design addition. The system for determining the phase at each point can be also directly implemented with optical mixers. Finally, there are significant advantages possible in remoting the antenna elements in the actual scanning device using optics. A picture of the millimeter linear array actually used in airports is shown in Figure 18.13.

Using a photonic feed system, it would be feasible to increase the number of elements and reduce the physical scanning. The use of optical phase measurement and optical interconnection would also permit much of this circuit to be fabricated in a monolithic configuration.

Finally, all of this processing to find phase and to analyze the signal becomes increasingly difficult at high frequencies. One of the solutions has been to mix down and to process the signal at the lower sideband frequency. Another approach that has recently gained support is the use of digital techniques such as used in digital radar. This has been limited by the A-to-D (analog-to-digital) systems at high

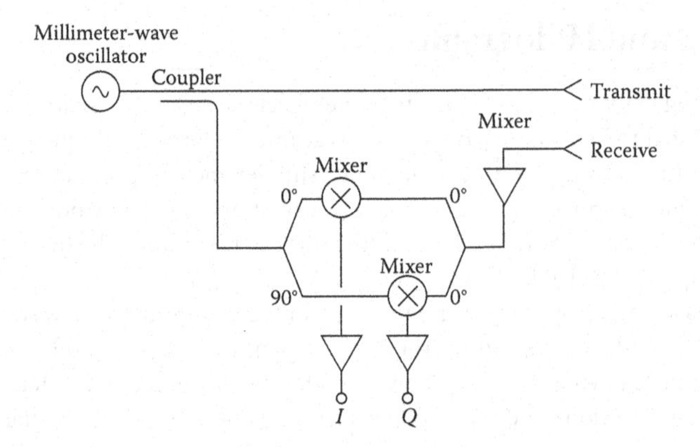


**FIGURE 18.12** Simplified holographic receiver using mixers to determine the effective phase and amplitude at each antenna element. The resulting signals are labeled as $I$ and $Q$ for in-phase and quadrature and contain the complex scattered waves amplitude and phase $I + jQ = Ae^{-j2kR}$, where $A$ is the amplitude of the scattered signal, $R$ is the distance to the target, and $k$ is the wavenumber. (Reproduced with permission from D. M. Sheen, D. L. McMakin, and T. E. Hall, *IEEE Trans. Microwave Theory Techniques*, vol. 49, no. 9, pp. 1581–1592, © 2001 IEEE.)

frequencies such as 100 GHz. In Figure 18.14, we show a modern photonic time stretching system [15] that can be used to transform the 100-GHz signals to 10 GHz for electronic A-to-Ds. Since the spectral width is conserved in linear propagation, the pulse width depends linearly on the fiber length. For time-stretching, the relevant stretch factor ($\equiv M$) is the width of the pulse exiting $L_2$ compared with that exiting $L_1$. If we denote by $\tau_0$ the pulse width exiting the laser, and $\delta\tau_1$, $\delta\tau_2$ the additional broadening from $L_1$, $L_2$ respectively, then the stretch factor can be described as

$$M = \frac{\tau_0 + \delta\tau_1 + \delta\tau_2}{\tau_0 + \delta\tau_1} = 1 + \frac{\delta\tau_2}{\tau_0 + \delta\tau_1} \approx 1 + \frac{L_2}{L_1} \quad if \quad \tau_0 \ll \delta\tau_0. \tag{18.8}$$

**FIGURE 18.13** The 1D linear array used in holographic scanning of millimeter waves in airports. The array is rotated about the subject to form a 3D image. (Reproduced with permission from D. M. Sheen, D. L. McMakin, and T. E. Hall, *IEEE Trans. Microwave Theory Techniques*, vol. 49, no. 9, pp. 1581–1592, © 2001 IEEE.)

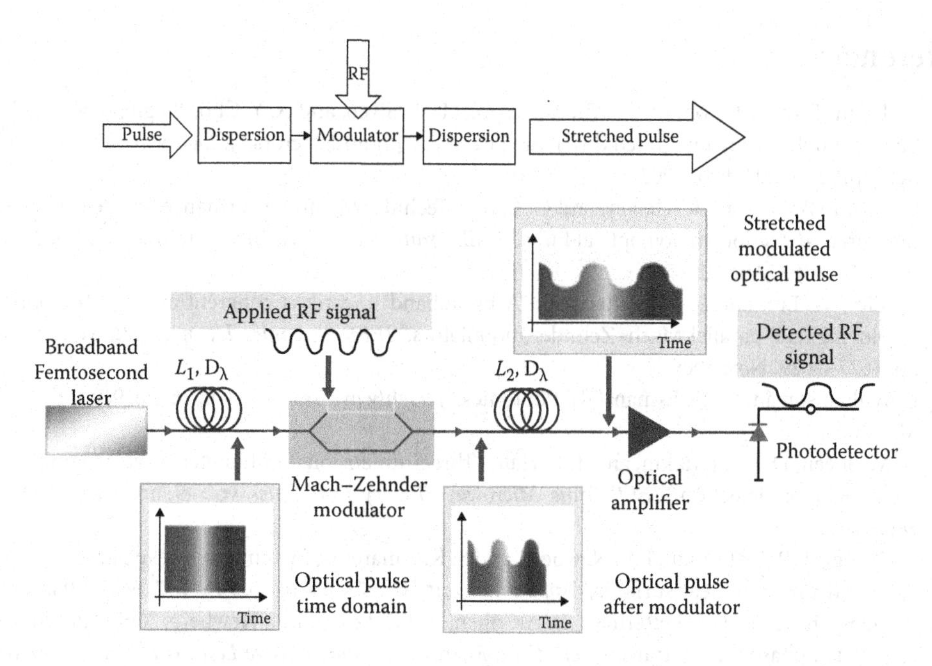


**FIGURE 18.14** Block diagram of the photonic time-stretch system showing the dispersive elements and the optical modulator.

For a modulating signal at the frequency $f_m$, the optical intensity on the photodetector $I_{out}(t)$ is proportional to

$$I_{out}(t) \propto f(t)\cos\left(\frac{2\pi f_m}{M}\right)\cos\left(2\pi^2 L_2 \beta'' \frac{L_1}{L_1 + L_2} f_m^2\right), \tag{18.9}$$

where $f(t)$ is the stretched pulse envelope and the last cosine term is a dispersion penalty distortion term that can be eliminated in a single sideband modulation system.

Essentially, in these systems, a dispersive element is used to generate a linear variation of wavelength with time from a broadband short pulse optical source. Then, using a high-frequency, single-sideband modulator, the signal to be digitized is used to amplitude modulate the light. Next, the signal is further dispersed in time to stretch the signal to a frequency as much as 15 times lower. After detection using a diode at the stretched lower frequency, the electronic signal can then be digitized using standard converters. This use of optics to convert the microwaves and millimeter-wave signals to digital represents a next generation in radar and holographic imaging systems.

## 18.5 Conclusion

In this chapter we have summarized a number of new optical techniques used in beam forming of microwave and millimeter-wave signals. We see this as the beginning of an entirely new area of development. The availability of new inexpensive lasers and low-cost high-frequency modulators now permits the creation of new systems for imaging, communications, and radars. The technology will make these devices available for scanning in airports, for prevention of autocollisions and to enable communications at extraordinary speeds. The use of OSP approaches and ultrahigh speed A-to-D converters will open new application areas that will lead future sensor systems for decades to come.

## References


1. M. Islam, T. Jung, T. Itoh, M. C. Wu, A. Nespola, D. L. Sivco, and A. Y. Cho, "High power and highly linear monolithically integrated distributed balanced photodetectors," *J. Lightwave Technol.*, vol. 20, no. 2, pp. 285–294, Feb. 2002.
2. C. Cox, E. Ackerman, R. Helkey, and G. Betts, "Techniques and performance of intensity modulation direct-detection analog optical links," *IEEE Trans. Microwave Theory Techniques*, vol. 45, no. 8, pp. 1375–1383, Aug. 1997.
3. G. Zhu, W. Liu, and H. R. Fetterman, "A broadband linearized coherent analog fiber-optic link employing dual parallel Mach–Zehnder modulators," *IEEE Photonics Technol. Lett.*, vol. 21, no. 21, pp. 1627–1629, Nov. 2009.
4. R. Williamson and R. D. Esman, "RF Photonics," *J. Lightwave Technol.*, vol. 26, no. 9, pp. 1145–1153, May 2008.
5. D. M. Sheen, D. L. McMakin, and T. E. Hall, "Three-dimensional millimeter-wave imaging for concealed weapon detection," *IEEE Trans. Microwave Theory Techniques*, vol. 49, no. 9, pp. 1581–1592, Sep. 2001.
6. Y. Chang, H. R. Fetterman, I. L. Newberg, and S. K. Panaretos, "Millimeter–wave phase conjugation using artificial nonlinear surfaces," *Appl. Phys. Lett.*, vol. 72, no. 6, pp. 745–747, Feb. 1998.
7. Y. Chang, B. Tsap, H. R. Fetterman, D. A.Cohen, A. F. J. Levi, and I. Newberg, "Optically controlled serially fed phased-array transmitter," *IEEE Microwave Guided Wave Lett.*, vol. 7, no. 3, pp. 69–71, Mar. 1997.
8. B. Tsap, Y. Chang, H. R. Fetterman, A. F. J. Levi, D. A. Cohen, and I. Newberg, "Phased-array optically controlled receiver using a serial feed," *IEEE Photonics Technol. Lett.*, vol. 10, no. 2, pp. 267–269, Feb. 1998.
9. D. B. Hunter, M. E. Parker, and J. L. Dexter, "Demonstration of a continuously variable true-time delay beamformer using a multichannel chirped fiber grating," *IEEE Trans. Microwave Theory Techniques*, vol. 54, no. 2, pp. 861–867, Feb. 2006.
10. Z. Fu, C. Zhou, and R. T. Chen, "Waveguide–hologram-based wavelength-multiplexed pseudo-analog true-time-delay module for wideband phased-array antennas," *Appl. Opt.*, vol. 38, no. 14, pp. 3053–3059, May 1999.
11. J. Han, B. Seo, S. Kim, H. Zhang, and H. Fetterman, "Single-chip integrated electro-optic polymer photonic RF phase shifter array," *J. Lightwave Technol.*, vol. 21, no. 12, pp. 3257–3261, Dec. 2003.
12. B. Seo, S. Kim, B. Bartnik, H. R. Fetterman, D. Jin, and R. Dinu, "Optical signal processor using electro-optic polymer waveguides," *J. Lightwave Technol.*, vol. 27, no. 15, pp. 3092–3106, Aug. 2009.
13. B. Seo and H. R. Fetterman, "True-time-delay element in lossy environment using EO waveguides," *IEEE Photnics Technol. Lett.*, vol. 18, no. 1, pp. 10–12, Jan. 2006.
14. Y. Chang, H. R. Fetterman, I. L. Newberg, and S. K. Panaretos, "Millimeter-wave phase conjugation using antenna arrays," *IEEE Trans. Microwave Theory Techniques*, vol. 46, no. 11, pp. 1910–1919, Nov. 1998.
15. Y. Han, B. Jalali, J. Han, B. Seo, and H. Fetterman, "Demonstration and analysis of single sideband photonic time-stretch system," *IEICE Trans. Electron.*, vol. E86-C, no. 7, pp. 1276–1280, Jul. 2003.
16. R. Y. Miyamoto, K. M. K. H. Leong, S.-S. Jeon, Y. Wang, Y. Qian, and T. Itoh, "Digital wireless sensor server using an adaptive smart-antenna/retrodirective array," *IEEE Trans. Vehicular Technol.*, vol. 52, no. 5, pp. 1181–1188, Sep. 2003.

# 19

# Electro-Optic Time Lenses for Shaping and Imaging Optical Waveforms



Brian H. Kolner
*University of California*

## 19.1 Introduction

The history of time lenses can be traced back to the development of chirp radar during World War II. Chirp radar arose out of a need to obtain better signal-to-noise ratios (SNRs) on returned microwave pulses and ushered in a new era of analog signal processing and network theory [1]. The sophistication of analog circuit signal processing techniques continued well after this period. It was P. Tournois who, apparently, first made the connection between chirp and dispersion and temporally magnifying electrical waveforms [2,3]. Caputi also realized these important relationships and independently developed a time-scaling system utilizing the same elements, which he called, "stretch" [4]. The tools of linear systems analysis were brought to bear on this technique and the analogs to optical processes were included in a textbook by Papoulis [5]. In a seminal paper on optical pulse compression, Treacy pointed out the analogy [6] and other parallels in optics soon began to emerge [7].

Early in the history of ultrashort light pulse techniques, the application of electro-optic modulators as a chirping mechanisms was explored by many researchers [8–11]. This technique of pulse chirping, combined with newer approaches to dispersive delay lines and the feverish pace of short pulse technology, renewed interest in the possibility of temporal imaging [12,13] and soon experiments were demonstrating pulse compression [14–16]. A comprehensive analysis of the mathematical duality between the spatial problem of diffraction and the temporal problem of dispersion [17] defined the temporal resolution of an imaging system and the magnification, the sign of which suggested the possibility of local time reversal (a feature suggested in a number of prior papers [18]).

What became clear during this period was that electro-optic time lenses, although conceptually satisfying, held great technical challenges to gain wide utility. This was the impetus for the development of the parametric time lens [19–26]. Today, the electro-optic approach to temporal imaging is more promising as we develop ever-more efficient modulators in integrated optical form, but the principles transcend the technological hurdles. These principles have gained wide acceptance and are starting to appear in textbooks [27–30]. This present contribution addresses electro-optic time lenses and temporal imaging in a comprehensive manner with the hope of stimulating further work in improving the capabilities of the electro-optic modulator as a time lens.

To justify the notion of time lenses and temporal imaging systems in general, it is first necessary to establish the analogous properties of the physics of paraxial diffraction and narrowband dispersion. This will then lead to a system of dual equations upon which lies the basis of temporal imaging. We will begin with a review of paraxial diffraction and then narrowband dispersion and cast the results in nearly identical forms. We will then proceed to a general discussion of space lenses and then the concept of the time lens. With these elements in hand, we will see how electro-optic modulators can be used as time lenses and establish their properties in accordance with their spatial counterparts. Finally, we discuss the application of time lenses in optical pulse compression and temporal imaging, which is a special case of waveform crafting.

## 19.2 Space–Time Duality: Diffraction and Dispersion

The usual approach to studying the space–time evolution of arbitrary electromagnetic wavefronts is to confine one's attention to monochromatic waves, which are spatially limited in the plane transverse to the general direction of propagation. Within this approximation, wave propagation is known as, "paraxial diffraction" [29]. We write the total field including time dependence as

$$E(x,y,z,t)=U(x,y,z)e^{i(\omega_0 t-k_0 z)}, \tag{19.1}$$

where $U(x,y,z)$ is an envelope function that varies slowly compared to $e^{-ik_0 z}$. The differential equation governing propagation is the paraxial wave equation for the envelope $U(x,y,z)$,

$$\frac{\partial U}{\partial z}=-\frac{i}{2k_0}\left(\frac{\partial^2 U}{\partial x^2}+\frac{\partial^2 U}{\partial y^2}\right), \tag{19.2}$$

or, using subscripts to indicate differentiation with respect to the indicated coordinate, gives the more compact form

$$U_z=-\frac{i}{2k_0}(U_{xx}+U_{yy}). \tag{19.3}$$

Spatial Fourier transformation of this equation in the transverse coordinates $(x,y)$ converts it into an ordinary differential equation

$$\frac{dU(k_x,k_y,z)}{dz}=\frac{i}{2k_0}\left(k_x^2+k_y^2\right)U\left(k_x,k_y,z\right), \tag{19.4}$$

where $U(k_x,k_y,z)$ is the Fourier transform of $U(x,y,z)$, and $k_x = 2\pi f_x$ and $k_y = 2\pi f_y$ are the transverse spatial frequencies. This step is facilitated by the derivative theorem from Fourier transform theory. The solution to Equation 19.4 is

$$U(k_x,k_y,z)=U(k_x,k_y,0)\exp\left[i\frac{z}{2k_0}\left(k_x^2+k_y^2\right)\right], \tag{19.5}$$

where $U(k_x,k_y,0) \equiv \Im\{U(x,y,0)\}$ (the Fourier transform of $U(x,y,0)$) is the spectrum at an arbitrary input plane, here labeled $z = 0$. To recover the electric field in the real space coordinates, we simply inverse Fourier transform Equation 19.5:

$$U(x,y,z)=\frac{1}{4\pi^2}\iint_{-\infty}^{\infty} U(k_x,k_y,0)\exp\left[i\frac{z}{2k_0}\left(k_x^2+k_y^2\right)\right]\exp\left[-i(k_xx+k_yy)\right]dk_x\,dk_y. \tag{19.6}$$

The key feature of this result is that, in the paraxial approximation, an incident wavefront $U(x,y,0)$ has its transverse Fourier spectrum filtered upon propagation by a phase-only transfer function that is quadratic in the spatial frequency variables and linear in the propagation distance $z$. This result is quite general, and we will see that it occurs in the narrowband dispersion problem (Section 19.3).

## 19.3 Narrowband Dispersion

In the dispersion problem, we assume fields with infinite constant transverse amplitude so that $\partial^2/\partial x^2 = \partial^2/\partial y^2 = 0$. In a material medium such as a dielectric with electromagnetic properties given by $\mu = \mu_0$ and $\varepsilon = \varepsilon(\omega)$, the wave equation for the electric field reduces to

$$\nabla^2E(x,y,z,t)=\frac{\partial^2E(z,t)}{\partial z^2}=\mu_0\varepsilon(\omega)\frac{\partial^2E(z,t)}{\partial t^2}. \tag{19.7}$$

A solution to this equation is the simple monochromatic plane wave:

$$E(z,t) = E_0\exp[i(\omega_0t - \beta_0z)], \tag{19.8}$$

where

$$\beta_0=\beta(\omega_0)=\omega_0\sqrt{\mu\varepsilon(\omega_0)} \tag{19.9}$$

and $E_0$ is an arbitrary amplitude.

Now, dispersion is a phenomenon that causes the broadening of pulses, or wavepackets, in media with a frequency-dependent propagation constant $\beta(\omega)$. It is an essential feature for exploiting the attributes of time lenses. Suppose we construct a wavepacket of carrier envelope form:

$$E(z,t)=A(z,t)e^{i(\omega_0t-\beta_0z)} \tag{19.10}$$

where $A(z,t)$ is an envelope function that confines the carrier plane wave in the direction of propagation. If we know the initial shape of the envelope in a plane $z$ = constant, then the evolution of the wavepacket upon propagation can be deduced by allowing each Fourier component of the spectrum to propagate according to Equation 19.8, where $\omega_0 \to \omega$. To make the superposition of all monochromatic wave constituents tractable, we expand the propagation constant in a Taylor series to second order about a nominal carrier frequency $\omega_0$,

$$\beta(\omega) = \beta(\omega_0) + (\omega - \omega_0)\left.\frac{d\beta}{d\omega}\right|_{\omega=\omega_0} + \frac{(\omega-\omega_0)^2}{2}\left.\frac{d^2\beta}{d\omega^2}\right|_{\omega=\omega_0} \tag{19.11}$$

$$= \beta_0 + (\omega - \omega_0)\beta' + \frac{(\omega-\omega_0)^2}{2}\beta''. \tag{19.12}$$

A new differential equation can then be constructed for the envelope $A(z,t)$:

$$\left(\frac{\partial}{\partial z} + \frac{1}{v_g}\frac{\partial}{\partial t}\right)A(z,t) = \frac{i\beta''}{2}\frac{\partial^2 A(z,t)}{\partial t^2} \tag{19.13}$$

where $v_g^{-1} = \beta'$. It is useful to transform this result to a traveling-wave coordinate system moving at the group velocity $v_g$. This is accomplished by setting

$$\begin{aligned} \tau &\equiv (t - t_0) - \frac{z - z_0}{v_g}, \\ \xi &\equiv z - z_0 \end{aligned} \tag{19.14}$$

where $t_0$ and $z_0$ are arbitrary reference points. The derivatives transform according to

$$\begin{aligned} \frac{\partial}{\partial z} &\to \frac{\partial}{\partial \xi} - \frac{1}{v_g}\frac{\partial}{\partial \tau}, \\ \frac{\partial}{\partial t} &\to \frac{\partial}{\partial \tau}, \\ \frac{\partial^2}{\partial t^2} &\to \frac{\partial^2}{\partial \tau^2}. \end{aligned} \tag{19.15}$$

Equation 19.13 in the traveling-wave coordinate system is now

$$\frac{\partial A(\xi,\tau)}{\partial \xi} = \frac{i\beta''}{2}\frac{\partial^2 A(\xi,\tau)}{\partial \tau^2}. \tag{19.16}$$

Using the compact notation, this can also be written

$$A_\xi = \frac{i\beta''}{2}A_{\tau\tau}. \tag{19.17}$$

Comparing Equation 19.17 with the paraxial wave equation (19.3) completes the space-time duality. The solution to the dispersion problem is therefore

$$A(\xi,\tau)=\frac{1}{2\pi}\int_{-\infty}^{\infty}A(0,\omega)\exp\left[-i\frac{\xi}{2}\beta''\omega^2\right]\exp[i\omega\tau]d\omega. \tag{19.18}$$

## 19.4 Space Lenses: General Properties

To appreciate time lenses and how they apply to waveshaping and imaging systems, it will help to review the physical principles of conventional space lenses. The physical action of a thin space lens on a propagating wavefront can be described in terms of an instantaneous spatial phase modulation or transformation, which in the $\exp[i(\omega_0 t - k_0 z)]$ basis is [29]

$$\begin{aligned} t_\iota(x,y) &= \exp\left[-ik_0 n\Delta_0 + i\frac{k_0}{2f_S}(x^2+y^2)\right], \\ &= \exp\left[i\phi_S(x,y)\right] \end{aligned} \tag{19.19}$$

where $\Delta_0$ is the lens thickness at its center, $k_0 = 2\pi/\lambda_0$, and $f_S$ is the focal length of the lens defined in terms of its physical properties (e.g., radii of curvature, $R_1$, $R_2$, and index, $n$, of the glass) and the medium in which it is immersed. When traveling through a vacuum,

$$f_s = \frac{1}{(n-1)\left(\frac{1}{R_1}-\frac{1}{R_2}\right)}. \tag{19.20}$$

The fields immediately past the lens then vary according to

$$\exp\left[i\left(\omega_0 t - \left[k_0 n\Delta_0 - \frac{k_0}{2f_S}\left(x^2+y^2\right)\right]\right)\right] = \exp\left[i\left(\omega_0 t + \phi_S(x,y)\right)\right]. \tag{19.21}$$

The conventional description of a lens's focal length can be interpreted in a broader context, which will have a dual in the development of the theory of time lenses if we expand the phase in a Maclaurin series,

$$\phi_S(x) = \phi_S(0) + x\left.\frac{\partial\phi_S}{\partial x}\right|_{x=0} + \frac{x^2}{2!}\left.\frac{\partial^2\phi_S}{\partial x^2}\right|_{x=0} + \ldots \tag{19.22}$$

$$= \phi_{S0} + x\phi_S' + \frac{x^2}{2}\phi_S'' + \ldots, \tag{19.23}$$

where, due to the usual spherical symmetry of the lens, we need only consider one coordinate. Comparing the actual lens phase of Equation 19.19 with the expansion (Equation 19.23) we see that

$$\phi_S'' = \frac{k_0}{f_S} \quad \text{or} \quad f_S = \frac{k_0}{\phi_S''}. \tag{19.24}$$

Thus, the phase has positive curvature for a lens with a positive focal length. Equivalently, it is the strength of the quadratic phase imparted by the lens that determines its focal length.

The instantaneous transverse spatial frequency or wavenumber in the $x$-direction is given by

$$k_{xi} = -\frac{\partial}{\partial x}\big(\omega_0 t + \phi_S(x,y)\big) = -\frac{\partial \phi_S(x,y)}{\partial x} = -\frac{k_0}{f_S}x. \tag{19.25}$$

We see that a positive thin lens introduces a negative frequency sweep, or chirp, on the wavefront. If we differentiate one more time,

$$\frac{\partial k_{xi}}{\partial x} = -\frac{k_0}{f_S} \tag{19.26}$$

which gives the slope of the transverse spatial frequency (or chirp rate) versus position. For a purely quadratic-phase lens, the slope is constant (see Figure 19.1a). Thus, the focal length can be interpreted as (minus) the ratio of the free-space wavenumber to the chirp rate of the instantaneous transverse spatial frequency [31]:

$$f_S = -\frac{k_0}{\partial k_{xi}/\partial x} = -\frac{k_0}{\partial k_{yi}/\partial y}. \tag{19.27}$$

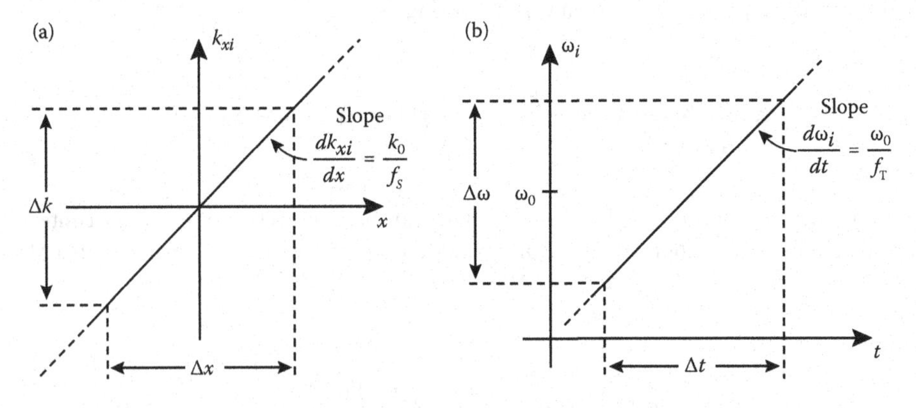


**FIGURE 19.1** Instantaneous frequency spectra of (a) spatial and (b) temporal waveforms. In (a) the spectrum is positively chirped and therefore corresponds to an expanding quadratic wavefront, which would be produced by either paraxial diffraction or a negative thin lens ($f_s < 0$). The temporal spectrum in (b) is positively chirped and corresponds to a wavepacket propagating in a medium with normal dispersion or a time lens that produces a positive quadratic phase curvature [31]. (Reproduced from B. H. Kolner, *J. Opt. Soc. Am. A*, vol. 11, no. 12, pp. 3229–3234, Dec. 1994. With permission of Optical Society of America.)

## 19.5 Time Lenses

The concept of a time lens develops directly from the analogy to a space lens [17]. Here, however, the quadratic phase modulation must take place in the traveling-wave time coordinate $\tau = t - z/v_g$, moving at the group velocity of the wavepacket. By analogy with the quadratic phase transformation produced by a thin lens (Equation 19.19), suppose a phase modulation

$$\phi_\tau(\tau) = \frac{\omega_0 \tau^2}{2 f_T} \tag{19.28}$$

is applied to the wavepacket in the traveling-wave reference frame. Here we have introduced the concept of a *focal time*, $f_T$, in direct analogy with the space lens. If we compare Equation 19.28 to a general Taylor series expansion of the phase function [17,31]

$$\phi_\tau(\tau) = \phi_\tau(\tau_0) + (\tau - \tau_0)\frac{d\phi_\tau}{d\tau}\bigg|_{\tau=\tau_0} + \frac{(\tau-\tau_0)^2}{2!}\frac{d^2\phi_\tau}{d\tau^2}\bigg|_{\tau=\tau_0} + \ldots \tag{19.29}$$

$$= \phi_{\tau 0} + (\tau - \tau_0)\phi'_\tau + \frac{(\tau-\tau_0)^2}{2}\phi''_\tau + \ldots \tag{19.30}$$

we immediately conclude that

$$f_T = \frac{\omega_0}{\phi''_\tau}. \tag{19.31}$$

Also, since the instantaneous angular frequency is defined as the time derivative of the phase,

$$\omega_i = \frac{d\phi_t}{d\tau} = \frac{\omega_0}{f_T}\tau \tag{19.32}$$

we conclude that a time lens with a positive focal time induces a positive frequency sweep, or chirp, on a waveform. This is opposite to the result from our analysis of space lenses described by Equation 19.25 where a positive lens produces a negatively chirped spatial frequency spectrum.

Differentiating Equation 19.32 once more,

$$\frac{d\omega_i}{d\tau} = \omega'_i = \frac{d^2\phi_\tau}{d\tau^2} = \phi''_\tau \tag{19.33}$$

Equation 19.31 can be written as

$$f_T = \frac{\omega_0}{\omega'_i}. \tag{19.34}$$

In words, the focal time is the ratio of the carrier frequency to the chirp rate induced by the lens. The one-to-one correspondence with the properties of the space lens are now established and summarized in Table 19.1.

**TABLE 19.1** Physical Characteristics and Definitions for Space and Time Lenses

| | Space | Time |
|---|---|---|
| Lens phase | $\phi_s(x,y)=\frac{k_0(x^2+y^2)}{2f_S}$ | $\phi_\tau(\tau)=\frac{\omega_0\tau^2}{2f_T}$ |
| Focal length | $f_S=\frac{k_0}{\phi''_x}=-\frac{k_0}{k'_{xi}}$ | $f_T=\frac{\omega_0}{\phi''_\tau}=\frac{\omega_0}{\omega'_i}$ |
| $f$-Number | $f_S^\#=\frac{k_0}{\Delta k}$ | $f_T^\#=\frac{\omega_0}{\Delta\omega}$ |

## 19.6 The Focal Length-to-Aperture Ratio: *f*-Number

The *f*-number, or ratio of focal length to diameter, of a conventional space lens is universally used to primarily specify the light-gathering power of the lens and to a lesser extent indicate its diffraction-limited resolution (where warranted). We can use the previously derived generalizations of the concept of focal length to arrive at a similarly generalized notion of *f*-number [31]. If we call the useful aperture of the lens $\Delta x$ and use Equation 19.27 for the generalized focal length, the *f*-number of a space lens is

$$f_S^\# \equiv \frac{|f_S|}{\Delta x}=\frac{k_0}{\Delta x\left|\partial k_{xi}/\partial x\right|} \tag{19.35}$$

where, without loss of generality, we use the chirp of the spatial frequency spectrum in the $x$-direction. The denominator of Equation 19.35 has a simple physical interpretation; it represents the total spatial frequency spectrum generated by the lens

$$\Delta k_x \equiv \Delta x\left|\frac{\partial k_{xi}}{\partial x}\right|. \tag{19.36}$$

For the usual case of circularly symmetric lenses, we need make no distinction between the $x$- and $y$-axes, and thus,

$$f_S^\# = \frac{k_0}{\Delta k}. \tag{19.37}$$

In words, the *f*-number of a space lens is the ratio of the wavenumber to the spread in transverse spatial frequencies induced by the lens.

We can apply the same reasoning to a time lens. Suppose we define the temporal *f*-number as the ratio of the focal time $f_T$ to the aperture time $\Delta t$, where $\Delta t$ is controlled by some mechanism that limits the available aperture of the time lens to a region of predominantly quadratic phase. Then,

$$f_T^\# \equiv \frac{|f_T|}{\Delta t}=\frac{\omega_0}{\Delta t\left|d\omega_i/d\tau\right|}. \tag{19.38}$$

Again, we recognize that the denominator describes the total bandwidth $\Delta\omega$ induced by the quadratic phase modulation of the time lens

$$\Delta\omega = \Delta t\left|\frac{d\omega_i}{d\tau}\right|. \tag{19.39}$$

Thus, the temporal *f*-number becomes

$$f_T^{\#} = \frac{\omega_0}{\Delta\omega} \tag{19.40}$$

which, in words, is the inverse of the fractional bandwidth induced by the time lens phase modulation process. The *f*-number definitions presented in Equations 19.37 and 19.40 can now be added to Table 19.1.

## 19.7 Electro-Optic Time Lenses

A conventional electro-optic phase modulator is perhaps the simplest mechanism for achieving the necessary quadratic phase required of a time lens. The general phase shift accumulated upon passage through any modulator for a $z$-propagating wave is

$$\phi(z,t) = \omega_0 t - k_0 z - \frac{\omega_0}{c}\int_0^z \Delta n(z',t)dz' \tag{19.41}$$

$$= \omega_0 t - k_0 z - \Gamma(z,t), \tag{19.42}$$

where $\Gamma(z,t)$ is the net retardation in the laboratory reference frame, due to a space and time-varying index perturbation $\Delta n(z,t)$. In the traveling-wave reference frame, this phase becomes

$$\phi(\xi,\tau) = \omega_0\tau + \xi\left(\frac{\omega_0}{v_{go}} - k_0\right) - \Gamma(\xi,\tau) \tag{19.43}$$

$$\approx \omega_0\tau - \Gamma(\xi,\tau), \tag{19.44}$$

where Equation 19.44 is obtained by assuming there is no significant optical dispersion in the modulator. The total optical field of a wavepacket, $A_i(0,\tau)e^{i\omega_0\tau}$, incident on a modulator will emerge as

$$E_0(\xi,\tau) = A_i(0,\tau)e^{i[\omega_0\tau - \Gamma(\xi,\tau)]} = A_0(\xi,\tau)e^{i\omega_0\tau}. \tag{19.45}$$

The relationship between the input and output envelope functions is therefore

$$A_0(\xi,\tau) = A_i(0,\tau)e^{-i\Gamma(\xi,\tau)}. \tag{19.46}$$

Since the electro-optic effect is linear in the applied electric field, we must develop a field that varies quadratically in space–time over the duration of the traveling optical waveform. We will study two drive fields that are strongly quadratic within a finite region; sinusoids and Gaussian pulses. We begin by deriving the phase retardation for an electrically short modulator with a homogeneous field distribution.

### 19.7.1 Short Homogeneous Field Modulators: Sinusoidal Drive

Consider an electro-optic modulator configured as a parallel-plate capacitor with a homogeneous, time-varying, field distribution (Figure 19.2a), which produces an index modulation:

$$\Delta n(z,t) = \Delta n_0 \cos(\omega_m t + \theta), \tag{19.47}$$

where $\theta$ is an arbitrary initial phase offset. The length of the optical propagation path is $L$. Since we are interested in casting the resulting retardation in the reference frame of the modulated wavepacket, let us transform the integral to traveling-wave coordinates using Equation 19.14:

$$\Gamma(\xi,\tau) = \frac{\omega_0}{c}\int_0^L \Delta n(\xi',\tau)d\xi' \tag{19.48}$$

$$= \frac{\omega_0 \Delta n_0}{c}\int_0^L \cos\left[\omega_m\left(\tau + \xi'/v_{go}\right) + \theta\right] d\xi' \tag{19.49}$$

$$= \frac{\omega_0 \Delta n_0}{c}\frac{v_{go}}{\omega_m}\left\{\sin\left[\omega_m\left(\tau + L/v_{go}\right) + \theta\right] - \sin(\omega_m \tau + \theta)\right\}. \tag{19.50}$$

Using the identity $\sin(\alpha) - \sin(\beta) = \cos[(\alpha + \beta)/2]\sin[(\alpha - \beta)/2]$, Equation 19.50 can be written in the more illustrative form:

$$\Gamma(\xi,\tau) = \Gamma_0 \frac{\sin(\Delta\phi/2)}{\Delta\phi/2}\cos(\omega_m \tau + \Delta\phi/2 + \theta), \tag{19.51}$$

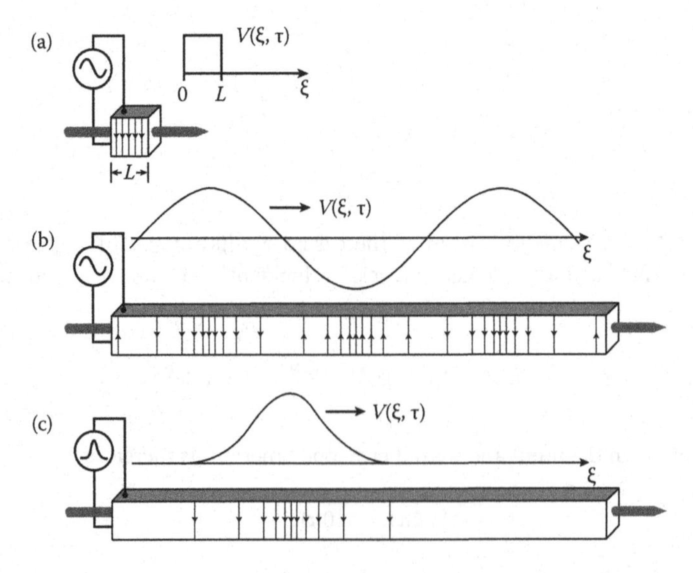


**FIGURE 19.2** Electro-optic phase modulators as time lenses. Voltage profile $V(\xi, \tau)$ shown as a function of spatial coordinate $\xi$ at an arbitrary time $\tau$. (a) Electrically short modulator with sinusoidal drive voltage and spatially homogeneous field distribution. (b) Electrically long, traveling-wave modulator with sinusoidal drive. (c) Electrically long, traveling-wave modulator with Gaussian pulse drive.

where

$$\Gamma_0 \equiv \frac{\omega_0 \Delta n_0 L}{c} = \pi \frac{V}{V_\pi} \tag{19.52}$$

is the peak phase retardation and

$$\Delta\phi \equiv \frac{\omega_m L}{v_{go}} \tag{19.53}$$

is a transit-time reduction, or walkoff, factor. This is the ratio of the transit time through the modulator to the period of the modulation. For long crystals or high modulation frequencies, the field in the crystal can change appreciably during propagation and this will reduce the overall strength of the retardation. It will also shift the symmetry point by $\Delta\phi/2$, but this can be precompensated by virtue of the phase offset $\theta$. Note that restoring this symmetry will not restore the amplitude.

From Equation 19.51, we see that in the traveling-wave reference frame of the optical wavepacket, a quadratic phase modulation is obtained in the vicinity of the center of the packet ($\tau \approx 0$) provided that the transit-time effect is not too large. However, to keep this transit time effect small, we must have a relatively short interaction length, $L$, which leads to a small peak phase retardation. The usual way to get around this problem is using a traveling-wave modulator, where the modulating field travels at, or near, the same velocity as the optical field.

## 19.7.2 Traveling-Wave Phase Modulators: Sinusoidal Modulation

In a traveling-wave electro-optic modulator, the driving field is guided through the crystal, usually by means of a transmission line, to produce an index of refraction perturbation that propagates as a traveling wave (Figure 19.2b). In the case where the driving field is a sinusoid at frequency $\omega_m$ with propagation constant $k_m$ propagating in the $+z$ direction,

$$\Delta n(z,t) = \Delta n_0 \cos(\omega_m t - k_m z + \theta), \tag{19.54}$$

where $\Delta n_0$ is the peak instantaneous index of refraction modulation and $\theta$ is a fixed phase offset.

We again transform the integral for the net retardation in Equation 19.41 to a traveling-wave coordinate system:

$$\Gamma(\xi,\tau) = \frac{\omega_0}{c} \int_0^{\xi} \Delta n(\xi',\tau) d\xi' \tag{19.55}$$

$$= \frac{\omega_0}{c} \int_0^{\xi} \Delta n_0 \cos\left[\omega_m \left(\tau + \xi' \left(\frac{1}{v_{go}} - \frac{1}{v_{pm}}\right) + \frac{\theta}{\omega_m}\right)\right] d\xi' \tag{19.56}$$

$$= \Gamma_0 \frac{\sin(\Delta\phi/2)}{\Delta\phi/2} \cos(\omega_m \tau + \Delta\phi/2 + \theta), \tag{19.57}$$

where

$$\Gamma_0 \equiv \frac{\omega_0 \Delta n_0 \xi}{c} = \pi \frac{V}{V\pi} \tag{19.58}$$

is the peak phase retardation and

$$\Delta\phi \equiv \omega_m \xi \left( \frac{1}{v_{go}} - \frac{1}{v_{pm}} \right) \tag{19.59}$$

is the normalized walkoff between the optical wavepacket and the modulating field.

Although similar in appearance to the homogeneous-field result of Equation 19.51, the traveling-wave retardation of Equation 19.57 has a substantial difference due to the nature of the transit-time, or walkoff, factor $\Delta\phi$. In the case of the simple homogeneous-field modulator, increasing the interaction length eventually causes a reduction in net retardation while in the traveling-wave modulator we have the additional degree of freedom of matching the optical group velocity with the modulating field phase velocity. In fact, once they are matched, the modulator length can be arbitrarily long (neglecting losses and beam diffraction effects).

Figure 19.3a (left) shows the retardation of Equation 19.57, normalized to the maximum $\Gamma_0$ for three values of walkoff ($\Delta\phi = 0, \pm\pi/4$) and no initial phase offset ($\theta = 0$). Included in the figure is a Gaussian pulse, which represents the intensity of a hypothetical optical wavepacket. When the wavepacket and modulating field are synchronized with a modulation extremum, the accumulated retardation is cosinusoidal. And if the duration of the wavepacket is less than about $\pi/4$ in modulation phase, the retardation is predominantly quadratic.

The introduction of velocity walkoff in the time lens produces a new feature: displacement and reduction of the resulting retardation, which is a new feature. This is immediately obvious in Figure 19.3a for the cases of $\Delta\phi = \pm\pi/4$. The reduction in amplitude arises because the wavepacket slides through the cosinusoidal field

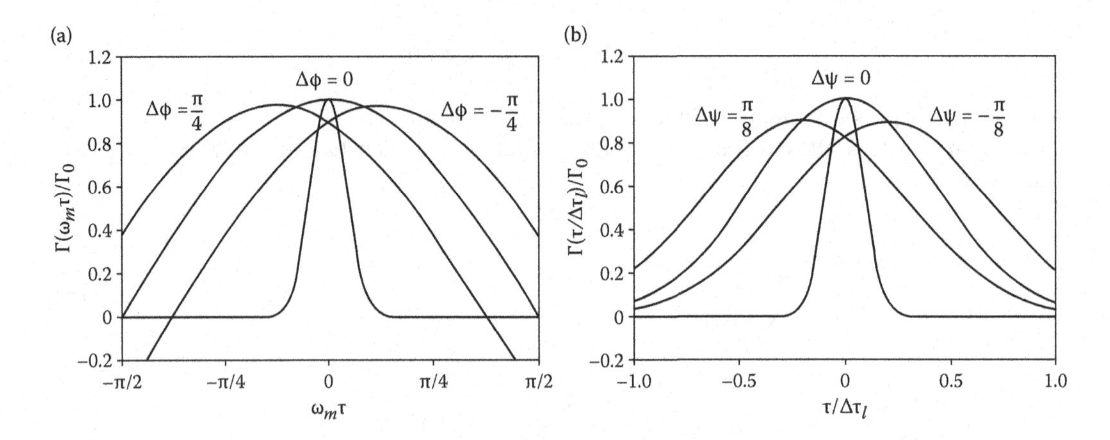


**FIGURE 19.3** Normalized time lens retardation after propagation as a function of the local time variable. (a) Retardation produced by a cosinusoidal phase modulation with walkoff between local optical wavepacket (shown as a Gaussian pulse) and modulating field of $\Delta\phi = 0, \pm\pi/4$. (b) Retardation produced by a Gaussian modulation with walkoff between local wavepacket and field of $\Delta\psi = 0, \pm\pi/8$. Notice that velocity walkoff generates a linear phase term, which corrupts the pure quadratic response required for a time lens. Notice too that a wavepacket of long duration will incur higher order phase terms on the wings of the retardation.

rather than staying synchronized with it. The shift in the resulting retardation arises for the same reason. The net effect is that the extremum of the retardation is no longer centered on the wavepacket, its displacement introduces a linear phase term onto the center of the wavepacket. From the Fourier shift theorem, we know that a linear phase in one domain results in a shift in the other. Therefore, the linear phase across the center of the wavepacket will cause a time shift of the packet with respect to the traveling-wave coordinate system. This is exactly what happens when an optical beam is parallel to the axis of a space lens but displaced from that axis. As the beam exits the lens it is progressively displaced laterally from the lens axis.

### 19.7.3 Traveling-Wave Phase Modulators: Gaussian Pulse Modulation

Next, we consider a traveling-wave modulator with a driving field in the shape of a Gaussian pulse (Figure 19.2c). This type of pulse can also create a quadratic phase retardation and has certain advantages over the sinusoidal drive. Let the traveling-wave index be given by

$$\Delta n(z,t)=\Delta n_0 \exp\left[-4\ln 2\left(\frac{t-z/v_{gm}-\tau_0}{\Delta\tau_\iota}\right)^2\right], \tag{19.60}$$

where $\Delta\tau_\iota$ is the full-width at half-maximum (FWHM) pulse width and $\tau_0$ is a fixed time delay. Transforming to the traveling-wave coordinate system moving at the group velocity of the optical waveform gives the net retardation as

$$\Gamma(\xi,\tau)=\frac{\omega_0}{c}\int_0^{\xi}\Delta n_0 \exp\left[-4\ln 2\left(\frac{\tau-\tau_0+\xi'\left(1/v_{go}-1/v_{gm}\right)}{\Delta\tau_\iota}\right)^2\right]d\xi'. \tag{19.61}$$

This integral can be related to the error function

$$\mathrm{erf}(v)=\frac{2}{\sqrt{\pi}}\int_o^{v} e^{-u^2}\,du \tag{19.62}$$

if we let

$$u=\sqrt{4\ln 2}\left(\frac{\tau-\tau_0+\xi'\left(1/v_{go}-1/v_{gm}\right)}{\Delta\tau_\iota}\right). \tag{19.63}$$

The retardation integral Equation 19.61 then becomes

$$\Gamma(\xi,\tau)=\frac{\omega_0\Delta n_0}{c}\frac{\Delta\tau_\iota}{\sqrt{4\ln 2}\left(1/v_{go}-1/v_{gm}\right)}\int_{u_1}^{u_2} e^{-u^2}\,du \tag{19.64}$$

$$=\frac{\omega_0\Delta n_0}{c}\frac{\Delta\tau_\iota}{\sqrt{4\ln 2}\left(1/v_{go}-1/v_{gm}\right)}\frac{\sqrt{\pi}}{2}\left[\mathrm{erf}(u_2)-\mathrm{erf}(u_1)\right]. \tag{19.65}$$

The resulting properties of the retardation of Equation 19.65 for a Gaussian phase-modulation pulse are not immediately clear. We can gain considerable insight into this situation, however, by assuming that there will be good velocity matching between the optical wavepacket and the modulating pulse and therefore examine the behavior of $\Gamma(\xi,\tau)$ in the limit ($v_{go} \rightarrow v_{gm}$). This is equivalent to requiring that after propagating a distance $\xi$ in the modulator, the walkoff between the optical wavepacket and the modulating pulse is small compared to the pulsewidth:

$$\xi\left(\frac{1}{v_{go}}-\frac{1}{v_{gm}}\right) << \Delta\tau_\iota. \tag{19.66}$$

The integral to be evaluated in Equation 19.64 can be simplified by noting that as $v_{go} \rightarrow v_{gm}$, $u_1 \rightarrow u_2$, and therefore the integral returns the area of a very narrow slice of the Gaussian function $\exp(-t^2)$. The area can then be approximated by

$$\lim_{u_1 \rightarrow u_2} \int_{u_1}^{u_2} e^{-u^2}\,du \approx (u_2-u_1)\left[\frac{e^{-u_2^2}+e^{-u_1^2}}{2}\right], \tag{19.67}$$

where

$$e^{-u_2^2} = \exp\left[-4\ln 2\left(\frac{\tau-\tau_0+\xi\left(1/v_{go}-1/v_{gm}\right)}{\Delta\tau_\iota}\right)^2\right] \tag{19.68}$$

$$= \exp\left[-4\ln 2\left(\frac{\tau-\tau_0}{\Delta\tau_\iota}+\Delta\psi\right)^2\right] \tag{19.69}$$

and

$$e^{-u_1^2} = \exp\left[-4\ln 2\left(\frac{\tau-\tau_0}{\Delta\tau_\iota}\right)^2\right]. \tag{19.70}$$

It was notationally convenient in Equation 19.69 to define a normalized walkoff time as we did with the sinusoidal modulating field in Equation 19.59:

$$\Delta\psi \equiv \frac{\xi\left(1/v_{go}-1/v_{gm}\right)}{\Delta\tau_\iota}. \tag{19.71}$$

Substituting these into Equation 19.67, we have

$$(u_2-u_1)\left[\frac{e^{-u_2^2}+e^{-u_1^2}}{2}\right] = \sqrt{\ln 2}\,\Delta\psi \exp\left[-4\ln 2\left(\frac{\tau-\tau_0}{\Delta\tau_\iota}\right)^2\right] \times \left\{1+\exp\left[-4\ln 2\left(\frac{2(\tau-\tau_0)\Delta\psi}{\Delta\tau_\iota}+\Delta\psi^2\right)\right]\right\}. \tag{19.72}$$

Finally, using this result in Equation 19.64 gives

$$\Gamma(\xi,\tau) \approx \frac{\Gamma_0}{2}\exp\left[-4\ln 2\left(\frac{\tau-\tau_0}{\Delta\tau_\iota}\right)^2\right] \times \left\{1+\exp\left[-4\ln 2\left(\frac{2(\tau-\tau_0)\Delta\psi}{\Delta\tau_\iota}+\Delta\psi^2\right)\right]\right\} \tag{19.73}$$

valid for small walkoff and where, again,

$$\Gamma_0 = \frac{\omega_0 \Delta n_0 \xi}{c}. \tag{19.74}$$

Figure 19.3b (right) shows the retardation expressed in Equation 19.73 normalized to the maximum $\Gamma_0$ for three values of walkoff ($\Delta\psi = 0, \pm\pi/8$) and no initial phase offset ($\tau_0 = 0$). We see the same effect as with the cosinusoidal modulation. That is, when there is walkoff, the center of quadratic phase retardation slips away from the center of the wavepacket, which is equivalent to entering a space lens off of, and parallel to, the lens axis.

## 19.8 Drive Waveform Comparisons

Suppose the envelope waveform $A_i(0,\tau)$ enters a time lens. The output waveform, $A_0(\xi,\tau)$, is given in the most general case by

$$A_0(\xi,\tau) = A_i(0,\tau)H_\iota(\xi,\tau), \tag{19.75}$$

where $H_\iota(\xi,\tau)$ is the time lens phase transformation. Neglecting losses,

$$H_\iota(\xi,\tau) = \exp[i\phi_\tau(\xi,\tau)]. \tag{19.76}$$

In the situation considered here, the electro-optic time lens produces a retardation that governs the time lens phase and therefore:

$$\phi_\tau(\xi,\tau) = -\Gamma(\xi,\tau). \tag{19.77}$$

As we saw in Section 19.5, it is the second derivative of the phase function that gives rise to the quadratic phase necessary to create a time lens. To compare the efficacy of the two types of drive waveforms considered, we will assume the best-case scenario of perfect velocity matching ($\Delta\phi = \Delta\psi = 0$) and no offset ($\theta = 0$, $\tau_0 = 0$). Then

$$\phi_\tau(\xi,\tau) = -\Gamma_0\cos(\omega_m\tau) \qquad \text{COSINE DRIVE} \tag{19.78}$$

$$= -\Gamma_0\exp\left[-4\ln 2\left(\frac{\tau}{\Delta\tau_\iota}\right)^2\right]. \qquad \text{GAUSSIAN DRIVE} \tag{19.79}$$

The second derivatives define the strength of the time lens:

$$\frac{d^2\phi_\tau}{d\tau^2} = \Gamma_0\omega_m^2\cos(\omega_m\tau) \qquad \text{COSINE DRIVE} \tag{19.80}$$

$$= \Gamma_0 \frac{8\ln 2}{\Delta\tau_\iota^2}\left[1 - 8\ln 2\left(\frac{\tau}{\Delta\tau_\iota}\right)^2\right] e^{-4\ln 2(\tau/\Delta\tau_\iota)^2}. \qquad \text{GAUSSIAN DRIVE} \tag{19.81}$$

We wish to compare the behavior of the quadratic phases above as a function of the local time coordinate, $\tau$. This will help define the effective size of the time lens window. That is, the available time over which the lens is predominantly quadratic. An ideal time lens would, after all, have a constant quadratic phase for all $\tau$. To compare these two functions, we first assume that the peak retardations, $\Gamma_0$, are the same in both cases and then set the peak values of the second derivatives equal to each other at $\tau = 0$. This gives us an equivalence between the period of the cosine drive, $T = 2\pi/\omega_m$, and the Gaussian pulsewidth $\Delta\tau_1$. We find

$$\frac{\Delta\tau_\iota}{T} = \frac{\sqrt{8\ln 2}}{2\pi}. \tag{19.82}$$

When this condition is satisfied, the maximum values of Equations 19.80 and 19.81 become

$$\phi''_{\tau\text{MAX}} = \Gamma_0\omega_m^2 = \Gamma_0 \frac{8\ln 2}{\Delta\tau_\iota^2}. \tag{19.83}$$

Figure 19.4 shows plots of the second derivatives of the cosine and Gaussian drive waveforms defined by Equations 19.80 and 19.81 as a function of the local traveling-wave coordinate, $\tau$, normalized to the Gaussian pulsewidth (FWHM), $\Delta\tau_1$. It is clear that a window of strong quadratic phase occurs only over a duration of roughly ±20% of the Gaussian pulsewidth. Also of interest is the fact that, for the situation described here (equal peak retardation and second derivative), the Gaussian pulse waveform produces a more consistent time lens over a longer time window than the cosine waveform. Thus, it will be a faster lens in the traditional sense that the useful aperture is wider for the same focal length (time), which is,

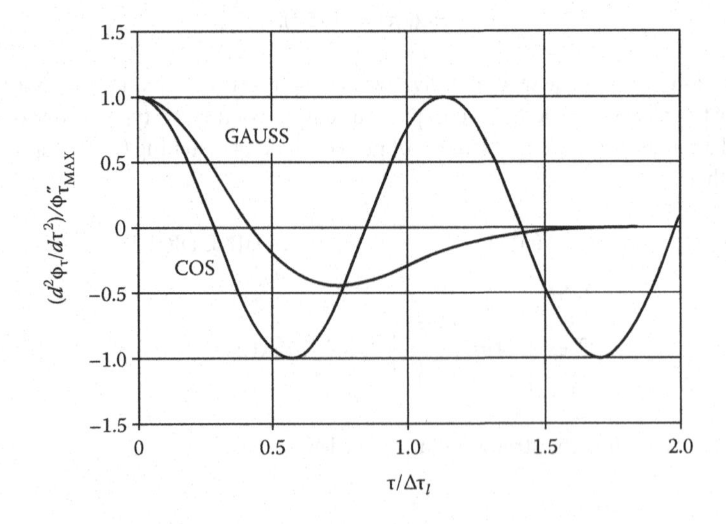


**FIGURE 19.4** Strength of quadratic phase in cosine and Gaussian phase modulation as a function of the traveling-wave coordinate normalized to the Gaussian pulse width (FWHM).

we recall, given by Equation 19.31 $f_T = \omega_0/\phi''_\tau$. From Equations 19.80 and 19.81, the focal times for the two cases are given by

$$f_T = \frac{\omega_0}{d^2\phi_\tau/dr^2\big|_{\tau=0}} \tag{19.84}$$

$$= \frac{\omega_0}{\Gamma_0\omega_m^2} \qquad \text{COSINE DRIVE} \tag{19.85}$$

$$= \frac{\omega_0\Delta\tau_t^2}{\Gamma_0 8\ln 2}. \qquad \text{GAUSSIAN DRIVE} \tag{19.86}$$

## 19.9 Pulse Compression

The compression of optical pulses by first chirping with electro-optic phase modulators and then compressing in a dispersive delay line has been studied and reported for quite some time [8–10]. For a transform-limited pulse, this is the time-domain equivalent of focusing a beam from a waist. Assuming an ideal time lens and neglecting a constant phase shift through the lens, the output envelope given a transform-limited, unit-amplitude Gaussian input pulse is, according to Equation 19.75,

$$A_0(\xi,\tau) = \exp\left[-2\ln 2\left(\frac{\tau}{\Delta\tau_p}\right)^2\right]\exp\left[i\frac{\omega_0\tau^2}{2f_T}\right]$$

$$= \exp\left[-\left(\frac{2\ln 2}{\Delta\tau_p^2} - i\frac{\omega_0}{2f_T}\right)\tau^2\right], \tag{19.87}$$

where $\Delta\tau_p$ is the FWHM of the pulse intensity. If this pulse is now passed through a dispersive delay line with exclusively quadratic dispersion, the pulse will expand or compress depending on the signs of the time lens phase and the dispersion. To see how this pulse evolves, we must Fourier transform it and then apply the integral solution given by Equation 19.18. Since Equation 19.87 becomes the input to the dispersion problem, we reset the distance variable $\xi \to 0$. The Fourier transform of Equation 19.87 is easily found using the basic transform pair:

$$\Im\left\{e^{-(a+ib)t^2}\right\} = \left[\frac{\pi^2}{a^2+b^2}\right]^{\frac{1}{4}}\exp\left[-\frac{a-ib}{4(a^2+b^2)}\omega^2\right], \tag{19.88}$$

where

$$a \equiv \frac{2\ln 2}{\Delta\tau_p^2},$$
$$b \equiv -\frac{\omega_0}{2f_T}. \tag{19.89}$$

Thus,

$$A(0,\omega)=\left[\frac{\pi^2}{\left(2\ln 2/\Delta\tau_p^2\right)^2+\left(\omega_0/2f_T\right)^2}\right]^{\frac{1}{4}}\exp\left[-\frac{2\ln 2/\Delta\tau_p^2+i\omega_0/2f_T}{4\left(\left(2\ln 2/\Delta\tau_p^2\right)^2+\left(\omega_0/2f_T\right)^2\right)}\omega^2\right]. \tag{19.90}$$

Now, upon propagating a distance $\xi$ through the dispersive medium, the spectrum acquires an additional quadratic spectral phase according to Equation 19.18. For convenience, we will use the notation indicated in Equation 19.89:

$$A(\xi,\omega)=A(0,\omega)\exp\left[-i\frac{\xi}{2}\beta''\omega^2\right] \tag{19.91}$$

$$=A(0,0)\exp\left[-\left(\frac{a+i\left[2(a^2+b^2)\xi\beta''-b\right]}{4(a^2+b^2)}\right)\omega^2\right]. \tag{19.92}$$

The possibility of eliminating the quadratic phase in this spectrum implies a minimum pulsewidth, which we will show shortly. This can be accomplished by setting the focal time:

$$f_T=-\omega_0\xi\beta'' \tag{19.93}$$

provided that $\xi\beta'' << \Delta\tau_p^2$. Before applying this condition, consider the time domain waveform found by inverse Fourier transformation of Equation 19.92. This yields the rather formidable-looking result:

$$A(\xi,\tau)=\left[\frac{a^2+b^2}{a^2+\left[2(a^2+b^2)\xi\beta''-b\right]^2}\right]^{\frac{1}{4}}\times\exp\left[-\frac{a-i\left[2(a^2+b^2)\xi\beta''-b\right]}{a^2+\left[2(a^2+b^2)\xi\beta''-b\right]^2}(a^2+b^2)\tau^2\right]. \tag{19.94}$$

This complex Gaussian pulse will have a minimum duration when the real part of the argument in the exponent is maximized. For a fixed focus time lens and a fixed input pulsewidth, $f_T$ and $\Delta\tau_p$ are fixed and thus so are $a$ and $b$. The only variable left to adjust is the net group delay dispersion, $\xi\beta''$. The argument is thus maximized when

$$2(a^2+b^2)\xi\beta''=b \tag{19.95}$$

which is exactly the same condition that we previously found to eliminate the quadratic phase in the spectrum of the waveform in Equation 19.92. Notice that it also eliminates the quadratic phase in the time domain. From Equation 19.95, we can express the optimum group delay dispersion for pulse compression in terms of the focal time or chirp rate of the time lens as

$$\xi_f\beta''\equiv-\frac{f_T}{\omega_0}=-\left(1/\omega_i'\right)\ \text{s/Hz}, \tag{19.96}$$

where we have now labeled the propagation distance $\xi_f$ in the dispersive medium as that with which we associate compression of an initially unchirped Gaussian pulse due to the action of a time lens. It is, in effect, the focal distance in the dispersive medium.

By satisfying the condition of Equation 19.96, we simultaneously minimize the compressed pulse width and eliminate the quadratic temporal and spectral phase. In words, a group delay dispersion of $\xi_f\beta''$ s/Hz is required to exactly cancel the time lens chirp rate of $-\omega_i'$ Hz/s. The presence of the minus sign can be explained as follows: a time lens that produces a positive chirp, $\omega_i' > 0$, generates a waveform whose lower frequency components arrive before the higher frequency components. This is why there is a negative sign at the beginning of the term. Therefore, to achieve compression of the waveform, the dispersive medium must cause the lower frequency components to travel slower than the higher frequency components so that they overlap in time at the end of the medium. This can only be achieved with negative group delay dispersion, $\beta'' < 0$. Thus, a positively chirped pulse can only be compressed with, for example, an anomalously dispersive dielectric medium or a pair of parallel diffraction gratings [6].

When the condition expressed in Equation 19.96 is satisfied, the output waveform becomes

$$A(\xi,\tau) = \left[1+(b/a)^2\right]^{\frac{1}{4}} \exp\left[-\frac{a^2+b^2}{a}\tau^2\right]. \tag{19.97}$$

Again, we will define the compressed pulsewidth ($\Delta\tau_{pc}$) in terms of the FWHM of the pulse's intensity and therefore let

$$\frac{2\ln 2}{\Delta\tau_{pc}^2} = \frac{a^2+b^2}{a}. \tag{19.98}$$

Substituting from Equation 19.89 gives the compressed pulsewidth in terms of the input pulsewidth and time lens focal time or chirp rate:

$$\Delta\tau_{pc} = \frac{\Delta\tau_p}{\sqrt{1+\left(\dfrac{\omega_0\Delta\tau_p^2}{4\ln 2 f_T}\right)^2}} = \frac{\Delta\tau_p}{\sqrt{1+\left(\dfrac{\omega_i'\Delta\tau_p^2}{4\ln 2}\right)^2}}. \tag{19.99}$$

Thus, we see that the pulse compression ratio depends on a short focal time, $f_T$, or, equivalently, a strong chirp rate, $\omega_i'$, compared to the input pulsewidth, $\Delta\tau_p^2$.

## 19.10 Temporal Imaging

We now extend the use of time lenses to a more interesting situation, that of temporal imaging. By this we mean the expansion or contraction of the envelope of an optical waveform, in the time domain, in a way that preserves the shape of the original waveform. This is the temporal equivalent of spatial imaging, which is traditionally understood as a relationship between input and output field intensities, and therefore, phase is not an issue. Figure 19.5a shows a conventional spatial imaging system upon which we base our analogous temporal imaging system. The input and output diffraction produce a quadratic phase filtering of the transverse spatial frequency spectrum, while the lens introduces a quadratic phase transformation in the real-space coordinates $x$ and $y$.

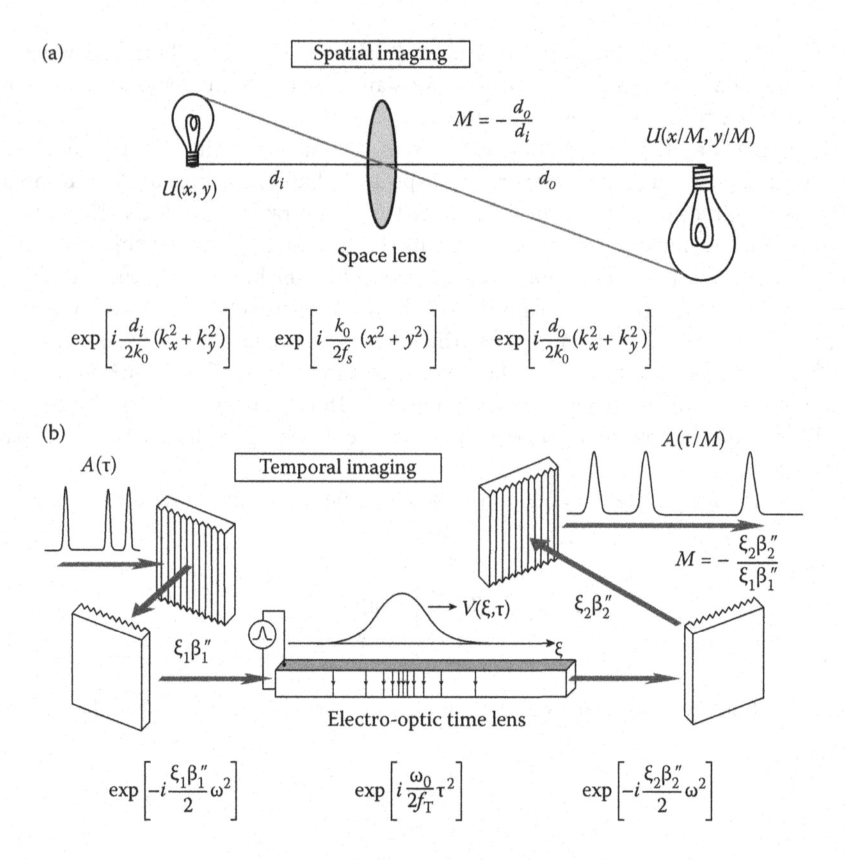


**FIGURE 19.5** (a) Elementary single-lens spatial imaging system. $U(x, y)$ is the slowly varying transverse electric field amplitude. (b) Analogous temporal imaging system. $A(\tau)$ is the slowly varying temporal electric field envelope. Dispersion, represented here as a pair of diffraction gratings, plays the role of the spatial diffraction in the Fresnel approximation and a quadratic phase modulator acts as a time lens (shown here as an electro-optic modulator). The temporal imaging system preserves the overall envelope shape, produces a magnification $M$ and even local time reversal, which is the temporal analogue of image inversion.

The temporal dual of the spatial imaging system is configured by cascading input dispersion, a time lens, and output dispersion as shown in Figure 19.5b. The analysis of the previous sections specifies how we deal with these effects. Dispersion is handled by introducing a quadratic phase filter in the temporal frequency domain, while the time lens produces a quadratic phase modulation in the time domain. We simply apply repeated Fourier transforms to the input waveform and apply the appropriate filter or modulation function. Recall, however, that since we cast the entire space–time theory in terms of envelope functions, we need only deal with them in our system analysis; the carrier is absent.

Consider the effects upon an arbitrary input waveform, $A(0,\tau)$, of a sequence of input dispersion, a time lens, and output dispersion. Mathematically, the effects are described by

Input dispersion $$A(\xi_1,\omega) = A(0,\omega)\exp[-ia\omega^2] \qquad a \equiv (\xi_1/2)\beta_1'' \tag{19.100}$$

Time lens $$A(\xi_1+\varepsilon,\tau) = A(\xi_1,\tau)\exp\left[i\tau^2/4c\right] \qquad c \equiv \left(f_T/2\omega_0\right) \tag{19.101}$$

Output dispersion $$A(\xi_2,\omega) = A(\xi_1+\varepsilon,\omega)\exp[-ib\omega^2] \qquad b \equiv (\xi_2/2)\beta_2''. \tag{19.102}$$

Skipping the intermediate steps (for details, see [17]) and neglecting an overall multiplicative phase factor, the waveform that exits the system becomes

$$A(\xi_2,\tau)=\frac{1}{2\pi}\int_{-\infty}^{\infty}A(0,\omega')\times\exp\left[-i\left(a+c-\frac{c^2}{b+c}\right)\omega'^2\right]\exp\left[i\left(\frac{c}{b+c}\right)\omega'\tau\right]d\omega', \tag{19.103}$$

where $A(0,\omega')$ is the spectrum of the input waveform. To eliminate the quadratic phase in the integrand we set $a + c = c_2/(b + c)$, which is equivalent to

$$\frac{1}{a}+\frac{1}{b}=-\frac{1}{c}\xrightarrow{or}\frac{1}{\xi_1\beta_1''}+\frac{1}{\xi_2\beta_2''}=-\frac{\omega_0}{f_T}. \tag{19.104}$$

Using the results from the pulse compression analysis, Equation 19.96 allows this to be written in the more intuitive form,

$$\frac{1}{\xi_1\beta_1''}+\frac{1}{\xi_2\beta_2''}=\frac{1}{\xi_f\beta_2''}, \tag{19.105}$$

where $\xi_f$ is the physical propagation distance in the second dispersive medium that will compress an object waveform from the first medium when it has infinite dispersion. This is equivalent to an object at infinity in the spatial case.

The similarity between Equation 19.105 and its spatial counterpart is unmistakable. Another important similarity can be found in the output waveform expression Equation 19.103. Notice that the frequency domain variable $\omega'$ in the kernel has a scale factor $(b + c)/c$. From the scaling theorem, we know that the time-domain waveform that is obtained upon the inverse Fourier transform operation has a scale factor $c/(b + c)$. This rescaling of the time coordinate defines a *temporal magnification factor* M. Substituting for b and c, and using the imaging condition of Equation 19.105, we find

$$M_T=-\frac{\xi_2\beta_2''}{\xi_1\beta_1''} \tag{19.106}$$

which has its counterpart in the spatial imaging environment. All of the important functions and relationships relevant to imaging in both the space and time domains are compiled in Table 19.2.

**TABLE 19.2** Synopsis of Important Functions and Relationships Governing Spatial and Temporal Imaging

| | Space | Time |
|---|---|---|
| Input spectral filter | $\exp\left[i\frac{d_i}{2k_0}(k_x^2+k_y^2)\right]$ | $\exp\left[i\frac{\xi_1\beta_1''}{2}\omega^2\right]$ |
| Lens function | $\exp\left[i\frac{k_0}{2f_S}(x^2+y^2)\right]$ | $\exp\left[i\frac{\omega_0}{2f_T}\tau^2\right]$ |
| Output spectral filter | $\exp\left[i\frac{d_o}{2k_0}(k_x^2+k_y^2)\right]$ | $\exp\left[-i\frac{\xi_2\beta_2''}{2}\omega^2\right]$ |
| Imaging condition | $\frac{1}{d_o}+\frac{1}{d_i}=\frac{1}{f}$ | $\frac{1}{\xi_1\beta_1''}+\frac{1}{\xi_2\beta_2''}=\frac{1}{\xi_f\beta_2''}$ |
| Magnification | $M_S=-\frac{d_i}{d_o}$ | $M_T=-\frac{\xi_2\beta_2''}{\xi_1\beta_1''}$ |

## 19.11 Apertures, Impulse Response, and Resolution

Up to this point, we have said nothing about the window in time through which a waveform flows into a time lens. Without such a window, or aperture, several possibilities arise: (1) The quadratic phase necessary for time lens action must continue indefinitely (clearly an unattainable situation). (2) For a finite-duration quadratic phase, field energy passing through the remaining static, or linear, phase would simply be dispersed by the following dispersive media and create a bright background of interfering light. This is very similar to placing a lens on a sheet of plate glass. (3) If the quadratic phase does not come to an abrupt transition to the linear phase, there will be higher-order phase terms that will effectively become aberrations [32]. Therefore, some form of temporal window, or gate, is necessary. This presents some problems when dealing with electro-optic time lenses because these are strictly phase modulation mechanisms and do not inherently window out the waveform past the lens boundary time, as happens in parametric time lenses [23,24,33]. Nevertheless, suppose that some mechanism such as an electro-optic amplitude modulator or Kerr-shutter is used to limit the time duration of an input waveform to a prescribed window. If we call this function a pupil, $P(\tau)$, then the total transmission function of a time lens becomes $P(\tau)H_l(\xi,\tau)$, where $H_l(\xi,\tau)$ describes the phase modulation.

To understand the effect that a finite time window has on an imaged temporal waveform, we can apply the tools of linear systems theory and find the impulse response of the temporal imaging system [17]. The output waveform is then found to be a scaled replica of the input waveform convolved with the impulse response of the temporal imaging system, which is the Fourier transform of the temporal pupil function. For example, consider a rectangular time window of width $\tau_a$. This will yield an impulse response referenced to the input of the imaging system,

$$\tilde{h}(M_T\tau) = \frac{\tau_a}{4\pi b\sqrt{M_T}} \exp\left[i\frac{\omega_0 M_T}{2f_T}\tau^2\right] \text{sinc}\left(\frac{\tau_a M_T \tau}{4\pi b}\right), \tag{19.107}$$

where $\text{sinc}(x) \equiv \sin(\pi x)/(\pi x)$. The width of the sinc function can be approximated by the time to the first zero, and thus the input resolution is

$$\delta\tau_{in} = \left|\frac{4\pi b}{M_T \tau_a}\right| = \left|\frac{4\pi a}{\tau_a}\right| = \left|\frac{2\pi\xi_1\beta_1''}{\tau_a}\right|. \tag{19.108}$$

One of the principal applications of temporal imaging is microscopy in the time domain. That is, stretching waveforms out onto a time scale where conventional high-speed photodetectors and electronics can be employed to resolve them. In this case, the input dispersion is close to the temporal equivalent of one focal length. Or, equivalently, the output dispersion is much greater than the input dispersion. From Equation 19.104 or Equation 19.105, this implies that $b \approx -c$ or

$$\xi_1\beta_1'' \approx -\frac{f_t}{\omega_0} = \xi_f\beta_2'' \tag{19.109}$$

and therefore the input resolution can be written

$$\delta\tau_{in}\left|\frac{2\pi f_T}{\omega_0\tau_a}\right| = T_0\left|\frac{f_T}{\tau_a}\right| = T_0 f_T^{\#}, \tag{19.110}$$

where $T_0 = 2\pi/\omega_0$ is the period of the carrier and the last step is obtained from the development of the temporal *f*-number in Section 19.6. This is similar to the case of spatial imaging, where the resolution in the transverse dimension is $\delta r \approx \lambda f_S^{\#}$.

## 19.12 Practical Issues

There are a number of practical issues that arise when contemplating temporal imaging systems based on electro-optic time lenses. First and foremost is the severe requirement on the degree of phase modulation necessary to obtain a useful lens. We can make some estimates by invoking the results from the previous section on resolution. Equation 19.110 tells us that the key is the temporal *f*-number. From Equation 19.38, we know that in terms of a useful aperture time, $\tau_a$, we have $f_T^{\#} = |f_T|/\tau_a$. Suppose that we are using a cosinusoidal drive to the electro-optic time lens and that a shutter limits the aperture time to $\tau_a \approx 1/\omega_m$, where $\omega_m$ is the modulation frequency. This allows the waveform to enter the modulator over a duration of phase that is predominantly quadratic. The focal time for a cosinusoidal drive waveform is given by Equation 19.85, and therefore the *f*-number is

$$f_T^{\#} = \frac{|f_T|}{1/\omega_m} = \frac{\omega_0}{\Gamma_0 \omega_m}. \quad \text{COSINE DRIVE} \tag{19.111}$$

Substituting into Equation 19.110 gives the input resolution:

$$\delta\tau_{in} = \frac{1}{\Gamma_0 f_m}. \quad \text{COSINE DRIVE} \tag{19.112}$$

This is a useful expression. For example, suppose we wish to build a time microscope with input resolution of 100 fs. Using an electro-optic modulator with a drive frequency of $f_m = 20$ GHz, it would have to produce a peak phase retardation of $\Gamma_0 = 500$ radians.

We can apply the same analysis to the case of the Gaussian drive waveform. We saw in Section 19.8 that the Gaussian phase modulator produces a useful quadratic phase over about ±20% of the FWHM of the drive pulse. A conservative value for the aperture time could then be set at $\tau_a = \Delta\tau_1/4$. Using this in the expression for the input resolution and the focal time for the Gaussian drive Equation 19.86 gives

$$\delta\tau_{in} = \frac{\pi\Delta\tau_1}{\Gamma_0 \ln 2}, \quad \text{GAUSSIAN DRIVE} \tag{19.113}$$

where, again, $\Delta\tau_1$ is the FWHM of the Gaussian drive pulse. For example, suppose we drive the modulator with a 100-ps pulse and wish to obtain 100-fs time resolution. Then, the modulator must produce a peak phase retardation of ≈4500 radians.

The peak phase retardation calculated in the above examples are quite formidable. A reasonable approach to obtaining very large peak retardations is to employ resonant phase modulators, but this precludes the use of anything but a narrowband sinusoidal drive. The use of resonant phase modulators has a long history and has recently gained renewed interest for generating optical frequency combs for metrology [34–41] and for intracavity simultaneous AM and FM modelocking [42]. Details about the design considerations can be found in these references.

One interesting consequence of using resonant modulators is the possibility of coupling the optical waveform to the backward-propagating traveling-wave field that is a necessary component of a standing-wave in the resonator. We can augment the analysis of Section 19.7.2 by simply adding a second term to the traveling-wave index of refraction (Equation 19.54) with a propagation constant representing the backward wave. Thus

$$\Delta n(z,t) = \Delta n_0[\cos(\omega_m t - k_m z + \theta) - \cos(\omega_m t + k_m z + \theta)], \tag{19.114}$$

where the minus sign is included to satisfy the boundary condition $\Delta n(0, \tau) = 0$. Carrying out the integral in Equation 19.55 yields

$$\begin{aligned}\Gamma(\xi,\tau) &= \Gamma_0\left\{\frac{\sin(\Delta\phi/2)}{\Delta\phi/2}\cos\left(\omega_m\tau+\Delta\phi/2+\theta\right)-\frac{\sin(\eta/2)}{\eta/2}\cos\left(\omega_m\tau+\eta/2+\theta\right)\right\}\\ &= \Gamma_+(\xi,\tau)+\Gamma_-(\xi,\tau)\end{aligned} \tag{19.115}$$

where

$$\eta \equiv \omega_m\xi\left(\frac{1}{v_{go}}+\frac{1}{v_{pm}}\right). \tag{19.116}$$

The backward propagating wave produces the contribution $\Gamma_-(\xi,\tau)$ to the total retardation. The term $\eta$ counts the number of radians of phase that the optical waveform passes through in the backward-propagating modulating field. For long modulators, and thus large $\eta$, the optical waveform passes through many cycles of backward-propagating field. As such, the positive and negative lobes of the field eventually cancel each other out in the integral. This is reflected in the $\sin(\eta/2)/(n/2)$ coefficient. If there are an equal number of positive and negative lobes, they cancel identically [17].

## 19.13 Conclusions

The basics of the duality between the equations of paraxial diffraction and narrowband dispersion have been presented. In conjunction with a quadratic time-phase modulation, which plays the role of a time lens, they form the basis of the principles of temporal imaging. Electro-optic phase modulators, in particular, have been analyzed in terms of their applicability as time lenses. We found that the performance of a time lens can be characterized by its focal time, which is the equivalent of the focal length of a space lens. The feature of the focal time that is of consequence is the chirp rate, $d\omega/d\tau$. Just as in spatial imaging systems, the *f*-number of the time lens determines the resolution. According to our definition, the *f*-number of a time lens is given by $f_T^{\#} = \omega_0/\Delta\omega$, which is the inverse of the fractional bandwidth induced by the time lens. We found that for the same peak phase retardation, $\Gamma_0$, a Gaussian pulse produces a quadratic phase modulation over a longer duration than the cusp of a cosine. The disadvantage of this approach is that one cannot use a microwave resonant structure to support a Gaussian pulse, so for the same peak microwave power, a high-Q resonator should outperform any single-pass structure supporting a single Gaussian pulse. As quickly becomes obvious, the design of high-efficiency time lenses becomes a multiple-parameter optimization problem. It is hoped that this brief introduction to the principles regarding the operation and performance of electro-optic time lenses will generate more enthusiasm for exploring this parameter space.

There is a considerable body of literature on topics related to, and based upon, the concepts presented here, but which we do not have the space to present. They still merit mention, however, for the interested reader. These include temporal filtering and signal processing [43,44], temporal self-imaging, and the temporal Talbot effect [45–47], the self Fourier-transforming properties of dispersive media [48], compound temporal lens designs [49,50], and issues of aberrations [32].

## Acknowledgments

This work has been supported by the David and Lucile Packard Foundation, the National Science Foundation (grants ECS-9110678, ECS-9521604, and ECS-9900414), the U.S. Department of Defense Air Force Office of Scientific Research (ATRI Program F30602-94-2-0001), the U.S. Department of Energy's Lawrence Livermore National Laboratory (contract W-7405Eng48, LLNL Photonics Group Grant 98-ERD-027), the Hewlett-Packard Company, Agilent Technologies, and the University of California at Los Angeles.

## References

1. J. R. Klauder, A. C. Price, S. Darlington, and W. J. Albersheim, "The theory and design of chirp radars," *Bell Sys. Tech. J.*, vol. 39, no. 4, pp. 745–808, Jul. 1960.
2. P. Tournois, "Analogie optique de la compression d'impulsion," *C. R. Acad. Sci.*, vol. 258, pp. 3839–3842, Apr. 1964.
3. P. Tournois, J.-L. Verner, and G. Bienvenu, "Sur l'analogie optique de certains montages électroniques: Formation d'images temporelles de signaux électriques," *C. R. Acad. Sci.*, vol. 267, pp. 375–378, 1968.
4. W. J. Caputi, "Stretch: A time transformation technique," *IEEE Trans. Aerosp. Electron. Syst.*, vol. AES-7, no. 2, pp. 269–278, Mar. 1971.
5. A. Papoulis, *Systems and Transforms with Applications in Optics*. New York: McGraw-Hill, 1968.
6. E. B. Treacy, "Optical pulse compression with diffraction gratings," *IEEE J. Quantum Electron.*, vol. QE-5, no. 9, pp. 454–458, Sep. 1969.
7. S. A. Akhmanov, A. P. Sukhorukov, and A. S. Chirkin, "Nonstationary phenomena and spacetime analogy in nonlinear optics," *Soviet Phys. JETP*, vol. 28, pp. 748–757, Apr. 1969.
8. J. A. Giordmaine, M. A. Duguay, and J. W. Hansen, "Compression of optical pulses," *IEEE J. Quantum Elect.*, vol. QE-4, no. 5, pp. 252–255, May 1968.
9. D. R. Grischkowsky, "Optical pulse compression," *Appl. Phys. Lett.*, vol. 25, no. 10, pp. 566–568, Nov. 1974.
10. J. E. Bjorkholm, E. H. Turner, and D. B. Pearson, "Conversion of cw light beam into a train of subnanosecond pulses using frequency modulation and the dispersion of a near resonant atomic vapor," *Appl. Phys. Lett.*, vol. 26, no. 10, pp. 564–566, May 1975.
11. J. K. Wigmore and D. R. Grischkowsky, "Temporal compression of light," *IEEE J. Quantum Electron.*, vol. QE-14, no. 4, pp. 310–315, Apr. 1978.
12. B. H. Kolner and M. Nazarathy, "Temporal imaging with a time lens," *Opt. Lett.*, vol. 14, no. 12, pp. 630–632, Jun. 1989.
13. B. H. Kolner and M. Nazarathy, "Temporal imaging with a time lens: erratum," *Opt. Lett.*, vol. 15, no. 11, p. 655, Jun. 1990.
14. B. H. Kolner, "Active pulse compression using an integrated electro-optic phase modulator," *Appl. Phys. Lett.*, vol. 52, no. 14, pp. 1122–1124, Apr. 1988.
15. A. A. Godil, B. A. Auld, and D. M. Bloom, "Time lens producing 1.9 ps optical pulses," *Appl. Phys. Lett.*, vol. 62, no. 10, pp. 1047–1049, Mar. 1993.
16. R. P. Scott, C. V. Bennett, and B. H. Kolner, "Picosecond laser source with single knob adjustable pulsewidth," in *Proc. Lasers for RF Guns*, Anaheim, CA, 1994.
17. B. H. Kolner, "Space–time duality and the theory of temporal imaging," *IEEE J. Quantum Electron.*, vol. 30, no. 8, pp. 1951–1963, Aug. 1994.
18. L. S. Telegin and A. S. Chirkin, "Reversal and reconstruction of the profile of ultrashort light pulses," *Sov. J. Quantum Electron.*, vol. 15, no. 1, pp. 101–102, 1985.

19. C. V. Bennett, R. P. Scott, and B. H. Kolner, "Temporal magnification and reversal of 100 Gb/s optical data with an up-conversion time microscope," *Appl. Phys. Lett.*, vol. 65, no. 20, pp. 2513–2515, Nov. 1994.
20. C. V. Bennett and B. H. Kolner, "Upconversion time microscope demonstrating 103X magnification of femtosecond waveforms," *Opt. Lett.*, vol. 24, no. 11, pp. 783–785, Jun. 1999.
21. C. V. Bennett and B. H. Kolner, "Parametric temporal imaging," in *Trends in Optics and Photonics, Ultrafast Electronics and Optoelectronics*, Washington, DC: Optical Society of America, 1999.
22. C. V. Bennett and B. H. Kolner, "Subpicosecond single-shot waveform measurement using temporal imaging," in *Proc. IEEE Lasers Electro-optics Society Annu. Meeting, LEOS '99*, San Francisco, CA, 1999, paper ThBB1.
23. C. V. Bennett and B. H. Kolner, "Principles of parametric temporal imaging—Part I: System configurations," *IEEE J. Quantum Electron.*, vol. 36, no. 4, pp. 430–437, Apr. 2000.
24. C. V. Bennett and B. H. Kolner, "Principles of parametric temporal imaging—Part II: System performance," *IEEE J. Quantum Electron.*, vol. 36, no. 6, pp. 649–655, Jun. 2000.
25. A. M. Shaw, R. N. Zare, C. V. Bennett, and B. H. Kolner, "Bounce-by-bounce cavity ring-down spectroscopy: Femtosecond temporal imaging," *Chemphyschem*, vol. 2, no. 2, pp. 118–121, Feb. 2001.
26. M. A. Foster, R. Salem, D. F. Geraghty, A. C. Turner-Foster, M. Lipson, and A. L. Gaeta, "Silicon-chip–based ultrafast optical oscilloscope," *Nature*, vol. 456, pp. 81–85, Nov. 2008.
27. A. Yariv, *Optical Electronics in Modern Communications*, 5th ed. New York: Oxford University Press, 1997.
28. J.-C. Diels and W. Rudolf, *Ultrashort Laser Pulse Phenomena*. San Diego: Academic Press, 1996.
29. J. W. Goodman, *Introduction to Fourier Optics*, 3rd ed. Englewood, CO: Roberts and Company, 2005.
30. A. M. Weiner, *Ultrafast Optics*. Hoboken, NJ: John Wiley and Sons, 2009.
31. B. H. Kolner, "Generalization of the concepts of focal length and f-number to space and time," *J. Opt. Soc. Am. A*, vol. 11, no. 12, pp. 3229–3234, Dec. 1994.
32. C. V. Bennett and B. H. Kolner, "Aberrations in temporal imaging," *IEEE J. Quantum Electron.*, vol. 37, no. 1, pp. 20–32, Jan. 2001.
33. C. V. Bennett, R. P. Scott, and B. H. Kolner, "Up-conversion time lens demonstrates 12x magnification of 100 Gb/s data," in *Proc. Seventh Annu. Meeting IEEE Lasers Electro-optics Society; LEOS '94*, Boston, MA, 1994, paper UO4.2.
34. K. Amano, T. Kobayashi, H. Yao, A. Morimoto, and T. Sueta, "Generation of 0.64-THz-wide optical sidebands by a novel electrooptic modulator for the purpose of forming ultrashort optical pulses," *J. Lightwave Technol.*, vol. LT-5, no. 10, pp. 1454–1458, Oct. 1987.
35. M. Kourogi, K. Nakagawa, and M. Ohtsu, "Wide-span optical frequency comb generator for accurate optical frequency difference measurements," *IEEE J. Quantum Electron.*, vol. 29, no. 10, pp. 2693–2701, Oct. 1993.
36. L. R. Brothers, D. Lee, and N. C. Wong, "Terahertz optical frequency comb generation and phase locking of an optical parametric oscillator at 665 GHz," *Opt. Lett.*, vol. 19, no. 4, pp. 245–247, Feb. 1994.
37. A. S. Bell, G. M. Mcfarlane, E. Riis, and A. I. Ferguson, "Efficient optical frequency-comb generator," *Opt. Lett.*, vol. 20, no. 12, pp. 1435–1437, Jun. 1995.
38. G. M. Macfarlane, A. S. Bell, and A. I. Ferguson, "Optical comb generator as an efficient short-pulse source," *Opt. Lett.*, vol. 21, no. 7, pp. 534–536, Apr. 1996.
39. A. Huber, Th. Udem, B. Gross, J. Reichert, M. Kourogi, K. Pachucki, M. Weitz, and T. W. Hansch, "Hydrogen–deuterium 1S–2S isotope shift and the structure of the deuteron," *Phys. Rev. Lett.*, vol. 80, no. 3, pp. 468–471, Jan. 1998.
40. J. L. Hall, L.-S. Ma, M. Taubman, B. Tiemann, F.-L. Hong, L. Pfister, and J. Ye, "Stabilization and frequency measurement of the I2-stabilized Nd:YAG laser," *IEEE Trans. Instrum. Meas.*, vol. 48, no. 2, pp. 583–586, Apr. 1999.

41. S. A. Diddams, L. Ma, J. Ye, and J. L. Hall, "Broadband optical frequency comb generation with a phase-modulated optical parametric oscillator," *Opt. Lett.*, vol. 24, no. 23, pp. 1749–1751, Dec. 1999.
42. R. P. Scott, C. V. Bennett, and B. H. Kolner, "AM and high-harmonic FM modelocking," *Appl. Opt.*, vol. 36, no. 24, pp. 5908–5912, Aug. 1997.
43. A. W. Lohmann and D. Mendlovic, "Temporal perfect shuffle optical processor," *Opt. Lett.*, vol. 17, no. 11, pp. 822–824, Jun. 1992.
44. A. W. Lohmann and D. Mendlovic, "Temporal filtering with time lenses," *Appl. Opt.*, vol. 31, no. 29, pp. 6212–6219, Oct. 1992.
45. T. Jannson and J. Jannson, "Temporal self-imaging effect in single-mode fibers," *J. Opt. Soc. Am.*, vol. 71, no. 11, pp. 1373–1376, Nov. 1981.
46. P. A. Bélanger, "Periodic restoration of pulse trains in a linear dispersive medium," *IEEE Photon. Technol. Lett.*, vol. 1, no. 3, pp. 71–72, Mar. 1989.
47. D. Duchesne, R. Morandotti, and J. Azaña, "Temporal Talbot phenomena in high-order dispersive media," *J. Opt. Soc. Am. B*, vol. 24, no. 1, pp. 113–125, Jan. 2007.
48. T. Jannson, "Real-time Fourier transformation in dispersive optical fibers," *Opt. Lett.*, vol. 8, no. 4, pp. 232–234, Apr. 1983.
49. I. P. Christov, "Design of a compound time lens," *J. Modern Opt.*, vol. 36, no. 8, pp. 1027–1030, Aug. 1989.
50. I. P. Christov, "Theory of a time telescope," *Opt. Quantum Electron.*, vol. 22, no. 5, pp. 473–480, Sep. 1990.

# 20

# RF Photonics-Optical Pulse Synthesizer




Hiroshi Murata
*Osaka University*


## 20.1 Introduction

Ultrashort optical pulses are important and indispensable in many science and engineering application fields. A lot of theoretical and experimental studies on the generation of ultrashort optical pulses have been explored and developed since the invention of the laser. Today, ultrashort optical pulses of several femtoseconds can be obtained by use of a commercially available mode-locked laser. However, there are some weaknesses in the mode-locked laser; it is difficult to control the parameters of the optical pulses: the pulse width, the pulse shape, the pulse position in a time slot, the pulse repetition frequency, and the wavelength. It can also be difficult to synchronize the pulse repetition frequency with other electrical signals although it is required in many applications. The available wavelength ranges of the optical pulses are restricted by the intrinsic characteristics of the gain medium used in the mode-locked laser. Therefore, the development of a compact easy-to-use optical pulse source in arbitrary optical wavelength ranges still remains an issue.

Another candidate for ultrashort optical pulse generation is based on electro-optic (EO) phase modulation and optical synthesis [1–7]. This method, which is called the "EO modulation method," has a lot of advantages compared with optical pulse generation by the mode-locked laser. This method utilizes high-speed (typically >10 GHz) EO phase modulation by use of a sinusoidal single-tone modulation signal with a large amplitude, so that many discrete optical sideband components are generated on both sides of the input light spectrum according to the modulation index of the phase modulation. The optical sidebands generated by the EO phase modulation are completely phase-locked to each other and have in-phase or out-of-phase relationships. Therefore, by changing the phases and amplitudes of the generated optical sidebands with a group-delay circuit and an optical synthesizer, a train of ultrashort optical pulses can be obtained. The pulse repetition frequency is easily controlled by tuning the modulation frequency. The other pulse parameters are also easily controlled by tuning the amplitude of the modulation signal and the optical synthesis conditions.

In the following sections, the principle, required optical devices, and demonstrated pulse generation using this method are presented.

## 20.2 Principle of Optical Pulse Generation

### 20.2.1 Basic Scheme

The basic block diagram of the optical pulse generation in the EO modulation method is shown in Figure 20.1. A CW laser operating at any wavelength is applicable as the light source. The lightwave with a narrow spectrum from the CW laser is fed to the EO phase modulator and converted to a deeply phase-modulated lightwave with a large modulation index in the EO phase modulator, which is driven by a single-tone sinusoidal signal. A typical modulation frequency range is 10 GHz–30 GHz (quasi-millimeter-waves). The modulation frequency determines the pulse repetition frequency. To obtain a large modulation index at the quasimillimeter-wave frequency range, high-speed EO modulators using traveling-wave electrodes with velocity-matching technology, standing-wave resonant electrodes, or optical resonator structures are mainly used with a large-amplitude modulation signal.

The generated phase-modulated lightwave with a large modulation index contains many optical sidebands whose phases are completely locked to each other. From the other point of view, the phase-modulated lightwave is periodically chirped, that is, its instantaneous frequency is periodically increased and decreased by the same amount at the modulation frequency [1]. The maximum deviation of the instantaneous frequency corresponds to the index of the phase modulation.

The phase-modulated lightwave generated by the EO modulator is fed to a group-delay dispersion circuit or an optical synthesizer where the amplitudes and phases of the optical sidebands in the phase-modulated lightwave are controlled. As a result, a train of optical pulses is generated. For example, if the all phases of the sidebands are set to be in-phase, the shortest possible optical pulses in the Fourier-transform–limited condition are generated.

With optical pulse generation by the EO modulation method, it is easy to control the parameters of the optical pulse. The pulse width is inversely proportional to the optical spectrum width generated by the EO phase modulator when the pulse is under the Fourier-transform–limited condition. The pulse

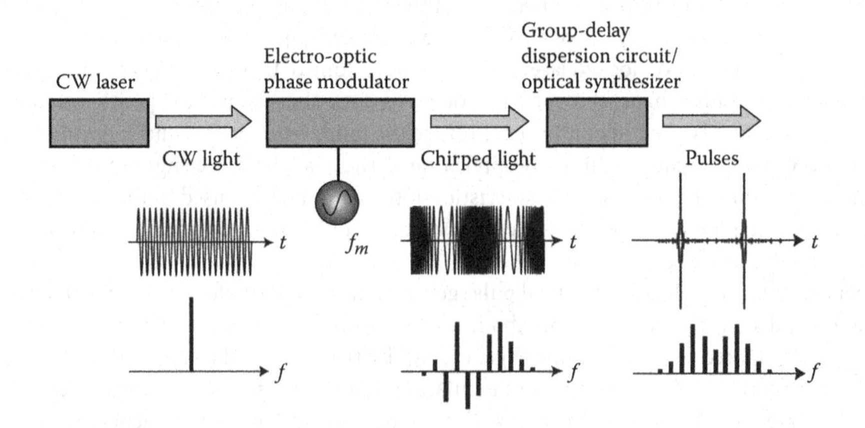


**FIGURE 20.1** Optical pulse generation based on EO phase modulation. (Reprinted with permission from H. Murata, A. Morimoto, T. Kobayashi, and S. Yamamoto, *IEEE J. Selected Top. Quantum Electron.*, vol. 6, no. 6, pp. 1325–1331, © 2000 IEEE.)

position in a time slot is related to the temporal phase of the modulation signal supplied to the modulator. Therefore, these are electrically controllable by adjusting the amplitude and phase of the modulation signal. It is also easy to synchronize the pulse repetition frequency with another electronic signal, since the repetition frequency is determined by the modulation frequency supplied to the EO phase modulator. In addition, not only simple Gaussian-like optical pulses but also optical signals with an arbitrary envelope are obtainable by controlling the amplitudes and phases of the optical sidebands using the optical synthesizer. Multiplication of the pulse repetition frequency is also possible by selecting optical sidebands.

It should be noted that the optical spectrum width of the phase-modulated lightwave can become much larger (over several hundred times) than the modulation frequency value due to cascading modulation. For example, by using a 10-GHz single-tone modulation signal, optical sidebands over a THz can be obtained by using a modulation signal with an extremely large amplitude [3–5] or by using an optical cavity-like optical frequency comb generator [8,9]. Therefore, using only 10-GHz-order sinusoidal signals with a period of ~100-ps, ultrashort optical pulses that have subpicosecond temporal widths can be generated with the EO modulation method [3–5].

### 20.2.2 Generation of Optical Sidebands by Phase Modulation

The phase-modulated lightwave can be expressed by use of the Bessel functions assuming that the modulation signal is a single tone at frequency $f_m$ and that the input lightwave has a line spectrum with an amplitude $E_0$ at frequency $\nu$.

$$\begin{aligned} E &= E_0 \exp[j\{2\pi\nu t + \Delta\theta \sin(2\pi f_m t)\}] \\ &= E_0 \sum_{n=-\infty}^{\infty} J_n(\Delta\theta) \exp\{j2\pi(\nu + nf_m)t\}, \end{aligned} \tag{20.1}$$

where $J_n$ is the $n$-th order Bessel function and $\Delta\theta$ is the modulation index of the optical phase modulation. Therefore, many optical sideband components with a frequency spacing of $f_m$ can be generated through the phase modulation process. The generated optical sideband components are completely phase-locked (in-phase or out-of-phase) in pure phase modulation. A typical example of the spectrum of the phase-modulated lightwave is shown in Figure 20.2a. Each modulation sideband is completely phase-locked in the phase-modulated lightwave. However, their magnitudes, which are described by the Bessel functions, are not completely the same.

The spectral width $\Delta\nu$ of the phase-modulated lightwave is approximately expressed by the following equation when the modulation index $\Delta\theta$ is much larger than 1 radian:

$$\Delta\nu = 2f_m\Delta\theta. \tag{20.2}$$

This means that the modulation index of $\Delta\theta$ over several dozen radians is necessary to generate the optical sideband width of $\Delta\nu$ ~1 THz using a modulation signal of 10 GHz–30 GHz.

### 20.2.3 Pulse Compression

The deeply phase-modulated CW lightwave with wide optical sidebands from the EO modulator is fed to an optical group-delay dispersion circuit, for example, a pair of diffraction gratings with a wide aperture, an optical fiber of an adjusted length, an asymmetric directional coupler, a distributed Bragg reflector (DBR) structure, or an optical synthesizer. Then, the chirped CW light can be compressed and converted to a train of ultrashort optical pulses.

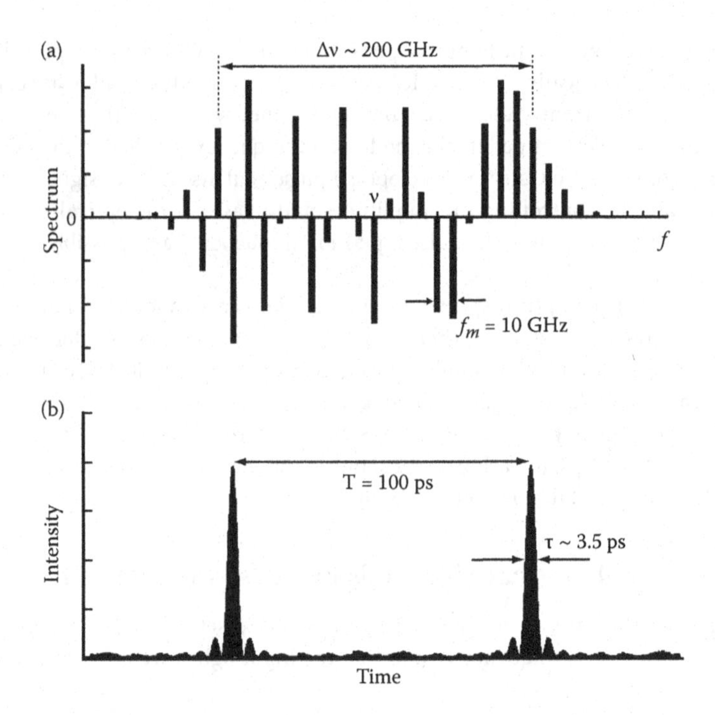


**FIGURE 20.2** The spectrum of a deeply phase-modulated lightwave. (a) The temporal profile of the optical pulses obtained with the simple pulse compression condition ($f_m$ = 10 GHz, $\Delta\theta$ = 10 rad, $\nu$ = 200 THz). (b) The pulse width is $\tau \sim 3.5$ ps when all optical sideband components are set in phase.

The obtainable pulse width by pulse compression, $\tau$ is expressed by

$$\tau \approx \frac{0.7}{\Delta\nu}, \tag{20.3}$$

under the condition for optimum pulse compression where the phases of the generated optical sidebands are all in phase as shown in Figure 20.2b [1]. It should be noted that the peak power of the generated pulse becomes much larger than the CW light power owing to the temporal compression of the lightwave energy at the pulse position. For example, the peak power of the optical pulses in Figure 20.2b is about 20 times of the input CW light power. The residual optical power between adjacent optical pulses remains as shown in Figure 20.2b since only the up-chirped, or down-chirped components can be compressed and contribute to the pulses. To remove the residual optical power, amplitude modulation driven by the same frequency is rather effective [1,10].

## 20.3 EO Phase Modulator

### 20.3.1 Phase Modulation

The EO phase modulator is a key device for pulse generation using the EO modulation method. The modulator is required to operate with a large modulation index (typically over 10 rad) at a high frequency (10 GHz–30 GHz) to generate wide optical sidebands of ~1 THz and to obtain very short optical pulses, whose pulse widths are below 1 ps [2–4].

It should be noted that an optical intensity modulator is not suitable for optical pulse generation since the output power from an optical intensity modulator is smaller than the input power, while essentially there is no power dissipation in the optical phase modulation process. The optical pulse width generated by an intensity modulator is usually proportional to the temporal width of the driving electrical signal. Therefore, to obtain short optical pulses below 1 ps by using an intensity modulator, an extremely high-speed driving signal of ~1 THz and an extremely high-speed intensity modulator operated at ~1 THz are necessary.

EO phase modulation is a kind of nonlinear optical process with cascaded wavelength conversion. The generated upper and lower optical sideband components correspond to sum and difference frequency components in the optical wavelength conversion, respectively. Namely, two optical sidebands of frequency $\nu + f_m$ and $\nu - f_m$ are generated through nonlinear coupling between the input lightwave of frequency $\nu$ and the modulation wave of frequency $f_m$. When the modulation index is rather small ($\Delta\theta \ll 1$) and the nonlinear coupling is weak, the phase-modulated lightwave consists of the input lightwave (carrier) and +1st and −1st order optical sidebands only. As the modulation index becomes larger, the cascading modulation process becomes more significant so that the lower order optical sidebands are converted to the higher order ones successively. As a result, many optical sidebands with a frequency separation of $f_m$ are generated through EO phase modulation.

The phase matching conditions for the generation of the upper and lower optical sidebands are described below in a traveling-wave electrode modulator.

$$\begin{aligned}\Delta\beta &= \beta(\nu + f_m) - \beta(\nu) - \beta(f_m)\\ &= \beta(\nu) - \beta(\nu - f_m) - \beta(f_m)\\ &\cong \frac{\partial\beta}{\partial\omega}\{2\pi(\nu + f_m) - 2\pi\nu\} - \beta(f_m)\\ &= \frac{2\pi f_m}{v_g} - \frac{2\pi f_m}{v_m}\\ &= \frac{2\pi f_m}{c}(n_g - n_m),\end{aligned} \tag{20.4}$$

where $\beta(\nu)$, $\beta(\nu + f_m)$, $\beta(\nu - f_m)$, and $\beta(f_m)$ are the phase constants of the input lightwave (carrier), the +1st sideband, the −1st sideband and the modulation microwave, respectively. $n_g$ and $n_m$ are the group index of the lightwave and the effective index of the modulation microwave, respectively. Therefore, the phase matching conditions in the nonlinear coupling between successive optical sidebands can be satisfied simultaneously in velocity-matched traveling-wave electrode modulators ($n_g = n_m$), since the frequency differences between the successive sidebands are much smaller than their own frequencies. This means that the optical sideband width, obtainable through optical phase modulation, is limited by the group-velocity dispersion of the EO material or the optical waveguide.

The power of the phase-modulated lightwave is essentially the same as the input light power as long as the EO material is transparent and the optical propagation loss in the waveguide is negligible.

In the following subsections, typical EO phase modulators for optical pulse generation are described.

## 20.3.2 Traveling-Wave Electrode EO Modulator

$LiNbO_3$ EO modulators using traveling-wave electrodes and Ti-diffused optical waveguides are standard devices for high-speed optical modulation in long-haul optical fiber communication systems. High-performance EO modulators with a low operational voltage ($V_\pi < 5$ V), good stability, high reliability, and advanced functions have been developed and are commercially available at the wavelength ranges around O-, S-, C-, and L-bands [11,12]. For optical pulse generation in these wavelength ranges, $LiNbO_3$ EO phase modulators using traveling-wave electrodes are applicable. Optical amplifiers (EDFA, SOA, Raman amplifier) are also available in the wavelength ranges around the same bands, although relatively large noise may increase the background dark levels surrounding the optical pulses.

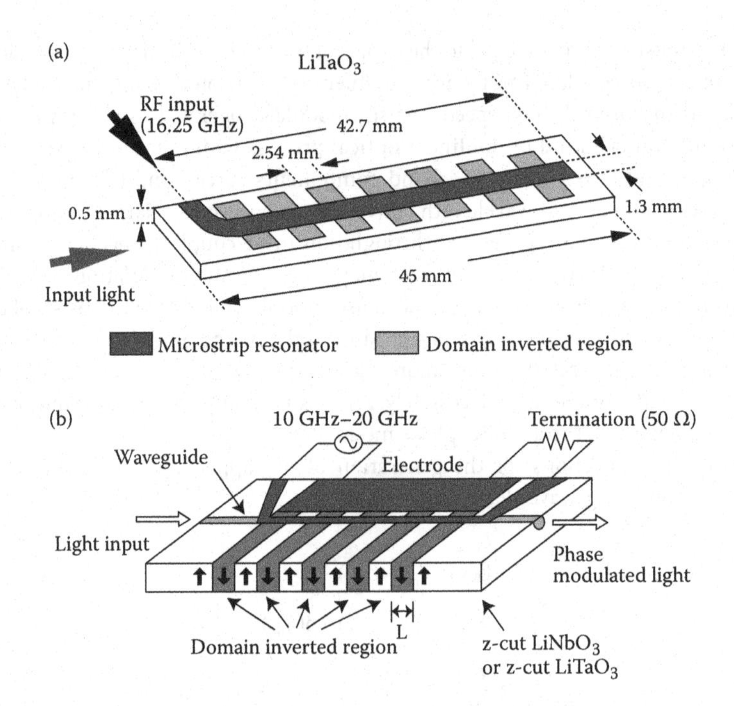


**FIGURE 20.3** Traveling-wave electrode EO phase modulator with periodic polarization-reversal (domain inversion) structures. (a) Bulk-type. (D.-S. Kim, M. Arisawa, A. Morimoto, and T. Kobayashi, *IEEE J. Selected Top. Quantum Electron.*, vol. 2, no. 3, pp. 493–499, © 1996 IEEE.) (b) Guided-wave type. (Reprinted with permission from H. Murata, A. Morimoto, T. Kobayashi, and S. Yamamoto, *IEEE J. Selected Top. Quantum Electron.*, vol. 6, no. 6, pp. 1325–1331, © 2000 IEEE.)

For operation in the visible wavelength range, excellent traveling-wave electrode EO modulators have been proposed and developed using the technology of the polarization reversal of ferroelectric optical crystals [3–7]. Both bulk-type and guide-wave type EO phase modulators have been developed using $LiTaO_3$. The generation of wide optical sidebands over 1 THz and ultrashort optical pulses below 1 ps has also been demonstrated. Utilizing the guided-wave structure, the operational power for EO modulation can be lowered drastically compared to that for the bulk-type one. It is also possible to integrate the guided-wave modulator with a CW laser and a guided-wave optical group-delay dispersion circuit for pulse compression and thus to construct integrated optical pulse generators (Figure 20.3).

## 20.3.3 Resonant Standing-Wave Electrode EO Modulator

A resonant standing-wave electrode EO modulator is attractive for operation in the quasi-millimeter- and millimeter-wave frequency ranges. Especially in the millimeter-wave frequency range over 30 GHz, the decay of the modulation signal in a traveling-wave electrode modulator may become significant. This decay shortens the effective interaction length in EO modulation and thus causes a degradation of the optical phase modulation in the traveling-wave electrode modulator. By utilizing the resonance effect of the modulation signal in a standing-wave electrode, the modulation electric field is significantly enhanced and high-efficiency modulation can be obtained while using a compact resonant electrode [7,13,14].

The resonant standing-wave electrode modulator is essentially a band modulator. Its operational bandwidth becomes narrower as the Q-factor in the resonant electrode becomes larger. However, unlike digital data encoding in optical fiber communication systems, a base-band modulation frequency

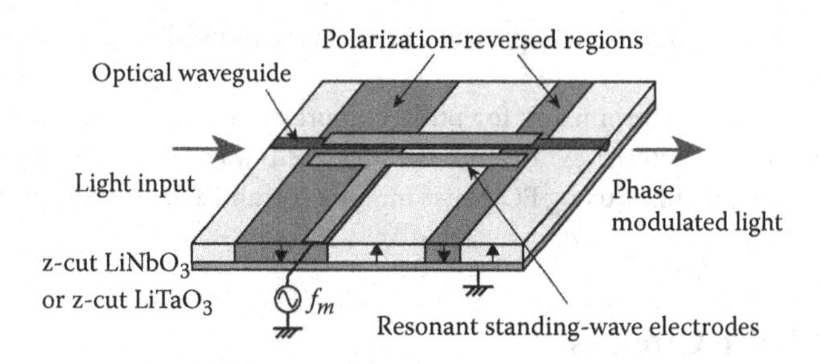


**FIGURE 20.4** Resonant standing-wave electrode EO phase modulator with polarization-reversal structures. (Reprinted with permission from S. Matsunaga, H. Murata, and Y. Okamura, *J. Lightwave Technol.*, vol. 24, no. 9, pp. 3334–3340, © 2006 IEEE.)

response is not necessary for a phase modulator with the EO modulation method. High-efficiency modulators with a 3-dB operational bandwidth of a few gigahertz have been reported [7,14] (Figure 20.4).

By utilizing a high-quality optical cavity like optical frequency comb generators [8,9], the modulation signal power can be lowered drastically although the light wavelength and the modulation frequency must be precisely tuned to the designed values.

## 20.4 Group-Delay Dispersion Circuit

The optical group-delay dispersion circuit is another key device for the generation of optical pulses. Several optical group-delay dispersion circuits and optical synthesizers are available for the compression of a phase-modulated lightwave with many optical sidebands to generate a train of optical pulses. Typical optical circuits are presented in the following.

### 20.4.1 Bulk Optics

It is well-known that several bulk optical components are applicable for the compensation of optical dispersion effects. Pairs of diffraction gratings or prisms have been used for the compensation of the dispersion effect in pulsed lasers and the compression of chirped lightwaves to generate optical pulses. The application of a pair of diffraction gratings for pulse compression in the EO modulation method has been reported since the method's proposal [1–3].

Optical synthesizers, which are composed of a pair of gratings and Fourier-transforming optics with a spatial modulator, are also attractive not only to compress the chirped lightwave but also to manipulate and synthesize output optical pulses [15,16].

### 20.4.2 Optical Fiber

A silica optical fiber is a good optical delay circuit since it possesses extremely low propagation loss (~0.2 dB/km at $\lambda$ = 1.55 μm), negligible cross-talk, good stability, and is low-cost. However, the group-delay dispersion in silica fibers is small compared with other optical circuits. In a standard single-mode fiber, the group-delay dispersion value is ~16 ps/(nm km) at $\lambda$ = 1.55 μm. Despite this, it is possible to utilize a long fiber (>1 km) due to the extremely low propagation loss so that enough optical group-delay dispersion for pulse compression can be obtained.

Compression of a phase-modulated lightwave and optical pulse generation by use of a silica single-mode fiber has been reported [17]. Specific optical fibers such as dispersion shifted types using double cladding structures and photonic crystal fibers are also applicable. The nonlinearity in optical fibers is attractive for further optical pulse compression because it can produce a broadening of the optical spectrum.

A fiber Bragg grating is also promising for pulse compression since it can have large and tailored group-delay dispersion characteristics around the designed wavelength. The generation of high-repetition-frequency optical pulses using EO phase modulation and a fiber Bragg grating has been demonstrated [18].

### 20.4.3 Guided-Wave Circuits

A guided-wave type group-delay dispersion circuit is attractive to integrate with an EO phase modulator and a CW semiconductor laser to construct integrated optical pulse generators. Several guided-wave group-delay dispersion circuits are applicable, such as an asymmetric directional coupler composed of a single-mode guide and a multimode guide [19,20], a distributed Bragg reflector/distributed feed-back waveguide [21,22], and an arrayed waveguide grating (AWG).

An AWG is a powerful device to separate the generated optical sideband components, which enables us to control the amplitude and phase of the generated optical sidebands independently with the combination of an array of optical intensity/phase modulators. It is also possible to integrate an AWG with a guided-wave EO phase modulator for the generation of optical sidebands and to construct an integrated optical pulse generator. Experimental demonstrations of optical pulse generation using an AWG have been reported [23].

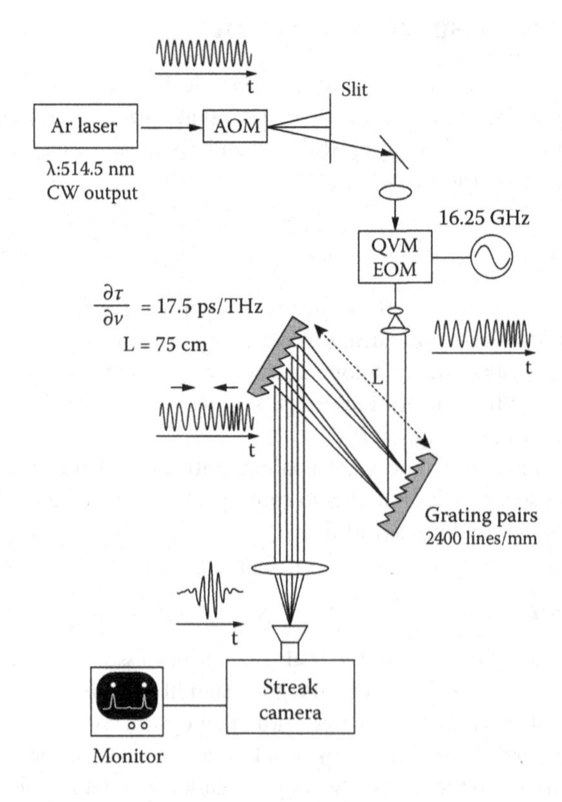


**FIGURE 20.5** Experimental setup for optical pulse generation using a pair of gratings. (Reprinted with permission from D.-S. Kim, M. Arisawa, A. Morimoto, and T. Kobayashi, *IEEE J. Selected Top. Quantum Electron.*, vol. 2, no. 3, pp. 493–499, © 1996 IEEE.)

## 20.5 Optical Pulse Generator and Optical Frequency Comb

### 20.5.1 Generation by Discrete Devices

Several experimental studies on the demonstration of optical pulse generation by the EO modulation method have been reported. Typical examples are shown below.

Figure 20.5 shows an experimental setup for optical pulse generation using a bulk-type EO phase modulator and a pair of diffraction gratings for the pulse compression. In this experiment, the EO phase modulator with a traveling-wave electrode and periodic polarization-reversed structures shown in Figure 20.3a was used for the optical phase modulation. This modulator was driven at the operational frequency of 16.25 GHz, and many optical sidebands were generated. The 3-dB bandwidth of the generated optical sidebands was over 2 THz [3]. With the compression of the deeply phase-modulated lightwave by use of the pair of diffraction gratings, ultrashort optical pulses of ~560 fs were generated at the repetition frequency of 16.25 GHz (Figure 20.6).

Figure 20.7 shows an experimental setup for optical pulse generation using a guided-wave type EO phase modulator and an optical synthesizer. In this experiment, the EO phase modulator with a resonant standing-wave electrode shown in Figure 20.4 was used for the optical phase modulation. The obtained bandwidth of the optical sidebands was over 100 GHz. Each sideband component of the modulated lightwave was spatially separated using an optical synthesizer composed of a diffraction grating and a focal mirror. For compression of the optical sidebands, the lower sideband components were blocked with a special mask in the frequency plane of the optical synthesizer. While the other sideband components passed through the frequency plane without any disturbance and recombined by going through the reverse optical path of the grating and the Fourier transforming mirror, and thus the trains

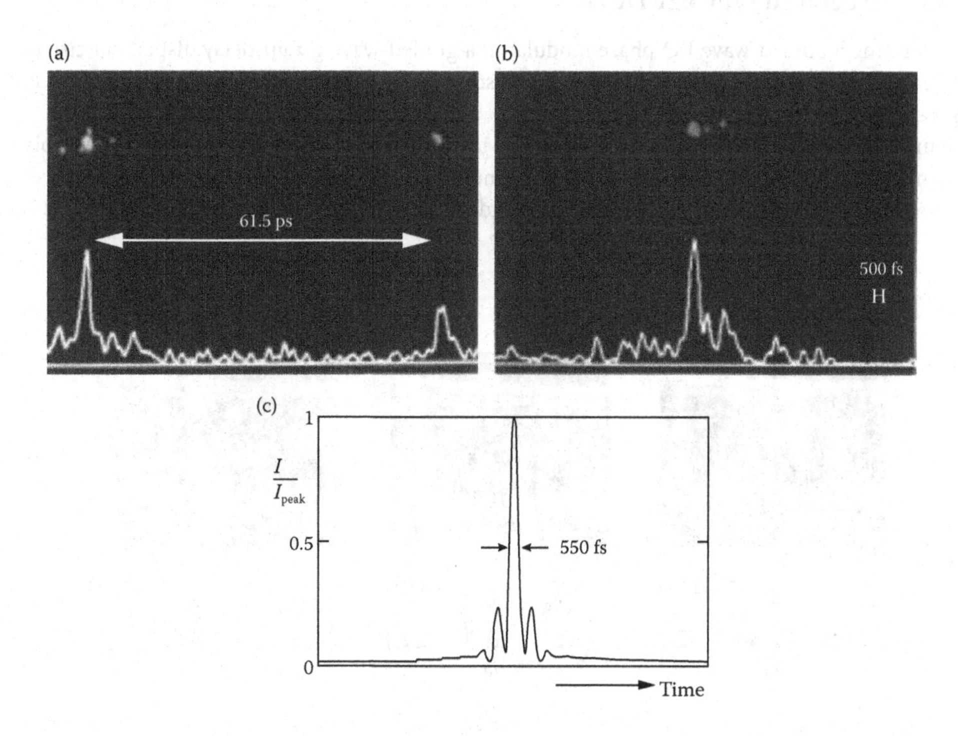


**FIGURE 20.6** Optical pulse produced by the configuration given in Figure 20.5 as measured using an optical streak camera. (Reprinted with permission from D.-S. Kim, M. Arisawa, A. Morimoto, and T. Kobayashi, *IEEE J. Selected Top. Quantum Electron.*, vol. 2, no. 3, pp. 493–499, © 1996 IEEE.)

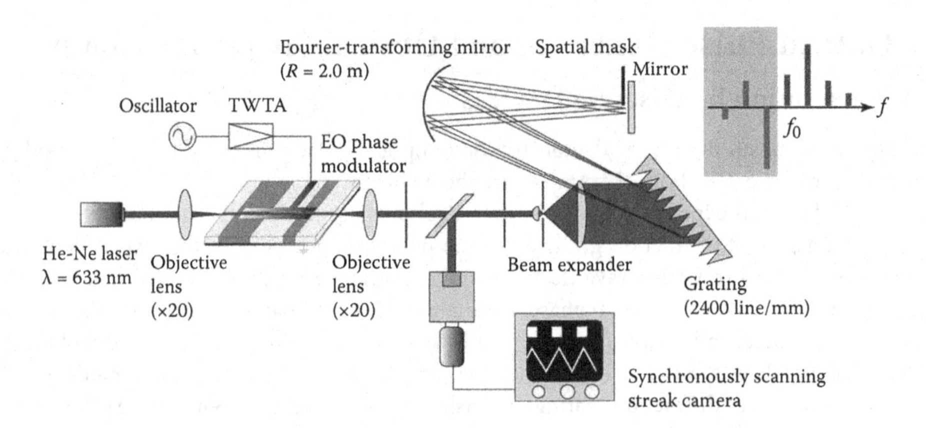


**FIGURE 20.7** Experimental setup for optical pulse generation using an optical synthesizer. (Reprinted with permission from S. Matsunaga, H. Murata, and Y. Okamura, *J. Lightwave Technol.*, vol. 24, no. 9, pp. 3334–3340, © 2006 IEEE.)

of optical pulses were generated. An example of the observed envelope pattern of the optical pulse trains is shown in Figure 20.8.

## 20.5.2 Integrated Optical Device

By integrating a guided-wave EO phase modulator, a guided-wave group-delay dispersion circuit, and a CW laser on the same substrate, a compact and stable module for a light source of ultrashort optical pulses can be constructed.

Figure 20.9 shows the basic structures of several integrated optical pulse generators [6]. Combining a CW semiconductor laser, a guided-wave EO phase modulator and (1) a guided-wave group-delay dispersion circuit of the asymmetric directional waveguide, (2) a DBR waveguide, or (3) a pair of AWGs with

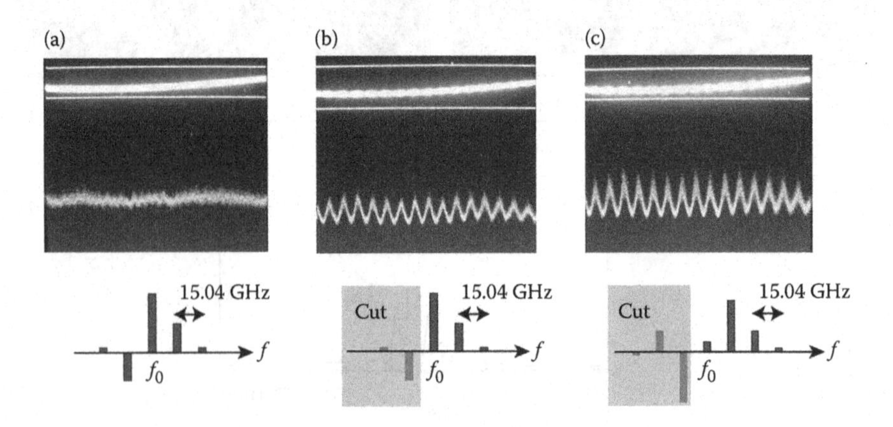


**FIGURE 20.8** Optical pulse train produced by the configuration given in Figure 20.7 as measured using an optical streak camera. (Reprinted with permission from S. Matsunaga, H. Murata, and Y. Okamura, *J. Lightwave Technol.*, vol. 24, no. 9, pp. 3334–3340, © 2006 IEEE.)

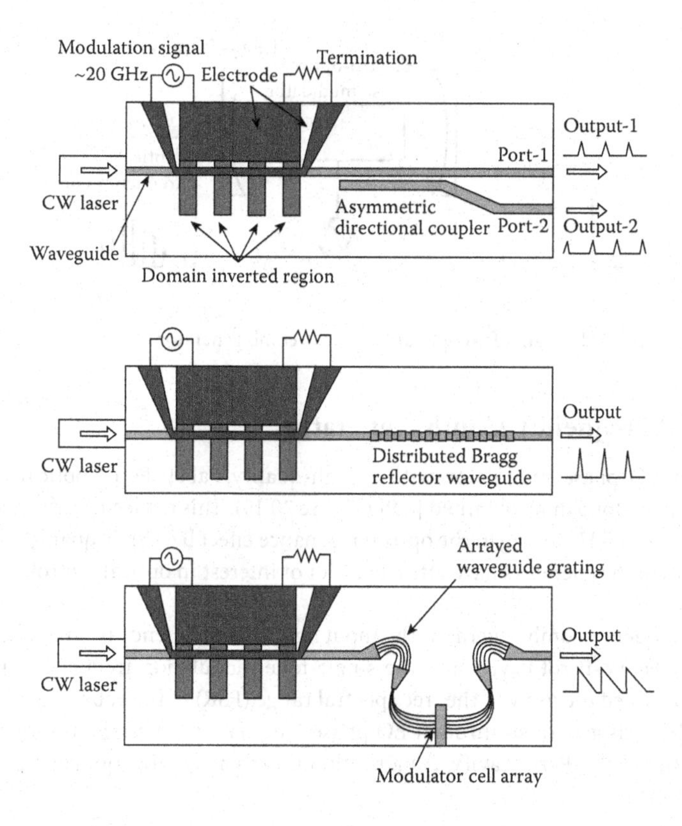


**FIGURE 20.9** Integrated optical pulse generators. (Reprinted with permission from H. Murata, A. Morimoto, T. Kobayashi, and S. Yamamoto, *IEEE J. Selected Top. Quantum Electron.*, vol. 6, no. 6, pp. 1325–1331, © 2000 IEEE.)

modulator cell arrays as a guided-wave optical synthesizer, integrated optical pulse generators can be obtained.

It is interesting to note that in the integrated device using an asymmetric directional coupler, simultaneous pulse compression of both up-chirped light and down-chirped light is possible because the asymmetric directional coupler waveguide considered here has two guided modes of positive and negative group-delay dispersion [19]. Therefore, if the modulated light from the EO phase modulator is incident upon the directional coupler such that it couples equally to the two modes of positive and negative dispersion, two trains of optical pulses are obtained from the two output ports, where the pulse positions within a time slot are shifted by a half period from each other, as shown in Figure 20.9. This splitting occurs because the lightwave produced by a sinusoidal phase modulation is composed of two frequency-chirped parts in the time period of the modulation signal; one part is up-chirped, while the rest is down-chirped.

The integration of a guided-wave grating with an EO modulator is attractive for obtaining integrated optical pulse generators. The fabrication and measurement of $LiNbO_3$ waveguides with 1D grating structures or 1D photonic crystal structures has been reported [21,22].

In the integrated device using an AWG, not only simple optical pulses but also optical signals of an arbitrary envelope can be obtained, such as a signal with two or more pulses in a time slot, a saw-toothed envelope signal, and triangle-like or rectangle-like envelope signals. These optical pulses with arbitrary envelopes are generated by controlling the amplitudes and phases of the optical sidebands using the modulator cell array.

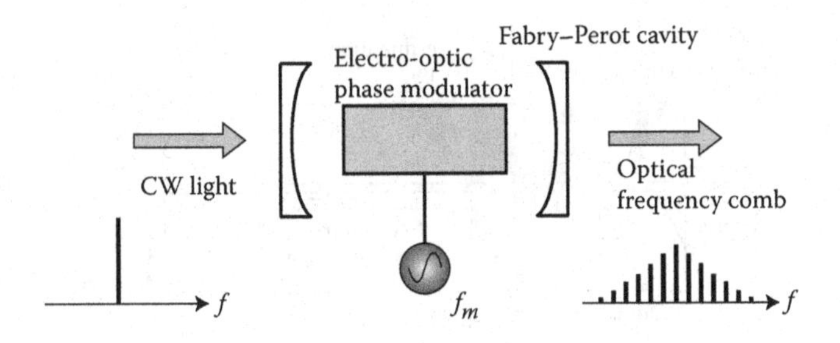


**FIGURE 20.10** Basic block diagram of an optical frequency comb generator.

### 20.5.3 Optical Frequency Comb Generator

By combining an EO phase modulator with a high-quality Fabry–Perot optical cavity, an optical frequency comb generator can be obtained [8,9] (Figure 20.10). This device is very useful for generating optical sidebands over 1 THz owing to the optical resonance effect by a high-quality Fabry–Perot cavity. Optical frequency comb generators have attracted a lot of interest in optical metrology and many other fields [24].

In the optical frequency comb generator, the input lightwave frequency is set at one of the resonance frequencies of the Fabry–Perot cavity and the single-tone modulation frequency supplied to the EO phase modulator is tuned precisely to the free spectral range (FSR) of the Fabry–Perot cavity. Therefore, all the optical sidebands generated through EO phase modulation are in the resonance condition and are enhanced in the Fabry–Perot cavity. As a result, an extremely wide optical frequency comb over 1 THz can be obtained.

### 20.5.4 Mach–Zehnder Modulator-Based Optical Frequency Comb

Recently, a new method for the generation of a flat optical frequency comb using a guided-wave Mach–Zehnder EO modulator has been proposed and demonstrated [25,26]. This method utilizes a pair of EO phase modulations with the same modulation frequency but different modulation index values with an appropriate optical bias in a Mach–Zehnder waveguide. By tuning the modulation index values in the two arms of the Mach–Zehnder waveguide and the optical bias between the two arms, flat optical frequency comb signals can be obtained.

Figure 20.11 shows the basic operational diagram. The required condition for the generation of the flat optical frequency comb is expressed using the simple equation:

$$\Delta\theta_{12} + \Delta\phi_{12} = \pi, \tag{20.5}$$

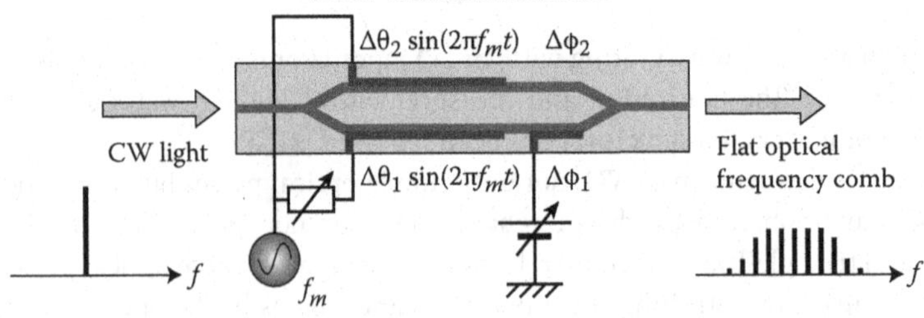


**FIGURE 20.11** Basic block diagram of the Mach–Zehnder modulator-based optical frequency comb generator.

where $\Delta\theta_{12}$ is the difference between the modulation index values in the two arms of the Mach–Zehnder waveguide and $\Delta\phi_{12}$ is the difference of the optical bias between the two arms. Demonstrations of the generation of flat optical frequency combs and optical pulses have been reported [26].

## 20.6 Conclusions

Optical pulse generation methods using EO phase modulation are powerful techniques for generating tunable and controllable optical pulse trains. Recently, optical pulse sources based on EO phase modulation have become commercially available. Optical frequency comb generators have also become important in many application fields. Further developments of high-speed EO modulators for advanced functions are expected with the utilization of new photonic technologies like the photonic crystal structure, the meta-material structure, and the polarization reversal technology.

## References


1. T. Kobayashi, H. Yao, K. Amano, Y. Fukushima, A. Morimoto, and T. Sueta, "Optical pulse compression using high-frequency electrooptic phase modulation," *IEEE J. Quantum Electron.*, vol. 24, no. 2, pp. 382–387, Feb. 1988.
2. K. Amano, T. Kobayashi, H. Yao, A. Morimoto, and T. Sueta, "Generation of 0.64 THz width optical sidebands by a novel electrooptic modulator for the purpose of forming ultrashort optical pulses," *J. Lightwave Technol.*, vol. LT-5, no. 10, pp. 1454–1458, Oct. 1987.
3. D.-S. Kim, M. Arisawa, A. Morimoto, and T. Kobayashi, "Femtosecond optical pulse generation using quasi-velocity-matched electrooptic phase modulator," *IEEE J. Selected Top. Quantum Electron.*, vol. 2, no. 3, pp. 493–499, Sep. 1996.
4. D.-S. Kim, T. Khayim, A. Morimoto, and T. Kobayashi, "Ultrashort optical pulse shaping by electrooptic synthesizer," *IEICE Trans. Electron.*, vol. E81-C, no. 2, pp. 260–263, Feb. 1998.
5. T. Khayim, M. Yamauchi, D.-S. Kim, and T. Kobayashi, "Femtosecond optical pulse generation from a CW laser using an electrooptic phase modulator featuring lens modulation," *IEEE J. Quantum Electron.*, vol. 35, no. 10, pp. 1412–1418, Oct. 1999.
6. H. Murata, A. Morimoto, T. Kobayashi, and S. Yamamoto, "Optical pulse generation by electrooptic modulation method and its application to integrated ultrashort pulse generators," *IEEE J. Selected Top. Quantum Electron.*, vol. 6, no. 6, pp. 1325–1331, Nov. 2000.
7. S. Matsunaga, H. Murata, and Y. Okamura, "Optical pulse generation using guided-wave electrooptic modulator with resonant electrodes and polarization reversal," *J. Lightwave Technol.*, vol. 24, no. 9, pp. 3334–3340, Sep. 2006.
8. M. Kourogi, K. Nakagawa, and M. Ohtsu, "Wide-span optical frequency comb generator for accurate optical frequency difference measurement," *IEEE J. Quantum Electron.*, vol. 29, no. 10, pp. 2693–2701, Oct. 1993.
9. T. Saitoh, M. Kourogi, and M. Ohtsu, "A waveguide-type optical-frequency comb generator," *IEEE Photon. Technol. Lett.*, vol. 7, no. 2, pp. 197–199, Feb. 1995.
10. T. Otsuji, M. Taita, T. Nagatsuma, and E. Sano, "10–80-Gb/s highly extinctive electrooptic pulse pattern generation," *IEEE J. Selected Topics Quantum Electron.*, vol. 2, no. 3, pp. 643–649, Sep. 1996.
11. E. L. Wooten, K. M. Kissa, A. Yi-Yan, E. J. Murphy, D. A. Lafaw, P. F. Hallemeier, D. Maack, D. V. Attanasio, D. J. Fritz, G. J. McBrien, and D. E. Bossi, "A Review of lithium niobate modulators for fiber-optic communication systems," *IEEE J. Selected. Top. Quantum Electron.*, vol. 6, no. 1, pp. 69–82, Jan. 2000.
12. T. Kawanishi, T. Sakamoto, and M. Izutsu, "High-speed control of lightwave amplitude, phase, and frequency by use of electro-optic effect," *IEEE J. Selected. Top. Quantum Electron.*, vol. 13, no. 1, pp. 79–91, Jan. 2007.

13. M. Izutsu, H. Murakami, and T. Sueta, "Guided-wave light modulator using a resonant coplanar electrode," *Trans. IEICE*, vol. J71-C, no. 5, pp. 653–658, May 1988.
14. A. Enokihara, H. Furuya, H. Yajima, M. Kosaki, H. Murata, and Y. Okamura, "60 GHz guided-wave electro-optic modulator using novel electrode structure of coupled microstrip line resonator," in *Proc. IEEE MTT-S International Microwave Symposium (IMS2004)*, Fort Worth, TX, IFTH-59, vol. 3, pp. 2055–2058, June. 2004.
15. A. M. Weiner, "Femtosecond pulse shaping using spatial light modulators," *Rev. Sci. Instrum.*, vol. 71, no. 5, pp. 1929–1960, May 2000.
16. Z. Jiang, D. E. Leaird, and A. M. Weiner, "Optical processing based on spectral line-by-line pulse shaping on a phase-modulated CW laser," *IEEE J. Quantum Electron.*, vol. 42, no. 7, pp. 657–666, Jul. 2006.
17. D.-S. Kim, M. Matsuda, A. Morimoto, and T. Kobayashi, "Electro-optic femtosecond pulse generation using a dispersive optical fiber," *Jpn. J. Appl. Phys.*, vol. 36, no. 8, pp. 5125–5129, Aug. 1997.
18. T. Komukai, T. Yamamoto, and S. Kawanishi, "Optical pulse generator using phase modulator and linearly chirped fiber Bragg gratings," *IEEE Photon. Technol. Lett.*, vol. 17, no. 8, pp. 1746–1748, Aug. 2005.
19. U. Peshel, T. Peshel, and F. Lederer, "A compact device for highly efficient dispersion compensation in fiber transmission," *Appl. Phys. Lett.*, vol. 67, no. 15, pp. 2111–2113, Oct. 1995.
20. Y. Lee, "Pulse compression using coupled-waveguide structures as highly dispersive elements," *Appl. Phys. Lett.*, vol. 73, no. 19, pp. 2715–2717, Nov. 1998.
21. A. Enokihara, A. Suzuki, J. Adachi, T. Iwamoto, H. Murata, and Y. Okamura, "Fabrication and evaluation of $LiNbO_3$ periodic waveguide with etched grooves," *Electron. Lett.*, vol. 43, no. 11, pp. 629–630, May 2007.
22. A. Suzuki, T. Iwamoto, A. Enokihara, H. Murata, and Y. Okamura, "Fabrication of Bragg gratings with deep grooves in $LiNbO_3$ ridge optical waveguide," *Microelectronic Engineering*, vol. 85, no. 5–6, pp. 1417–1420, May 2008.
23. H. Tsuda, Y. Tanaka, T. Shioda, and T. Kurokawa, "Analog and digital optical pulse synthesizers using arrayed-waveguide gratings for high-speed optical signal processing," *J. Lightwave Technol.*, vol. 26, no. 6, pp. 670–677, Mar. 2008.
24. S. T. Cundiff and J. Ye, "Femtosecond optical frequency combs," *Rev. Mod. Phys.*, vol. 75, no. 1, pp. 325–342, Jan. 2003.
25. T. Sakamoto, T. Kawanishi, and M. Izutsu, "Asymptotic formalism for ultraflat optical frequency comb generation using a Mach–Zehnder modulator," *Opt. Lett.*, vol. 32, no. 11, pp. 1515–1517, Jun. 2007.
26. T. Sakamoto, T. Kawanishi, and T. Tsuchiya, "10 GHz, 2.4 ps pulse generation using a single-stage dual-drive Mach–Zehnder modulator," *Opt. Lett.*, vol. 33, no. 8, pp. 890–892, Apr. 2008.

# 21

# Optoelectronic Oscillator


Lute Maleki
*OEwaves Inc.*

Danny Eliyahu
*OEwaves Inc.*

Andrey B. Matsko
*OEwaves Inc.*




## 21.1 Introduction

The information age in which we live is powered by data flowing at ever-increasing rates across a growing number of channels. Generation, transmission, and processing of this expanding data volume is regulated by the beat of the output signal of RF, microwave, and as of late, optical oscillators. Oscillators produce signals that are the carriers on which data are modulated. They set the "clock speed" in any data system. In radar systems, oscillator performance determines the extent of capability for clutter rejection, and detection of small signals. In communications systems, oscillator performance determines the channel capacity and channel density.

Since the range of applications of the oscillator is varied, so are the types of oscillators that serve these applications. For example, the proliferation of small, hand-held communication devices and their associated frequency bands, has created a demand for small, high performance oscillators operating in the 1-GHz to 6-GHz band, with future systems likely to require higher performance at higher frequencies. On the other extreme, advanced radars that detect slow moving targets in severe cluttered environment require extremely high spectral purity at frequencies ranging from about 10 GHz–35 GHz to cover radar frequency bands.

The technological areas of the applications mentioned above have their counterparts in scientific applications. Aside from being an integral part of physical sensors that measure various parameters of interest, oscillators have many applications in metrology and fundamental scientific investigations. These range from the measurement of the size of particles in planetary rings to tests of Einstein's general relativity.

### 21.1.1 Photonic RF Oscillators

The quality of the signal produced by an oscillator generally depends on the energy storage element in the oscillator loop, which is typically a resonator or a delay line. Various physical realizations of RF and microwave resonators have been used to address the specific needs of each application. Quartz resonators, surface acoustic wave (SAW) devices, microwave cavities, and lumped element electronic circuits, are examples of such resonators.

The photonic oscillator is a new class of device that generates stable RF signals, using optical energy storage elements. This class of oscillator became feasible as optical communications technology evolved, and as commercially available components were developed. Typically, the photonic oscillator is comprised of the following stages: (1) upconversion of the RF signal to the optical frequency domain using a modulator, (2) translation of the modulated laser light through the optical storage element, (3) conversion of the modulated light back to an RF frequency using a fast photodiode, and (4) closing the oscillator loop by feeding the RF signal back to the modulator. More stages, such as amplification of the demodulated RF signal, RF filtering, or phase shifting can also be added. Such oscillators, known as optoelectronic oscillators (OEOs), possess characteristics and performance parameters unmatched by their electronic counterparts.

It is worth noting that an optical oscillator (laser) can also be used as a photonic RF oscillator. In such an oscillator, several harmonics are generated in a purely optical way (no modulators are involved). This kind of a device can be produced, for example, by locking the frequencies of two lasers to two different modes of a single optical resonator. In this scheme, harmonics are generated due to nonlinear optical processes, such as parametric or hyperparametric, or due to optomechanical processes. We do not discuss those devices in this chapter because they do not include the use of a modulator.

### 21.1.2 General Architecture of OEO and COEO

The OEO is a unique oscillator in that the equivalent quality factor ($Q$) in the oscillator loop can be selected to be very large, which is done by using an optical storage element, such as a length of fiber, or an optical resonator. This is possible, since optical storage elements, unlike their electronic counterparts, dissipate very little energy, ensuring that losses are small. The OEO is also distinct from other oscillators, since the frequency of oscillation can be any frequency within the bandwidth of its components, set by the center frequency of an RF filter. As such, the OEO can serve numerous applications in communications and radar at frequencies limited only by the bandwidth of the modulator; currently, all other components in the OEO loop can achieve higher bandwidths.

In a conventional photonic link, an RF signal modulated on a laser carrier produces sidebands that are conveyed in a fiber waveguide and then recovered at the output of a photodetector. If the signal is a pure tone and the modulation is not of purely phase type, the modulated sidebands beat with the carrier on a photodiode, to reproduce the tone. This simple link can be transformed into a feedback loop by applying the output of the photodiode to the modulator. This configuration functions as an oscillator if gain is introduced in the loop to overcome all losses. The oscillation, initiated by noise in the system, is self-sustaining if there is enough gain in the loop. This configuration, shown in Figure 21.1, is an OEO. The oscillator in Figure 21.1 also includes a filter and a phase shifter. The phase shifter ensures that all waves propagating in the loop add-in phase. Oscillation will be sustained for waves going around the loop one time, twice,..., $m$ times—the number of which is determined by the bandwidth of components in the loop. The filter selects a single frequency corresponding to a single mode of operation, otherwise this becomes a multimode oscillator.

The OEO architecture described above is quite generic and can be extended to other configurations by the choice of its elements. For example, the gain element and the filter shown in the electronic segment of the loop can be replaced with optical counterparts. A tunable filter in the loop produces a tunable oscillator. Multiple loops can replace a single fiber loop to produce added filtering to suppress any

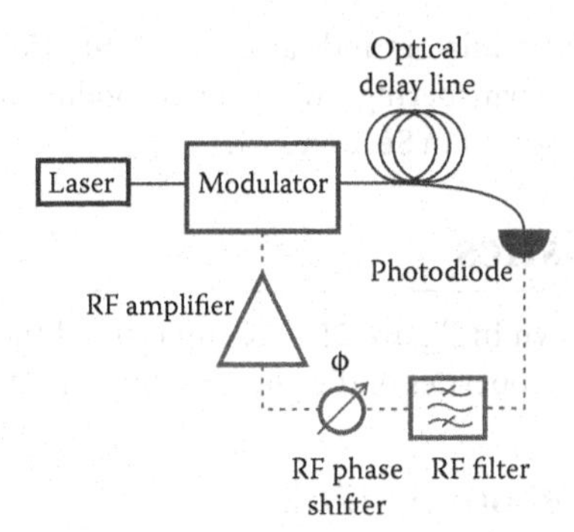


**FIGURE 21.1** A schematic of an OEO.

additional modes that fall within the bandwidth of the filter. While an intensity modulator is used in most applications, a phase modulator can also be used, if the optical segment includes a dispersive element to convert phase modulation to amplitude modulation.

The light source, which produces the energy driving the oscillation, and shown to be outside the feedback loop in Figure 21.1, can be brought inside the loop. This is accomplished by forming a second loop, this time around an optical amplifier, to oscillate and produce laser light. Such a crude laser is then mode-locked using the modulator in the loop driven by oscillation in the electrical segment of the loop. This dual optical and electrical loop configuration is known as the coupled optoelectronic oscillator (COEO), shown in Figure 21.2. Like the OEO, the COEO has an electrical output and an optical output, but the COEO produces a pulsed optical output, as in a conventional mode-locked laser, while the OEO's optical output simply carries the modulated tone.

The performance of the OEO and the COEO depends on a number of parameters, including characteristics of optical and electronic components in the oscillator circuit. This dependence has been previously analyzed in detail and described fairly widely in the literature. To be consistent with the theme of this book, we focus primarily on the role of the modulator on the performance of the OEO and the signals it produces, rather than reproducing a detailed analysis of every parameter here. We still describe the basic OEO features to show how the efficiency of the modulator, modulation bandwidth, and response time, noise, and insertion loss, and optical power handling properties impact the oscillator performance.

This chapter is organized as follows. In Section 21.2, we present the basic equations describing OEO properties using a generic model of the modulator. In Section 21.3, we discuss performance of OEOs and

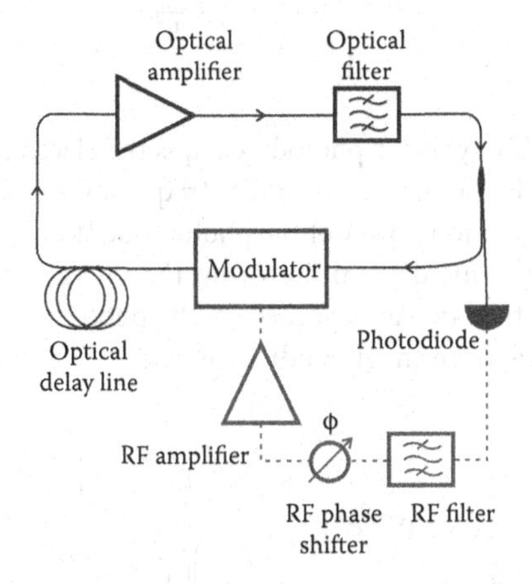


**FIGURE 21.2** A schematic of a COEO.

COEOs based on various types of modulators and various configurations of optical fiber delay lines. The performance of OEOs based on the whispering gallery mode modulators is studied in Section 21.4. The discussion and conclusion are presented in Section 21.7.

## 21.2 OEO Characteristics

We consider an OEO scheme shown in Figure 21.1. To understand the behavior and characteristics of the OEO, we have to describe the properties of the photonic link the OEO is based on.

### 21.2.1 Optical Gain of a Photonic Link

The photonic link operates in the most efficient way if the modulator included to the link generates a single sideband (SSB). For an ideal SSB modulator, the power of the output carrier ($P_{Cout}$) and sideband ($P_{Sout}$) optical waves is given by

$$\frac{P_{Cout}}{P_{Cin}} = \left( \frac{1 - \frac{P_{RFin}}{P_{RF0}}}{1 + \frac{P_{RFin}}{P_{RF0}}} \right)^2, \tag{21.1}$$

$$\frac{P_{Sout}}{P_{Cin}} = \frac{4\frac{P_{RFin}}{P_{RF0}}}{\left(1 + \frac{P_{RFin}}{P_{RF0}}\right)^2}, \tag{21.2}$$

where $P_{RF0}$ is the characteristic RF power of the modulator, $P_{RFin}$ in is the input RF power, and $P_{Cin}$ is the input optical power. These expressions determine the most efficient modulation process irrespective of the nature of the modulation, as they require conservation of the optical power $P_{Cout} + P_{Sout} = P_{Cin}$.

To describe the photodiode in the OEO loop (Figure 21.1) we use the following phenomenological model for the photocurrent:

$$I = \frac{RP}{\left[1 + (P/P_{sat})^2\right]^{1/2}}, \tag{21.3}$$

where $R = \eta q/\hbar\,\omega_0$ is the responsivity of the photodiode, $q$ is the electron charge, $\eta$ is the quantum efficiency of the photodiode, $\omega_0$ is the carrier wave angular frequency, $\hbar$ is Planck's constant, $P$ is the input optical power, and $P_{Sat}$ is the saturation power of the photodiode [1,2].

The saturation power $P_{Sat}$ is limited from above by the energy conservation law. According to Equations 21.1 and 21.2, the optical power absorbed by the photodiode is $P_{Cin}$. The time-averaged RF power generated in and emerging from the photodiode is

$$P_{RFout} = \rho R^2 P_{Cin}^2 \frac{4\frac{P_{RFin}}{P_{RF0}}\left(1 - \frac{P_{RFin}}{P_{RF0}}\right)^2}{\left(1 + \frac{P_{Cin}^2}{P_{sat}^2}\right)\left(1 + \frac{P_{RFin}}{P_{RF0}}\right)^4}, \tag{21.4}$$

where ρ is the resistance of the RF transmission line (we assume for now that the resistance of the photodiode is the same as the resistance of the RF transmission line). According to the energy conservation law,

$$P_{Cin} \geq P_{RFout}. \tag{21.5}$$

It is convenient to parameterize the inequality (21.5) by introducing two parameters: $P_{Cin}/P_{Sat}$ and $P_{RFin}/P_{RF0}$. It is possible to show that Equation 21.5 is satisfied for any parameter value if

$$P_{sat} \leq \frac{8\left(3-2\sqrt{2}\right)}{\rho R^2}. \tag{21.6}$$

It is easy to see that $P_{sat} < 43$ mW for $R = 0.8$ A/W and $\rho = 50\ \Omega$. Another useful expression that can be derived from Equation 21.5 under the assumption that the modulator operates far from its saturation is

$$\frac{8}{\rho R^2 P_{RF0}} > \frac{P_{RFout}}{P_{RFin}} \equiv G_{ph}, \tag{21.7}$$

which imposes a restriction on the gain $G_{ph}$ of the photonic link. The photonic gain fundamentally cannot exceed unity if $P_{RF0} > 250$ mW, for $R = 0.8$ A/W and $\rho = 50\ \Omega$.

To understand the meaning of characteristic RF power of a modulator, $P_{RF0}$, it is useful to consider a phase modulator. The power in the lowest order modulation sidebands of the modulator is given by

$$\frac{P_{S\pm out}}{P_{Cin}} = \left(\frac{\pi}{2}\frac{V}{V_\pi}\right)^2, \tag{21.8}$$

where $V$ is the voltage of the RF signal fed into the modulator and $V_\pi$ is the half-wave voltage of the modulator. Keeping that in mind,

$$P_{RFin} = \frac{V^2}{2\rho_M}, \tag{21.9}$$

where $\rho_M$ is the resistance of the modulator. Comparing Equation 21.8 with Equation 21.2 and assuming that $P_{S+out} + P_{S-out} = P_{Sout}$, we define for the phase modulator:

$$P_{RF0} = \frac{V_\pi^2}{\rho_M}\left(\frac{2}{\pi}\right)^2. \tag{21.10}$$

It is easy to find that $P_{RF0} \approx 200$ mW for a comparably large half-wave voltage $V_\pi = 5$ V and $\rho_M = 50\ \Omega$. This means that if phase modulation is transferred to amplitude modulation without loss, the quite ordinary modulator-based photonic link can have a gain larger than unity. However, the lower $V_\pi$ is, the higher the gain can be.

One way to increase the value of the photonic gain is by using a transimpedance amplifier that allows increasing the effective value of ρ at the expense of a reduction in the reception bandwidth of the photodiode (larger ρ results in a larger photonic gain, in accordance with Equation 21.4). This is an appropriate approach for a generic OEO operating at low optical power. The photodiode in the OEO circuit should be sensitive to only a single RF tone, so bandwidth reduction is not a problem, but rather an advantage.

However, this approach is useless if one employs a high power laser in the OEO circuit because the use of a transimpedance amplifier reduces the saturation power of the photodiode (Equation 21.6) and the maximal achievable photonic gain (Equation 21.7).

It is important to highlight the generality of Equations 21.6 and 21.7. The boundary values they describe do not depend on the optical power falling on a photodiode. Hence, $P_{sat}$ and $G_{ph}$ stay unchanged if one introduces optical gain into the optical part of the OEO loop. Only the properties of the modulator and the photodiode determine the maximal value of the photonic gain, $G_{ph}$. It is worth noting that the experimental achievement of a photonic gain exceeding unity ($G_{ph} > 1$) has been reported in several earlier works, e.g., [3–5].

There is a way to increase $P_{Sat}$ using a DC-biased photodiode. Our theoretical model is not valid for biased photodiodes because the energy conservation law should include the energy that the diode consumes from the DC current. Consideration of biased photodiodes does not change the results of our study significantly because the saturation power is small enough in fast, biased photodiodes, and is generally less than 10 mW. Moreover, a biased photodiode is equivalent to an unbiased photodiode and an RF amplifier. Therefore, such a photodiode introduces more noise into the RF signal.

### 21.2.2 RF Power Generated by an OEO

To find the RF power circulating in the oscillator we have to require unity gain in the closed OEO loop, i.e.,

$$G_{ph}(P_{RFin})G_{RF}\left(1+\frac{G_{ph}\left(P_{RFin}\right)P_{RFin}}{P_{GRF}}\right)^{-1}=1, \tag{21.11}$$

where $G_{RF}$ and $P_{GRF}$ are the net gain and saturation parameters of the RF loop. The RF field circulating inside the optoelectronic loop is determined by saturation of either the optical or electronic parts of the loop. Practice shows that making the RF loop saturate faster than the optical loop results in better OEO performance. In this case, $G_{ph}(P_{RFin}) \approx$ constant and the power leaving the fast photodiode is given by

$$P_{RFout} = P_{GRF}(G_{ph}G_{RF} - 1). \tag{21.12}$$

As a generic rule, to have a reasonable amount of power in the loop, one needs $G_{ph}G_{RF}$ to be as large as 2 dB–3 dB. Usually such a trivial formula results in a reasonably good description of the experimental results. It is also worth noting that $G_{ph}$ and $G_{RF}$ are frequency dependent, so Equation 21.11 should also be used to determine the frequency of the oscillation. The equation is insufficient, though, to predict the frequency. It should be used together with Equation 21.13.

Modulators are important elements for setting the OEO oscillation power as they determine the value as well as the RF saturation level of the photonic gain, $G_{ph}$. Low $V_\pi$ modulators have low saturation levels, which eventually results in small RF power generated by the OEO. Since RF power from the photodiode, $P_{RFout}$, is usually less than the RF saturation power of the optical circuit, the output power of the oscillator, being a fraction of $P_{RFout}$, can be impractically low. For this reason, a significant post-amplification is required in the OEO.

OEOs without RF amplification in the loop have been demonstrated. The basic drive for the development of such systems results from the assumption that RF amplifiers introduce $1/f$ noise that significantly degrades the phase noise of an oscillator at low (Fourier) frequencies. The power-efficient OEO of this kind requires a very efficient modulator as well as a high power photodiode [6]. Initially, amplifierless OEOs operating at 1 GHz [7] and 1.25 GHz [6] were demonstrated. Very recently, a 10-GHz OEO was also produced [8]. It was shown that the amplifierless OEO has better phase noise close to the carrier, as compared to the conventional OEO. However, the overall amplifierless oscillator phase noise was significantly worse compared to the phase noise of conventional OEOs operating at a similar frequency [9]. In our opinion, there is no solid proof that an amplifierless OEO is better than a conventional OEO and this subject requires a more comprehensive study.

### 21.2.3 Oscillation Frequency

A convolution of the group delay of the optical circuit, $\tau_f$, and the RF phase delay of the electronic circuit determines the possible oscillation frequencies of an OEO:

$$\omega_{RF}(\tau_f + \tau_e) = 2\pi m, \tag{21.13}$$

where $m$ is an integer number indicating an RF supermode of an OEO (half integer mode numbers are allowed, $m \to m + 1/2$, depending on the bias of the RF loop [10]) and $\omega_{RF}$ is the oscillation frequency. The delays $\tau_f$ and $\tau_e$ are connected to the eigenfrequencies of the optical, as well as RF, circuits. $\omega_f$ is an eigenfrequency of the optical circuit (found from $\omega_f\tau_f = 2\pi k$, where $k$ is an integer) and $\omega_e$ is an eigenfrequency of the electronic circuit (found from $\omega_e\tau_e = 2\pi l$, where $l$ is integer). Equation 21.13 describes the usual frequency pulling effect. The basic achievement of RF photonics is the realization of oscillators with very large $\tau_f$ (i.e., $\tau_f >> \tau_e$), so that the optical loop determines the oscillation frequency.

An OEO can generate multiple RF frequencies, in accordance with Equation 21.13. However, a joint solution of Equations 21.11 and 21.13 generally specifies a unique frequency at which the OEO oscillates [10]. Other RF modes, determined by Equation 21.13, result in the appearance of multiple spurious noise peaks (supermodes) in the phase noise spectrum of the oscillator. In an OEO based on a long fiber loop, the supermode spectrum is very dense and any change in the temperature of the fiber results in a change in optical length and concomitant jumps in the oscillator frequency from one supermode to another. The only means for eliminating supermodes is by using an RF filter that is narrow enough to pass only one mode. This, of course, is impractical with microwave filters in oscillators containing long lengths of fiber, which require sub-MHz linewidths centered at tens of gigahertz.

### 21.2.4 Phase Noise and Frequency Stability

The single-sideband, low-frequency (from the carrier up to the value of the start of the supermode free spectral range) phase noise of an OEO can be found from the simplified generic equation [10,11]:

$$L(f) = \left[\frac{2q\rho RP + Fk_BT}{2P_{RFout}} + \frac{\delta F_{f^{-1}}}{f}\right]\left[1 + \left(\frac{\delta v_{S21}}{2f}\right)^2\right], \tag{21.14}$$

where $P$ is the optical DC power absorbed by the photodiode, $P_{RFout}$ is the demodulated RF power at the photodiode, $q$ is the charge of an electron, $F$ is the integral noise figure of the RF amplifiers used in the OEO loop, $k_B$ is the Boltzmann constant, $T$ is the physical temperature of the OEO, $\delta v_{S21}$ is the full-width at the half-maximum (FWHM) of $S_{21}$ of the entire photonic system, $\delta F_{f^{-1}}$ is the dimensionless constant determined by the flicker noise of the RF amplifiers in the circuit at 1 Hz, and $f$ is the linear frequency.

Several more generic expressions for the phase noise of fiber-based OEOs, valid in a broader frequency range compared to a single RF FSR (free spectral range), were presented in [12]. For instance, for a single optical fiber loop-based OEO:

$$L(f) = \left[\frac{2q\rho RP + Fk_BT}{2P_{RFout}} + \frac{\delta F_{f^{-1}}}{f}\right] \times \left|1 - \frac{\exp(2\pi i f\tau)}{1 - if/\delta v_{RF}}\right|^{-2}, \tag{21.15}$$

where $\delta\nu_{RF}$ is the HWHM of the RF bandpass filter and $\tau_f = n_f L_f/c$ is the phase delay of the optical delay line.

At Fourier frequencies that fall within the bandwidth of the filter in the loop, the noise floor of an OEO significantly exceeds the noise of low frequency RF oscillators, as optical shot noise is generally much larger than thermal noise (Equation 21.14). This happens because the signal-to-noise ratio of an optoelectronic loop is determined by the number of optical photons and not by the (much larger) number of RF photons circulating in the loop. The major advantage of the OEO over its electronic competitors is twofold: (1) it can operate at high frequencies where RF amplifiers have large noise figures and good electronic oscillators are unavailable, and (2) it can have low phase noise in the vicinity of the carrier. It is the large RF frequency dispersion of the optical loop that allows for the realization of OEOs with low noise close to the carrier, since the dispersion of the RF loop is generally low. The photonic link delay time $\tau_f$ can be made on the order of milliseconds or larger, depending on the length of the optical fiber loop and/or quality factor of the optical resonator used in the loop.

## 21.3 Characteristics of OEOs and COEOs Based on Various Modulator Types

There have been multiple OEO schemes based on Mach–Zehnder lithium niobate electro-optic modulators (MZI EOMs), electroabsorption modulators (EAMs), and whispering gallery mode modulators. In what follows, we briefly summarize our best results obtained with MZI EOMs and EAMs and devote the next section to a detailed description of OEO's based on whispering gallery mode electro-optical modulators (WGM EOMs).

### 21.3.1 The OEO Based on an MZI EOM

A high-performance OEO was demonstrated based on the conventional OEO architecture (Figure 21.1) [9,13]. Light from a high power YAG laser (200 mW of 1319 nm light at the output) was sent into a lithium

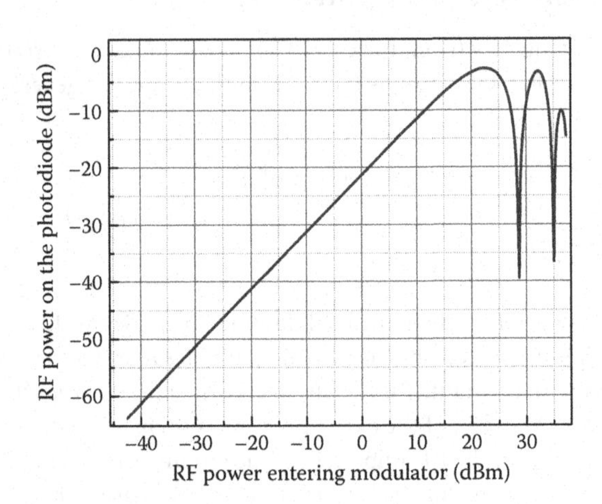


**FIGURE 21.3** This plot illustrates the 10-GHz RF output power at the end of a microwave photonic optical link (output of the photodiode), as a function of the RF modulation power fed into the MZI modulator. This behavior is linear at low modulation RF power levels, however, it reaches saturation at about 24 dBm, and then drops dramatically at certain power levels. It is a direct result of the interferences between the two RF arms of a Mach–Zehnder modulator (the modulator includes two interfering phase modulators with π-shifted RF output). The resonant decrease in the modulation efficiency corresponds to suppression of the modulation optical sidebands at 10 GHz, so that 20-GHz modulation sidebands prevail.

niobate Mach–Zehnder EOM (Figure 21.3). The modulator's half wave voltage and optical insertion loss (IL) were $V_\pi = 5$ V and 4 dB, respectively. Light from the modulator was subsequently sent through a 16-km-long SMF-28 fiber (11 dB total loss), and then to a photodiode (responsivity $R = 0.8$ A/W, $\rho \approx 50\ \Omega$ resistance, and $\delta F_{f^{-1}} = -122$ dB flicker). The total power reaching the photodiode was 10 dBm. The fiber was placed in a thermally stabilized box to reduce thermal drift (the box had 2- to 4-day thermal time constant) and the thermal frequency drifts of the fiber delay line were smaller than 1 Hz/min. The RF output of the photodiode (–4 dBm) was amplified (the amplifier had $G_{RF} = 54$ dB gain, $\delta F_{f^{-1}} = -128$ dB flicker, and $F = 6$ dB noise figure), then sent through a bandpass filter (2 MHz FWHM and 12 dB insertion loss) and a voltage controlled phase shifter (~10 dB insertion loss) centered at 10 GHz and finally to the modulator, completing the feedback loop.

The phase noise performance of the OEO is shown in Figure 21.4. To achieve this result, the laser frequency was modulated at 18 kHz to reduce noise effects resulting from an interferometric interaction of the scattered light (where the light comes from Rayleigh scattering). This kind of noise would otherwise degrade the performance at Fourier frequencies below a few kilohertz [13]. Flicker noise of the photodiode, being higher than the flicker of the high performance amplifiers used in this unit, is the basic source of the noise at low frequencies. While the photodiode flicker limits the phase noise of the OEO below a few kilohertz, it can be reduced by using multiple photodiodes, or a photodiode array, as discussed in [13]. The optical power handling capability of the photodiode, which was limited to about 10 mW, is another factor limiting the phase noise of the device. This limiting factor is directly related to the white noise level obtained and will be improved when a higher power photodiode is utilized.

The demonstrated oscillator proves the utility of OEOs for generation of RF signals with ultrahigh spectral purity. We note that the single optical loop OEO exhibits supermode noise at frequencies that correspond to multiples of the $\approx L_f n_f / c$ mode separation frequency, where $L_f$ is the length of the fiber, $n_f$ the index of refraction, and $c$ the speed of light. Various approaches, such as the use of multiple loops [12,14] can be applied to reduce these noise peaks. We did not attempt any other approaches with this unit, as the goal was to demonstrate the lowest achievable phase noise with the OEO architecture. The phase noise of a 10-GHz dual loop OEO based on a lithium niobate MZI EOM [12] is illustrated in Figure 21.5.

Fiber-based OEOs that employ MZI EOMs are rather large (2 ft$^3$) and power hungry (power consumption ~110 W). The size can be reduced by about 30% with tighter packaging. Fiber-based COEOs generally have better characteristics than OEOs.

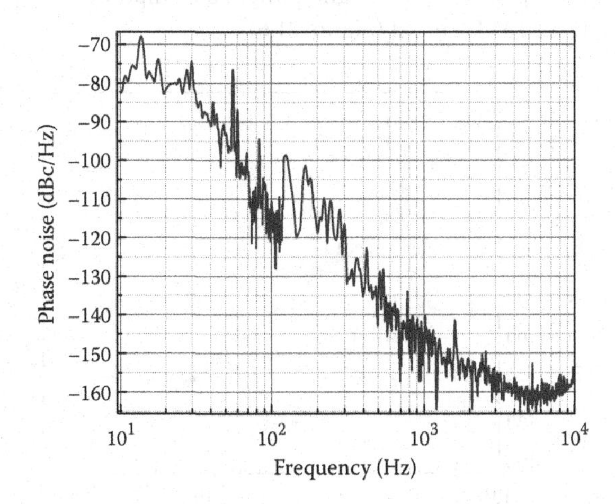


**FIGURE 21.4** Phase noise of the fiber-based OEO with 16 km optical fiber delay line and lithium niobate EOM. (Reproduced with permission from D. Eliyahu, D. Seidel, and L. Maleki, in *2008 IEEE Int. Freq. Cont. Symp.*, Honolulu, HI, pp. 811–814, © 2008 IEEE.)

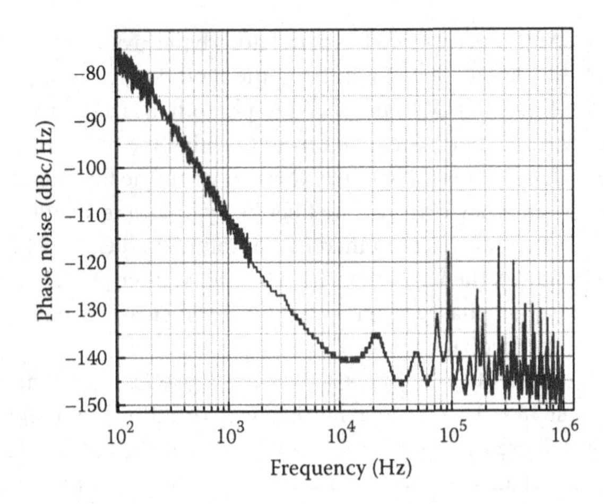


**FIGURE 21.5** Phase noise of the dual loop OEO. (Reproduced with permission from D. Eliyahu and L. Maleki, *Proc. 2003 IEEE Int. Freq. Cont. Symp. PDA Exhibition*, vol. 1, pp. 405–410, © 2003 IEEE.)

## 21.3.2 An OEO Based on an EAM

Since the conventional fiber-based OEO is usually large in size and many applications require small oscillators, we examined various ways of reducing the size of the OEO. An X-band mini-OEO was also demonstrated based on the conventional OEO architecture (Figure 21.1), with relatively short fiber and small components, wherever possible. The fiber laser and the relatively long Mach–Zehnder modulator were replaced by a miniature integrated DFB laser with an electoabsorbtion modulator (EAM), having drive voltage of 2.6 V (10-dB extinction ratio). Small, low-power amplifiers were used instead of the bulky high performance amplifiers in the large OEOs. Light from the modulator (2 mW) was coupled to a 2.2-km single-mode fiber (microwave Q of about 350,000) and a fast photodetector, before entering the RF section. The fiber temperature was actively controlled inside a thermally isolated box in order to reduce thermal drift [15,16].

The RF output of the photodiode (–15 dBm) was amplified (the amplifier is characterized with $P_{GRF}$ = 20 dBm saturation power with respect to the output signal, $\delta F_{f^{-1}} = -103$ dB flicker at 1 Hz offset, and $F = 4$ dB noise figure), and sent through a bandpass filter (4 MHz FWHM and 10 dB insertion loss). This mini-OEO had a volume of only 4100 cm$^3$, and power consumption of <10 W. The phase noise was approximately –132 dBc/Hz at 10 kHz offset (Figure 21.6).

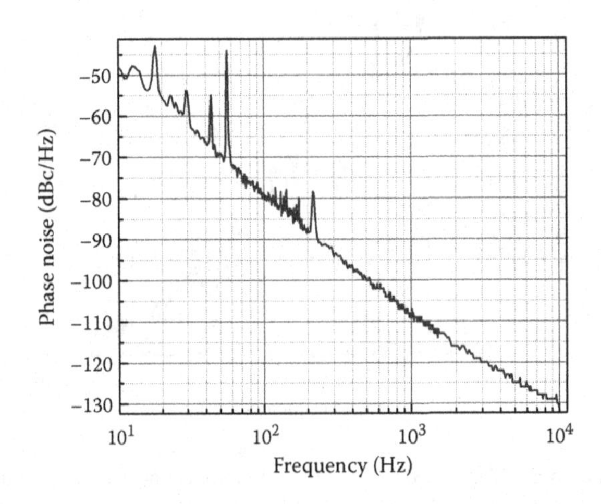


**FIGURE 21.6** Phase noise of the fiber-based OEO with 2.2-km optical fiber delay line and EAM.

### 21.3.3 A Tunable Fiber-Based OEO

A high-performance tunable OEO was constructed and demonstrated using a multiloop OEO configuration with a Mach–Zehnder lithium niobate modulator and a tunable RF YIG filter [17]. The tunable YIG filter was used to select the desired mode (frequency) of operation among the many possible modes of the fiber-based OEO. The demonstrated device had tunability from 6 GHz to 12 GHz (limited by the bandwidth of amplifiers) in steps of 3 MHz and exhibited phase noise of about –128 dBc/Hz at 10 kHz offset (illustrated in Figure 21.7), limited by the amplifier phase noise and fiber length, as well as by the YIG filter noise. The measured phase noise level was about 30 dB better than the phase noise of conventional tunable YIG oscillators. The noise characteristics can be further improved if one uses a less noisy driver for the YIG filter. RF tunability based on dispersive feedback was also achieved in a dual-loop OEO [18].

### 21.3.4 Coupled Optoelectronic Oscillator

A small, high-frequency, and stable RF source is generally required for realization of an actively mode-locked laser with high pulse repetition rate. This requirement can be eliminated if the stable RF signal is generated in the same system. A COEO is an architecture for realization of such a system [19,20].

A generic COEO consists of two photonic loops generating light as well as RF radiation (Figure 21.2). These loops are connected by means of an EOM and a photodiode. The laser light energy is converted directly to spectrally pure RF signals, using an electro-optic feedback loop containing a high-$Q$ optical element, at a frequency limited only by the available optical modulation and detection elements. This frequency is the repetition frequency of the optical pulses generated in the system.

The optical loop is similar to a conventional ring fiber laser with either an erbium or a semiconductor optical amplifier. If the RF photonic loop of the COEO is open, the ring laser generates several independent optical modes. The number of modes is determined by the loop length and the linewidth, as well as the gain of the optical amplifier. If the RF photonic loop is closed and sufficient microwave amplification is available to ensure RF oscillation in the system, as in an OEO, the optical modes become phase-locked.

The laser radiation propagates through a modulator and an optical energy storage element (delay line) before it is converted, with a photodiode, to electrical energy. The electrical signal at the output of the photodiode is amplified and filtered before it is fed back to the modulator, thereby completing a feedback loop with gain, which generates sustained oscillation. Since the noise performance of an oscillator is

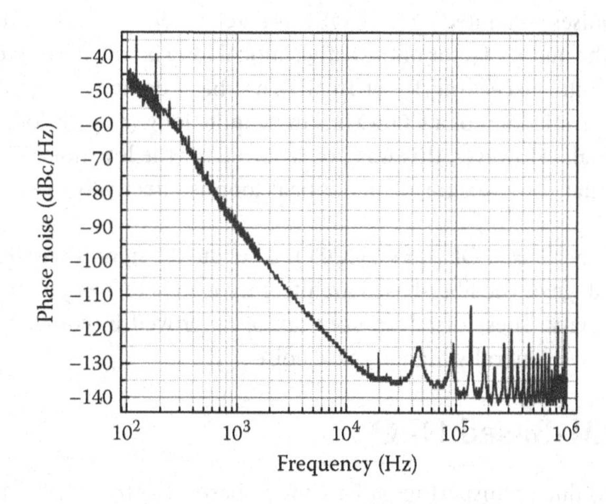


**FIGURE 21.7** Phase noise of the tunable fiber-based OEO. (Reproduced with permission from D. Eliyahu and L. Maleki, *Proc. IEEE MTT-S Int. Microwave Symp.*, vol. 3, pp. 2185–2187, © 2003 IEEE.)

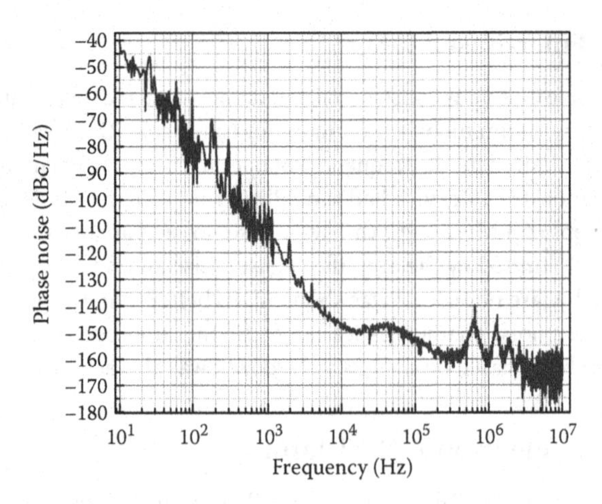


**FIGURE 21.8** Phase noise of the fiber-based COEO. (Reproduced with permission from D. Eliyahu, D. Seidel, and L. Maleki, in *2008 IEEE Int. Freq. Cont. Symp.*, Honolulu, HI, pp. 811–814, © 2008 IEEE.)

determined by the energy storage time, or quality factor Q, the use of optical storage elements provide extremely high Qs, and thus spectrally pure signals.

The EOM is one of the main sources of power consumption in the COEO because of the large power required to drive conventional modulators. Broadband Mach–Zender modulators as well as EAMs used in COEOs typically require one to a few watts of RF power to achieve significant modulation. This means that either the photocurrent in the COEO system should be amplified significantly or the laser loop of the COEO should operate much above the laser threshold to produce enough optical radiation for the drive power. If the RF power sent to the modulator is small, the information about the RF signal will not be transduced to light through the EOM.

Compact COEOs offer extremely low-phase noise and vibration, as well as acceleration sensitivity, for signal source modules required in high-frequency, high-performance applications. Multiple X-band oscillators possessing similar performance and frequencies ranging from 10 GHz to 12 GHz were built. The compact COEO offers typical phase noise performance levels of better than –140 dBc/Hz at 10 kHz offset from the carrier (see Figure 21.8).

The jitter in optical pulses generated by the COEO is determined by spectral characteristics of the RF signal, and vice versa. The quality factor of the active optical loop is transferred to the RF quality factor, and an effective RF Q-factor exceeding $10^6$ at 10 GHz has been demonstrated [21,22]. It was expected in [19] that the optoelectronic loop of a COEO has to be quite long (a few kilometers) for generation of low-phase-noise RF signals. However, this was not required for the low-noise COEO reported recently [21–23]. The reason is that the Q-factor of the optical loop is effectively enhanced due to the mode-locking process [24].

We have shown that fiber based OEOs and COEOs generate RF signals with extremely high spectral purity. A clear disadvantage of the fiber-based architecture is its size, weight, power consumption, and the low vibration sensitivity. These disadvantages can be improved in compact WGM OEOs, though these resonant devices have generally higher phase noise.

## 21.4 WGM EOM-Based OEO

In this section, we introduce whispering gallery mode-based electro-optical modulators and OEOs based on these modulators. WGM EOMs are of particular interest since they allow production of chip scale, fiberless, and continuously tunable OEOs. To create a miniature fiberless OEO one ultimately needs a modulator characterized by (1) small size, (2) monolithic architecture, (3) ultimately, single sideband (SSB) modulation of light, (4) narrow spectral width of $S_{21}$, and (5) high efficiency (the maximal magnitude of $S_{21}$ should be comparable to the saturation power of the modulator).

The efficiency of a modulator can be significantly enhanced if the electro-optical material is placed in or formed into a resonant structure. The resonant interaction increases the interaction length of light and RF radiation and increases the interaction efficiency by confining the optical and/or RF photons into small volumes. Excellent properties of experimentally demonstrated resonant WGM modulators represent a compelling proof of this statement.

It was pointed out a decade ago that WGM resonators can be made out of electro-optical crystals and that such resonators can be used for efficient modulation of light. The resonant interaction of several optical WGMs and an RF signal can be achieved by engineering the shape of the RF resonator (electrode) as well as the WGM resonator [25–27]. A WGM-based EOM and a photonic link (initially called a microwave receiver) were suggested and realized shortly after the discovery [28–35]. Quadratic [36–38] and coherent [39,40] WGM-based receivers were introduced and demonstrated shortly afterwards. Very recently, an approach for the realization of WGM-based single-sideband (SSB) modulators, and tunable OEOs and receivers based on these modulators, was found [41–46].

The first experimental implementation of the WGM-based OEO was discussed in [47]. In what follows, we report on the results of our very recent experiments with the system (Figure 21.9) [48]. The OEO is based on a resonator made out of Z-cut stoichiometric lithium tantalate (SLT). The resonator has a radius of 1.27 mm and a thickness of 100 μm; the extremity of the resonator is shaped such that WGMs are mode-matched with the prism coupler.

The resonator is interrogated with 1550 nm light, emitted from a DFB laser that is self-injection locked to a selected WGM. The measured indices of refraction of the material are $n_0 = 2.16823$ and $n_e = 2.16273$ at this wavelength. The unloaded quality factor of a WGM is $Q = Q_{e0} = Q_{o0} = 10^8$ (2 MHz FWHM), the loaded $Q$-factor used in the oscillator is $Q_e = 2 \times 10^7$ (10 MHz FWHM).

The laser emits 7 mW of light, the optical power entering the resonator is $P_{in} = 5$ mW, and the optical power at the fast photodiode is $P = 2$ mW. The laser is locked to the slope of the optical resonance to achieve amplitude modulation. However, sometimes amplitude modulation is achieved even for resonant tuning of the laser due to the interference effect of the light entering the resonator and the light reflected from the coupling point.

The RF resonator has $Q_M = 40$ quality factor and $\nu_{RF} \approx 35$ GHz carrier frequency. The WGM EOM saturates when interrogated with RF signal of $P_M = 3$ mW power. The photonic gain of the modulator itself (here it is the ratio of the demodulated RF power in the absence of RF amplifier and the input RF power) is $G = -15$ dB.

The fast photodiode has responsivity $R = 0.8$ A/W, and the RF line has resistance $\rho = 50\ \Omega$. The RF power at the photodiode output is 10 μW. The RF circuit includes two RF amplifiers and two RF attenuators/couplers. The first amplifier in the array has gain $G_1 = 21$ dB, output saturation power $P_{sat1} = 15$ dBm, and noise figure $F_1 = 2.8$ dB. An attenuator with $A_1 = 2$ dB attenuation is placed after the first amplifier.

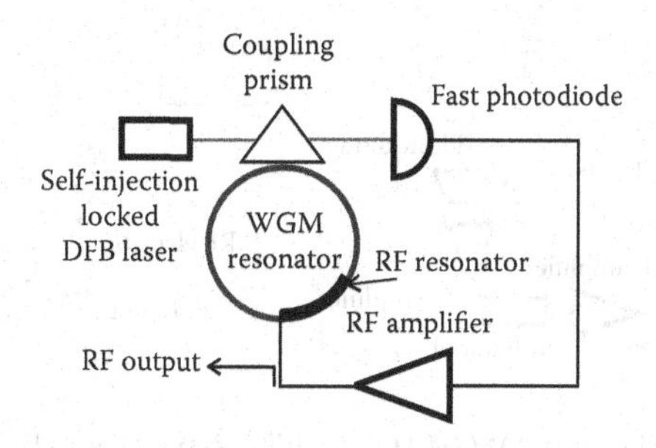


**FIGURE 21.9** Schematic of the WGM-based resonant OEO.

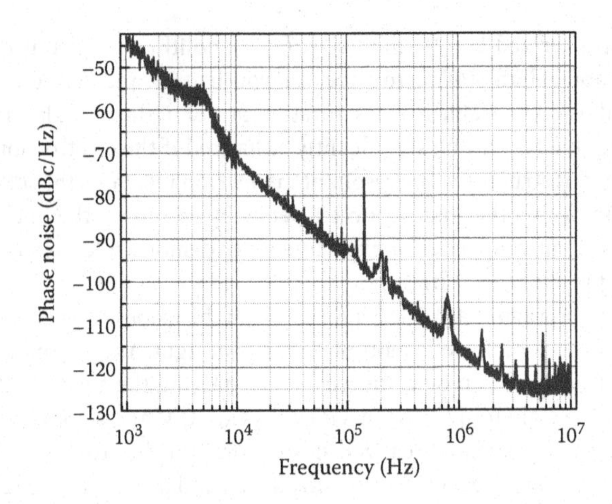


**FIGURE 21.10** Phase noise of the oscillator. The oscillation frequency is ≈35 GHz. To measure the phase noise, the signal was divided by four. The noise floor of the phase noise is determined by the technical noise due to this division. (Reproduced with permission from A. A. Savchenkov et al., *Proc. IEEE Int. Freq. Cont. Symp.*, vol. 1, pp. 554–557, © 2010 IEEE.)

The second amplifier in the array has gain $G_2 = 21$ dB, output saturation power $P_{sat2} = 10$ dBm, and noise figure $F_2 = 2.8$. An attenuator and coupler with total $A_2 = 15$ dB attenuation is placed after the second amplifier. The integral gain of the amplifier chain is $G = 25$ dB and the noise figure found from the expression

$$F = F_1 + \frac{F_2 - 1}{G_1 A_1}, \tag{21.16}$$

is $F \cong 2.8$.

The phase noise of the OEO is shown in Figure 21.10. The theoretical values (Equation 21.14) are significantly better for the parameters indicated above: $L(f = 100 \text{ kHz}) \cong -109$ dBc/Hz and the noise floor is –149 dBc/Hz. The sources of noise should be analyzed more rigorously to find the reason for the observed discrepancy with the expected noise floor.

## 21.4.1 WGM-Based Tunable OEO

Tunable OEOs involve SSB WGM EOMs [42] in their circuits (Figure 21.11). Tunable OEOs [45] discussed in this section are based on a stoichiometric lithium tantalate WGM resonator with diameter

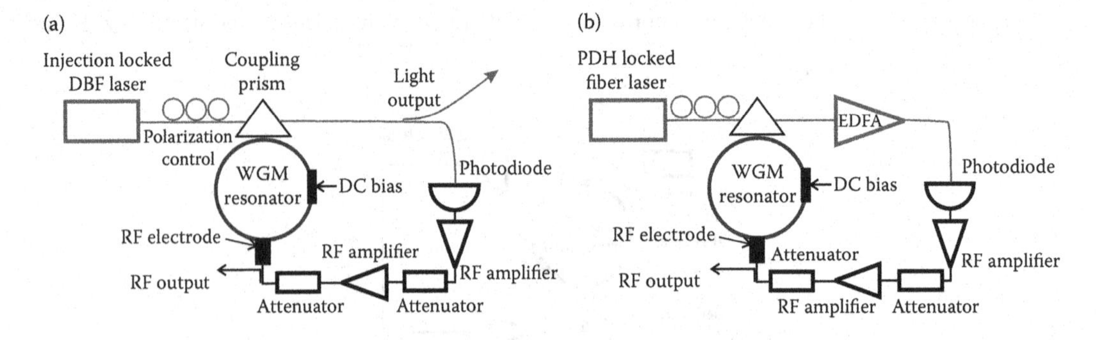


**FIGURE 21.11** WGM TOEO schemes: (a) WGM TOEO with self-injection locked DFB laser and (b) WGM TOEO with PDH-locked fiber laser.

of 1 mm and thickness of 50 μm. The resonator is pumped with 1550 nm light emitted by a laser that is locked to a selected WGM. The measured indices of refraction of the material are $n_0$ = 2.16864 and $n_e$ = 2.16516 at this wavelength, at room temperature. The unloaded quality factor of the WGMs is $Q = Q_{e0} = Q_{o0} = 5.7 \times 10^8$ (350 kHz unloaded bandwidth), the loaded $Q$-factor of the mode, pumped optically, is $Q_e = 6.7 \times 10^6$ (30 MHz bandwidth). The quality factor of the sideband mode is $Q_0 = 1.1 \times 10^8$ (1.8 MHz bandwidth). The free spectral range (FSR) of the resonator is approximately the same for both mode families, $\nu_{FSR}$ = 46 GHz.

We consider two distinct schemes for the tunable oscillator (Figure 21.11). Initially, we built a table-top model of an oscillator containing a fiber laser and an erbium-doped fiber amplifier (EDFA). The fiber laser was locked to the selected WGM via the well-known Pound–Drever–Hall (PDH) technique (Figure 21.11b). Later, we made several packaged oscillators utilizing semiconductor DFB lasers (Figure 21.11a). The packaged oscillators have better performance compared to the table-top unit.

## 21.5 Tunable OEO with a Fiber Laser

In this setup, we use a 30-mW fiber laser to pump the WGM resonator. The power at the resonator input is about 24 mW. The coupling is inefficient so the power at the EDFA input port is 72 μW (EDFA has 12 dB amplification), and the power on the photodiode is 1.4 mW.

The RF power at the photodiode output is 30 μW and the total RF amplification and attenuation is 26 dB (7 dB noise figure). The RF white noise level (Figure 21.12) is about −133 dBc/Hz. Below the Leeson frequency (2 MHz), the noise has $f^{-2}$ behavior (down to about 300 Hz). The noise floor is apparently determined by the amplified spontaneous emission of the EDFA, because the theoretical limit of the noise floor is −152 dBc/Hz.

## 21.6 Packaged Tunable OEO

We have produced a brassboard model of the packaged tunable OEO. The 4 × 5.5 × 1.2-cm = 26.4-cm$^3$ package contains all the optical parts of the oscillator. A semiconductor DFB laser is self-injection locked to a selected WGM. The RF feed circuit has a nearly flat frequency response in the 1-GHz to

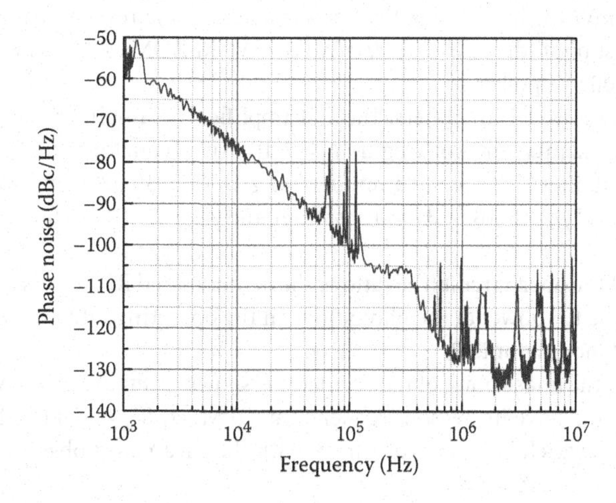


**FIGURE 21.12** Phase noise of the WGM-based tunable OEO with EDFA in the optical loop. The oscillator frequency is selected to be 5.28 GHz. Note that the peak at 1.54-MHz offset and its harmonics are related to the length (130 m) of the fiber delay line used in our phase noise measurement system (though, there is an additional peak from the oscillator noise just below 5 MHz offset). (Reproduced with permission from A. A. Savchenkov et al., *Proc. IEEE Int. Freq. Cont. Symp.*, vol. 1, pp. 554–557, © 2010 IEEE.)

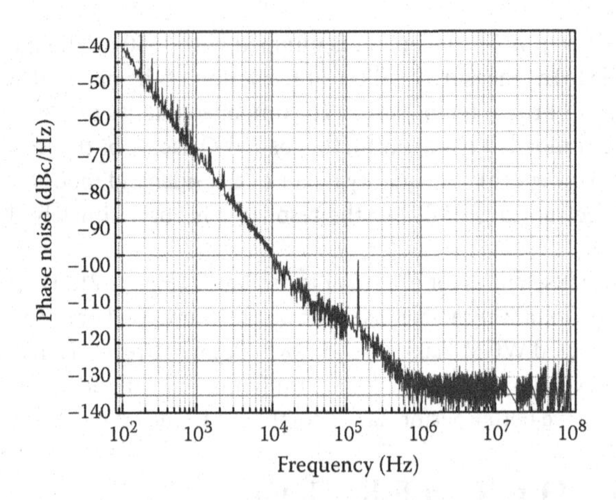


**FIGURE 21.13** Phase noise of the WGM-based tunable OEO. Oscillation frequency is 9.8 GHz. The peak at 1.5 × $10^5$ Hz results from external noise coming from the laboratory environment. (Reproduced with permission from A. A. Savchenkov et al., *Proc. IEEE Int. Freq. Cont. Symp.*, vol. 1, pp. 554–557, © 2010 IEEE.)

25-GHz frequency range (RF reflection $S_{11}$ is smaller than 10 dB in this frequency range). The RF field enhancement within the WGM localization is ultimately $|E_{SLTWGM}|/|E_{stripline}| \approx 5.6$, according to our HFSS numerical model (the RF line operates as a transimpedance amplifier). The RF electrode has a 250-μm length and the geometrical overlap parameter of the optical and RF fields is $\zeta = 0.08$. The total length of the RF cables is 27 cm, so the Leeson frequency is determined by the spectral width of the $S_{21}$ function of the SSB EOM. The fast photodiode has responsivity $R = 0.6$ A/W and the RF line has resistance $\rho = 50\ \Omega$. The tunability of the oscillation frequency is achieved by applying a DC voltage to the electrodes attached to the top and bottom surfaces of the resonator, and the tuning slope is 69 MHz/V. The origin of the tunability is the same as in the resonant modulator used in the OEO [42].

The laser emits 3.2 mW of light. The optical power entering the resonator is $P_{in} = 1.9$ mW and the optical power at the fast photodiode is $P = 0.9$ mW. The WGM EOM saturates when interrogated with an RF signal of $P_{sat} \approx 100$ μW power.

The RF circuit includes two RF amplifiers. The first amplifier (preamplifier) has gain $G_1 = 30$ dB, input saturation power $P_{sat1} = 0.2$ mW, and noise figure $F_1 = 7$ dB. An attenuator with $A_1 = 24$ dB attenuation is placed after the first amplifier. The second amplifier has gain $G_2 = 30$ dB, input saturation power $P_{sat2} = 0.2$ mW, and noise figure $F_2 = 7.5$ dB. A 3-dB RF splitter and $A_2 = 27$-dB attenuation are placed after the second amplifier.

The photonic gain of the open circuit (the ratio of the demodulated RF power and the input RF power) is kept at $G \cong -5$ dB. The RF power at the photodiode in the operating OEO is $P_{RF} \cong 10$ μW, so that the input power of the SSB modulator is $P_{RFin} \cong 31.6$ μW.

The power density of the phase noise of the oscillator is shown in Figure 21.13. According to Equation 21.14, the oscillator should have its Leeson frequency at 0.9 MHz, $L(f = 100\text{ kHz}) \cong -122$ dBc/Hz, and −148 dBc/Hz noise floor, which has good agreement with the parameters observed experimentally.

## 21.7 Summary and Conclusion

Besides its direct application as an RF oscillator, an OEO can be used as a self-oscillating atomic clock [49,50], a magnetometer [2,51–53], or a gyroscope [54]. In what follows, we briefly discuss OEO applications in clocks and magnetometers.

We reported on a new scheme for a magnetometer with potentially high sensitivity as well as accuracy [2,51]. We proposed to combine OEO technology with an atomic vapor cell filter [49] and to stabilize the OEO using the effect of coherent population trapping (CPT) [55]. The CPT resonances are applicable for construction of all-optical miniature atomic clocks [56] and magnetometers [57].

It is possible to produce a stable RF clock or stable magnetic field-tunable RF signal using the same OEO, simply by locking the oscillator frequency to the magneto-insensitive or magneto-sensitive atomic transitions. In this manner, both goals of sensitivity and accuracy could be achieved.

In optical, atomic magnetometers, light is used to detect shifts of magneto-sensitive atomic energy levels, by tracking the frequencies of corresponding magneto-optical resonances. As a general rule, better measurement sensitivity is achieved for narrower resonances. Magneto-optical resonances of Hz widths allow for subfemtotesla magnetic field measurement sensitivities [58–60].

Optical atomic magnetometers can operate in open- as well as closed-loop modes [61]. Open-loop magnetometers respond to applied magnetic fields through changes of the probe light parameters such as phase shift, polarization rotation, or transmission. For example, passive magnetometers sometimes involve external oscillators as "fly wheels" [62] and use atoms to stabilize those oscillators. Self-oscillating, or closed-loop, optical magnetometers [63] are electronic self-excited oscillators with an atomic resonance filter in their feedback loop. They can also be considered as masers with electronic amplifiers. The output frequency of these devices is approximately proportional to the measured magnetic field.

The main drawback of these atomic devices is the unreliability of the electronic oscillator segment and the uncertainty of the shape of the resonance used for stabilization. The sensitivity of these devices has been demonstrated to be on the order of 10 pT [64], lower than in the passive devices. However, active atomic magnetometers have a larger dynamic range and superior accuracy compared to passive ones.

The OEO-based magnetometer is self-oscillating; hence, it differs from the passive CPT magnetometer [57]. On the other hand, the OEO-based magnetometer is different from other active magnetometers, where stability is achieved through the RF field stored in an RF resonator or an RF coil containing an atomic vapor cell. There is no need for an RF resonator in the OEO because the RF energy is carried as the sidebands of modulated light. This allows for minimizing the size and reducing the magnetometer power consumption without performance loss.

Use of the SSB EOM based on a WGM resonator would result in a significant improvement of properties of both active and passive optical clocks based on the effect of CPT [41]. The advantage arises because the SSB EOM generates an optical sideband polarized orthogonally with respect to the optical pump polarization and because locking of the carrier sideband optical pair to the atomic transition stabilizes the entire WGM spectrum.

The effect of CPT suffers from a small signal contrast resulting from optical pumping. Interrogation of atomic transitions with crossed linear polarizations has been proposed to overcome this limitation [65]. Implementation of this method is complicated because existing EOMs, EAMs, and modulatable lasers produce modulation sidebands having the same polarization as the carrier. The SSB EOM creates a modulation sideband with the polarization orthogonal to the polarization of the incoming optical pumping. Hence, the SSB EOM significantly simplifies the realization of high contrast CPT resonances and also is promising in optical clock/magnetometer applications.

To conclude, we have discussed our recent achievements in research, development, and fabrication of fiber-based as well as optical resonator-based OEOs and COEOs. The oscillators include lithium niobate EOMs, semiconductor EAMs, as well as WGM-based EOMs. We have demonstrated both fixed frequency and tunable OEOs having good RF phase noise characteristics. We have shown that COEOs are also attractive as low jitter, high repetition rate mode-locked lasers. Though WGM-based oscillators have worse noise characteristics compared to fiber-based devices, they demonstrate unique frequency tunability. The performance of WGM-based oscillators can be further improved and the packaged devices can have a small form factor. Such potential makes them attractive for a variety of applications

**TABLE 21.1** Summary of OEO Characteristics

| Description | Modulator Type | Fiber Length (km) | Phase Noise (dBc/Hz) | Volume (cm$^3$) | Power Consumption (W) |
|---|---|---|---|---|---|
| Single-loop OEO, 10 GHz | EAM | 2.2 | −131 at 10 kHz | 4,100 | <10 |
| Single-loop OEO, 10 GHz | LN MZI | 4.4 | −141 at 10 kHz | 6,500 | 10 |
| Single-loop OEO, 10 GHz | LN MZI | 16 | −164 at 10 kHz | 65,000 | 110 |
| Dual-loop OEO, 10 GHz | LN MZI | 8.8 | −140 at 10 kHz | 26,000 | 24 |
| Triple-loop tunable OEO, 6–12 GHz | LN MZI | 8.8 | −128 at 10 kHz | 26,000 | 28–32 |
| COEO, 10 GHz | EAM | <0.2 | −145 at 10 kHz | 345 | 30 |
| WGM OEO, 35 GHz | LT WGM | 0 | −93 at 100 kHz | 1 | 1.5 |
| WGM TOEO, 5–8 GHz | LT SSB WGM | <0.1 (EDFA) | −100 at 100 kHz | BB | BB |
| WGM TOEO, 8–11.8 GHz | LT SSB WGM | 0 | −100 at 10 kHz | 26 | 1.5 |
| WGM TOEO, 9.4 GHz | LT SSB WGM | 4.4 | −140 at 10 kHz | BB | BB |

*Note:* BB = breadboard demonstration.

where, in addition to phase noise, size, and power consumption are important. The parameters of variety of the OEOs created at OE waves are presented in Table 21.1.

# References


1. J. F. Holmes and B. J. Rask, "Optimum optical local-oscillator power levels for coherent detection with photodiodes," *Appl. Opt.*, vol. 34, no. 6, pp. 927–933, Feb. 1995.
2. A. B. Matsko, D. Strekalov, and L. Maleki, "Magnetometer based on the opto-electronic microwave oscillator," *Opt. Commun.*, vol. 247, no. 1–3, pp. 141–148, Mar. 2005.
3. C. H. Cox, III, G. E. Betts, and L. M. Johnson, "An analytic and experimental comparison of direct and external modulation in analog fiber-optic links," *IEEE Trans. Microwave Theory Techniques*, vol. 38, no. 5, pp. 501–509, May 1990.
4. C. H. Cox, III, E. I. Ackerman, G. E. Betts, and J. L. Prince, "Limits on the performance of RF-over-fiber links and their impact on device design," *IEEE Trans. Microwave Theory Techniques*, vol. 54, no. 2, pp. 906–920, Feb. 2006.
5. V. J. Urick, M. S. Rogge, F. Bucholtz, and K. J. Williams, "Wideband (0.045–6.25 GHz) 40 km analogue fibre-optic link with ultra-high (>40 dB) all-photonic gain," *Electron. Lett.*, vol. 42, no. 9, pp. 552–553, Apr. 2006.
6. C. W. Nelson, A. Hati, D. A. Howe, and W. Zhou, "Microwave optoelectronic oscillator with optical gain," in *IEEE Int. Freq. Cont. Symp.*, Geneva, 2007, pp. 1014–1019.
7. P. Devgan, V. Urick, J. McKinney, and K. Williams, "A low-jitter master–slave optoelectronic oscillator employing all-photonic gain," in *2007 IEEE International Topical Meeting on Microwave Photonics*, Victoria, BC, pp. 70–73.
8. V. J. Urick, M. E. Godinez, P. S. Devgan, J. D. McKinney, and F. Bucholtz, "Analysis of an analog fiber-optic link employing a low-biased Mach–Zehnder modulator followed by an erbium-doped fiber amplifier," *J. Lightwave Technol.*, vol. 27, no. 12, pp. 2013–2019, Jun. 2009.
9. D. Eliyahu, D. Seidel, and L. Maleki, "Phase noise of a high performance OEO and an ultra low noise floor cross-correlation microwave photonic homodyne system," in *2008 IEEE Int. Freq. Cont. Symp., Honolulu, HI*, pp. 811–814.
10. X. S. Yao and L. Maleki, "Optoelectronic microwave oscillator," *J. Opt. Soc. Am. B*, vol. 13, no. 8, pp. 1725–1735, Aug. 1996.
11. E. Rubiola, *Phase Noise and Frequency Stability in Oscillators*, Cambridge, England: Cambridge University Press, 2008.

12. D. Eliyahu and L. Maleki, "Low phase noise and spurious level in multiloop opto-electronic oscillators," *Proc. 2003 IEEE Int. Freq. Cont. Symp. PDA Exhibition,* vol. 1, pp. 405–410, May 2003.
13. D. Eliyahu, D. Seidel, and L. Maleki, "RF amplitude and phase-noise reduction of an optical link and an opto-electronic oscillator," *IEEE Trans. Microwave Theory Techniques,* vol. 56, no. 2, pp. 449–456, Feb. 2008.
14. X. S. Yao and L. Maleki, "Multi-loop optoelectronic oscillator," *IEEE J. Quant. Electron.,* vol. 36, no. 1, pp. 79–84, Jan. 2000.
15. D. Eliyahu, K. Sariri, A. Kamran, and M. Tokhmakhian, "Improving short and long term frequency stability of the opto-electronic oscillator," *Proc. IEEE Int. Freq. Cont. Symp. PDA Exhibition,* vol. 1, pp. 580–583, 2002.
16. D. Eliyahu, K. Sariri, J. Taylor, and L. Maleki, "Optoelectronic oscillator with improved phase noise and frequency stability," *Proc. SPIE,* vol. 4998, pp. 139–147, Jul. 2003.
17. D. Eliyahu and L. Maleki, "Tunable, ultra low phase noise YIG based optoelectronic oscillator," *Proc. IEEE MTT-S Int. Microwave Symp.,* vol. 3, pp. 2185–2187, Jun. 2003.
18. S. Poinsot, H. Porte, J.-P. Goedgebuer, W. T. Rhodes, and B. Boussert, "Continuous radio-frequency tuning of an optoelectronic oscillator with dispersive feedback," *Opt. Lett.,* vol. 27, no. 15, pp. 1300–1302, Aug. 2002.
19. X. S. Yao and L. Maleki, "Dual microwave and optical oscillator," *Opt. Lett.* vol. 22, no. 24, pp. 1867–1869, Dec. 1997.
20. X. S. Yao, L. Davis, and L. Maleki, "Coupled optoelectronic oscillators for generating both RF signal and optical pulses," *J. Lightwave Technol.,* vol. 18, no. 1, pp. 73–78, Jan. 2000.
21. N. Yu, E. Salik, and L. Maleki, "Ultralow-noise mode-locked laser with coupled optoelectronic oscillator configuration," *Opt. Lett.,* vol. 30, no. 10, pp. 1231–1233, May 2005.
22. D. Eliyahu and L. Maleki, "Modulation response (S21) of the coupled opto-electronic oscillator," *Proc. 2005 IEEE Int. Frequency Control Symposium and Exposition,* vol. 1, pp. 850–856, Aug. 2005.
23. E. Salik, N. Yu, and L. Maleki, "An ultralow phase noise coupled optoelectronic oscillator," *IEEE. Photon. Technol. Lett.,* vol. 19, no. 6, pp. 444–446, Mar. 2007.
24. A. B. Matsko, D. Eliyahu, P. Koonath, D. Seidel, and L. Maleki, "Theory of coupled optoelectronic microwave oscillator I: expectation values," *J. Opt. Soc. Am. B,* vol. 26, no. 5, pp. 1023–1031, May 2009.
25. V. S. Ilchenko, X. S. Yao, and L. Maleki, "Microsphere integration in active and passive photonics devices," *Proc. SPIE,* vol. 3930, pp. 154–162, Jan. 2000.
26. V. S. Ilchenko and L. Maleki, "Novel whispering-gallery resonators for lasers, modulators, and sensors," *Proc. SPIE,* vol. 4270, pp. 120–130, Jan. 2001.
27. L. Maleki, V. S. Ilchenko, A. A. Savchenkov, and A. B. Matsko, "Crystalline Whispering Gallery Mode Resonators in Optics and Photonics," in *Practical Applications of Microresonators in Optics and Photonics,* A. B. Matsko, Ed., Boca Raton, FL: CRC Press, 2009.
28. L. Maleki, A. F. J. Levi, S. Yao, and V. Ilchenko, "Light modulation in whispering-gallery-mode resonators," U.S. patent 6 473 218, 2002.
29. D. A. Cohen and A. F. J. Levi, "Microphotonic millimetre-wave receiver architecture," *Electron. Lett.,* vol. 37, no. 1, pp. 37–39, Jan. 2001.
30. D. A. Cohen, M. Hossein-Zadeh, and A. F. J. Levi, "Microphotonic modulator for microwave receiver," *Electron. Lett.,* vol. 37, no. 5, pp. 300–301, Mar. 2001.
31. D. A. Cohen and A. F. J. Levi, "Microphotonic components for a mm-wave receiver," *Solid State Electron.,* vol. 45, no. 3, pp. 495–505, Mar. 2001.
32. D. A. Cohen, M. Hossein-Zadeh, and A. F. J. Levi, "High-Q microphotonic electro-optic modulator," *Solid State Electron.,* vol. 45, no. 9, pp. 1577–1589, Sep. 2001.
33. V. S. Ilchenko, A. B. Matsko, A. A. Savchenkov, and L. Maleki, "High-efficiency microwave and millimeter-wave electro-optical modulation with whispering-gallery resonators," *Proc. SPIE,* vol. 4629, pp. 158–163, Jan. 2002.

34. V. S. Ilchenko, A. A. Savchenkov, A. B. Matsko, and L. Maleki, "SubmicroWatt photonic microwave receiver," *IEEE Photon. Tech. Lett.*, vol. 14, no. 11, pp. 1602–1604, Nov. 2002.
35. V. S. Ilchenko, A. A. Savchenkov, A. B. Matsko, and L. Maleki, "Whispering gallery mode electro-optic modulator and photonic microwave receiver," *J. Opt. Soc. Am. B*, vol. 20, no. 2, pp. 333–342, Feb. 2003.
36. M. Hossein-Zadeh and A. F. J. Levi, "Self-homodyne RF-optical LiNbO microdisk receiver," *Solid State Electron.*, vol. 49, no. 8, pp. 1428–1434, Aug. 2005.
37. M. Hossein-Zadeh and A. F. J. Levi, "14.6-GHz LiNbO3 microdisk photonic self-homodyne RF receiver," *IEEE Trans. MTT*, vol. 54, no. 2, pp. 821–831, Feb. 2006.
38. V. S. Ilchenko, A. A. Savchenkov, I. Solomatine, D. Seidel, A. B. Matsko, and L. Maleki, "Ka-band all-resonant photonic microwave receiver," *IEEE Photon. Tech. Lett.*, vol. 20, no. 19, pp. 1600–1602, Oct. 2008.
39. V. S. Ilchenko, J. Byrd, A. A. Savchenkov, D. Seidel, A. B. Matsko, and L. Maleki, "Coherent resonant Ka band photonic microwave receiver," lanl.arXiv.org>physics>arXiv:0806.3239.
40. A. B. Matsko, V. S. Ilchenko, P. Koonath, J. Byrd, A. A. Savchenkov, D Seidel, and L. Maleki, "RF photonic receiver front-end based on crystalline whispering gallery mode resonators," *Proc. 2009 IEEE Radar Conf.*, vol. 1, pp. 1–6, May 2009.
41. A. A. Savchenkov, W. Liang, V. S. Ilchenko, A. B. Matsko, D. Seidel, and L. Maleki, "RF photonic signal processing components: From high order tunable filters to high stability tunable oscillators," *Proc. 2009 IEEE Radar Conf.*, vol. 1, pp. 1–6, May 2009.
42. A. A. Savchenkov, W. Liang, A. B. Matsko, V. S. Ilchenko, D. Seidel, and L. Maleki, "Tunable optical single-sideband modulator with complete sideband suppression," *Opt. Lett.*, vol. 34, no. 9, pp. 1300–1302, May 2009.
43. A. A. Savchenkov, V. S. Ilchenko, L. Maleki, A. B. Matsko, D. Seidel, and W. Liang, "Tunable resonant single-sideband electro-optical modulator," *Proc. of 2009 IEEE/LEOS Summer Topical Meeting*, vol. 1, pp. 63–64, Jul. 2009.
44. A. A. Savchenkov, A. B. Matsko, V. S. Ilchenko, D. Seidel, and L. Maleki, "Single-sideband electro-optic modulators and their application in tunable opto-electronic oscillators," *Proc. of 2009 Int. Topical Meeting on Microwave Photonics*, vol. 1, pp. 1–4, Oct. 2009.
45. A. A. Savchenkov, V. S. Ilchenko, W. Liang, D. Eliyahu, A. B. Matsko, D. Seidel, and L. Maleki, "Voltage controlled photonic oscillator," *Opt. Lett.*, vol. 35, no. 10, pp. 1572–1574, May 2010.
46. A. A. Savchenkov, A. B. Matsko, W. Liang, V. S. Ilchenko, D. Seidel, and L Maleki, "Single sideband electro-optical modulator and tunable microwave photonic receiver," *IEEE Trans. MTT and J. Lightwave Technol. Special Issue on Microwave Photonics*, vol. 58, no. 11, pp. 3167–3174, Nov. 2010.
47. A. B. Matsko, L. Maleki, A. A. Savchenkov, and V. S. Ilchenko, "Whispering gallery mode based optoelectronic microwave oscillator," *J. Mod. Opt.*, vol. 50, no. 15, pp. 2523–2542, Oct. 2003.
48. A. A. Savchenkov, V. S. Ilchenko, J. Byrd, W. Liang, D. Eliyahu, A. B. Matsko, D. Seidel, and L. Maleki, "Whispering-gallery mode based optoelectronic oscillators," *Proc. of IEEE Int. Freq. Cont. Symp.*, vol. 1, pp. 554–557, Aug. 2010.
49. D. Strekalov, D. Aveline, N. Yu, R. Thompson, A. B. Matsko, and L. Maleki, "Stabilizing an optoelectronic microwave oscillator with photonic filters," *J. Lightwave Technol.*, vol. 21, no. 12, pp. 3052–3061, Dec. 2003.
50. D. Strekalov, A. B. Matsko, N. Yu, A. A. Savchenkov, and L. Maleki, "Application of vertical cavity surface emitting laser in self-oscillating atomic clocks," *J. Mod. Opt.*, vol. 53, no. 16, pp. 2469–2484, Nov. 2006.
51. A. B. Matsko, D. Strekalov, and L. Maleki, "Magnetometer based on the opto-electronic oscillator," *Mater. Res. Soc. Symp. Proc.*, vol. 906E, art. 0906-HH03-06.1, 2006.
52. W. Gawlik, L. Krzemien, S. Pustelny, D. Sangla, J. Zachorowski, M. Graf, A. O. Sushkov, and D. Budker, "Nonlinear magneto-optical rotation with amplitude modulated light," *Appl. Phys. Lett.*, vol. 88, no. 13, art. 131108, Mar. 2006.

53. J. M. Higbie, E. Corsini, and D. Budker, “Robust, high-speed, all-optical atomic magnetometer,” *Rev. Sci. Instrum.*, vol. 77, no. 11, art. 113106, Nov. 2006.
54. V. N. Konopsky, “A new type of optical gyro via electro-optical oscillator,” *Opt. Commun.*, vol. 126, no. 4, pp. 236–239, May 1996.
55. E. Arimondo, “Coherent population trapping in laser spectroscopy,” in *Progress in Optics,* vol. 35, E. Wolf, Ed., Amsterdam, The Netherlands: Elsevier, 1996, ch. 5, pp. 257–354.
56. J. Kitching, S. Knappe, and L. Hollberg, “Miniature vapor-cell atomic frequency references,” *Appl. Phys. Lett.*, vol. 81, no. 3, pp. 553–555, Jul. 2002.
57. A. Nagel, L. Graf, A. Naumov, E. Mariotti, V. Biancalana, D. Meschede, and R. Wynands, “Experimental realization of coherent dark-state magnetometers,” *Europhys. Lett.*, vol. 44, no. 1, pp. 31–36, Oct. 1998.
58. D. Budker, V. Yashchuk, and M. Zolotorev, “Nonlinear magneto-optic effects with ultranarrow widths,” *Phys. Rev. Lett.*, vol. 81, no. 26, pp. 5788–5791, Dec. 1998.
59. J. C. Allred, R. N. Lyman, T. W. Kornack, and M. V. Romalis, “Highsensitivity atomic magnetometer unaffected by spin-exchange relaxation,” *Phys. Rev. Lett.*, vol. 89, no. 13, art. 130801, Sep. 2002.
60. I. K. Kominis, T. W. Kornack, J. C. Allred, and M. V. Romalis, “A subfemtotesla multichannel atomic magnetometer,” *Nature,* vol. 422, no. 6932, pp. 596–599, Apr. 2003.
61. E. B. Alexandrov, “Recent progress in optically pumped magnetometers,” *Phys. Scr.,* vol. T105, pp. 27–30, 2003.
62. G. Bison, R. Wynands, and A. Weis, “A laser-pumped magnetometer for the mapping of human cardiomagnetic fields,” *Appl. Phys. B.*, vol. 76, no. 3, pp. 325–328, Mar. 2003.
63. A. L. Bloom, “Principles of operation of the rubidium vapor magnetometer,” *Appl. Opt.*, vol. 1, no. 1, pp. 61–68, Jan. 1962.
64. E. Pulz, K.-H.Jackel, and H.-J. Linthe, “A new optically pumped tandem magnetometer: principles and experiences,” *Meas. Sci. Technol.*, vol. 10, no. 11, pp. 1025–1031, Nov. 1999.
65. T. Zanon, S. Guerandel, E. de Clercq, D. Holleville, N. Dimarcq, and A. Clairon, “High contrast Ramsey fringes with coherent-population trapping pulses in a double lambda atomic system,” *Phys. Rev. Lett.*, vol. 94, no. 19, art. 193002, May 2005.

# 22

# Electro-Optic Analog-to-Digital Converters

Minyu Yao
*Tsinghua University*

Hongming Zhang
*Tsinghua University*



## 22.1 Introduction

An analog-to-digital converter (ADC) is a device that converts an analog signal to a corresponding digital signal. It is the bridge connecting analog systems and digital systems. Digital systems, compared to analog systems, have obvious advantages in terms of stability, antijamming capability, precision, and integration. This is why in recent years digital technology has undergone rapid development. It has penetrated into various areas of technology. However, most physical signals are analog in nature, so it is necessary to convert analog signals to digital signals before any processing is performed on them. As bridges between the digital and analog worlds, ADCs have become increasingly important, and there is a continuous demand for higher performance devices. Areas such as ultrawideband (UWB) communication [1], software radio [2], navigation, aircraft identification, electronic warfare, UWB radar receivers, ultrabandwidth signal acquisition, and digital processing require ADCs with high capacities.

However, traditional electronic ADCs have performance bottlenecks such as timing jitter of the sampling clock, relaxation of the sample-and-hold circuits, and different turn-on voltages of the transistors. These bottlenecks have constrained development of electronic ADCs, especially when high sampling rates and high quantization precision are required [3].

Looking at the history of the electronic ADC's development [4] (see Figure 22.1), we find that the bandwidth of most existing ADCs is below 1 GHz and that improvements in bandwidth and quantization precision has been slow.

Although there are very high-speed electronic ADCs that have 10-GHz input bandwidths, their quantization precision is only about 2 bits. Because of their fundamental limitations, electronic ADCs with bandwidths of more than 10 GHz and more than 2 bits of quantization precision are difficult to build. Since the development of analog-to-digital conversion technology has not kept pace with the advances of the rest of the digital world, ADC limitations are more and more responsible for restricting improvements in information processing technology.

In recent years, the rapid development of photonics technology has expanded to include research into techniques that promise to overcome the bottlenecks of electronic analog-to-digital conversion schemes. These techniques have the potential to enable simultaneous ultrahigh-speed and high-precision analog-to-digital conversion [5,6]. There are advantages to photonic analog-to-digital conversion. First, mode-locked lasers can generate optical sampling pulses that have subpicosecond widths [7,8]. These pulses

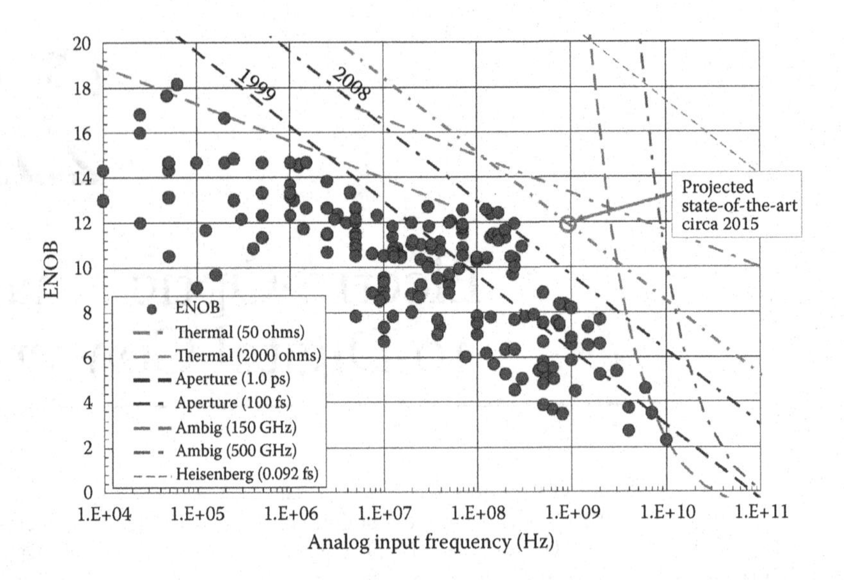


**FIGURE 22.1** ENOB (effective number of bits) as a function of analog input frequency. (R. H. Walden, "Analog-to-digital converters and associated IC technologies," *IEEE Compound Semiconductor Integrated Circuits Symposium* (CSICS) 2008, © 2008 IEEE.)

can be used to achieve sampling rates of more than 100 GS/s (samples per second) [9–12]. Second, the timing jitter of mode-locked lasers is very small. The jitter can be several femtoseconds (fs = $10^{-15}$ s), which is two orders of magnitude below the timing jitter of the best electrical clocks [13,14]. So, using mode-locked lasers as the sampling clock sources can enable higher quantization accuracy. Therefore, optical analog-to-digital conversion technology has advantages compared to traditional electronic analog-to-digital conversion technology for applications where high sampling rates, ultrawide input bandwidths, and high quantization accuracy are needed. This research field is receiving the attention of many of the world's research groups, as well as widespread emphasis at national defense research institutions [15–22].

## 22.2 The Development of Optical Analog-to-Digital Conversion

As early as 1970, A. E. Siegman and D. J. Kuizenga proposed the idea of using optical pulses to sample analog signals [23]. Although their original intention was to measure picosecond ultrashort optical pulses, this idea of sampling with optical pulses can be considered the beginning of optical analog-to-digital conversion technology. In 1974, S. Wright first proposed an optical quantization scheme [24]. In this scheme, an optical diffraction grating was modulated by an input analog signal and the power of the light diffracted by the grating would change accordingly. By measuring the power of the light diffracted in different directions, the analog signal could be quantized. These early ideas of optical sampling and optical quantization laid the foundation for later studies of optical analog-to-digital conversion technology.

In 1975, H. F. Taylor proposed an optical analog-to-digital conversion scheme where the input analog signal was sampled and quantized by a series of electro-optic modulators with different half-wave voltages [25]. Taylor's approach was very attractive and over the next 10 years, optical analog-to-digital conversion studies mainly focused on the realization and improvement of his scheme. After the 1990s, optical analog-to-digital conversion technology entered a period of rapid progress, and many new optical analog-to-digital conversion schemes were proposed.

Optical analog-to-digital conversion schemes can be divided into two categories. The first category is an optoelectronic hybrid analog-to-digital conversion, in which either the sampling process or the quantization process or both are still achieved with electronic devices. Here, photonic techniques play an auxiliary role in the conversion process. The second category is all-optical analog-to-digital conversion, in which both the sampling process and the quantization process are achieved using photonic techniques. Current optical analog-to-digital conversion technology research is focused on converting analog electronic signals to digital electronic signals. Therefore, the photoelectric conversion process is inevitable. If the two required processes of analog-to-digital conversion, sampling and quantization, are accomplished with photonic techniques, the whole process can be called all-optical analog-to-digital conversion.

### 22.2.1 Optoelectronic Hybrid Analog-to-Digital Conversion

In optoelectronic hybrid analog-to-digital conversion, the sampling or quantization is still achieved by electronic techniques. According to the different roles that photonic techniques play in the conversion process, optoelectronic hybrid analog-to-digital conversion can be divided into three categories: photonic-assisted analog-to-digital conversion, optical sampling analog-to-digital conversion, and electrical sampling and photonic quantization analog-to-digital conversion.

In the photonic-assisted analog-to-digital conversion scheme, the sampling and quantization processes are done using electronic devices, while being assisted by photonic techniques. With a scheme that used optical pulses as the clock source for the sample-and-hold step, a 1-GS/s, 10-bit effective number of bits (ENOB) conversion method was reported [26]. Another scheme using an electron-beam triggered by optical pulses to achieve conversion was reported [27]. The most promising conversion scheme that has been reported in recent years is the time-stretched optical analog-to-digital conversion [10,28–31]. Chapter 19 of this book has a detailed description of time-stretching techniques.

The UCLA research team, led by B. Jalali, proposed a scheme to achieve stretching of analog signals in the time domain by stretching optical pulses that were carrying the waveform of the analog signals. The principle of time-stretched optical analog-to-digital conversion is shown in Figure 22.2. First, a wide-spectrum optical pulse passes through a dispersive medium, such as a dispersion-compensation fiber (DCF), of length $L_1$. Thus, the time-domain width of the pulse is stretched to the order of a few nanoseconds (i.e., chirp is introduced in the pulse). Then, the analog signal to be converted is mapped onto the intensity envelope of the chirped optical pulse with an electro-optic modulator. The pulse then passes through a dispersive medium of length $L_2$, resulting in further broadening of the optical pulse in the time domain. The analog signal modulated onto the pulse is broadened at the same time. After photoelectric conversion of the modulated, stretched optical pulse, the electronic signal is then sampled and quantized by a traditional, low-speed, electronic analog-to-digital converter. If the amount of dispersion of the first dispersive medium is $D_1$ and that of the second dispersive medium is $D_2$, then the broadening factor is $(D_1 + D_2)/D_1$. If the two dispersive media are made of the same material (e.g., using the

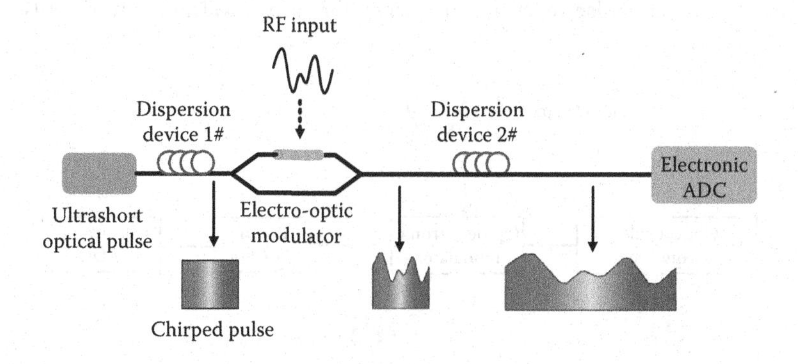


**FIGURE 22.2** Time-stretched optical analog-to-digital conversion.

same kind of dispersion-compensating fiber), then the broadening factor will be $(L_1 + L_2)/L_1$. With this approach, the group achieved a broadening factor of 16 and ADC bandwidth of 18 GHz with a 45-dB signal-to-noise ratio (SNR), which is equivalent to an ENOB of 7 [29]. The group has also reported 10 TS/s (1 TS = $10^{12}$ samples) hybrid optical analog-to-digital conversion experiments, which is the highest sampling rate ever reported [30].

The main difficulty of time-stretched optical analog-to-digital conversion is how to linearly broaden the optical pulse. Since the analog signal to be converted has been mapped onto the temporal intensity envelope of the optical pulse, stretching the modulated optical pulse is equivalent to reducing the bandwidth of the analog signal. The nonlinear character of the dispersive medium results in irregularities in the stretched optical pulse. Ultimately, these irregularities decrease the SNR of this conversion scheme. At the same time, the long optical fibers (used as the dispersive medium) can lead to instability and hamper efforts to reduce the size of systems based on this technique. In addition, this time-stretched conversion scheme is suitable for high-speed transient signal capture, but there are many problems that need to be resolved for continuous signal capture and broadening [31].

To achieve high-speed optical analog-to-digital conversion for continuous electrical signals, we can directly sample the electronic signal using a high-repetition-frequency optical pulse sequence (i.e., modulate the optical pulse sequence with the electrical signal) and then process the modulated pulse sequence with an electronic ADC as shown in Figure 22.3. This is the basic idea of optical sampling analog-to-digital conversion.

As can be seen from Figure 22.3, using this scheme does not reduce the requirements on the electronic ADC's sampling rate, $f_{REQ}$, although the scheme can be modified to achieve higher effective sampling rates and lower sampling timing jitter. If we place a demultiplexer after the modulator to separate the modulated pulse sequence into $N$ channels then, at the end of each channel, an electronic ADC capable of sampling at $f_{REQ}/N$ can be used. Bell first reported the use of time division multiplexing (TDM) in optical sampling analog-to-digital conversion and achieved a 2-GS/s sampling rate with an ENOB of 2.8 [32]. Williamson reported using time-division multiplexing and a dual-output Mach–Zehnder electro-optical modulator to inhibit nonlinear modulation and noise to achieve a 505-MS/s sampling rate and a (rather high) ENOB of 9.8 [33].

In a time-division multiplexing approach, the demultiplexing step will be more complex and difficult to achieve as the number of channels is increased. To solve this problem, in 1998, Yariv proposed a wavelength division multiplexing (WDM) combined with time-division multiplexing optical analog-to-digital conversion approach [16], as shown in Figure 22.4.

This scheme is characterized by interleaving ultrashort optical pulses of different wavelengths. Compared to time-division multiplexing, wavelength and time interleaved multiplexing can achieve a high sampling rate with a very simple demultiplexing approach. Therefore, this approach has been extensively researched since it was proposed [12,17,34–36]. In 2004, Hughs Research Labortories reported a 4 × 10-GS/s WDM/TDM optical pulse sampling analog-to-digital conversion experiment (four wavelengths, 10-GHz pulse repetition frequency for each wavelength, 16-ps pulse width). The

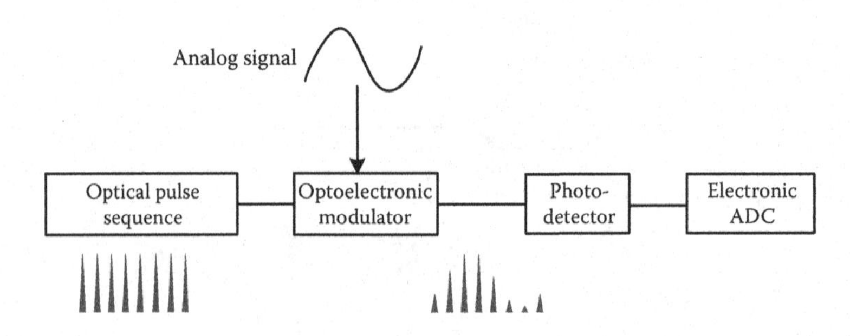


**FIGURE 22.3** The principle of optical sampling analog-to-digital conversion.

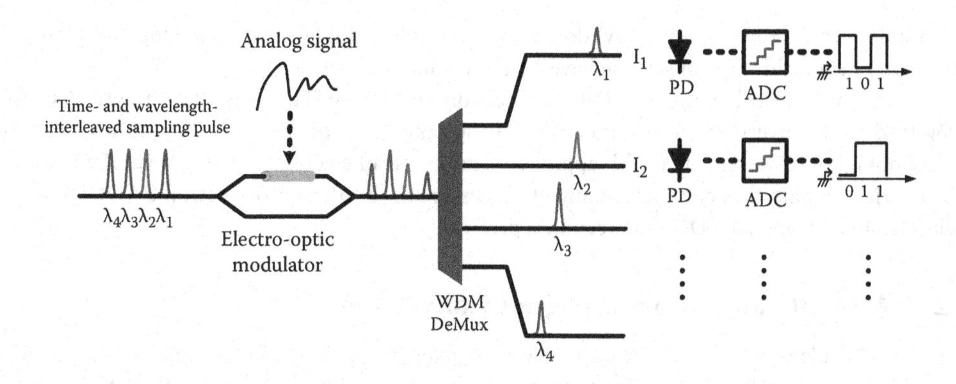


**FIGURE 22.4** The principle of wavelength-division multiplexing combined with time-division multiplexing analog-to-digital conversion.

group used specially designed electronics (10 GS/s ADCs with integrated photoelectric converters) to reach a nearly 40 GS/s sampling rate and an ENOB of 8.3 [37], which is the highest ENOB at that sampling rate, although the conversion bandwidth was less than 2 GHz. In both the time-division multiplexing approach and the WDM/TDM interleaved approach, high sampling rates are achieved by multiplexing. Therefore, the electronic ADCs at the end of each demultiplexed channel can sample and quantize the optical pulses at a lower speed. The number of the electronic ADCs is equal to the number of demultiplexed channels. For example, in a 40-GS/s optical sampling analog-to-digital conversion system, if the electronic ADCs' sampling and quantizing rate is 1.25 GS/s, we need 32 parallel channels for demultiplexing the sampling optical pulse sequence and 32 electronic ADCs (one for each channel). A large number of channels and electronic ADCs would lead to increased complexity and cost of the system, while balancing control between the channels would become difficult.

There are also schemes using electronic techniques to sample the signal and photonic techniques to quantize the signal. For example, in 2002, Zmuda proposed a scheme, as shown in Figure 22.5 [38]. In this scheme, an electronic sample-and-hold circuit generates a ladder-shaped voltage, according to the input analog signal voltage, and a wavelength tunable semiconductor laser is modulated by the ladder-shaped voltage. As the laser wavelength is changed, the direction of the light diffracted by the grating will change. A detector array will record this change of direction and thus we know the change of the original analog signal. It is clear that the bandwidth of this scheme is subject to the wavelength tunable semiconductor laser's response speed. The scheme's accuracy relies on the tuning linearity of the semiconductor laser and the detection accuracy. More importantly, the advantages of optical sampling

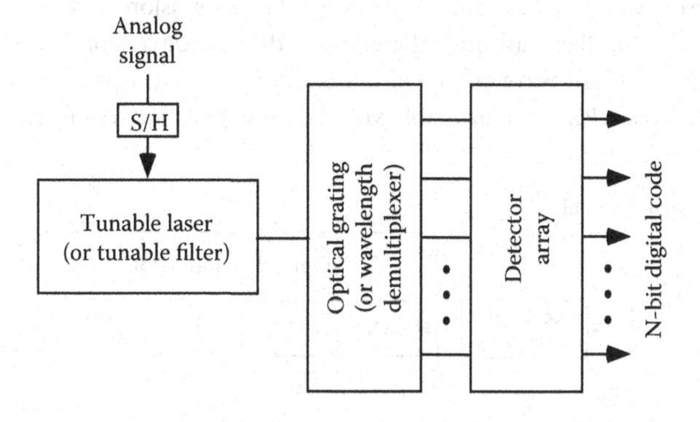


**FIGURE 22.5** The principle of electronic sampling and photonic quantization.

(e.g., sampling rate, sampling accuracy) do not play any role in this scheme. Therefore, few conversion methods relying on this type of scheme have been reported recently.

Time-stretched optical analog-to-digital conversion and WDM/TDM interleaved optical sampling analog-to-digital conversion are two more promising optoelectronic hybrid analog-to-digital conversion technologies. However, since both approaches make use of electronic ADCs to get the final quantized electrical signals, they cannot fundamentally resolve the bottleneck of electronic ADCs. Therefore, development of all-optical ADCs has become a priority.

## 22.2.2 All-Optical Analog-to-Digital Conversion

The basic principle of all-optical analog-to-digital conversion is shown in Figure 22.6. A high-repetition-frequency optical sampling pulse sequence first samples the input analog signal in an electro-optic modulator. Then, the sampled pulses are quantized and encoded and, after the threshold decision, the digital signal corresponding to the input analog signal is acquired. With this technique, the sampling and quantization processes are achieved with photonic technology, which can essentially avoid the bottlenecks of electronic ADCs.

The best-known all-optical analog-to-digital conversion scheme was proposed by H. F. Taylor in 1975 [25]. Figure 22.7 shows a 3-bit, all-optical analog-to-digital conversion scheme. There are three Mach–Zehnder electro-optical modulators used to modulate the optical pulse sequence that samples the analog signal. Each modulator has a transmission curve that is cyclical with respect to the applied voltage, as shown in Figure 22.8. If each modulator's half-wave voltage is half that of the previous one (e.g., the first modulator's half-wave voltage is $V_\pi$, the second modulator's is $V_\pi/2$, and the third modulator's is $V_\pi/4$, as Figure 22.8 shows) then, after the threshold decision at the end of each optical channel, the quantized and encoded digital signal is acquired. Therefore, $N$ modulators with such a relationship between their half-wave voltages can achieve an $N$-bit, all-optical analog-to-digital conversion.

Taylor's scheme has the advantages of a simple quantization principle and a direct binary code output. The number of modulators and optical output channels is equal to the number of bits of output. Therefore, in recent decades, many studies have been carried out based on Taylor's idea [39–44]. In 1984, Becker reported a 1-GS/s, all-optical analog-to-digital conversion experiment based on Taylor's scheme [43]. However, in Taylor's approach, the half-wave voltages of the modulators must be half of each other exactly, this means that modulators with ultralow $V_\pi$ are needed. For 5-bit, all-optical analog-to-digital conversion, the lowest $V_\pi$ should be 1/16 of the highest $V_\pi$. Commercially available modulators with bandwidths greater than 10 GHz have $V_\pi$ larger than 2 V. This means that 5-bit quantization can only be achieved for analog signals that have at least 32-V peak-to-peak voltages. Currently, signal sources with bandwidths greater than 10 GHz are not able to provide such high voltages. Using techniques based on Taylor's idea, K. Ikeda demonstrated all-optical analog-to-digital conversion by using nonlinear effects in an optical fiber and has reported 10-GS/s, 3-bit conversion. In that experiment, several nonlinear optical fiber loop mirrors were used to achieve the cyclical transmission curves [9].

Jalali has proposed folding-flash, all-optical analog-to-digital conversion [45]. This approach avoids the requirement of ultralow half-wave voltage modulators by cascading modulators to achieve many-cycled transmission curves within a reasonable voltage range; such a system (with many modulators

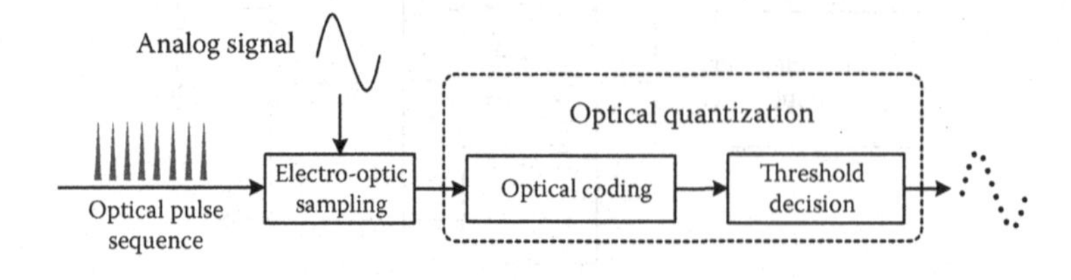


**FIGURE 22.6** The basic principle of all-optical analog-to-digital conversion.

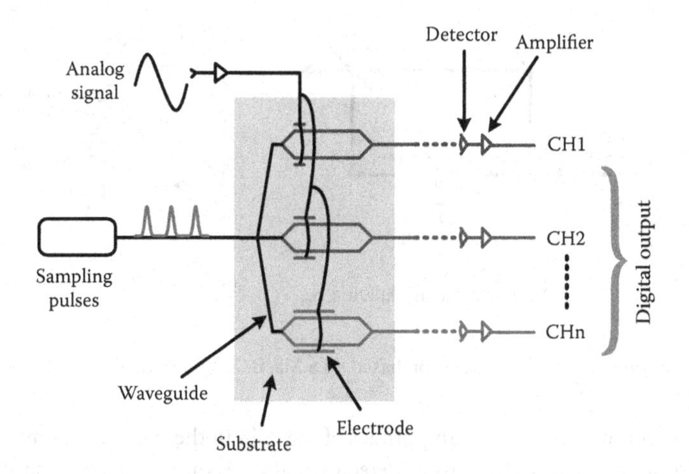


**FIGURE 22.7** A 3-bit all-optical analog-to-digital conversion scheme.

and Y-splitters) is very complex. Additionally, the time delays and insertion losses in the waveguides between modulators limits the system's bandwidth and processing speed.

To avoid the complex structure of Taylor's scheme, in 2005, Stigwall proposed a novel all-optical analog-to-digital conversion approach called phase-shifted optical quantization [21]. As Figure 22.9 shows, a phase modulator is placed in one arm of a Mach–Zehnder interferometer. The input analog signal modulates the optical phases of the pulses in the modulator. At the output of the interferometer, the pulses from the two arms interfered and the interference fringes were detected at a screen behind the interferometer. $N$ optoelectronic detectors are placed on the screen at certain points to detect changes in the optical power. The optical power along the $x$-axis and the phase that is modulated onto the optical pulse, $\varphi_m$, has the relationship:

$$I(x,\varphi_m) = a^2 + 2ab\cos(xk_x+\varphi_m) + b^2, \tag{22.1}$$

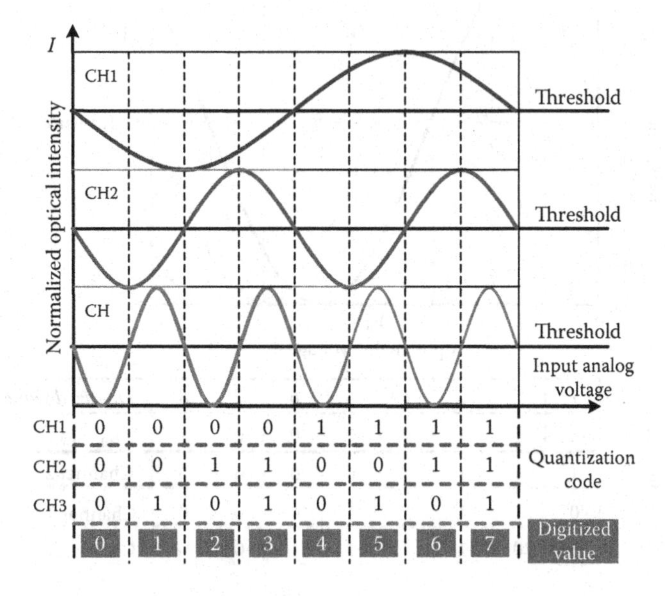


**FIGURE 22.8** All-optical quantization principle.

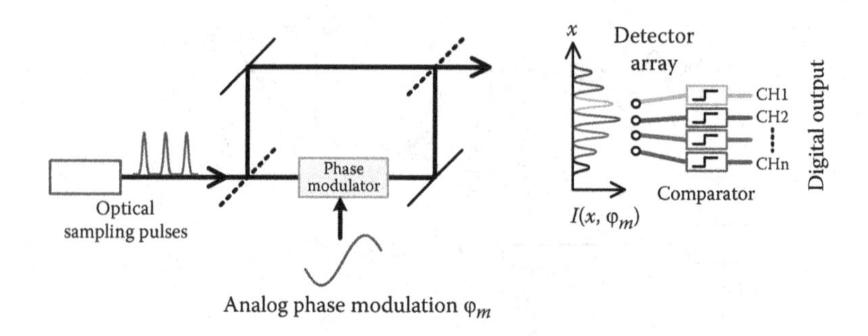


**FIGURE 22.9** Phase shifted optical quantization based on a Mach–Zehnder interferometer.

where $a$ and $b$ represent the electric field amplitudes of the light in the two arms of interferometer, $k_x$ is a constant, $x$ is the location of the optoelectronic detector, and $xk_x$ is the spatial phase shift due to the detector's location. By specific placement of the $n$ detectors, the phase shift difference between two adjacent detectors can be just $\pi/n$. For example, if $n = 4$, then the phase difference between the transmission curves of adjacent (and properly positioned) detectors will be $\pi/4$ (see Figure 22.10). After threshold detection, the combined outputs of the four detectors can achieve 8-level quantization with the digital code shown in Figure 22.10. Therefore, the use of four optoelectronic detectors allows 8-level quantization of an input analog voltage that is in the range 0–$2V_\pi$ ($V_\pi$ is the half-wave voltage of the phase modulator). $N$ detectors can achieve $2n$ levels of quantization, and the corresponding bit number is $\log_2(2n)$ bits.

The appeal of phase-shifted optical quantization is that only one modulator is used. In 2006, Stigwall reported all-optical analog-to-digital conversion results in which a 1.25-GHz microwave signal was sampled at 40 GS/s with a 3.6 ENOB [46]. Compared to Taylor's scheme, phase-shifted optical quantization

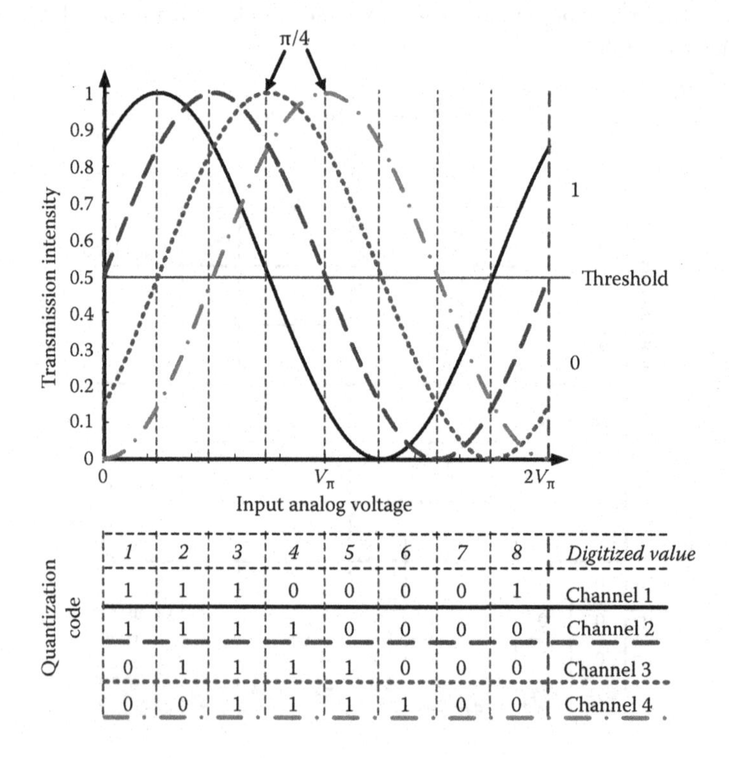


| 1 | 2 | 3 | 4 | 5 | 6 | 7 | 8 | *Digitized value* |
|---|---|---|---|---|---|---|---|---|
| 1 | 1 | 1 | 0 | 0 | 0 | 0 | 1 | Channel 1 |
| 1 | 1 | 1 | 1 | 0 | 0 | 0 | 0 | Channel 2 |
| 0 | 1 | 1 | 1 | 1 | 0 | 0 | 0 | Channel 3 |
| 0 | 0 | 1 | 1 | 1 | 1 | 0 | 0 | Channel 4 |

**FIGURE 22.10** Quantization principle of a 3-bit optical analog-to-digital converter.

has obvious advantages when a high sampling rate and high accuracy are needed. Phase-shifted optical quantization has good prospects for future development and application.

In 2007, Li improved Stigwall's approach by recognizing that the individual interferometer arms could be replaced by the two orthogonal optical axes of the phase modulator [47]. This principle is shown in Figure 22.11. The difference between the indices of refraction of the two axes vary linearly with the voltage of the input analog signal. A linear polarizer is placed such that its transmission axis is 45° between the two orthogonal axes. Thus, light from the two orthogonal directions is projected onto the same axis and will interfere after the polarizer.

The sampling pulses that exit the modulator are divided into $n$ channels, as shown in Figure 22.11. Next, the optical pulses in each channel pass through a tunable optical phase shifter (PSM) to acquire a fixed phase difference between the two orthogonal polarization directions. The phase shifters are set to make the phase difference between the two polarizations equal to $i \times \pi/n$ ($i = 1,2,\ldots,n$) in the $i$-th channel. The transmission curves of each channel will have fixed $\pi/n$ phase differences between each other, as shown in Figure 22.10. Therefore, phase-shifted optical quantization is achieved. Tunable optical phase-shifters can be built by physically stretching a polarization maintaining fiber. Applying different strains to the fiber will induce different optical phase shifts between the two polarizations axes of the fiber.

Using an approach based on the scheme just outlined, the research team from Tsinghua University demonstrated an ENOB of 3.45 at 40 GS/s [48]. Figure 22.12 shows their 40-GS/s, all-optical analog-to-digital conversion apparatus. In this setup, a time and wavelength-interleaved optical pulse train with a repetition rate of 40 GHz was the sampling pulse source. Two polarization states with the same amplitude exited the phase modulator. After the phase-modulated sampling pulses were demultiplexed into 8 wavelength channels, the optical fiber of each channel was pressed by a fiber squeezer to induce the desired extra phase difference between the two polarization states. Interference between the polarization states occurred after the sampling pulses passed through the in-line analyzer array. The sampling pulses were then detected and digitized with electronic comparators. Each wavelength channel had a 10-GS/s, 4-bit optical ADC. By combining the four, time-interleaved channels, a sampling rate of 40 GS/s with 4-bit resolution can be achieved. If more optical channels are phase-shifted, higher sampling rates can be obtained.

In addition to Taylor's approach and the phase-shifted optical quantization approach, a spectrum-encoded all-optical analog-to-digital conversion scheme has been proposed [19,49,50]. The core idea

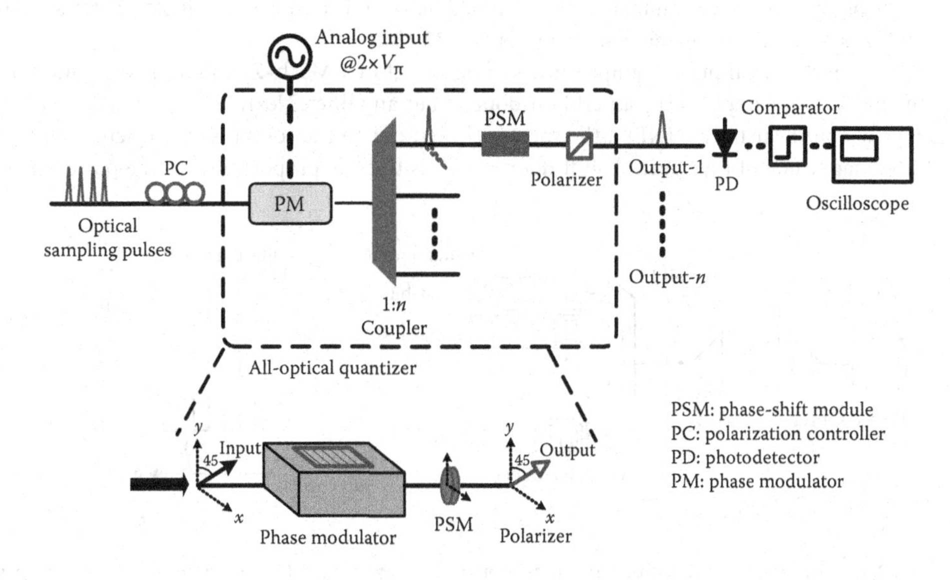


**FIGURE 22.11** Interference of the two orthogonal polarizations in a Mach–Zehnder phase modulator.

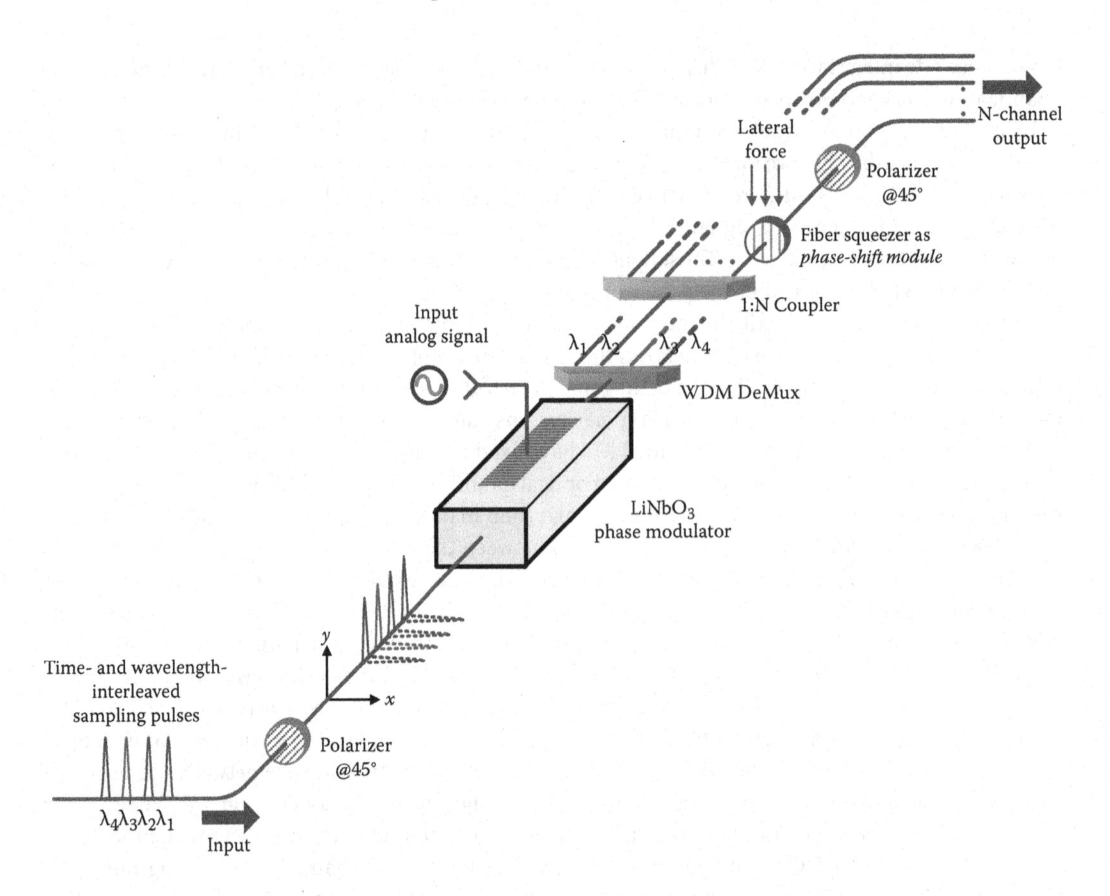


**FIGURE 22.12** Schematic illustration of a 40-GS/s optical ADC system.

behind this conversion method is that nonlinear optical effects will convert intensity modulations of the sampling pulses into frequency modulations of their spectra. The best known such scheme, spectrum-encoded all-optical analog-to-digital conversion based on the self-frequency shift of a soliton, was proposed by Chris Xu in 2003 [19] and is shown in Figure 22.13.

First, ultrashort optical pulses sampled the analog signal in a Mach–Zehnder intensity modulator. Then, the pulses were amplified by an erbium-doped fiber amplifier. Next, as the light pulses traveled in the highly nonlinear fiber (HNLF), the spectra shifted due to the soliton self-frequency shift effect (SSFS); the magnitude of the spectral shift due to SSFS is linearly proportional to the pulse intensity.

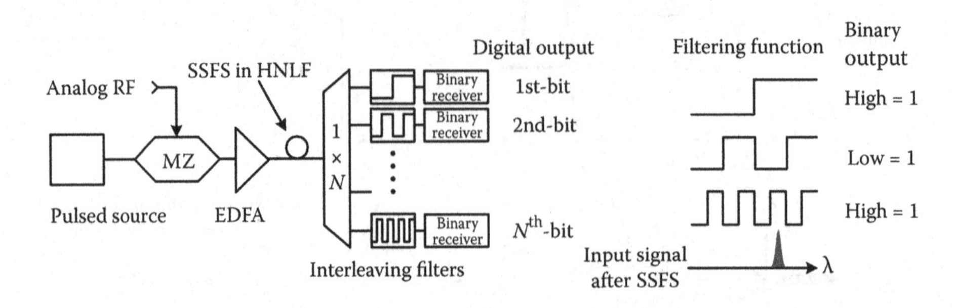


**FIGURE 22.13** The spectrum encoding in an all-optical analog-to-digital conversion scheme based on self-frequency shift of a soliton.

Then, a set of interleavers was used to filter the spectra of the pulses. Each interleaver has a transmission cycle half of its previous one. Finally, the quantized result of the analog signal was acquired after the interleavers, as shown in Figure 22.13. The number of quantization bits is equal to the number of interleavers. This scheme avoids the need for multiple modulators in Taylor's scheme. However, increasing the efficiency of SSFS in the highly nonlinear fiber and making a set of interleavers with the filter cycles shown in Figure 22.13 are difficulties that need to be overcome.

In addition to the use of interleavers, Nishitani from Osaka University also proposed an optical interconnect structure to achieve spectrum-encoded all-optical analog-to-digital conversion [49]. There is also a scheme using the supercontinuum spectrum broadening effect and filters with specific transmission features to achieve spectrum-encoded all-optical analog-to-digital conversion [50]. All these schemes rely on the spectra of the sampling pulses changing as a function of sampling pulse intensity. The efficiency of the nonlinear effect relies critically on pulse quality and energy. The amount by which the pulse spectra can change limits the quantization accuracy of this method. Therefore, high accuracy is difficult to achieve, although these schemes can work at very high sampling rate.

# References


1. L. Yang and G. B. Giannakis, "Ultra-wideband communications: An idea whose time has come," *IEEE Signal Processing Magazine,* vol. 21, pp. 26–54, 2004.
2. S. P. Reichart, B. Youmans, and R. Dygert, "The software radio development system," *IEEE Personal Communications,* vol. 6, pp. 20–24, 1999.
3. R. H. Walden, "Analog-to-digital converter survey and analysis," *IEEE Journal on Selected Areas in Communications,* vol. 17, pp. 539–550, 1999.
4. R. H. Walden, "Analog-to-digital converters and associated IC technologies," *IEEE Compound Semiconductor Integrated Circuits Symposium (CSICS) 2008.*
5. G. V. Valley, "Photonic analog-to-digital converters," *Optics Express,* vol. 15, pp. 1955–1982, 2007.
6. B. L. Shoop, "Photonic analog-to-digital conversion," Berlin: Springer, 1984.
7. E. Yoshida, N. Shimizu, and M. Nakazawa, "A 40-GHz 0.9-ps regeneratively mode-locked fiber laser with a tuning range of 1530–1560 nm," *IEEE Photonics Technology Letters,* vol. 11, pp. 1587–1589, 1999.
8. M. Nakazawa and E. Yoshida, "A 40-GHz 850-fs regeneratively FM mode-locked polarization-maintaining erbium fiber ring laser," *IEEE Photonics Technology Letters,* vol. 12, pp. 1613–1615, 2000.
9. K. Ikeda, J. M. Abdul, H. Tobioka, T. Inoue, S. Namiki, and K. Kitayama, "Design considerations of all-optical A/D conversion: Nonlinear fiber-optic Sagnac-loop interferometer-based optical quantizing and coding," *Journal of Lightwave Technology,* vol. 24, pp. 2618–2628, 2006.
10. A. S. Bhushan, P. V. Kelkar, B. Jalali, O. Boyraz, and M. Islam, "130-GSa/s photonic analog-to-digital converter with time stretch pre-processor," *IEEE Photonics Technology Letters,* vol. 14, pp. 684–686, 2002.
11. Thomas R. Clark, Jr. and Michael L. Dennis, "Toward a 100-GSample/s photonic A/D converter," *IEEE Photonics Technology Letters,* vol. 13, pp. 236–238, 2001.
12. P. Jiang, Y. Chai, I. White, R. Penty, J. Heaton, A. Kuver, S. Clements, et al., "80 GSPS photonic analogue to digital conversion system using broadband continuous wave source," *Conference on Lasers and Electro-optics (CLEO),* pp. 874–876, 2005.
13. S. Gee, S. Ozharar, F. Quinlan, J. J. Plant, P. W. Juodawlkis, and P. J. Delfyett, "Self-stabilization of an actively mode-locked semiconductor-based fiber-ring laser for ultralow jitter," *IEEE Photonics Technology Letters,* vol. 19, pp. 498–500, 2007.
14. M. Yoshida, T. Hirayama, N. Masataka, H. Ken, and I. Takeshi, "Regeneratively mode-locked fiber laser with a repetition rate stability of $4.9 \times 10^{-15}$ using a hydrogen master phase-locked loop," *Optics Letters,* vol. 32, pp. 1827–1829, 2007.
15. H. F. Taylor, "An optical analog-to-digital converter—Design and analysis," *IEEE Journal of Quantum Electronics,* QE-15, pp. 210–216, 1979.

16. A. Yariv and R. G. M. P. Koumans, "Time interleaved optical sampling for ultra-high speed A/D conversion," *Electronics Letters,* vol. 34, pp. 2012–2013, 1998.
17. T. R. Clark, J. U. Kang, and R. D. Esman, "Performance of a time- and wavelength-interleaved photonic sampler for analog–digital conversion," *IEEE Photonics Technology Letters,* vol. 11, pp. 1168–1170, 1999.
18. J. C. Twichell and R. Helkey, "Phase-encoded optical sampling for analog-to-digital converters," *IEEE Photonics Technology Letters,* vol. 12, pp. 1237–1239, 2000.
19. C. Xu and X. Liu, "Photonic analog-to-digital converter using soliton self-frequency shift and interleaving spectral filters," *Optics Letters,* vol. 28, pp. 986–988, 2003.
20. K. Ikeda, J. M. Abdul, S. Namiki, and K. Kitayama, "Optical quantizing and coding for ultrafast A/D conversion using nonlinear fiber-optic switches based on Sagnac interferometer," *Optics Express,* vol. 13, pp. 4296–4302, 2005.
21. J. Stigwall and S. Galt, "Interferometric analog-to-digital conversion scheme," *IEEE Photonics Technology Letters,* vol. 17, pp. 468–470, 2005.
22. S. Yu, S. Koo, and N. Park, "Coded output photonic A/D converter based on photonic crystal slow-light structures," *Optics Express,* vol. 16, pp. 13752–13757, 2008.
23. A. E. Siegman and D. J. Kuizenga, "Proposed method for measuring picosecond pulsewidths and pulse shapes in cw mode-locked lasers," *IEEE Journal of Quantum Electronics,* vol. 6, pp. 212–215, 1970.
24. S. Wright, I. M. Mason, and M. G. F. Wilson, "High-speed electro-optic analogue–digital conversion," *Electronics Letters,* vol. 10, pp. 508–509, 1974.
25. H. F. Taylor, "An electro-optic analog-to-digital converter," *Proceedings of the IEEE,* vol. 63, pp. 1524–1525, 1975.
26. E. W. Jacobs, J. B. Sobti, V. F. Vella, R. Nguyen, D. J. Albares, R. B. Olsen, C. T. Chang, et al., "Optically clocked track-and-hold for high-speed high-resolution analog-to-digital conversion," *IEEE International Topical Meeting on Microwave Photonics (MWP),* pp. 190–192, 2004.
27. R. F. Pease, K. Ioakeimidi, R. Aldana, and R. Leheny, "Photoelectronic analog-to-digital conversion using miniature electron optics: Basic design considerations," *Journal of Vacuum Science and Technology,* vol. 21, pp. 2826–2829, 2003.
28. A. S. Bhushan, F. Coppinger, and B. Jalali, "Time-stretched analogue-to-digital conversion," *Electronics Letters,* vol. 34, pp. 1081–1083, 1998.
29. A. Nuruzzaman, O. Boyraz, and B. Jalali, "Time-stretched short-time Fourier transform," *IEEE Transactions on Instrumentation and Measurement,* vol. 55, pp. 598–602, 2006.
30. Y. Han, O. Boyraz, and B. Jalali, "Tera-sample per second real-time waveform digitizer," *Applied Physics Letters,* vol. 87, p. 241116, 2005.
31. Y. Han, and B. Jalali, "Continuous-time time-stretched analog-to-digital converter array implemented using virtual time gating," *IEEE Transactions on Circuits and Systems,* vol. 52, pp. 1502–1507, 2005.
32. J. A. Bell, M. C. Hamilton, and D. A. Leep, "Optical sampling and demultiplexing applied to A/D conversion," *Proceedings of SPIE,* vol. 1562, pp. 276–280, 1991.
33. R. C. Williamson, R. D. Younger, P. W. Juodawlkis, J. J. Hargreaves, and J. C. Twichell, "Precision calibration of an optically sampled analog-to-digital converter," *Digest of IEEE LEOS Summer Topical Meeting on Photonic Time/Frequency Measurement and Control,* pp. 22–23, 2003.
34. F. Coppinger, A. S. Bhushan, and B. Jalali, "12 Gsample/s wavelength division sampling analogue-to-digital converter," *Electronics Letters,* vol. 36, pp. 316–318, 2000.
35. R. C. Williamson, P. W. Juodawlkis, J. L. Wasserman, G. E. Betts, and J. C. Twichell, "Effects of crosstalk in demultiplexers for photonic analog-to-digital converters," *Journal of Lightwave Technology,* vol. 19, pp. 230–236, 2001.
36. M. P. Fok, K. L. Lee, and C. Shu, "4 × 2.5 GHz repetitive photonic sampler for high-speed analog-to-digital signal conversion," *IEEE Photonics Technology Letters,* vol. 16, pp. 876–878, 2004.

37. W. Ng, L. Luh, D. Persechini, D. Le, Y. M. So, M. Mokhtari, C. H. Fields, D. Yap, and J. E. Jensen, "Ultra-high speed photonic analog-to-digital conversion technologies," *Proceedings of SPIE,* vol. 5435, pp. 171–177, 2004.
38. H. Zmuda, M. J. Hayduk, R. J. Bussjager, and E. N. Toughlian, "Wavelength-based analog-to-digital conversion," *Proceedings of SPIE,* vol. 4547, pp. 134–145, 2002.
39. K. Takizawa and M. Okada, "Analog-to-digital converter: A new type using an electrooptic light modulator," *Applied Optics,* vol. 18, pp. 3148–3151, 1979.
40. R. A. Becker and F. J. Leonberger, "2-bit 1 Gsample/s electrooptic guided-wave analog-to-digital converter," *IEEE Journal of Quantum Electronics,* vol. 18, pp. 1411–1413, 1982.
41. F. J. Leonberger, C. E. Woodward, and R. A. Becker, "4-bit 828-megasample/s electro-optic guided-wave analog-to-digital converter," *Applied Physics Letters,* vol. 40, pp. 565–568, 1982.
42. C. L. Chang and C. S. Tsai, "Electro-optic analog-to-digital converter using channel waveguide Fabry-Perot modulator array," *Applied Physics Letters,* vol. 43, pp. 22–24, 1983.
43. R. A. Becker, C. E. Woodward, F. J. Leonberger, and R. C. Williamson, "Wideband electrooptic guidedwave analog-to-digital converters," *Proceedings of the IEEE,* vol. 72, pp. 802–819, 1984.
44. P. E. Pace and D. D. Styer, "High-resolution encoding process for an integrated optical analog-to-digital converter," *Optical Engineering,* vol. 33, pp. 2638–2645, 1994.
45. B. Jalali and Y. M. Xie, "Optical folding-flash analog-to-digital converter with analog encoding," *Optics Letters,* vol. 20, pp. 1901–1903, 1995.
46. J. Stigwall and S. Galt, "Demonstration and analysis of a 40 gigasample/s interferometric analog-to-digital conversion," *Journal of Lightwave Technology,* vol. 24, pp. 1247–1256, 2006.
47. W. Li, H. Zhang, Q. Wu, and M. Yao, "All-optical analog-to-digital conversion based on polarization-differential interference and phase modulation," *IEEE Photonics Technology Letters,* vol. 19, no. 8, pp. 625–627, 2007.
48. Q. Wu, H. Zhang, Y. Peng, X. Fu, and M. Yao, "40 GS/s Optical analog-to-digital conversion system and its improvement," *Optics Express,* vol. 17, pp. 9252–9257, 2009.
49. T. Nishitani, T. Konishi, and K. Itoh, "Optical coding scheme using optical interconnection for high sampling rate and high resolution photonic analog-to-digital conversion," *Optics Express,* vol. 15, pp. 15812–15817, 2007.
50. S. Oda and A. Maruta, "A novel quantization scheme by slicing supercontinuum spectrum for all-optical analog-to-digital conversion," *IEEE Photonics Technology Letters,* vol. 17, pp. 465–467, 2005.

# 23

# Fiber-Optic Gyroscopes

Karl Kissa
*JDSU*

Jeffrey E. Lewis
*Honeywell International Inc.*



## 23.1 Introduction

Fiber-optic gyroscopes (FOGs) belong to a class of rotation sensors that employ an optical interferometer to measure rotation rate by means of the Sagnac effect. The ring laser gyro also belongs to the family of rate sensors that employ the Sagnac effect to measure rotation. The interferometric fiber-optic gyroscope (IFOG) is a ring interferometer that uses a multiturn optical fiber coil to enhance the Sagnac effect induced by rotation with respect to inertial space. A significant advantage of the FOG is that it has no moving parts and hence the resulting solid state architecture ensures high reliability. Other benefits include scalability, high precision and accuracy, and low power consumption.

Applications requiring knowledge of attitude and heading, such as inertial navigation, employ gyroscopes to measure angular displacement with respect to an inertial frame of reference. Inertial navigation and attitude and heading knowledge are important for many applications. Some of these applications include aircraft inertial navigation, marine inertial navigation, aircraft and spacecraft attitude and heading, and missile guidance.

The performance of gyros is divided into three grades in accordance to the angular drift per hour (bias drift) caused by ancillary effects such as temperature, pressure, etc. These three categories are tactical grade, navigation grade, and strategic grade. Tactical grade gyros have a bias error of approximately 1°/hr, Navigation grade gyros have a bias error better than 0.01°/hr, and strategic grade gyros have a bias error that is better than 0.005°/hr.

In recent years, IFOGs have become widely used in various applications such as missile guidance, satellite inertial attitude reference, and submarine navigation. The IFOG's scalability makes it a versatile sensor with diverse applications that has demonstrated performance that can meet or exceed the performance of other gyroscope technologies. IFOG's high frequency response to the angle position makes it an excellent sensor for platform stabilization and pointing. IFOGs are found in applications ranging from terrestrial navigation to spacecraft and planetary rovers as a consequence of its high reliability, light weight, and low power consumption.

## 23.2 Principle of Operation

The Sagnac effect was first demonstrated in 1913 by Georges Sagnac [1,2]. Sagnac constructed a closed optical path on a revolving platform with mirrors. He then split a beam of light and sent the two beams around the closed optical path in opposite directions. Concomitant with recombining the two beams, Sagnac was able to observe an interference pattern that changed as the rotation rate of the platform changed.

Even though the origin of the Sagnac effect lies in special relativity, it is instructive to look at a classical derivation to understand how the Sagnac effect works. The classical derivation will lead to insight as to the origin of the optical phase shift between the two counter propagating light waves [3]. Consider a circular optical path (i.e., a loop of optical fiber) with a radius $r$ rotating with an angular velocity of $\omega$ radians/sec (see Figure 23.1). Let there be a point in this optical path where light can enter and exit the optical loop. If the entry/exit point starts at point A in Figure 23.1, then at a time $t$ later the entry/exit point has moved to point B. Light is injected into the loop and split such that half the light is traveling clockwise in the loop and half of the light is traveling counterclockwise in the loop.

Now, consider the time required for light to propagate around the loop in each direction. The speed of light in the optical fiber is given by the speed of light in a vacuum, $c$, divided by the index of refraction of the fiber encountered by the light wave traveling in the clockwise direction, $n_{cw}$, or $v_{cw} = c/n_{cw}$. The light traveling in the clockwise direction will travel a distance equivalent to the length of the loop, $L$, plus the additional distance that the entry/exit point has traveled during the time the light traversed the loop. The time required to travel the total distance $L + \Delta L$ is given by $\tau_{cw} = (L + \Delta L)/v_{cw}$, or:

$$\tau_{cw} = \frac{n_{cw}(L+\Delta L)}{c}.$$

Similarly, the light traveling in the counterclockwise direction travels a distance $L$ while the entry/exit point advances a distance $\Delta L$, hence the total distance the light travels will be $L - \Delta L$. The time required for the light to traverse the loop in the counterclockwise direction is given by

$$\tau_{ccw} = \frac{n_{ccw}(L-\Delta L)}{c}.$$

The two indices of refraction, $n_{cw}$ and $n_{ccw}$, are the indices seen by the clockwise and counterclockwise optical waves in the fiber, which are slightly different due to the motion of the medium.

The distance that the entry/exit point moves during the time the light traverses the loop is given by $\Delta L = r\omega t$ where $r = D/2$, $t = L/v$, and $v = n/c$. Putting this together we get:

$$\Delta L = \frac{D}{2}\left(\frac{Ln}{c}\right)\omega = \frac{nLD}{2c}\omega.$$

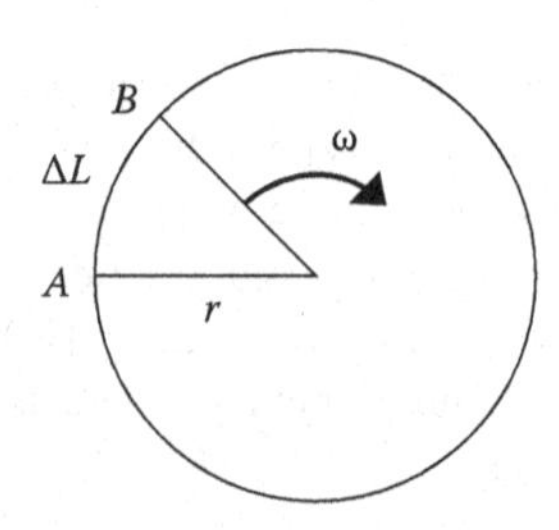


**FIGURE 23.1** Closed optical loop rotating clockwise with angular velocity ω.

Now we can find the phase shift between the two counter propagating beams, induced by rotating the optical fiber, by first finding the difference between the time it took for the two beams to traverse the optical path:

$$\tau_{cw} - \tau_{ccw} = \frac{L}{c}(n_{cw} - n_{ccw}) + \frac{n^2 LD}{c^2}\omega. \tag{23.1}$$

To find $n_{cw}$ and $n_{ccw}$, we need to find the velocities of the counterpropagating waves in the medium by relativistic addition of the velocity of the fiber with the velocity of the light waves in the fiber. To perform this addition, we will use the composition law for velocities from the theory of special relativity, given by

$$v_1 \oplus v_2 = \frac{v_1 + v_2}{\left(1 + \frac{v_1 v_2}{c^2}\right)}.$$

Combining the velocity of light in the optical fiber at rest, which is given by $v_1 = n/c$, with the velocity of the rotating optical fiber, given by $v_2 = \pm r\omega = \pm D\omega/2$, yields:

$$v_{cw} = \frac{v_1 + v_2}{\left(1 + \frac{v_1 v_2}{c^2}\right)}$$

and

$$v_{ccw} = \frac{v_1 + v_2}{\left(1 - \frac{v_1 v_2}{c^2}\right)}.$$

Using the binomial expansion to expand the denominator, $(1 \pm v_1 v_2/c^2)^{-1}$ and keeping only terms that are first order in $\omega$ gives

$$v_{cw} = \frac{c}{n_{cw}} = \frac{c}{n} + \left(1 - \frac{1}{n^2}\right)\frac{\omega D}{2} + \ldots$$

and

$$v_{ccw} = \frac{c}{n_{ccw}} = \frac{c}{n} - \left(1 - \frac{1}{n^2}\right)\frac{\omega D}{2} + \ldots$$

Solving for $n_{cw}$ and $n_{ccw}$ gives

$$n_{cw} = \frac{c}{\frac{c}{n} + \left(1 - \frac{1}{n^2}\right)\frac{D\omega}{2}}$$

$$n_{ccw} = \frac{c}{\frac{c}{n} - \left(1 - \frac{1}{n^2}\right)\frac{D\omega}{2}}.$$

Taking the difference between $n_{cw}$ and $n_{ccw}$ yields

$$n_{cw} - n_{ccw} = (1 - n^2)\frac{D\omega}{c}.$$

Substituting into Equation 23.1 gives

$$\tau_{cw} - \tau_{ccw} = \left\{(1 - n^2)\frac{LD}{c^2}\omega + \frac{n^2 LD}{c^2}\omega\right\}. \tag{23.2}$$

The first term in Equation 23.2 is referred to as the Fresnel drag term and the second term is the delay caused by the motion of the entry/exit point. The terms that depend on the index of refraction cancel, leaving

$$\tau_{cw} - \tau_{ccw} = \frac{LD}{c^2}\omega.$$

The phase difference between the two returning optical waves will then be given by

$$\Delta\varphi = \frac{2\pi}{\lambda}\Delta L = \frac{2\pi c}{\lambda}(\tau_{cw} - \tau_{ccw}) = \frac{2\pi LD}{\lambda c}\omega. \tag{23.3}$$

The net result of the math is that the observed phase shift is proportional to the area enclosed in the optical loop, LD. Hence, increasing the LD product by increasing the number of turns of fiber or by increasing the diameter of the fiber loop will increase the sensitivity of the fiber-optic gyro.

## 23.3 Device Topologies

### 23.3.1 Minimum Reciprocal Configuration

The preferred architecture of FOGs is the minimum reciprocal configuration [3–5]. Optical phase differences between the counterpropagating waves that are not directly caused by the rotation of the gyro are indistinguishable from the phase shifts induced by rotation. Hence, the nonrotation induced phase shifts need to be minimized. One source of phase shifts that are not caused by rotation occurs when the clockwise optical path is different than the counterclockwise optical path. To minimize the phase shifts brought on by differences in the optical path, the two counterpropagating beams must follow identical paths through the gyro. Two beam splitters (or fiber-optic couplers) are required to ensure that the two counter propagating beams see the loop splitter the same way.

To illustrate, consider the nonreciprocal configuration illustrated in Figure 23.2. The light that travels in the clockwise direction (dashed arrows) in the loop will have passed straight through the loop coupler both when entering the fiber loop then again when exiting the loop. The light traveling in the counterclockwise direction (solid arrows) will have crossed the coupler both when entering the fiber loop and when exiting the loop. Thus, the clockwise and counterclockwise paths are not the same and could be a major source of optical phase shift.

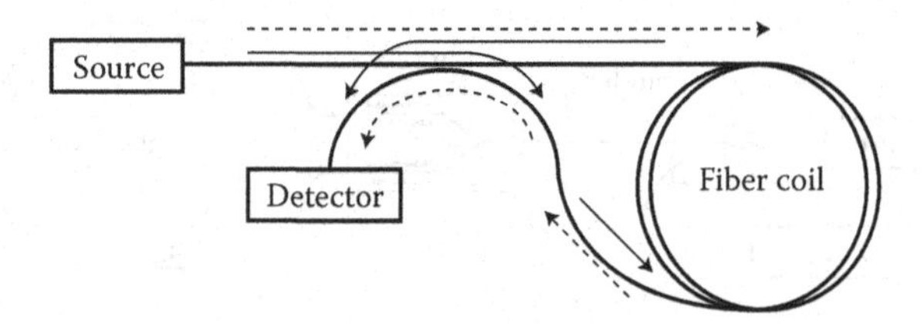


**FIGURE 23.2** FOG configuration that is not reciprocal. Dashed arrows represent the clockwise path and solid arrows represent the counterclockwise path.

The optical path can be made reciprocal by adding another fiber-optic coupler so that the detector is not attached to the loop coupler (Figure 23.3). In this configuration, both of the counterpropagating beams will cross through the loop coupler once and pass straight through the loop coupler once before reaching the detector. Therefore, both the counterpropagating light waves see the loop coupler the same way and the configuration is said to be reciprocal.

The loop splitter may be implemented either using an optical fiber splitter, as depicted in Figure 23.3, or by using an integrated optic Y-junction, as in Figure 23.4. Using an integrated optic Y-junction offers several advantages. One of the advantages is that multiple functions can be integrated into a single device, reducing the number of components. The integrated optic circuit commonly used in fiber-optic gyros integrates a Y-junction, a polarizer, and a phase modulator into a single compact device.

The optical fiber developed for the telecommunications industry was designed to have low dispersion and low loss in the 1.3-μm and 1.55-μm wavelength ranges. The wavelength of the optical source is designed to take advantage of these properties of the fiber. Also, short coherence length light sources are used in fiber-optic gyros to reduce spurious signals due to light backscattered from imperfections in the loop, to reduce interference formed by polarization cross-coupling points, to reduce backscatter from Rayleigh scattering, and to minimize nonlinear effects such as the Kerr effect. Typical broadband sources that are employed are superluminescent diodes or amplified spontaneous emission sources.

Light from the source passes through the first coupler and is directed to the integrated optics circuit (IOC), where it is split into two counterpropagating light waves. Each wave travels around the fiber coil in either a clockwise (CW) or a counterclockwise (CCW) direction. The CW and CCW waves then exit the coil, are recombined in the integrated optics circuit's Y-junction, and are directed to the signal photodetector by means of the first coupler.

Any phase difference in the recombined CW and CCW waves will result in an intensity change at the detector. The intensity at the detector is given by a raised cosine function:

$$I_d = \frac{I_0}{2}\left\{1+\cos(\Delta\phi_{rotation})\right\}.$$

Here $I_0$ is the intensity of the light reaching the detector in the absence of any rotation, and $\Delta\phi$ is the phase difference due to rotation as given by Equation 23.3. When the intensity of the interference pattern

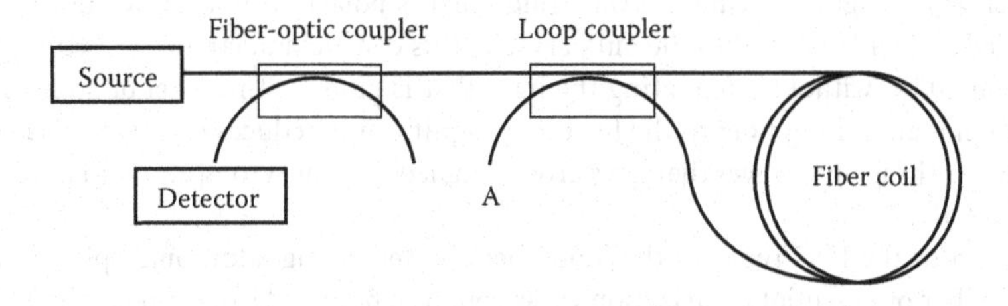


**FIGURE 23.3** Architecture of a fiber-optic gyro in the minimum reciprocal configuration.

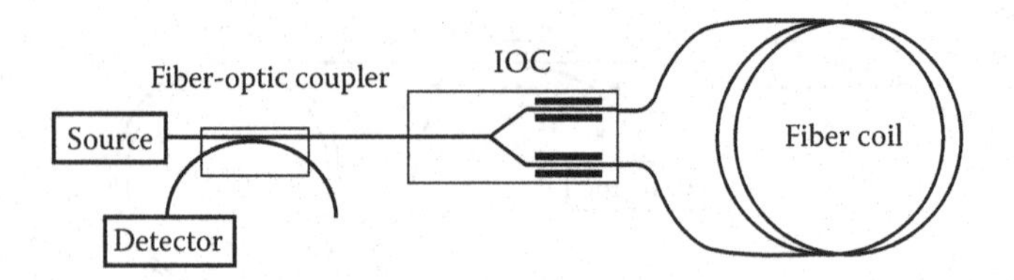


**FIGURE 23.4** Architecture of a fiber-optic gyro in the minimum reciprocal configuration employing an IOC.

on the detector is used directly to determine the rotation rate, the gyro is said to be operating open loop. In order to overcome direction ambiguity an additional time dependent phase can be injected via the phase modulator on the IOC. The simplest approach is to inject a sinusoidal phase via a phase modulator and synchronously demodulate the output signal on the odd harmonics of the injected phase. The output will have a maximum sensitivity near zero rotation rate and no directional ambiguity [4]. This injected phase is referred to as bias modulation and bias modulation is not just limited to sinusoidal waveforms [3,6] but can take other forms (cf. [3, pp. 32–36] for an alternate form of bias modulation). The primary disadvantages to open loop operation are that the intensity of the output signal, being a raised cosine function, is not linear over a large range of input rotation rates, and that the rate range of operation is limited.

To address the issues of poor linearity and limited rate range a closed-loop (or phase nulling) configuration can be employed. In closed-loop operation, a phase shift that is equal and opposite to the rotation induced phase shift is generated electronically and added to the bias modulation and applied by means of an electro-optic modulator such as the phase modulators of the IOC. The form of the phase nulling voltage time waveform can take any number of forms. Typical waveform patterns used include sawtooth waves such as the serrodyne waveform, square waves, or triangle waves used in "dual-ramp" waveform [4,3, Chapter 8]. Each waveform has its own advantages. The magnitude of this additional applied voltage is used to determine the rotation rate of the instrument. Operating a FOG in a closed-loop mode will result in liner rate output over a broad range of input rates, which results in improved gyro accuracy. Another advantage is the increased range of rotation rates that the gyro can measure. Closed-loop operation is made possible by the high-fidelity and high-bandwidth electro-optic modulation made possible by the phase modulators on the IOC.

## 23.4 Integrated Optics Circuit

The integrated optic circuits currently used for fiber gyros offer several distinct advantages. The first advantage is that the device combines three critical functions into a single device. The benefits are reduced size and weight, improved performance, reduced number of splices, increased reliability, and reduced assembly effort.

Single-mode waveguides fabricated on lithium niobate substrates using annealed proton exchange (APE) are inherently polarizing. APE waveguides in X-cut lithium niobate will only guide the TE (transverse electric) mode, the TM (transverse magnetic) mode will not be guided and disperse into the substrate. In other words, light entering the waveguide that is polarized normal to the crystal surface is totally unguided and will be lost into the bulk crystal. This can yield polarization extinction ratios that are greater than 60 dB without attenuating the light that is guided. This level of attenuation, coupled with polarization maintaining fiber in the fiber loop significantly reduces errors associated with polarization by extinguishing light waves that have cross-coupled into the wrong axis and have taken a spurious path in the coil.

The Y-junction in the IOC replaces the loop fiber splitter saving additional splices, which in turn reduces the number of potential polarization cross coupling points. Also, forming the Y-junction with the waveguides forms a more stable and compact splitter.

The phase modulators of the IOC enable the closed-loop gyro operation. High-bandwidth voltage waveforms applied to the phase modulators are converted into optical phase delays. An electric waveform is applied by means of the phase modulators. The applied voltage modifies the index of refraction of the waveguide, thus imparting a phase shift to light traveling through the waveguide. The voltage required to shift the optical phase by 180° is referred to as $V_\pi$. The bias modulation and the feedback ramp used to null the phase shifts caused by rotation are generated electronically and applied to the phase modulators of the IOC, which convert the electrical signal into optical phase shifts in the optical waves guided in the IOC. In addition to the bias modulation, error suppression modulation, and modulation used to measure the IOC's $V_\pi$ are other applications that employ the optical phase modulators on the IOC.

The design and performance of the IOC are critical to the gyro performance. The important optical properties include low-loss single-mode waveguides, high polarization extinction ratio, balanced split at the Y-junction, and low optical backscatter from the interface between the waveguides and the fiber pigtails. The important electrical properties include low $V_\pi$, high bandwidth electric-to-optic (E–O) conversion, linear response (low harmonic distortion), low hysteresis, and low residual intensity modulation.

Any effects that occur in addition to the simple functions described above generally create aberrations in the measured response at the photodiode, thereby degrading performance. For example, optical reflections and/or polarization effects in the system cause multiple paths for the light to travel, causing the resultant interference pattern at the photodiode to deviate from the ideal [7]. Often the undesired aberrations cannot be separated easily from the main desired effect, causing errors in the detected rotation rate. Most of the design challenges are related to suppressing these undesired aberrations, in order to maximize system performance.

Many of the design considerations that maximize performance for the gyro application are similar to those that maximize performance for communications applications; however, the gyro application is more sensitive to certain aberrations, hence great care needs to be paid to measurement of chip performance, and optimization of the chip design. Some of the design considerations are evident by closer examination of Figure 23.5.

First, the chip endfaces are typically angled 10° in the horizontal plane at the optical input/output to reduce back-reflections from being captured by the optical fibers [8]. The optical fiber carrying light from the optical source is connected to the optical input and aligned to the TE polarization in the chip. Due to mode-mismatch between the optical fiber and waveguide, some optical power radiates into the substrate. In addition, if APE waveguides are used in the chip, any light having the unwanted polarization state (TM) is radiated into the substrate as well. A short length of waveguide, the input spatial filter,

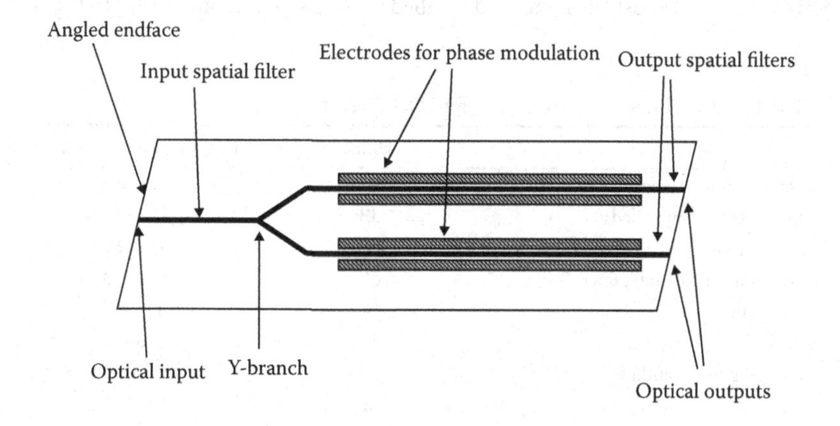


**FIGURE 23.5** Architecture of a fiber-optic gyro integrated optics circuit.

allows all of the radiation modes (TE and TM) to dissipate before the main beam reaches the Y-branch. Hence, the light is spatially filtered and well polarized before reaching the Y-branch.

The waveguides are designed to be single-mode as any light traveling in the higher-order modes may introduce spurious signals at the detector, much in the same way as light in the TM polarization. The higher order modes are split into two paths by the Y-junction [9]. Coherent beating between the split higher-order modes and fundamental mode manifests itself as an imbalance in the split ratio of the Y-junction. Careful attention must be paid to limit the number of guided modes in the waveguide to one.

Once the light is split into two beams, phase modulation is applied. The beams will experience phase modulation again upon returning from the fiber coil. Typically, the phase modulators also create a small amount of residual intensity modulation (IM) due to several effects within the chip. First, the electric field from the electrodes modifies the refractive index profile within the waveguide slightly, causing the mode shape to change slightly. Propagation loss may change slightly due to the new mode shape. This effect will be much larger if there is significant overlap of the mode with the metal electrode on the surface of the chip. In this case, the optical loss due to metal loading becomes a function of the mode shape, and hence the applied voltage. The voltage-dependent optical loss creates intensity modulation that is nearly linear with applied voltage. If a coherent optical source is used, other stray components of optical field can create additional components of IM. For example, a portion of the light radiated within the substrate is reflected within the substrate, and carried to the optical outputs, where a tiny portion can couple into the output fibers. The coherent interference of the main beam with the stray light component creates IM. Fortunately, the incoherent source used in the system has a short enough coherence length to eradicate the IM caused by substrate light. Note that IM is typically measured with an LED or other incoherent source. The light having TM polarization also travels this same path, and may be more deleterious, even with an incoherent source.

After phase modulation, an output spatial filter strips out any modal aberrations introduced by the phase modulators. If the phase modulators are designed properly, these aberrations are small; hence, the length of output spatial filter can be relatively short.

Table 23.1 lists the most critical performance parameters. The list is not in any prioritized order, as the importance of various parameters depends on the particular gyro application. Ranges of typical values are also listed. Note that FOG circuits are designed to work over a narrow wavelength range, e.g., 820–840, or 1280–1320, or 1520–1550 nm. Back-reflection occurs at the $LiNbO_3$-to-fiber interface.

## 23.5 Improvements to the Integrated Optic Circuit

There are several chip design improvements that improve performance of the gyro. First of all, designs to absorb both TE and TM substrate light are described in the patent literature [10,11]. Figure 23.6 shows

**TABLE 23.1** Important Parameters for FOG Circuit

| Parameter | Units | Range of Typical Values |
|---|---|---|
| Operating wavelength | Nm | 820–1550 |
| Insertion loss (pigtailed) | dB | 3.0–5.0 |
| Polarization extinction ratio, chip | dB | 55 to >65 |
| Polarization crosstalk, fiber | dB | −24 to <−30 |
| Split ratio | % | 45–55 |
| Back-reflection | dB | −60 |
| Optical power handling | mW | 5–200 |
| Half-wave voltage, $V_\pi$ | Volts | 3.1–4.5 |
| Intensity modulation | % | 0.02–0.2 |

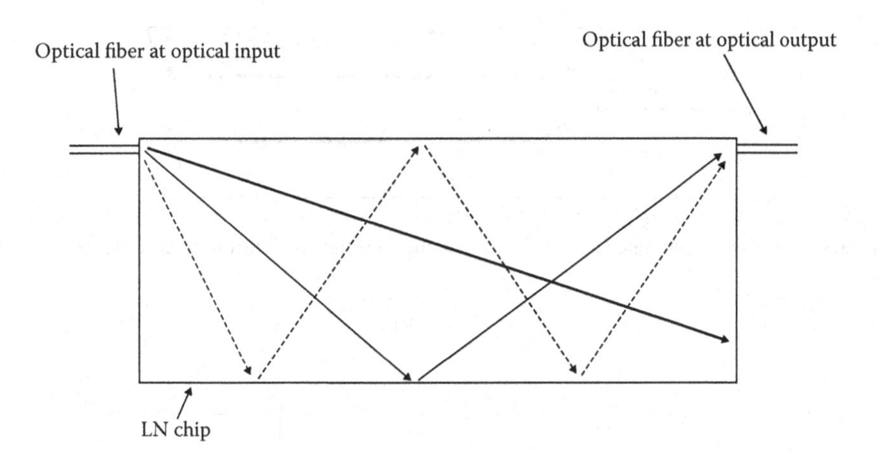


**FIGURE 23.6** Vertical cross section along optical propagation direction showing stray light path through substrate.

a vertical cross section of the chip. The chip and optical fibers are shown while the optical waveguide in the chip is not shown for sake of clarity. Also shown are several rays emanating from the optical input, which represent the TE or TM light that is radiated into the substrate. The heavy line, thin line, and dashed line represent different paths the radiated light takes within the substrate, experiencing one or more reflections before reaching the end of the chip. Note that due to reciprocity, the light within the substrate enters the optical fiber at the output. The acceptance of substrate light back into the system degrades gyro performance. A design to mitigate reflection off the bottom of the chip is described [10] and shown in Figure 23.7. Slots that are filled with optical absorber extend inward into the substrate. The absorbers help to remove any TE and TM light that is radiated into the substrate, in order to prevent introduction of the light back into the system at the optical output. Note that the substrate light absorption also works in reverse for light entering the output port and exiting the input port.

Slots can also be introduced from the side of the chip [11]. Figures 23.8 and 23.9 describe the improvement from side slots. Figure 23.8 shows a top view of a chip having angled endfaces, along with rays representing light radiated into the substrate. The path taken by these rays is shown in the horizontal plane. The solid line shows a path having only one reflection off the side of the chip, while the dashed line shows a path with three reflections. Figure 23.9 shows a design with slots in the side of the chip that absorbs light radiated into the substrate. The larger slot midway down the length of the device absorbs light having the primary single-bounce reflection while the smaller slots help to absorb light having secondary multiple-bounce reflections. The slots can be filled with absorptive material, as described in the patent literature [10].

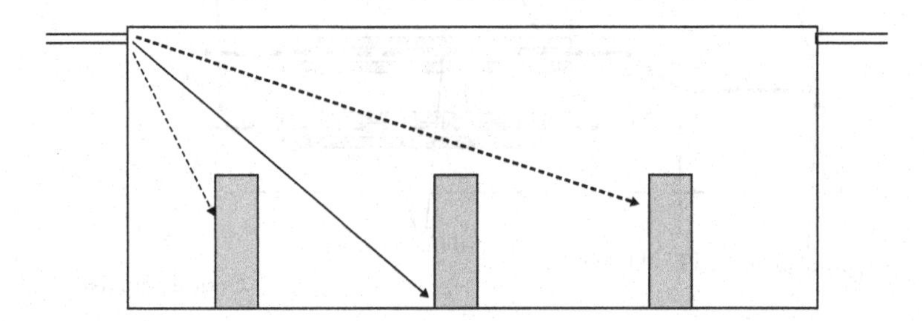

**FIGURE 23.7** Vertical cross section along optical propagation direction showing stray light path through substrate with slots to absorb stray light.

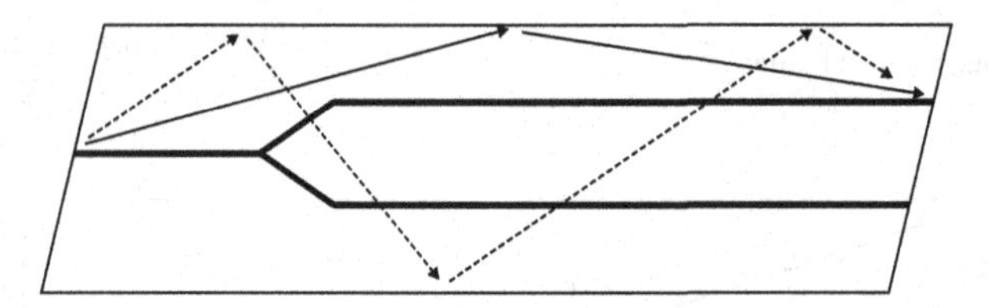

**FIGURE 23.8** Top view of integrated optics circuit showing stray light path through substrate in horizontal plane.

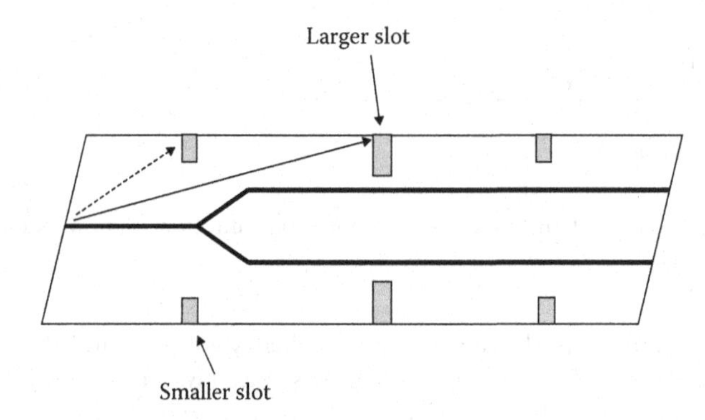


**FIGURE 23.9** Top view of integrated optics circuit showing stray light path through substrate in horizontal plane with slots to absorb stray light.

Environmental effects on these parameters need to be considered, as well. Strategies to suppress environmental sensitivity have been developed [12]. The FOG circuit is cleaned and then exposed to a reactive oxide such as carbon monoxide (CO) to passivate the surface. Finally, a sealing material such as silicon dioxide ($SiO_x$) or a material from the poly-*para*-xylylene family is deposited onto the surface.

Figure 23.10 describes a further refinement where two different types of waveguides are integrated in the device. APE waveguides described earlier have excellent polarization extinction, however, the electro-optic response changes in the presence of a vacuum [13]. The effect is mitigated by stitching in a Ti in-diffused waveguide in the section where modulation occurs. The Ti-waveguide has poorer

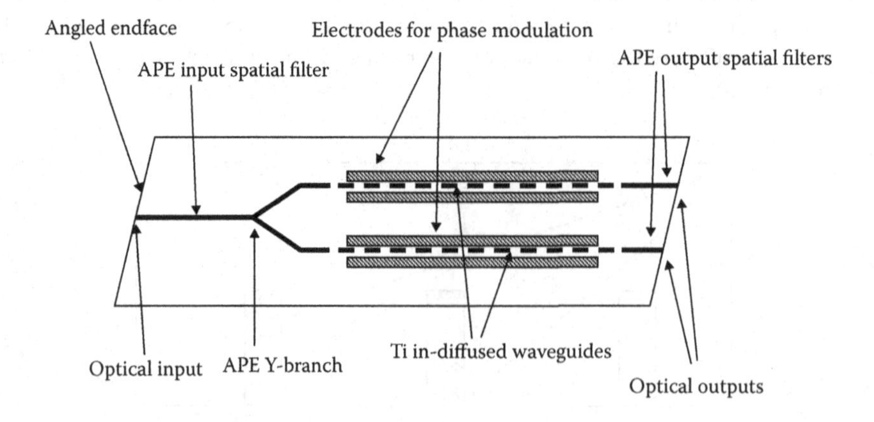


**FIGURE 23.10** Architecture of a fiber-optic gyro-integrated optics circuit having Ti in-diffused waveguides stitched in between sections of APE waveguides.

polarization extinction, but does not exhibit the change in electro-optic behavior in vacuum. The overall degradation in polarization extinction ratio of the chip is small.

In summary, adequate performance of the lithium niobate fiber-optic gyro circuit is possible by paying attention to all the mechanisms that degrade gyro performance. A number of features that improve performance have been described.

## References


1. G. Sagnac, "L'éther lumineux démontré par l'effet du vent relative d'éther dans un interféromètre en rotation uniforme," *Comptes Rendus l'Académie Sciences*, vol. 157, pp. 708–710, 1913.
2. G. Sagnac, "Sur la prevue de la réalité de l'éther lumineux par l'expérience de l'interférographe tournant." *Comptes Rendus l'Académie Sciences*, vol. 157, pp. 1410–1413, 1913.
3. H. Lefevre, *The Fiber-Optic Gyroscope*, Norwood, MA: Artech House, 1993.
4. J. Blake, "Fiber optic gyroscopes," in *Optical Fiber Sensor Technology*, vol. 2, K. T. V. Gratten and B. T. Meggitt, Eds., London: Chapman & Hall, 1998, pp. 303–328.
5. S. Ezekiel, S. P. Smith, and F. Zarinetchi, "Basic principles of fiber-optic gyroscopes," in *Optical Fiber Rotation Sensing*, W. K. Burns, Ed., Boston: Academic Press, 1994, pp. 1–25.
6. B. Y. Kim, "Signal Processing Techniques," in *Optical Fiber Rotation Sensing*, W. K. Burns, Ed., Boston: Academic Press, 1994.
7. S. Ezekiel and H. J. Arditty, *Fiber-Optic Rotation Sensors*, 1st ed., Berlin, Germany: Springer-Verlag, 1982, pp. 23, 52–81, 102–110, 124–135.
8. *UTP Catalog of Integrated Optical Circuits*, UTP Electro-optics Product Division (now JDSU), Bloomfield, CT, 1997, pp. 21–22.
9. Z. Weismann, A. Hardy, and E. Marom, "Mode-dependent radiantion loss in y-junctions and directional couplers," *IEEE Journal of Quantum Electronics*, vol. 25, no. 6, pp. 1200–1208, Jun. 1998.
10. K. M. Kissa, "Optical substrate with light absorbing segments," U.S. Patent 5 321 779, Jun. 14, 1994.
11. J. E. Lewis, "Devices and methods for spatial filtering," U.S. Patent 7 711 216, May 4, 2010.
12. L. M. Hendry, J. E. Lewis, J. C. Grooms, and C. B. Gray, "Environmentally stable electro-optic device and method for making same," U.S. Patent 7 228 046, Jun. 5, 2007.
13. J. Feth, "Stitched waveguide for use in a fiber-optic gyroscope," U.S. Patent application publication 2009/0 219 545, Sep. 3, 2009.

# 24

# Polarization Control


Fred Heismann
*JDSU Optical Networks Research Lab*

Suwat Thaniyavarn
*EOSPACE, Inc.*

David Moilanen
*EOSPACE, Inc.*




## 24.1 Introduction

Single-mode optical fibers employed in long-distance telecommunication systems usually do not preserve the state of polarization (SOP) of the launched optical signal. Thus, the SOP of an optical signal emerging from a fiber-optic cable is usually unknown and may even fluctuate randomly with time [1]. Optical components at the receiving end of a fiber-optic link are therefore required to operate with all possible polarization states. However, certain optical elements, such as polarization demultiplexers or polarization-mode dispersion compensators [2,3], are inherently sensitive to the signal's SOP. If such an element is employed at the receiving end of a fiber, then the signal emerging from the fiber most likely needs to be transformed into the preferred input SOP of the polarization-sensitive element. This transformation may be accomplished by inserting an adjustable polarization controller before the polarization-sensitive element [4]. Moreover, if the input SOP to the controller fluctuates randomly with time, the polarization transformation needs to be continuously adjusted to compensate for the changes in the input SOP. Such adaptive transformation of a time-varying input SOP into a fixed output SOP is known as automatic polarization stabilization [5]. Figure 24.1 shows a typical setup for an automatic polarization stabilization system, in which the polarization controller is adjusted via a feedback signal received from an optical monitor detector that is connected to the output of the polarization-sensitive element [6].

In other applications, a polarization controller may be used to transform a fixed (and sometimes unknown) input SOP into a time-varying output SOP. This feature is often desired in test and measurement equipment to synthesize a certain SOP or to sweep the SOP of an optical test signal through a predetermined range of states. Moreover, fast polarization modulation (or "scrambling") is sometimes employed to depolarize the output light on the timescale of the modulation frequency, which may range from nanoseconds to milliseconds [7].

Aside from cost and physical size, the two most important attributes of a polarization controller are the range of polarization transformations it can perform and the speed at which it can track changes in

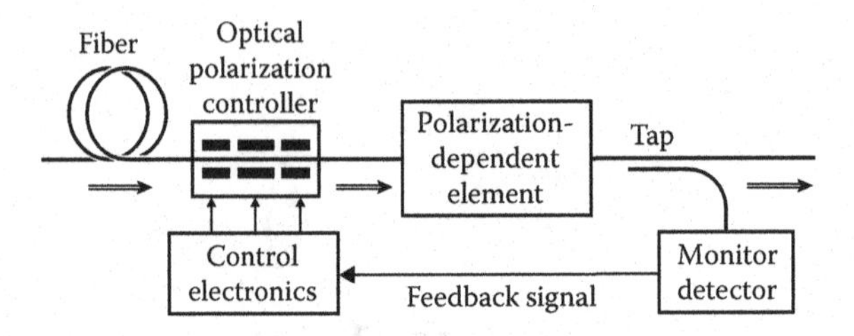


**FIGURE 24.1** Typical setup of an automatic polarization control system.

the input or output SOP. Some applications may also need high accuracy and repeatability in obtaining the desired transformation.

Most applications require a polarization controller that is capable of transforming an arbitrary input SOP into the desired output SOP, where the latter may either be constant and well-defined (e.g., a linear SOP at fixed angular orientation) or arbitrary and even time-varying. It is well known that transformation of an arbitrary input SOP to an arbitrary or fixed output SOP requires at least two independently adjustable control parameters or "degrees of freedom." Polarization controllers, therefore, typically comprise two or more transformation stages or require adjustment of at least two independent control parameters. Three-stage polarization controllers composed of rotatable fiber-optic loops, for instance, are widely used in laboratory setups to convert arbitrary input SOPs into any desired output SOP [8].

Polarization controllers may be realized with bulk-optic components as well as with fiber-optic or integrated-optic waveguides. Some of the more popular control schemes and their main features are summarized in Table 24.1. Mechanically rotated wave plates or fiber-optic coils, for example, can provide highly accurate polarization transformations but exhibit inherently slow control speeds [9,10]. These controllers are frequently employed in optical test and measurement equipment as well as in experimental setups, where input and output SOPs vary only slowly. Electro-optic polarization controllers, on the other hand, offer very fast control speeds, in particular when implemented with single-mode waveguides [11–14]. These integrated-optic devices require much lower drive voltages than their bulk-optic counterparts and, hence, allow high-speed operation. In fact, integrated-optic polarization controllers in $LiNbO_3$ are among the fastest devices currently available and are predominantly employed

**TABLE 24.1** Implementations of Optical Polarization Controllers and Their Main Attributes

| Principle of Operation | Implementation | Adjustment Type | Control Speed | Reference |
|---|---|---|---|---|
| Rotatable wave plates | Bulk crystals | Mechanical | Very slow | 9 |
| | Fiber loops | Mechanical | Slow | 10 |
| | Bulk crystals | Electro-optic | Fast | 11 |
| | Liquid crystals | Electro-optic | Medium | 12 |
| | Micro-optic PLZT | Electro-optic | Fast | 13 |
| | I/O waveguides | Electro-optic | Very fast | 14 |
| Phase retarders at fixed orientation | Fiber squeezers | Mechanical | Fast | 15 |
| | Fiber stretchers | Mechanical | Fast | 8 |
| | Liquid crystals | Electro-optic | Medium | 16 |
| | I/O Waveguides | Electro-optic | Very fast | 17 |
| Faraday rotators | Fiber coils | Magneto-optic | Fast | 18 |
| | Micro-optic crystal | Magneto-optic | Very fast | 19 |

for fast polarization scrambling as well as for fast automatic polarization stabilization. In the following section, we describe the design and operation of these polarization controllers.

## 24.2 Lithium Niobate Polarization Controllers

The ease of optical waveguide fabrication in lithium niobate ($LiNbO_3$) as well as its electro-optic properties make it an ideal candidate for high speed polarization control over a broad range of wavelengths. Early $LiNbO_3$ polarization controllers utilized a periodic series of electrodes aligned orthogonally to the Ti in-diffused waveguide on Z-cut, X-propagating $LiNbO_3$ [20]. Due to the periodic electrode configuration, the devices inherently operated efficiently over a narrow band of wavelengths. Thaniyavarn demonstrated the first broadband $LiNbO_3$ polarization controller in 1985 [14,21]. This device implemented three electrodes running parallel to the Ti in-diffused waveguide in X-cut, Z-propagating $LiNbO_3$. This design is widely used in modern high-speed polarization controllers, and this section will describe the basic device operation and demonstrate how to determine device parameters in a typical commercially available polarization controller device.

### 24.2.1 Device Description

High-speed $LiNbO_3$ waveguide–based polarization control is achieved through the linear electro-optic effect. Figure 24.2 is an illustration of the device, its electrode arrangement, and the coordinate system for a standard polarization controller in $LiNbO_3$, along with the $LiNbO_3$ electro-optic tensor.

The device consists of a Z-(optical axis) propagating single-mode waveguide fabricated on an X-cut $LiNbO_3$ substrate. The waveguide supports one TM- ($x$-polarized) and one TE ($y$-polarized) mode. The three-electrode structure provides adjustable control for both vertical and horizontal electrical fields inside the optical waveguide. The horizontal electric field ($E_y$) causes an optical phase shift by electro-optically inducing index changes, $\Delta n_{\text{TM}}$ and $\Delta n_{\text{TE}}$ via the electro-optic coefficients $r_{12}$ and $r_{22}$ (= $-r_{12}$), respectively (see the electro-optic tensor in Figure 24.2). This device design is relatively independent of the optical wavelength because the waveguide propagation direction is along the optical Z-axis so that both TM- and TE-polarized modes experience the same ordinary index ($n_o$) and are very nearly phase-matched. With only small adjustments to the horizontal electrical field strength, the (TE/TM) phase matching condition can be reached. To achieve polarization control, the vertical electrical field ($E_x$) provides the electro-optic induced TE/TM mode coupling via the off-diagonal $r_{61}$ coefficient of the electro-optic tensor [21]. Since both TE/TM polarization coupling (via $E_x$), and differential optical phase shifting (via $E_y$) between the two polarization states can be precisely controlled with the applied voltages, an arbitrary output polarization state can be achieved for a relatively broad wavelength range.

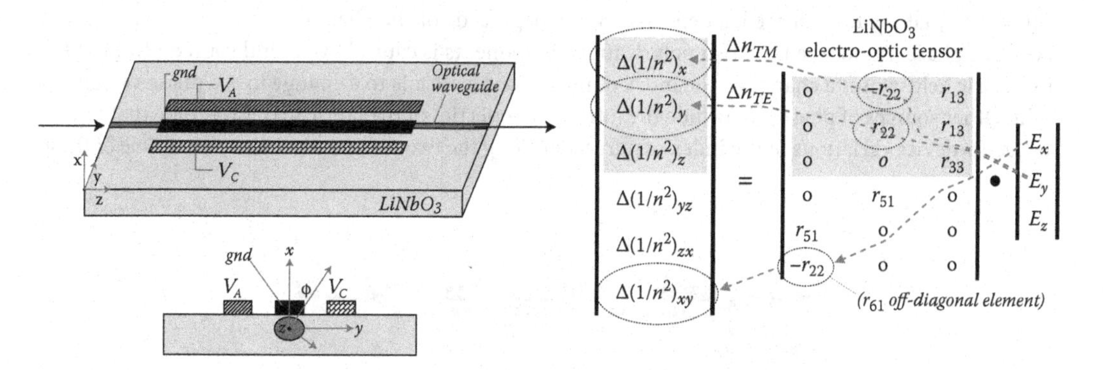


**FIGURE 24.2** Basic $LiNbO_3$ polarization controller device and $LiNbO_3$ electro-optic tensor.

### 24.2.2 Example of a Commercially Available $LiNbO_3$ Polarization Controller Device and Derivation of Device Parameters

A photograph of a commercially available high-speed electro-optic LiNbO3 waveguide polarization controller device manufactured by EOSPACE Inc. is shown in Figure 24.3.

In general, the module consists of multiple stages (1, 3, 4, 5, 6, or 8 stages) of the basic polarization controller devices, equivalent to a series of independent voltage-adjustable waveplates. Multiple staged devices are used to accommodate various electronic control algorithms for a variety of applications, including reset-free polarization control. The module is 4″ × 0.35″ × 0.24″ in dimension. At a wavelength of 1.55 μm, a typical device has an optical insertion loss of ~2 dB (including FC/PC connector mating loss) and a polarization-dependent loss (PDL) <0.1 dB.

To characterize a typical polarization controller device and its control voltage parameters, we analyze the behavior of polarization states as a function of the control voltages [22]. Polarized light may always be expressed as a linear combination of two orthogonal polarization components. If polarized light is propagating in the *z*-direction in X-cut, Z-propagating $LiNbO_3$ with the two polarization components in the xy plane making an angle of $\phi$ and $\phi + \pi/2$ with the *x* axis, the angle $\phi$ is given by [23]:

$$\phi = \frac{1}{2}\arctan\left(\frac{-\tau_1}{\tau_2}\right). \tag{24.1}$$

Here, $\tau_1$ and $\tau_2$ are the phase retardations for light polarized at $\phi$ and $\phi + \pi/2$ with respect to the *x* axis. The total phase lag between the two polarization components is given by

$$\tau = \tau_2\cos(2\phi) - \tau_1\sin(2\phi). \tag{24.2}$$

This is clearer when viewed from a geometric standpoint. $\tau_2$ follows the *y*-direction and $-\tau_1$ follows the $-x$ direction. The projection of the polarization components onto the vertical and horizontal axes determines the magnitude of the phase delay, as shown in Figure 24.4.

We can rewrite cos(2$\phi$) as $\cos(2\phi) = \frac{\tau_2}{\sqrt{\tau_1^2+\tau_2^2}}$ and $\sin(2\phi) = \frac{-\tau_1}{\sqrt{\tau_1^2+\tau_2^2}}$.

Then,

$$\tau = \frac{\tau_2^2}{\sqrt{\tau_1^2+\tau_2^2}} + \frac{\tau_1^2}{\sqrt{\tau_1^2+\tau_2^2}} = \sqrt{\tau_1^2+\tau_2^2}. \tag{24.3}$$

This shows that the total phase lag depends on the magnitude of the phase delay along each of the two axes. The phase lag between the two polarization components is caused by several sources. In $LiNbO_3$, an electric field causes a change in the index of refraction that leads to a change in the phase velocity of light. Other sources of phase lag include the intrinsic properties of the fabricated waveguide that lead to device-to-device variations in the index of refraction along the two polarization directions. One of these

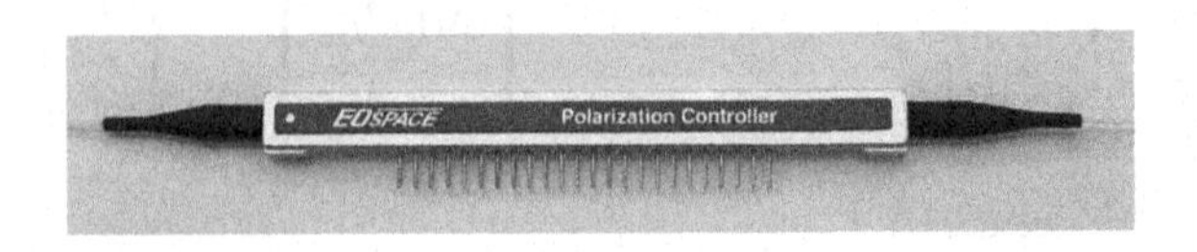


**FIGURE 24.3** Electro-optic $LiNbO_3$-based polarization controller module.

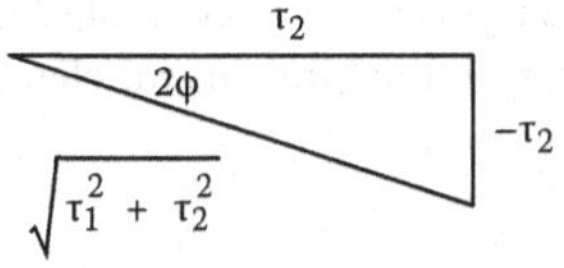

$\tau_2$

$2\phi$

$-\tau_2$

$\sqrt{\tau_1^2 + \tau_2^2}$

**FIGURE 24.4** Projection of polarization components.

sources depends on the voltage applied to the electrodes around the waveguide, the other does not. This relationship can be expressed as:

$$\begin{bmatrix} \tau_1 \\ \tau_2 \end{bmatrix} = \begin{bmatrix} t_{11} & t_{12} \\ t_{21} & t_{22} \end{bmatrix} \cdot \begin{bmatrix} U_S \\ U_{AS} \end{bmatrix} + \begin{bmatrix} t_{1i} \\ t_{2i} \end{bmatrix}. \tag{24.4}$$

Here, $t_{1i}$ and $t_{2i}$ are the intrinsic phase retardation along the $x$- and $y$-axes, respectively. The components of the matrix $T$ describe the response of the $LiNbO_3$ to the applied voltages with

$$T = \begin{bmatrix} t_{11} & t_{12} \\ t_{21} & t_{22} \end{bmatrix}.$$

The voltages are expressed in terms of a symmetric, $U_S$, and an asymmetric, $U_{AS}$, term. The utility of this form has to do with the arrangement of the electrodes around the waveguide. Figure 24.2 shows a schematic diagram of the electrodes and the waveguide in a polarization controller. The electrode directly above the waveguide is held at ground while the two outer electrodes have voltages applied, denoted $V_A$ and $V_C$.

When both outer electrodes have the same voltage, the electric field in the waveguide is predominately in the vertical, $x$-direction, terminating on the central ground electrode. This is the "symmetric" configuration. When the outer electrodes have equal and opposite voltages the electric field in the waveguide is predominately in the horizontal, $y$-direction. This is the "asymmetric" configuration. It is convenient to define $U_S$ and $U_{AS}$ in terms of the voltages on the two outer electrodes. The symmetric term is defined as the average of the two voltages:

$$U_S = \frac{V_C + V_A}{2}. \tag{24.5}$$

The asymmetric term is the difference of the two voltages:

$$U_{AS} = -V_C + V_A. \tag{24.6}$$

This can be conveniently expressed in matrix form:

$$\begin{bmatrix} U_S \\ U_{AS} \end{bmatrix} = \begin{bmatrix} \frac{1}{2} & \frac{1}{2} \\ -1 & 1 \end{bmatrix} \cdot \begin{bmatrix} V_C \\ V_A \end{bmatrix}. \tag{24.7}$$

The task of determining the response of a device to the applied voltages, $V_A$ and $V_C$ is equivalent to understanding how varying these two voltages alters the input state of polarization as it passes through the device. The theory describing the transformation of an input state of polarization (SOP) to an output

SOP in a birefringent element such as a polarization controller can be expressed in terms of a matrix of coefficients based on the phase lag, $\tau$, in the birefringent element that operates on the input state of polarization vector [24].

$$\vec{S}_{out} = \begin{bmatrix} \cos^2 2\phi + \sin^2 2\phi\cos\tau & \frac{1}{2}(\cos\tau - 1)\sin 4\phi & \sin\tau\sin 2\phi \\ \frac{1}{2}(\cos\tau - 1)\sin 4\phi & \sin^2 2\phi + \cos^2 2\phi\cos\tau & -\sin\tau\cos 2\phi \\ -\sin\tau\sin 2\phi & \sin\tau\cos 2\phi & \cos\tau \end{bmatrix} \cdot \vec{S}_{in}. \tag{24.8}$$

Essentially, the goal is to determine how $\tau$ depends on the applied voltages, $V_A$ and $V_C$ because this implies knowledge of the matrix $T$, which fully describes the intrinsic properties of the $LiNbO_3$ polarization controller. Equation 24.8 is intimidating, and determining the device properties in a manufacturing setting requires significant simplification. In order to deal with the simplest form of this matrix, only one term is considered. This is accomplished by adjusting the input and output SOP so that $\vec{S}_{in} = \vec{S}_{out} = \begin{pmatrix} 0 \\ 0 \\ 1 \end{pmatrix}$. $\vec{S}$ is known as a Stokes vector in the Poincaré sphere representation and it describes the state of polarization in terms of several basis states [25].

A Stokes vector of the form $\vec{S}_{in} = \vec{S}_{out} = \begin{pmatrix} 0 \\ 0 \\ 1 \end{pmatrix}$ describes circularly polarized light. Devices are characterized after fiber pigtailing with single-mode fiber so that setting $\vec{S}_{in} = \vec{S}_{out} = \begin{pmatrix} 0 \\ 0 \\ 1 \end{pmatrix}$ for the stage of the polarization controller under test (DUT) means that the polarization of the input light must be adjusted by means of an additional polarization controller and single-mode fiber so that it is circularly polarized. If the light from the laser is linearly polarized, the input polarization controller and fiber should amount to an effective quarter wave plate. After passing through the polarization controller, the output state of polarization is given by

$$\vec{S}_{out} = \begin{pmatrix} \sin\tau\sin 2\phi \\ -\sin\tau\cos 2\phi \\ \cos\tau \end{pmatrix}. \tag{24.9}$$

This is still more complicated than we would like, so the polarization of the light must be rotated by the output single-mode fiber and another polarization controller so that it is linearly polarized. In the case where $\tau = 0$ (no phase lag between the two polarization components), the output light remains circularly polarized so that another quarter wave plate is required to return the light to linear polarization. A fixed, linear polarizer in front of the photodiode is used to determine when the output polarization controller and fiber act as an effective quarter wave plate so that only the third component of $\vec{S}$ is retained:

$$\vec{S}_{detected} = \begin{pmatrix} 0 \\ 0 \\ \cos\tau \end{pmatrix}. \tag{24.10}$$

Equation 24.10 provides some insight into the intensity of the transmitted light. When the phase lag is zero, $\tau = 0$, the transmitted intensity is maximized. With increasing phase lag, the transmitted intensity varies periodically with $\tau$. However, $\tau$ depends on the phase lag induced along both the $x$- and $y$-axes so the problem is inherently 2D. For any given $\tau$,

$$\begin{aligned} \tau_1 &= -\tau \sin(2\phi) \\ \tau_2 &= \tau \cos(2\phi). \end{aligned} \tag{24.11}$$

Equations 24.4, 24.7, and 24.11 can be combined to give

$$\begin{bmatrix} -\tau \sin(2\phi) \\ \tau \cos(2\phi) \end{bmatrix} = \begin{bmatrix} t_{11} & t_{12} \\ t_{21} & t_{22} \end{bmatrix} \cdot \begin{bmatrix} \frac{1}{2} & \frac{1}{2} \\ -1 & 1 \end{bmatrix} \cdot \begin{bmatrix} V_C \\ V_A \end{bmatrix} + \begin{bmatrix} t_{1i} \\ t_{2i} \end{bmatrix}. \tag{24.12}$$

Equation 24.12 shows how the total phase lag, $\tau$, depends on the voltages applied to the electrodes, $V_C$ and $V_A$, as well as the intrinsic device parameters, $t_{11}$, $t_{12}$, $t_{21}$, $t_{22}$, $t_{1i}$, and $t_{2i}$. Our goal is to determine the intrinsic device parameters of a typical polarization controller, and to do this, we use the 2D cos($\tau$) dependence of the intensity on the voltages $V_C$ and $V_A$.

Figure 24.5 shows a typical pattern achieved for a single stage of a four-stage polarization controller manufactured at EOSPACE Inc. The two voltages $V_C$ and $V_A$ are the two independent variables, lying on the horizontal and vertical axis, respectively. The measured intensity is the dependent variable, shown by the contours. White corresponds to higher transmitted intensity, black to lower transmitted intensity. The peak at the center of Figure 24.5 shows the voltages at which $\tau = 0$. As $\tau$ increases at all voltages away from this point, the transmitted intensity decreases until it reaches its first null corresponding to $\tau = \pi$. As $\tau$ continues to increase away from the point of $\tau = 0$, the intensity varies periodically.

Based on the theoretical formalism developed above, it is possible to fit the 2D pattern to extract the device parameters, $t_{11}$, $t_{12}$, $t_{21}$, $t_{22}$, $t_{1i}$, and $t_{2i}$. These parameters can be used to derive the voltage

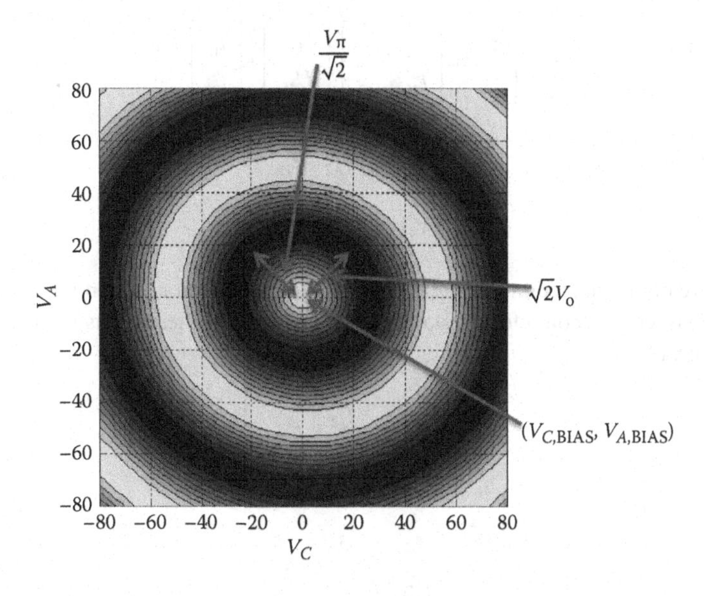


**FIGURE 24.5** Typical data acquired by raster scanning the voltages $V_C$ and $V_A$ with circular input and output polarizations.

requirements for device operation. These are the DC bias required to achieve zero phase lag, the voltage required to cause a $\pi$ phase shift between the TM and TE modes in the waveguide, $V_\pi$, and the voltage, $V_0$, required to rotate the polarization direction by 90° so that TM becomes TE and vice versa. The voltages can be derived from the $T$ matrix values, but it is also possible to roughly estimate them from the 2D pattern in Figure 24.5.

First, we find the DC bias required to set $\tau = 0$. This set of voltages is simply given by the position of the central peak in Figure 24.5, ($V_{C,\text{BIAS}}, V_{A,\text{BIAS}}$). In terms of the $T$ matrix and $V_C$ and $V_A$ this can be written as

$$0 = \begin{bmatrix} t_{11} & t_{12} \\ t_{21} & t_{22} \end{bmatrix} \cdot \begin{bmatrix} \frac{1}{2} & \frac{1}{2} \\ -1 & 1 \end{bmatrix} \cdot \begin{bmatrix} V_{C,\text{BIAS}} \\ V_{A,\text{BIAS}} \end{bmatrix} + \begin{bmatrix} t_{1i} \\ t_{2i} \end{bmatrix}. \tag{24.13}$$

Rearranging and solving for the voltages give

$$-\begin{bmatrix} \frac{1}{2} & \frac{1}{2} \\ -1 & 1 \end{bmatrix}^{-1} \cdot \begin{bmatrix} t_{11} & t_{12} \\ t_{21} & t_{22} \end{bmatrix}^{-1} \cdot \begin{bmatrix} t_{1i} \\ t_{2i} \end{bmatrix} = \begin{bmatrix} V_{C,\text{BIAS}} \\ V_{A,\text{BIAS}} \end{bmatrix}. \tag{24.14}$$

The voltage required to cause a $\pi$ phase shift between the TM and TE modes in the waveguide, $V_\pi$, and the voltage, $V_0$, required to rotate the polarization direction by 90° so that TM becomes TE and vice versa are determined by the symmetric and asymmetric voltages required to reach the first null away from the central peak in the pattern shown in Figure 24.5.

In X-cut, Z-propagating $LiNbO_3$ an electric field along the $y$-axis causes a phase lag between the TM and TE modes whereas an electric field along the $x$-axis causes a rotation of the polarization. When $U_S = 0$ (only asymmetric voltage) there is nominally no electric field in the $x$-direction so that only phase lag should occur. This occurs along the –1 slope diagonal from the central peak when $V_C + V_{C,\text{BIAS}} = -V_A + V_{A,\text{BIAS}}$.

$$\begin{bmatrix} \tau_1 \\ \tau_2 \end{bmatrix} = \begin{bmatrix} t_{11} & t_{12} \\ t_{21} & t_{22} \end{bmatrix} \cdot \begin{bmatrix} 0 \\ U_{AS} \end{bmatrix} + \begin{bmatrix} t_{1i} \\ t_{2i} \end{bmatrix} \tag{24.15}$$

$$\tau_1 = t_{12} U_{AS} + t_{1i} \tag{24.16}$$

$$\tau_2 = t_{22} U_{AS} + t_{2i}. \tag{24.17}$$

In practice, we are not interested in the static phase lag terms, $t_{1i}$ and $t_{2i}$, when we determine $V_\pi$ and $V_0$ because the DC bias terms account for the static phase lag. Ignoring these terms and using Equation 24.3 leads to a value of $\tau$ of

$$\tau = U_{AS}\sqrt{t_{12}^2 + t_{22}^2}. \tag{24.18}$$

To have $\tau = \pi$, we need

$$\frac{\pi}{\sqrt{t_{12}^2 + t_{22}^2}} = U_{AS} = V_\pi. \tag{24.19}$$

This is the most general expression for $V_\pi$ but in practice, the off-diagonal $T$ matrix elements ($t_{12}$ and $t_{21}$) are two or more orders of magnitude smaller than the diagonal elements ($t_{11}$ and $t_{22}$) so it is a good approximation to neglect them and write

$$V_\pi = \frac{\pi}{t_{22}}. \tag{24.20}$$

A similar result is obtained when $U_{AS} = 0$. This is the symmetric voltage case where there is nominally no electric field in the $y$-direction so that only polarization rotation should occur. This situation occurs along the +1 slope diagonal from the central peak in Figure 24.5 when $V_C + V_{C,\mathrm{BIAS}} = V_A + V_{A,\mathrm{BIAS}}$. A slice along this diagonal is shown in Figure 24.6. The central peak appears close to 0 V. As the symmetric voltage is increased, the intensity decreases, reaching a minimum at ~25 V.

$$\tau = U_S\sqrt{t_{11}^2 + t_{21}^2} \tag{24.21}$$

$$\frac{\pi}{\sqrt{t_{11}^2 + t_{21}^2}} = U_S = V_0 \tag{24.22}$$

$$V_0 = \frac{\pi}{t_{11}}. \tag{24.23}$$

In Figure 24.5, the voltage from the central peak to the first null along the −1 diagonal is $V_\pi/\sqrt{2}$. This is due to the definition of $U_{AS} = -V_C + V_A$. Similarly, the voltage to the first null along the +1 diagonal is $\sqrt{2}V_o$ due to the definition of $U_S = (V_C + V_A)/2$. Using this result and the data in Figure 24.6, it is clear that $V_0$ for this four-stage device is ~17.5 V. The 2D pattern shown in Figure 24.5 provides a simple, intuitive visual aid to illustrate the key device parameters of a $LiNbO_3$ polarization controller.

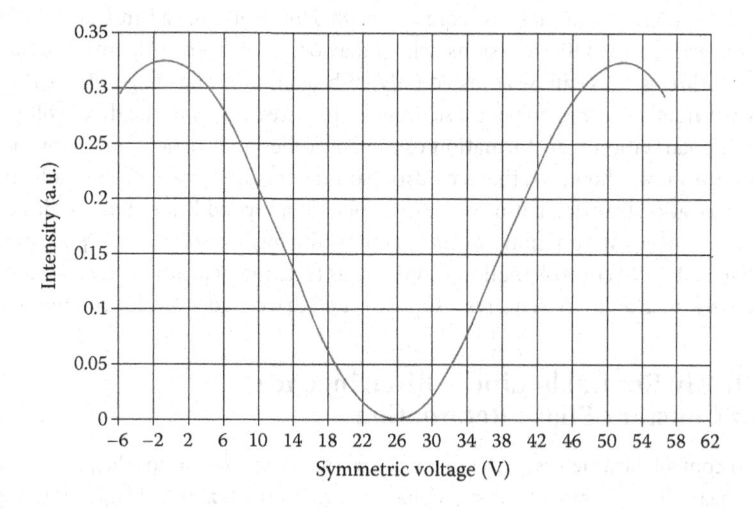


**FIGURE 24.6** Slice through the data shown in Figure 24.5 along the +1 diagonal. The central peak appears near 0 V. The first null appears near 25 V.

## 24.3 Operation of Lithium Niobate Polarization Controllers

Lithium niobate polarization controllers are well suited for automatic polarization stabilization because of their fast control speed and because several transformer stages can be readily combined in a single compact device, as shown in Figure 24.2. Moreover, the three-electrode structure employed in these devices is extremely versatile in that it can generate variable combinations of differential TE/TM phase shifting and TE/TM mode conversion. In this section, we describe three basic modes of operating a single electrode section of the controller. In all three cases, the control voltages applied to the electrode section are of the general form

$$\begin{aligned} V_A(t) &= a(t)V_o + b(t)V_\pi/2 + V_{A,\text{BIAS}} \\ V_C(t) &= a(t)V_o - b(t)V_\pi/2 + V_{C,\text{BIAS}}, \end{aligned} \tag{24.24}$$

wherein $V_{A,\text{BIAS}}$ and $V_{C,\text{BIAS}}$ are the bias voltages determined in Section 24.2.2, $V_o$ and $V_\pi$ are the voltages to generate complete mode conversion and a TE/TM phase shift of $\pi$, respectively, and $a(t)$ and $b(t)$ are two adjustable control parameters.

### 24.3.1 Pure TE/TM Phase Shifting and Pure TE/TM Mode Conversion

In the most basic mode of operation, a single electrode section generates either pure TE/TM phase shifts or pure TE/TM mode conversion. With $a(t) \equiv 0$ and $b(t)$ adjustable in the range $-1 \le b(t) \le 1$, for example, the electrode section generates pure TE/TM phase shifting, i.e., linear birefringence at 0°, which is described by a transfer matrix of the form Equation 24.8 wherein $\phi \equiv 0$ and $\tau = b(t)\pi$. On the other hand, with $b(t) \equiv 0$ and $a(t)$ adjustable in the range $-1 \le a(t) \le 1$, it generates pure TE/TM mode conversion, i.e., linear birefringence at 45°, which is described by Equation 24.8 when $\phi \equiv \pi/4$ and $\tau = a(t)\pi$. In both cases, however, the electrode section can perform only a very limited range of polarization transformations. In fact, there exist two well-defined SOPs, called eigenstates, which cannot be altered by the controller stage, no matter what value we choose for $a(t)$ or $b(t)$. It is easily verified, for example, that the eigenstates of a differential TE/TM phase shifter are pure TE or TM polarization states. Arbitrary polarization transformations from any general input SOP into any general output SOP can be obtained with a three-stage controller comprising alternating TE/TM phase shifters (PC) and mode converters (MC), i.e., by cascaded PS-MC-PS or MC-PS-MC controllers [8]. However, such a three-stage controller cannot compensate arbitrarily large SOP excursions, which may occur in fiber-optic links, because the control parameters in the three stages can be adjusted only within a finite range (e.g., $-1 \le a(t), b(t) \le 1$). Thus, when a control parameter reaches one of these limits, it requires a reset of the drive voltages, which may affect the overall polarization transformation of the controller and, hence, interrupt the signal transmission. It has been shown, however, that a control parameter can be reset without affecting the overall (end-to-end) polarization transformation if the controller employs additional (i.e., redundant) electrode sections, which take over the function of other sections during a reset cycle [17]. A cascaded PS-MC-PS-MC or (MC-PS-MC-PS) controller allows continuous tracking of arbitrarily varying input or output SOPs [26]. However, such a device requires a sophisticated drive algorithm for proper operation [27].

### 24.3.2 Endlessly Rotatable Linear Birefringence with Constant Phase Retardation

With sinusoidal control parameters of the form $a(t) = \sin\theta(t)$ and $b(t) = \cos\theta(t)$, a single electrode section generates linear birefringence of constant phase retardation $\tau \equiv \pi$, but at adjustable angular orientation $\phi \equiv \theta/2$, which leads to a polarization transformation that is identical to that of a rotatable half-wave plate (HWP) [24]. An important feature of this mode of operation is that the angular orientation of the induced birefringence can be endlessly rotated, because the electrical phase θ in the control parameters may be increased (or decreased) without limits. Similarly, when driven with control parameters $a(t) = 0.5 \sin \theta(t)$ and $b(t) = 0.5 \cos \theta(t)$, the electrode section can emulate the polarization transformation of an endlessly rotatable quarter-wave plate (QWP) with $\tau \equiv \pi/2$. However, just like with pure TE/TM phase shifting or mode conversion, the transformation range of a single rotatable HWP or QPW is very limited. A HWP, for instance, always converts a circular input SOP into the orthogonally polarized circular state, independent of its angular orientation [28]. Again, arbitrary polarization transformation can be accomplished with multistage controllers that employ, for example, two or three cascaded QWPs or a QWP–HWP combination [8]. Moreover, certain combinations of rotating-wave-plate controllers exhibit an inherently unlimited transformation range and, hence, never require a reset cycle. A three-stage cascade of an endlessly rotatable QWP followed by an endlessly rotatable HWP and another endlessly rotatable QWP, for example, allows polarization transformations between any two arbitrarily varying SOPs [24]. Such a controller can be operated with a relatively simple drive algorithm and, thus, is capable of fast control speeds that can compensate SOP fluctuations of up to 4000 rad/s [29].

### 24.3.3 Endlessly Rotatable Linear Birefringence with Variable Phase Retardation

It is also possible to combine the two modes of operation described above and generate rotatable linear birefringence of variable phase retardation with just a single electrode section. This is readily accomplished by introducing two independent control parameters of the form $a(t) = c(t)\sin \theta(t)$ and $b(t) = c(t) \cos \theta(t)$, wherein $-1 \le c(t) \le 1$. The corresponding polarization transformation is similar to that of a Soleil–Babinet compensator and described by Equation 24.8 with $\tau \equiv c\pi$ and $\phi \equiv \theta/2$ [27]. This type of controller may be used as a single-stage device to transform an arbitrarily varying input SOP into a fixed linear output SOP or as a double-stage device for general polarization transformations [27]. Because of its simplicity, a single-stage controller of this type may track polarization changes at speeds of up to 15,000 rad/s [30].

## 24.4 Automatic Polarization Control in Fiber-Optic Systems

Automatic polarization stabilization in fiber-optic communication systems is one of the most challenging applications for optical polarization controllers because the output SOP of a long fiber link may fluctuate rapidly with time, at rates of up to several 1000 rad/s and by large amounts [31]. It is critically important in these applications that the controller does not—at any time—impair or interrupt transmission of the optical signal. Thus, it must be quick enough to follow even the fastest polarization changes occurring in the fiber and, furthermore, it must exhibit an effectively unlimited transformation range, so that it can continuously compensate for arbitrarily large SOP fluctuations in the fiber.

In most applications, the actual input SOP to the polarization controller is not known. The controller is therefore operated in a closed feedback loop, where the desired polarization transformation is obtained with the help of an electrical feedback signal, which is generated by a detector downstream from the controller, as shown schematically in Figure 24.1. Such an arrangement is frequently used for automatic polarization demultiplexing of two polarization multiplexed signals or for adjusting the input SOP to an optical PMD compensator [2,3]. The feedback signal is usually proportional to the optical power in the desired SOP and, hence, assumes maximal amplitude when the desired output SOP is obtained. To generate the necessary error signals for these "blind" adjustments, the controller periodically dithers the output SOP slightly about its current setting and registers the corresponding variations on the feedback signal, which are then analyzed by a suitable "hill-climbing" algorithm to maximize the feedback signal.

It is often overlooked that such an automatic control loop may get trapped occasionally and, thus, provide suboptimal polarization transformations, even when employing a polarization controller with unlimited transformation range. To analyze the specific conditions under which these trappings may occur, we consider in the following a three-stage lithium niobate QWP–HWP–QWP controller with independently adjustable electrical control phases $\theta_i(t)$, $i = 1, 2, 3$. To generate the dither in the output SOP, the three phases are varied by some small amounts $\Delta\theta_i$ about their current values, either separately in different time periods, or together in certain combinations. The required magnitude $\Delta\theta_i$ depends on the sensitivity of the feedback signal to the corresponding SOP excursions and is typically of the order of 0.1 rad [29].

Ideally, the phase dither should vary the Stokes vector of the output SOP, $\vec{S}_{out}$, symmetrically along at least two orthogonal directions, so that the control circuit can evaluate the local dependencies of the feedback signal on $\theta_i(t)$ and determine whether the signal is maximal or not. However, the dither-induced variations of $\vec{S}_{out}$ strongly depend on the actual values of the phases $\theta_i(t)$ as well as on the current input and output SOPs. In most cases, $\vec{S}_{out}$ moves along two sufficiently different directions, but under certain conditions, it essentially varies along only one direction, because it requires very large changes in $\theta_i(t)$ to move $\vec{S}_{out}$ into a different direction [24]. As a result, the control circuit may not be able to adjust the controller for maximal feedback signal and, hence, becomes trapped. The particular conditions at which these trappings may occur can be determined from the overall transformation matrix of the controller, $\mathbf{R}(\theta_1,\theta_2,\theta_3) = \mathbf{R}_3(\theta_3)\mathbf{R}_2(\theta_2)\mathbf{R}_1(\theta_1)$, wherein $\mathbf{R}_1$ denotes the transfer matrix Equation 24.8 of the first QWP section, $\mathbf{R}_2$ that of the HWP section, and $\mathbf{R}_3$ that of the second QWP section. With $\vec{S}_{in} = \mathbf{R}^{-1}\,\vec{S}_{out}$, one obtains for the sensitivity of $\vec{S}_{out}$ to a phase dither with $\Delta\theta_1 << 1$ to first order,

$$\vec{S}_{out}(\theta_i+\Delta\theta_i) \approx \left[\mathbf{I}+\Delta\theta_i\left(\frac{\partial \mathbf{R}}{\partial \theta_i}\right)\mathbf{R}^{-1}\right]\vec{S}_{out}(\theta_i), \tag{24.25}$$

wherein $\mathbf{I}$ denotes the identity matrix and $i = 1, 2, 3$. Thus, if $\vec{S}_{out}$ is an eigenstate of $(\partial\mathbf{R}/\partial\theta_m)\mathbf{R}^{-1}$, then dithering $\theta_m$ has no significant effect on the output SOP, at least to first order in $\Delta\theta_m$. Specifically, the QWP–HWP–QWP controller exhibits the following three dither eigenstates [32]:

$$\vec{S}_{\Delta\theta_1} = \begin{bmatrix} \sin(2\tilde{\theta}_2)\cdot\cos(\theta_3) \\ \sin(2\tilde{\theta}_2)\cdot\sin(\theta_3) \\ -\cos(2\tilde{\theta}_2) \end{bmatrix}, \quad \vec{S}_{\Delta\tilde{\theta}_2} = \begin{bmatrix} -\sin(\theta_3) \\ \cos(\theta_3) \\ 0 \end{bmatrix}, \quad \vec{S}_{\Delta\theta_3} = \begin{bmatrix} 0 \\ 0 \\ 1 \end{bmatrix}, \tag{24.26}$$

wherein $\tilde{\theta}_2(t) = \theta_2(t) + \left[\theta_1(t) - \theta_3(t)\right]/2$ and $\Delta\tilde{\theta}_2$ represents the dither combination $[\theta_1 \pm \Delta\theta_1/2, \theta_2 \pm \Delta\theta_2, \theta_3 \mp \Delta\theta_3/2]$. It turns out that dithering $\tilde{\theta}_2(t)$ has certain advantages over dithering just $\theta_2(t)$ by itself [24]. The eigenstates Equation 24.26 define three axes in Stokes space, about which $\vec{S}_{out}$ is rotated when a dither $\Delta\theta_i$ is applied. These three axes do not share a common plane as long as $2\tilde{\theta}_2 \neq m\pi$, where $m = 0, \pm1, \pm2, \ldots$. Thus, even if $\vec{S}_{out}$ is equal to one of these eigenstates Equation 24.26, it still can be dithered symmetrically along two significantly different directions, which become orthogonal at $2\tilde{\theta}_2 = \pm\pi/2, \pm3\pi/2, \ldots$. However, when $2\tilde{\theta}_2 = m\pi$, with $m = 0, \pm1, \pm2, \ldots$, two of the three axes are identical, i.e., $\vec{S}_{\Delta\theta_1} = \pm\vec{S}_{\Delta\theta_3}$, so that $\vec{S}_{out}$ can be rotated about only two (albeit orthogonal) axes. Therefore, when $\vec{S}_{out}$ happens to be equal to either $\vec{S}_{\Delta\theta_1}$ (circular polarized) or $\vec{S}_{\Delta\tilde{\theta}_2}$ (linear polarized), it can be dithered along only one direction.

This condition may be readily visualized on the Poincaré sphere, as shown in Figure 24.7a, which displays the polarization transformation of a controller that became trapped at $\tilde{\theta}_2 = \pi$ and $\vec{S}_{out} \approx S_{\Delta\theta_1} = S_{\Delta\theta_3}$. The diagram depicts the evolution of the SOP in the three electrode sections as well as the excursions in the output SOP that result from dithering $\theta_1$, $\tilde{\theta}_2$, and $\theta_3$. All three phase dithers move $\vec{S}_{out}$ along only

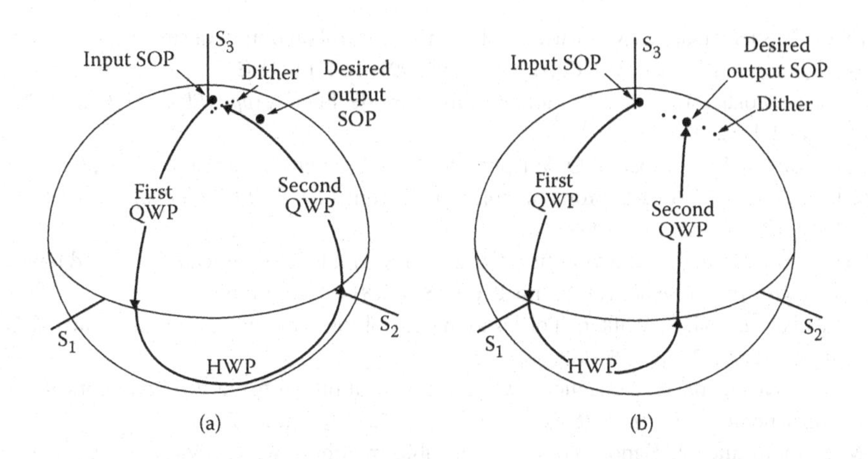


**FIGURE 24.7** Poincaré sphere representation of the polarization transformation performed by a QWP-HWP-QWP controller after becoming trapped at $\tilde{\theta}_2 = \pi$ and $\vec{S}_{out} \approx S_{\Delta\theta_1} = S_{\Delta\theta_3}$ (a), and after readjustment of the control parameters (b).

one direction, which in the case of Figure 24.7 is orthogonal to the direction of the desired output SOP. Hence, the control loop is temporarily trapped until the input SOP changes sufficiently to move $\vec{S}_{out}$ away from $S_{\Delta\theta_3}$.

A similar analysis for a QWP–HWP–QWP polarization controller with only two independent control parameters, having also an unlimited transformation range [24], shows that it becomes trapped even more often than the three-control-parameter device considered above because it exhibits just two different dither eigenstates.

Conversely, one finds that undesired trapping occurs much less frequently with polarization controllers that comprise four or more independently adjustable stages, such as a QWP–HWP–QWP–HWP–QWP combination or even a cascade of two independent QWP–HWP–QWP polarization controllers. However, adding extra stages to the controller only reduces the probability of trapping but does not completely solve the problem. A more effective solution is to deliberately force a change in the three control parameters, which can be done without affecting the overall polarization transformation, because there always exists more than one combination of $\theta_1$, $\theta_2$, and $\theta_3$ that yields the desired transformation [24]. Such forced variations of the control parameters give rise to corresponding changes in two of the dither eigenstates (via $\tilde{\theta}_2$ and $\theta_3$) and may even move them away from $\vec{S}_{out}$, so that the loop can resume normal operation, as shown in Figure 24.7b. However, the feedback loop and the controller have to be fast enough to introduce sufficiently large changes in the control parameters before trapping occurs. Finally, it should be noted that similar trapping problems may be encountered with other types of endless polarization controllers including cascaded MC–PS–MC–PS controllers [33,34].

# References


1. T. Imai and T. Matsumoto, "Polarization fluctuations in a single-mode optical fiber," *J. Lightwave Technol.*, vol. 6, no. 9, pp. 1366–1375, Sep. 1988.
2. F. Heismann, P. B. Hansen, S. K. Korotky, G. Raybon, J. J. Veselka, and M. S. Whalen, "Automatic polarization demultiplexer for polarization multiplexed transmission systems," *Electron. Lett.*, vol. 29, no. 22, pp. 1965–1966, Oct. 1993.
3. F. Heismann, D. A. Fishman, and D. L. Wilson, "Automatic compensation of first-order polarization mode dispersion in a 10-Gb/s transmission system," in *Proc. 24th European Conf. Optical Communication*, Madrid, Spain, 1998, vol. 1, pp. 529–530.

4. Y. Kidoh, Y. Suematsu, and K. Furuya, "Polarization control on output of single-mode optical fibers," *IEEE J. Quantum Electron.*, vol. QE-17, no. 6, pp. 991–994, Jun. 1981.
5. R. Ulrich, "Polarization stabilization on single-mode fiber," *Appl. Phys. Lett.*, vol. 35, no. 11, pp. 840–842, Dec. 1979.
6. F. Heismann, A. F. Ambrose, T. O. Murphy, and M. S. Whalen, "Polarization-independent photonic switching system using fast automatic polarization controllers," *IEEE Photon. Technol. Lett.*, vol. 5, no. 11, pp. 1341–1343, Nov. 1993.
7. F. Heismann, "Compact electro-optic polarization scramblers for optically amplified lightwave systems," *J. Lightwave Technol.*, vol. 14, no. 8, pp. 1801–1814, Aug. 1996.
8. N. G. Walker and G. R. Walker, "Polarization control for coherent communications," *J. Lightwave Technol.*, vol. 8, no. 3, pp. 438–458, Mar. 1990.
9. T. Imai, K. Nosu, and H. Yamaguchi, "Optical polarization control utilizing an optical heterodyne detection scheme," *Electron. Lett.*, vol. 21, no. 2, pp. 52–53, Jan. 1985.
10. T. Matsumoto and H. Kano, "Endlessly rotatable fractional-wave devices for single-mode fibre optics," *Electron. Lett.*, vol. 22, no. 2, pp. 78–79, Jan. 1986.
11. H. Shimazu and K. Kaede, "Endless polarization controller using electro-optic waveplates," *Electron. Lett.*, vol. 24, no. 7, pp. 412–413, Mar. 1988.
12. T. Chiba, Y. Ohtera, and S. Kawakami, "Polarization stabilizer using liquid crystal rotatable waveplates," *J. Lightwave Technol.*, vol. 17, no. 5, pp. 885–890, May 1999.
13. K. Hirabayashi and C. Amano, "Variable and rotatable waveplates of PLZT electrooptic ceramic material on planar waveguide circuits," *IEEE Photon. Technol. Lett.*, vol. 14, no. 7, pp. 956–958, Jul. 2002.
14. S. Thaniyavarn, "Wavelength independent, optical damage immune $Z$-propagation $LiNbO_3$ waveguide polarization converter," *Appl. Phys. Lett.*, vol. 47, no. 7, pp. 674–677, Oct. 1985.
15. H. Shimizu, S. Yamazaki, T. Ono, and K. Emura, "Highly practical fiber squeezer polarization controller," *J. Lightwave Technol.*, vol. 9, no. 10, pp. 1217–1224, Oct. 1991.
16. S. H. Rumbaugh, M. D. Jones, and L. W. Casperson, "Polarization control for coherent fiber-optic systems using nematic liquid crystals," *J. Lightwave Technol.*, vol. 8, no. 3, pp. 459–464, Mar. 1990.
17. H. Heidrich, C. H. von Hemholt, D. Hoffmann, H.-J. Hensel, and A. Kleinwächter, "Polarisation transformer on Ti:$LiNbO_3$ with reset-free optical operation for heterodyne/homodyne receivers," *Electron. Lett.*, vol. 23, no. 7, pp. 335–336, Mar. 1987.
18. J. Prat, J. Comellas, and G. Junyent, "Experimental demonstration of an all-fiber endless polarization controller based on Faraday rotation," *IEEE Photon. Technol. Lett.*, vol. 7, no. 12, pp. 1430–1432, Dec. 1995.
19. T. Saitoh and S. Kinugawa, "Magnetic field rotating-type Faraday polarization controller," *IEEE Photon. Technol. Lett.*, vol. 15, no. 10, pp. 1404–1406, Oct. 2003.
20. R. C. Alferness, "Efficient waveguide electro-optic TE↔TM mode converter/wavelength filter," *Appl. Phys. Lett.*, vol. 36, no. 7, pp. 513–515, Apr. 1980.
21. S. Thaniyavarn, "Wavelength-independent, optical-damage-immune $LiNbO_3$ TE–TM mode converter," *Opt. Lett.*, vol. 11, no. 1, pp. 39–41, Jan. 1986.
22. Derivation and device characterization data provided by David Moilanen, EOSPACE Inc.
23. A. J. P. van Haasteren, J. J. G. M. van der Tol, M. O. van Deventer, and H. J. Frankena, "Modeling and characterization of an electrooptic polarization controller on $LiNbO_3$", *J. Lightwave Technol.*, vol. 11, no. 7, pp. 1151–1157, Jul. 1993.
24. F. Heismann, "Analysis of a reset-free polarization controller for fast automatic polarization stabilization in fiber-optic transmission systems," *J. Lightwave Technol.*, vol. 12, no. 4, pp. 690–699, Apr. 1994.
25. E. Collett, *Polarized Light—Fundamentals and Applications*, New York: Marcel Dekker, 1993.
26. P. Oswald and C. K. Madsen, "Deterministic analysis of endless tuning of polarization controllers," *J. Lightwave Technol.*, vol. 7, no. 7, pp. 2932–2939, Jul. 2006.

27. R. Noé, H. Heidrich, and D. Hoffmann, "Endless polarization control systems for coherent optics," *J. Lightwave Technol.*, vol 6, no. 7, pp. 1199–1208, Jul. 1988.
28. W. Shieh and H. Kogelnik, "Dynamic eigenstates of polarization," *IEEE Photon. Technol. Lett.*, vol. 13, no. 1, pp. 40–42, Jan. 2001.
29. F. Heismann and M. S. Whalen, "Fast automatic polarization control system," *IEEE Photon. Technol. Lett.*, vol. 4, no. 5, pp. 503–505, May 1992.
30. A. Hidayat, B. Kock, H. Zhang, V. Mirvoda, M. Lichtinger, D. Sandel, and R. Noé, "High-speed endless optical polarization stabilization using calibrated waveplates and field-programmable gate array-based digital controller," *Optics Express*, vol. 16, no. 23, pp. 18984–18991, Nov. 2008.
31. P. M. Krummrich, E.-D. Schmidt, W. Weiershausen, and A. Mattheus, "Field trial results on statistics of fast polarization changes in long haul WDM transmission systems," in *Optical Fiber Communication Conf.*, Anaheim, CA, 2005, paper OThT6.
32. A. Chen and F. Heismann, "Polarization control," in *Encyclopedic Handbook of Integrated Optics*, Boca Raton, FL: CRC Press, 2006, pp. 306–318.
33. W. H. J. Aarts and G.-D. Khoe, "New endless polarization control method using three fiber squeezers," *J. Lightwave Technol.*, vol. 7, no. 7, pp. 1033–1043, Jul. 1989.
34. M. Martinelli and R. A. Chipman, "Endless polarization control algorithm using adjustable linear retarders with fixed axes," *J. Lightwave Technol.*, vol. 21, no. 9, pp. 2089–2096, Sep. 2003.

# Index

Page numbers followed by f and t indicate figures and tables, respectively.

## E

## F

## G



## H

## M

## N



## O

## P

## Q



## R



## S

## T



## U



## V

Printed in the United States
by Baker & Taylor Publisher Services